AF735437

SUPERCRITICAL FLUID TECHNOLOGY IN MATERIALS SCIENCE AND ENGINEERING

SUPERCRITICAL FLUID TECHNOLOGY IN MATERIALS SCIENCE AND ENGINEERING

SYNTHESES, PROPERTIES, AND APPLICATIONS

EDITED BY

YA-PING SUN

Clemson University
Clemson, South Carolina

MARCEL DEKKER, INC. NEW YORK • BASEL

ISBN: 0-8247-0651-X

This book is printed on acid-free paper.

Headquarters
Marcel Dekker, Inc.
270 Madison Avenue, New York, NY 10016
tel: 212-696-9000; fax: 212-685-4540

Eastern Hemisphere Distribution
Marcel Dekker, Inc.
Hutgasse 4, Postfach 812, CH-4001 Basel, Switzerland
tel: 41-61-261-8482; fax: 41-61-261-8896

World Wide Web
http://www.dekker.com

The publisher offers discounts on this book when ordered in bulk quantities. For more information, write to Special Sales/Professional Marketing at the headquarters address above.

Copyright © 2002 by Marcel Dekker, Inc. All Rights Reserved.

Neither this book nor any part may be reproduced or transmitted in any form or by any means, electronic or mechanical, including photocopying, microfilming, and recording, or by any information storage and retrieval system, without permission in writing from the publisher.

Current printing (last digit):
10 9 8 7 6 5 4 3 2 1

PRINTED IN THE UNITED STATES OF AMERICA

Preface

Supercritical fluid technology has attracted the attention of both scientists and engineers. In the last 20 years or so, applications of supercritical fluid technology have been primarily in extraction and chromatography. Extensive experimental and theoretical investigations have been aimed toward an understanding of the properties of supercritical fluid systems, particularly intermolecular interactions (solute–solvent, solvent–solvent, and solute–solute) in supercritical fluid solutions. Recently, however, significant progress has been made in the use of supercritical fluids and mixtures as reaction media for chemical syntheses and polymer preparations and as alternative solvent systems for materials processing. In fact, materials-related applications have emerged as a new frontier in the development of supercritical fluid technology. I hope that this book will be a timely contribution to this emerging research field by serving at least two purposes. One is to provide interested readers with a rich source of information on the current status of supercritical fluid technology as related to materials research. The second is to stimulate more interest within the multidisciplinary supercritical fluid research community for the further development of the technology in materials-related applications.

I would like to thank all the contributors. I also thank my students and postdoctoral associates; together we have had a lot of fun in the pursuit of many interesting and exciting projects in this research field. I am grateful for financial support from the National Science Foundation and the U.S. Department of Energy during my editing of this book.

On a more personal note, I want to credit Professor Wen-Hsing Yen, on the occasion of his 95th birthday celebration, for introducing me to the world of chemical thermodynamics and the critical phenomenon, at Zhejiang University in China many years ago. Credit is also due my postdoctoral mentor Professor Marye Anne Fox. It was her collaboration with Professor Keith Johnston at the University of Texas at Austin that introduced me to the field of supercritical fluid research.

Ya-Ping Sun

Contents

Contributors

Tadafumi Adschiri, Ph.D. Department of Chemical Engineering, Tohoku University, Sendai, Japan

Kunio Arai, Ph.D. Department of Chemical Engineering, Tohoku University, Sendai, Japan

Jayasundera Bandara, Ph.D. Department of Chemistry, Clemson University, Clemson, South Carolina

Eric J. Beckman, Ph.D. Department of Chemical Engineering, University of Pittsburgh, Pittsburgh, Pennsylvania

Christopher E. Bunker, Ph.D. Propulsion Directorate, Air Force Research Laboratory, Wright-Patterson Air Force Base, Ohio

Can Erkey, Ph.D. Department of Chemical Engineering, University of Connecticut, Storrs, Connecticut

Markus M. Hoffmann, Ph.D. Department of Chemistry, State University of New York–Brockport, Brockport, New York

Linda J. Holm, Ph.D. School of Chemical Engineering, Georgia Institute of Technology, Atlanta, Georgia

Keith W. Hutchenson, Ph.D. Central Research and Development, DuPont Company, Wilmington, Delaware

Andreas Kordikowski, Dr.rer.nat. Technology Development, Bradford Particle Design plc, Bradford, West Yorkshire, England

Frederick S. Mandel, Ph.D. Department of Chemical Engineering, Virginia Commonwealth University, Richmond, Virginia

Mark A. McHugh, Ph.D. Department of Chemical Engineering, Virginia Commonwealth University, Richmond, Virginia

Jaouad M. Meziani, Ph.D. Department of Chemistry, Clemson University, Clemson, South Carolina

Srinivas Palakodaty, Ph.D. Process Engineering, Bradford Particle Design plc, Bradford, West Yorkshire, England

Janice L. Panza, Ph.D. Department of Chemical and Petroleum Engineering, University of Pittsburgh, Pittsburgh, Pennsylvania

Harry W. Rollins, Ph.D. Chemistry Department, Idaho National Engineering and Environmental Laboratory, Idaho Falls, Idaho

Raymond Sloan, Ph.D. Bioprocessing Department, Bradford Particle Design plc, Bradford, West Yorkshire, England

Ya-Ping Sun, Ph.D. Department of Chemistry, Clemson University, Clemson, South Carolina

Amyn S. Teja, Ph.D. School of Chemical Engineering, Georgia Institute of Technology, Atlanta, Georgia

Mark C. Thies, Ph.D. Department of Chemical Engineering, Clemson University, Clemson, South Carolina

Chien M. Wai, Ph.D. Department of Chemistry, University of Idaho, Moscow, Idaho

J. Don Wang, Ph.D. Consultant, Supercritical Fluid Development, Cleveland, Ohio

Kimberly F. Webb, Ph.D. School of Chemical Engineering, Georgia Institute of Technology, Atlanta, Georgia

Markus Weber, Dr.sc.techn. Department of Chemical Engineering, Clemson University, Clemson, South Carolina

Clement R. Yonker, Ph.D. William R. Wiley Laboratory, Pacific Northwest National Laboratory, Richland, Washington

Peter York, Ph.D., F.R.S.C., C.Chem. School of Pharmacy, University of Bradford, Bradford, West Yorkshire, England

1
Fundamental Properties of Supercritical Fluids

Christopher E. Bunker
Wright-Patterson Air Force Base, Ohio

Harry W. Rollins
Idaho National Engineering and Environmental Laboratory, Idaho Falls, Idaho

Ya-Ping Sun
Clemson University, Clemson, South Carolina

I. INTRODUCTION

Supercritical fluids* have been studied extensively for the past two decades in attempts to gain accurate and detailed knowledge of their fundamental properties. Such knowledge is essential to the utilization and optimization of supercritical fluid technology in materials preparation and processing. Among the most important properties of a supercritical fluid are the low and tunable densities that can be varied between those of a gas and a normal liquid and the local density effects observed in supercritical fluid solutions (most strongly associated with near-critical conditions). A supercritical fluid may be considered macroscopically homogeneous but microscopically inhomogeneous, consisting of clusters of solvent molecules and free volumes. That a supercritical fluid is macroscopically homogeneous is obvious—the fluid at a temperature above the critical temperature exists as a single phase regardless of pressure. As a consequence,

*A supercritical fluid is defined loosely as a solvent above its critical temperature because under those conditions the solvent exists as a single phase regardless of pressure. It has been demonstrated that a thorough understanding of the low-density region of a supercritical fluid is required to obtain a clear picture of the microscopic properties of the fluid across the entire density region from gas-like to liquid-like (1–3).

extremely wide variations in the solvent properties may be achieved. The microscopic inhomogeneity of a supercritical fluid is a more complex issue and is probably dependent on the density of the fluid. The microscopic properties and their effects on and links to the macroscopic properties have been the focus of numerous experimental investigations, many of which employed molecular spectroscopic techniques. The main issues have been the existence and extent of local density augmentation (or solute–solvent clustering) and solvent-facilitated solute concentration augmentation (or solute-solute clustering) in supercritical fluid solutions.

Solute–solvent clustering is typically defined as a local solvent density about a solute molecule that is greater than the bulk solvent density in a supercritical fluid solution. Initially, local density augmentation was proposed to explain the unusual density dependence of the basic solvent parameters (i.e., polarity, dielectric constant, refractive index, viscosity, etc.). These early studies tended to demonstrate significant discrepancies between experimental results and those predicted by continuum theory. It is now known that for different supercritical fluids a common pattern exists for the density dependence of the solute–solvent interactions. The pattern is characterized by different spectroscopic (or other) responses in the three density regions: (a) a rapid increase in response in the low-density region; (b) a plateau-like response in the near-critical density region; and (c) a further increase in response in the high-density region (Figure 1) (1–3). This characteristic pattern is a reflection of the specific solute–solvent interactions occurring in the three density regions. Thus, an empirical

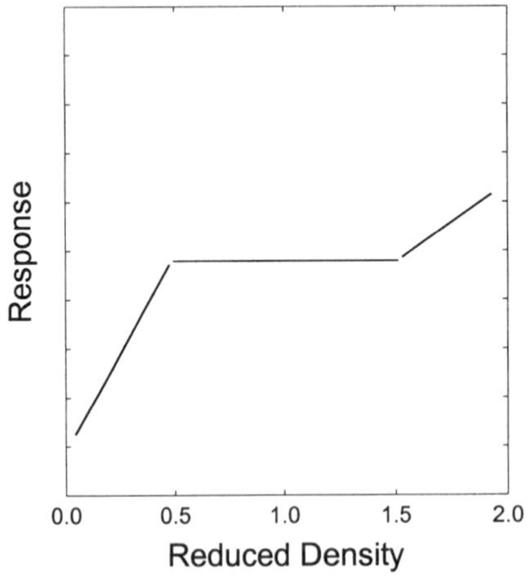

Figure 1 Cartoon representation of typical spectroscopic and other responses in the three density regions in a supercritical fluid.

three-density-region solvation model has been developed to serve as a baseline in the interpretation of supercritical fluid properties (1–3).

Solute-solute clustering is somewhat less well defined. As an extension of the concept of solute–solvent clustering, the type of solute-solute clustering commonly discussed in the literature may be defined loosely as local solute concentrations that are greater than the bulk solute concentration. An important consequence of solute-solute clustering is the enhancement of bimolecular reactions in supercritical fluid solutions. Thus, well-established bimolecular probes (most commonly intermolecular reactions or intramolecular reactive molecules) have been used in the study of the clustering phenomenon. Experimental results that confirm and others that deny the existence of significant solute–solute clustering in supercritical fluid solutions have been presented, and some interpretations remain controversial. That solute–solute clustering is probably system dependent makes the issue more complex. Nevertheless, a critical review of the available evidence and various opinions on the issue is warranted.

On the topics of solute–solvent and solute-solute clustering, there is a significant number of publications by research groups from around the world, demonstrating the tremendous interest of the international research community. This chapter is a review of representative literature results, especially those based on molecular spectroscopy and related experimental techniques. Discussion of the fundamental properties of supercritical fluids will be within the context of enhanced solute–solvent and solute–solute interactions in supercritical fluid solutions, and the current understanding of the reasonably well-established solute–solvent clustering model and the somewhat controversial solute-solute clustering concept will be presented.

II. SOLUTE–SOLVENT INTERACTIONS

Numerous experimental studies have been conducted on solute–solvent interactions in supercritical fluid solutions. In particular, issues such as the role of characteristic supercritical solvent properties in solvation and the dependence of solute–solvent interactions on the bulk supercritical solvent density have been extensively investigated. Results from earlier experiments showed that the partial molar volumes υ_2 became very large and negative near the critical point of the solvent (4–12). The results were interpreted in terms of a collapse of the solvent about the solute under near-critical solvent conditions, which served as a precursor for the solute–solvent clustering concept. Molecular spectroscopic techniques, especially ultraviolet-visible (UV-vis) absorption and fluorescence emission, have since been applied to the investigation of solute–solvent interactions in supercritical fluid solutions. Widely used solvent environment–sensitive molecular probes include Kamlet–Taft π^* scale probes for polarity/polarizability

(13,14), pyrene (Py scale) (15,16), solvatochromic organic dyes, and molecules that undergo twisted intramolecular charge transfer (TICT) in the photoexcited state (17).

A. Kamlet-Taft π^* Scale for Polarity/Polarizability

The π^* scale of solvent polarity/polarizability is based on the correlation between the experimentally observed absorption or emission shifts (ν_{max} values) of various nitroaromatic probe molecules and the ability of the solvent to stabilize the probe's excited state via dielectric solute–solvent interactions (18). Since π^* values are known for many commonly used liquid solvents, the π^* scale allows comparison of the solvation strength of supercritical fluids and normal liquid solvents. Several research groups have utilized the π^* probes to investigate solvent characteristics for a series of supercritical fluids (19–34). For example, Hyatt (19) employed two nitroaromatic dyes and the penta-*tert*-butyl variation of the Riechardt dye (18) to determine the π^* values in liquid and supercritical CO_2 (0.7 reduced density at 41°C). The experimental results were also used to calculate the $E_T(30)$ solvent polarity scale (19), which is similar to the π^* scale.*

The π^* values obtained in both liquid (−0.46) and supercritical CO_2 (−0.60) were much lower than that of liquid hexane (−0.08), whereas the $E_T(30)$ value (33.8 kcal/mol) compared well with those of simple aromatic hydrocarbons such as toluene (33.9 kcal/mol). Sigman et al. determined the π^* values for 10 different nitroaromatic dyes in supercritical CO_2 at several densities (20,21). For temperatures between 36°C and 42°C, the π^* values varied between −0.5 and −0.1 over the CO_2 density range ~0.4–0.86 g/mL (reduced density 0.87–1.87). These π^* values place the solvent strength of high-density supercritical CO_2 near that of liquid hexane (−0.08). The results also show that the solvent strength of supercritical CO_2 increases with increasing density. Hyatt's results for the infrared absorption spectral shifts of the C=O stretch of acetone and cyclohexanone and the N-H stretch of pyrrole in liquid and supercritical CO_2 are also consistent with the conclusion that supercritical CO_2 is near to liquid hexane in solvent strength (19).

A more detailed examination of the density dependence of the π^* values was performed by Yonker et al. and Smith et al. using primarily 2-nitroanisole as probe in sub- and supercritical CO_2, N_2O, $CClF_3$, NH_3, ethane, Xe, and

*The $E_T(30)$ solvent polarity scale is based on the spectral shift of a betaine dye (Riechardt dye) in a large number of solvents and correlates the dye's spectral shift to the ability of the solvent to stabilize the probe molecule via dielectric solute–solvent interactions (18). The $E_T(30)$ scale has found limited application in the investigation of supercritical fluids, mainly because of solubility issues.

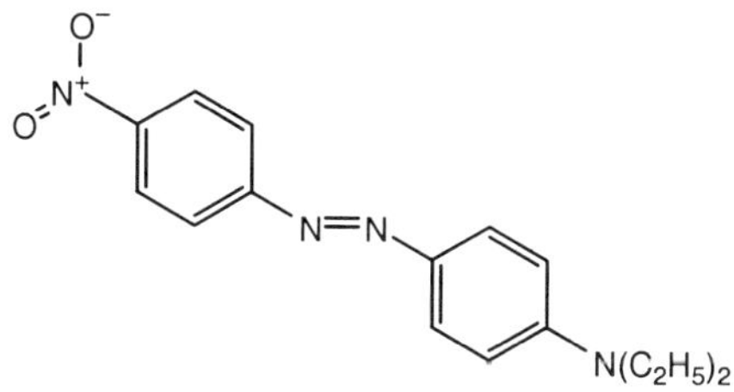

Diethyl-[4-(4-nitro-phenylazo)-phenyl]-amine

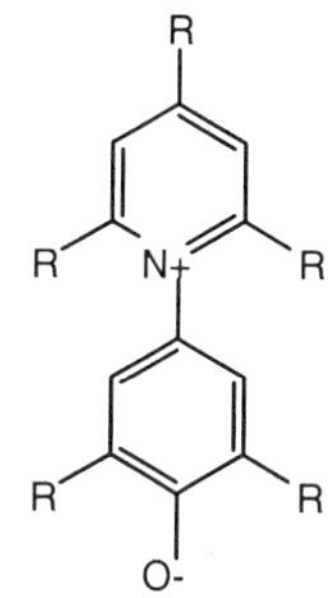

Reichardt's Dye
R = C_6H_5 or
tert-butylphenyl

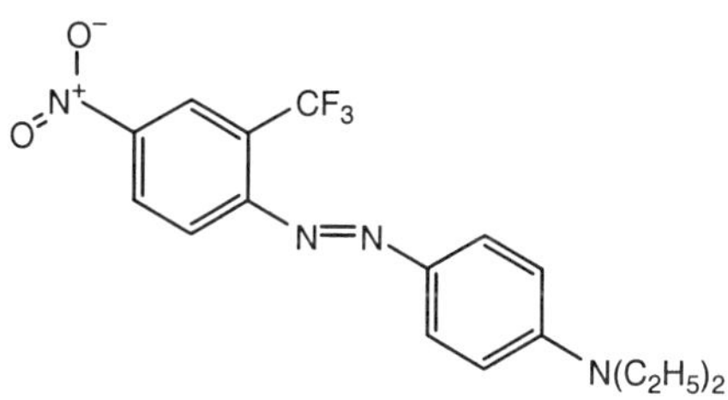

Diethyl-[4-(4-nitro-2-trifluoromethyl-phenylazo)-phenyl]-amine

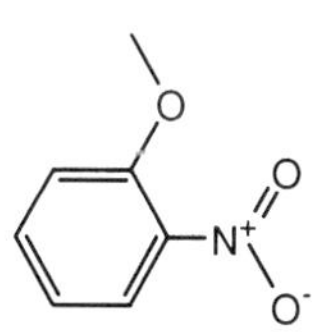

2-Nitroanisole

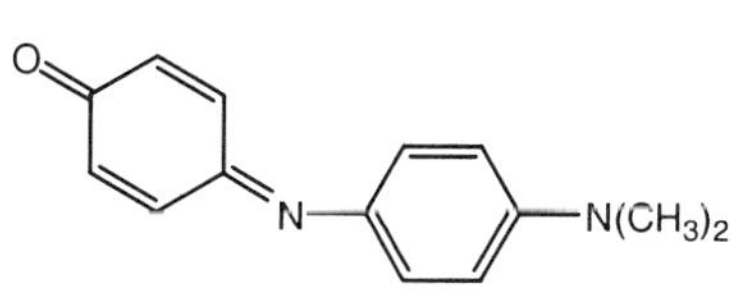

Phenol Blue

Structure 1

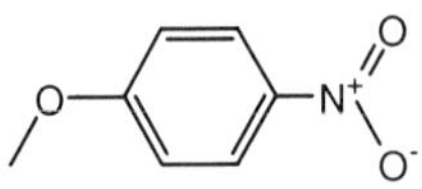

4-Nitroanisole

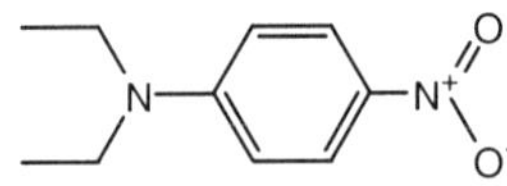

N,N-Diethyl-4-nitroaniline

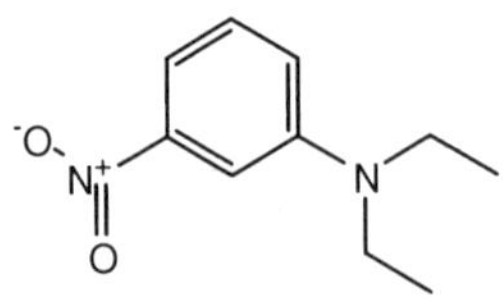

N,N-Diethyl-3-nitroaniline

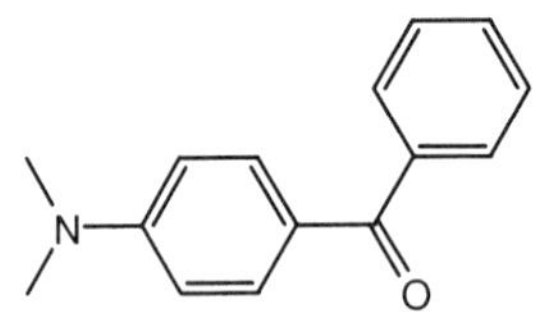

4-(Dimethylamino)benzophenone

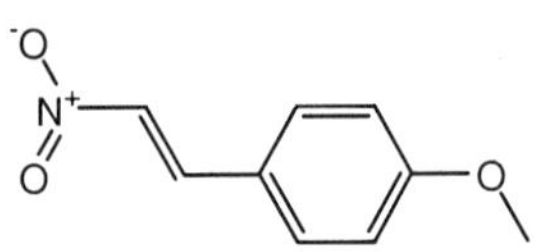

4-Methoxy-β-nitrostyrene

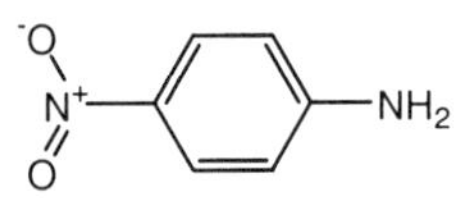

4-Nitroaniline

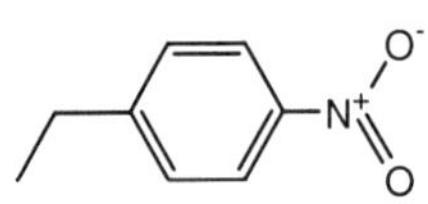

4-Ethyl-nitrobenzene

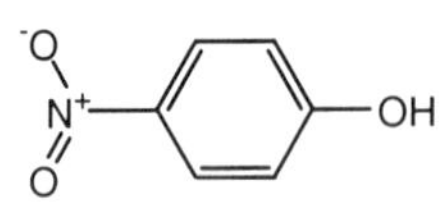

4-Nitrophenol

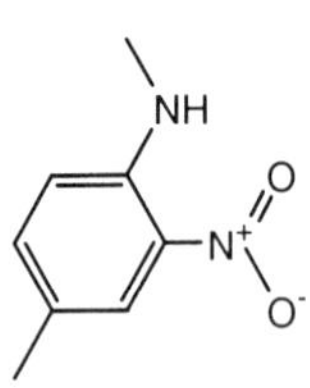

N-Methyl-2-nitro-p-toluidine

Structure 1 *(Continued)*

SF_6 (22–25). Under subcritical (liquid) conditions, a wide variation in π^* was found among the solvents: 0.8 (NH_3), 0.04 (CO_2), −0.03 (N_2O), −0.21 ($CClF_3$), −0.22 (ethane), and −0.36 (SF_6) (22,23). These π^* values correlate well with the Hildebrand solubility parameters of the solvents. The same variation in π^* was observed for the solvents under supercritical conditions when compared at a single reduced density (Figure 2) (22). For a given supercritical fluid, the π^* values were again found to increase with increasing fluid density; however, the solvent strength was clearly nonlinear with density, especially in the low-density region (Figure 2). This was particularly true for supercritical CO_2, ethane, and Xe, for which characteristic three-density-region solvation model behavior was observed. The apparent linear dependence of the π^* values on fluid density in supercritical NH_3 and SF_6 was attributed to specific solute–solvent interactions that represent the two extremes—unusually high polarity in NH_3 and a general lack of sensitivity due to the nonpolar nature of SF_6 (22).

Kim and Johnston made a similar observation of nonlinear density dependence for the shift in the absorption spectral maximum of phenol blue in

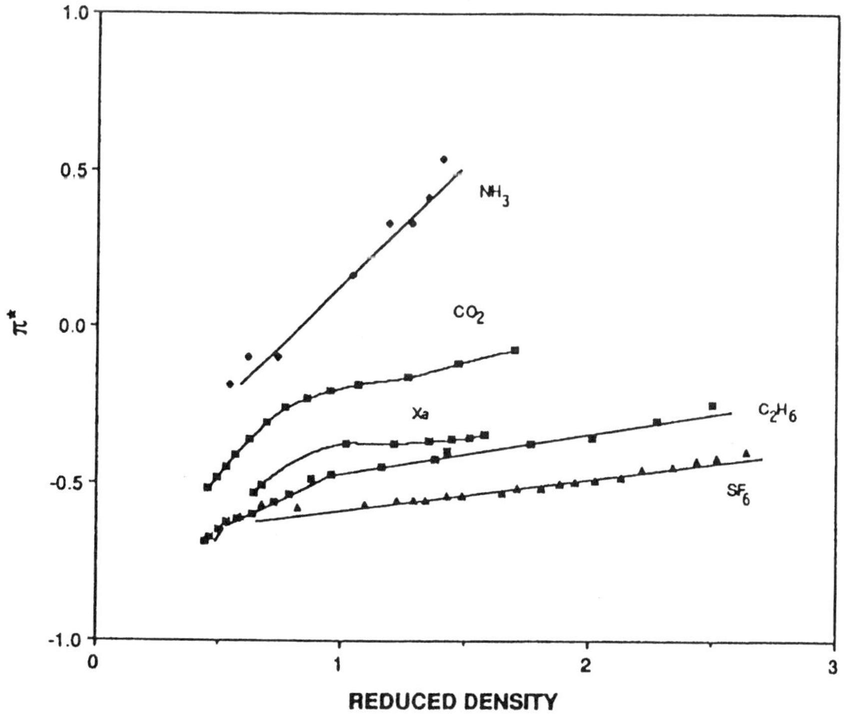

Figure 2 Plot of π^* vs. reduced density (ρ/ρ_c) for the five fluids. (From Ref. 22.)

supercritical ethylene, $CClF_3$, and fluoroform (26). Quantitatively, the stabilization of the photoexcited probe molecule in solution is linearly related to the intrinsic solvent strength, $E^0{}_T$.

$$E^0{}_T = A[(n^2 - 1)/(2n^2 + 1)] + B[(\varepsilon - 1)/(\varepsilon + 2) - (n^2 - 1)/(n^2 + 2)] + C \quad (1)$$

where A, B, and C are constants specific to the solvent, n is the solvent refractive index, and ε is the solvent dielectric constant. According to Kim and Johnston (26), the plot of the absorption spectral maximum of phenol blue vs. $E^0{}_T$ deviates from the linear relationship [Eq. (1)] in the near-critical density region; this deviation can be attributed to the clustering of solvent molecules about the solute probe (Figure 3).

A similar deviation was observed by Yonker et al. in the plot of π^* values as a function of the first term in Eq. (1), $(n^2 - 1)/(2n^2 + 1)$; the deviation was also discussed in terms of solute–solvent clustering (Figure 4) (23–25).

The use of similar molecular probes in various supercritical fluids has been reported (27–34), e.g., 9-(α-perfluoroheptyl-β,β-dicyanovinyl)julolidine dye for supercritical ethane, propane, and dimethyl ether (27); nile red dye for 1,1,1,2-tetrafluoroethane (28); 4-nitroanisole and 4-nitrophenol for ethane and fluorinated ethanes (29); 4-aminobenzophenone for fluoroform and CO_2 (30); phenol blue for CO_2, CHF_3, N_2O, and ethane (31); and coumarin-153 dye for CO_2,

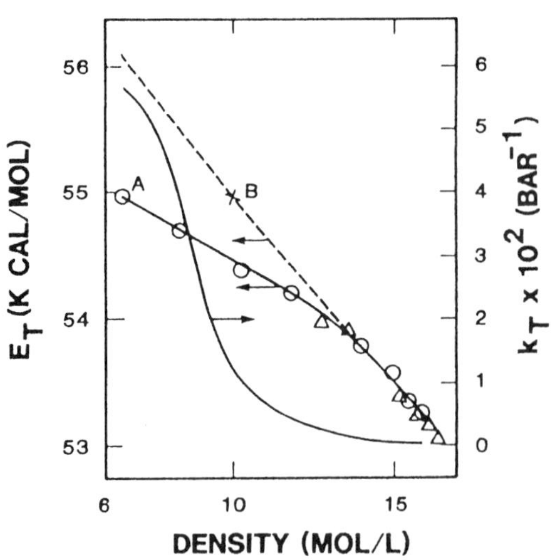

Figure 3 Transition energy (E_T) and isothermal compressibility vs. density for phenol blue in ethylene: (○) 25°C, (△) 10°C, (---) calculated $E^0{}_T$. (From Ref. 26.)

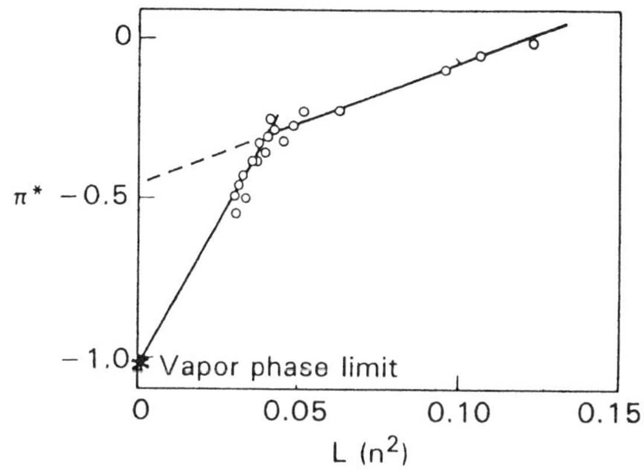

Figure 4 π* vs. Onsager reaction field function ($L(n^2)$) for CO_2 at 50°C. (From Ref. 23.)

fluoroform, and ethane (32,33). The results of these studies showed the characteristic density dependence of solvation in supercritical fluids, supporting the solute–solvent clustering concept.

B. Pyrene and the Py Scale

The molecular probe pyrene is commonly employed to elucidate solute solvent interactions in normal liquids (18,35). Because of the high molecular symmetry, the transition between the ground and the lowest excited singlet state is only weakly allowed, subject to strong solvation effects (36–39). As a result, in the fluorescence spectrum of pyrene the relative intensities of the first (I_1) and third (I_3) vibronic bands vary with changes in solvent polarity and polarizability. The ratio I_1/I_3 serves as a convenient solvation scale, often referred to as the Py solvent polarity scale. Py values for an extensive list of common liquid solvents have been tabulated (15,16).

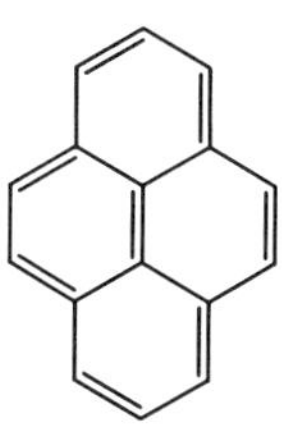

Pyrene

Structure 2

Several research groups have used pyrene as a fluorescent probe in the study of supercritical fluid properties (2,3,40–48). In particular, the density dependence of the Py scale has been examined systematically in a number of supercritical fluids such as CO_2 (2,3,40–43,45,46), ethylene (40,41,47), fluoroform (3,40,41,43,47), and CO_2-fluoroform mixtures (43). The Py values obtained in various supercritical fluids correlate well with the polarity or polarizability parameters of the fluids (3,40,41,43,47). For example, Brennecke et al. (40) found that the Py values obtained in fluoroform were consistently larger than those obtained in CO_2, which were, in turn, consistently larger than those found in ethylene over the entire density region examined. In addition, the Py values obtained in the liquid-like region (reduced density ~1.8) indicate that the local polarity of fluoroform is comparable to that of liquid methanol, CO_2 with xylenes, and ethane with simple aliphatic hydrocabons (15,16).

For the density dependence of solute–solvent interactions in supercritical fluids, the Py values were found to increase with increasing density in a nonlinear manner (2,3,40–43). For example, Sun et al. reported Py values in supercritical CO_2 over the reduced density (ρ_r) range 0.025–1.9 at 45°C (Figure 5) (2). At low densities ($\rho_r < 0.5$), the Py values are quite sensitive to density changes, increasing rapidly with increasing density. However, at higher densities, the Py values exhibit little variation with density over the ρ_r range ~0.5–~1.5, followed by slow increases with density at $\rho_r > 1.5$. The nonlinear density dependence was attributed to solvent clustering effects in the near-critical region of the

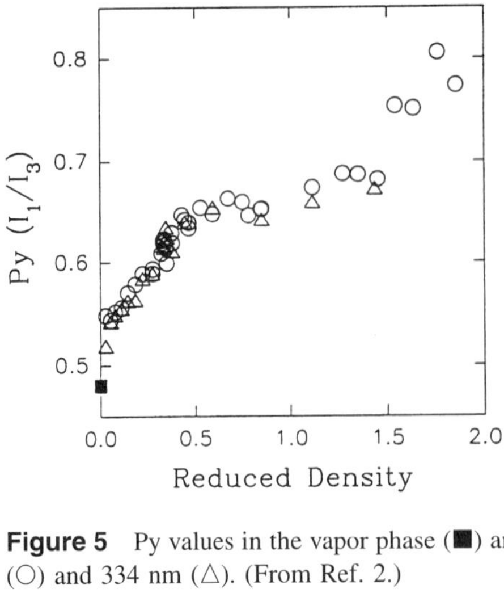

Figure 5 Py values in the vapor phase (■) and CO_2 at 45°C with excitation at 314 nm (○) and 334 nm (△). (From Ref. 2.)

supercritical fluid. Quantitatively, the clustering effects were evaluated using the dielectric cross-term $f(\varepsilon, n^2)$ (2):

$$f(\varepsilon, n^2) = [(\varepsilon - 1)/(2\varepsilon + 1)] * [(n^2 - 1)/(2n^2 + 1)] \quad (2)$$

Extrapolation of the data obtained in the liquid-like region to the gas-phase values confirmed that significant deviation of the experimental data from the prediction of Eq. (2) for the low-density region of supercritical CO_2 was occurring (Figure 6). The results are consistent with those obtained from investigations using other polarity-sensitive molecular probes. It appears that the largest deviation (or the maximum clustering effect) occurs at a reduced density of about 0.5 rather than at the critical density, as was naturally assumed (40,42,43).

The investigation of high-critical-temperature supercritical fluids is a more challenging task. One of the significant difficulties associated with these studies is probe-molecule thermal stability; many molecular probes commonly used with ambient supercritical fluids decompose at the temperatures required by these high-critical-temperature fluids. Fortunately, pyrene can be employed for such tasks. Several reports have been made of the use of pyrene as a molecular probe to investigate solute–solvent interactions in high-critical-temperature supercritical fluids (e.g., pentane, hexane, heptane, octane, cyclohexane, methcyclohexane, benzene, toluene, and water) (44,48,49). In supercritical hexane

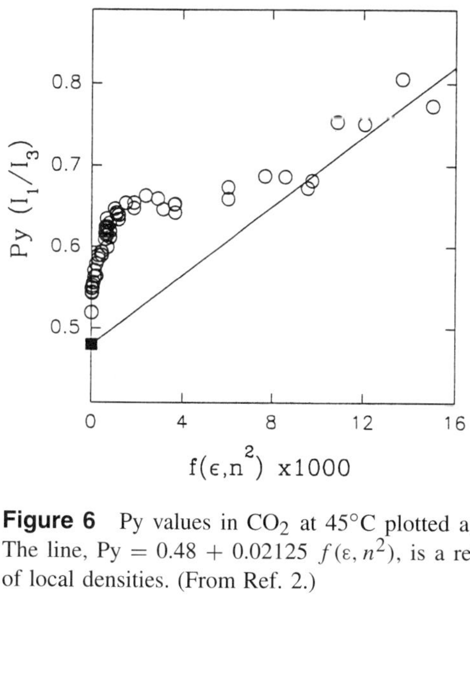

Figure 6 Py values in CO_2 at 45°C plotted against a dielectric cross term $f(\varepsilon, n^2)$. The line, Py = 0.48 + 0.02125 $f(\varepsilon, n^2)$, is a reference relationship for the calculation of local densities. (From Ref. 2.)

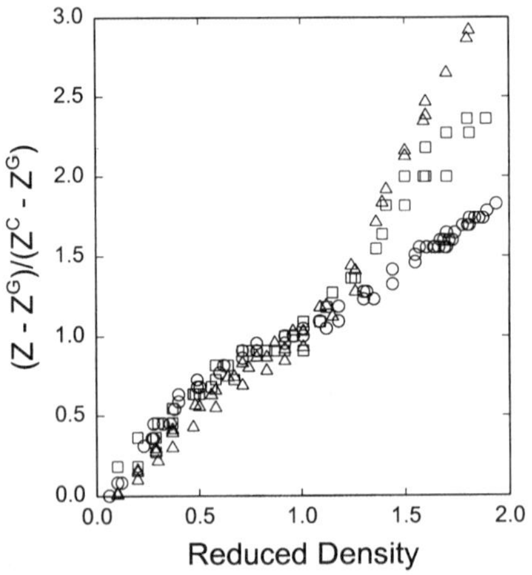

Figure 7 Pyrene fluorescence excitation spectral shifts (○), and hexane C-H Raman shifts (□) and Raman intensities (▽) in supercritical hexane at 245°C. The *y* axis represents normalized spectral responses, with Z^G being the spectral response obtained in the gas phase, Z^C the spectral response at the critical density, and Z the observed responses. (From Ref. 49).

the pyrene fluorescence spectrum is very broad, lacking the characteristic structural detail observed in the room-temperature spectrum (49). The fluorescence spectrum for low and high densities is essentially the same; however, the fluorescence excitation spectrum maintains its characteristic vibronic structure and displays a small but measurable red shift with increasing fluid density (49). A plot of the fluorescence excitation spectral maximum as a function of the reduced density of supercritical hexane (Figure 7) shows the same characteristic pattern observed for pyrene in supercritical CO_2 (2,3); and the results can be explained in terms of the three-density-region solvation model (1–3). It appears that even in the high-temperature supercritical fluids, solute–solvent clustering is prevalent. This is supported by results obtained from the investigation of supercritical hexane using Raman spectroscopy, where the spectral shifts and relative intensities of the C-H stretch transition of hexane were measured at different densities (Figure 7) (49).

C. TICT State Probes

Molecules that form a TICT state serve as excellent probes to elucidate solute–solvent interactions in condensed media (17). Upon photoexcitation, the excited-state processes of TICT molecules in polar solvents are characterized by a thermodynamic equilibrium between the locally excited (LE) singlet state and the TICT state (Figure 8) (50). Because of the two excited states, TICT molecules often exhibit dual fluorescence, with the fluorescence band due to the TICT state being extremely sensitive to solvent polarity. The spectral shifts of the TICT emission band can be used to establish a polarity scale similar to the Py and π^* scales.

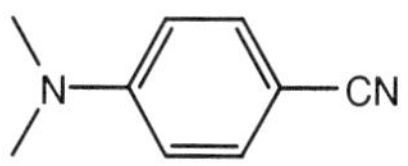

DMABN

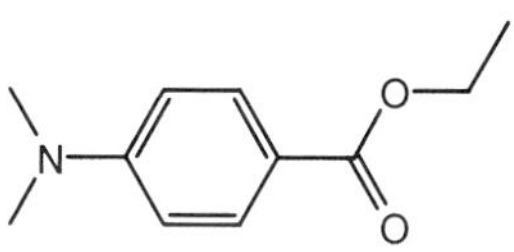

DMAEB

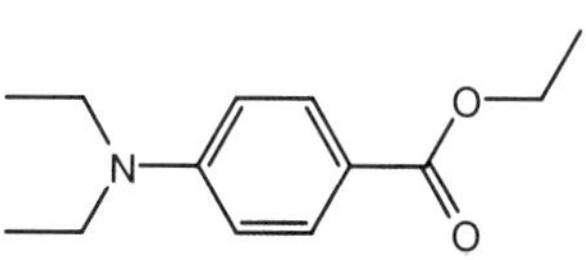

DEAEB

Structure 3

Kajimoto et al. used the classic TICT molecule *p*-(*N*,*N*-dimethylamino) benzonitrile (DMABN) to investigate solute–solvent interactions in supercritical fluoroform (51–54) and ethane (55). In fluoroform, the TICT emission was readily observed. The emission band shifted to the red with increasing fluoroform density. The shift was accompanied by an increase in the relative contribution of the TICT emission to the observed total fluorescence (Figure 9) (51). The solvent effects were evaluated by plotting the shift in the TICT band maximum as a function of the dielectric cross-term [Eq. (2)]:

$$P = [(\varepsilon - 1)/(\varepsilon + 2)] - [(n^2 - 1)/(n^2 + 2)] \qquad (3)$$

The shifts of the TICT band maximum in normal liquid solvents correlated well with those of P, confirming the linear relationship predicted by classical continuum theory. However, the results in supercritical fluoroform deviated

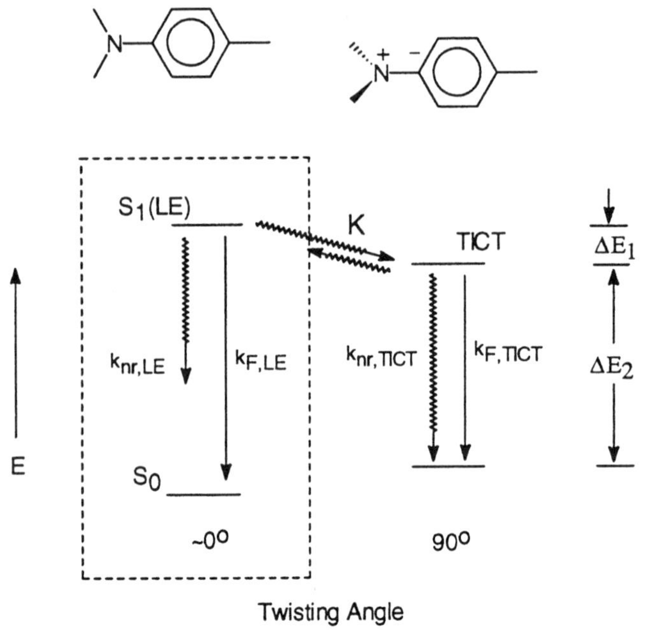

Figure 8 Energy diagram for the formation and decay of a TICT state in DEAEB and related molecules. The coordinate is for the twisting of amino-phenyl linkage. The diagram represents a mechanism in which fast and slow emission processes are considered. The fast process is restricted in the region surrounded by dashed lines. (From Ref. 50.)

significantly from the relationship, indicating that the effective polarities in supercritical fluoroform were significantly larger than expected (Figure 10) (51). According to Kajimoto et al. (51), the deviation may be attributed to unusual solute–solvent interactions (or solute–solvent clustering) in supercritical fluid solutions. From the results at low fluid densities, they were able to determine the number of solvent molecules about the solute using a simple model with solute–solvent interaction potentials (51,52,54,55).

Sun et al. carried out a more systematic investigation of the TICT molecules DMABN and ethyl *p*-(*N*,*N*-dimethylamino)benzoate (DMAEB) in supercritical fluoroform, CO_2, and ethane as a function of fluid density (1). They found that the absorption and TICT emission spectral maxima shifted to the red with increasing fluid density. The results were comparable to those reported by Kajimoto et al. (51–55). More importantly, the spectral shifts and the fractional contribution of the TICT state emission changed with fluid density following the characteristic three-density-region pattern (Figures 11 and 12) (1). In fact, these results furnished the impetus for the development of the three-density-region solvation model for solute–solvent interactions in supercritical fluid solutions (2,3).

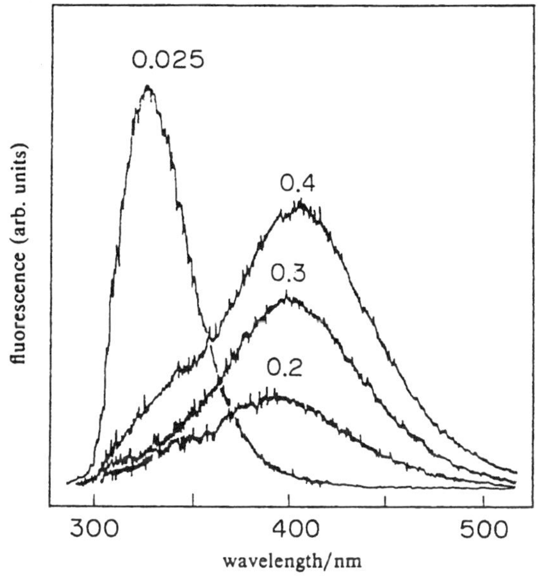

Figure 9 Dependence of the relative intensity of the CT emission of DMABN on the density of the supercritical solvent, CF_3H (in g/mL). (From Ref. 51.)

Another TICT molecule, ethyl *p*-(*N*,*N*-diethylamino)benzoate (DEAEB), was used to probe solute–solvent interactions in supercritical ethane, CO_2, and fluoroform (3,50,56). Unlike DMABN and DMAEB, DEAEB forms a TICT state even in nonpolar solvents (Figure 13) (50), resulting in dual fluorescence emissions. Because of the excited-state thermodynamic equilibrium, the relative intensities (or fluorescence quantum yields) of the LE-state (x_{LE}) and TICT-state (x_{TICT}) emissions may be correlated with the enthalpy (ΔH) and entropy (ΔS) differences between the two excited states:

$$K = (x_{TICT}/x_{LE})(k_{F,LE})/(k_{F,TICT}) \quad (4)$$

$$\ln(x_{TICT}/x_{LE}) = -\Delta H/RT + \Delta S/R + \ln[(k_{F,TICT})/(k_{F,LE})] \quad (5)$$

where $k_{F,LE}$ and $k_{F,TICT}$ are the radiative rate constants of the two excited states. If solvent effects on the entropy difference are assumed to be negligible, the relative contributions of the LE-state and TICT-state emissions are dependent primarily on the enthalpy difference ΔH. The energy gap between the two excited states is obviously dependent on solvent polarity because the highly polar TICT state is more favorably solvated than the LE state in a polar or polarizable solvent environment. Thus, $\ln(x_{TICT}/x_{LE})$ serves as a sensitive measure for the solvent-induced stabilization of the TICT state (Figure 14) (3). For DEAEB in the supercritical fluids (ethane in particular), the LE and TICT emission bands

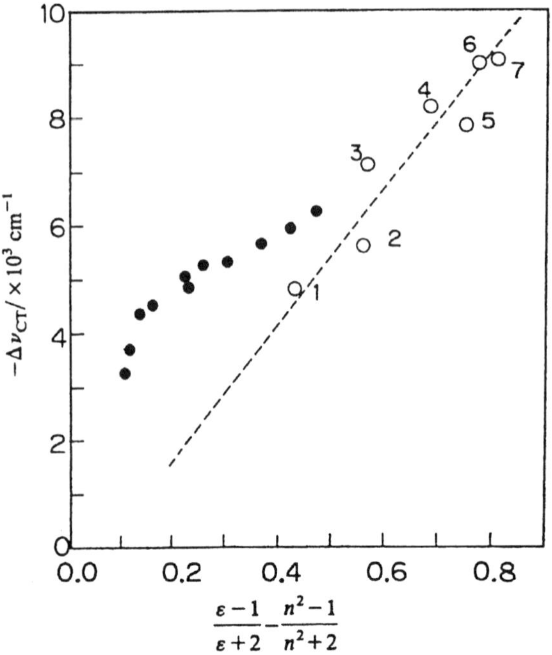

Figure 10 Shift of the maximum of the CT emission as a function of the polar parameter of the solvent. The open circle shows the data obtained in the liquid solvent: (1) bromobenzene, (2) *n*-butyl chloride, (3) THF, (4) butylnitrile, (5) cyclohexanol, (6) ethanol, and (7) methanol. The solid circles represent the results of the supercritical experiments. The polar parameters for the supercritical fluid were calculated based on the reported dielectric constants. (From Ref. 51.)

overlap significantly. A quantitative determination of the x_{TICT}/x_{LE} ratio as a function of the fluid density requires the separation of overlapping fluorescence spectral bands. In the work of Sun et al. (50,56), the spectral separation was aided by the application of a chemometric method known as principal component analysis—self-modeling spectral resolution (57–62). As shown in Figure 14, the plot of $\ln(x_{TICT}/x_{LE})$ as a function of reduced density in supercritical ethane again shows the characteristic three-density-region pattern, which validates the underlying concept of the three-density-region solvation model for solute–solvent interactions in supercritical fluid solutions.

Other investigations of supercritical fluid systems have been conducted using TICT and TICT-like molecules as probes. For example, DMABN and DMAEB were used to study solvation in two-component supercritical fluid mixtures (63). Another popular probe has been the highly fluorescent molecule

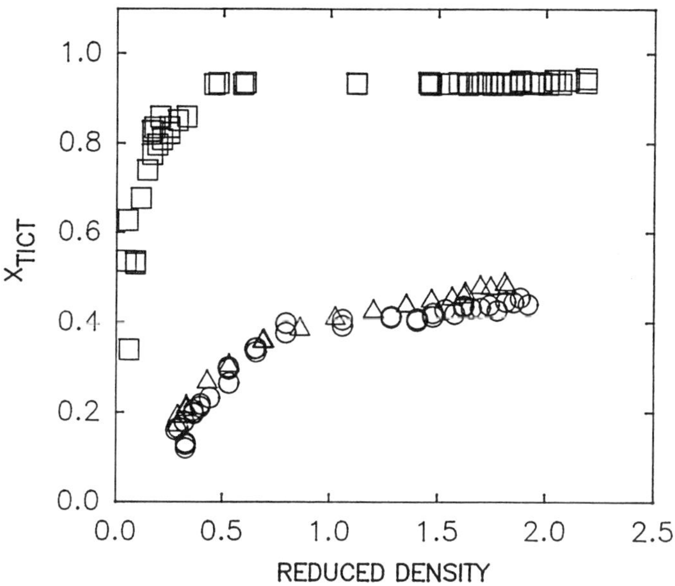

Figure 11 Bathochromic shifts of ν^{TICT}_{max} (relative to the LE band maximum in the absence of solvent, 330 nm) of DMAEB as a function of the reduced solvent density in CHF_3 at 28.0°C (□), in CO_2 at 33.8°C (○), and in CO_2 at 49.7°C (△). (From Ref. 1.)

6-propionyl-2-(dimethylamine)naphthalene (PRODAN). Although it shares the structural features of the TICT molecules discussed above, PRODAN apparently forms no TICT state upon photoexcitation; however, the fluorescence spectrum of PRODAN does undergo extreme solvatochromic shifts. The shifts also correlate well with those of the TICT emissions (Figure 15), implying that the emissive excited state of PRODAN is similar to a typical TICT state (64). The strong solvatochromism of PRODAN was the basis for its use in the study of solute–solvent interactions in supercritical CO_2 and fluoroform and other supercritical fluid systems (3,65). In addition, PRODAN was also used as probe for rotational reorientation in supercritical N_2O through fluorescence anisotropy measurements (66).

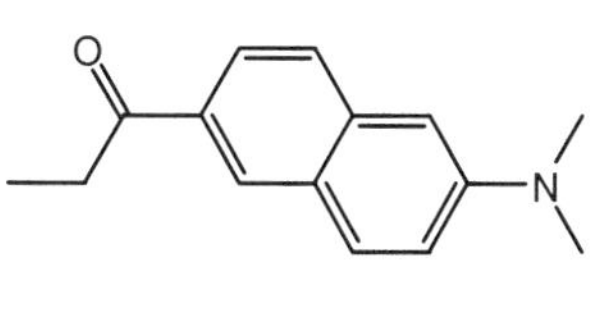

PRODAN

Structure 4

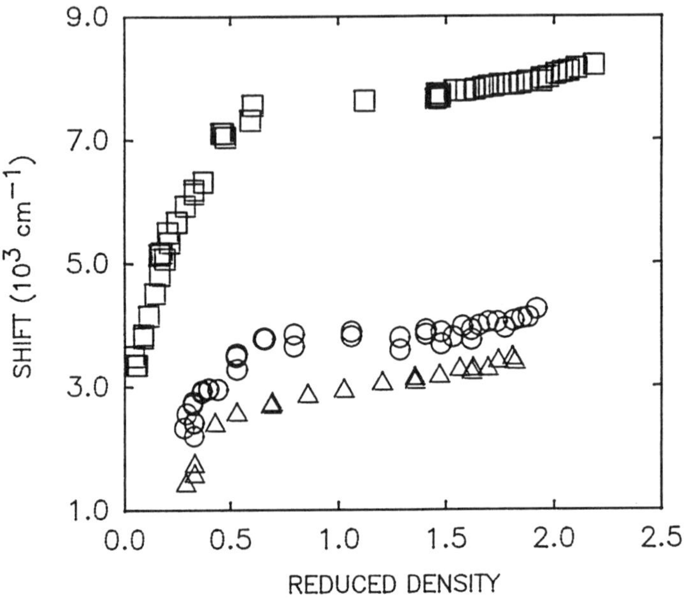

Figure 12 Fractional contribution of the TICT state emission of DMAEB as a function of the reduced solvent density in CHF_3 at 28.0°C (□), in CO_2 at 33.8°C (○), and in CO_2 at 49.7°C (△). (From Ref. 1.)

D. Other Systems and Methods

The π*, Py, and TICT solvation scales discussed above have been the basic techniques used in the investigation of solute–solvent interactions in supercritical fluid solutions. In addition, other methods have been applied for the same purpose, including the use of unimolecular reactions and vibrational spectroscopy and the probing of rotational diffusion; the results obtained have been important to the understanding of the fundamental properties of supercritical fluids.

1. Unimolecular Reactions

Unimolecular reactions that have been used to investigate the solvation properties of supercritical fluids include tautomeric reactions (67–71), rotational isomerization reactions (72–78), and radical reactions (79–83). O'Shea et al. used the tautomeric equilibrium of 4-(phenylazo)-1-naphthol to examine the solvent strength in supercritical ethane, CO_2, and fluoroform and to determine its dependence on density (67). The equilibrium is strongly shifted to the azo tautomer in supercritical ethane and the hydrazone tautomer in supercritical chloroform; and

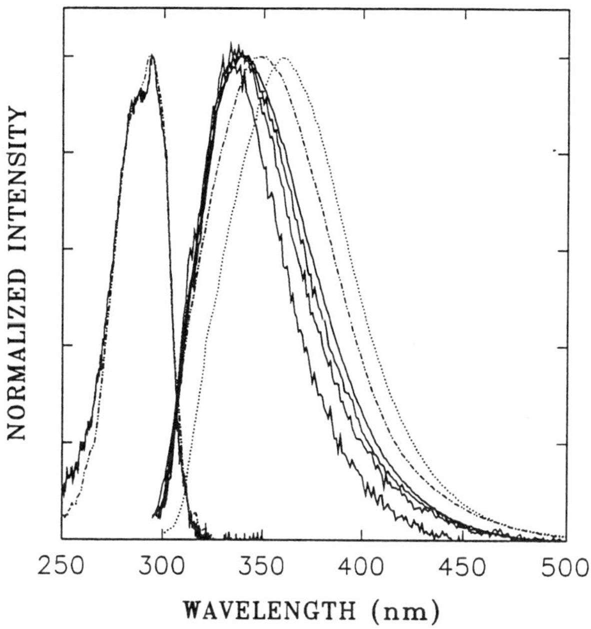

Figure 13 Absorption and fluorescence spectra of DEAEB in supercritical ethane (—) and CO_2 (-···-). Absorption in ethane: 580 psia and 53°C. Absorption in CO_2: 800 psia and 50°C. Fluorescence in ethane (in the order of increasing band width): the vapor phase, 340, 470, and 750 psia at 45°C. Fluorescence in CO_2: 600 psia and 50°C. The fluorescence spectrum in room-temperature hexane (···) is also shown for comparison. (From Ref. 50.)

the equilibrium is inert to density changes in both fluids. In supercritical CO_2 neither extreme applies; therefore, the equilibrium is strongly density dependent, favoring the azo tautomer at low densities and the hydrazone tautomer at high densities. Using the equilibrium between the azo and hydrazone tautomers as a solvation scale, the authors concluded that the solvent strength of supercritical CO_2 is similar to that of liquid benzene and that the solvent strength of supercritical fluoroform is similar to that of liquid chloroform. The results are consistent with the findings based on the π^* and Py scales. (See Scheme 1.)

Lee et al. investigated the photoisomerism of *trans*-stilbene in supercritical ethane to observe the so-called Kramer's turnover region where the solvent effects are in transition from collisional activation (solvent-promoting reaction) to viscosity-induced friction (solvent-hindering reaction) (76). In the experiments the Kramer's turnover was observed at the pressure of about 120 atm at 350 K. (See Scheme 2.)

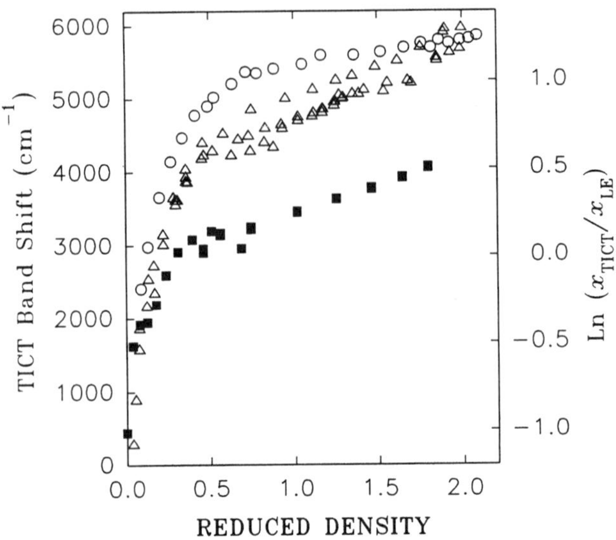

Figure 14 Solvatochromatic shifts of the TICT band maximum for DEAEB in supercritical CHF_3 at 35°C (○) and 50°C (△), and the relative contributions of the TICT and LE emissions, $\ln(x_{TICT}/x_{LE})$, for DEAEB in supercritical ethane at 50°C (■) as a function of reduced density. (From Ref. 3.)

Randolph and coworkers (79,80) used electron paramagnetic resonance (EPR) spectroscopy to determine the hyperfine splitting constants A_N for di-*t*-butylnitroxide radicals in supercritical ethane, CO_2, and fluoroform. Plots of A_N as a function of reduced density clearly revealed the three-density-region pattern. The solute–solvent clustering issue was evaluated using the $[(\varepsilon - 1)/(2\varepsilon + 1)]$ term as a measure of solvent polarity. Again, it was found that the maximum clustering effects occurred at a reduced density around 0.5.

2. Vibrational Spectroscopy

A number of investigations of supercritical fluids have been conducted using vibrational spectroscopy methods, including infrared absorption (19,84–89), Raman scattering (90–100), and time-resolved vibrational relaxation and collisional deactivation (101–112). The results of these investigations have significantly aided the understanding of solute–solvent interactions in supercritical fluid systems. For example, Hyatt used infrared absorption to examine the spectral shifts of the C=O stretch mode for acetone and cyclohexanone and those of the N-H stretch mode for pyrrole in liquid and supercritical CO_2 to determine the solvent strength of CO_2 relative to normal liquid solvents (19). Blitz et al.

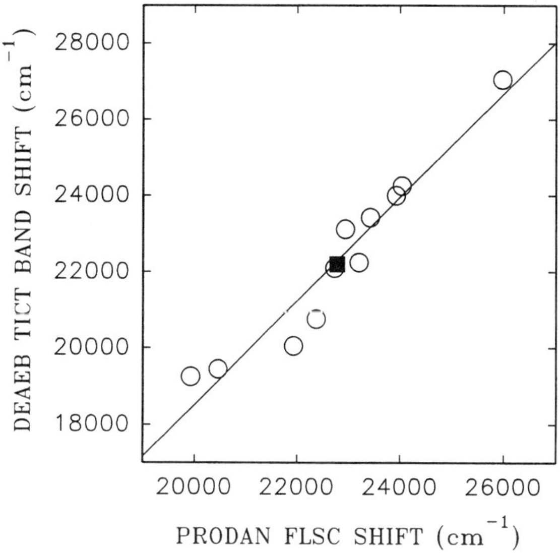

Figure 15 A plot of the DEAEB TICT band maxima vs. the PRODAN fluorescence spectral maxima in a series of room-temperature solutions. The result in CHF_3 at the reduced density of 2 and 35°C (■) follows the empirical linear relationship closely. (From Ref. 3.)

utilized infrared and near-infrared absorption to study CO_2 under supercritical conditions in both neat CO_2 and CO_2–cosolvent mixtures (84). For neat CO_2 at 50°C, plots of the frequency shifts and the absorption bandwidths as a function of fluid density were clearly nonlinear, similar to the plots made using data obtained with the π^* polarity probes (22–25). Ikushima et al. used

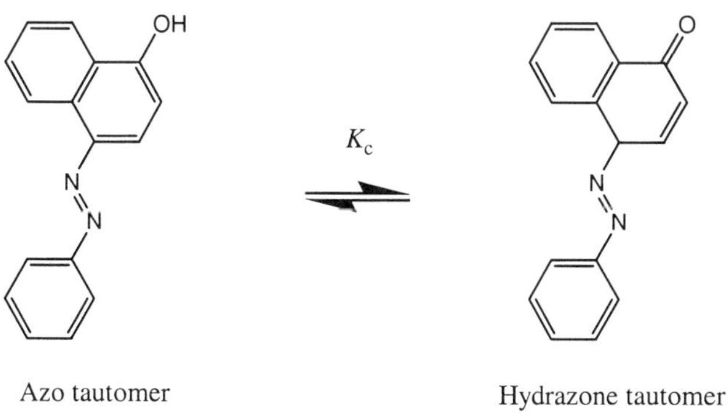

Scheme 1

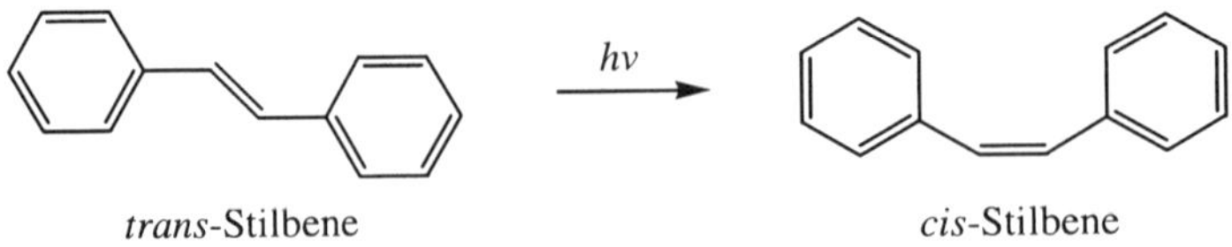

Scheme 2

frequency shifts of the C=O stretch mode in cyclohexanone, acetone, *N,N*-dimethylformamide, and methyl acetate to probe the solvent strength in supercritical CO_2 (85); Wada et al. used the molar absorptivity changes of the C-C ring stretch and the substituent deformation stretch in several substituted benzenes to study solvation effects in supercritical CO_2 (89). Both investigations yielded results that are characteristic of solute–solvent clustering. The results of Wada et al. again suggest that the maximum clustering effects occur at a reduced density of around 0.5 (89).

The collisional deactivation of vibrationally excited azulene was recently investigated in several supercritical fluids for a series of fluid densities (106, 108,109). Theoretically, the rate constant of collisional deactivation k_c should be proportional to the coverage of azulene by the collision (solvent) molecules, and thus k_c should be a function of the local solvent density in a supercritical fluid. A plot of k_c as a function of reduced density in propane shows the characteristic three-density-region solvation behavior (Figure 16). The results correlate well with the observed shifts in the absorption maximum of azulene under the same solvent conditions (106). Similarly, Fayer and coworkers (101–103) examined the vibrational relaxation of tungsten hexacarbonyl $W(CO)_6$ in supercritical ethane, CO_2, and fluoroform as a function of fluid density. Their results show that the lifetime of the T_{1u} asymmetric C=O stretch mode decreases with increasing fluid density in the characteristic three-density-region pattern. A concept similar to the solute–solvent clustering, "local phase transitions," was introduced by these authors to explain the experimental results. The results were also discussed in terms of a mechanistic scheme in which the competing thermodynamic forces may cancel out the density dependence of the lifetimes of the vibrational modes in the near-critical density region. However, the validity of such a scheme remains open to debate (113,114).

3. Rotational Diffusion

Another important topic in the study of supercritical fluids is viscosity effects. Several research groups have used well-established probes to examine the effect of viscosity on rotational diffusion in supercritical fluid systems.

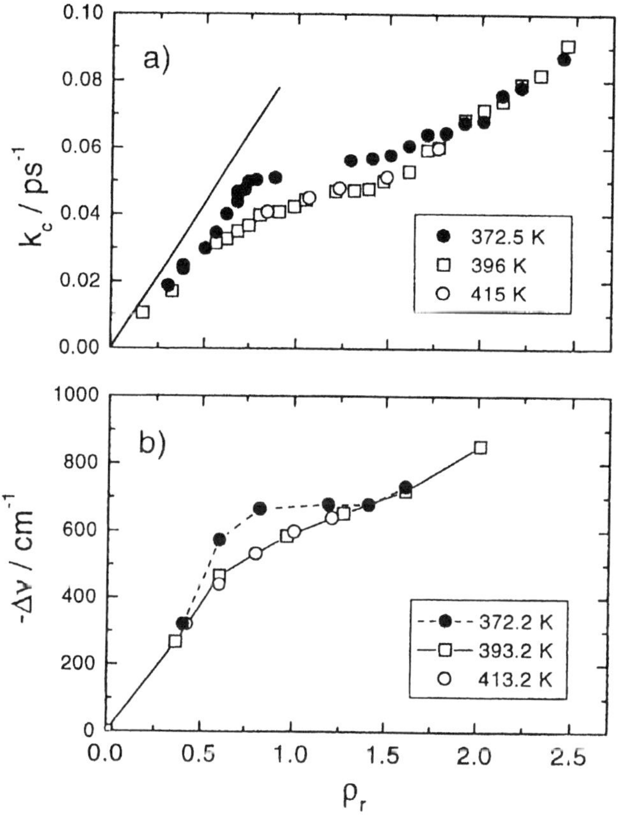

Figure 16 (a) Density dependence of collisional deactivation rate constants of azulene in propane at various temperatures (full line: extrapolation from dilute gas phase experiments). (b) Density dependence of the shift of the azulene $S_3 \leftarrow S_0$ absorption band in propane at various temperatures. (From Ref. 106.)

The time for rotational diffusion τ_{rot} can be related to the viscosity η using the modified Stokes–Debye–Einstein equation (115):

$$\tau_{rot} = (\eta V_p / k_B T) f C \tag{6}$$

where V_p is the volume of the probe molecule, k_B is the Boltzmann constant, T is the temperature in K, and f and C are correction factors. The factor f corrects for the shape of the probe molecule, whereas the factor C takes into account variations in hydrodynamic boundary conditions. In the absence of these corrections (both factors being unity), the rotational diffusion time τ_{rot} is linearly dependent on the viscosity (115). Experimentally, rotational diffusion times of the probes in supercritical fluids have been determined via various spectroscopic techniques, including infrared absorption and Raman scattering (116–125), NMR (126–133), fluorescence depolarization (66,115,134,135), and EPR (136). For example, Betts et al. used the fluorescence depolarization method to obtain rotation reorientation times of PRODAN in supercritical N_2O (66). The results show that, contrary to the behavior predicted by Eq. (6), τ_{rot} actually increases with decreasing pressure and density (lower bulk viscosity of the fluid). As unusual as it seems, the observation that rotation reorientation times increase with decreasing density in supercritical fluids has been reported in other investigations. Heitz and Bright (135) reported similar behavior for the rotational diffusion of *N*,*N*′-bis-(2,5-*tert*-butylphenyl)-3,4,9,10-perylenecarboxodiimide (BTBP) in supercritical ethane, CO_2, and fluoroform; and deGrazia and Randolph (136) made similar

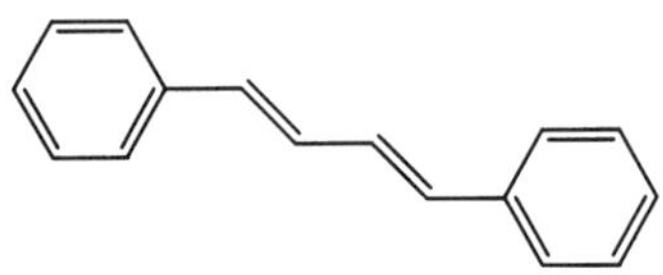

DPB

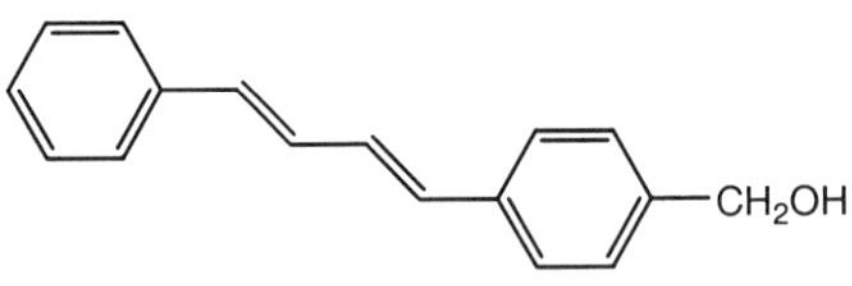

HMS

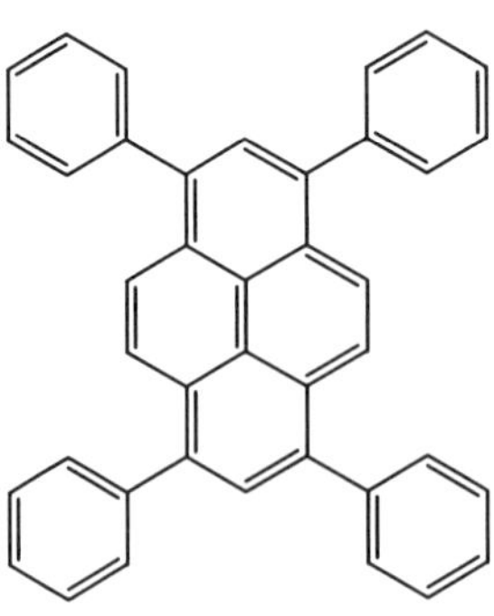

TPP

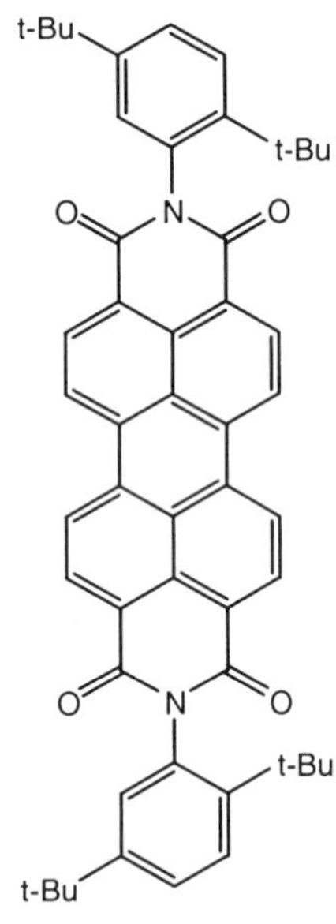

BTBP

Structure 5

observations in their EPR (electron paramagnetic resonance) study of copper 2,2,3-trimethyl-6,6,7,7,8,8,8-heptafluoro-3,5-octanedionate in supercritical CO_2. These rotational diffusion results are somewhat controversial, partially due to the fact that the probes involved are complicated and subject to other effects beyond viscosity-controlled rotational diffusion. deGrazia and Randolph suggested that solute–solute interactions might be responsible for the anomalous density dependence of τ_{rot} in supercritical CO_2 (136). Heitz and Maroncelli (115) repeated the rotational reorientation study of BTBP in supercritical CO_2 and also added two more probes, 1,2,6,8-tetraphenylpyrene (TPP) and 9,10-bis(phenylethynyl)anthracene (PEA). They found that for all three probes, the τ_{rot} values actually increase with increasing fluid density (115). More quantitatively, the PEA results clearly deviate from the prediction of Eq. (6). The deviations were discussed in terms of significant solute–solvent clustering in the near-critical density region, namely, that local solvent density augmentation results in locally enhanced viscosities. Anderton and Kauffman (134) studied the rotational diffusion of *trans,trans*-1,4-diphenylbutadiene (DPB) and *trans*-4-(hydroxymethyl)stilbene (HMS) in supercritical CO_2 and found that the τ_{rot} values increase with increasing fluid density for both probes. The debate concerning the density dependence of rotational diffusion in supercritical fluids is likely to continue.

E. The Three-Density-Region Solvation Model

The wealth of data characterizing solute–solvent interactions in supercritical fluids show a surprisingly characteristic pattern for the density dependence. Even more incredible is the fact that the same density dependence pattern has been observed in virtually all supercritical fluids (from nonpolar to polar and from ambient to high temperature) with the use of numerous molecular probes that are based on drastically different mechanisms. These results suggest that three distinct density regions are present in a supercritical fluid: gas-like, near-critical, and liquid-like. The density dependence of the molecular probe response in a supercritical fluid differs in each of the three density regions (Figure 1): strong in the gas-like region, increasing significantly with increasing density; plateau-like in the near-critical density region, beginning at $\rho_r \sim 0.5$ and extending to $\rho_r \sim 1.5$; and again increasing in the liquid-like region, in the manner predicted by the dielectric continuum theory.

To account for the characteristic density dependence of the spectroscopic (and other) responses in supercritical fluids, a three-density-region solvation model was proposed, reflecting the different solute–solvent interactions in three distinct density regions (Figure 17) (1–3). According to the model, the three density-region solvation behavior in supercritical fluid solutions is determined

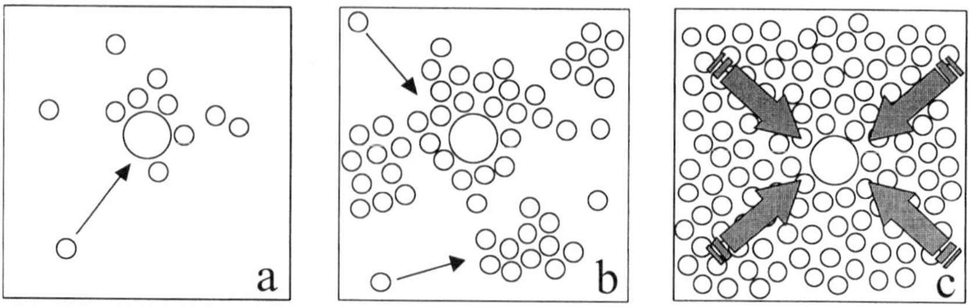

Figure 17 Cartoon representation of the empirical three-density-region solvation model depicting molecular level interactions for the three density regions: (a) low-density region; (b) near-critical density region; (c) liquid-like region.

primarily by the intrinsic properties of the neat fluid over the three density regions.

The behavior in the gas-like region at low densities is probably dictated by short-range interactions in the inner solvation shell of the probe molecule. The strong density dependence of the spectroscopic and other responses is probably associated with a process of saturating the inner solvation shell. Before saturation of the shell, microscopically the consequence of increasing the fluid density is the addition of solvent molecules to the inner solvation shell of the probe, which produces large incremental effects (Figure 17a). In the near-critical region, where the responses are nearly independent of changes in density, the microscopic solvation environment of the solute probe undergoes only minor changes. Such behavior is probably due to the microscopic inhomogeneity of the near-critical fluid—a property sheared by all supercritical fluids. As discussed in the introduction, a supercritical fluid may be considered macroscopically homogeneous (remaining one phase regardless of pressure) but microscopically inhomogeneous, especially in the near-critical density region. Although the solvent environment is highly dynamic, on the average the fluid in the near-critical region can be viewed as consisting of solvent clusters and free volumes that possess liquid-like and gas-like properties, respectively. Changes in bulk density through compression primarily correspond to decreases in the free volumes, leaving solute–solvent interactions in the solvent clusters largely unaffected (Figure 17b). This explains the insensitivity of the responses of the probe molecules to changes in bulk density in the near-critical region. Above a reduced density of about 1.5, the free volumes become less significant (consumed), and additional increases in bulk density again affect the microscopic solvation environment of the probe. The solvation in the liquid-like region at high densities should be similar to that in a compressed normal liquid solvent (Figure 17c).

The three-density-region solvation model provides a generalized view of the solvation behavior in supercritical fluid solutions, providing a qualitative but global explanation of the available experimental results; however, a theoretical basis for the model remains to be explored and established.

III. SOLUTE–SOLUTE INTERACTIONS

An important topic in supercritical fluid research is the effect of solvent local density augmentation on solute solute interactions in a supercritical fluid solution. The most important question seems to be whether the supercritical solvent environment facilitates solute-solute clustering, which may be loosely defined as local solute concentrations that are greater than the bulk solute concentration. Unlike solute–solvent clustering discussed in the previous section, solute-solute clustering in supercritical fluid solutions is a more complex and somewhat controversial issue. Following is a summary of the available experimental results and a review of the various explanations and mechanistic proposals on the clustering of solute molecules in supercritical fluid solutions.

A. Entrainer Effect in Mixtures

In early investigations of supercritical fluid extraction and chromatography, it was found that the addition of a small quantity of a polar cosolvent could dramatically improve the solubility of organic analytes in a nonpolar supercritical fluid, such as CO_2. This is commonly referred to as the *entrainer effect* in supercritical fluid mixtures. In many studies attempts have been made to quantify the entrainer effect. For example, Dobbs and coworkers examined the solubility of phenanthrene, hexamethylbenzene, and benzoic acid in supercritical CO_2 mixtures with simple alkanes (pentane, octane, and undecane) as cosolvent (137). Solubility enhancements of up to 3.6 times the solubility in neat CO_2 were observed in mixtures containing 3.5 mol % cosolvent. The enhancements were found to increase with cosolvent concentration over the range 3.5–7.0 mol % and with increasing chain length (and polarizability) of the cosolvent; however, no differences were observed in the solutes, with all exhibiting similar levels of enhancement (137). On the other hand, the addition of polar cosolvents led to solubility enhancements that were solute specific, with more dramatic solubility increases for polar solutes. As an example, the addition of methanol to supercritical CO_2 (3.5 mol %) resulted in a solubility enhancement of 6.2 times for 2-aminobenzoic acid, although no effect on the solubility of hexamethylbenzene was observed (138). The ability to selectively enhance the solubility of polar solutes (over that of nonpolar solutes) in supercritical fluid–cosolvent mixtures was further demonstrated for several quaternary systems, each consisting of a supercritical fluid, a cosolvent, and a nonpolar and a polar solute (139).

Mechanistically, the entrainer effect has been explained in terms of a higher than bulk population of the cosolvent molecules in the vicinity of the solute molecule. It may be argued that the "clustering" of cosolvent molecules about a solute is a consequence of the local density augmentation in supercritical fluid solutions and that the observation of the entrainer effect is a precursor to the solute-solute clustering concept. Specific solute–cosolvent interactions such as hydrogen bonding may also play a significant role in the observed entrainer effect in some systems (140).

Several investigations of supercritical fluid–cosolvent systems have focused on the effects of hydrogen bonding and the role of specific intermolecular interactions in solubility enhancements. Walsh et al. used infrared absorption results to show that the entrainer effect in supercritical fluid–cosolvent mixtures is due to various types of hydrogen bonding interactions (141–143). Infrared absorption spectra have also been employed to estimate the extent of hydrogen bonding between solutes such as benzoic acid and salicylic acid and alcohol cosolvent molecules in supercritical CO_2 (144). Bennet et al. used a supercritical fluid chromatography technique to determine the solubilities of 17 solutes in three supercritical fluids (ethane, CO_2, and fluoroform) with eight cosolvents (145). Their results showed that solubility enhancements are present in the supercritical fluid–cosolvent mixtures and that the enhancements become more significant at higher densities. More quantitatively, the solubility enhancement observed for anthracene in an ethane-ethanol mixture was predominantly due to the change in density that occurs on going from the neat fluid to the mixture. However, for carbazole and 2-naphthol in the same mixture, the solubility enhancements were considerably higher than those predicted on the basis of the density change, suggesting the involvement of specific intermolecular interactions (145). Ting et al. investigated the solubility of naproxen [(*S*)-6-methoxy-α-methyl-2-naphthaleneacetic acid] in supercritical CO_2–cosolvent mixtures (six different polar cosolvents at concentrations up to 5.25 mol %) at different temperatures (146). The solubility enhancements differ for the various cosolvents—in the order of increasing enhancement, ethyl acetate, acetone, methanol, ethanol, 2-propanol, and 1-propanol. For example, the solubility of naproxen in the supercritical CO_2-1-propanol (5.25 mol %) mixture at 125 bar and 333.1 K is about 50 times higher than that in neat CO_2 under the same conditions (146). It was estimated that the density increase from neat CO_2 to the mixtures could account for 30–70% of the observed solubility enhancements at low cosolvent concentrations (1.75 mol %) but be less significant at higher cosolvent concentrations. It was suggested that the observed solubility enhancements in the supercritical CO_2–cosolvent mixtures were consistent with a solute-solute clustering mechanism and were also strongly influenced by hydrogen bonding interactions

(146,147). Foster and coworkers measured the solubility of hydroxybenzoic acid in supercritical CO_2 with 3.5 mol % methanol or acetone as a cosolvent and found enhancements that were beyond the effects of the density increases from neat CO_2 to the mixtures (148). They attributed the solubility enhancements to a higher local concentration of cosolvent molecules around the solute and even estimated the local mixture compositions in terms of the experimental solubility data.

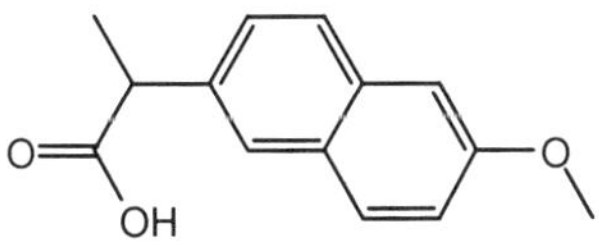

Naproxen

Structure 6

Molecular spectroscopy methods have also been applied to the study of the entrainer effect in supercritical fluid–cosolvent mixtures. Again, the molecular probes employed for absorption and fluorescence measurements include the Kamlet–Taft π^* polarity/polarizability scale probes (13,14), pyrene (15,16), and TICT molecules (17).

Kim and Johnston used phenol blue as a probe to investigate the local compositions of octane, acetone, ethanol, and methanol in CO_2 at 35°C over the entire bulk composition range (mole fraction from 0 to 1) (149). Their results show that the cosolvent local concentrations calculated on the basis of absorption spectral shifts are higher than the corresponding bulk concentrations over the entire mixture composition range. In addition, the local concentration enhancement is more significant at low cosolvent mole fractions, although the absolute local concentration increases with increasing bulk concentration of the cosolvent.

Nitroanisoles have achieved popularity as probes in the study of supercritical fluid–cosolvent mixtures (150–153). For example, Yonker and Smith used 2-nitroanisole to determine local concentrations of the cosolvent 2-propanol in supercritical CO_2 at different temperatures (150). Their results are similar to those of Kim and Johnston (149); the difference between the local and bulk cosolvent concentrations is more significant at low pressures and decreases with increasing pressure, approaching the bulk concentration at high pressures (Figure 18) (150). Also, results obtained in supercritical CO_2 with methanol and tetrahydrofuran (THF) as cosolvents are similar (151,152). Eckert and coworkers investigated supercritical ethane with several cosolvents using the solvatochromatic shifts of 4-nitroanisole and 4-nitrophenol (153). When the cosolvent is basic, the spectral shifts of 4-nitrophenol are larger than those of 4-nitroanisole

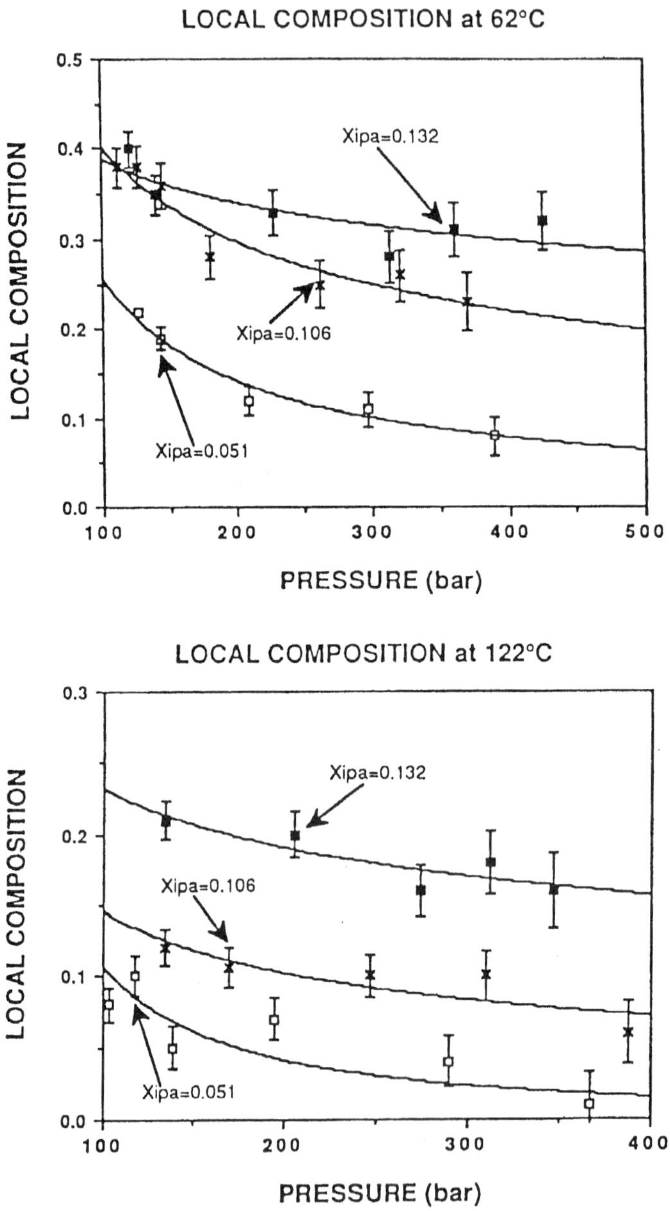

Figure 18 Local composition vs. pressure for constant temperature at 62°C and 122°C at (□) 0.051, (×) 0.106, and (■) 0.132 bulk mole fraction compositions. (From Ref. 150.)

because 4-nitrophenol can participate in hydrogen bonding. In addition, for 4-nitrophenol in the supercritical ethane–basic cosolvent mixtures, the spectral shifts correlate well with the Kamlet–Taft solvent basicity parameters (153).

Many other probes have been used to study supercritical fluid–cosolvent mixtures, including the charge transfer complexes Fe^{II}(1,10-phenanthroline)$_3{}^{2+}$ and Fe^{III}(2,4-pentadionate)$_3$ (for CO_2-methanol mixtures) (154), Nile red dye (for Freon-13, Freon-23, and CO_2 with the cosolvents methanol, THF, acetonitrile, and dichloromethane) (155), benzophenone (for ethane with the cosolvents 2,2,2-trifluoroethanol, ethanol, chloroform, propionitrile, 1,2-dibromoethane, and 1,1,1-trichloroethane) (156), 4-amino-*N*-methylphthalimide (for CO_2–2-propanol mixtures) (157), and other molecular probes such as 2-naphthol, 5-cyano-2-naphthol, and 7-azaindole for a variety of supercritical fluid–cosolvent mixtures (158,159).

As expected, pyrene has also been used to characterize supercritical fluid–cosolvent mixtures. For example, Zagrobelny and Bright used the Py polarity scale and pyrene excimer formation to study supercritical CO_2–methanol and CO_2–acetonitrile mixtures (160). Their results suggest the clustering of cosolvent molecules around pyrene. Similarly, Brennecke and coworkers measured Py values in CO_2, CHF_3, and CO_2-CHF_3 mixtures (43).

TICT molecules are also excellent probes for the study of supercritical fluid–cosolvent mixtures. Sun et al. carried out a systematic investigation of supercritical CO_2-CHF_3 mixtures using DMABN and DMAEB as probes (63,161). In their experiments, shifts of the LE and TICT emission bands and TICT emission fractional contributions were determined for the probe molecules in the neat fluids and mixtures of various CHF_3 compositions (6% and 11%). The data indicate that the solute is preferentially solvated by the polar component CHF_3 in the mixtures. The preferential solvation can be observed for pyrene in the same supercritical fluid mixtures, according to Brennecke and coworkers (43). The results of Sun et al. also suggest that the local composition effect is more significant at lower reduced densities (161). In another experiment, DMABN was used by Sun and Fox to determine the microscopic solvation effects in CO_2-THF and CHF_3-hexane mixtures (162). Schulte and Kauffman have also used TICT molecules [bis(aminophenyl)sulfone and bis(4,4′-dimethylaminophenyl)sulfone] to characterize supercritical CO_2-ethanol mixtures (163,164). Their results, based on the shifts of the LE and TICT emission bands, suggest that the local ethanol concentrations are an order of magnitude higher than the bulk concentrations.

Dillow et al. investigated the tautomeric equilibrium of the Schiff base 4-(methoxy)-1-(*N*-phenylforminidoyl)-2-naphthol in supercritical ethane with acetone, chloroform, dimethylacetamide, ethanol, 2,2,2-trifluoroethanol, and 1,1,1,3,3,3-hexafluoro-2-propanol as cosolvents (165). Their results show that the polar cosolvents acetone, chloroform, and dimethylacetamide have little effect on the keto-enol equilibrium but that the cosolvents capable of hydrogen bonding

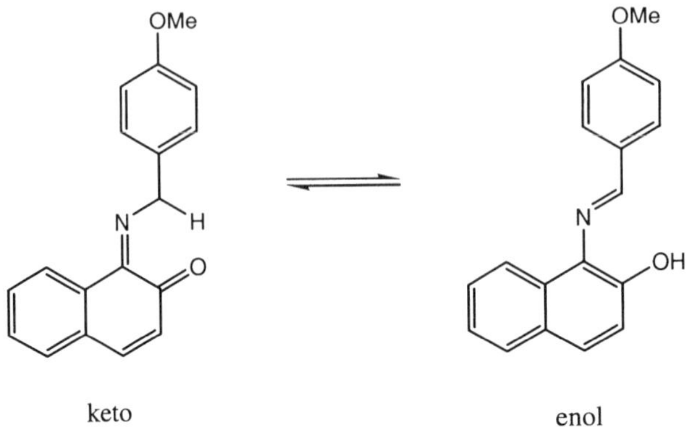

Scheme 3 4-(Methoxy)-1-(*N*-phenylforminidoyl)-2-naphthol.

shift the equilibrium toward the keto tautomer. For ethanol and trifluoroethanol as cosolvents, the equilibrium was found to shift back toward the enol form with increasing density. It was also found that the position of the keto-enol equilibrium in the near-critical region of the solvent was more toward the keto form than what would be predicted on the basis of the bulk cosolvent concentration. It was concluded that the clustering of cosolvent molecules about the Schiff base was responsible for these results. (See Scheme 3.)

B. Bimolecular Reactions

Studies of the entrainer effect discussed above demonstrate that the solute in supercritical fluid–cosolvent mixtures is, in many cases, surrounded preferentially by the cosolvent molecules. Since the cosolvent may be regarded as a second solute, the solute–cosolvent clustering may be considered as a special case of solute-solute clustering. An important consequence of the entrainer effect is enhancement in solute–cosolvent interactions or reactions. Similarly, solute-solute clustering in supercritical fluid solutions may enhance bimolecular reactions between the solute molecules. Extensive investigation of the solute-solute clustering phenomenon by many research groups has been prompted by the prospect of being able to influence bimolecular interactions and reactions under supercritical fluid conditions and, as a result, increase reaction yields and alter product distributions. Spectroscopic and other instrumental techniques combined with molecular probes that undergo well-characterized bimolecular processes or reactions (such as the formation of an excimer or exciplex, photodimerization, and fluorescence quenching) have been used.

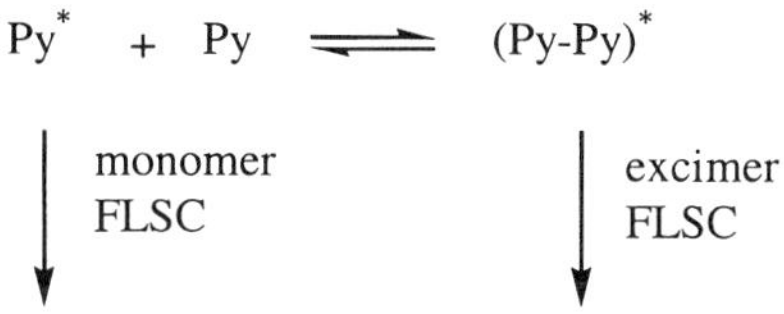

Scheme 4

1. Excimer and Exciplex

Formation of pyrene excimer (a complex between a photoexcited and a ground-state pyrene molecule; Scheme 4) is an extensively characterized and well-understood bimolecular process (35). Because the process is known to be diffusion controlled in normal liquid solutions, it serves as a relatively simple model system for studying solvent effects on bimolecular reactions. In fact, it has been widely employed in the probing of the solute-solute clustering in supercritical fluid solutions (40–42,46,47,160,166–168). (See Scheme 4.)

Eckert's group was the first to report pyrene-excimer formation in supercritical fluids at pyrene concentrations significantly below those required in normal liquid solutions (Figure 19) (40,41). Taking into account the difference in viscosity and molecular diffusion in supercritical CO_2 (150 bar and 35°C) as opposed to normal liquid cyclohexane, they concluded that the observed yield for excimer formation in CO_2 exceeded what might be expected from the higher

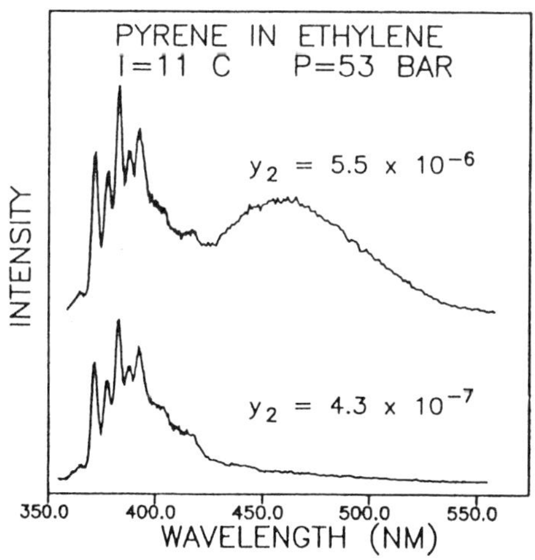

Figure 19 Excimer formation in dilute supercritical fluid solutions. (From Ref. 40.)

diffusivity. Thus, enhanced solute–solute interactions in a supercritical fluid became a possibility. According to Eckert, Brennecke, and coworkers (40,41,166), similarly efficient pyrene-excimer formation takes place in nonpolar and polar supercritical fluids such as ethylene and fluoroform.

Bright and coworkers investigated pyrene-excimer formation in supercritical fluids from a somewhat different angle using not only steady-state but also time-resolved fluorescence techniques (47,167). They measured fluorescence lifetimes of the pyrene monomer and excimer at a pyrene concentration of 100 μM in supercritical ethane, CO_2, and fluoroform at reduced densities higher than 0.8. Since the kinetics for pyrene-excimer formation was found to be diffusion controlled in ethane and CO_2 and less than diffusion controlled in fluoroform, they concluded that there was no evidence for enhanced pyrene–pyrene interactions in supercritical fluids. The less efficient excimer formation in fluoroform was discussed in terms of the influence of solute–solvent clustering on excimer lifetime and stability. Experimentally, their fluorescence measurements were influenced by extreme inner-filter (self-absorption) effects due to the high pyrene concentration in the supercritical fluid solutions (35).

Sun and Bunker performed a more quantitative analysis of the photophysical results of pyrene in supercritical CO_2 (46). In their experiments absolute and relative fluorescence quantum yields of the pyrene monomer and excimer were determined in supercritical CO_2 at 35°C and 50°C over the CO_2 reduced-density range of about 0.5–2 (Figures 20 and 21). Although the pyrene concentrations were between 2×10^{-6} and 7×10^{-5} M in these supercritical CO_2 solutions, significant pyrene excimer fluorescence was observed. In an attempt to quantitatively model the experimental results in terms of the classical photophysical mechanism established for pyrene in normal liquid solutions, they found that the results deviate significantly from the classical mechanism. The disagreement could be reconciled by replacing the pyrene concentration in the photophysical model with a local pyrene concentration (the actual concentration of ground-state pyrene molecules in the vicinity of a photoexcited pyrene molecule). In the sense that the local concentration of pyrene is higher than the bulk concentration—up to a factor of 9, assuming diffusion-controlled conditions—pyrene-pyrene clustering enhances excimer formation in supercritical CO_2 (Figure 22) (46).

An excimer is a special case of exciplex—a complex between an excited-state molecule and a ground-state molecule, where the two molecules have different identities. Exciplex formation has been used as a model bimolecular process in the study of solute-solute clustering in supercritical fluid solutions. Brennecke et al. reported the investigation of naphthalene-triethylamine exciplex formation in supercritical CO_2 at 35°C and 50°C (166). Their results show that the exciplex emission can be observed, even at low triethylamine concentrations (5×10^{-3}–5×10^{-2} M). Similarly, Inomata et al. investigated the formation of pyrene-dimethylaniline excimer in supercritical CO_2 at 45°C (169). They

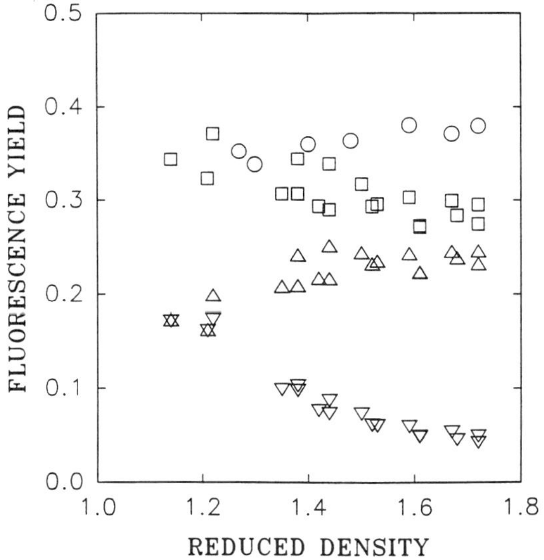

Figure 20 Fluorescence quantum yields of pyrene in supercritical CO_2 (35°C) at concentrations of 2×10^{-6} M (○) and 6×10^{-5} M (total, □: monomer, △: and excimer, ▽) as a function of CO_2 reduced densities. (From Ref. 46.)

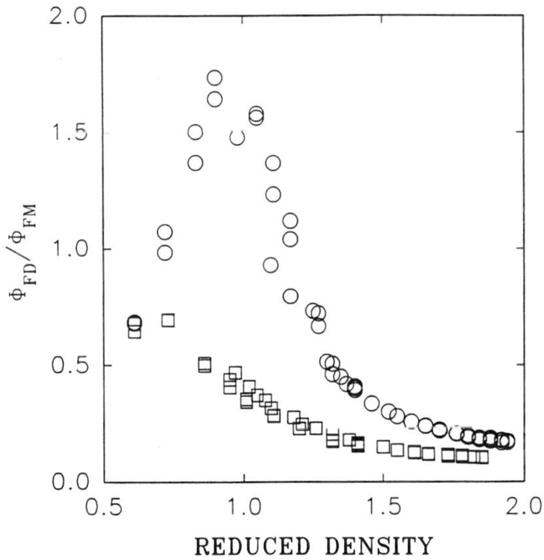

Figure 21 Ratios of pyrene excimer and monomer fluorescence quantum yields as a function of CO_2 reduced densities at 35°C (6.2×10^{-5} M, ○) and 50°C (5.9×10^{-5} and 6.8×10^{-5} M, □). (From Ref. 46.)

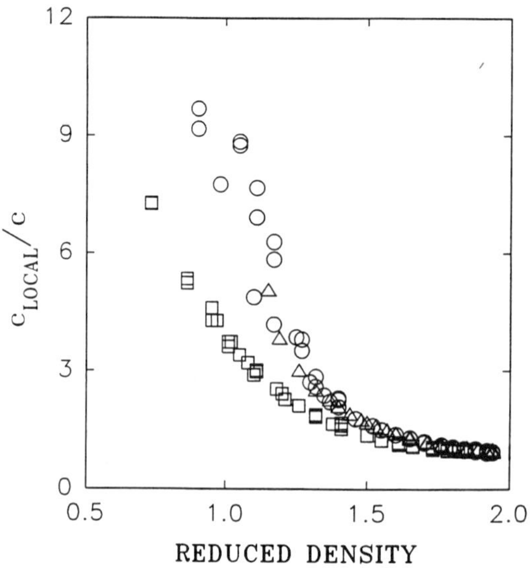

Figure 22 Ratios of the local and bulk pyrene concentrations as a function of CO_2 reduced densities at 35°C (2.8 × 10^{-5} M, △; 6.2 × 10^{-5} M, ○) and 50°C (5.9 × 10^{-5} and 6.8 × 10^{-5} M, □). (From Ref. 46.)

found unusually efficient exciplex formation and attributed the enhancement to preferential clustering of dimethylaniline molecules about pyrene.

Molecules capable of forming an intramolecular exciplex have also been used in the probing of solute-solute clustering in supercritical fluid solutions (170–172). These systems are fundamentally different from their intermolecular counterparts because intramolecular exciplex formation is independent of both bulk and local concentration as a result of the two participating pieces of the complex being linked by a tether. Okada et al. investigated the intramolecular exciplex formation of *p*-(*N*,*N*-dimethylaminophenyl)-$(CH_2)_2$-9-anthryl (DMAPA) in supercritical ethylene and fluoroform at 30°C (170). No exciplex formation was observed in the nonpolar fluid ethylene; however, in supercritical fluoroform two emission bands (normal and exciplex) were detected. Similarly, Rice et al. investigated the intramolecular excimer formation of 1,3-bis(1-pyrenyl)propane in supercritical ethane and fluoroform (171). They found that the ratio of excimer emission to monomer emission increases with increasing fluid density and that the excimer formation is at least partially dynamic in nature. Quantitative interpretation of their results was complicated by the existence of multiple ground-state species of the probe at all fluid densities.

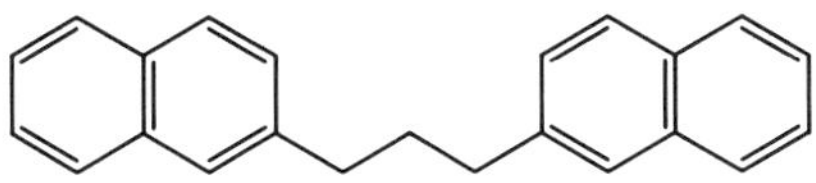

1,3-Di(2-naphthyl)propane

Structure 7

Rollins et al. investigated the intramolecular excimer formation of 1,3-di(2-naphthyl)propane in supercritical CO_2 (172) and compared the results with intermolecular pyrene-excimer formation recorded under similar conditions (46). Their results show that the ratio of excimer emission to monomer emission decreases gradually with increasing CO_2 density (Figure 23), in a pattern that agrees well with that predicted from viscosity changes in terms of the classical photophysical model for excimer formation (35). In a comparison of 1,3-di(2-naphthyl)propane and pyrene in the same fluid, the ratio of excimer emission to

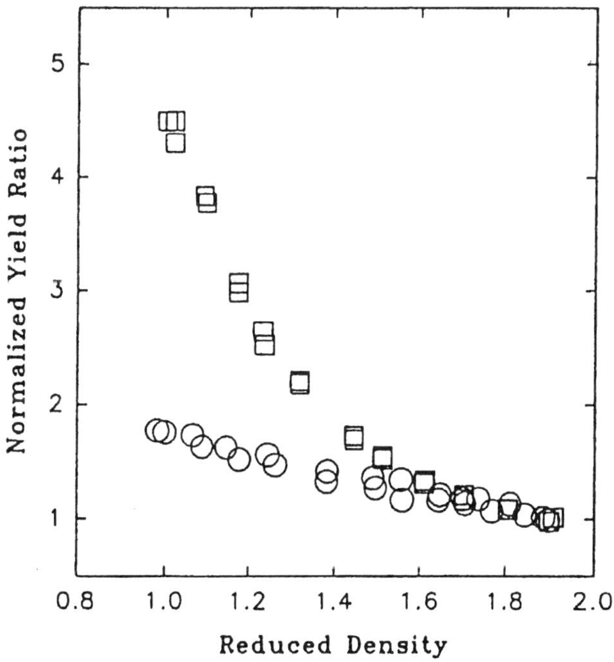

Figure 23 The Φ_{FD}/Φ_{FM} ratios (normalized at the reduced density of 1.9) for the intramolecular excimer formation in 1,3-di(2-naphthyl)propane (○) and the intermolecular excimer formation in pyrene [□] in supercritical CO_2 at 40°C. (From Ref. 172.)

monomer emission is considerably less sensitive to changes in fluid density for the tethered system, which seems to support the conclusion that the formation of intermolecular pyrene excimer is affected by solute-solute clustering.

2. Photodimerization

Photodimerization reactions in supercritical fluid solutions have been used to probe the effects of possible solute-solute clustering. Kimura et al. investigated the dimerization of 2-methyl-2-nitrosopropane in CO_2, chlorotrifluoromethane, fluoroform, argon, and xenon (173–176). Their results show that the density dependence of the dimerization equilibrium constant is rather complex, probably due to the existence of various dimerization mechanisms in different density regions.

Hrnjez et al. evaluated the product distribution of the photodimerization of isophorone in supercritical fluoroform and CO_2 (177). The reaction typically produces a mixture of various regioisomers and stereoisomers. Relative yields of the regioisomers are fluid density dependent in the polar fluid fluoroform but exhibit little or no change with fluid density in CO_2. On the other hand, relative yields of the stereoisomers are affected by changes in the fluid density in both fluoroform and CO_2. The results were discussed in terms of solvation and various degrees of solvent reorganization required for the various products. (See Scheme 5.)

Tsugane et al. used Fourier transform infrared absorption spectroscopy to investigate the dimerization reaction of benzoic acid in saturated supercritical

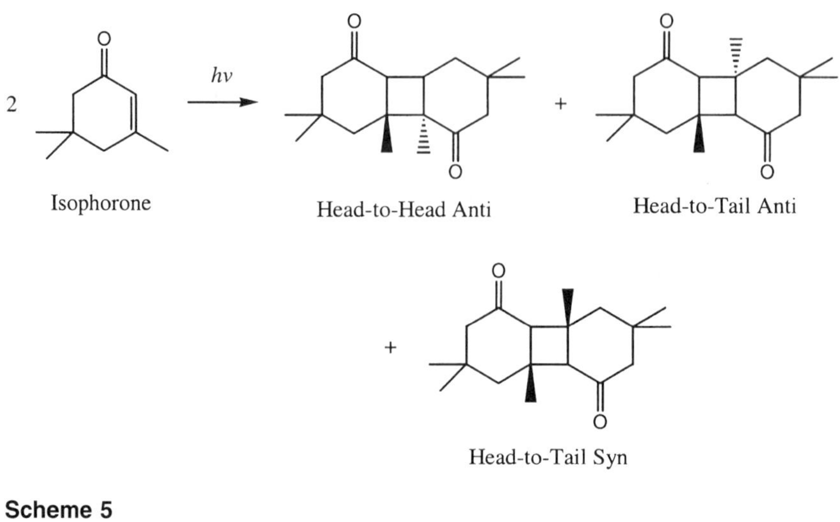

Scheme 5

CO_2 solutions at 45°C (178). The ratio of dimer absorption to monomer absorption was found to be a strong function of fluid density, with a clear maximum in the near-critical region. In addition, the dimer formation was observed at benzoic acid mole fractions of as low as 10^{-4}; this was attributed to significant solute–solute interactions in the dilute supercritical fluid solutions.

Bunker et al. studied the photodimerization reaction of anthracene in supercritical CO_2 at 35°C (179). They found that the reaction quantum yields are up to an order of magnitude higher in supercritical CO_2 (35°C, $\rho_r = 1.9$) than in liquid benzene at the same anthracene concentrations; however, for the fluid density dependence, the yields obtained at different densities agree well with the yields calculated on the basis of experimentally determined viscosities (Figure 24). Since the results provided no evidence of solute-solute clustering effects, the higher photodimerization yields in the supercritical fluid were attributed to more efficient anthracene diffusion associated with the lower viscosity. (See Scheme 6.)

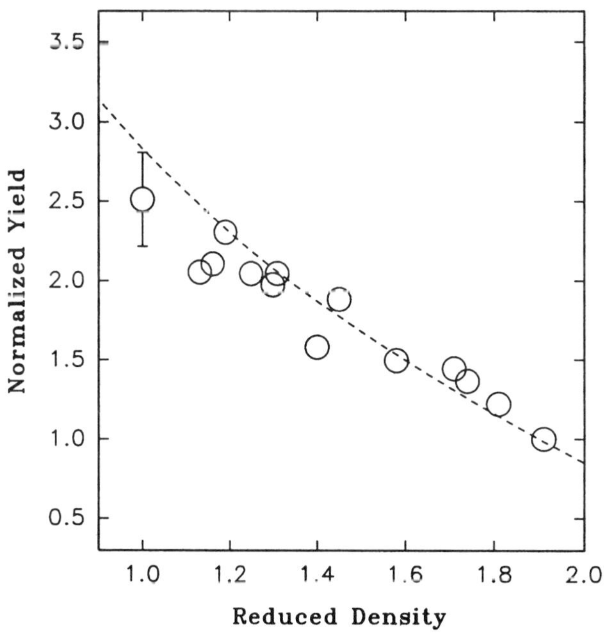

Figure 24 Photodimerization yields of anthracene in supercritical CO_2 at 35°C as a function of CO_2 reduced density compared with the values calculated from viscosities in terms of Debye equation. All results are normalized against those at the reduced density of 1.9. (From Ref. 179.)

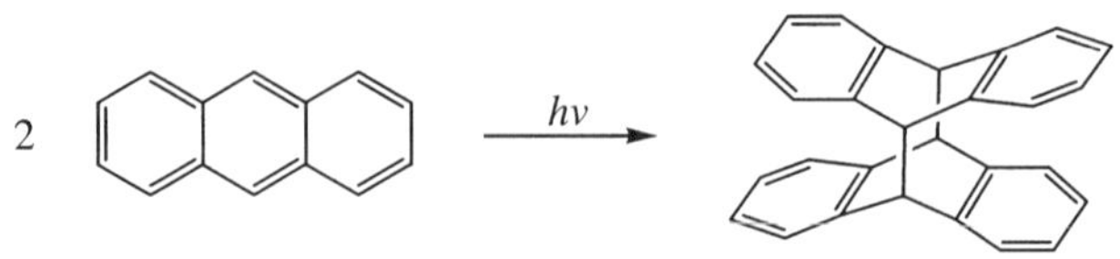

Scheme 6

3. Fluorescence Quenching

The formation of an excimer or exciplex is a fluorescence quenching process in which the monomer excited state is quenched by the ground-state molecule to form an excited-state complex. However, the fluorescence quenching discussed here is somewhat different in that the quenching results in no complex between the molecule being quenched and the quencher. The absence of excimer or exciplex formation in these systems that undergo bimolecular fluorescence quenching eliminates some of the complications in the probing of solute–solute interactions in supercritical fluid solutions (180).

Bunker and Sun studied the quenching of 9,10-bis(phenylethynyl)anthracene (BPEA) fluorescence by carbon tetrabromide (CBr_4) in supercritical CO_2 at 35°C using time-resolved fluorescence methods (180). The bimolecular reaction of the photoexcited anthracene derivative BPEA with CBr_4 is known to be diffusion controlled in normal liquid solutions (35). Because fluorescence is the only decay pathway of the excited BPEA in the absence of quenchers (fluorescence yield of unity), the bimolecular fluorescence quenching process is clean and simple, involving no competing reaction processes and no formation of an emissive excited-state complex (35). For the quenching of the fluorescence lifetime, the Stern–Volmer equation is as follows:

$$\tau_f{}^0/\tau_f = 1 + K_{SV}[CBr_4] = 1 + k_q\tau_f{}^0[CBr_4] \tag{7}$$

where $\tau_f{}^0$ and τ_f are fluorescence lifetimes of BPEA in the absence and presence of quenchers, respectively, K_{SV} is the Stern–Volmer quenching constant, and k_q is the quenching rate constant. When the process is diffusion controlled, k_q should be equal to k_{diff}. The diffusion rate constant k_{diff} is typically estimated from the Smoluchowski equation with a correction factor f.

$$k_{diff} = fk_{SE} \tag{8}$$

$$k_{SE} = (4 \times 10^{-3})\pi N(r_{BPEA} + r_{CBr4})(D_{BPEA} + D_{CBr4}) \tag{9}$$

where r_{BPEA} and r_{CBr4} are the molecular radii of BPEA and CBr_4, respectively, and D represents the diffusion coefficients.

$$D_i = kT/6\pi\eta r_i \tag{10}$$

For the quenching of BPEA fluorescence by CBr_4, the k_q values obtained from the Stern–Volmer equation are larger than the k_{diff} values obtained from Eqs. (8) and (9) (180); and the difference between k_q and k_{diff} is more significant at lower fluid densities (Figure 25). The results were interpreted in terms of the local quencher CBr_4 concentration in the vicinity of the excited BPEA being higher than the bulk concentration in the supercritical fluid solutions (180).

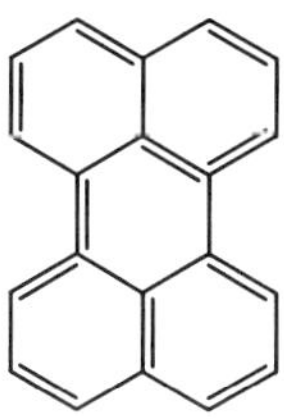

Perylene

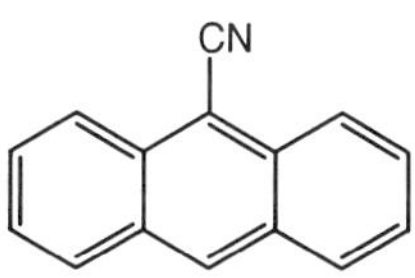

9-Cyanoanthracene

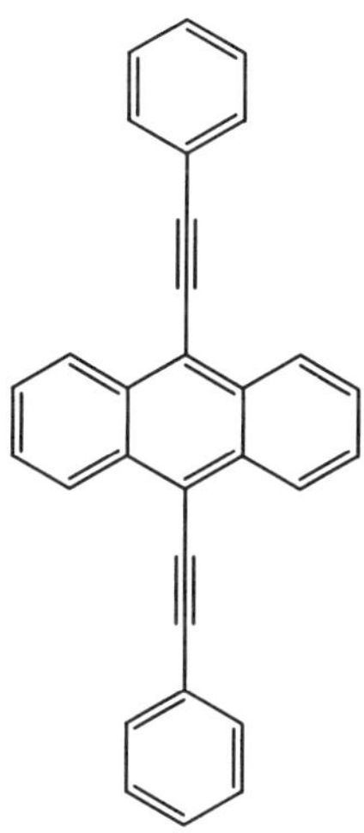

BPEA

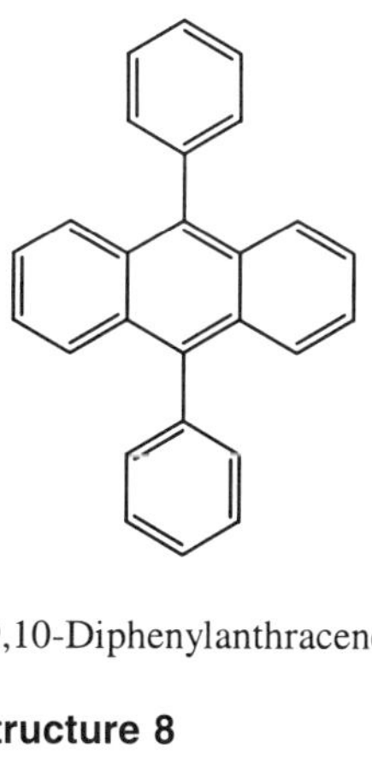

9,10-Diphenylanthracene

Structure 8

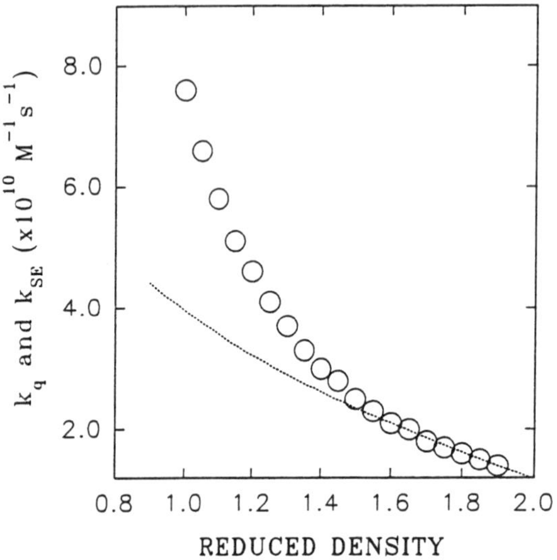

Figure 25 Observed quenching rate constants as a function of CO_2 reduced densities at 35°C. The dashed line represents the density dependence of the Smoluchowski diffusion rate constants. (From Ref. 180.)

The same fluorescence quenching study was expanded to other fluorophores, including anthracene, perylene, 9-cyanoanthracene, and 9,10-diphenylanthracene (181). The results show that the solute-solute clustering in the form of higher local CBr_4 concentration is dependent on the fluorescent molecule being quenched. Enhanced quenching effects are present in the 9-cyanoanthracene-CBr_4 and 9,10-diphenylanthracene-CBr_4 systems but not in the anthracene-CBr_4 and perylene-CBr_4 systems (Figure 26). In more recent studies of similar fluorescence quenching processes (fluorophores anthracene, 1,2-benzanthracen, and naphthalene with quenchers CBr_4 and C_2H_5Br) in supercritical ethane and CO_2 (182,183), Brennecke and coworkers found the same system dependence for the quencher. For example, enhanced fluorescence quenching was observed in the anthracene-C_2H_5Br system but, again, not in the anthracene-CBr_4 system.

4. Other Bimolecular Reactions

Brennecke, Chateauneuf, and coworkers used laser flash photolysis to investigate the excited triplet-state reactions of benzophenone, including triplet-triplet annihilation and hydrogen abstraction reactions with a variety of hydrogen donors in supercritical fluids (184–191). For example, when 2-propanol and 1,4-cyclohexadiene were used as hydrogen donors, the hydrogen abstraction reactions of the triplet benzophenone in supercritical CO_2 were found to be

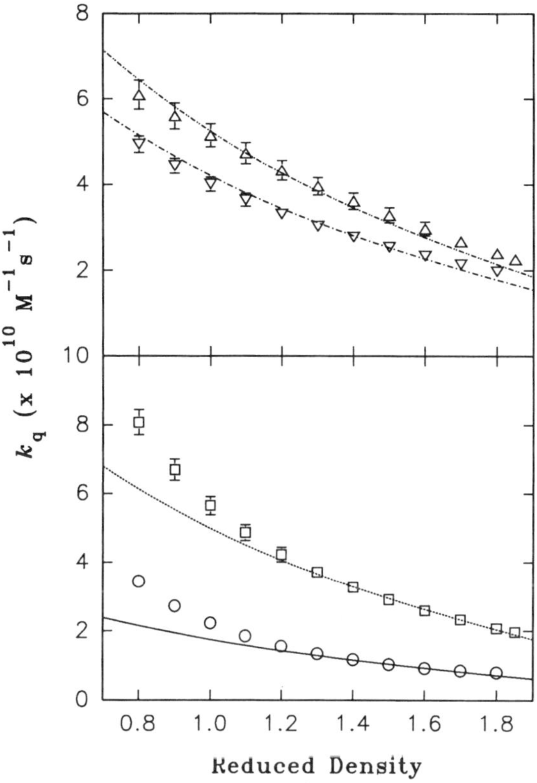

Figure 26 Observed quenching rate constants k_q at different reduced densities for anthracene-CBr_4 (top, ▽), perylene-CBr_4 (top, △), 9CA-CBr_4 (bottom, ○), and DPA-CBr_4 (bottom, □) in supercritical CO_2 at 35°C. The lines represent the CO_2 density dependence of the Debye–Smoluchowski diffusion rate constants adjusted with the f factors. (From Ref. 181.)

particularly efficient in the near-critical density region (Figure 27) (184). The enhancement in the reactions was attributed to the clustering of hydrogen donor molecules around the solute benzophenone, conceptually similar to the entrainer effect. The same reactions were also carried out in supercritical ethane and fluoroform, yielding similar results (185); however, it is difficult to understand why no clustering-related enhancements were observed in the same reactions of benzophenone with triethylamine and 1,4-diazabicyclo[2.2.2]octane (186). Also no solute-solute effect on the triplet-triplet annihilation reaction of benzophenone in several supercritical fluids and mixtures was observed (187,188). On the other hand, the results for the triplet-triplet annihilation of anthracene in supercritical water may invoke a solute-solute clustering explanation (189).

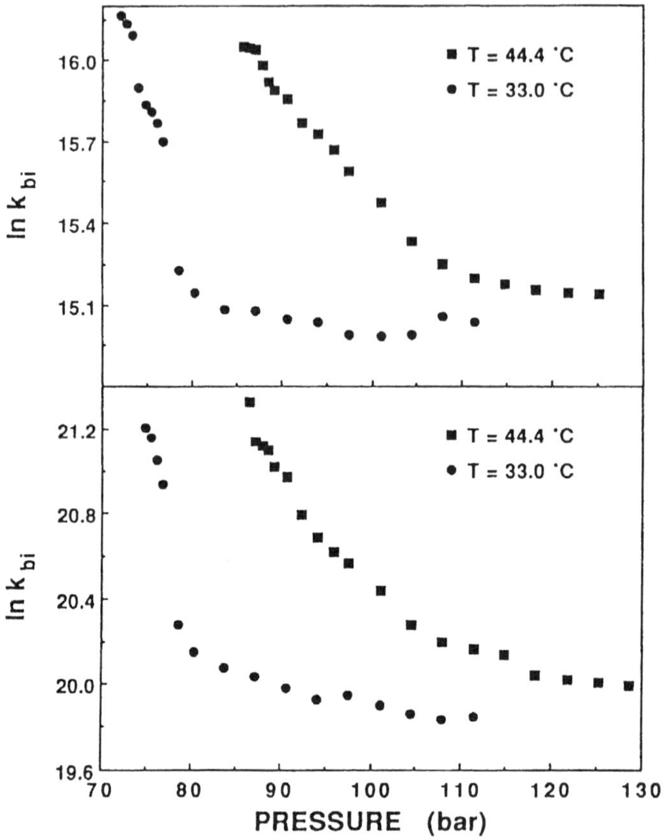

Figure 27 Pressure dependence on the bimolecular rate constant k_{bi} ($M^{-1}s^{-1}$), at 33.0°C (●) and 44.4°C (■) for the reaction of ^{3}BP with 2-propanol (top) and 1,4-cyclohexadiene (bottom). (From Ref. 184.)

Electron transfer reactions have also been used in the probing of solute–solute interactions in supercritical fluid solutions. For example, Takahashi and Jonah examined the electron transfer between biphenyl anion and pyrene in supercritical ethane (192). Worrall and Wilkinson studied triplet-triplet energy transfer reactions for a series of donor–acceptor pairs, including anthracene-azulene in supercritical CO_2-acetonitrile and supercritical CO_2-hexane and benzophenone-naphthalene in supercritical CO_2-acetonitrile (193). The high efficiency of the energy transfer reactions at low cosolvent concentrations was attributed to the effect of solute-solute clustering.

Randolph and Carlier used EPR spectroscopy to study the Heisenberg spin exchange reaction of nitroxide free radicals in supercritical ethane (194). The reaction rate constants were found to be pressure dependent, decreasing with increasing pressure and decreasing rapidly at temperatures nearer to the critical temperature. Despite the disagreement between experimental and predicted reaction rate constants (Figure 28), solute-solute clustering was considered to be highly unlikely because of the independence of the reaction rate constants on the solute concentration; instead, the enhanced reaction rates were explained in terms of the effects of solute–solvent clustering on the average reaction contact times and the conversion rates.

Tanko et al. examined cage effects on the free-radical chlorination of cyclohexane in supercritical CO_2 at 40°C and a series of pressures (195). The ratio of monochlorination to polychlorination was found to be linear with the diffusivity in CO_2—similar to the relationship in normal liquid solvents. Thus, apparently clustering has no effect on the reaction in supercritical CO_2 (195,196).

These studies show clearly the intense interest of the research community in the phenomenon of solute-solute clustering in supercritical fluid systems. The diverse and sometime inconsistent results demonstrate the difficulties associated with the issue. Obviously, additional investigations, especially those based on novel approaches and intrinsically more accurate experimental techniques, are required.

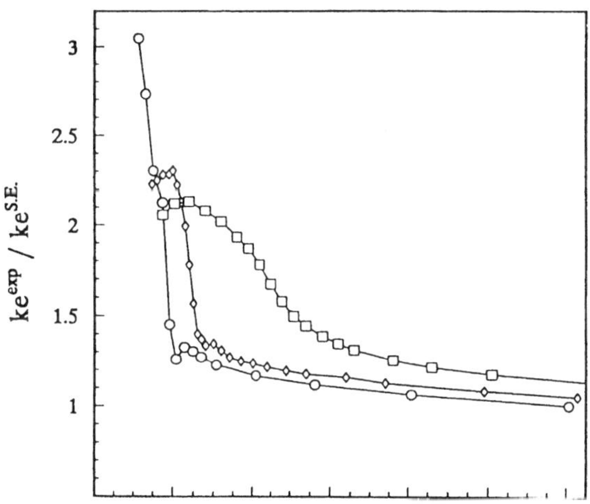

Figure 28 Ratio of observed bimolecular rate constant for spin exchange in ethane to the rate constant predicted from the Stokes–Einstein relationship as a function of pressure. Temperatures are 308 K (circles), 313 K (diamonds), and 331 K (squares). (From Ref. 194.)

IV. SUMMARY

Significant progress has been made in our understanding of the fundamental properties of supercritical fluids as a result of the extensive experimental investigations carried out over the last two decades. This understanding has prompted widespread applications of supercritical fluid technology, including in particular the recent proliferation for the use of supercritical fluids in materials preparation and processing. It may also be expected that such applications will stimulate further development of the technology.

ACKNOWLEDGMENTS

We thank M. Whitaker, R. Martin, and B. Harruff for assistance in the preparation of the manuscript. This work was made possible by the support of Dr. J. Tishkoff and the Air Force Office of Scientific Research (C.E.B.), the Department of Energy under Contracts DE-AC07-99ID13727 (H.W.R.) and DE-FG02-00ER45859 (Y.-P.S.), and the National Science Foundation through CHE-9729756 and the Clemson Center for Advanced Engineering Fibers and Films (Y.-P.S).

REFERENCES

1. Y-P Sun, MA Fox, KP Johnston. Spectroscopic studies of *p*-(*N*,*N*-dimethylamino) benzonitrile and ethyl *p*-(*N*,*N*-dimethylamino)benzoate in supercritical trifluoromethane, carbon dioxide, and ethane. J Am Chem Soc 114:1187, 1992.
2. Y-P Sun, CE Bunker, NB Hamilton. Py scale in vapor phase and in supercritical carbon dioxide. Evidence in support of a three-density-region model for solvation in supercritical fluids. Chem Phys Lett 210:111, 1993.
3. Y-P Sun, CE Bunker. Solute and solvent dependencies of intermolecular interactions in different density regions in supercritical fluids. A generalization of the three-density-region solvation mechanism. Ber Bunsenges Phys Chem 99:976, 1995.
4. P Ehrlich, RH Fariss. Negative partial molar volumes in the critical region. Mixtures of ethylene and vinyl chloride. J Phys Chem 73:1164, 1969.
5. P Ehrlich, PC Wu. Volumetric properties of supercritical ethane-n-heptane mixtures: the isothermal compressibility in the critical region. AIChE J 19:540, 1973.
6. PC Wu, P Ehrlich. Volumetric properties of supercritical ethane-n-heptane mixtures: molar volumes and partial molar volumes. AIChE J 19:533, 1973.
7. CA Eckert, DH Ziger, KP Johnston, S Kim. Solute partial molar volumes in supercritical fluids. J Phys Chem 90:2738, 1986.
8. KP Johnston, C Haynes. Extreme solvent effects on reaction rate constants at supercritical fluid conditions. AIChE J 33:2017, 1987.

9. DG Peck, AJ Mehta, KP Johnston. Pressure tuning of chemical reaction equilibria in supercritical fluids. J Phys Chem 93:4297, 1989.
10. Q Ji, EM Eyring, R van Eldik, KP Johnston, SR Goates, ML Lee. Laser flash photolysis studies of metal carbonyls in supercritical CO_2 and ethane. J Phys Chem 99:13461, 1995.
11. PG Debenedetti. Clustering in dilute, binary supercritical mixtures: a fluctuation analysis. Chem Eng Sci 42:2203, 1987.
12. GM Simmons, DM Mason. Pressure dependency of gas phase reaction rate coefficients. Chem Eng Sci 27:89, 1972.
13. MJ Kamlet, J-LM Abbound, RW Taft. Solvatochromic comparison method. 6. Pi-Star scale of solvent polarities. J Am Chem Soc 98:6027, 1977.
14. MJ Kamlet, TN Hall, J Boykin, RW Taft. Linear solvation energy relationships. 6. Additions to and correlations with the Pi-Star scale of solvent polarities. J Org Chem 44:2599, 1979.
15. DC Dong, MA Winnik. The Py scale of solvent polarities—solvent effects on the vibronic fine-structure of pyrene fluorescence and empirical correlations with ET-value and Y-value. Photochem Photobiol 35:17, 1982.
16. DC Dong, MA Winnik. The Py scale of solvent polarities. Can J Chem 62:2560, 1984.
17. W Rettig. Charge separation in excited-states of decoupled systems—TICT compounds and implications regarding the development of new laser-dyes and the primary processes of vision and photosynthesis. Angew Chem Int Ed Eng 25:971, 1986.
18. C Reichardt. Solvents and Solvent Effects in Organic Chemistry. 2nd ed. New York: VCH, 1990.
19. JA Hyatt. Liquid and supercritical carbon dioxide as organic solvents. J Org Chem 49:5097, 1984.
20. ME Sigman, SM Lindley, JE Leffler. Supercritical carbon dioxide: behavior of Pi* and beta solvatochromic indicators in media of different densities. J Am Chem Soc 107:1471, 1985.
21. ME Sigman, JE Leffler. Supercritical carbon dioxide. 2. Pi* and the dielectric function for supercritical CO_2 media at various densities. J Phys Chem 90:6063, 1986.
22. RD Smith, SL Frye, CR Yonker, RW Gale. Solvent properties of supercritical Xe and SF_6. J Phys Chem 91:3059, 1987.
23. CR Yonker, SL Frye, DR Kalkwarf, RD Smith. Characterization of supercritical fluid solvents using solvatochromic shifts. J Phys Chem 90:3022, 1986.
24. CR Yonker, RD Smith. Solvatochromism: a dielectric continuum model applied to supercritical fluids. J Phys Chem 92:235, 1988.
25. CR Yonker, RD Smith. Thermochromic shifts in supercritical fluids. J Phys Chem 93:1261, 1989.
26. S Kim, KP Johnston. Molecular interactions in dilute supercritical fluid solutions. Ind Eng Chem Res 26:1206, 1987.
27. RM Lemert, JM DeSimone. Solvatochromic characterization of near- and supercritical ethane, propane, and dimethyl ether using 9-(α-perfluoroheptyl-β,β-dicyanovinyl)julolidine. J Supercrit Fluids 4:186, 1991.

28. AP Abbott, CA Eardley. Solvent properties of liquid and supercritical 1,1,1,2-tetrafluoroethane. J Phys Chem B 102:8574, 1998.
29. AF Lagalante, RL Hall, TJ Bruno. Kamlet–Taft solvatochromic parameters of the sub- and supercritical fluorinated ethane solvents. J Phys Chem B 102:6601, 1998.
30. K Takahashi, K Abe, S Sawamura, CD Jonah. Spectroscopic study of 4-aminobenzophenone in supercritical CF_3H and CO_2: local density and Onsager's reaction cavity radius. Chem Phys Lett 282:361, 1998.
31. T Yamaguchi, Y Kimura, M Hirota. Solvent and solvent density effects on the spectral shifts and the bandwidths of the absorption and the resonance raman spectra of phenol blue. J Phys Chem 101:9050, 1997.
32. R Biswas, JE Lewis, M Maroncelli. Electronic spectral shifts, reorganization energies, and local density augmentation of coumarin 153 in supercritical solvents. Chem Phys Lett 310:485, 1999.
33. K Takahashi, K Fujii, S Sawamura, CD Jonah. Density dependence of the Stokes shift and solvent reorganization energy in supercritical fluids. Rad Phys Chem 55:579, 1999.
34. JD Wechwerth, PW Carr. Study of interactions in supercritical fluids and supercritical fluid chromatography by solvatochromic linear solvation energy relationships. Anal Chem 70:1404, 1998.
35. JB Birks. Photophysics of Aromatic Molecules. London: Wiley-Interscience, 1970.
36. A Nakajima. Solvent effect on the vibrational structures of the fluorescence and absorption spectra of pyrene. Bull Chem Soc Jpn 44:3272, 1971.
37. A Nakajima. Solvent enhancement in first singlet–singlet transition of pyrene-D_{10}. Spectrochim Acta Part A 30:860, 1974.
38. A Nakajima. Effects of isomeric solvents on vibronic band intensities in fluorescence spectrum of pyrene. J Mol Spectrosc 61:467, 1976.
39. A Nakajima. Fluorescence spectra of pyrene in chlorinated aromatic solvents. J Lumin 11:429, 1976.
40. JF Brennecke, DL Tomasko, J Peshkin, CA Eckert. Fluorescence spectroscopy studies of dilute supercritical solutions. Ind Eng Chem Res 29:1682, 1990.
41. JF Brennecke, CA Eckert. Fluorescence spectroscopy studies of intermolecular interactions in supercritical fluids. ACS Symp Series 14:406, 1989.
42. JK Rice, ED Niemeyer, RA Dunbar, FV Bright. State-dependent solvation of pyrene in supercritical CO_2. J Am Chem Soc 117:5832, 1995.
43. J Zhang, LL Lee, JF Brennecke. Fluorescence spectroscopy and integral equation studies of preferential solvation in supercritical fluid mixtures. J Phys Chem 99:9268, 1995.
44. ED Neimeyer, FV Bright. Determination of the local environment surrounding pyrene in supercritical alkanes: a first step toward solvation in supercritical aciation fuels. Energy Fuels 12:823, 1998.
45. S-H Chen, VL McGuffin. Temperature effect on pyrene as a polarity probe for supercritical fluid and liquid solutions. Appl Spectrosc 48:596, 1994.
46. Y-P Sun, CE Bunker. Quantitative spectroscopic investigation of enhanced excited state complex formation in supercritical carbon dioxide under near-critical conditions: inconsistency between experimental evidence and classical photophysical mechanism. J Phys Chem 99:13778, 1995.

47. J Zagrbelny, FV Bright. Influence of solute–fluid clustering on the photophysics of pyrene emission in supercritical C_2H_4 and CF_3H. J Am Chem Soc 114:7821, 1992.
48. ED Niemeyer, RA Dunbar, FV Bright. On the local environment surrounding pyrene in near- and supercritical water. Appl Spectrosc 51:1547, 1997.
49. CE Bunker, JR Gord. unpublished results.
50. Y-P Sun, TL Bowen, CE Bunker. Formation and decay of the ethyl *p*-(*N*,*N*-diethylamino)benzoate twisted intramolecular charge-transfer state in the vapor phase, supercritical fluids, and room-temperature solutions. J Phys Chem 98:12486, 1994.
51. O Kajimoto, K Yamasaki, K Honma. Intramolecular charge-transfer reactions studied in a supercritical fluid of varying densities and in a molecular beam. Faraday Disc Chem Soc 85:65, 1988.
52. O Kajimoto, M Futakami, T Kobayashi, K Yamasaki. Charge-transfer-state formation in supercritical fluid: (*N*,*N*-dimethylamino)benzonitrile in CF_3H. J Phys Chem 92:1347, 1988.
53. O Kajimoto, T Nayuki, T Kobayashi. Picosecond dynamics of the twisted intramolecular charge-transfer state formation of 4-(*N*,*N*-dimethylamino)benxonitrile (DMABN) in supercritical fluid solvent. Chem Phys Lett 209:357, 1993.
54. O Kajimoto, K Sekiguchi, T Nayuki, T Kobayashi. Dynamics of charge-transfer state formation in supercritical fluid solvent. Ber Bunsenges Phys Chem 101:600, 1997.
55. A Morita, O Kajimoto. Solute–solvent interaction in nonpolar supercritical fluid: a clustering model and size distribution. J Phys Chem 94:6420, 1990.
56. Y-P Sun, CE Bunker. Twisted intramolecular charge transfer of ethyl *p*-(*N*,*N*-diethylamino)benzoate in the gas phase and in the low-density non-polar supercritical fluids. a quantitative spectral resolution using principal component analysis and self modeling. J Chem Soc, Chem Commun 5, 1994.
57. ER Malinowski. Factor Analysis in Chemistry. 2nd ed. New York: Wiley Interscience, 1991.
58. DW Osten, BR Kowalski. Multivariate curve resolution in liquid chromatography. Anal Chem 56:991, 1984.
59. Y-P Sun, DF Sears, J Saltiel, FB Mallory, CW Mallory, CA Buser. Principal component–three component self-modeling analysis applied to trans-1,2-di(2-naphthyl)ethene fluorescence. J Am Chem Soc 110:6974, 1988.
60. CE Bunker, NB Hamilton, Y-P Sun. Quantitative application of principal component analysis and self-modelling spectral resolution to product analysis of tetraphenylethylene photochemical reactions. Anal Chem 65:3460, 1993.
61. LS Ramos, KR Beebe, WP Carey, EM Sanchez, BC Erickson, BE Wilson, LE Wangen, BR Kowalski. Chemometrics. Anal Chem 58:294R, 1986.
62. LB McGown, IM Warner. Molecular fluorescence, phosphorescence and chemiluminescence spectrometry. Anal Chem 62:255R, 1990.
63. Y-P Sun, G Bennett, KP Johnston, MA Fox. Quantitative resolution of dual fluorescence spectra in molecules forming twisted intramolecular charge-transfer states. Toward establishment of molecular probes for medium effects in supercritical fluids and mixtures. Anal Chem 64:1763, 1992.

64. CE Bunker, TL Bowen, Y-P Sun. A photophysical study of molecular probe 6-propionyl-2-(*N*,*N*-dimethylamino)-naphthalene (PRODAN) in aqueous and nonaqueous solutions. Photochem Photobiol 58:499, 1993.
65. TA Betts, J Zagrobelny, FY Bright. Investigation of solute–fluid interactions in supercritical CF_3H: a multifrequency phase and modulation fluorescence study. J Supercrit Fluids 5:48, 1992.
66. TA Betts, J Zagrobelny, FV Bright. Spectroscopic determination of solute–fluid cluster size in supercritical N_2O. J Am Chem Soc 114:8163, 1992.
67. KE O'Shea, KM Kirmse, MA Fox, KP Johnston. Polar and hydrogen-bonding interactions in supercritical fluids. Effects on the tautomeric equilibrium of 4-(phenylazo)-1-naphthol. J Phys Chem 95:7863, 1991.
68. K Akao, Y Yoshimura. Keto-enol tautomeric equilibrium of acetylacetone in trifluoromethane near the critical temperature. J Chem Phys 94:5243, 1991.
69. K Yamasaki, O Kajimoto. Solvent effect in supercritical fluids: keto-enol equilibria of acetylacetone and ethyl acetoacetate. Chem Phys Lett 172:271, 1990.
70. SG Kazarian, M Poliakoff. Can conformational equilibria be "tuned" in supercritical fluid solution? An IR spectroscopic study of trans/gauche isomerism of hexafluoropropan-2-ol in supercritical SF_6 and CHF_3 solutions. J Phys Chem 99:8624, 1995.
71. Y Yagi, S Saito, H Inomata. Tautomerization of 2,4-pentanedione in supercritical CO_2. J Chem Eng Jpn 26:116, 1993.
72. K Hara, N Ito, O Kajimoto. High pressure studies of the Kramers turnover behavior for the excited-state isomerization of 2-alkenylanthracene in alkene. J Chem Phys 110:1662, 1999.
73. K Hara, H Kiyotani, O Kajimoto. High-pressure studies on the excited-state isomerization of 2-vinylanthracene: experimental investigation of Kramers turnover. J Chem Phys 103:5548, 1995.
74. C Gehrke, J Schroeder, D Schwarzer, J Troe, F Voss. Photoisomerization of diphenylbutadiene in low-viscosity nonpolar solvents: experimental manifestations of multidimensional Kramers behavior and cluster effects. J Chem Phys 92:4805, 1990.
75. J Schoeder, D Schwarzer, J Troe, F Voss. Cluster and barrier effects in the temperature and pressure dependence of the photoisomerization of trans-stilbene. J Chem Phys 93:2393, 1990.
76. M Lee, GR Holtom, RM Hochstrasser. Observation of the Kramers turnover region in the isomerization of trans-stilbene in fluid ethane. Chem Phys Lett 118:359, 1985.
77. ME Sigman, JE Leffler. Supercritical carbon dioxide. The cis to trans relaxation and π,π^* transition of 4-(diethylamino)-4′-nitroazobenzene. J Org Chem 52:3123, 1987.
78. AK Dillow, JS Brown, CL Liotta, CA Eckert. Supercritical fluid tuning of reactions rate: the cis-trans isomerization of 4-4′-disubstituted azobenzenes. J Phys Chem A 102:7609, 1998.
79. C Carlier, TW Randolph. Dense-gas solvent–solute clusters at near-infinite dilution: EPR spectroscopic evidence. AIChE J 39:876, 1993.

80. S Ganapathy, C Carlier, TW Randolph, JA O'Brien. Influence of local structural correlations of free-radical reactions in supercritical fluids: a hierarchical approach. Ind Eng Chem Res 35:19, 1996.
81. SN Batchelor, B Henningsen, H Fischer. EPR of transient free radicals during photochemical reactions in high temperature and pressure gasses. J Phys Chem A 101:2969, 1997.
82. SN Batchelor. Free radical motion in super critical fluids probed by EPR spectroscopy. J Phys Chem B 102:615, 1998.
83. KE Dukes, EJ Harbron, MDE Forbes. Flow system and 9.5 Ghz microwave resonators for time-resolved and steady-state electron paramagnetic resonance spectroscopy in compressed and supercritical fluids. Rev Sci Instrum 68:2505, 1997.
84. JP Blitz, CR Yonker, RD Smith. Infrared spectroscopic studies of supercritical fluid solutions. J Phys Chem 93:6661, 1989.
85. Y Ikushima, N Saito, M Arai, K Arai. Solvent polarity parameters of supercritical carbon dioxide as measured by infrared spectroscopy. Bull Chem Soc Jpn 64:2224, 1991.
86. Y Ikushima, N Soito, M Arai. Supercritical carbon dioxide as reaction medium: examination of its solvent effects in the near-critical region. J Phys Chem 96:2293, 1992.
87. JC Meredith, KP Johnston, JM Seminario, SG Kazarian, CA Eckert. Quantitative equilibrium constants between CO_2 and Lewis bases from FTIR spectroscopy. J Phys Chem 100:10837, 1996.
88. SG Kazarian, MF Vincent, FV Bright, CL Liotta, CA Eckert. Specific intermolecular interactions of carbon dioxide with polymers. J Am Chem Soc 118:1729, 1996.
89. N Wada, M Saito, D Kitada, RL Smith Jr, I Inomata, K Arai, S Saito. Local excess density about substituted benzene compounds in supercritical CO_2 based on FT-IR spectroscopy. J Phys Chem B 101:10918, 1997.
90. CH Wang, RB Wright. Effect of density on the Raman scattering of molecular fluids. I. A detailed study of the scattering polarization, intensity, frequency shift, and spectral shape in gaseous N_2. J Chem Phys 59:1706, 1973.
91. TW Zerda, X Song, J Jonas. Raman study of intermolecular interactions in supercritical solutions of naphthalene in CO_2. Appl Spec 40:1194, 1986.
92. Y Garrabos, V Chandrasekharan, MA Echargui, F Marsault-Herail. Density effect on the raman fermi resonance in the fluid phases of CO_2. Chem Phys Lett 160:250, 1989.
93. 91-1 D Ben-Amotz, F LaPlant, D Shea, J Gardecki, D List. Raman studies of molecular potential energy surface changes in supercritical fluids. ACS Symp Ser 488:18, 1991.
94. F Marsault-Herail, M Echargui. Isotopic dilution effects on the isotropic Raman spectra of methane in supercritical fluid. J Molecular Liquids 48:211, 1991.
95. S Akimoto, O Kajimoto. Solvent-induced shift of the Raman spectra in supercritical fluids. Chem Phys Lett 209:263, 1993.
96. F Marsault-Herail, F Salmoun, J Dubessy, Y Garrabos. Isotropic Raman spectra of H_2S in supercritical fluid. J Mol Liquids 62:251, 1994.

97. T Yamaguchi, Y Kimura, N Hirota. Solvation state selective excitation in resonance Raman spectroscopy. 1. Experimental study of the C=N and C=O stretching modes of phenol blue. J Chem Phys 109:9075, 1998.
98. JNM Hegarty, JJ McGarvey, SEJ Bell, AHR Al-Obaidi. Time-resolved resonance Raman scattering of triplet state anthracene in supercritical CO_2. J Phys Chem 100:15704, 1996.
99. AS Zinn, D Schiferi, MF Nicol. Raman spectroscopy and melting of nitrogen between 290 and 900 K and 2.3 and 18 Gpa. J Chem Phys 87:1267, 1987.
100. MA Echargui, F Marsault-Herail. Critical effects on vibrational dephasing in CH_4 diluted in CO_2. Chem Phys Lett 179:317, 1991.
101. RS Urdahl, KD Rector, DJ Myers, PH Davis, MD Fayer. Vibrational relaxation of a polyatomic solute in a polyatomic supercritical fluid near the critical point. J Chem Phys 105:8973, 1996.
102. RS Urdahl, DJ Myers, KD Rector, PH Davis, BJ Cherayil, MD Fayer. vibrational lifetimes and vibrational line positions in polyatomic supercritical fluids near the critical point. J Chem Phys 107:3747, 1997.
103. BJ Cherayil, MD Fayer. Vibrational relaxation in supercritical fluids near the critical point. J Chem Phys 107:7642, 1997.
104. DJ Myers, RS Urdahl, BJ Cherayil, MD Fayer. Temperature dependence of vibrational lifetimes at the critical density in supercritical mixtures. J Chem Phys 107:9741, 1997.
105. DJ Myers, S Chen, M Shigeiwa, BJ Cherayil, MD Fayer. Temperature dependent vibrational lifetimes in supercritical fluids near the critical point. J Chem Phys 109:5971, 1998.
106. D Schwarzer, J Troe, M Zerezke. The role of local density in the collisional deactivation of vibrationally highly excited azulene in supercritical fluids. J Chem Phys 107:8380, 1997.
107. R Kroon, M Baggen, A Lagendijk. Vibrational dephasing in liquid nitrogen at high densities studied with time-resolved stimulated raman gain spectroscopy. J Chem Phys 91:74, 1989.
108. D Schwarzer, J Troe, M Votsmeier, M Zerezke. Collisional deactivation of vibrationally highly excited azulene in compressed liquids and supercritical fluids. J Chem Phys 105:3121, 1996.
109. D Schwarzer, J Troe, M Zerezke. Preferential solvation in the collisional deactivation of vibrationally highly excited azulene in supercritical xenon/ethane mixtures. J Phys Chem A 102:4207, 1998.
110. J Benzler, S Linkersdörfer, K Luther. Density dependence of the collisional deactivation of highly vibrationally excited cycloheptatriene in compressed gases, supercritical fluids, and liquids. J Chem Phys 106:4992, 1997.
111. BM Ladanyi, RM Stratt. Short-time dynamics of vibrational relaxation in molecular fluids. J Phys Chem A 102:1068, 1998.
112. A Moustakas, E Weitz. A comparison between the vibrational relaxation rate of HCl in liquid versus supercritical xenon. Chem Phys Lett 191:264, 1992.
113. SC Tucker, MW Maddox. The effect of solvent density inhomogeneities on solute dynamics in supercritical fluids: a theoretical perspective. J Phys Chem B 102:2437, 1998.

114. SC Tucker. Solvent density inhomogeneities in supercritical fluids. Chem Rev 99:391, 1999.
115. MP Heitz, M Maroncelli. Rotation of aromatic solutes in supercritical CO_2: are rotation times anomalously slow in the near critical regime? J Phys Chem A 101:5852, 1997.
116. SM Howdle, VN Bagratashvili. The effects of fluid density on the rotational Raman spectrum of hydrogen dissolved in supercritical carbon dioxide. Chem Phys Lett 214:215, 1993.
117. LD Ziegler, R Fan. The resonance fluorescence polarization of free rotors: methyl iodide in methane and carbon dioxide. J Chem Phys 105:3984, 1996.
118. S Okazaki, M Matsumoto, I Okada, K Maeda, Y Kataoka. Density dependence of rotational relaxation of supercritical CF_3H. J Chem Phys 103;8594, 1995.
119. S Okazaki, N Terauchi, I Okada. Raman spectroscopic study of rotational and vibrational relaxation of CF_3H in the supercritical state. J Mol Liquids 65/66:309, 1995.
120. X Pan, RA MacPhail. A Raman study of cyclopentane-d_9 pseudorotation dynamics as a function of pressure. Chem Phys Lett 212:64, 1993.
121. H Versmold. Depolarized Rayleigh scattering, molecular reorientation of CO_2 in a wide density range. Mol Phys 43:383, 1981.
122. TW Zerda, J Schroeder, J Jonas. Raman band shapes and dynamics of molecular motion for SF_6 in the supercritical dense fluid region. J Chem Phys 75:1612, 1981.
123. Y Garrabos, R Tufeu, B Le Neindre, G Zalczer, D Beysens. Rayleigh and Raman scattering near the critical point of carbon dioxide. J Chem Phys 72:4637, 1980.
124. J Eastoe, BH Robinson, AJWG Visser, DC Steytler. Rotational dynamics of AOT reversed micelles in near-critical and supercritical alkanes. J Chem Soc Faraday Trans 87:1899, 1991.
125. LE Bowman, BJ Palmer, BC Garrett, JL Fulton, CR Yonker, DM Pfund, SL Wallen. Infrared and molecular dynamics study of D_2O rotational relaxation in supercritical CO_2. J Phys Chem 100:18327, 1996.
126. CJ Jameson, AK Jameson, NC Smith. ^{15}N spin-relaxation studies of N_2 in buffer gases. Cross sections for molecular reorientation and rotational energy transfer. J Chem Phys 86:6833, 1987.
127. DM Lamb, ST Adamy, KW Woo, J Jonas. Transport and relaxation of naphthalene in supercritical fluids. J Phys Chem 93:5002, 1989.
128. RF Evilia, JM Robert, SL Whittenburg. NMR studies of rotational motion at low viscosity. J Phys Chem 93:6550, 1989.
129. CL Jameson, AK Jameson, MA ter Horst. ^{14}N Spin relaxation studies of N_2 in buffer gases. Cross sections for molecular reorientation and rotational energy transfer. J Chem Phys 95:5799, 1991.
130. P Etesse, AM Ward, WV House, R Kobayashi. Spin-lattice relaxation and self-diffusion near the critical point of carbon dioxide. Physica B 183:45, 1993.
131. S Bai, CMV Taylor, F Liu, CL Mayne, RJ Pugmire, DM Grant. CO_2 Clustering of 1-decanol and methanol in supercritical fluids by ^{13}C nuclear spin-lattice relaxation. J Phys Chem B 101:2923, 1997.

132. CMV Taylor, S Bai, CL Mayne, DM Grant. Hydrogen bonding of methanol in supercritical CO_2 studied by ^{13}C nuclear spin-lattice relaxation. J Phys Chem B 101:5652, 1997.
133. S Bai, CR Yonker. Pressure and temperature effects on the hydrogen-bond structure of liquid and supercritical fluid methanol. J Phys Chem 102:8641, 1998.
134. RM Anderton, JF Kauffman. Rotational relaxation in the compressible region of CO_2: Evidence for solute-induced clustering in supercritical fluid solutions. J Phys Chem 99:13759, 1995.
135. MP Heitz, FV Bright. Probing the scale of local density augmentation in supercritical fluids: a picosecond rotational reorientation study. J Phys Chem 100:6889, 1996.
136. JL deGrazia, TW Randolph, JA O'Brien. Rotational relaxation in supercritical carbon dioxide revisited: a study of solute-induced local density augmentation. J Phys Chem A 102:1674, 1998.
137. JM Dobbs, JM Wong, KP Johnston. Nonpolar co-solvents for solubility enhancement in supercritical fluid carbon dioxide. J Chem Eng Data 31:303, 1986.
138. JM Dobbs, JM Wong, RJ Lahiere, KP Johnston. Modification of supercritical fluid phase behavior using polar cosolvents. Ind Eng Chem Res 26:56, 1987.
139. JM Dobbs, KP Johnston. Selectivities in pure and mixed supercritical fluid solvents. Ind Eng Chem Res 26:1476, 1987.
140. DK Joshi, JM Prausnitz. Supercritical fluid extraction with mixed solvents. AIChE J 30:522, 1984.
141. JM Walsh, GD Ikonomou, MD Donohue. Supercritical phase behavior: the entrainer effect. Fluid Phase Equil 33:295, 1987.
142. JM Walsh, MD Donohue. Hydrogen bonding in entrainer cosolvent mixtures: a parametric analysis. Fluid Phase Equil 52:397, 1989.
143. JM Walsh, ML Greenfield, GD Ikonomou, MD Donohue. Hydrogen-bonding competition in entrainer cosolvent mixtures. Chem Eng Comm 86:125, 1989.
144. J Ke, S Jin, B Han, H Yan, D Shen. Hydrogen bonding of some organic acid in supercritical CO_2 with polar cosolvents. J Supercrit Fluids 11:53, 1997.
145. MP Ekart, KL Bennett, SM Ekart, GS Gurdial, CL Liotta, CA Eckert. Cosolvent interactions in supercritical fluid solutions. AIChE J 39:235, 1993.
146. SS Ting, SJ Macnaughton, DL Tomasko, NR Foster. Solubility of naproxen in supercritical carbon dioxide with and without cosolvents. Ind Eng Chem Res 32:1471, 1993.
147. SST Ting, DL Tomasko, SJ Macnaughton, NR Foster. Chemical-physical interpretation of cosolvent effects in supercritical fluids. Ind Eng Chem Res 32:1482, 1993.
148. GS Gurdial, SJ Macnaughton, DL Tomasko, NR Foster. Influence of chemical modifiers on the solubility of o- and m-hydroxybenzoic acid in supercritical CO_2. Ind Eng Chem Res 32:1488, 1993.
149. S Kim, KP Johnston. Clustering in supercritical fluid mixtures. AIChE J 33:1603, 1987.
150. CR Yonker, DR Smith. Solvatochromic behavior of binary supercritical fluids: the carbon dioxide/2-propanol system. J Phys Chem 92:2374, 1988.

151. ML O'Neill, P Kruus, RC Burk. Solvatochromic parameters and solubilities in supercritical fluid systems. Can J Chem 71:1834, 1993.
152. DS Bulgarevich, T Sako, T Sugeta, K Otake, M Sato, M Uesugi, M Kato. Microscopic solvent structure of supercritical carbon dioxide and its mixtures with methanol in the cybotactic region of the solute molecule. J Chem Phys 108:3915, 1998.
153. KP Hafner, FLL Pouillot, CL Liotta, CA Eckert. Solvatochromic study of basic cosolvents in supercritical ethane. AIChE J 43:847, 1997.
154. JM Tingey, CR Yonker, RD Smith. Spectroscopic studies of metal chelates in supercritical fluids. J Phys Chem 93:2140, 1989.
155. JF Deye, TA Berger, AG Anderson. Nile red as a solvatochromic dye for measuring solvent strength in normal liquids and mixtures of normal liquids with supercritical and near critical fluids. Anal Chem 62:615, 1990.
156. BL Knutson, SH Sherman, KL Bennett, CL Liotta, CA Eckert. Benzophenone as a probe of local cosolvent effects in supercritical ethane. Ind Eng Chem Res 36:854, 1997.
157. TA Betts, FV Bright. Reversible excited-state transient solvation in binary supercritical fluids revealed by multifrequency phase and modulation fluorescence. Appl Spectrosc 44:1203, 1990.
158. DL Tomasko, BL Knutson, F Pouillot, CL Liotta, CA Eckert. Spectroscopic study of structure and interactions in cosolvent-modified supercritical fluids. J Phys Chem 97:11823, 1993.
159. 93-3 DL Tomasko, BL Knuston, JM Coppom, W Windson, B West, CA Eckert. Fluorescence spectroscopy study of alcohol–solute interactions in supercritical carbon dioxide. ACS Symp. Series 514:220, 1993.
160. J Zagrobelny, FV Bright. Probing solute–entrainer interactions in matrix modified supercritical CO_2. J Am Chem Soc 115:701, 1993.
161. Y-P Sun, G Bennett, KP Johnston, MA Fox. Studies of solute–solvent interactions in mixtures of supercritical fluids using fluorescence spectroscopy. J Phys Chem 96:10001, 1992.
162. Y-P Sun, MA Fox. Picosecond transient absorption study of the twisted excited singlet state of tetraphenylethylene in supercritical fluids. J Am Chem Soc 115:747, 1993.
163. RD Schulte, JF Kauffman. Solvation in mixed supercritical fluids: TICT spectra of bis(4,4′-aminophenyl) sulfone in ethanol/CO_2. J Phys Chem 98:8793, 1994.
164. RD Schulte, JF Kauffman. Fluorescence from the twisted intramolecular charge transfer compound bis(4,4′-dimethylaminophenyl)sulfone in ethanol/CO_2: a probe of local solvent composition. Appl Spectrosc 49:31, 1995.
165. AK Dillow, KP Hafner, SLJ Yun, F Deng, SG Kazarian, CL Liotta, CA Eckert. Cosolvent tuning of tautomeric equilibrium in supercritical fluids. AIChE J 43:515, 1997.
166. JF Brennecke, DL Tomasko. Naphthalene/triethylamine exciples and pyrene excimer formation in supercritical fluid solutions. J Phys Chem 94:7692, 1990.
167. J Zagrobelny, TA Betts, FV Bright. Steady-state and time-resolved fluorescence investigations of pyrene excimer formation in supercritical CO_2. J Am Chem Soc 114:5249, 1992.

168. Y-P Sun. Excitation wavelength dependence of pyrene fluorescence in supercritical carbon dioxide. Evidence for a supercritical solvent-assisted solute–solute clustering mechanism. J Am Chem Soc 115:3340, 1993.
169. H Inomata, H Hamatani, N Wada, Y Yagi, S Saito. Intermolecular exciplex of pyrene/*N*,*N*-dimethylaniline in supercritical carbon dioxide. J Phys Chem 97:6332, 1993.
170. T Okada, Y Kobayashi, H Yamasa, N Mataga. Intramolecular exciplex fluorescence in supercritical fluids. Chem Phys Lett 128:583, 1986.
171. JK Rice, SJ Christopher, U Narang, WR Peifer, FV Bright. Evidence for changes in the conformation of flexible solutes dissolved in supercritical solvents. Analyst 119:505, 1994.
172. HW Rollins, R Dabestanni, Y-P Sun. Spectroscopic investigations of intramolecular and intermolecular excimers of 1,3-di(2-naphthyl)propane and methylnaphthalenes in supercritical carbon dioxide. Chem Phys Lett 268:187, 1997.
173. Y Kimura, Y Yoshimura. Chemical equilibrium in fluids from the gaseous to liquid states: solvent density dependence of the dimerization equilibrium of 2-methyl-2-nitrosopropane in carbon dioxide, chlorotrifluoromethane, and trifluoromethane. J Chem Phys 96:3085, 1992.
174. Y Kimura, Y Yoshimura, M Nakahara. Chemical reactions in the medium density fluid. anomaly in the volume profile of the dimerization reaction of 2-methyl-2-nitrosopropane in carbon dioxide. Chem Lett 617, 1987.
175. Y Kimura, Y Yoshimura, M Nakahara. Chemical reaction in medium density fluid. solvent density effects on the dimerization equilibrium of 2-methyl-2-nitrosopropane in carbon dioxide. J Chem Phys 90:5679, 1989.
176. Y Kimura, Y Yoshimura. Chemical equilibrium in simple fluids: solvent density dependence of the dimerization equilibrium of 2-methyl-2-nitrosopropane in argon and xenon. J Chem Phys 96:3824, 1992.
177. BJ Hrnjez, AJ Mehta, MA Fox, KP Johnston. Photodimerization of isophorone in supercritical trifluoromethane and carbon dioxide. J Am Chem Soc 111:2662, 1989.
178. H Tsugane, Y Yagi, H Inomata, S Saito. Dimerization of benzoic acid in saturated solutions of supercritical carbon dioxide. J Chem Eng Japan 25:351, 1992.
179. CE Bunker, HW Rollins, JR Gord, YP Sun. Efficient photodimerization reaction of anthracene in supercritical carbon dioxide. J Org Chem 62:7324, 1997.
180. CE Bunker, Y-P Sun. Evidence for enhanced bimolecular reactions in supercritical CO_2 at near-critical densities from a time-resolved study of fluorescence quenching of 9,10-bis(phenylethynyl)anthracene by carbon tetrabromide. J Am Chem Soc 117:10865, 1995.
181. CE Bunker, Y-P Sun, JR Gord. Time-resolved studies of fluorescence quenching in supercritical carbon dioxide: system dependence in the enhancement of bimolecular rates at near-critical densities. J Phys Chem A 101:9233, 1997.
182. J Zhang, DP Roek, JE Chateauneuf, JF Brennecke. A steady-state and time-resolved fluorescence study of quenching reactions of anthracene and 1,2-benzanthracene by carbon tetrabromide and bromoethane in supercritical carbon dioxide. J Am Chem Soc 119:9980, 1997.

183. DP Roek, JE Chateauneuf, JF Brennecke. A fluorescence lifetime and integral equation study of the quenching of naphthalene fluorescence by bromoethane in super- and subcritical ethane. Ind Eng Chem Res 39:3090, 2000.
184. CB Roberts, JE Chateaunuef, JF Brennecke. Unique pressure effects on the absolute kinetics of triplet benzophenone photoreduction in supercritical CO_2. J Am Chem Soc 114:8455, 1992.
185. CB Roberts, JF Brennecke, JE Cheteauneuf. Solvation effects on reactions of triplet benzophenone in supercritical fluids. AIChE J 41:1306, 1995.
186. DP Roek, MJ Kremer, CB Roberts, JE Cheteauneuf, JF Brennecke. Spectroscopic studies of solvent effects on reactions in supercritical fluids. Fluid Phase Equilibria 158:713, 1999.
187. CB Roberts, J Zhang, JF Brennecke, JE Chateauneuf. Laser flash photolysis investigations of diffusion-controlled reactions in supercritical fluids. J Phys Chem 97:5618, 1993.
188. CB Roberts, J Zhang, JE Chateauneuf, JF Brennecke. Diffusion-controlled reactions in supercritical CHF_3 and CO_2/acetonitrile mixtures. J Am Chem Soc 115:9576, 1993.
189. M Kremer, KA Connery, MM DiPippo, J Feng, JE Chateauneuf, JF Brennecke. Laser flash photolysis investigation of the triplet-triplet annihilation of anthracene in supercritical water. J Phys Chem A 103:6591, 1999.
190. CB Roberts, J Zhang, JE Chateauneuf, JF Brennecke. Laser flash photolysis and integral equation theory to investigate reactions of dilute solutes with oxygen in supercritical fluids. J Am Chem Soc 117:6553, 1995.
191. J Zhang, KA Connery, JF Brennecke, JE Chateauneuf. Pulse radiolysis investigations of solvation effects on arylmethyl cation reactivity in supercritical fluids. J Phys Chem 100:12394, 1996.
192. K Takahashi, CD Jonah. The measurement of an electron transfer reaction in a non-polar supercritical fluid. Chem Phys Lett 264:297, 1997.
193. DR Worrall, FJ Wilkinson. Photochemistry in modified supercritical carbon dioxide. Effect of modifier concentration of diffusion probed by triplet-triplet energy transfer. Chem Soc Faraday Trans 92:1467, 1996.
194. TW Randolph, C Carlier. Free-radical reactions in supercritical ethane: a probe of supercritical fluid structure. J Phys Chem 96:5146, 1992.
195. JM Tanko, NK Suleman, B Fletcher. Viscosity-dependent behavior of geminate caged-pairs in supercritical fluid solvent. J Am Chem Soc 118:11958, 1996.
196. B Fletcher, NK Suleman, JM Tanko. Free radical chlorination of alkanes in supercritical carbon dioxide: the chlorine atom cage effect as a probe for enhanced cage effects in supercritical fluid solvents. J Am Chem Soc 120:11839, 1998.

2
NMR Investigation of High-Pressure, High-Temperature Chemistry and Fluid Dynamics

Clement R. Yonker
Pacific Northwest National Laboratory, Richland, Washington

Markus M. Hoffmann
State University of New York–Brockport, Brockport, New York

I. INTRODUCTION

Supercritical fluids hold special promise as novel solvents in current and future industrial applications. The favorable phase behavior, solubility, and mass transport characteristics in supercritical fluids has lead to a growing interest in exploiting supercritical fluids as a medium for chemical reactions. Presently, supercritical CO_2 is being used in industrial processes involving the extraction of hop essence, in dry cleaning of clothes, and in paint spraying. Recently, DuPont announced to build a facility to evaluate supercritical CO_2 as a solvent for the industrial production of fluoropolymers. Most of industry's supercritical fluid applications involve either bulk extractions or their use as reaction solvents. In the drug industry there is the need for conversion of pharmaceuticals into nanometer-size particles for injectable use. This materials commutation can be accomplished using a supercritical fluid procedure [Rapid Expansion of Supercritical Fluid Solutions (RESS)]. All these industrial processes are built on a fundamental understanding of supercritical fluids. The need still exists for a molecular-level understanding of both the kinetic and thermodynamic behavior of these compressible solvents [a recent excellent review of supercritical fluid fundamentals and applications has been published (1)].

The rationale for using pressure as an experimental variable is to gain an enhanced understanding, on a molecular level, of the nature of the chemical or dynamic process occurring in solution which could be hidden or unobtainable by temperature variation alone. Some of the effects of pressure in solution are briefly described as follows. First, using variable-pressure measurements the activation volume for a reaction can be obtained that can be used for mechanistic assignment in a reaction sequence. Second, changing temperature at atmospheric pressure produces a concurrent change in the thermal energy and volume of the system. To separate this thermal effect from the volume effect, a high-pressure experiment must be performed. From numerous experimental efforts, it has been determined that volume effects can determine the mechanism of a dynamic process, whereas temperature changes the frequency of molecular motion usually without affecting the mechanism. Third, the solvent environment (density, viscosity, dielectric) about a solute molecule can be continuously changed as a function of pressure without having to alter the solvent composition. This is a particularly important reason for using supercritical fluid solvents in physicochemical investigations of reaction mechanisms, chemical equilibria, or solution dynamics because of the wide range of solvent conditions that can be sampled as a function of pressure (and temperature). Therefore, from high-pressure NMR experiments, one could expect to obtain fundamental molecular-level information about a wide range of chemical and dynamic processes in solutions for both liquids and supercritical fluids.

There are numerous experimental techniques that have been used to investigate supercritical fluids. These range from Fourier transform infrared (FTIR), UV-visible (UV-vis), fluorescence, electron spin resonance (ESR), and x-ray spectroscopies (2). NMR is a technique that has seen limited application to supercritical fluid solvents due to the specialized need for the design of a high-pressure, nonmagnetic probe and its associated electronics. There have been different successful solutions to a functioning high-pressure NMR probe (3–13) and each of these probe designs has its own strengths and weaknesses. Overall, NMR is an information-rich spectroscopic technique that can describe the solvent environment about a solute molecule, determine self-diffusion coefficients, ascertain molecular structure, measure hydrogen bonding in solution, and describe molecular clustering as a function of density. NMR can provide important molecular-level information about the density dependence of rotational and translational dynamics in supercritical fluid solutions. Similarly, high-pressure kinetics and chemical equilibria can be investigated by the use of NMR.

We hope this chapter encourages and inspires researchers to use high-pressure NMR to directly investigate chemical reactions and solution dynamics in supercritical fluids. An overview of this type, by its nature, cannot be considered totally inclusive; it is hoped that practitioners will gain a better appreciation of the type of molecular-level information that NMR can provide for both materials and solution chemistries in supercritical fluids.

II. FUNDAMENTAL SCIENCE

This section describes the use of high-pressure NMR to obtain an understanding on a molecular level of the physicochemical processes occurring in a supercritical fluid solution. NMR is a powerful technique not only for determining the chemical structure of a molecule but also for determining solution dynamics through measurement of self-diffusion coefficients and molecular relaxation rates. Due to the sensitivity of a nuclei to its local solvation environment, it is possible to investigate this environment and determine the effect of pressure and temperature on its composition and structure. The investigations described below have mainly been undertaken in our laboratory using a capillary high-pressure NMR cell. Other research efforts have been illustrated, where appropriate, to enhance the discussion covering critical areas of high-pressure NMR research as applied to the physicochemical determination of solution behavior as a function of pressure and temperature.

A. Chemical Shifts: Hydrogen Bonding

Aggregation and association in alcohols have typically been used to study hydrogen bonding dynamics in solutions. Methanol and *tert*-butanol can associate through hydrogen bonding, and details about the dynamics of this interaction in solution have been investigated for both liquid and supercritical conditions (14–16). Temperature is typically the thermodynamic variable used to investigate changes in hydrogen bonding in self-associating molecules such as methanol, but pressure can be used in a similar manner to study the extent of hydrogen bonding. High-pressure NMR investigations of hydrogen-bonding alcohols have provided information about the solution structure in such solvents (17–22), but they have been limited in number due to the complexity of the experimental system.

The nuclear shielding constant (σ) is an absolute measure of the electronic distribution about the nucleus and its effect on the observed magnetic moment of that nuclei in the applied magnetic field, which is sensitive to a molecule's chemical structure and local solvation environment. The nuclear shielding for a molecule can be related to (23)

$$\sigma_{(total)} = \sigma_B + \sigma_A + \sigma_W + \sigma_E + \sigma_{EX} + \sigma_S \tag{1}$$

where σ_B is the contribution from the bulk magnetic susceptibility, σ_A is the contribution from the anisotropy of the magnetic susceptibility of the molecule, σ_W is due to the van der Waals dispersion interactions, σ_E arises from the polarization of the solvent due to a permanent dipole moment in the molecule, σ_{EX} represents the effective short-range exchange interactions, and σ_S is the contribution from specific interactions such as hydrogen bonding.

For the molecules investigated, the $(CH_3)_3$, CH_3, and OH groups will each experience their own shielding environment as seen in Eq. (1). One assumes that changes in pressure or temperature affect the nonspecific contributions to the nuclear shielding in a similar manner for all the different groups' resonances. Thus, the difference between the shielding of the groups can be related to the specific interactions in solution, σ_S, which is due to hydrogen bonding of the OH group. The use of the chemical shift difference [$\Delta\delta$ (ppm)] between either the OH and $(CH_3)_3$ groups or the OH and CH_3 groups for *tert*-butanol and methanol, respectively, eliminates the nonspecific contributions that augment the nuclear shielding as a function of pressure and temperature. Therefore, through the investigation of the chemical shifts of the $(CH_3)_3$, CH_3, and OH groups of pure *tert*-butanol and methanol over an extended temperature and pressure range, $\Delta\delta$ can be used to qualitatively estimate changes in the hydrogen bond network in solution as a function of pressure and temperature.

Figure 1 is a plot of $\Delta\delta$ (ppm) vs. pressure at various temperatures for *tert*-butanol and methanol. Focusing on *tert*-butanol, at constant temperature, $\Delta\delta$ increases as pressure increases. At constant pressure, $\Delta\delta$ decreases with increasing temperature. The slope $[(\partial\Delta\delta/\partial P)T]$ for the three temperatures shown in Figure 1 for *tert*-butanol (50, 100, and 150°C) increases with increasing tempera-

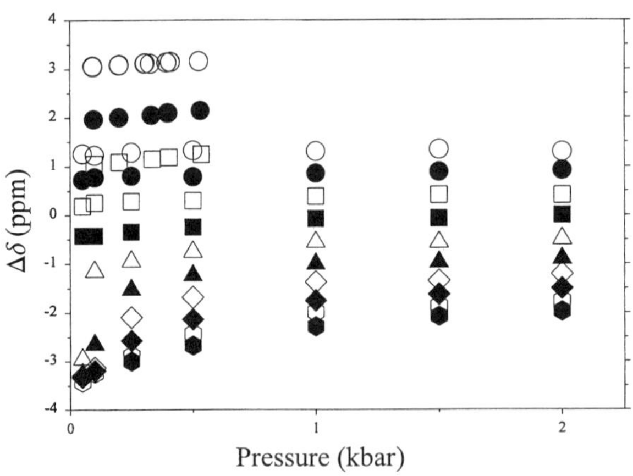

Figure 1 Plot of the chemical shift difference $\Delta\delta$ (ppm) vs. pressure for *tert*-butanol and methanol. *tert*-Butanol: (○, 50°C), (●, 100°C), (□, 150°C), pressure to 0.5 kbar. Methanol: (○, 50°C), (●, 100°C), (□, 150°C), (■, 200°C), (△, 250°C), (▲, 300°C), (◇, 350°C), (◆, 400°C), (⬡, 450°C), (⬢, 500°C), pressure to 2.0 kbar.

ture. These observations can be explained within the framework of the hydrogen bonding occurring in solution. Hydrogen bonding removes electron density from the vicinity of the ^1H nucleus contributing to the deshielding of the proton. Qualitatively, an increase in $\Delta\delta$ correlates with an increase in the deshielding of the OH proton relative to that of the $(CH_3)_3$ group and thus an increase in hydrogen bonding in solution. The results in Figure 1 demonstrate that increasing temperature at constant pressure tends to decrease the extent of hydrogen bonding in *tert*-butanol, whereas increasing pressure at constant temperature increases hydrogen bonding in solution. One would anticipate that increasing temperature would more readily disrupt hydrogen bonds in solution. Increasing pressure at high temperatures should have a large effect on the solutions' hydrogen bond network, contributing to the larger slope $[(\partial\Delta\delta/\partial P)_T]$ seen at the higher temperatures. This behavior is more readily apparent for methanol as discussed in the following section.

For methanol, the $\Delta\delta$ data were obtained over a much wider range of pressure and temperature (50–500°C and 2 kbar). Similar behavior is seen for $\Delta\delta$ in methanol as a function of pressure and temperature when compared to *tert*-butanol. A dramatic change of $\Delta\delta$ in the vicinity of the methanol critical point (methanol T_c is 239.4°C) at low pressure is observed (note the 250°C to 350°C isotherm). This is related to the large changes in density and thus hydrogen bonding of solution in this region. As pressure is increased through this temperature region, hydrogen bonding increases, which contributes to a change in shielding of the nucleus and thus to a change in $\Delta\delta$. If one focuses on the pressure region of 0.5–2.0 kbar, it is more readily discernible that the slope $[(\partial\Delta\delta/\partial P)_T]$ increases at higher temperatures. This could be due to a change in both the extent and strength of the hydrogen bond network at high temperatures as one changes pressure as compared to a change in hydrogen bond strength alone at low (50°C) temperatures with increasing pressure (22). However, the NMR chemical shift data presented in Figure 1 clearly indicate that significant hydrogen bond interactions exist for methanol at high temperatures and pressures and in the critical region. The results reported here using the capillary high-pressure NMR cell are in good agreement with earlier measurements (18–20) at comparable temperatures and pressures.

Further support for hydrogen bonding interactions remaining important in methanol in the near-critical region is provided by a comparison with gaseous methanol. Ultimately, at the limit of infinitely high temperatures and low densities any substance approaches the ideal gas limit where there are no intermolecular interactions and hence no hydrogen bonding. Hoffmann and Conradi measured the hydroxyl proton chemical shifts of ethanol, methanol (19), and water (24) over pressure and temperature ranges that cover all three phases: vapor, liquid, and supercritical. These measurements established for all three solvents the hydroxyl proton chemical shift value for the low-density, high-temperature ideal gas limit. In each case, an estimate of the extent of hydrogen bonding near the critical point was obtained by applying a simple two-state model, where each site is either bonded or nonbonded. The chemical shift measurement is an average value over all sites, which leads to a linear relationship between the chemical shift σ and the extent of hydrogen bonding θ. Setting $\theta = 0$ at the ideal gas limit and $\theta = 1$ at ambient conditions, the extent of hydrogen bonding at the critical point of water, ethanol, and methanol compared to room temperature was 0.25, 0.3, and 0.4, respectively. This analysis further supports the premise that hydrogen bonding remains important at the critical point for these three solvents. Hoffmann and Conradi also provide a comparison between the solvents on the basis of reduced thermodynamic variables (19). The density-reduced extent of hydrogen bonding (θ/ρ^*) was calculated for the alcohols where the density ρ^* was scaled to ambient conditions. The density dependence of this quantity reveals a near constancy at liquid-like densities but a strong density dependence at gas-like densities. The observed strong increase of the θ/ρ^* isotherms with density at gas-like densities was related to the formation of hydrogen bonded oligomers with indications that dimer formation is limited in favor of trimers and possibly higher ordered oligomers. The NMR evidence for the formation of hydrogen-bonded oligomers in water and the alcohols at low, gas-like densities is in keeping with a large body of experimental gas phase results, as reviewed in reference (25).

B. Chemical Shifts: Solution Behavior

High-pressure NMR has been used to study the solution behavior of polymers in supercritical CO_2. The polymers investigated by Dardin et al., using a folded capillary design for the high-pressure NMR cell, were poly(1,1-dihydroperfluorooctyl acrylate) and poly(1,1-dihydroperfluorooctyl acrylate-block-stryene) (26). The proton chemical shifts of the polymers were determined as a function of temperature and pressure. The upper critical solution temperature and the upper critical solution density were determined from a transition in the chemical shift of the proton resonances of the polymers with CO_2 density. A coexistence region was determined at intermediate densities of CO_2, in which the polymers existed in two distinct solution environments. In the two-phase region, the chemical shift of the polymer-rich phase was used to estimate the amount of CO_2 in that phase. The block copolymer is known to form micelles in supercritical CO_2. From the NMR studies it was inferred that at high CO_2 densities the styrene units close to the core–shell interface were highly solubilized in CO_2.

In a second series of experiments by Dardin et al. (27), using a folded-capillary high-pressure NMR cell, the authors investigated the ^1H and ^{19}F chemical shifts of hexane, perfluorohexane, and 1,1-dihydroperfluorooctylpropionate in supercritical CO_2 as a function of pressure and temperature. The nuclear shielding for a molecule as described by Eq. (1) is related to specific and nonspecific intermolecular interactions. These interactions will all contribute to the experimental chemical shift of the nuclei. The ^1H chemical shifts of hexane in solution were determined to be solely governed by the contribution to the nuclear shielding due to the bulk susceptibility, σ_B, which is a function of CO_2 density. For the ^{19}F chemical shift, the van der Waals dispersion interaction term, σ_W, was also needed to explain the experimental chemical shifts of perfluorohexane as a function of temperature and pressure. These findings were interpreted as indicating nonspecific van der Waals interactions between the fluorinated sites on the solute molecule and CO_2.

Overall, high-pressure capillary NMR investigations are proving to be very useful for studying solute–solvent interactions in solution and other solvent effects (i.e., hydrogen bonding) as a function of solution pressure and temperature.

C. Relaxation Time (T_1) Measurements

NMR relaxation measurements provide information about the rotational reorientation and spatial reorientation (translational motion) of molecules in solution. A recent review of the density dependence on rotational and translational molecular dynamics was published in 1993 (28). Using high-pressure capillary NMR spectroscopy, the determination of ^{19}F, ^1H, and ^2H relaxation times (T_1) of perfluorobenzene, benzene, and perdeuterobenzene were measured in carbon dioxide as a function of pressure and temperature to address the role of potential CO_2/F intermolecular interactions in solution (29). The pressure range for the relaxation time measurement was between 0.4 and 2.33 kbar over the temperature limits of 25–150°C. The density of the solvent, carbon dioxide, over these conditions was between 0.55 and 1.27 g/cm^3. Over these conditions, the contributions to the molecular relaxation processes for both ^1H and ^{19}F in CO_2 could be determined. From the comparison of the relaxation processes for ^{19}F and ^1H in CO_2, especially at high densities, the occurrence of specific molecular interactions between CO_2 and fluorine could be addressed.

The spin-lattice relaxation of a molecule is governed by its interactions with the surrounding solvent bath through complex processes, some of which are composed of internal reorientation, spatial translation, and changes in angular momentum. The spin-lattice relaxation mechanism can be described as being composed of dipole–dipole interactions (DD), spin-rotation interactions (SR), quadrupolar interactions (Q), chemical shift anisotropy (CSA), and scalar coupling (SC). These represent the more common nuclear relaxation processes encountered in liquids and gases. These magnetic interactions DD, SR, CSA, SC and magnetic/electric field interactions (Q), will contribute to different degrees in the reequilibration of the nuclei after excitation by the radiofrequency (rf) field pulse.

The DD relaxation mechanism is combined of both intramolecular and intermolecular processes. A detailed account of the derivation of T_1 from the various intermolecular and intramolecular dipole–dipole interactions has been described by Bloembergen et al. (30). The intramolecular relaxation process is governed by the angular reorientation of the vector connecting the spin one-half ($\frac{1}{2}$) nuclei—in this case either ^{1}H or ^{19}F in benzene or perfluorobenzene. The relaxation rate ($1/T_1$) is

$$1/T_1(\mathrm{DD-intra}) = 9/10\gamma_{\mathrm{H}}{}^4\hbar^2 r^{-6}{}_{(\mathrm{H-H})}\tau_{\mathrm{c}} \quad (2)$$

here γ is the magnetogyric ratio for the proton (or ^{19}F), $\hbar$ is Planck's constant over 2π, $r_{(\mathrm{H-H})}$ is the proton–proton distance in benzene (or fluorine–fluorine distance in perfluorobenzene), and τ_c is the rotational correlation time of the molecule. All of these molecules have a C_6 symmetry axis for in plane rotational orientation and a C_2 symmetry axis of tumbling about the molecular plane. It is impossible from this investigation to determine the difference between these two types of molecular motion. The intermolecular relaxation process has a complex dependence on angular position and spatial reorientation. This has been simplified by expressing the dependence of the relaxation rate in terms of the self-diffusion coefficient (D):

$$1/T_1(DD-\mathrm{inter}) = (3\pi/10)N_{\mathrm{o}}\gamma_{\mathrm{H}}{}^4\hbar^2/aD \quad (3)$$

here N_o is the number density and a is the distance of closest approach of the nuclei.

The spin-rotation relaxation process becomes important in gases or supercritical fluids at low densities and high temperatures (16). This relaxation rate is expressed as

$$1/T_1(SR) = (2/3)kT\hbar^{-2}\mathrm{I}(2c_{\perp}{}^2 + c_{\parallel}{}^2)\tau_{\mathrm{J}} \quad (4)$$

where I is the moment of inertia for the molecule, k is Boltzmann's constant, T is temperature, $c_{\perp}$ and $c_{\parallel}$ are the spin-rotation coupling constants, and τ_J is the angular momentum correlation time for the molecule. The spin-rotation relaxation rate is greater for ^{19}F as compared to ^{1}H because the moment of inertia and the coupling constants are larger for perfluorobenzene (31). It should be noted that τ_J and τ_c have opposite dependence on temperature, i.e., as the gas temperature increases τ_c decreases whereas τ_J increases. The density dependence is opposite also, i.e., at high density/low temperatures τ_c is long and τ_J is short, whereas for high temperatures/low density, τ_c is short and τ_J is long. In fact, for liquids $\tau_J \ll \tau_c$, such that spin rotation does not play a role in relaxation.

The quadrupolar relaxation mechanism is the dominant process for nuclei with spins greater than $\frac{1}{2}$. Quadrupolar relaxation efficiency is determined by the magnitude of the nuclear quadrupole and the electric field gradient at the nucleus. This interaction is modulated by molecular rotation in a similar manner as for dipole–dipole relaxation. The quadrupolar relaxation rate is

$$1/T_1(\mathrm{Q}) = (3\pi^2/10)(2I + 3/(I^2(2I - 1)))(1 + \eta_s{}^2/3)\chi^2\tau_c \qquad (5)$$

where η_s is the measure of asymmetry of the quadrupolar nuclei, I is the spin quantum number, and χ is the nuclear quadrupole coupling constant (which is the product of the electric field gradient and the nuclear quadrupole moment). The other relaxation processes, CSA and SC, are assumed to play a negligible role in molecular relaxation for the CO_2 solution (32).

The effect of density on the relaxation time for benzene and perfluorobenzene is shown in Figure 2 for the two temperature extremes. At high densities/low temperatures the relaxation times for both C_6F_6 and C_6H_6 are similar. At density values $\geq$ 0.8 g/cm^3 the relaxation times for C_6F_6 over the temperature range studied (25–150°C) were very similar. Benzene has a large variation in T_1 as a function of temperature and density. The T_1 values for C_6F_6 and C_6H_6 in CO_2 are similar to those of the pure liquids for the lower temperature (31,33). The relaxation time for C_6D_6 was much faster than the other two solute molecules due to quadrupolar relaxation. Using C_6D_6 allows one to separate the intermolecular from the intramolecular dipole–dipole relaxation contributions for this series of solute molecules. As the molecular reorientation correlation time in Eq. (5) is the same as the molecular reorientation correlation time in Eq. (2). For benzene the dipole–dipole intramolecular relaxation time

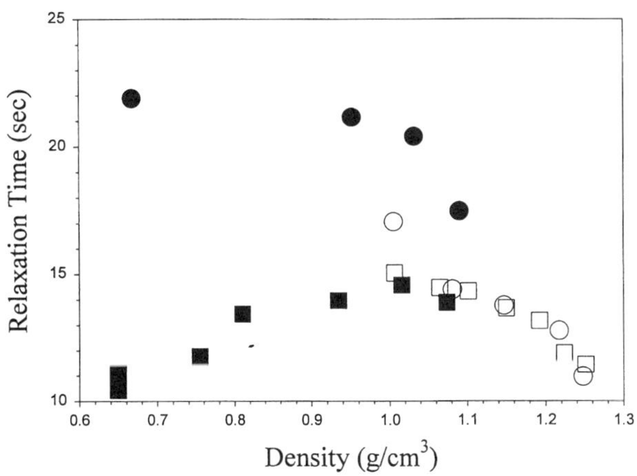

Figure 2 Plot of relaxation time for C_6H_6 and C_6F_6 vs. CO_2 density; C_6H_6: (○, 30°C), (●, 150°C), and C_6F_6: (□, 25°C), (■, 150°C).

calculated from Eq. (2) is dominant throughout most of the density range. At high density, intermolecular dipole–dipole relaxation [determined from Eq. (3)] begins to play a role in relaxation as the diffusion coefficient decreases. This is similar to the T_1 results reported for methanol as a function of pressure and temperature (16). For C_6F_6 at low temperatures the relaxation mechanism is similar to that determined for benzene. On the other hand, for high temperatures, the spin-rotation relaxation mechanism [determined from Eq. (4)] effects nuclear relaxation. This relaxation process is related to the number of molecular collisions in solution. At high temperature/low density spin rotation becomes a major factor in molecular relaxation as reported for benzene near its critical temperature (31) and methanol (16). The difference between the T_1 values for C_6H_6 and C_6H_6 at 150°C with decreasing density seen in Figure 2 is due to spin-rotation relaxation. ^{19}F is affected to a much greater degree than ^{1}H since C_6H_6 has a larger moment of inertia then benzene ($I_{C6F6}/I_{C6H6} = 5.6$) and the spin-rotation coupling constants are larger for ^{19}F than ^{1}H and appear squared in Eq. (4).

As apparent in these measurements of the relaxation times for C_6H_6 and C_6H_6 in CO_2 over similar pressures and temperatures, there is no experimental manifestation of a specific intermolecular interaction between CO_2 and fluorine. These interactions, if prevalent, would be expected to be seen in a change in relaxation rate or mechanism at high densities where the intermolecular distance between the CO_2 molecule and the fluorine group would be the smallest and their potential specific interaction the greatest. It appears that at high densities, solution viscosity dominates the relaxation process, and the relaxation mechanism for both ^{19}F and ^{1}H are similar. Therefore, there is no experimental evidence for a specific CO_2-F interaction that impacts on the relaxation of these two molecules, which supports the calculations of Diep et al. (34) and the experimental efforts of Yee et al. (35).

D. Vapor Liquid Equilibrium Measurements

NMR can be used to investigate the phase behavior of complex, multicomponent solvent systems as a function of pressure and temperature, with the molar composition of the different phases being determined simultaneously, in situ, using a high-pressure capillary NMR cell (36). In a similar manner, the hydrogen bonding behavior of the polar modifier can be determined and provides important physicochemical information regarding solvent interactions occurring in both the liquid and vapor phase. Typically, the vapor–liquid equilibrium (VLE) for a solution is determined using variable-volume high-pressure view cells, with remote sampling and off-line analysis to determine phase composition. In this way, the phase behavior of the system with regard to pressure, temperature, and composition can be determined. However, these techniques are labor and equipment intensive and are not commonly available in most typical laboratories. For most supercritical fluid solvent systems of interest the solvent molecules are hydrocarbons (a notable exception is supercritical CO_2). This presents an opportunity for the use of high-pressure NMR to determine the pressure–temperature–composition behavior of binary hydrocarbon solvents because of the ease of proton detection on the solvent molecules in both the vapor and liquid phase simultaneously. In practice, the application of high-pressure NMR for the determination of VLE phase behavior could be extended to any spin one-half ($\frac{1}{2}$) nuclei of adequate sensitivity. A hydrocarbon-containing solvent system, ethylene/methanol, was investigated demonstrating the advantages and limitations of high-pressure NMR for VLE determinations.

The VLE experimental data for the ethylene/methanol binary solvent system at 140°C is shown in Figure 3. The initial mole fraction of methanol at the starting conditions of the NMR experiment was 0.54. This was determined from the peak areas in the single-phase liquid region of the VLE phase diagram. Liquid phase equilibrium data determined by McHugh et al. (37) is shown for comparison in Figure 3. At pressures above the two-phase region only a single liquid phase was detected in the capillary cell. As pressure decreased the two-phase region was entered, resulting in an NMR spectrum containing both liquid and vapor phase. Figure 4A and 4B shows the two-phase and single-phase NMR spectra for the ethylene/methanol binary system at the pressures of 130.0 and 269.5 bar, respectively. The vapor and liquid phases are readily distinguished due to the differences in the chemical shifts between the two sepa-

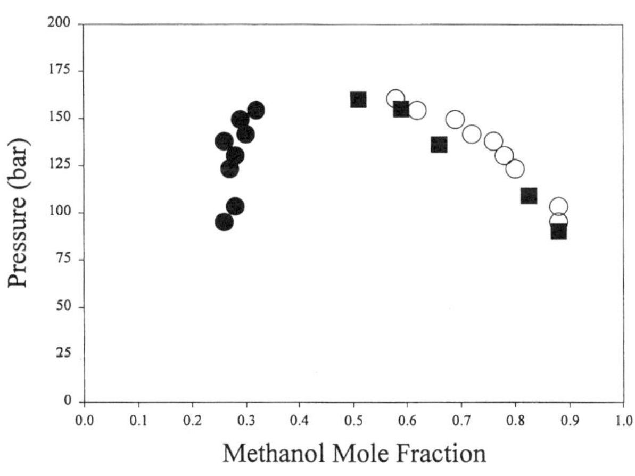

Figure 3 Plot of the experimental phase behavior for ethylene/methanol at 140°C; vapor phase (●), liquid phase (○), and liquid phase data (■) reported from McHugh et al. (From Ref. 37.)

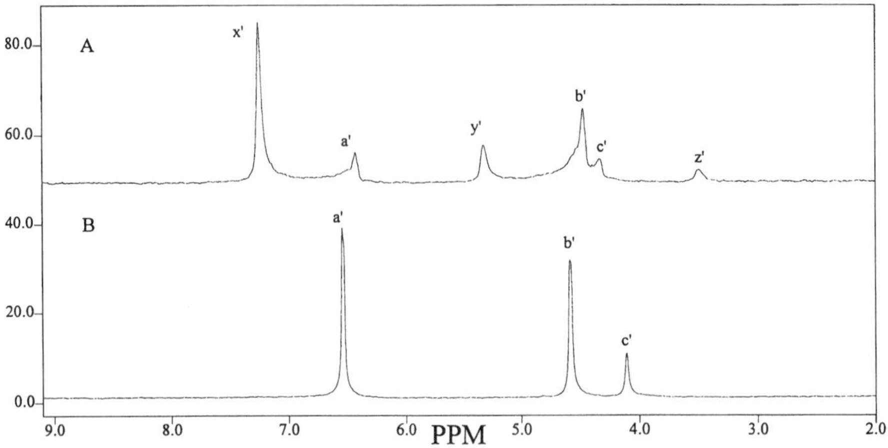

Figure 4 The high-pressure NMR spectra of the binary phase behavior of ethylene/methanol at 140°C. Spectra A—ethylene/methanol at 130.0 bar; liquid phase: a′, ethylene; b′, CH_3 group in methanol; and c′, OH group in methanol. Vapor phase: x′, ethylene; y′, CH_3 group in methanol, and z′, OH group in methanol. Spectra B—ethylene/methanol at 269.5 bar in the single-phase region: a′, ethylene, b′, CH_3 group in methanol, and c′, OH group in methanol.

rate environments. The separately resolved peaks of the vapor and liquid phases allow the simultaneous quantitation of the mole fraction of methanol in the two phases.

A unique advantage of the NMR technique is the ability to determine the behavior of the methanol protons in the two phases as a function of pressure. The change in chemical shift, $\Delta\delta$ (ppm), was used to describe the dynamics and extent of methanol–hydrogen bonding in the two-phase system as a function of pressure. For the ethylene/methanol binary solvent system at 140°C, methanol in the liquid phase exhibits a decrease in the extent of hydrogen bonding with increasing pressure, but in the vapor phase methanol demonstrates an increase in hydrogen bonding with increasing pressure (36). This is due to the change in density of the two phases as one approaches the critical pressure. The density of the liquid phase decreases, while the density of the vapor phase increases with a concomitant decrease and increase, respectively, in the extent of methanol–hydrogen bonding in the two phases. As pressure increases in the two-phase region, the physical characteristics of the two phases become more similar (density, viscosity, and composition) by definition. As the liquid and vapor phases merge to a single phase, the extent of hydrogen bonding and the chemical shifts will merge to a single value. If the mole fraction of the binary modifier is near the critical composition for the temperature and pressure under investigation, then the change in the extent of hydrogen bonding in the two phases with pressure can be used to determine the critical pressure of the binary solvent.

E. Diffusion Coefficient Measurements

Diffusion coefficients measured by the spin-echo technique provide a means of investigating the translational motion of molecules under the extremes of temperature and pressure. There have been numerous studies of the self-diffusion coefficients of high-pressure liquids and supercritical fluids by NMR. As an illustration of the potential of these physicochemical measurements, we will focus on CO_2 (3,28,33,38,39). The availability of a wide range of diffusion coefficients and viscosities allows one to test the Stokes–Einstein equation at the molecular level. From hydrodynamic theory,

$$D = (k_B T)/(\kappa \pi a \eta) \tag{6}$$

where k_B is Boltzmann's constant, T is temperature, κ is a constant equal to 4 for slip boundary conditions and 6 for stick boundary conditions, a is the molecular radius, and η is viscosity. Therefore, Eq. (6) relates the self-diffusion coefficient to the viscosity of the solution. This equation is valid for a macroscopic sphere moving in a solvent continuum and should apply only to solutions where the solute is large in comparison with the solvent (stick condition). If the solute and solvent are of comparable size, the slip condition should apply. It is interesting to note that in past studies Eq. (6) has demonstrated its ability to provide reasonable estimates of the diffusion coefficient for simple molecules (40). For CO_2, using the slip condition and the assumption that the packing structure is arranged as a cubic lattice with all CO_2 molecules just touching, the predicted self-diffusion coefficients were within $\pm 5\%$ of the experimentally determined values for reduced densities greater than 1.5 (38). In their high-pressure CO_2 diffusion coefficient measurements, Lüdemann et al. (39) have shown that using a rough-hard-sphere approximation to describe a in Eq. (6), results in the fluid becoming more "sticky" as temperature increases. This is caused by the deviation of the CO_2 molecule from spherical symmetry as assumed in Eq. (6).

From Eq. (6), the self-diffusion coefficient can be related to the viscosity of the solution in a simple manner. The measurement of solution viscosities at high pressures and temperatures is a difficult, time-consuming task. Using high-pressure NMR measurements of the diffusion coefficient under these extreme conditions, the solution viscosity can be easily estimated from Eq. (6). Therefore, high-pressure NMR not only provides a very good estimate of solution viscosity under extreme conditions of pressure and temperature but also provides an experimental method to test the different microscopic physicochemical models of translational dynamics in solutions.

F. Quadrupolar NMR Measurements

The linewidth of a NMR absorption peak is dependent on the relaxation processes of the nuclei in solution. From the uncertainty principle,

$$\Delta E \Delta t \sim \hbar \tag{7}$$

since $\Delta E = h\Delta\nu$, and Δt can be related to the lifetime of the excited state T_2, then $\Delta\nu \sim 1/T_2$. T_2 represents all of the factors influencing the linewidth (relaxation processes) and $\Delta\nu$ is the linewidth at half-height. When the relaxation contributions are from spin-lattice effects, then $\mathrm{T}_2 = \mathrm{T}_1$. Therefore, in units of Hz the peak width at half-height ($\Delta\nu_{1/2}$) of a NMR absorption is $1/\pi\ T_1$. From Eq. (5) for the quadrupolar relaxation rate of nuclei,

$$\Delta\nu_{1/2} = 1/\pi T_1 = (3\pi/10)(2I + 3/(I^2(2I - 1)))(1 + \eta_s{}^2/3)\chi^2\tau_c \tag{8}$$

where the parameters are the same as described in Sec. II.B. It is important to note that $\Delta\nu_{1/2}$ is proportional to the rotational correlation time (τ_c) of the molecule containing the nuclei of interest. The molecular reorientation time in a polar liquid has been described by Debye as

$$\tau = 4\pi\eta a^3/k_B T \tag{9}$$

where the molecule is treated as a sphere of radius a in the viscous liquid and τ involves the same motions as τ_c. Therefore, the rotational correlation time depends on both the solution viscosity and temperature. From Eqs. (8) and (9), it can be seen that the peak width at half height of the quadrupolar nuclei is directly related to the viscosity of the solution. Thus, a decrease in the linewidth of an NMR resonance undergoing a quadrupolar relaxation process can be accomplished either through an increase in temperature or a decrease in solution viscosity. This is critical, as the linewidth for a quadrupolar nucleus can be very broad due to short relaxation times, and when coupled with low NMR receptivity, peak detection is essentially impossible.

The advantage of supercritical fluids lies in their low viscosities as compared to liquids and in their variable solvating powers by changing density. The study of quadrupolar nuclei naturally benefits from use of a supercritical fluid solvent in terms of the decrease in viscosity, which contributes to a decrease in linewidth. Studies have been reported investigating organic molecules containing ^{14}N and ^{17}O in supercritical fluids where the decrease in the linewidth for the quadrupolar nuclei was substantial (41–43). In homogeneous catalysis the organometallic compound can contain a transition metal that is a quadrupolar nuclei. The first report of the NMR investigation of such an organometallic compound (methylmanganesepentacarbonyl, ^{55}Mn) in supercritical ethylene was by Jonas et al. (43). The line narrowing between the supercritical fluid and a liquid solution for ^{55}Mn was reported to be a factor of 5.8. Rathke and coworkers (44) were the first to exploit the line narrowing of quadrupolar nuclei in supercritical fluids to study in situ chemical reactions with high-pressure NMR. Their work on the homogeneous catalytic activity of cobalt carbonyl complexes in supercritical fluids is covered in Sec. III.D. While the number of investigations of quadrupolar nuclei in supercritical fluid solvents have been few (41–45), the distinct advantages of supercritical fluids for the investigation of homogeneous catalysis and chemical synthesis in such systems should expand in the future.

III. REACTIONS IN SUPERCRITICAL FLUIDS

NMR spectroscopy is routinely used in today's organic synthesis laboratories to identify and structurally characterize reaction products. Yet despite the enormous structural information content of NMR and the availability of a variety of high-pressure NMR techniques (3–13), it is little used for studying in situ chemical reactions and solvent effects on chemical reactions in supercritical media.

Some advantages that are inherent to in situ NMR are as follows: (a) the linewidth narrowing of quadrupolar nuclei in supercritical fluids allows NMR measurements on nuclei with $I \geq 1$; (b) in situ NMR conveniently provides information on the phase behavior in the solution system; and (c) the nuclear spin may be regarded as a label for reactive species. NMR spectroscopy can use a variety of "labeling" techniques, such as a magnetization transfer experiment, that manipulates the sample's spin system with suitable pulse sequences. These capabilities, which are unique to NMR spectroscopy, provide a wide array of powerful means for studying chemical reactions. We hope the following sections will capture some of the flavor of NMR's ability toward the in situ investigation of reactions in supercritical fluids.

A. Deuteration and Hydrogen Exchange Reactions

In supercritical water extremely weak acidic protons, such as aliphatic or aromatic protons, undergo substitution reactions. Hydrogen exchange of this class of molecules in deuterium oxide can be studied by following the extent of deuteration as a function of time. Evilia et al. (46–48) have studied deuteration reactions of various organic molecules in supercritical D_2O using ex situ NMR. Their results have established that (a) the exchange reactions are base catalyzed, (b) for aromatic systems the inductive electron withdrawing of oxygen- or nitrogen-containing functional groups may be stronger than resonance effects resulting in preferential ortho deuteration selectivity, and (c) the substitution reaction most likely does not occur through hydride abstraction and the formation of carbocations.

Hoffmann and Conradi (49) studied the deuteration of 4-ethylphenol in supercritical D_2O at 460°C and 250 bar using high-pressure NMR. The methyl protons were found to be stable under these conditions. The temperature was raised to 500°C, and the deuteration of the methyl protons in conjunction with decomposition reactions was monitored. From this investigation, the selective deuteration of the different protons in 4-ethylphenol in supercritical D_2O could be demonstrated.

Another investigation of the deuteration of resorcinol in supercritical D_2O used a high-pressure NMR capillary cell as an in situ flow reactor (50). The temperatures covered in these measurements ranged from 200°C to 450°C at a pressure of about 400 bar. The deuterium exchange in resorcinol under these conditions was observed using both proton and deuterium NMR as a function of the resorcinol residence time in the capillary tubular reactor. The NMR results indicate that H/D exchange in resorcinol for the ring protons was observed at temperatures as low as 200°C. The activation energy of the H/D exchange of the ortho/para ring protons on resorcinol was reported. The microvolume and low thermal impedance of the capillary tubing contributes to the rapid temperature equilibration in the flow-through reactor region in the NMR probe.

If the hydrogen exchange rate $1/\tau_{\mathrm{life}}$ is comparable to the chemical shift frequency difference, $\Delta\omega_{\mathrm{shift}}$, of the spin exchanging sites, then lifetime broadening effects are visible in the NMR spectrum. Specifically, if hydrogen exchange is fast, such that $\Delta\omega_{\mathrm{shift}}\tau_{\mathrm{life}} \ll 1$, then one observes one sharp resonance. A gradual slowing of the chemical exchange rate, to the regime of $\Delta\omega_{\mathrm{shift}}\tau_{\mathrm{life}} \approx 1$, results in resonance broadening. This eventually leads to the emergence of two broad lines, each representing one of the two hydrogens. Finally, when hydrogen exchange is fast, such that $\Delta\omega_{\mathrm{shift}}\tau_{\mathrm{life}} \gg 1$, two sharp lines are present in the NMR spectrum. This was observed for a supercritical aqueous solution of methanol when the density at a constant temperature of 400°C was gradually decreased (49). Methanol in comparison with aliphatic or aromatic hydrocarbons is a much stronger acid. At high densities and 400°C, hydrogen exchange is fast, and one resonance line for the hydroxyl and water protons is observed. However, with decreasing density the solvation characteristics of supercritical water and methanol change. Hence, polar intermediates during hydrogen exchange, such as H_3O^+, are less stabilized, which in turn slows the hydrogen exchange rate. As a result, the hydrogen exchange rate at low, gas-like densities shows two distinct resonances for the hydroxyl and water protons. This example demonstrates one of the major promises of supercritical fluids: reaction rates can be smoothly altered at a fixed temperature by pressure-tuning (density-tuning) the solvation characteristics of the supercritical fluid solvent.

Similar to lifetime effects based on chemical shift differences, the J coupling can be exploited to study changes in reaction rates as well (49). Using this approach, the collapse of the J-coupling multiplets in ethanol (from the hydroxyl proton coupling with the CH_2 protons) was monitored as a function of pressure at several constant temperatures. Interestingly, the J-coupling multiplet collapse was observed in gaseous ethanol, indicating that even at low, gas-like densities intermolecular hydrogen exchange is important. With the J-coupling constant being 5 Hz, the exchange rate is $2\pi \times 5 \approx 30\ \text{sec}^{-1}$ at these conditions.

B. Keto-enol Equilibrium

The keto-enol tautomeric equilibrium of acetylacetone is an intramolecular hydrogen exchange process. High-pressure NMR was used to study changes in this equilibrium over a pressure range to 2.5 kbar and temperatures to 145°C (51). With an increase in temperature at constant pressure, the equilibrium distribution shifted to the keto tautomer. An increase in pressure did not change the keto-enol distribution at any temperature. From the high-pressure experiments as a function of temperature the reaction enthalpy, ΔH, and entropy, ΔS, were determined to be 2.80 ± 0.02 kcal/mol and 7.2 ± 0.3 cal/K mol, respectively.

A subsequent study investigated the effect of fluorine substitution on the tautomeric equilibrium of acetylacetonate β-diketones (acetylacetone, trifluoroacetylacetone, and hexafluoroacetylacetone) (52). The equilibrium between the keto and enol tautomers was studied as a function of pressure and temperature for both the pure compounds and those dissolved in supercritical CO_2. Similarly, no pressure dependence was observed on the tautomeric equilibrium. However, the degree of fluorination was found to have a dramatic stabilizing effect on the enol tautomer. This is because the electron-withdrawing fluorine further stabilizes the enol form through enhanced electron delocalization in the intramolecular resonance-assisted hydrogen bond. The stability of the enol tautomer in hexafluoroacetylacetone to temperature also points to the magnitude of this enol stabilization. The ΔH values indicate that the enol tautomer is enthalpically favored in both acetylacetone and trifluoroacetylacetone. The small experimental difference seen in the value of ΔS for acetylacetone and trifluoroacetylacetone can be rationalized on the basis of the differing hydrogen bond strength in the enol form of the two compounds with trifluoroacetylacetone having a stronger intramolecular hydrogen bond. These findings imply that changes in temperature should have dramatic consequences on supercritical fluid extraction involving β-diketones, whereas pressure should play a minor role. In contrast to the increase in the intermolecular hydrogen bond strength of simple alcohols with pressure, it is interesting to note that increasing pressure either weakens or does not affect the intramolecular hydrogen bond strength for the three molecules investigated.

C. Photolysis Reactions

An advantage of using the high-pressure capillary NMR cell is the ease of access to optical light of the supercritical fluid solution. Therefore, the range of chemical reactions in supercritical fluids that can be studied in situ by high-pressure NMR can be broadened to include photolysis reactions. The benchmark experiment that demonstrated the capability of this new technique was an investigation of the photoreversible fulgide, Aberchrome-540, as a function of pressure and temperature to 2.0 kbar and 120°C (53). The interconversion of the two Aberchrome-540 species (ring open and ring closed) was monitored during continuous photolysis. One could readily determine the decrease in the initial structure (ring open) and a concomitant increase in the ring-closed structure with time. For the ring-closed molecule, the equatorial methyl group in the fulgide is deshielded by the carbonyl on the anhydride group as compared to the axial methyl group. Thus, during photoconversion the largest chemical shift occurs on ring closure for the methyl ^{1}H resonances due to the change in conjugation of the fulgide structure upon ring closure. The temperature dependence of fulgide photoconversion was investigated at 25, 60, and 120°C at a constant pressure of 2.0 kbar. Interestingly, at 120°C there was no photoconversion observed at 2.0 kbar. It was hypothesized that during photolysis at higher temperatures the back reaction of the photocyclization (ring closed to ring open) was promoted.

For organometallic chemistry in supercritical fluids, high-pressure NMR has the advantage of describing the physicochemical environment and molecular structure of the ligand and complex directly coupled with the investigation of the central metal atom (dependent on the spin of the nuclei under investigation). The in situ photolytic substitution of ethylene and hydrogen for carbon monoxide on cymantrene [$CpMn(CO)_3$] and methylcymantrene [$MeCpMn(CO)_3$] dissolved in subcritical and supercritical solvents (CO_2 and ethylene) was investigated by high-pressure NMR over the temperature range −40°C to 100°C and the pressure range 35 to 2000 bar (54). These in situ photolysis investigations of organometallic species involved the direct detection of reaction products and the observation of the substituted ligand attached to the metal center. Photolytic substitution of ethylene for CO proceeded to completion under all conditions investigated, but only one ethylene was observed to add to the manganese complexes even in neat ethylene under extreme conditions of pressure and temperature. Small amounts of dihydrogen were observed to substitute for CO at 35°C in a binary mixture of CO_2/H_2. Hydrogen is a very poor solvent for these organometallic complexes, and small amounts in either CO_2 or ethylene can precipitate the metal complex from solution or cause phase separation. The photolysis of pentamethylcyclopentadienyl rhenium tricarbonyl [$Cp^*Re(CO)_3$], in H_2/CO_2 at 35°C and 490 bar was observed in which the hydride was formed on displacement of one of the carbonyl ligands (51). Similarly, the photolysis reaction chemistry of $Cp^*Re(CO)_3$ at 26.5°C and 540 bar in ethylene was investigated in which the mechanism involves the formation of a diethylene-substituted complex $Cp^*Re(CO)(\eta^2\text{-}C_2H_4)_2$ (51). The diethylene-substituted rhenium complex was studied using time-dependent high-pressure NMR, to investigate the slow, thermal loss of ethylene for CO. This reaction sequence is

$$\begin{aligned}Cp^*Re(CO)(\eta^2\text{-}C_2H_4)_2 + CO &\leftrightarrows Cp^*Re(CO)_2(\eta^2\text{-}C_2H_4) + CO \\ &\leftrightarrows Cp^*Re(CO)_3 \qquad (10)\end{aligned}$$

From a kinetic fit to the time-resolved data shown in Figure 5, the rate constant for the two thermal loss reactions could be determined. Since the overall reaction is first order in $Cp^*Re(CO)(\eta^2\text{-}C_2H_4)_2$ with no observed dependence upon $Cp^*Re(CO)_3$, the source of the CO appears to be the free CO in the supercritical ethylene solution liberated during the initial photolysis of $Cp^*Re(CO)_3$. In this investigation, advantage was taken of the fluid solution homogeneity to investigate rates of reactions involving small molecules (CO and ethylene) that are normally only obtainable in the gas phase. These in situ high-pressure NMR

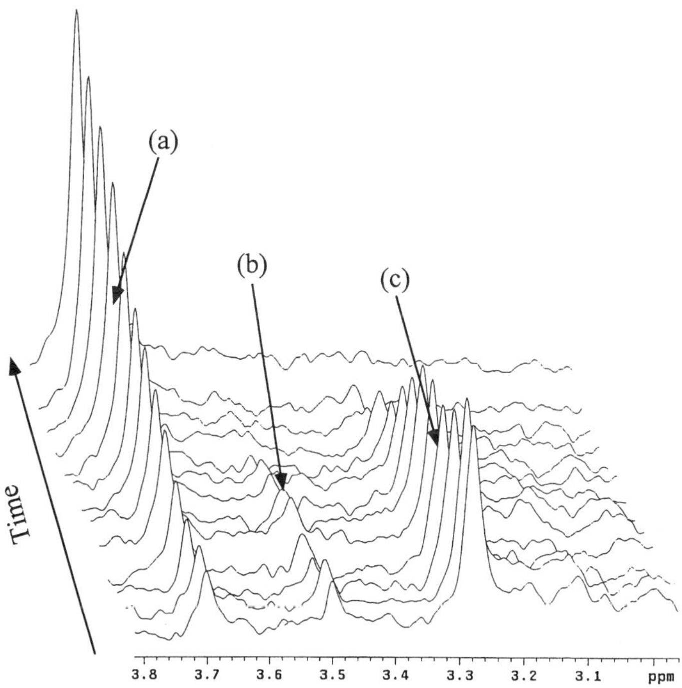

Figure 5 The time-resolved high-pressure NMR spectra of the back reactions for $Cp^*Re(CO)(\eta^2\text{-}C_2H_4)_2$ and CO at 345 bar and 30°C in supercritical ethylene; (A) $Cp^*Re(CO)_3$, (B) $Cp^*Re(CO)_2(\eta^2\text{-}C_2H_4)$ and (C) $Cp^*Re(CO)(\eta^2\text{-}C_2H_4)_2$. The spectra are spaced about 80 min apart from bottom to top.

studies have provided direct mechanistic insight into the photoinduced ligand substitution reactions of these organometallic compounds in supercritical fluids.

D. Homogeneous Catalysis

$CO_2(CO)_8$-catalyzed hydroformylation of olefins is an important reaction in the field of organometallics and is probably the first homogeneous catalysis reaction that has been studied by in situ NMR using supercritical carbon dioxide as the reaction solvent (55). The significantly reduced line width of the quadrupolar nuclei ^{59}Co in the supercritical fluid made it possible to resolve all of the catalytic intermediate species [$RC(O)Co(CO)_4$, $HCo(CO)_4$, and $Co_2(CO)_8$], and the high solubility of hydrogen in supercritical CO_2 resulted in a single-phase reaction mixture, thus eliminating the need for agitation. Hence, the concentrations of all the intermediate species were followed directly in real time with ^{59}Co NMR. The olefin decline and aldehyde production were followed by ^1H NMR over the course of the hydroformylation reaction at 80°C (88% product yield).

One important intermediate step in the catalytic reaction of $Co_2(CO)_8$ involves the chemical equilibrium:

$$CO_2(CO)_8 + H_2 \leftrightarrows 2HCO(CO)_4 \tag{11}$$

This equilibrium is established in supercritical CO_2 at 80°C in less than 2 hr for both the forward and reverse reactions (55). This equilibrium was subsequently studied as a function of temperature and the reaction enthalpy (4.7 ± 0.2 kcal/mol) and entropy [4.4 ± 0.5 cal/(mol K)] were determined (56). $Co_2(CO)_{10}$ was also observed to efficiently promote the hydrogenation of $Mn_2(CO)_{10}$ (57). Using in situ ^{55}Mn NMR, this finding was exploited to establish the reaction enthalpy (8.7 ± 0.3 kcal/mol) and entropy [8.5 ± cal/(K mol)] for the hydrogenation of $Mn_2(CO)_{10}$.

A second concurrent reaction involving the catalytic activity of $Co_2(CO)_8$ is the carbonyl exchange reaction of free and coordinated carbon monoxide. An earlier investigation of temperatures up to 80°C used a magnetic transfer technique for labeling the nuclear spins to track the chemical reaction (58). In this technique, one selectively inverts the spin population of one NMR signal and follows the transfer of the inverted population through the chemical reaction sequence with time. With knowledge of the individual T_1 relaxation times, one can separate out the relaxation contributions and obtain the reaction rates. At temperatures exceeding 80°C, the CO exchange rate was too fast to use the magnetization transfer technique. Another study was carried out at temperatures up to 180°C using lineshape analysis (44). It was observed that the ^{13}CO chemical shift showed a large temperature dependence. This unusual chemical shift dependence was interpreted as a contact shift from the interaction of the paramagnetic radical metal center $\cdot(CO)_4$ with the ^{13}CO ligand. The temperature dependence of the contact shift was used to estimate the homolytic bond dissociation energy of $Co_2(CO)_2$ (19 ± 2 kcal/mol). This value compared favorable with the experimental results from magnetic susceptibility and linewidth measurements (44), as well as the value predicted from the Marcus theory. In contrast to $Co_2(CO)_8$, the CO exchange rate in the corresponding manganese carbonyl compounds was too small to be measured (44).

The temperature–pressure behavior, as well as the solvent dependence of the equilibrium species of phosphine-substituted and unsubstituted cobalt carbonyl oxo catalysts, was investigated using in situ NMR techniques (59). The solvent polarity was found to have a dramatic effect on the equilibrium concentrations of species present and a new, as yet unidentified species was observed to be present preferentially in nonpolar solvents.

Another investigation describes a high-pressure NMR flow cell for the in situ study of homogeneous catalysis (60). This flow cell was constructed from a sapphire tube and reaction intermediates from a ruthenium-rhodium organometallic complex with CO were detected for the first time.

IV. CONCLUSIONS AND FUTURE DIRECTIONS

NMR is a very useful and versatile technique for the investigation of supercritical fluids, gases, and liquids under extreme conditions of pressure and temperature. Some of the examples discussed in this chapter demonstrate the potential for studying chemical reactions and solution dynamics in supercritical fluids as a function of density using high-pressure NMR. With pressure as a variable one gains an understanding of the solution process that is unobtainable through temperature variation alone. Spectroscopic studies of the physicochemical properties of supercritical fluid solutions and reactions are still at an initial stage of growth. There are areas of current application of NMR in the arena of materials and solution chemistry that a chapter of this nature allows one to explore for their potential impact in the future on high-pressure, high-temperature NMR studies in fluids and liquids. A synopsis of these areas is included in the following discussion.

A notable recent impetus for the growing interest in using supercritical fluids as a reaction medium has come from the efforts to explore chemical routes using carbon dioxide as a carbon source for the synthesis of organic compounds. The exploitation of CO_2 as an inexpensive, nonhazardous C_1 building block for organic reactions has long been a research topic of wide interest. However, only since the early 1990s has it been realized that the reactant may also be used as the supercritical solvent with beneficial miscibility and transport properties for these reactions (61,62). Researchers in this area have traditionally used NMR not only to identify and structurally characterize the reaction products but also to follow reaction kinetics. However, with the exception of one early investigation (63), these studies were exclusively conducted with ex situ NMR methods. Only recently have in situ NMR methods been adopted in this research field (64,65). This research area will in particular benefit from in situ high-pressure NMR techniques because, as pointed out by Burgemeister et al., "every step of the catalytic cycle for the rhodium-catalyzed hydrogenation of CO_2 to formic acid can be monitored by ^{1}H-NMR spectroscopy" (66).

A new development in high-pressure NMR probe design is a multipurpose high-pressure autoclave made from the thermoplastic polyetheretherketone (PEEK) (67). This NMR autoclave was used for in situ NMR imaging of a compressed gas system, namely, the exchange of methanol for liquid CO_2 in nanoporous silica-alcogels, reported for the first time.

Magic-angle spinning (MAS) solid-state NMR spectroscopy has for a number of years provided a means to study heterogeneous catalysis reactions by directly probing the chemical species present on the catalyst surface. Some of these experiments have been conducted at temperatures in excess of 200°C (68–71) and up to 400°C (72). By application of laser (73) or rf heating (74), fast transient sample heating (temperature jump) can be achieved. However, the most interesting development is the very recent construction of an isolated flow MAS NMR probe (75). This development has brought solid-state NMR much closer for studying heterogeneous catalysis in supercritical fluids.

Another area of potential impact on high-pressure NMR in the future may come from recent advances in enhancing NMR sensitivity by use of laser-polarized noble gases. The pioneering theoretical and experimental work by Happer (76) laid the foundation for understanding the physics involved. The major research efforts in this area have focused on (a) applying the laser-polarized noble gases directly, as in magnetic resonance imaging, or (b) transferring the ^{129}Xe polarization to another nuclei. Recently, significant progress has been made in transferring the polarization from laser-polarized ^{129}Xe to other nuclei. Signal enhancements of 70-fold in ^{13}C NMR have been achieved by cross-relaxation from laser-polarized liquid xenon (77). A further breakthrough was recently reported by the Pines et al. (78), who can routinely polarize supercritical xenon to enhancements of about 1000 that last for hundreds of seconds. These reports suggest numerous applications to supercritical fluids, i.e., the study of the efficiency of cross-relaxation from polarized supercritical xenon to dissolved solute molecules as a function of temperature and density (pressure). Second, a spin-labeled reactive functional group may be useful for studying chemical reactions in supercritical xenon (or eventually in other supercritical fluids).

These are a few of many potential areas were new techniques in combination with high-pressure NMR could make an important contribution to the fundamental understanding of solution chemistry and physics in supercritical and high-pressure liquid solvents.

ACKNOWLEDGMENTS

I express my appreciation to my former colleagues, Drs. S. L. Wallen and S. Bai, whose contributions to the work in my laboratory (CRY) is represented in this chapter. The work performed at the Pacific Northwest National Laboratory (PNNL) was supported by the Office of Science, Office of Basic Energy Sciences, Chemical Sciences Division of the U. S. Department of Energy, under Contract DE-AC076RLO 1830.

REFERENCES

1. R Noyori. Supercritical fluids. Chem Rev 99:353-634, 1999.
2. CR Yonker, JC Linehan, JL Fulton. UV, EPR, x-ray and related spectroscopic techniques. In: P Jessop, W Leitner, eds. Chemical Synthesis Using Supercritical Fluids. Weinheim, Germany: Wiley-VCH, 1999, pp 195–212.
3. J Jonas. High Pressure NMR, NMR Basic Principles and Progress. New York: Springer-Verlag, 1991.
4. K Woelk, JW Rathke, RJ Klingler. The toroid cavity NMR detector. J Magn Res A 109:137–146, 1994.
5. CR Yonker, TS Zemanian, SL Wallen, JC Linehan, JA Franz. A new apparatus for the convenient measurement of NMR spectra in high-pressure liquids. J Magn Res A 113:102–107, 1995.
6. U Matenaar, J Richter, MD Zeidler. High-temperature–high-pressure NMR probe for self-diffusion measurements in molten salts. J Magn Res A 122:72–75, 1996.
7. S Funahashi, K Ishihara, S Aizawa, T Sugata, M Ishii, Y Inada, M Tanaka. High-pressure stopped-flow nuclear magnetic resonance apparatus for the study of fast reactions in solution. Rev Sci Instrum 64:130–134, 1993.
8. S Bai, Taylor C. M., CL Mayne, RJ Pugmire, DM Grant. A new high pressure sapphire nuclear magnetic resonance cell. Rev Sci Instrum 67:240–243, 1996.
9. L Ballard, C Reiner, J Jonas. High-resolution NMR probe for experiments at high pressures. J Magn Res A 123:81–86, 1996.
10. MM Hoffmann, MS Conradi. Nuclear magnetic resonance probe for supercritical water and aqueous solutions. Rev Sci Instrum 68:159–164, 1997.
11. S-H Lee, MS Conradi, RE Norberg. Improved NMR resonator for diamond anvil cells. Rev Sci Instrum 63:3674–3676, 1992.
12. A Zahl, A Neubrand, S Aygen, R van Eldik. A high-pressure probehead for measurements at 400 MHz. Rev Sci Instrum 65:882–886, 1994.
13. H Yamada. Pressure-resisting glass cell for high pressure, high resolution NMR measurements. Rev Sci Instrum 45:640–642, 1974.
14. SL Wallen, BJ Palmer, BC Garrett, CR Yonker. Density and temperature effects on the hydrogen bond structure of liquid methanol. J Phys Chem 100:3959–3964, 1996; J Phys Chem 100:20173, 1996.
15. CR Yonker, SL Wallen, BJ Palmer, BC Garrett. Effects of pressure and temperature on the dynamics of liquid tert-butyl alcohol. J Phys Chem A 101:9564–9570, 1997.

16. S Bai, CR Yonker. Pressure and temperature effects on the hydrogen-bond structures of liquid and supercritical fluid methanol. J Phys Chem A 102:8641–8647, 1998.
17. JG Oldenziel, NJ Trappeniers. High resolution nuclear magnetic resonance spectroscopy in liquids and gases at pressures up to 2500 bar, III. Density dependence of the proton magnetic shielding constants in the polar liquids methanol and ethanol. Physica A 83:161–172, 1976.
18. EM Schulman, DW Dwyer, Doetschman. Temperature and pressure dependence of hydrogen bonding in liquid methanol studied by nuclear magnetic resonance. J Phys Chem 94:7308–7312, 1990.
19. MM Hoffmann, MS Conradi. Are there hydrogen bonds in supercritical methanol and ethanol. J Phys Chem B 102:263–271, 1998.
20. N Asahi, Y Nakamura. Chemical shift study of liquid and supercritical methanol. Chem Phys Lett 290:63–67, 1998.
21. N Asahi, Y Nakamura. Nuclear magnetic resonance and molecular dynamics study of methanol up to the supercritical region. J Chem Phys 109:9879–9887, 1998.
22. C Czeslik, J Jonas. Pressure and temperature dependence of hydrogen-bond strength in methanol clusters. Chem Phys Lett 302:633–638, 1999.
23. WT Raynes, AD Buckingham, HJ Bernstein. Medium effects in proton magnetic resonance. I. Gases. J Chem Phys 36:3481–3488, 1962.
24. MM Hoffmann, MS Conradi. Are there hydrogen bonds in supercritical water? J Am Chem Soc 119:3811–3817, 1997.
25. LA Curtiss, M Blander. Thermodynamic properties of gas-phase hydrogen-bonded complexes. Chem Rev 88:827–841, 1988.
26. A Dardin, JB Cain, JB DeSimone, CS Johnson, Jr., ET Samulski. High-pressure NMR of polymers dissolved in supercritical carbon doxide. Macromolecules 30: 3593–3599, 1997.
27. A Dardin, JM DeSimone, ET Samulski. Fluorocarbons dissolved in supercritical carbon dioxide. NMR evidence for specific solute-solvent interactions. J Phys Chem B 102:1775–1780, 1998.
28. EW Lang, H-D Lüdemann. Density dependence of rotational and translational molecular dynamics in liquids studied by high pressure NMR. Prog Nucl Magn Reson Spectrosc 25:507–633, 1993.
29. CR Yonker. Solution dynamics of perfluorobenzene, benzene, and perdeuterobenzene in carbon dioxide as a function of pressure and temperature. J Phys Chem A 104:685–691, 2000.
30. N Bloembergen, EM Purcell, RV Pound. Relaxation effects in nuclear magnetic resonance absorption. Phys Rev 73:679–712, 1948.
31. DK Green, JG Powles. Nuclear spin-lattice relaxation, including the spin-rotation interaction, in liquid benzene and several benzene derivatives up to the critical temperature. Proc Phys Soc 85:87–102, 1965.
32. MA Suhm, KJ Müller, H Weingärtner. A nuclear magnetic relaxation study of the structure of binary liquid mixtures of hexafluorobenzene with benzene and cyclohexane. Z Phys Chem, Neue Folge 155:101–119, 1987.
33. N Asahi, Y Nakamura. NMR study of liquid and supercritical benzene. Ber Bunsen-Ges Phys Chem 101:831–836, 1997.

34. P Diep, KD Jordan, JK Johnson, EJ Beckman. CO_2-fluorocarbon and CO_2-hydrocarbon interactions from first-principles calculations. J Phys Chem A 102:2231–2236, 1998.
35. GG Yee, JL Fulton, RD Smith. Fourier transform infrared spectroscopy of molecular level interactions of heptafluoro-1-butanol or 1-butanol in supercritical carbon dioxide and supercritical ethane. J Phys Chem 96:6172–6181, 1992.
36. CR Yonker, JC Linehan, JL Fulton. The use of high pressure NMR for the determination of phase behavior for selected binary solvent systems. J Supercrit Fluids 14:9–16, 1998.
37. BM Hasch, EJ Maurer, LF Ansanelli, MA McHugh. (Methanol + ethene): phase behavior and modeling with the SAFT equation of state. J Chem Thermodyn 26: 640–652, 1994.
38. P Etesse, JA Zega, R Kobayashi. High pressure nuclear magnetic resonance measurement of spin-lattice relaxation and self-diffusion in carbon dioxide. J Chem Phys 97:2022–2029, 1992.
39. T Gross, J Buchhauser, H-D Lüdemann. Self-diffusion in fluid carbon dioxide at high pressures. J Chem Phys 109:4518–4522, 1998.
40. J Jonas. High pressure NMR studies of the dynamics in liquids and complex systems. In: J Jonas, ed. High Pressure NMR: NMR Basic Principles and Progress. New York: Springer-Verlag, 1991, Vol. 24, pp 85–128.
41. JM Robert, RF Evilia. Nuclear magnetic resonance studies of nitrogen-14-containing species in supercritical fluids. Anal Chem 60:2035–2040, 1988.
42. JM Robert, RF Evilia. High-resolution nuclear magnetic resonance spectroscopy of quadrupolar nuclei: nitrogen-14 and oxygen-17 examples. J Am Chem Soc 107: 3733–3735, 1985.
43. DM Lamb, DG Vander Velde, J Jonas. Line Narrowing for solutes containing quadrupolar nuclei using supercritical fluid solvents. J Magn Reson 73:345–348, 1987.
44. RJ Klingler, JW Rathke. High-pressure NMR investigation of hydrogen atom transfer and related dynamic processes in oxo catalysis. J Am Chem Soc 116:4772–4785, 1994.
45. S Gaemers, CJ Elsevier. Reducing the NMR line widths of quadrupole nuclei by employing supercritical solvents. Chem Soc Rev 28:135–141, 1999.
46. J Yao, RF Evilia. Deuteration of extremely weak organic acids by enhanced acid–base reactivity in supercritical deuteroxide solution. J Am Chem Soc 116:11229–11233, 1994.
47. MR Hibbs, J Yao, RF Evilia. Reactivity of activated phenyl groups in supercritical water. High Temp Mater Sci 36:9–14, 1996.
48. Y Yang, RF Evilia. Deuteration of hexane by ^2HCl in supercritical deuterium oxide. J Supercrit Fluids 15:165–172, 1999.
49. MM Hoffmann, MS Conradi. Hydrogen exchange reactions in supercritical media monitored by in situ NMR. J Supercrit Fluids 14:31–40, 1998.
50. S Bai, BJ Palmer, CR Yonker. Kinetics of deuterium exchange of resorcinol in D_2O at high pressures and high temperatures. J Phys Chem A 104:53–58, 2000.
51. CR Yonker, SL Wallen, JC Linehan. Reaction chemistries in supercritical fluid solutions. J Microcolumn Sep 10:153–160, 1998.

52. SL Wallen, CR Yonker, CL Phelps, CM Wai. Effect of fluorine substitution, pressure and temperature on the tautomeric equilibria of acetylacetonate β-diketones. J Chem Soc, Faraday Trans 93:2391–2394, 1997.
53. CR Yonker, SL Wallen. High-pressure on-line photolysis with NMR detection. Appl Spectrosc 50:781–784, 1996.
54. JC Linehan, SL Wallen, CR Yonker, TE Bitterwolf, JT Bays. In situ NMR observations of the photolysis of cymantrene and methylcymantrene in supercritical fluids: a new technique using high-pressure NMR. J Am Chem Soc 119:10170–10177, 1997.
55. JW Rathke, RJ Klingler, TR Krause. Propylene hydroformylation in supercritical carbon dioxide. Organometallics 10:1350–1355, 1991.
56. JW Rathke, RJ Klingler, TR Krause. Thermodynamics for the hydrogenation of dicobalt octacarbonyl in supercritical carbon dioxide. Organometallics 11:585–588, 1992.
57. RJ Klingler, JW Rathke. Thermodynamics for the hydrogenation of dimanganese decacarbonyl. Inorg Chem 31:804–808, 1992.
58. DC Roe. High-pressure NMR studies of CO exchange with cobalt carbonyl species. Organometallics 6:942–946, 1987.
59. KW Kramarz, RJ Klinger, DE Fremgen, JW Rathke. Toroid NMR probes for the in situ examination of homogeneous cobalt hydroformylation catalysts at high pressures and temperatures. Catal Today 49:339–352, 1999.
60. JA Iggo, D Shirley, NC Tong. High pressure NMR flow cell for the in situ study of homogeneous catalysis. N J Chem 1043–1045, 1998.
61. MT Reetz, W Könen, T Strack. Supercritical carbon dioxide as a reaction medium and reaction partner. Chimia 47:493, 1993.
62. PG Jessop, T Ikariya, R Noyori. Homogeneous catalytic hydrogenation of supercritical carbon dioxide. Nature 368:231–233, 1994.
63. J-C Tsai, KM Nicholas. Rhodium-catalyzed hydrogenation of carbon dioxide to formic acid. J Am Chem Soc 114:5117–5124, 1992.
64. S Kainz, D Koch, W Leitner. Homogeneous catalysis in supercritical carbon dioxide: a "better solution?" In: H Werner, W Schreier, eds. Selective Reactions of Metal-Activated Molecules. Wiesbaden, Germany: Vieweg, 1998, pp 151–156.
65. S Kainz, A Brinkmann, W Leitner, A Pfaltz. Iridium-catalyzed enantioselective hydrogenation of imines in supercritical carbon dioxide. J Am Chem Soc 121: 6421–6429, 1999.
66. T Burgemeister, F Kastner, W Leitner. $[(PP)_2RhH]$ and $[(PP)_2Rh][O_2CH]$ complexes as models for the catalytically active intermediates in the Rh-catalyzed hydrogenation of CO_2 to HCOOH. Angew Chem, Int Ed Engl 32:739–741, 1993.
67. W Behr, A Haase, G Reichenauer, J Fricke. High-pressure autoclave for multipupose nuclear megnetic resonance measurements up to 10 MPa. Rev Sci Instrum 70:2448–2453, 1999.
68. AG Stepanov. In situ NMR identification of the intermediates and the reaction products in alcohols and hydrocarbons conversion on zeolites. Catal Today 24: 341–348, 1995.
69. EJ Munson, JF Haw. Reaction tuning in zeolites: an in situ MAS NMR study of acetaldehyde on HZSM-5. Angew Chem, Int Ed Engl 32:615–617, 1993.

70. MW Anderson, J Klinowski. Monitoring organic products of catalytic reactions on zeolites by two-dimensional J-resolved solid-state NMR. Chem Phys Lett 172: 275–278, 1990.
71. T Xu, EJ Munson, JF Haw. Toward a systematic chemistry of organic reactions in zeolites: in situ NMR studies of ketones. J Am Chem Soc 116:1962–1972, 1994.
72. FG Oliver, EJ Munson, JF Haw. High-temperature in situ magic angle spinning NMR studies of chemical reactions on catalysts. J Phys Chem 96:8106–8111, 1992.
73. H Ernst, D Freude, T Mildner, I Wolf. Laser supported high temperature MAS NMR. A new method for time resolved in site studies of reaction steps in heterogeneous catalysis. Stud Surf Sci Catal 94:413–418, 1995.
74. DB Ferguson, JF Haw. Transient methods for in situ NMR of reactions on solid catalysts using temperature jumps. Anal Chem 67:3342–3348, 1995.
75. PK Isbester, A Zalusky, DH Lewis, MC Douskey, MJ Pomije, KR Mann, EJ Munson. NMR probe for heterogeneous catalysis with isolated reagent flow and magic-angle spinning. Catal Today 49:363–375, 1999.
76. W Happer, E Miron, S Schaefer, D Schriber, WA van Wijngaarden, X Zeng. Polarization of the nuclear spins of noble-gas atoms by spin exchange with optically pumped alkali-metal atoms. Phys Rev A 29:3092–3110, 1984.
77. RJ Fitzgerald, KL Sauer, W Happer. Cross-relaxation in laser-polarized liquid xenon. Chem Phys Lett 284:87–92, 1998.
78. M Haake, BM Goodson, DD Laws, E Brunner, MC Cyrier, RH Havlin, A Pines. NMR of supercritical laser-polarized xenon. Chem Phys Lett 292:686–690, 1998.

3
Organic Chemical Reactions and Catalysis in Supercritical Fluid Media

Keith W. Hutchenson
DuPont Company, Wilmington, Delaware

I. INTRODUCTION

The development of more efficient and economical chemical transformation processes remains an important challenge. Organic synthesis reactions in the polymer, chemical, and pharmaceutical industries are increasingly required to be highly selective, economical, and environmentally benign. As one means of developing such processes, the use of supercritical fluids (SCFs) as reaction solvents has been the subject of increasing investigation over the last two decades because of the host of potential advantages afforded by these media over conventional liquid solvents and gaseous diluents. Specific applications in the materials area include polymer synthesis as well as the synthesis of monomers and key intermediates. This chapter focuses on the "small-molecule" end of this scale and includes a comprehensive survey of the major organic reaction classes currently under investigation in SCF media.

SCF technology has found widespread use in a number of industrial-scale processes, primarily in separations through SCF extraction. Examples of commercial implementation of SCF extraction applications include coffee and tea decaffeination, flavors from hops, cholesterol and fat from eggs, nicotine from tobacco, acetone from antibiotics, and organics from water (1,2). Other commercial applications include the CO_2-based dry cleaning facilities that are in direct competition with conventional perchloroethylene systems and Union Carbide's Unicarb technology for CO_2-based spraying of paint and other coatings (1,3).

Commercial implementation of SCF technology for conducting chemical reactions has been much more restricted, although limited applications have been in operation for decades. Jessop and Leitner (4) provide an excellent history of the early industrial applications. These include the well-known ammonia synthesis (1913), methanol synthesis (1923), oxidation of light alkanes (1920s), and the synthesis of low-density polyethylene (1940s). These early successful applications of SCF technology in the manufacture of bulk chemicals and polymers demonstrated that the technical challenges associated with large-scale operation at high temperatures and pressures could be resolved. However, despite the precedence established with these early applications, broad implementation of SCF technology for conducting chemical reactions has been limited to a handful of specialized applications. For example, the supercritical water oxidation (SCWO) process for the total oxidative destruction of hazardous organic compounds in aqueous waste streams has been commercialized (5,6), although in general these plants are limited to waste streams that minimize the impact of the two significant technical challenges facing this technology—scaling and corrosion (7). The Japanese firm Idemitsu Petrochemical Co. commercialized a 40,000 metric ton per year integrated reaction and separation process utilizing SCF butene in 1985 (2,4,8). The acid-catalyzed reaction of 1- and 2-butene to 2-butanol occurs in an aqueous phase, and the product is extracted into the SCF butene phase to drive the reversible reaction forward. Cooling and depressurizing the SCF extract phase then isolates the 2-butanol. The Danish firm Paul Møller Consulting, in conjunction with the Chalmers University of Technology (Göteborg, Sweden), is developing SCF processes at pilot plant scale for the hydrogenation of fatty acid methyl esters to fatty alcohols and the synthesis of hydrogen peroxide (4,9,10). The Swiss company Hoffmann-La Roche has reported the development of an 800 ton per year continuous pilot plant utilizing a 40-liter reactor for the heterogeneously catalyzed hydrogenation of vitamin precursors (4,11–13). Thomas Swan & Co., a fine-chemicals manufacturer in the United Kingdom, has announced the development of a commercial-scale continuous hydrogenation facility using an SCF solvent (1,14). This facility, which is being designed in conjunction with the Swedish engineering firm Chemature, is slated for startup during the second half of 2001 with an annual capacity of 500–1000 tons of an unspecified product. DuPont has announced (15) the construction of a 2.5 million pound per year market development facility at its Fayetteville, North Carolina, site that will evaluate supercritical CO_2 ($scCO_2$) as a reaction solvent for the production of tetrafluoroethylene-based fluoropolymers and copolymers. This work was initiated by DeSimone and coworkers (16–18) and further developed by DuPont (19). Successful demonstration of this technology in the market development facility will be followed by construction of a commercial-scale plant. These examples show that SCF reaction technology is beginning to gain broad commercial acceptance.

The topic of reactions in SCF media has been the subject of a number of reviews and surveys in recent years (7,8,12,20–30). Of note is a thorough review by Savage et al. (31) that provides a comprehensive analysis of this literature from 1985 to 1994. Several recent reviews have been more narrowly focused on specific topics. For example, Savage (32) and Baiker (33) present reviews of heterogeneous catalysis applications in SCF media, and Jessop and coworkers (34,35) focus on the homogeneous catalysis literature. Baiker's paper (33) includes a brief discussion on the various batch and continuous reactors that have been employed in much of the work discussed in these various reviews. Jessop and Leitner (36) recently edited a comprehensive monograph devoted exclusively to chemical synthesis in SCF media with contributions from a number of researchers in the field.

This chapter surveys the literature on small-molecule organic chemical transformations in SCF media. The scope is intended to be comprehensive, but the focus is on studies published since the extensive review by Savage (31), which covers the literature up through 1994. The chapter is organized to first present a fundamental understanding of the features of SCFs and the corresponding potential advantages for their utilization as reaction media. This is followed by a survey of the published literature on a variety of applications in various stages of research and development. This latter section is generally organized by major reaction classes for ease of reference.

As in conventional catalysis, catalytic chemical transformations utilizing SCF media can be conducted as both heterogeneous and homogeneous systems. Heterogeneous solid catalysts offer high reaction rates and simple separation of the catalyst from products. Hence, use of heterogeneous catalysis is by far the more common industrial practice in conventional solvents [approximately 85% of all known commercial catalytic processes use heterogeneous catalysts (37)]. However, the use of homogeneous molecular catalysis is increasingly becoming a viable alternative because these catalysts offer high selectivity, tunability, and even chirality for the production of a range of small to large molecules (38). Both types of systems have been investigated in SCF media, and the system type will be distinguished in the various cited studies.

The coverage of the patent literature is not exhaustive, although an attempt has been made to include the English language patents for SCF reactions dating back to 1990. Prior patent literature is covered in two primary sources. McHugh and Krukonis (39,40) provide a compilation of patent summaries in the area of SCF technology with an emphasis on extraction applications. Bruno (41) provides a listing and brief summary of patents in the field issued between 1982 and 1989. The appendix A to this chapter summarizes major patents issued from 1990 to 1999 in the area of chemistry and catalysis in SCFs within the restrictions noted below. This listing is believed to be representative, if not comprehensive, of the US, EP, and WO patents for this period along with a selection of other national patents. Selected patents from this summary are also cited in the text.

Three important areas within this general field that are beyond the scope of the current review include the topics of organic synthesis in supercritical water, polymer applications, and enzyme-catalyzed reactions. Shaw et al. (42) present an early review on the use of supercritical water as a reaction medium, and Savage (43) has updated this in a recent comprehensive review focusing on organic chemical reactions in supercritical water. Polymerizations and polymer modifications in SCF media have seen widespread interest in recent years, and this trend will likely accelerate as industrial motivations, such as solvent replacement, become increasingly important. DuPont's recent announcement (15) regarding the building of a $scCO_2$-based development facility for fluoropolymerizations is an example. Comprehensive reviews of this area are provided by Scholsky (44), Kiran (45), and, more recently, DeSimone and coworkers (46). Another rapidly emerging area of importance is that of enzyme-catalyzed reactions in SCFs. Randolph et al. (47), Clifford (23), and Savage et al. (31) provide reviews of this literature, and general overviews of the field are provided by Aaltonen and Rantakylä (48), Russell et al. (49), and Nakamura (50). Mesiano et al. (51) provide the most recent review available on this topic.

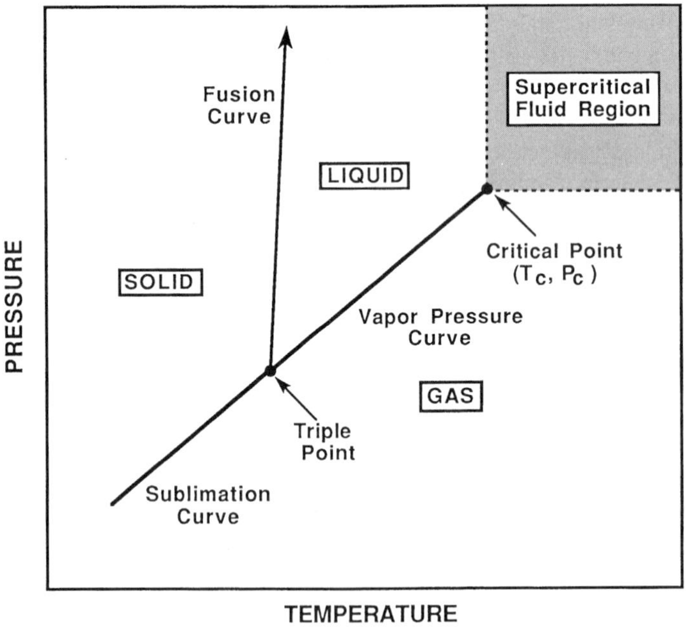

Figure 1 Pressure–temperature phase diagram for a pure fluid.

II. FUNDAMENTALS

A. Properties of SCFs

Figure 1 shows the SCF region for a pure component on a pressure–temperature diagram. By definition, a fluid is in the SCF state when the system temperature and pressure exceed the corresponding critical point values defined by the critical temperature (T_c) and pressure (P_c). Most useful applications of SCFs that take advantage of the unusual physical properties in this region occur in the range $T_R(= T/T_c) \approx 1.0$–1.1 and $P_R \approx 1$–2 (26). However, in the literature, some so-called SCF reaction systems actually are conducted at conditions slightly subcritical in temperature or pressure where some of the potential benefits afforded by SCF media are also present.

To a first approximation, the solvent strength of an SCF can be related to the solution density (52). One of the primary advantages of SCF reaction media is that the density can be varied continuously from liquid-like to gas-like values by varying either the temperature or the pressure. Figure 2 illustrates this unique feature of an SCF by showing the variation in density as a function

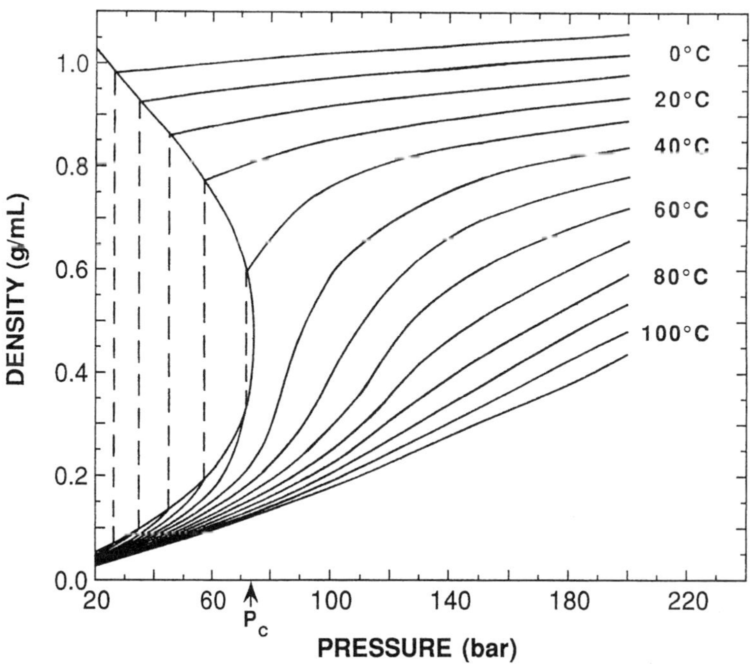

Figure 2 Density–pressure projection of the phase diagram for pure carbon dioxide (From Ref. 53).

of pressure for pure carbon dioxide along various isotherms. Note the dramatic variation in density in the immediate vicinity of the critical point. The various density-dependent physical properties also exhibit similar dramatic and continuous variation in this region. For example, Figure 3 shows significant changes in naphthalene solubility in carbon dioxide in the immediate vicinity of the critical point. Figure 4 shows similar variation in the dielectric constant of the polar solvent fluoroform, with an almost 1:1 correspondence with the density curves of Figure 2. Thus, in general, an SCF in the vicinity of its critical point has a liquid-like density and solvent strength, but transport properties (mass, momentum, and thermal diffusivities) that are intermediate to those of gases and liquids. Table 1 further illustrates this by listing order-of-magnitude density and transport properties for an SCF in this near-critical region.

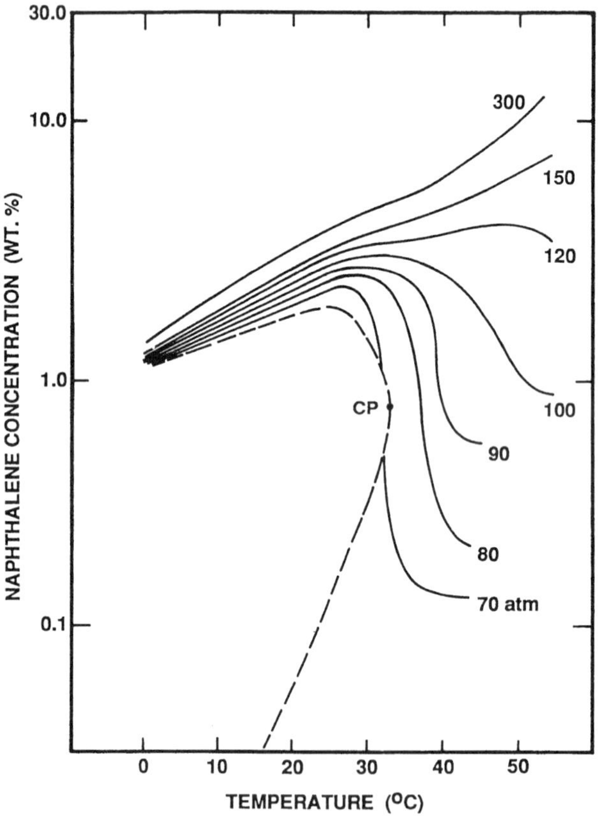

Figure 3 Solubility behavior of naphthalene in $scCO_2$. (Reprinted with permission from Ref. 39. Copyright 1986 Butterworth-Heinemann.)

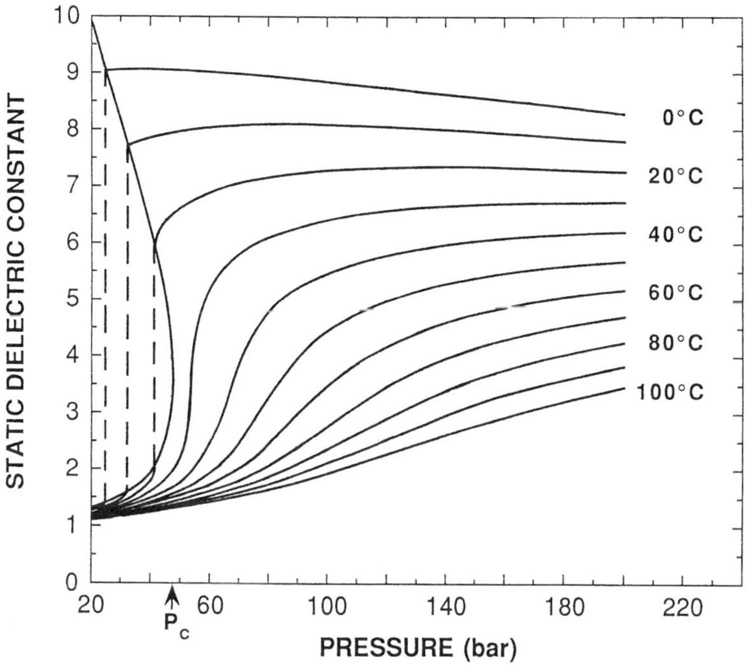

Figure 4 Variation of the dielectric constant of pure fluoroform in the vicinity of the critical point. [Calculated from published density and dielectric constant correlations (54–56), as adapted from Ref. 35.]

In addition to typical factors such as chemical inertness, cost, toxicity, and the like, one must carefully consider the critical temperature when selecting a potential solvent for conducting chemical transformations in the SCF regime. For practical applications, thermal and catalytic chemical reactions can only be conducted in a relatively narrow temperature range. Lower temperatures result in unacceptable reaction rates, and higher temperatures can result in significant

Table 1 Order-of-Magnitude Comparison of the Physical Properties of Gases, SCFs, and Liquids (57)

Physical property	Gases	Supercritical fluids	Liquids
Density (g/ml)	0.001	0.2–1.0	0.6–1.6
Dynamic viscosity (Pa·s)	10^{-5}	10^{-4}	10^{-3}
Diffusivity (cm^2/s)	10^{-1}	10^{-3}	10^{-5}

Table 2 Critical Properties[a] for Selected Supercritical Fluids Used in Chemical Reactions (58)

Solvent	Formula	T_c (°C)	P_c (bar)	ρ_c (g/cm^3)
Ethylene	C_2H_4	9.3	50.4	0.22
Xenon	Xe	16.6	58.4	1.11
Fluoroform	CHF_3	26.2	48.6	0.53
Carbon dioxide	CO_2	31.0	73.8	0.47
Ethane	C_2H_6	32.2	48.8	0.20
Nitrous oxide	N_2O	36.5	72.4	0.45
Sulfur hexafluoride	SF_6	45.6	37.6	0.73
Difluoromethane	CH_2F_2	78.4	58.3	0.43
Propylene	C_3H_6	91.8	46.0	0.23
Propane	C_3H_8	96.7	42.5	0.22
Dimethyl ether	C_2H_6O	126.9	52.4	0.26
Ammonia	NH_3	132.4	113.5	0.23
n-Pentane	C_5H_{12}	196.6	33.7	0.24
Isopropanol	$CH_3CH_2(OH)CH_3$	235.1	47.6	0.27
Methanol	CH_3OH	239.5	80.9	0.27
Ethanol	CH_3CH_2OH	240.7	61.4	0.28
Water	H_2O	374.2	221.2	0.32

[a] T_c, critical temperature; P_c, critical pressure; ρ_c, critical density.

selectivity and yield losses as well as catalyst deactivation. To obtain practical solvent densities and the corresponding density-dependent properties, this temperature optimization must be balanced against a general desire to operate in the vicinity of the mixture critical point of the reaction system to fully exploit the potential advantages afforded by SCF operation. The phase behavior of the reaction mixture, which is strongly influenced by the solvent critical temperature, is fundamentally important in defining this operating window. Table 2 lists the pure-component critical properties for selected solvents that have been considered as SCF reaction media. As shown, these solvents cover a relatively broad range of critical temperatures as well as pressures.

B. Potential Advantages of SCF Reaction Processes

SCFs are potentially very attractive media for conducting chemical transformations (4), primarily because the solvent and transport properties of a single solution can be varied appreciably and continuously with relatively minor changes in temperature or pressure. The density variation in an SCF also influences the chemical potential of solutes and, thus, reaction rates and equilibrium constants (59). Therefore, the solvent environment can be optimized for a specific reaction application by tuning the various density-dependent fluid properties (60). As an example, by pressure-tuning the reaction mixture, Kim and Johnston (61) demonstrated continuous control of the selectivity of two competing parallel Diels–Alder additions of methyl acrylate and cyclopentadiene to produce both the endo and exo products.

Solubility effects can influence reaction processes in several ways. For example, the solubilities of various reactants, products, and, in some cases, homogeneous catalysts can be manipulated to process advantage (e.g., homogenizing a previously heterogeneous reaction mixture). The enhanced solubility of catalyst-fouling components has been exploited to minimize heterogeneous catalyst deactivation through extraction of these fouling products or contaminants and prevention of subsequent coking (62). In addition, solubility effects can potentially be utilized in an SCF-based process through integration of the reaction step with product isolation to synthesize a novel reactive separation process.

Since gaseous reactants are completely miscible with SCFs, their concentrations in SCF reaction media are significantly higher than that obtainable in conventional liquid solvents, even at appreciable pressures (63). These higher reactant concentrations in SCF media combined with increased component diffusivities and relatively low system viscosities (see Table 1) can result in mass transfer rates that are appreciably higher than in liquid solvents. This can potentially shift a chemical reaction rate from mass transfer control to kinetic control in the reactor. For example, Noyori and coworkers (64) reported the hydrogenation of $scCO_2$ to formic acid at rates of up to 18 times that in liquid tetrahydrofuran, which they attribute to the enhanced mass transfer characteristics and high hydrogen miscibility afforded by the SCF mixture. The solubility of gaseous reactants in liquid solvents can also be enhanced by a volume expansion of the solvent with a dense SCF, which likewise results in increased mass transfer rates (65). Improved mass transport can also result in enhanced removal of residual solvents (26).

A key density-dependent property of SCFs that is sometimes overlooked is the heat capacity, which is relatively high in the vicinity of the critical point compared with gases (42,66). This high heat capacity produces effective heat transfer relative to gas-phase reactions. Thus, highly exothermic reactions such as hydrogenations can be conducted in SCF media with accurate temperature control (67).

As will be described below, both pressure and temperature effects can be used to influence chemical transformations. For example, reaction selectivity can be influenced indirectly through a pressure-dependent dielectric constant for a polar SCF solvent (68), and equilibrium constants can be shifted to favor desired products. Combining this manipulation of reaction characteristics through pressure effects with the use of solvents having moderate critical temperatures can be used for economical processing of temperature-sensitive materials at milder reaction conditions (26). Viscosity effects can be exploited to tune reaction selectivity. For example, Aida and Squires (69) have demonstrated a pronounced pressure dependence on the photoisomerization of *trans*-stilbene in $scCO_2$ at 40°C, which they attribute to significant changes in the solvent viscosity over the pressure range investigated.

A reason often cited for considering SCF-mediated reaction processes is the potential for utilizing a reaction solvent that exhibits improved safety, health, and environmental impact relative to typical organic solvents. Carbon dioxide, in particular, is generally considered environmentally benign, nontoxic, nonflammable, and inexpensive, and it is suitable for use as an SCF solvent at relatively moderate temperatures. However, as illustrated in Table 2, there are a variety of other practical SCF solvents that may have better solubility characteristics than CO_2 as well as beneficial impact relative to conventional organic solvents.

C. Phase Behavior

The dramatic property changes that occur in the vicinity of the critical point that result in these various potential advantages of conducting chemical reactions in SCF media have been illustrated for the simple case of a pure fluid in Figures 2–4. These property variations as well as the underlying mixture critical curve behavior are much more complex for the multicomponent systems that will be encountered in all practical applications for conducting SCF-mediated chemical transformations. One must know the location of phase boundaries and the magnitude of these property variations to fully exploit these potential advantages as well as to robustly control operating processes in the vicinity of a critical point where density fluctuations are significant (7,33,70–73). Thus, the importance of accurate measurement and modeling of solubility data and the corresponding phase behavior for the reactant–product–solvent systems is fundamental to the accurate interpretation of experimental reaction rate and selectivity data as well as the reliable scaling to commercial processes. Presentation of the appropriate phase equilibrium thermodynamics and calculational techniques for correlating experimental measurements and estimating multicomponent phase behavior is beyond the scope of this chapter, but a number of excellent sources are available in the literature for reference (e.g., see Refs. 40,70,74–77).

Utilization of SCF media for conducting chemical reactions results in several unique phase behavior features and challenges. A number of these are summarized here to better understand the phase behavior implications in the applications reviewed below. Some of these extend directly from the above discussion regarding potential advantages of using SCF media for conducting chemical reactions. For example, by conducting a reaction in the SCF-phase regime, all of the reactants can exist in a single homogeneous phase. This eliminates interfacial mass transfer resistances that would otherwise result in diffusional control in carrying out a specific reaction. For homogeneous catalysis applications, the catalyst can likewise exist in this same homogeneous phase with corresponding reductions of mass transfer resistances. Alternatively, in some applications, an SCF phase can be used to expand a separate liquid reaction phase and still provide substantial benefit in improved reactant mass transfer rates (65). Since the critical point of a multicomponent mixture is composition dependent, the critical point of a reaction mixture will change with the extent of reaction, and thus, with time in a batch reactor or with location in a continuous fixed-bed reactor (33). As illustrated previously, the various density-dependent physical properties can be manipulated with density changes for an SCF, resulting in a corresponding effect on reaction rates and selectivity (e.g., 68,78–80). For example, Combes et al. (81) have demonstrated control of both regio- and stereoselectivity through tuning of SCF solvation effects. Thus, manipulation of the phase behavior can be used to optimize the effect of these various parameters on specific reaction applications. The phase behavior of SCF systems can also be manipulated to control the number and composition of coexisting phases (79), thus controlling both reaction effects as well as the separation of products or homogeneous catalysts from the reaction mixture. Finally, the addition of cosolvents can be effectively utilized to exploit specific solute interactions such as enhancing solute solubilities (e.g., 70) and influencing reaction selectivities (79) and equilibria (82).

D. Thermodynamic Pressure Effects on Reaction Rates

Pressure is a fundamental physical property that affects various thermodynamic and kinetic parameters. Pressure dependence studies of a process reveal information about the volume profile of a process in much the same way as temperature dependence studies illuminate the energetics of the process (83). Since chemical transformations in SCF media require relatively high operating pressures, pressure effects on chemical equilibria and rates of reactions must be considered in evaluating SCF reaction processes (83–85). The most pronounced effect of pressure on reactions in the SCF region has been attributed to the thermodynamic pressure effect on the reaction rate constant (86), and control of this pressure dependency has been cited as one means of selecting between parallel reaction pathways (87). This pressure effect can be conveniently evaluated within the thermodynamic framework provided by transition state theory, which has often been applied to reactions in solutions (31,84,88–90). This theory assumes a true chemical equilibrium between the reactants and an activated transition state species that has the required energy and conformation corresponding to the internal energy barrier for chemical reaction. This transition state complex then proceeds directly to products, and the rate of the chemical reaction is governed by the rate constant for this decomposition from the activated state. This is illustrated for a bimolecular reaction between reactants A and B forming the transition state $M^{\ddagger}$ by

$$A + B \rightleftharpoons M^{\ddagger} \rightarrow \text{products}$$

The pressure dependence of the reaction rate constant is given by the following relation in terms of partial molar volumes and isothermal compressibility (90):

$$\left(\frac{\partial \ln k_{bm}}{\partial P}\right) = -\frac{\Delta\nu^{\ddagger}}{RT} - k_T$$

where k_{bm} = bimolecular rate constant (mol/L-min), $\Delta\nu^{\ddagger} = \nu_{M^{\ddagger}} - \nu_A - \nu_B$ = activation volume (the difference between the partial molar volumes of the transition state species and that of the reactants), ν_i = partial molar volume of component i at reaction conditions, k_T = mixture isothermal compressibility, and R = universal gas constant. Note that the isothermal compressibility term in the above equation accounts for changes in the reactant concentrations with pressure. This term is not included if the rate constant is expressed in pressure-independent units, such as mole fraction or molality (84,91). A more general expression for the pressure effect on the rate constant that accounts for the number of reactant species is given by

$$\left(\frac{\partial \ln k}{\partial P}\right) = -\frac{\Delta\nu^{\ddagger}}{RT} + (1 - n)k_T$$

where n is defined as the sum of the stoichiometric coefficients of the reactants (78,91).

Note from these expressions that a chemical reaction is accelerated by pressure if its activation volume is negative. This is generally the case for most addition reactions, and as an example, this effect has been exploited advantageously to accelerate cycloaddition reactions by pressure (92). In addition, dissociation reactions can be favored by pressure if charged species are formed through electrostriction effects. This causes an ordering of charged (ions) and uncharged (e.g., solvent) species, which results in a significant decrease in molar volume (93). The partial molar volumes and isothermal compressibility of the reaction mixture can be estimated from an equation of state. Brennecke and coworkers (90,94) present the appropriate thermodynamic correlations to estimate these values, and they have applied these calculations using the Peng–Robinson equation of state (95).

E. Activation Volumes

As noted previously, the activation volume can be defined as the difference between the partial molar volumes of the activated intermediate, or transition state, species and that of the reactants. Activation volumes for liquid-phase reactions are typically on the order of ±50 cm^3/mol (31,83,96), whereas apparent activation volumes of greater than ±1000 cm^3/mol have been reported for SCF reactions (96). For example, Johnston and Haynes (78) report an activation volume of −6000 cm^3/mol for the unimolecular thermal decomposition of α-chlorobenzylmethylether in 1,1-difluoroethane near the critical point. Eyring and coworkers (97,98) have observed activation volumes of as high as +7000 cm^3/mol at just above the critical point for the ring closure reaction of a metal carbonyl in both supercritical CO_2 and ethane. Such large positive-to-negative variations have been explained by considering the activation volume to be the sum of two contributing terms (26,78,97): (a) a repulsive or intrinsic component resulting from the change in occupied volume between the reactants and the transition state, and (b) an attractive contribution due to solute–solvent intermolecular forces. The repulsive part can be significant depending on changes in molecular volume (i.e., bond lengths) resulting from the breakage or formation of bonds. The attractive contribution is significant, for example, when there are large changes in polarity between the reactants and the transition state. Johnston and Haynes (78) attributed their large negative activation volume to this latter attractive contribution, which resulted from a more ordered solvent structure about a proposed highly polar transition state than about the reactant due to interactions of the strong dipole with the dielectric solvent (i.e., electrostriction). Conversely, Eyring and coworkers (97,98) attributed their large and positive activation volumes to a large repulsive (or intrinsic) contribution resulting from the dissociation of CO during the ring closure reaction.

The extreme divergences in activation volumes observed in SCF media are restricted to the region in the immediate vicinity of the critical point and approach liquid-like values at conditions removed from this critical transition (31,96). For example, the value of −6000 cm^3/mol reported by Johnston and Haynes (78) occurred at reduced conditions of about $T_r = 1.04$ and $P_r \approx 1.0$. They report a corresponding value of −71.6 cm^3/mol at $T_r = 1.09$ and $P_r \approx 6.1$. This large variation in activation volume is attributed to the pronounced pressure dependence of the compressibility of the SCF media. This same correspondence with compressibility has also been reported for partial molar volume data for several solutes in supercritical CO_2 and ethylene (99). Johnston and Haynes (78) provide an excellent summary of the thermodynamic arguments for these significant variations in activation volume and partial molar volume with pressure. To simplify and summarize these arguments, the dramatic variations in these parameters result from the sharp divergence of the isothermal compressibility toward infinity at the solvent critical point. The practical application of these observations is that the thermodynamic pressure effect on chemical reaction rates can be substantially higher in SCF media than in liquids and is most significant in the vicinity of the mixture critical point.

F. Clustering in SCF Media

There is a large and growing body of published experimental, theoretical, and simulation reports (56,81,90,100–124) that suggest that the time-averaged solvent density and composition in the immediate vicinity of a solute molecule may be significantly different from the bulk solution values for dilute SCF solutions. This phenomenon has generally been termed "solvent-solute clustering" (101,125). This clustering effect has also been reported for cosolvents or solutes about another solute molecule within an SCF phase (56,81,86,90,94,100,102, 106,109,126–130). These local density augmentations are reported to have typical values on the order of two to four times the bulk value (113,119,123), and as demonstrated in examples of cosolvent clustering around a solute molecule, they can result in local composition enhancements as high as seven times the bulk value (100,131). These local density and composition enhancements are generally believed to result primarily from specific short-range solvation effects associated with this molecular disparity and not long-range solvent critical phenomena that underlie the large compressibilities characteristic of these systems (26,90,101,107,108,122). In addition, this clustering phenomenon is believed to be a very dynamic process with rapid exchange of the clustering molecules with the bulk fluid occurring on the picosecond time scale (94,132). Spectroscopic experiments and molecular dynamics simulations have shown that these geometrically defined clusters may persist on the order of 100 ps (108). This topic of solvent-solute and solute-solute clustering has recently been reviewed in more detail by Brennecke and Chateauneuf (122) and Tucker (123).

This clustering phenomenon is important for consideration of organic chemical transformations in SCF media because cluster formation may influence the chemical reactivity (56,81,90,94,109–112,115,118,120–122,128–130, 133–136). For example, an enhanced local solvent density in the cluster around a solute molecule can significantly change the local values of density-dependent properties such as the dielectric constant (31,78,79), diffusivity (137), and viscosity (118,136). Such changes in the local solvent environment can consequently affect the reaction pathway through, for example, influencing the stability of a polar transition state species (122). The molecular solvent cluster can also enhance solvent cage effects around reactants or corresponding activated complexes, thus inhibiting the mass transfer of reactive species (81,86,103,118,133). However, for solvent cage effects to influence the chemical reactivity, the time scale of the reaction must be within that for which the cluster maintains its structural integrity (86,94,104,122). Finally, the local composition enhancements associated with the clustering phenomenon can result in an increased local concentration of reactive species which, in turn, can influence reaction rates through simple concentration effects (90).

In summary, the considerable "clustering" literature for SCF-mediated reactions suggests that the bulk physical properties of the SCF solvent media are the primary factors affecting chemical reactivity and selectivity (122). However, local molecular phenomenon can also influence these parameters in certain applications, so that the specific reaction mechanism must be considered to determine the relative influence of these factors.

III. APPLICATIONS

A. Carbon Dioxide as a C_1 Building Block

Utilization of $scCO_2$ as both a reaction solvent and a C_1 building block affords particular opportunities in CO_2-mediated reactions. Carbon dioxide is an abundant natural carbon source (138) that is relatively benign with regard to both environmental and health effects. As a result, CO_2 has found widespread use in a variety of industrial applications, including as a protective gas for sensitive foods, a source of beverage "carbonation," a fire-extinguishing agent, and an extraction solvent (139). Largely as a result of this abundance and benign nature, considerable efforts have been made in recent years to utilize CO_2 as a feedstock for the synthesis of valuable chemicals and fuels (139–149). Various intermediates and products that have been suggested include formic acid, alkyl formates, formamides, amines, methane and other hydrocarbons, methanol and higher alcohols, isocyanates, organic carbonates, carbamates, and even polycarbonate-based polymers. A number of these products are manufactured industrially using carbon monoxide and phosgene as reactants; hence the toxicological and physiological effects, corrosion, and environmental risks associated with these feedstocks make CO_2 a very attractive alternative for consideration (143,149,150). In fact, there are at least four important commercial-scale processes in which CO_2 is used for organic syntheses, including the manufacture of urea, cyclic carbonates, salicylic acid (the Kolbe–Schmitt process), and methanol (140).

Due to the relative inertness, readily attainable critical properties, and minimal environmental impact of CO_2 that was described previously, the vast majority of research and development in conducting chemical transformations in SCF media has utilized $scCO_2$ as the reaction medium (31,33,35). Thus, activation of CO_2 for use as a reactant for organic synthesis combined with the simultaneous use of $scCO_2$ as the reaction medium is an obvious synergy. A number of such studies have been reported in recent years.

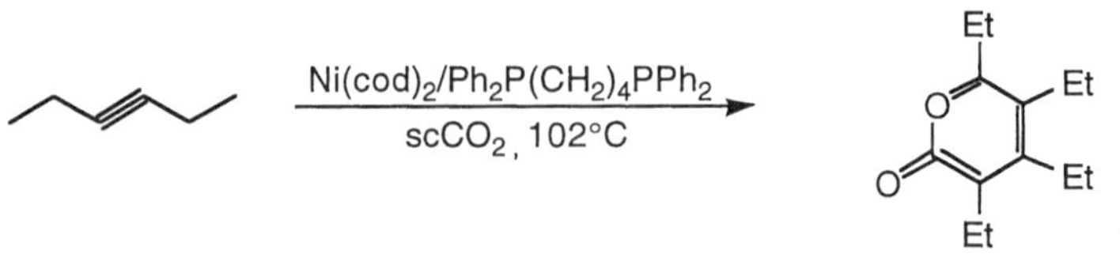

Scheme 1

Perhaps the earliest report of using $scCO_2$ as both reactant and reaction medium is that of Reetz et al. (151) who report the reaction of CO_2 with 3-hexyne to produce tetraethyl-2-pyrone using $[Ni(cyclooctadiene)_2]/Ph_2P(CH_2)_4PPh_2$ as catalyst (Scheme 1). The product distribution is reported to be similar to that in liquid benzene at 120°C.

Catalytic hydrogenation of CO_2 is one of the most promising approaches to CO_2 fixation in organic synthesis (143). The synthesis of formic acid by the catalytic reduction of CO_2 is of particular interest due to the widespread industrial consumption of formic acid (approximately 300,000 ton/y) and derivatives such as methyl formate, ethyl formate, and dimethylformamide (DMF) (141). Jessop, Ikariya, and Noyori (64,152–154) have reported an efficient synthesis of formic acid in an SCF mixture containing CO_2 and hydrogen using a ruthenium(II) phosphine complex as the catalyst, triethylamine, and a trace amount of water (Scheme 2). Addition of the basic tertiary amine and the use of high pressure in $scCO_2$ successfully shifted the equilibrium in favor of the formic acid product and led to a high initial reaction rate of 1400 mol of formic acid per mole of catalyst per hour. This is about 5 times faster than a similar synthesis in water at room temperature reported by Gassner and Leitner (155), and about 18 times faster than in the conventional liquid solvent tetrahydrofuran at the same temperature. The authors attribute such remarkable catalytic efficiency to the enhanced mass transfer characteristics and high hydrogen miscibility afforded by the SCF solvent/reactant. In subsequent papers, Jessop et al. (35,146) provide further details on this reaction, including results from catalyst screening experiments as well as the effects of the base, water and other additives, hydrogen pressure, and temperature. The $RuH_2[P(CH_3)_3]_4$ complex noted above and $RuCl_2[P(CH_3)_3]_4$ were found to be the most active catalysts, although an induction period of about 1 h was noted with the latter. The reaction rate was

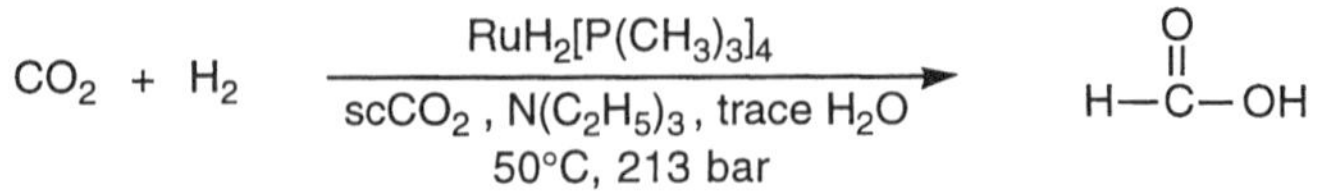

Scheme 2

observed to be highly dependent on the reagent concentrations, but this seemed to be more a function of the phase behavior than stoichiometry. In particular, the addition of trace amounts of water, methanol, or dimethylsulfoxide (DMSO) dramatically accelerated the initial rates. For example, use of methanol or DMSO rather than water in the above reaction increased the turnover frequency (TOF) to more than 4000 mol of formic acid per mole of catalyst per hour. The presence of the base is necessary for favorable thermodynamics (143) and has a strong effect on the reaction rate. Under the experimental conditions employed in this work, the triethylamine concentration had a strong optimum at about 100 mmol/L concentration.

Pomelli et al. (156) reported a theoretical study using density functional methods to investigate how coordination of a CO_2 molecule can assist in the release of formic acid from the catalyst complex in the last step of the catalytic cycle for the hydrogenation of CO_2 with rhodium complexes. They find that the presence of a CO_2 molecule in their active site model thermodynamically favors the formic acid dissociation from the complex and enhances the reaction rate. This may provide some additional explanation of the dramatic rate increases observed by Noyori and coworkers.

At higher temperatures and upon using methanol instead of water as the additive, Jessop et al. (146,157) observed that the homogeneous ruthenium-catalyzed hydrogenation of $scCO_2$ also results in appreciable thermal esterification of the formic acid with methanol to produce methyl formate (Scheme 3). The maximum yield of methyl formate was obtained at about 80°C due to slow thermal esterification at lower temperatures and low catalytic activity for hydrogenation at higher temperatures. The authors report a yield of up to 3500 TON (turnover number, moles of methyl formate per mole of catalyst), or 55 TOF (moles of methyl formate per mole of catalyst per hour), which is about two orders of magnitude greater than a literature value of 40 TON. Baiker and coworkers (158) also report the formation of methyl formate from the hydrogenation of $scCO_2$, but using a sol-gel-derived hybrid catalyst derived from $RuCl_2[PMe_2(CH_2)_2Si(OEt)_3]_3$. Best results were obtained at 100°C, where they report 100% selectivity to methyl formate with a TOF of 115 h^{-1}. This is more than double the TOF of 55 h^{-1} noted above.

Jessop et al. (146,159) also report the synthesis of DMF during $scCO_2$ hydrogenation at 100°C using the $RuCl_2[P(CH_3)_3]_4$ catalyst and primary or

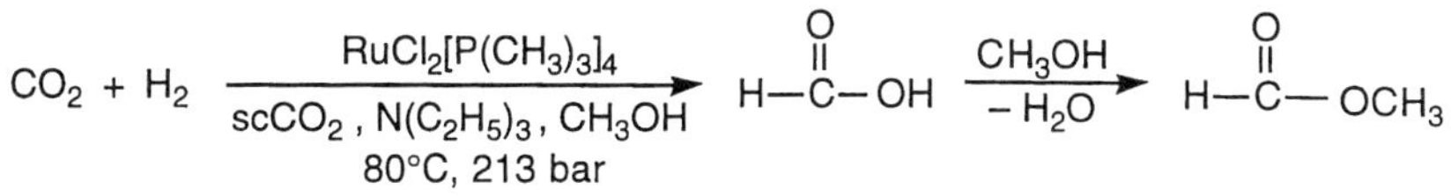

Scheme 3

secondary amines as the base (Scheme 4). For the case of dimethylamine, 99% conversion to DMF with 99% selectivity was observed for one reaction with a yield of 370,000 TON over 19 hours. This yield greatly exceeds a literature value of 3400 TON for liquid solvents that is cited by the authors. This reaction proceeds via a two-step mechanism: the Ru-catalyzed hydrogenation of CO_2 to formic acid followed by the slower thermal condensation of the formic acid and dimethylamine to produce DMF. Also, this reaction system is complicated by the fact that the secondary or primary amine forms a liquid carbamate salt on contact with CO_2 (35). Thus, two phases—the SCF phase and the liquid salt phase—were present in the vessel from the beginning of the reaction. Independent solubility tests showed that the catalyst was soluble in $scCO_2$ and not the liquid carbamate.

Subsequent studies by Baiker et al. (158,160,161) used sol-gel-derived and silica matrix–stabilized transition metal complexes as heterogeneous catalysts for synthesizing DMF by this reaction. These heterogeneous hybrid gel catalysts can easily be separated from the reaction mixture by pressure release and filtration, and they are stable under reaction conditions (33). A series of catalyst screening experiments with different transition metal catalysts showed a decreasing activity in the order Ru > Ir > Pt, Pd > Rh. The ruthenium-containing hybrid gel catalysts proved to be 100% selective and the most active for DMF synthesis, resulting in turnover numbers up to 110,800 with corresponding TOFs up to 1860 h^{-1}. This TOF exceeds previously reported values with heterogeneous catalysts by a factor of 600 (33). In subsequent work with homogeneous catalysts, Baiker et al. (162,163) showed that ruthenium complexes with bidentate phosphine ligands exhibit much higher activity for DMF synthesis in $scCO_2$ than any of these previous results. Using the complex $RuCl_2[Ph_2P(CH_2)_2PPh_2]$ as catalyst, they found that the synthesis proceeded rapidly with a TOF of 360,000 h^{-1} at 100°C with 99% yield and 100% selectivity to DMF.

Sakakura et al. (150) have investigated the use of $scCO_2$ for the selective synthesis of dimethyl carbonate (DMC) by reaction of CO_2 with trimethyl orthoacetate, which also produces methyl acetate (AcOMe) as a byproduct and dimethyl ether as a side product (Scheme 5). A primary incentive of this study was to evaluate the use of $scCO_2$ as an alternative to phosgene for this reaction. Various metal alkoxides with and without promoters were evaluated as catalytic systems, and the reaction was found to be strongly dependent on the alkox-

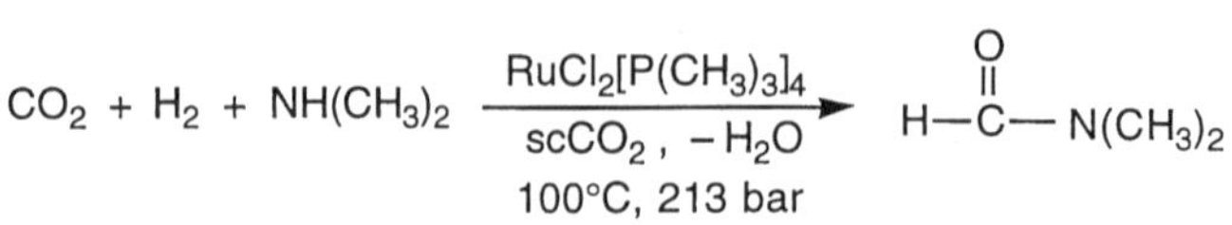

Scheme 4

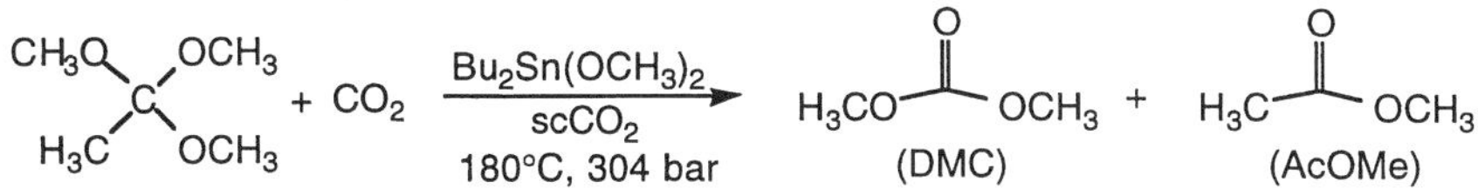

Scheme 5

ide structure. The example of $Bu_2Sn(OCH_3)_2$ with no promoter illustrated in Scheme 5 resulted in 20% yield of DMC after 24 h with 93% selectivity, and the reaction mixture was verified to be homogeneous by visual observation through sapphire windows. Evaluation of the influence of pressure showed a pronounced enhancement in catalytic activity in the vicinity of the critical pressure of CO_2, and selectivity to DMC was best at higher pressures. Another recent report of this group's efforts (164) describes a similar synthesis from acetal and $scCO_2$ in the presence of methanol using dibutyl tin dimethoxide as catalyst to produce DMC and acetone. An 80% yield with nearly 100% selectivity is achieved after 24 h at 80°C. Acetone is reconverted to acetal by dehydration with methanol in an SCF state at about 300°C and 150 bar pressure.

Vieville et al. (165) have examined the use of $scCO_2$ as a reaction medium and as a source of carbonate for the carbonatation of glycerol. Their results show that $scCO_2$ alone is not an adequate carbonate source for carbonatation of glycerol into glycerol carbonate. However, when utilized with an organic carbonate such as ethylene carbonate as a coreactant, the reaction proceeds to equilibrium in the presence of certain catalysts (Scheme 6). A catalyst screening study showed the zeolites Purosiv and 13X as well as the basic resin catalyst Amberlyst A26 in the hydoxyl form to be active for the production of glycerol carbonate with 1-h reaction yields of 32%, 25%, and 21%, respectively. It is not clear from the results presented whether or not the CO_2 actually participates as a reactant in this reaction.

Mizuno and coworkers (166,167) have explored a very different route to the hydrogenation of $scCO_2$ by using dispersed TiO_2 and Cu powders to photocatalytically reduce CO_2 to formic acid. The authors report that product formation requires a two-step procedure: (a) the CO_2 is first reduced by irradiating the TiO_2 powders dispersed in a $scCO_2$ reaction mixture, and (b) an aqueous solution is added to protonate reaction intermediates on the TiO_2 surface. Based

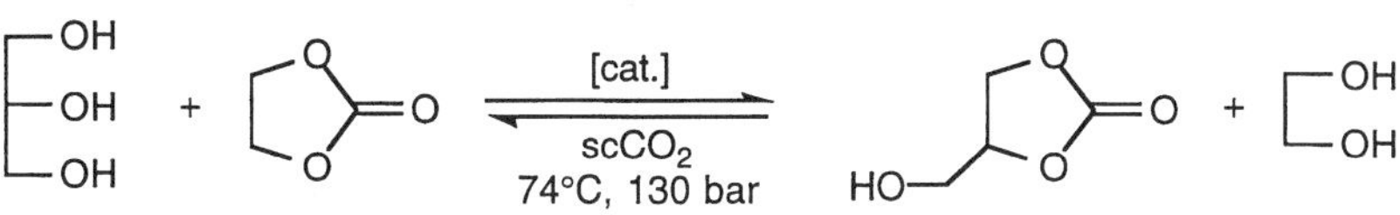

Scheme 6

on ESR measurements, the authors propose the formation of [$\cdot CO_2^-$] radicals as the reactive intermediates formed during irradiation, which are subsequently protonated to formic acid. The amount of formic acid produced increases with the pH of the aqueous solution when nitric, hydrochloric, and phosphoric acid solutions are used for the protonation step. Standard reaction conditions for the photocatalytic step were 35°C and 90 bar with an irradiation time of 5 h. Formic acid is apparently produced with high selectivity but at very low yields (on the order of 8.8 μmol/g-cat).

B. Hydrogenation

Many industrial hydrogenation processes are conducted in gas–liquid reactors with solid, heterogeneous catalysts (168). These are often relatively fast reactions, so the mass transfer resistances to the transport of the gaseous hydrogen through the liquid phase to the catalyst surface typically prove to be rate limiting. Indeed, Blackmond and coworkers (63,169) have shown in an elegant study that the key kinetic parameter affecting certain hydrogenation reaction rates is the actual molecular hydrogen concentration in the liquid phase, rather than the gas phase hydrogen partial pressure, which is typically taken as an indirect measurement of the liquid-phase solubility. They show that the actual liquid-phase hydrogen concentration can be significantly less than the equilibrium saturation value and result in limiting the reaction, depending on such factors as the inherent reaction rate and agitation rates. In fact, they were able to reproduce published data attributed to hydrogen pressure dependence by varying the gas–liquid mass transfer rate at constant pressure. This gas–liquid mass transfer resistance is eliminated in many SCF phase systems since the hydrogen is completely miscible with the SCF solvent [e.g., see Tsang and Streett (170) for the case of $scCO_2$]. As a result, the hydrogen concentration in an SCF phase can be an order of magnitude higher than in a comparable liquid-phase system at the same pressure (33), thus providing a significantly higher effective hydrogen concentration at the catalyst surface. This can lead to extremely fast reaction rates compared to liquid-phase systems and potentially affect product selectivities. This combination of increased hydrogen transport rates with the relatively high heat capacities afforded by SCF media (42,66) to facilitate removal of the substantial exothermic heat of reaction should prove beneficial for commercial application of SCF media for this class of reactions.

Several examples of research and development in conducting hydrogenations in SCF media are included herein. In one of the earliest papers on this topic (11), Pickel and Steiner report application of a continuous fixed-bed reactor using supercritical CO_2, ethane, and propane for an unspecified hydrogenation reaction of pharmaceutical interest. Other applications include the hydrogenation of fats and oils for the food industry, demonstration of hydrogenation of a variety of organic substrates, and enantioselective synthesis.

One application that has been reported by two research groups (10,171) in recent years is that of hydrogenating (or "hardening") of edible fats and oils. There is a significant market for hydrogenated oils with current annual production at about 25 million tons (10). Prior to use in the food industry for the manufacture of margarine or shortening, vegetable oils are selectively hydrogenated to increase both the melting point (hence the "hardening" terminology) and the oxidation stability as well as to provide the desired texture. Free fatty acids are generally hydrogenated completely for oleochemical applications, such as for detergents (171). The double bonds in natural oils and fatty acids are primarily cis bonds, which are preferred for edible oil applications due to the reported negative health impacts (e.g., high cholesterol and blood lipid levels) of the corresponding trans bonds. However, appreciable isomerization to these trans bonds typically occurs in conventional oil hydrogenation processes. This isomerization product is partially due to the nickel on kieselguhr catalyst used in the conventional process (172), but it has also been shown to be favored by low hydrogen concentrations at the catalyst surface. Hence, limited hydrogen solubility in the liquid oil phase of the conventional process and the corresponding mass transfer resistances in the gas–oil–catalyst system limit the hydrogenation reaction rates and result in significant formation of trans fatty acid bonds. Other disadvantages of the conventional process include discontinuous operation and low space–time yields (172).

Tacke and coworkers (171–173) describe the use of supercritical CO_2 and propane in a continuous fixed-bed reactor for both the selective and complete hydrogenation of fats and oils, free fatty acids, and fatty acid esters using De-Gussa's Deloxan polysiloxane-supported palladium and platinum catalysts. For one example, they report an increase in catalyst productivity by a factor of 18 (from 333 to 6086 kg fatty acid/kg catalyst) for a continuous fixed-bed process in $scCO_2$ using 1% Pd/Deloxan catalyst vs. the 25% Ni/kieselguhr catalyst of the conventional process. The authors attribute such increases in hydrogenation activity to the reduced viscosity of the $scCO_2$ reaction medium and the corresponding increase in the hydrogen mass transfer rates. Additional reported benefits include significantly improved selectivity to the desired cis isomer products and extended catalyst lifetimes.

Härröd and Møller (9,10) report hydrogenation of vegetable oils and fatty acid esters in a fixed-bed reactor packed with a commercial carbon-supported palladium catalyst using near-critical and supercritical propane as the solvent to address the hydrogen solubility and transport rate limitations noted previously. The authors state that the propane, oil, and hydrogen form an essentially homogeneous mixture under the reaction conditions studied, although the actual phase behavior of the reaction mixture is unclear from this report. Nevertheless, they report productivity increases up to 1000 times higher than with the conventional process as well as a considerable reduction in the formation of trans fatty acids for the same catalyst and degree of hydrogenation. The authors also report catalyst deactivation comparable to the conventional process at their reaction conditions. Opportunities to prevent this deactivation that are potentially afforded by use of the SCF media remain unresolved.

Bertucco, Zwahlen, and coworkers (174,175) report on the catalytic hydrogenation of the two double bonds of an unsaturated ketone that was identified only as a vitamin intermediate. The authors used a commercial alumina-supported palladium catalyst, and the reaction was run in a gradientless internal recycle reactor of a modified Berty type (176). The reaction mixture in this study was not homogeneous, consisting of both a liquid and an $scCO_2$-rich fluid phase. The benefit provided by the $scCO_2$ was apparently to expand the liquid phase and thus enhance the hydrogen transport. This was a detailed study including kinetic measurements in a statistically designed series of experiments to determine the effects of the temperature, pressure, and $scCO_2$ concentration on the reaction rate as well as the development of phase equilibrium thermodynamic calculations and kinetic models of the pertinent reaction chemistry. Phase equilibrium modeling using the Peng–Robinson equation of state (95) with mixture parameters fit to experimental binary data provided a satisfactory fit of the available binary and ternary vapor–liquid equilibrium data, so this phase behavior model was extended to multicomponent calculations. The simple homogeneous kinetic model developed for data interpretation satisfactorily reproduced the experimental results, but the kinetic calculations were reported to be very sensitive to the calculated reactant compositions in the liquid phase. Interestingly, the kinetic model results showed essentially zeroth order dependence on the hydrogen composition, suggesting the success of enhancing hydrogen transport by conducting the reaction in an $scCO_2$-expanded liquid phase.

Bertucco, Steiner, and coworkers (177) report further work on this application of using $scCO_2$ to expand the liquid phase in a catalytic hydrogenation reaction. The focus of this report was to present the development of a two-dimensional model describing the simulation of a high-pressure trickle-bed reactor that was validated with experimental results obtained on a pilot-scale process, and the specific application was the hydrogenation of apparently the same unsaturated ketone intermediate evaluated in the previous study (175). Several of the general conclusions of the study are pertinent to this review. First, echoing the results noted above, the authors stress the importance of an accurate evaluation of the phase equilibrium both to interpret kinetic data and to correctly understand the reactor behavior. Second, the authors note that this process allowed the use of a reaction volume significantly smaller than in conventional processes. This is an important conclusion considering the pilot scale of the reactor utilized in this work. Third, the authors note that the model was developed as a two-dimensional model primarily to account for radial temperature profiles due to the highly exothermic nature of the hydrogenation reaction, even in the presence of an evaporating liquid, as presumably occurs here. Thus, heat effects are critically important in scaling highly exothermic reactions, such as hydrogenations, and extreme care must be taken on this subject in interpreting results from small-scale reactors in SCF systems. Finally, the authors point out the possibility of performing such reactions in a homogeneous SCF phase to avoid interfacial mass transfer limitations, but extremely high temperatures and pressures may be needed to reach such conditions in specific applications. Thus, this compromise solution of expanding the liquid with an SCF phase to enhance reactant solubility in the liquid phase may prove invaluable in many potential commercial applications. In a recent further report on this work by Bertucco and coworkers (178), the authors report the detailed kinetic model development supporting the reactor modeling noted above. Here the authors show results that clearly demonstrate the positive effect of the $scCO_2$ expansion of the liquid phase on the reaction yield. The optimal reaction conditions are identified as 1:1 CO_2-to-liquid feed ratio, which result in a conversion of twice that with respect to the absence of the SCF solvent.

Hitzler and Poliakoff (179) report the hydrogenation of a variety of organic functionalities in supercritical CO_2 and propane using polysiloxane-supported noble metal catalysts in a small-scale (5 ml v) continuous flow reactor. Hydrogenation of cyclohexene to cyclohexane was used as a test reaction to establish the reactor performance. Using 5% palladium or platinum on polysiloxane, pressures of 120 bar for CO_2 and 60–80 bar for supercritical propane, cyclohexene rates of 0.005–0.20 mol/min and 2–4 times the stoichiometric hydrogen flow, the authors report fast hydrogenation rates and cyclohexane yields of 95–98%. However, the authors note substantial exothermic heating of the reactor to as high as 320°C. This graphically demonstrates the importance of temperature control in SCF-mediated hydrogenation reactions. Other compounds hydrogenated included nitro, epoxide, oxime, nitrile, hydroxyl, aldehyde, and ketone functionalities, reportedly with high conversions. One interesting example reported by the authors is that of varying the selectivity from acetophenone hydrogenation over at least four possible products depending on the reactor conditions of temperature, pressure, and hydrogen concentration (Scheme 7). Thus, the authors suggest that one can potentially tune reactor operating conditions to maximize the yield of a particular desired product without changing the catalyst. The

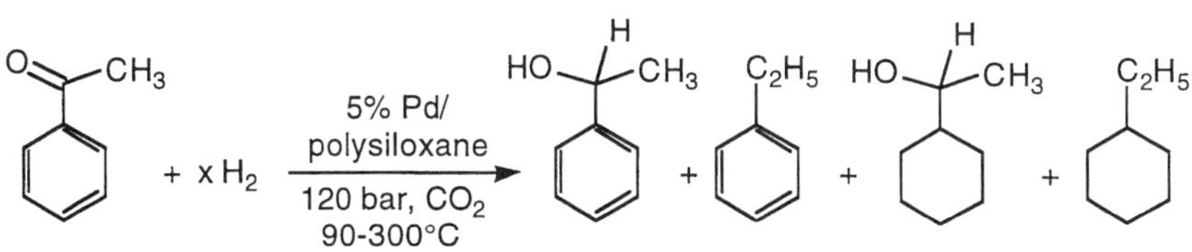

Scheme 7

authors attribute the apparent success of the SCF-phase hydrogenations to enhanced hydrogen solubility in the SCF solvent phase as well as reduced solvent viscosity and increased diffusion rates.

These authors and collaborators at Thomas Swan & Co. and Degussa report extension of this work and more details of the results in subsequent papers (172,180) and a process patent (181). Using small continuous-flow reactors of 5- and 10-ml volumes and heterogeneous noble metal catalysts on Degussa's Deloxan aminopolysiloxane supports, they report the hydrogenation of a wide range of organic compounds in both supercritical CO_2 and propane. These compounds include alkenes, alkynes, aliphatic and aromatic ketones and aldehydes, epoxides, phenols, cyclic ethers, oximes, nitrobenzenes, Schiff bases, and nitriles. The hydrogenolysis of aliphatic alcohols and ethers was also investigated. The authors report that several other catalysts were tried, including Pt/C and nickel catalysts, but they found the indicated Deloxan catalysts to be the most reliable. Propane was the favored SCF solvent for hydrogenating nitrogen-containing substrates, such as oximes, Schiff bases, nitriles, and nitrobenzenes to avoid the formation of insoluble carbamic acid salts formed with $scCO_2$ and the amine groups of the corresponding products. The authors note that propane has a lower critical pressure than CO_2, but the potential lower pressure process advantage is offset by the propane flammability. Several more examples of tuning product selectivity with reactor operating conditions are reported, including the hydrogenation of isophorone, a functionalized cyclohexene derivative of commercial interest in the fine chemicals industry (Scheme 8).

The authors report that the reaction conditions can be tuned to selectively hydrogenate the ring double bond and give quantitative conversion with high selectivity to the desired 3,3,5-trimethylcyclohexanone product. This is an item of commerce that is used as a solvent for vinyl resins, lacquers, varnishes, paints, and other coatings. Such high selectivity could significantly impact the capacity requirements for purification by distillation as is practiced in conventional processes. The authors note that one of the important findings of this work is that high substrate flow rates do not necessarily require high CO_2 flow rates, which can severely degrade the process economics due to the corresponding recycle costs. However, substantial exotherms are reported in the results, suggesting potential process limitations for accurate temperature control. Indeed, the authors

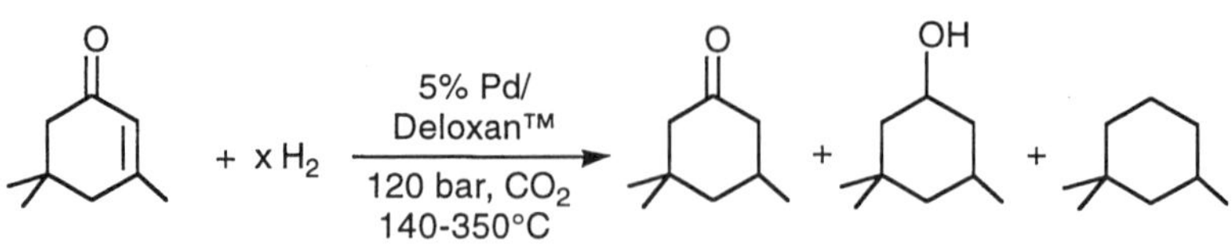

Scheme 8

report in a footnote (180) that scaling up to a 100-ml reactor by Thomas Swan & Co. necessitated the use of internal baffling to prevent excessively hot flows through the center of the catalyst bed.

Another report of hydrogenation under SCF conditions from the patent literature includes claims for a continuous heterogeneously catalyzed hydrogenation reaction process by Subramaniam and Said (182,183). The primary focus of these patents is the in situ mitigation of coke buildup in porous catalysts, but an SCF-mediated hydrogenation process is a cited application (and claim).

C. Asymmetrical Hydrogenation

Burk, Tumas, and coworkers (24,184) were the first to publish results demonstrating the feasibility of conducting asymmetrical catalytic hydrogenation reactions in $scCO_2$, and that, at least in certain cases, higher enantioselectivities can be achieved than in conventional liquid organic solvents. These researchers hydrogenated several prochiral α-enamides using cationic rhodium catalysts incorporating a chiral bidentate Et-DuPHOS ligand (Scheme 9). These cationic ET-DuPHOS-Rh complexes have been found to efficiently catalyze the hydrogenation of such α-enamide esters to the corresponding α-amino acid deriva-

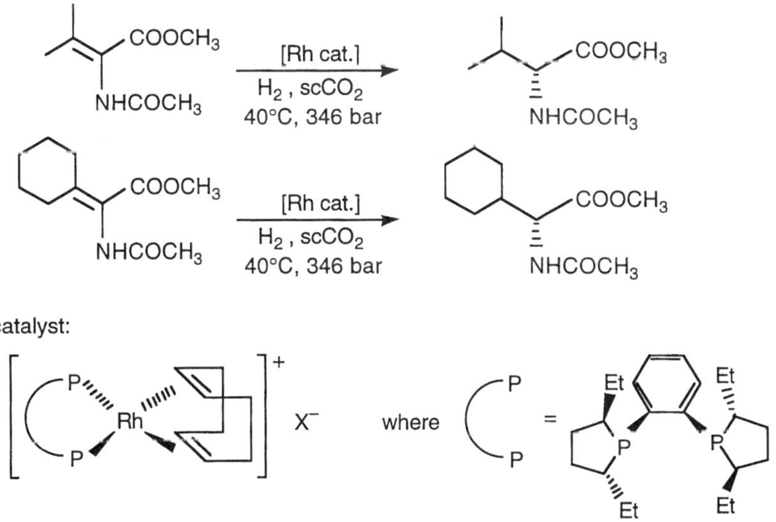

X1 = tetrakis(3,5-bis(trifluoromethyl)phenyl)borate (BARF)

X2 = trifluoromethanesulfonate (CF_3SO_3)

Scheme 9

tives with high enantioselectivities (≥98% ee) in conventional organic solvents. These alkyl-substituted diphosphine catalyst complexes with the indicated fluorous counterions ($CF_3SO_3^-$ or $BARF^-$) were found to be sufficiently soluble in $scCO_2$ for catalytic activity despite the ionic nature of the catalysts, which, as a general rule, results in poor solubility in $scCO_2$ (24). The solubility of the BARF complex was reported to be at least 0.030 mM at reaction conditions of 40°C and 346 bar. Hydrogenation of four β-monosubstituted α-enamide esters was conducted in $scCO_2$ as well as in liquid methanol and hexane. These reductions were reported to proceed cleanly and quantitatively to the corresponding α-amino acid derivatives with similar high enantioselectivity (90.9–99.7% ee) in each solvent. However, as listed in Table 3, reduction of the β,β-disubstituted α-enamide esters shown in Scheme 9 showed significantly higher enantioselectivity in $scCO_2$ than in methanol or hexane in three of the four cases illustrated. These high enantiomeric excesses are significant because these substrates are difficult to reduce with high enantioselectivity. In fact, the listed 88.4% ee of α-acetaminoisobutyric acid is the highest ever reported for reduction of the corresponding precursor. To verify that these results were not due to a pressure effect, the authors ran further experiments substituting nitrogen for the CO_2 at equivalent total pressures. These runs resulted in significantly lower enantiomeric excesses, suggesting that the reported selectivity enhancement is specifically associated with the use of $scCO_2$ as the reaction solvent and not a simple pressure effect.

Noyori et al. (185) report the asymmetrical hydrogenation of an olefinic substrate in $scCO_2$. They show that the chiral $[Ru(OCOCH_3)_2((S){-}H_8\text{-BINAP})]$ complex cleanly and efficiently catalyzes the α,β-unsaturated carboxylic acid

Table 3 Enantioselectivities for the Hydrogenation of β,β-Disubstituted α-Enamides Catalyzed by the Cationic Et-DuPHOS-Rh Complexes of Scheme 9 (184)

		% Enantiomeric excess		
Substrate	Catalyst	MeOH	Hexane	$scCO_2$
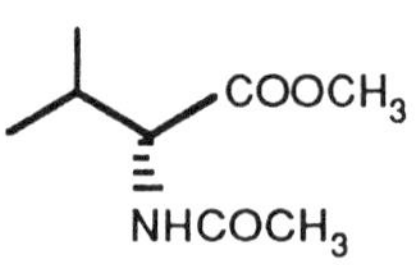	DuPHOS-Rh-X1	62.6	69.5	84.7
	DuPHOS-Rh-X2	67.4	70.4	88.4
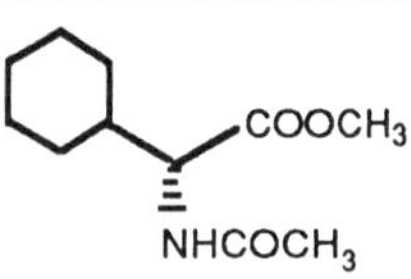	DuPHOS-Rh-X1	81.1	76.2	96.8
	DuPHOS-Rh-X2	95.0	91.2	92.5

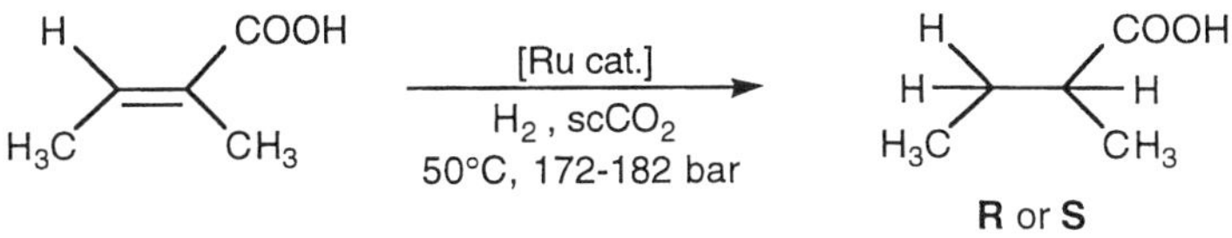

Scheme 10

tiglic acid to give (*S*)-2-methylbutanoic acid in over 99% yield and up to 81% enantiomeric excess (Scheme 10 and Table 4). This enantioselectivity is comparable to the 82% ee observed in methanol and exceeds the 73% ee seen in hexane. The analogous catalyst precursor [Ru(OCOCH$_3$)$_2$((R)-BINAP)] resulted in only 50% yield and 37% ee in a similar $scCO_2$-mediated reaction. The hydrogen partial pressure has been observed to influence the enantiomeric excess for this reaction in methanol (186). For example, lowering the hydrogen pressure from 30 to 5 bar in methanol increased the enantioselectivity from 82% ee to 95% ee (185). However, as shown in Table 4, lowering the hydrogen pressure from 33 to 7 bar in $scCO_2$ actually lowered the optical yield of the product from 81% ee to 71% ee.

As summarized elsewhere in this review, Kainz and Leitner (187) have demonstrated enantioselectivity of 66% ee (R) for the asymmetrical hydroformylation of styrene using [(CO)$_2$Rh(acac)]/(*R*,*S*)-BINAPHOS as the catalyst precursor in $scCO_2$ at low densities. In a more recent paper (188), this research group has reported the highly efficient iridium-catalyzed asymmetrical hydrogenation of prochiral imines in $scCO_2$. The test reaction selected was the enantioselective hydrogenation of *N*-(1-phenylethylidene)aniline to give *N*-phenyl-1-phenylethylamine (Scheme 11) using cationic iridium(I) complexes with chiral phosphinodihydrooxazoles modified with perfluoroalkyl groups in the ligand or in the anion to enhance solubility in $scCO_2$. Both the alkyl side chains and

Table 4 Asymmetrical Hydrogenation of Tiglic Acid with Chiral Ru(II) Catalyst Complexes in $scCO_2$ and Other Media (185)

Catalyst	Reaction medium	H_2 Pressure (bar)	Product	
			% Yield	% ee
Ru(OCOCH$_3$)$_2$((S)-H$_8$-BINAP)	liq. CO_2	30	0	—
Ru(OCOCH$_3$)$_2$((S)-H$_8$-BINAP)	$scCO_2$	33	99	81 (S)
Ru(OCOCH$_3$)$_2$((S)-H$_8$-BINAP)	$scCO_2$	7	23	71 (S)
Ru(OCOCH$_3$)$_2$((S)-H$_8$-BINAP)	methanol	30	100	82 (S)
Ru(OCOCH$_3$)$_2$((S)-H$_8$-BINAP)	hexane	30	100	73 (S)
Ru(OCOCH$_3$)$_2$((R)-BINAP)	$scCO_2$	33	50	37 (R)

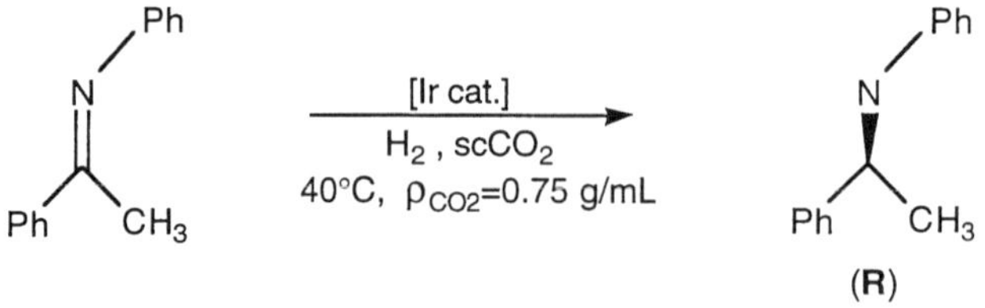

Scheme 11

the lipophilic anions increased the solubility, but the choice of the anion was also shown to have a dramatic effect on the enantioselectivity, with tetrakis-3,5-bis(trifluoromethyl)phenylborate (BARF) resulting in the highest asymmetrical induction. This was the same counterion used successfully by the Burk/Tumas group (184), as noted previously. In one example, the product was formed quantitatively within 1 h in $scCO_2$ with enantiomeric excesses of up to 81% (R). The homogeneous nature of the catalytically active species was demonstrated at the reaction conditions, and, interestingly, this was found to depend strongly on the presence of the substrate. Thus, the imine reactant essentially served as a cosolvent in the $scCO_2$ mixture to enhance the catalyst solubility.

The authors also demonstrated successful product recovery and isolation of the homogeneous catalyst by selectively extracting the product amine from the reactor with a $scCO_2$ purge following completion of the reaction. The iridium content of the recovered product was determined by atomic absorption spectroscopy to be less than 5 ppm. The authors have coined the acronym CESS for this "catalysis and extraction using supercritical solutions" process. Recharging the vessel with fresh reactant and repeating the reaction with no further addition of catalyst or ligand resulted in quantitative hydrogenation with similar enantioselectivity as the initial run for four catalysis cycles. Increased reaction times were required for quantitative conversion in subsequent cycles, but the enantiomeric excess of the product remained above 70%. These results are illustrated in Figure 5. The overall yield from this series of experiments corresponds to a total turnover number of 10,000 mol of product per mole of catalyst, with an average enantioselectivity of the isolated product of 76% ee. Another important result the authors emphasize in this study is that $scCO_2$ cannot be considered as a simple replacement solvent for conventional organic liquids, and that the successful "transfer" of reactions to this medium will require detailed knowledge about the physicochemical properties of the reaction mixture and about possible chemical interactions of the reaction components with the $scCO_2$ (or, for that matter, whichever SCF solvent is selected).

Baiker and coworkers (189,190) have investigated the asymmetrical hydrogenation of ethyl pyruvate in SCF solvents using a heterogeneous 5 wt % Pt/alumina catalyst modified with cinchonidine to promote the asymmetric in-

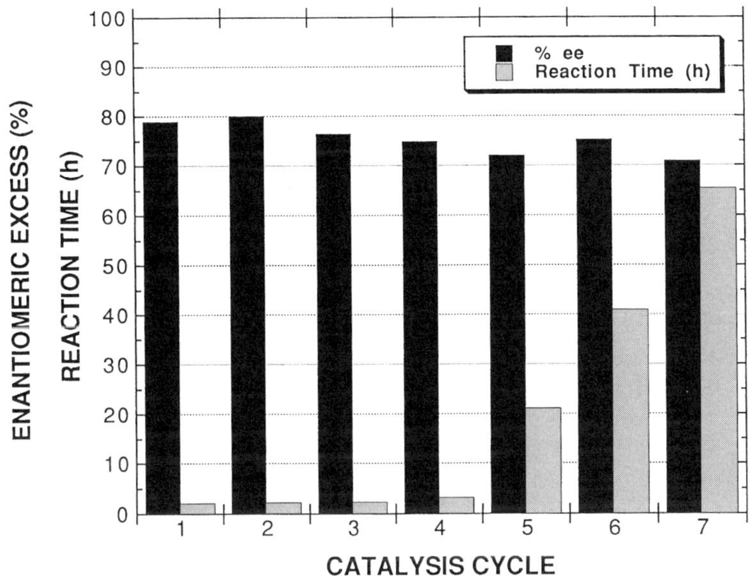

Figure 5 Catalyst recycling in the iridium-catalyzed enantioselective hydrogenation of imines in $scCO_2$. (Reprinted with permission from Ref. 188. Copyright 1999 American Chemical Society.)

duction (Scheme 12). They show that the reaction time in supercritical ethane can be reduced by a factor of 3.5 compared with that in toluene under similar conditions with similar enantioselectivity. In fact, the trend in enantioselectivity in the two apolar solvents ethane and toluene show very similar effects of reaction temperature, with both solvents having a pronounced decay of asymmetrical induction at temperatures greater than about 70°C. Figure 6 shows the change in

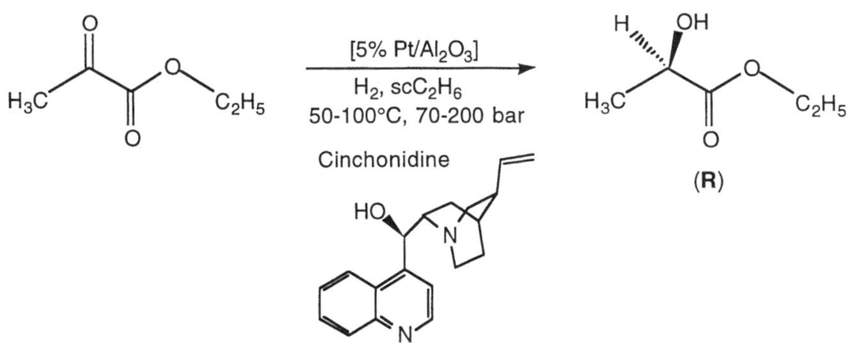

Scheme 12

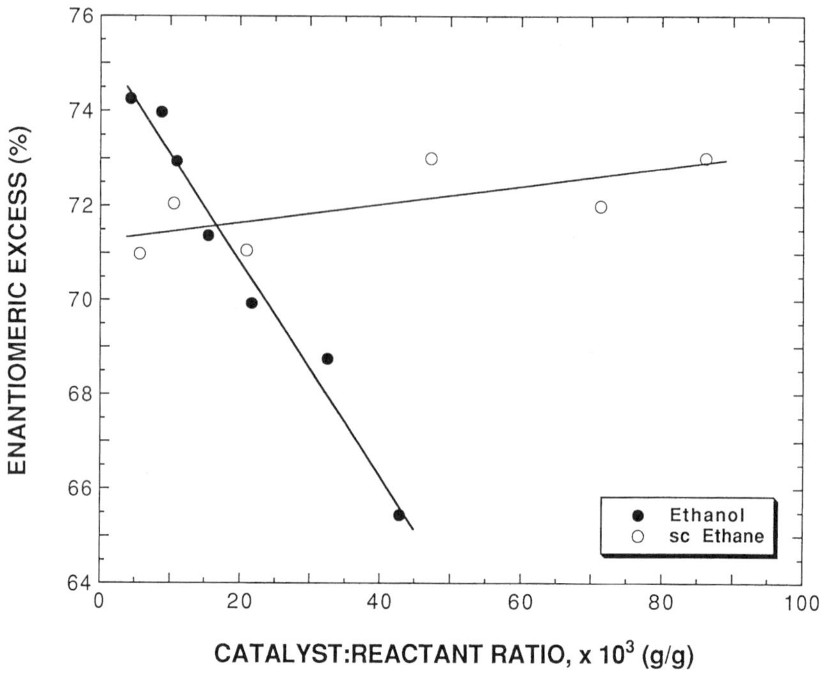

Figure 6 Enantiomeric excess for asymmetrical hydrogenation of ethyl pyruvate in liquid ethanol and supercritical ethane. (Reprinted with permission from Ref. 189. Copyright 1995 Baltzer Science Publishers.)

enantioselectivity as a function of the mass ratio of catalyst to reactant for experiments run in both supercritical ethane and liquid ethanol. As shown, increasing the catalyst loading has a small positive effect on the enantioselectivity in ethane but results in a sharply decreasing selectivity in liquid ethanol. This effect is believed to result from increasing mass transfer limitations in the liquid solvent on increased catalyst loading, whereas this is not a factor in the case of the SCF solvent. The authors point out that these results suggest that SCF solvents such as ethane are promising candidates for application in continuous fixed-bed reactors where the catalyst/reactant ratio is high and the mixing efficiency is well below that characteristic of slurry reactors. Table 5 summarizes other results reported in this study. As indicated, supercritical propane was also evaluated, but the enantioselectivity was poor at the higher temperature required for operating in the SCF phase with propane ($T_c = 96.6°C$) vs. ethane ($T_c = 32.2°C$). This points to the importance of tailoring the solvent choice for a particular reaction chemistry to match the solvent critical properties with the desired reaction temperature.

Table 5 Asymmetrical Hydrogenation of Ethyl Pyruvate over 5% Pt/Alumina Catalyst in Supercritical and Conventional Solvents (190)

Solvent	$P_{solvent}$ (bar)	P_{H2} (bar)	Temp. (°C)	Conversion (%)	ee (%)
Ethane	60	70	20	98	74
Ethane	60	70	50	96	74
Ethane	60	70	70	98	71
Ethane	60	70	100	68	33
Ethane	60	10	50	81	64
Ethane	60	30	50	99	69
Ethane	60	100	50	96	74
Ethane	60	140	50	92	74
Propane	50	70	100	40	34
CO_2	80	20	40	2	29
CO_2	80	70	40	3	28
CO_2	180	70	100	2	7
Toluene	—	70	50	100	75
Toluene	—	70	100	99	41

Table 5 also shows the effect of a strong catalyst deactivation that was observed in the enantioselective hydrogenation of ethyl pyruvate in $scCO_2$. The authors report that no conversion higher than 3% could be obtained in $scCO_2$ even after several hours reaction time. The authors investigated the source of this catalyst poisoning (190) and demonstrated with Fourier transform infrared (FTIR) spectral data the presence of adsorbed CO on the Pt/alumina catalyst surface after contacting it with CO_2, even at room temperature. One mechanism believed responsible for generating this CO, which is a strong poison for Pt-catalyzed hydrogenation of carbonyl compounds, is that of CO_2 reduction in the presence of the catalyst by the reverse water gas shift reaction. These results suggest a potential limitation to the use of $scCO_2$ as a solvent for commercial implementation of an SCF hydrogenation process. This potential limitation conflicts with reports noted previously of hydrogenations in $scCO_2$ that are being commercialized by Thomas Swan & Co (1).

D. Fischer–Tropsch Synthesis

The Fischer–Tropsch (FT) synthesis reaction is an important means of producing higher hydrocarbons (C_1–C_{20+}) from synthesis gas (CO and H_2), including those in the liquid fuel and chemical intermediates range. The reaction is typically heterogeneously catalyzed using supported cobalt, iron, or ruthenium catalysts, and

FT synthesis has been conducted in both the gas and liquid phases (67). The reaction is very exothermic, so adequate heat removal is essential to prevent localized overheating of catalyst surfaces. Gas-phase reactions exhibit higher reaction rates and reduced mass transport limitations relative to the liquid-phase alternative, but inadequate heat removal results in excessive methane formation (191) as well as high-molecular-weight waxes. The waxes are desirable for subsequent hydrocracking to diesel fuels (67), but they also tend to condense within the catalyst pores in gas-phase systems (192). The liquid-phase process exhibits superior heat removal capabilities and improved solubility of the high-molecular-weight waxes, but this reaction is restricted by inherent mass transfer limitations for transporting the gaseous reactants to the catalyst surfaces in the liquid solvent. Thus, SCF-mediated FT synthesis has been of considerable interest since optimal conditions should be attainable that combine gas-like mass transfer characteristics with liquid-like solubility and heat transfer capability.

Early work in this area was presented by Fujimoto and coworkers in a series of papers in the early 1990s (192–197). These reports have been reviewed by Savage et al. (31) and are only briefly summarized here. These studies included a side-by-side comparison of gas (N_2)–, SCF (*n*-hexane)–, and liquid (*n*-hexadecane)–mediated FT synthesis (192–194,196). The results effectively demonstrated removal of reaction heat and waxy products from the catalyst surfaces in the SCF case and produced a larger yield of higher hydrocarbons ($> C_{25}$) than in either the liquid- or gas-phase reactions. The rate of reaction and the diffusion of reactants were found to be slightly less than those in the gas-phase reaction. Due to diffusional differences in the reactants and products, the catalyst pore size was found to significantly impact both the reaction rate and the product distribution (195). For larger pore catalysts, the SCF solvent was found to be effective in extracting primary olefin products, which suppressed secondary hydrogenation to paraffins and olefin hydrocracking.

In more recent investigations, Fujimoto and coworkers extended their previous studies with an emphasis on developing the process to enhance the formation of heavy hydrocarbon waxes under optimized SCF conditions. Fujimoto et al. (198–200) show that the addition of a small amount (4 mol % on a CO basis) of long-chain primary olefins (e.g., 1-tetradecene or 1-hexadecene) to the SCF pentane reaction solvent significantly promotes carbon chain growth in FT reactions. Selectivity to waxy products is greatly enhanced with increased CO conversion, whereas the formation of methane, carbon dioxide, and light hydrocarbons is suppressed. An experimental and theoretical study of mass transfer effects in the FT synthesis reaction performed in the gas, SCF, and liquid phases was conducted (201) to compare the interplay of diffusion and reaction in these various reaction phases. Catalyst effectiveness factors calculated from a diffusion model applied to the reactant gases were in good agreement with experimentally measured values, and the relationships between these effectiveness factors and the catalyst particle size and reaction temperature were studied in all three phases. Model results suggested that the greater degree of carbon chain growth exhibited in catalysts calcined at higher temperatures could be partly attributed to rapid diffusion of the CO reactant and primary olefin products inside the catalyst pores. Mass transfer effects on secondary reactions of hydrogenation, cracking, and isomerization were also investigated. A subsequent study focused on the development of catalysts for selective wax production in SCF-mediated FT synthesis (200,202). Cobalt-based catalysts supported on silica were developed that exhibited high activity and wax selectivity even at relatively low reaction temperatures of 200°C. Addition of lanthanum and nickel to the catalysts enhanced the activity. Moderate calcination temperatures (300°C) and high catalyst cobalt loadings (20:100 weight ratio of Co to SiO_2) favored wax production by simultaneously increasing the CO conversion and chain growth probability. More recent evaluations extended earlier studies of the effect of cofeeding 1-tetradecene with synthesis gas over Co/SiO_2 (203) and Ru/Al_2O_3 (204) catalysts. Results showed that use of optimized SCF reaction conditions with 1-tetradecene as a chain initiator enhanced the formation of C_{14} and larger hydrocarbons at the expense of the corresponding shorter C_1–C_{13} chains.

Bukur and coworkers (205–207) have investigated FT synthesis in supercritical propane using a precipitated iron catalyst (Ruhrchemie LP 33/81 having a nominal composition of 100 Fe/5 Cu/4.2 K/25 SiO_2) which was originally used in commercial operations by Sasol in South Africa. The initial study (205) evaluated the performance of the iron catalyst in both supercritical propane and nitrogen at similar conditions to compare the results of SCF- and gas-phase reactions. The experiments were conducted for long run times (on the order of 300 h on-stream) in a continuous downflow fixed-bed reactor at 250°C and 70 bar using a 67% H_2/CO synthesis gas mixture with either nitrogen or propane. These results were compared with baseline conditions at 14.8 bar total pressure and with no added solvent or diluent. The synthesis gas partial pressure was the same in all cases. The study showed that operation at high pressure (70 bar) with nitrogen resulted in about a 10% decrease in FT activity relative to results at the base conditions with no effect on the product distribution. Operation with propane at the 70-bar pressure resulted in similar catalyst activity relative to the base conditions, but higher selectivity to 1-olefins. Bukur and coworkers next evaluated the effect of process conditions on olefin selectivity for similar SCF- vs. gas-phase operation (206,207). They observed that the total olefin content decreased for both modes of operation, but the 2-olefin content increased with an increase in conversion or H_2/CO molar feed ratio. However, olefin selectivities were essentially independent of reaction temperature. At high-synthesis gas conversions (~80%), the selectivity to high-molecular-weight 1-olefins during SCF operation were significantly higher than those obtained during gas-phase operation. Based on the results of both of these studies, the authors conclude that operation at SCF conditions results in higher diffusivities and more rapid desorption of the 1-olefins from the catalyst, which minimizes secondary hydrogenation and isomerization reactions relative to gas-phase synthesis. These results are consistent with those of Fujimoto and coworkers in the shorter duration tests described previously.

Bochniak and Subramaniam (67) investigated the effect of pressure-tuning the solvent and transport properties of the SCF reaction medium on FT reaction rates, catalyst activity, and product selectivity. FT synthesis was conducted over a Ruhrchemie iron catalyst (mass basis composition of 100 Fe/5.2 Cu/3.7 K/ 10.7 Si) in a continuous fixed-bed reactor using supercritical *n*-hexane as the reaction medium. Experiments were conducted over the pressure range of 35–70 bar (1.1–2.1 P_c of hexane) and at a fixed temperature of 240°C (1.01 T_c of hexane), space velocity of 50 cm^3/g cat, and H_2/CO ratio of 0.5. Synthesis gas conversion reached a steady state within a few hours and increased from 15% at 35 bar to 61% at 70 bar with virtually no catalyst deactivation during runs lasting up to 140 h. The 1-olefin selectivity at a given carbon number increased with pressure from 35 to 55 bar, but was virtually unaffected by a further pressure increase to 70 bar, whereupon the *n*-hexane density becomes less sensitive to pressure. Evaluation of the apparent rate constant (estimated by assuming a pseudo-first-order hydrogen dependence) showed that the catalyst effectiveness factor increases with pressure, suggesting the alleviation of pore diffusion limitations at the higher pressures. This increased accessibility of the pore volume was attributed to the enhanced extraction of the heavier hydrocarbon products from the catalyst pores through combination of the high density and mass transfer characteristics afforded by the SCF solvent. The enhanced desorption of the primary 1-olefin products at 55 and 70 bar inhibits secondary reactions such as hydrogenation, resulting in higher 1-olefin selectivities (~80%). This study is an excellent example of how operating conditions within the SCF regime can be manipulated to optimize the density and transport properties to maximize the utility of the SCF media for conducting chemical transformations.

E. Hydroformylation

The hydroformylation of olefins is a type of CO insertion reaction that is one of the most important industrial applications of homogeneous catalysis with transition metal complexes (208,209). Conventional industrial processes (e.g., the Oxo process) typically use either cobalt- or rhodium-based catalysts and conduct the reaction in two-phase gas–liquid reactors. Efficient transfer of the reactants from the gas phase into the liquid phase is of primary importance to minimize inherent mass transfer limitations (208). Reactor design thus focuses on optimizing this mass transfer rate by maximizing the interfacial area between phases. An SCF process eliminates this transport restriction since the hydrogen and carbon monoxide reactants are fully miscible in the homogeneous SCF phase (170,210).

Rathke, Klingler, and Krause (211–213) were the first to report an example of olefin hydroformylation in an SCF solvent, selecting the cobalt carbonyl catalyzed hydroformylation of propylene as a model reaction (Scheme 13). They demonstrated that the mass transfer limitations in the conventional liquid process can be reduced by operating in a single homogeneous SCF phase due to the higher concentration of synthesis gas (H_2/CO) reactants and enhanced transport rates. They demonstrated this by measuring the effect of the $scCO_2$ solvent on the linear-to-branched butyraldehyde product ratio, which is known to be dependent on both the reactant gas concentration and the agitation rate in conventional liquid solvents (211). Rathke et al. report that the reaction proceeds cleanly at 80°C with an enhanced n- to iso isomer product selectivity ratio of 88% relative to typical Oxo process selectivities of 75–80% (214). They also report that the hydroformylation rate and equilibrium concentrations of catalytic intermediates were found to be comparable to values for other linear-terminal olefins in nonpolar liquid media such as *n*-heptane and methylcyclohexane. Rathke et al. mention the potential for isolation of the catalyst and reaction products by a relatively simple pressure adjustment to control the density of the fluid mixture and, hence, the catalyst and product solubilities. This separation step is typically done by distillation in the conventional liquid process (208).

Broad claims were allowed in the process patent (213) resulting from this work, including the use of reactant substrates comprising C_2–C_{18} alkenes and carbonyl catalysts containing the group VIII metals cobalt, rhodium, ruthenium, and platinum as well as phosphine ligands. Specified reaction conditions included temperature (80–180°C), pressure (100–690 bar), and density (0.5–1.0 g/ml). Generic flowsheets defining the overall manufacturing process were also presented.

Rathke et al. monitored the progress of the reactions in their experiments using in situ high-pressure ^1H, ^{13}C, and ^{59}Co nuclear magnetic resonance (NMR) spectroscopy and a novel compact pressure probe design utilizing an efficient toroid detector. They emphasize the utility of NMR spectroscopy for SCF-mediated reactions because the linewidths of quadrupolar nuclei, such as ^{59}Co, are significantly decreased due to the low viscosity of the medium (211,215). One focus of the study was the determination of catalytically active

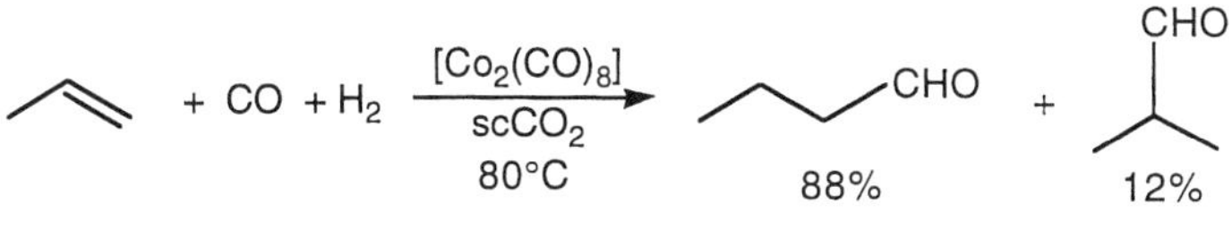

Scheme 13

intermediates utilizing this NMR technique. They report the rate and equilibrium constants as well as the enthalpy and entropy changes for the hydrogenation of the dicobalt octacarbonyl catalyst to form $HCo(CO)_4$ (Scheme 14). This is an important intermediate step in the overall reaction mechanism proposed by Heck and Breslow (216), since the $HCo(CO)_4$ is believed to be the actual catalyst in the reaction (217). Rathke et al. report that the kinetic and thermodynamic values measured in $scCO_2$ are in good agreement with values reported in conventional liquid hydrocarbon solvents (212,213). In a more recent study, Klingler and coworkers (218) have developed an improved high-pressure toroid detector NMR probe that can be operated at 300 bar and up to 250°C. This recent study reports application of this probe for in situ monitoring of phosphine-modified cobalt carbonyl hydroformylation catalysts in a variety of solvents.

Jessop et al. (219) present a mechanistic study for hydroformylation vs. hydrogenation of olefins in SCFs. Their chosen test reaction was that of the stoichiometric reaction (i.e., no added H_2 or CO) of 3,3-dimethyl-1,2-diphenylcyclopropene with hydridopentacarbonylmanganese(I) [$MnH(CO)_5$] catalyst. This reaction is known to give both hydroformylation and hydrogenation products (220), with the selectivity being dependent on the strength of the solvent cage (weaker cages favoring hydrogenation). Table 6 summarizes their results and shows the selectivity for hydrogenation in $scCO_2$ to be about 61–66%, which is similar to that in conventional liquid hydrocarbons solvents. This selectivity also seemed to be independent of solvent density. The authors conclude that the solvent cage strength in $scCO_2$ under these conditions is comparable to that of liquid alkanes, and thus, cage effects should not be an important factor in catalytic hydrogenations or hydroformylations by carbonyl catalysts in $scCO_2$. They also suggest that the aldehyde products of the hydroformylation reaction are primarily formed by nonradical pathways that are independent of solvent viscosity.

Guo and Akgerman (223,224) have also studied the hydroformylation of propylene in $scCO_2$ using the dicobalt octacarbonyl homogeneous catalyst. The focus of their study was to extend the work of Rathke and coworkers by using a well-mixed batch reactor operated over a range of conditions to evaluate the control of the reaction rate and product selectivity by tuning of the solvent density. They conducted their experiments at temperatures of 66–108°C and pressures ranging from 90.6 to 94.1 bar. They were able to provide accurate reaction times in these batch reactor experiments by isolating the catalyst in glass ampules, which were subsequently broken by switching on the reactor agitator once the desired reaction conditions were attained.

$$Co_2(CO)_8 + H_2 \rightleftharpoons 2\ HCo(CO)_4$$

Scheme 14

Table 6 Selectivity in the Stoichiometric Reaction of $MnH(CO)_5$ with 3,3-Dimethyl-1,2-diphenylcyclopropene (219)

Solvent	Gas/SCF (bar)	[Mn]/[olefin] (mM)	*P* (bar)	*T* (°C)	Time (h)	Hydrogenation (alkane) selectivity (%)	Hydroformylation (aldehyde) selectivity (%)
Micelle[a]	CO	8/2	—	50	15	8	92
Pentane[b]	Ar or CO	89/87	—	60	2–4	63	37
Hexane[c]	CO	3200/1100	—	55	5	66	34
None[d]	CO_2	20/6	5	60	4	66	34
$scCO_2$[d]	CO_2	20/6	203	60	3.5	66	34
$scCO_2$[d]	CO_2	18/6	239	35	16	61	39

[a]Matsui and Orchin (221).
[b]Nalesnik, Freudenberger, and Orchin (220).
[c]Nalesnik and Orchin (222).
[d]Jessop, Ikariya, and Noyori (219).

In the initial study (223), Guo and Akgerman assumed a pseudo-first-order rate expression for the hydroformylation reaction and fit their results to an empirical rate equation proposed by Cornils (217):

$$r = kC_{\mathrm{p}}W_{\mathrm{cat}}\frac{P_{\mathrm{H_2}}}{P_{\mathrm{CO}}}$$

where C_{p} is the propylene concentration, W_{cat} is the catalyst loading, $P_{\mathrm{H_2}}$ and P_{CO} are the partial pressures of hydrogen and carbon monoxide, respectively, and k is the reaction rate constant. They show this first-order rate assumption to be reasonably fit by the experimental data and report that the observed rate constant increases with increasing system pressure, more than doubling over the pressure range from 94 to 187 bar at 88°C . Based on Arrhenius fitting of these kinetic results at 111 and 166 bar, they also report a measured activation energy of 23 kcal/mol, which is comparable to values of 27–35 kcal/mol reported for this reaction in conventional organic solvents (225).

In the later study (224), Guo and Akgerman confirmed that the reaction mixture was in the single-phase region at the conditions studied by visual observation with a high-pressure view cell. They also confirmed a trend reported in the initial study that the density of the reaction mixture affects the product selectivity ratio of linear to branched butyraldehyde. The selectivity increased with pressure at constant temperature and decreased with temperature at constant pressure. For the former case, Figure 7 shows the selectivity ratio increasing from 1.6 (62%) to 2.6 (72%) at 88°C over the pressure range of 90.6–194.1 bar. Guo and Akgerman showed that these selectivity changes were not chemical equilibrium controlled; rather, the isothermal selectivity increase with pressure was attributed to differences of the partial molar volumes of the two isomers, and the isobaric selectivity decrease with temperature was explained in terms of differences in the isomeric partial molar enthalpies. This significant beneficial pressure effect on the product selectivity in an SCF medium is not realized in liquid solvents where the selectivity is only marginally affected by pressure (217,224).

Leitner and coworkers (226) demonstrated the use of a soluble rhodium complex catalyst modified with a highly fluorinated alkyl-substituted arylphosphane ligand for the single-phase hydroformylation of 1-octene to the corresponding isomeric aldehydes in scCO_2. As illustrated in Scheme 15 [adapted from Jessop et al. (35)], this gave a relatively high selectivity of 82% to the linear aldehyde isomer with a reported conversion of 92%. Running this reaction under identical conditions except substituting the analogous triphenylphosphine (TPP) ligand gave a significantly lower conversion of 26% with 78% selectivity to the linear aldehyde product. Leitner and coworkers attributed this dramatic difference to insufficient solubility of the active species in the case of the TPP ligand system, which lacked the fluorinated alkyl "tails" to promote CO_2 solu-

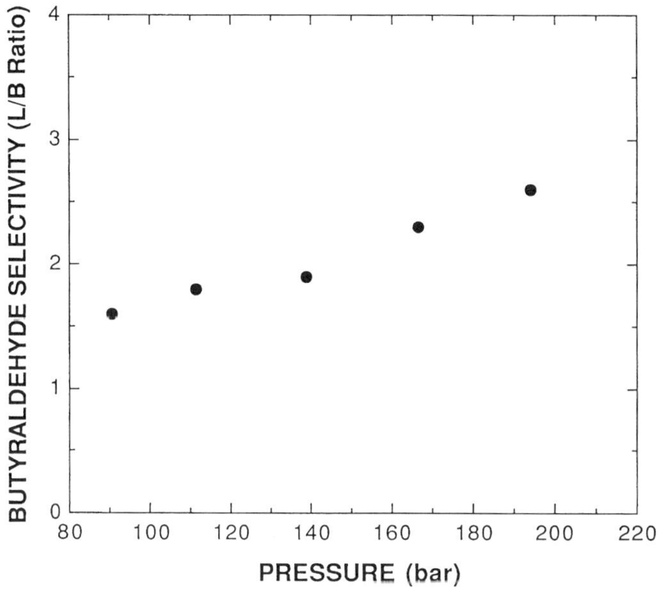

Figure 7 Butyraldehyde selectivity (*n*/iso ratio) in the hydroformylation of propylene in $scCO_2$ at 88°C. (Data from Ref. 224.)

bility. For both cases, they report that side reactions, such as hydrogenation or formation of further isomeric aldehydes, were not observed.

Koch and Leitner extended this work in a comprehensive study (227,228) comparing the hydroformylation of 1-octene using a rhodium complex catalyst without modifiers and with both phosphine and phosphite ligands. They report that $scCO_2$ is a generally applicable reaction medium for highly efficient

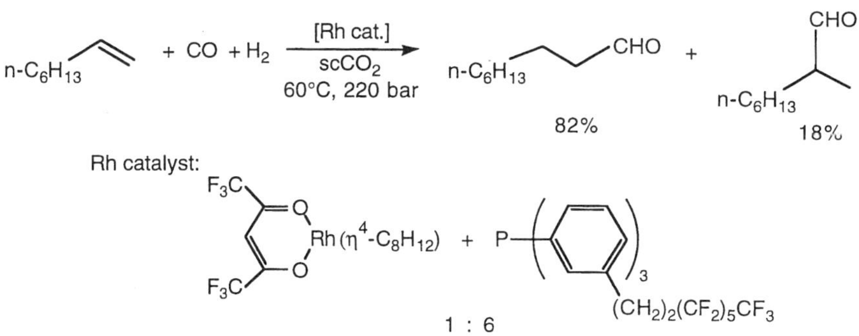

Scheme 15

rhodium-catalyzed hydroformylation reactions and that these rhodium-based catalysts offer particular advantages over the cobalt carbonyl catalysts, which typically require relatively high catalyst loadings and reaction temperatures (≥75°C). They demonstrate that a variety of olefinic substrates, including those with both terminal and internal double bonds, can be hydroformylated in $scCO_2$ at 40–65°C to the corresponding aldehydes in high yields. In one particular example, they demonstrate a fivefold increase in the rate of hydroformylation of *trans*-3-hexene in $scCO_2$ at 40°C relative to a control experiment conducted in liquid toluene at similar conditions.

The general 1-octene hydroformylation reaction and catalysts used in this work are shown in Scheme 16. Koch and Leitner report the key to the use of such triarylphosphorus ligands, particularly triarylphosphines, in $scCO_2$ media (where they are generally insoluble and hence inactive) is the attachment of perfluoroalkyl substituent groups as solubilizers (229). They report up to 99% conversion of the olefin to the indicated nonanal aldehydes at reaction condi-

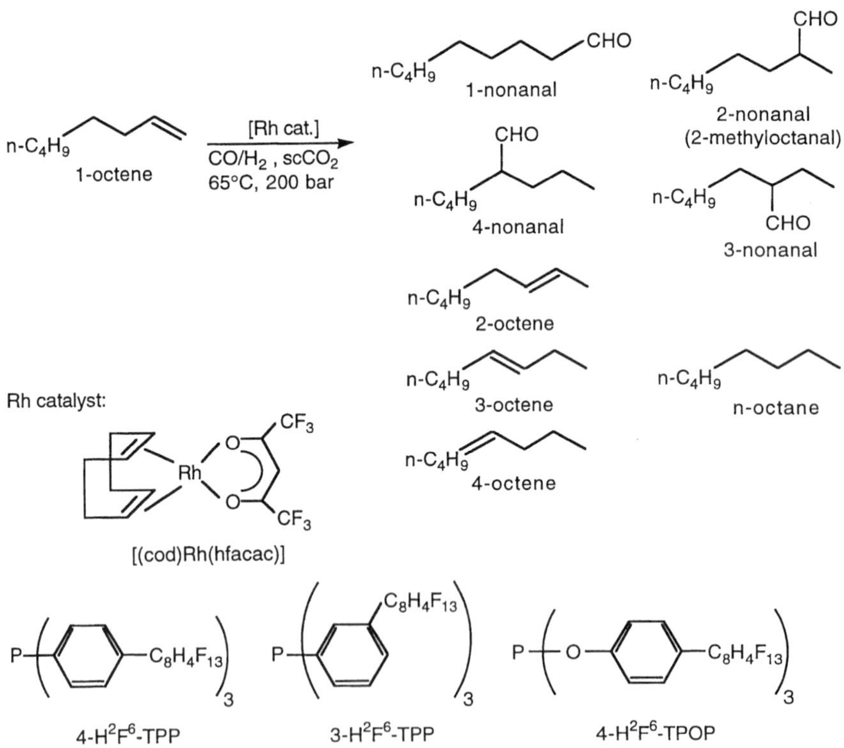

Scheme 16

tions of 65°C, 210 bar total pressure, reaction times on the order of 20 hours, and substrate/Rh ratios of up to 2650:1. Figure 8 illustrates the course of the reaction with the unmodified [(cod)Rh(hfacac)] catalyst in $scCO_2$ at 65°C and 20 bar initial synthesis gas pressure. Conversion of the olefin to aldehydes was over 75% after 3 h and almost reached completion within less than 20 h, giving a total catalytic turnover number (TON) for aldehyde formation of approximately 2400. The indicated catalyst TOF depends strongly on the course of the reaction and shows a distinct induction period during the early stages of reaction with a maximal value of 1375 h^{-1} after 1.1 h and olefin conversion of about 32%. Table 7 summarizes these results along with similar results for experiments with modified rhodium catalysts and with a control experiment in liquid toluene. These results, along with other data presented in this chapter, demonstrate rate increases of four- to fivefold for such olefin hydroformylations in an $scCO_2$ reaction medium relative to that in liquid toluene. These results also show the highest activity for the unmodified catalyst, with the TOF_{max} an order of magnitude larger than that for the phosphite-modified (TPOP) system, which was the least active of the catalysts investigated. However, Koch and Leitner

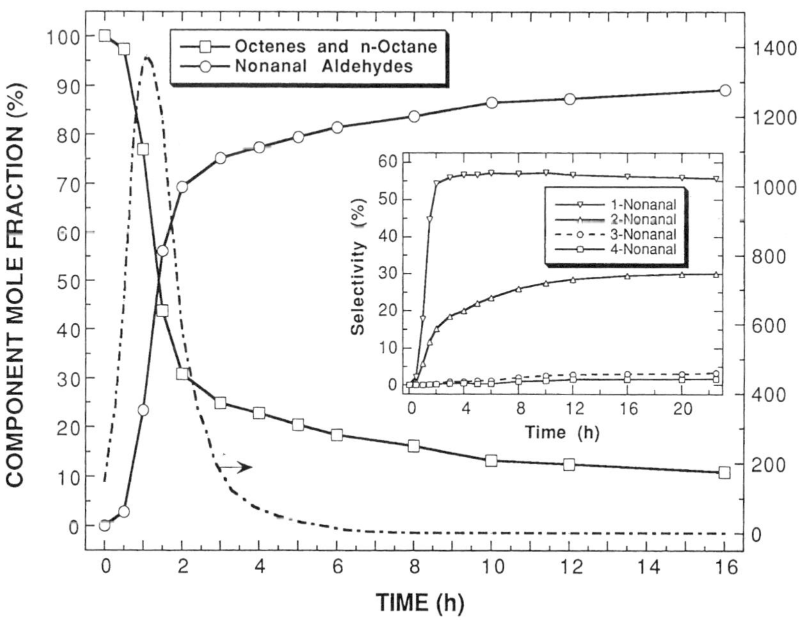

Figure 8 Product distribution and turnover frequency for 1-octene hydroformylation with [(cod)Rh(hfacac)]. (Reprinted with permission from Ref. 227. Copyright 1998 American Chemical Society.)

Table 7 Characteristic Activity Data for Various Rhodium Catalysts in the Hydroformylation of 1-Octene in $scCO_2$ and in Toluene (227)

Catalyst	1-Octene/Rh	[P]/Rh	TOF_{max} (h^{-1})	t_{max} (h)	$Conv._{max}$ (%)
[(cod)Rh(hfacac)]	2650	—	1375	1.1	32
[(cod)Rh(hfacac)]/TPP (in toluene)	2010	10:1	320	2.2	30
[(cod)Rh(hfacac)]/4-H^2F^6-TPP	2100	10:1	430	2.0	25
[(cod)Rh(hfacac)]/3-H^2F^6-TPP	2100	10:1	500	1.5	25
[(cod)Rh(hfacac)]/4- H^2F^6-TPOP	2175	10:1	115	7.0	26

report the intriguing result that the regioselectivity to the linear aldehyde isomer increases with increasing conversion to over 90% at 80% conversion, whereas the unmodified catalyst gave a maximum of 79% regioselectivity at about 55% conversion, and thereafter decreased with increasing conversion.

Palo and Erkey (230) also describe the use of a homogeneous rhodium catalyst with fluorinated arylphosphine ligands in the $scCO_2$ hydroformylation of 1-octene. They describe the primary limitation of conventional liquid process catalysts as being their limited solubility in an $scCO_2$ solution. For example, they report the maximum solubility of Wilkinson's catalyst [$RhCl(PPh_3)_3$] in $scCO_2$ at 45°C, 277 bar, and 0.88 g/ml to be no more than 0.02 mM as compared with typical catalyst concentrations utilized in conventional liquid homogeneous catalysis on the order of 1.0 mM. To increase the solubility of the active catalyst species in $scCO_2$, they synthesized a fluorinated analogue of the well-known catalyst *trans*-$RhCl(CO)(PPh_3)_2$. The novel complex, *trans*-$RhCl(CO)(P(p\text{-}CF_3C_6H_4)_3)_2$, incorporates *p*-trifluoromethyl groups into the phenyl rings of the phosphine ligands (Scheme 17). Palo and Erkey (230) report the solubility of this fluoromethyl-substituted catalyst in $scCO_2$ (70°C,

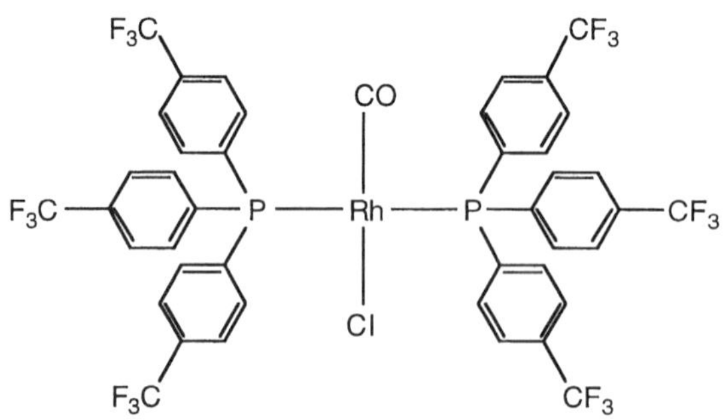

Scheme 17

277 bar, 0.77 g/ml) to be at least 5.5 mM, confirming the dramatic solubility enhancement afforded by the addition of small fluoroalkyl groups to the aryl rings of the phosphine. Palo and Erkey subsequently utilized this catalyst at a concentration of 2.0 mM under these conditions in the hydroformylation of 1-octene according to the general reaction of Scheme 15, producing the expected C_9 aldehydes with a selectivity of 71% to the linear 1-nonanal isomer relative to the branched isomer 2-nonanal. These results compared favorably with published values for 1-pentene hydroformylation in liquid benzene at similar conditions, so the researchers concluded that the fluoromethyl substituents impart substantial $scCO_2$ solubility to the rhodium complex without significantly affecting its catalytic activity.

In a subsequent study, Palo and Erkey (231) report an even more effective catalyst $[RhH(CO)(P(p\text{-}CF_3Ph)_3)_3]$ which exhibits high activity for conversion of 1-octene to C_9 aldehydes in $scCO_2$. The nonfluorinated analogue of this catalyst is used on a commercial scale for olefin hydroformylation (232). Palo and Erkey report the solubility of this catalyst in $scCO_2$ to be at least 7.61 mM at 277 bar and 50°C . For the selected reaction conditions (277 bar, 50°C, $[CO]_0 = [H_2]_0 = 1.05$ M, $[1\text{-octene}]_0 = 0.95$ M), they found the reaction to proceed cleanly with no observable isomerization or hydrogenation products. The n to iso selectivity was independent of the total pressure but increased monotonically from 75% to 79% as the catalyst concentration was increased from 0.63 to 7.61 mM. This behavior was reported to be similar to that of the nonfluorinated analogue of this catalyst in organic solvents (233). A kinetic study was conducted using this catalyst (234), and a kinetic rate expression was developed to fit the experimental data having the form:

$$r_{1-\text{octene}} = \frac{6.2[H_2]^{0.48}[\text{cat.}]^{0.84}[1-\text{octene}]^{0.50}}{1 + 0.69[CO]^{2.2}}$$

where $r_{1-\text{octene}}$ is the rate of 1-octene reaction in (mol/dm^3-min). This rate expression differed significantly from those obtained previously for the analogous nonfluorinated catalyst in organic media, specifically for the approximately one-half order rate dependence on $[H_2]$ (which was typically first order for the analogous catalyst), the lack of substrate inhibition, and the absence of a critical catalyst concentration. The authors attributed this altered kinetic behavior to several factors, including $scCO_2$ solvent effects, the modified phosphine ligands, and the increased H_2 and CO concentrations relative to the conventional systems.

In a recent study, Kainz and Leitner (187) demonstrated the use of the chiral phosphine/phosphite ligand (*R*,*S*)-BINAPHOS for the rhodium-catalyzed asymmetrical hydroformylation of styrene according to Scheme 18. They report that the reaction proceeds cleanly and almost quantitatively in $scCO_2$ at 60°C and gives appreciable enantiomeric excess [ee = 66% (*R*)] at densities close

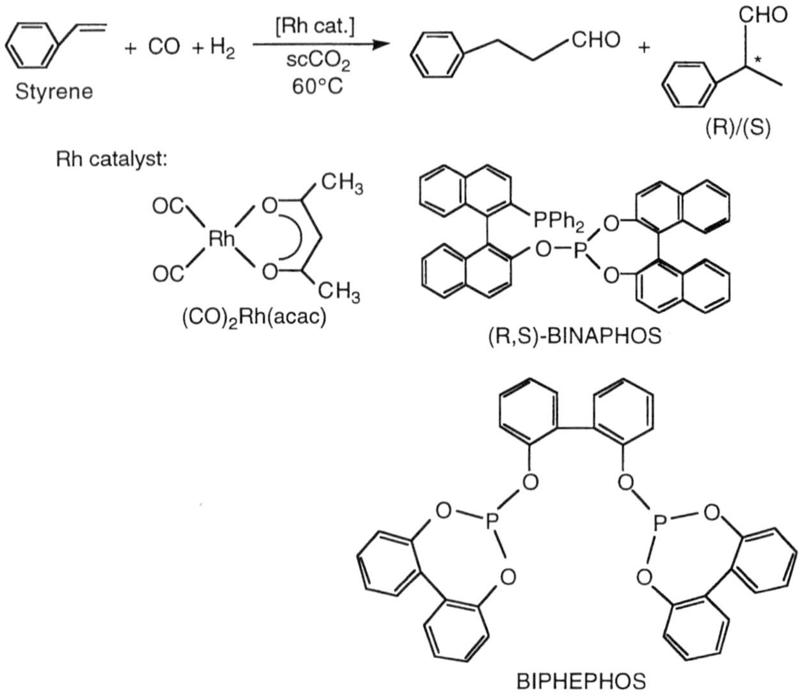

Scheme 18

to the critical density of CO_2 (0.47 g/ml). However, the enantiomeric excess decreases significantly with increasing solvent density as illustrated in Figure 9. At solvent densities between 0.70 and 0.82 g/ml, the enantiomeric excess is only 5–35% (R). However, they observe good selectivities of 86–89% to the chiral aldehyde product 3-phenylpropanal relative to the linear analogue. Furthermore, they observe that in the absence of CO_2, the hydroformylation of neat styrene proceeds quantitatively with similar regioselectivity to the chiral product and a high enantiomeric excess of 84% (R). Kainz and Leitner explain these intriguing results in this preliminary study as a complex interplay of density, phase behavior, and solubilities in such multicomponent reaction systems involving transition metal catalysts and $scCO_2$. They correctly note that such complicating factors can present significant obstacles to further development of such processes.

Bach and Cole-Hamilton (235) report that 1-hexene can be successfully hydroformylated in $scCO_2$ using triethylphosphine complexes of rhodium, and they point out that these catalysts would be significantly less expensive than the fluorinated analogues, such as those cited previously. For reaction conditions

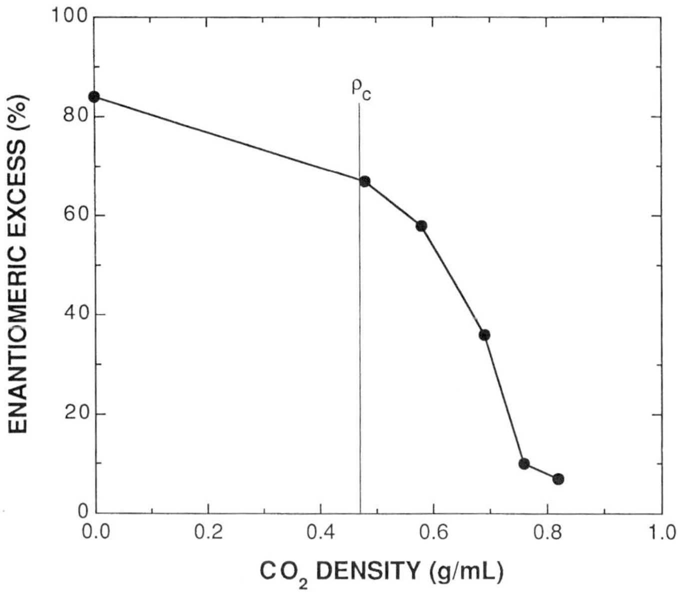

Figure 9 Influence of CO_2 density on the enantiomeric excess of 3-phenylpropanal obtained from styrene hydroformylation in $scCO_2$. (Reprinted with permission from Ref. 187. Copyright 1998 Baltzer Science Publishers.)

of 100°C and 105 bar, they report high conversion to C_7 aldehydes with about 71% selectivity to the linear isomer. However, they report measurable formation of C_7 alcohols. This selectivity was slightly improved over the 68% selectivity obtained in a control experiment in liquid toluene. In addition, a toluene control experiment resulted in significantly more C_7 alcohol formation.

Ojima et al. (236) report the hydroformylation of 1-octene, styrene, and vinyl acetate in $scCO_2$ using either BIPHEPHOS or (*R*,*S*)-BINAPHOS as the ligand with $[(CO)_2Rh(acac)]$ catalyst (see Scheme 16). Their results are summarized in Table 8 for run conditions of 65°C and a catalyst concentration of 0.067 mM $[(CO)_2Rh(acac)]$. They report a remarkable 99% selectivity to the linear aldehyde product for the hydroformylation of 1-octene in the presence of the Rh-BIPHEPHOS catalyst with 72 bar of $scCO_2$ loading. Increasing the CO_2 pressure to 86 bar resulted in a selectivity decrease to 86%, but increasing the ligand concentration by 50% resulted in a comparable selectivity of 99%. Branched aldehyde products were predominant for the indicated reactions of both styrene and vinyl acetate. However, the reported enantioselectivity to the (*R*)-2-phenylpropanal product resulting from styrene hydroformylation was found to be about 92% ee.

Table 8 Hydroformylation of Various Alkenes in $scCO_2$ at 65°C using $[(CO)_2Rh(acac)]$ Catalyst (236)

Substrate	Ligand conc. (mM)	Solvent	CO/H_2 loading/ratio (bar)	CO_2 loading (bar)	Time (h)	Conversion (%)	Linear aldehyde selectivity (%)
1-Octene	None	$scCO_2$	14 (1:1)	86	40	87	67
1-Octene	BIPHEPHOS (0.133 mM)	$scCO_2$	14 (1:1)	86	24	100	86
1-Octene	BIPHEPHOS (0.133 mM)	Hexane	14 (1:1)	86	12	100	94
1-Octene	BIPHEPHOS (0.133 mM)	$scCO_2$	14 (1:1)	72	12	88	99
1-Octene	BIPHEPHOS (0.200 mM)	$scCO_2$	14 (1:1)	86	12	69	99
Styrene	(R,S)-BINAPHOS (0.133 mM)	$scCO_2$	14 (1:1)	86	12	100	11[a]
Vinyl acetate	BIPHEPHOS (0.133 mM)	$scCO_2$	21 (1:2)	79	40	85	9
Vinyl acetate	(R,S)-BINAPHOS (0.133 mM)	$scCO_2$	21 (1:2)	79	43	74	11
Vinyl acetate	BIPHEPHOS (0.133 mM)	Et_2O	14 (1:1)	86	44	>95	10

[a] (*R*)-2-Phenylpropanal product obtained with 92% ee.

F. Selective Oxidations

An emerging technology in the environmental field that represents a more established application of SCFs is that of the total oxidative destruction of organic wastes in supercritical water, a process know as supercritical water oxidation (SCWO). Tester et al. (5) published a comprehensive review of ongoing fundamental research in this area as well as the technological issues facing the commercial-scale development of this process. Savage et al. (31) provided a more recent survey of the SCWO literature, and Ding et al. (237) present a more specialized review of catalytic oxidations in supercritical water. Schmieder et al. (6) recently summarized the status of current commercial developments. As mentioned previously, this important area is outside the scope of this chapter.

Another important class of industrially relevant reactions that has seen limited research in SCF media (7,31) is that of selective or partial oxidations for synthesis applications. These reactions are often heterogeneously catalyzed in gas- and liquid-phase reactors and as such are typically characterized by mass transfer limitations, low conversions, and poor selectivities. SCFs are attractive media for this class of reactions due to the potential advantages mentioned previously which address these limitations. For example, the rates of oxygen activation by homogeneous catalytic systems in organic solvents are often limited by low oxygen solubility in organic solvents and oxygen diffusion into the solvent (238). The use of an SCF solvent could potentially eliminate these mass transfer problems because gases are highly miscible in this medium.

Dooley and Knopf (239) present one of the earliest studies of selective oxidation reactions in an SCF media—the partial oxidation of toluene with air to benzaldehyde in $scCO_2$ in the presence of a supported cobalt oxide catalyst. This is a comprehensive study that includes phase equilibrium studies for the primary reaction components, a catalyst screening evaluation resulting in the selection of 5% CoO/Al_2O_3 as both active and selective for the partial oxidation of toluene, and initial kinetic measurements for the toluene oxidation reaction. The reaction experiments were conducted in a continuous fixed-bed reactor at 20–220°C and 81 bar using a feed composition of 1.5 wt % toluene and 6.5 wt % air in $scCO_2$. The toluene was oxidized with good selectivity to the partial oxidation products benzaldehyde, benzyl alcohol, and the cresol isomers, but with relatively low rates and conversions. The authors conclude that partial oxidation of alkyl aromatics with conventional supported metal oxide catalysts may be feasible at high pressures in SCF solvents due to these high selectivities relative to those characteristic of conventional processes. Subsequent work (240) included a study of the effect of total pressure on catalyst activity. The authors observed an approximately twofold increase in the partial oxidation rate of toluene with the supported CoO catalyst over a pressure range of about 81–140 bar at temperatures of 180–220°C. Only about 30% of this rate increase could be explained by detailed equation-of-state calculations of the effect of pressure on the concentration-based rate constant as provided by transition state theory. The authors suggest that the balance of the observed pressure effect could result from a pressure-dependent variation of oxygen concentration in the condensed phase in the catalyst pores. They indicate that the adsorbed phase in microporous materials at these conditions is essentially liquid-like in density. Hence, the pores become enriched in oxygen with increasing pressures, resulting in an increase in the apparent reaction rate.

Occhiogrosso and McHugh (241) and Suppes et al. (242) studied the partial oxidation of isopropyl benzene (cumene) in supercritical solvents including CO_2, Xe, and Kr at 110°C and 200–414 bar. Their results showed minimal effects of pressure, viscosity, and proximity to the mixture critical point on this reaction. However, they report a dramatic increase in the termination rate constant in the SCF regime, which they attribute to the presence of the metal reactor walls (316SS and gold-plated 316SS) functioning as catalyst sites. Their observed low selectivity to cumene hydroperoxide in an SCF solvent relative to that for the conventional process of autoxidation of neat liquid-phase isopropyl benzene is attributed to this metal catalysis in the large surface area-to-volume experimental reactor (a windowed, variable-volume view cell). The reported negligible pressure effect is notable in light of the relatively extreme pressure sensitivity on reaction rates observed by other researchers (78,97,98).

Gaffney and Sofranko (243–245) report the selective oxidation of olefins to the corresponding glycols by reaction of the olefin, oxygen, carbon dioxide, and water in an SCF reaction mixture. Heterogeneous catalysts included supported CaI_2, CuI, Cu_2O, and MnO_2. Examples are reported for propylene oxidation to propylene glycol in both batch and continuous fixed-bed reactors operating at 140–146°C and 139–154 bar. Optimal results were realized with a mixed $CuI/Cu_2O/MnO_2$ on alumina catalyst in the continuous-flow reactor. A productivity of 15 mmol propylene glycol/g cat.-h was achieved with 95% selectivity to propylene glycol at 140°C and 139 bar. However, an apparently unrealized increase in productivity of an order of magnitude was needed to warrant further development.

Srinivas and Mukhopadhyay (246) have investigated the selective thermal oxidation of cyclohexane in $scCO_2$ to produce cyclohexanone and cyclohexanol as the primary reaction products. Kinetic experiments were conducted at three temperatures (137°C, 150°C, and 160°C) and two pressures (170 and 205 bar), and the presence of a homogeneous SCF was verified experimentally for the initial reactor composition under these conditions (10 mol % cyclohexane, 10% O_2, and 80% CO_2). Kinetic results were interpreted assuming a free-radical reaction mechanism comparable to that observed in the conventional liquid-phase process, and the reaction was observed to be autocatalytic. Reported conversions are low relative to those in the liquid-phase process, which the authors attribute to the dilute reactant concentrations of the experiments. Cyclohexanone formation was favored over that of cyclohexanol, and the selectivity was found to increase with both pressure and temperature. The pressure increase from 170 to 205 bar resulted in a 50% reduction of the induction period, a significant increase in the activation energy from 13.0 to 22.6 kcal/mol, and an increase in the 160°C first-order rate constant by about 70%. The authors attribute the activation energy increase to cage effects where solvent clustering is more pronounced at lower pressures in the vicinity of the mixture critical point. The authors also report a variation in the estimated activation volume from 36 cm^3/mol at 137°C and 170 bar to -775 cm^3/mol at 160°C and 205 bar, suggesting that the reaction rate and conversion can be effectively manipulated by varying the reaction conditions in the $scCO_2$ medium.

Subsequent study of this reaction system by the authors included experimental measurement and modeling of the pertinent phase behavior of the system (247) and an investigation of additional effects on the reaction pathways, conversion, and rates (248). These effects included proximity to the mixture critical point, type of phase, concentration, temperature, and density of the initial reaction mixture. The authors observed the fastest reaction rates in a CO_2-expanded liquid cyclohexane phase but demonstrated that the reaction rate and conversion can be manipulated by varying the reaction temperature, pressure, and feed composition in the SCF phase.

Koda and coworkers (249) have reported the air oxidation of cyclohexane in $scCO_2$ to yield cyclohexanol and cyclohexanone in the presence of an iron porphyrin catalyst containing the *meso*-pentafluorophenyl group [FeCl(TPFPP), where TPFPP = 5,10,15,20-tetrakis(pentafluorophenyl)porphyrin]. The pentafluorophenyl groups were believed to enhance the catalyst solubility in the $scCO_2$ phase, although the solubility was not measured. The reaction required essentially stoichiometric quantities of acetaldehyde (250). The total yield of cyclohexanol and cyclohexanone was reported as 5% for 1 h reaction time at 70°C and 90 bar, which corresponded to a turnover number of 100. Selectivity to cyclohexanol and cyclohexanone was approximately equivalent at all conditions studied.

Tumas and coworkers (251) report the homogeneously catalyzed oxidation of cyclohexene in $scCO_2$ with aqueous $(CH_3)_3COOH$ oxidant and $Mo(CO)_6$ catalyst. They found that such unactivated alkenes can be effectively oxidized to their corresponding diols at temperatures of 95°C, but high Mo catalyst concentrations (12.5 mol %) are required for the reaction. Selectivity of 73% to 1,2-cyclohexanediol at 74% conversion was observed during a 12-h reaction with the other major products being 2-cyclohexen-1-one (10%) and 2-cyclohexen-1-ol (10%). They report that oxidations utilizing the aqueous hydroperoxide [90% $(CH_3)_3COOH/H_2O$] gave significantly higher conversions than those run with the anhydrous analogue (~15% conversion) for the $Mo(CO)_6$ system, suggesting a strong influence of the water. Presumably, the intermediate product was cyclohexene oxide, which then underwent hydrolysis to the diol product.

Fan et al. (252–254) reported the air oxidation of isobutane to *tert*-butyl alcohol (TBA), an important intermediate for the production of methyl *tert*-butyl ether (an important oxygenated fuel additive at that time) and isobutene. They report that TBA can be synthesized efficiently on selected heterogeneous catalysts if air is directly introduced to isobutane in the SCF phase, although the reported yields are relatively low. The study included evaluation of five types of catalysts, of which amorphous SiO_2-TiO_2 and Pd/C were found to be the most efficient. The authors have shown for both the catalyzed and noncatalytic reaction in a continuous fixed-bed reactor at 153°C that changing the isobutane state from the gas phase (44 bar) to the SCF phase (54 bar) gave remarkably enhanced conversion of isobutane and oxygen. For example, on 2.5% Pd/C catalyst, the isobutane conversion increased fivefold from 0.5% at 44 bar to 3.1% at 54 bar. Selectivity to the desired products (TBA and isobutene) was generally found to increase slightly on this state change. Increasing the air partial pressure under SCF reaction conditions further increased the TBA and isobutene selectivity. Catalyst deactivation was not observed after 30 h of continuous operation at the SCF conditions.

Suresh (255) has compiled and critically analyzed published data on the oxidation of isobutane to TBA, with particular emphasis on the data for oxidation in the SCF phase. The data for this latter case are taken from three patents assigned to Shell Oil (256–258) and do not include the data of Fan et al. (252–254) noted above. Suresh notes that the literature reports higher rates and yields for SCF-phase oxidation relative to liquid-phase oxidation, but it is not clear whether the reported effects result from the inherent properties of the SCF reaction mixture or merely the different conditions employed to reach the SCF state. Suresh's analysis was intended to clarify this point and suggested that the features of SCF oxidation extrapolate directly from those of liquid-phase oxidation. Calculations showed that the oxidation is first order in both converted and unconverted isobutane and zeroth or higher order in oxygen, depending on the oxygen concentration. These features are similar to those seen in the liquid-phase oxidation of cyclohexane (259), and a kinetic expression proposed for this latter reaction also performed well with both liquid- and SCF-phase isobutane oxidation. Suresh concluded that oxidation under SCF conditions offers higher rates and productivities than conventional liquid-phase oxidation because the liquid-like reaction behavior at higher temperatures is not accessible to liquid-phase oxidation.

Tumas and coworkers (238) have investigated the oxidation of cyclohexene using molecular oxygen catalyzed by halogenated iron porphyrins in $scCO_2$. The authors note that homogeneous oxidation catalysis in $scCO_2$ remains relatively unexplored, despite the attractive features afforded by this solvent, such as resistance to oxidation (CO_2 is already fully oxidized), nonflammability, and low critical temperature, which can lead to enhanced selectivities due to lower operating temperatures. Two fluorinated porphyrins were used in this study: tetrakis(pentafluorophenyl)porphyrinato iron(III) chloride [Fe(TFPP)Cl] and β-octabromo-tetrakis(pentafluorophenyl)porphyrinato iron(III) chloride [Fe(TFPPBr$_8$)Cl]. These catalysts are known to be active for alkene and alkane oxidation with molecular oxygen in organic solvents, and the fluorinated phenyl rings were expected to enhance the catalyst solubility in scCO_2. Solubility was confirmed using UV-vis spectroscopy for SCF conditions of 40°C and 345 bar, the mildest experimental oxidation conditions used in the study. Thc concentrations of Fe(TFPP)Cl and Fe(TFPPBr$_8$)Cl in scCO_2 were at least 18 and 10 μM, respectively, which is comparable to concentrations used in catalytic reactions in conventional organic solvents. Both halogenated metalloporphyrins were active catalysts for oxidation of cyclohexene to epoxide and radical-based allylic oxidation products in scCO_2 (Scheme 19). In 12 h at 80°C, up to 350 and 580 turnovers were observed for Fe(TFPP)Cl and FE(TFPPBr$_8$)Cl, respectively. Several organic solvent reactions at high temperature and pressure were also explored to benchmark relative activity and selectivity. Activity was higher in organic solvents but accompanied by substantial reaction with the solvent. Selectivity for epoxidation with Fe(TFPPBr$_8$)Cl was higher in scCO_2 than in organic solvents, with up to 34% cyclohexene oxide produced.

Wang and Willey (260) report the oxidation of methanol in scCO_2 over pure and mixed oxide (Fe, Si, and Mo) aerogel catalysts. Selective oxidation products of dimethyl ether, methyl formatc, and formaldehyde were found from 200°C to 300°C, and the selectivity depended on the catalyst used. Pure iron oxide favored dimethyl ether (80% yield), low levels of iron oxide on silica favored

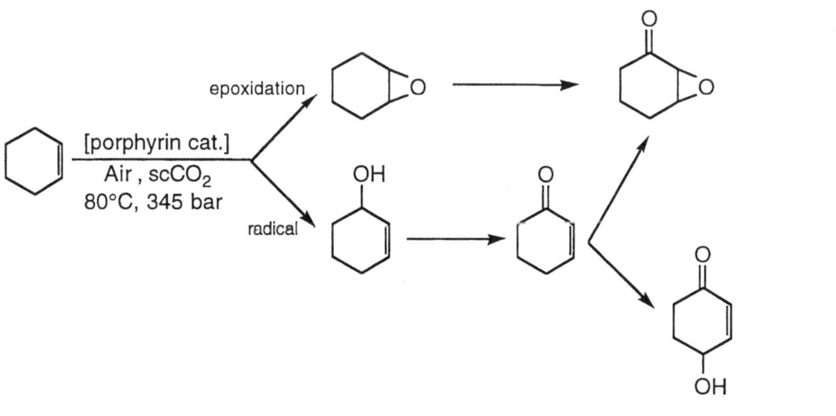

Scheme 19

methyl formate (60% yield), and iron oxide on molybdenum favored formaldehyde (90% yield). These high yields demonstrated the technical feasibility of aerogel catalysts for reactions in SCFs.

Haas and Kolis (261) report the diastereoselective epoxidation of a variety of allylic alcohols with *tert*-butyl hydroperoxide in the presence of a vanadyl(salen) epoxidation catalyst in $scCO_2$. In a series of 24- to 72-h batch scouting experiments at 40–50°C, the reaction of 11 olefinic alcohols was shown to proceed relatively cleanly with high yields (33–99%) and reasonable diastereoselectivity (60–99%) to the corresponding epoxides, which are comparable with those obtained in conventional organic solvents. Pressures were in excess of 221 bar (the room temperature CO_2 loading pressure), but were not reported at reaction temperature. Further oxidation of the allylic alcohol functionality to the allylic ketone was observed in most of the reactions but with less than 10% yield. Subsequent work on enantioselective epoxidations has also been presented (262).

Haas and Kolis (263) have also reported the oxidation of a variety of olefins to epoxides or diols using $Mo(CO)_6$ as the catalyst precursor and an organic peroxide [$(CH_3)_3COOH$] as the oxidant. Table 9 summarizes the results and reports the olefinic substrates and primary products. These runs were likewise initially charged with 221 bar CO_2 pressure at room temperature, but the pressure was monitored during the reaction and was typically about 517 bar at reaction temperature. Haas and Kolis report that these diol yields are generally better in $scCO_2$ than in conventional solvents and note that the epoxide products can be obtained under both wet and anhydrous conditions. These results demonstrate molybdenum as an effective oxygen transfer catalyst precursor under these conditions.

Oakes et al. (264) have reported the diastereoselective sulfoxidation of chiral sulfides derived from methionine and cysteine in $scCO_2$ as well as conventional organic liquid solvents. Scheme 20 shows the oxidation of Cbz methyl cysteine methyl ester and corresponding products. Oakes et al. report that use of *tert*-butyl hydroperoxide and Amberlyst-15 ion exchange resin is particularly effective for sulfoxide formation and shows a dramatic pressure-dependent increase in diastereoselectivity (up to >95% de) for such cysteine derivatives in $scCO_2$ compared with conventional solvents where essentially no diastereoselectivity is observed. The pressure dependence showed this optimum 95declining selectivity at higher and lower pressures. The authors report that they are continuing to investigate both the causes and generality of this dramatic effect, but these results clearly have tremendous potential impact in the pharmaceutical and fine-chemicals industry, where such stereoselective synthesis is fundamental.

Jia et al. (265) report the palladium(II)-catalyzed selective oxidation of methyl acrylate in $scCO_2$ to dimethyl acetal as a major product with excellent conversion and selectivity. Best results were obtained using $PdCl_2$ with $CuCl_2$ as

Table 9 Olefin Oxidation Products Using $Mo(CO)_6$ Catalyst in $scCO_2$ (263)

Olefin	Oxidant	Olefin/oxidant ratio	*T* (°C)	Time (h)	Yield/product
Cyclohexene	70% aq t-BuOOH	1:2	95	3	100% anti-1,2-cyclohexandiol
Cyclohexene	70% aq t-BuOOH	1:1	95	3	100% anti-1,2-cyclohexandiol
Cyclohexene	70% aq t-BuOOH	1:1	80	3	47% cyclohexene oxide
Cyclooctene	70% aq t-BuOOH	1:3	86	3	100% cyclooctene oxide
Cyclooctene	6M t-BuOOH/$C_{10}H_{12}$	1:3	86	3	100% cyclooctene oxide
Norbornylene	70% aq t-BuOOH	1:3	92	3	66% norbornanediol
1-Octene	70% aq t-BuOOH	1:3	103	4	80% 1,2-octanediol
cis-2-Heptene	6M t-BuOOH/$C_{10}H_{12}$	1:3	95	3	20% 2,3-heptane oxide
Vinylcyclohexane	6M t-BuOOH/$C_{10}H_{12}$	1:3	90	3	20% cyclohexenyl-ethylene oxide
2-Vinylnaphthalene	70% aq t-BuOOH	1:3	92	3	80% 2-vinylnaphthaldehyde
1-Vinylnaphthalene	70% aq t-BuOOH	1:3	92	3	80% 1-vinylnaphthaldehyde
trans-β-Methylstyrene	70% aq t-BuOOH	1:3	90	3	30% benzaldehyde
cis-Stilbene	70% aq t-BuOOH	1:3	95	3	50% benzaldehyde
trans-Stilbene	70% aq t-BuOOH	1:3	100	6	No reaction
trans-2-Heptene	70% aq t-BuOOH	1:3	94	3	No reaction

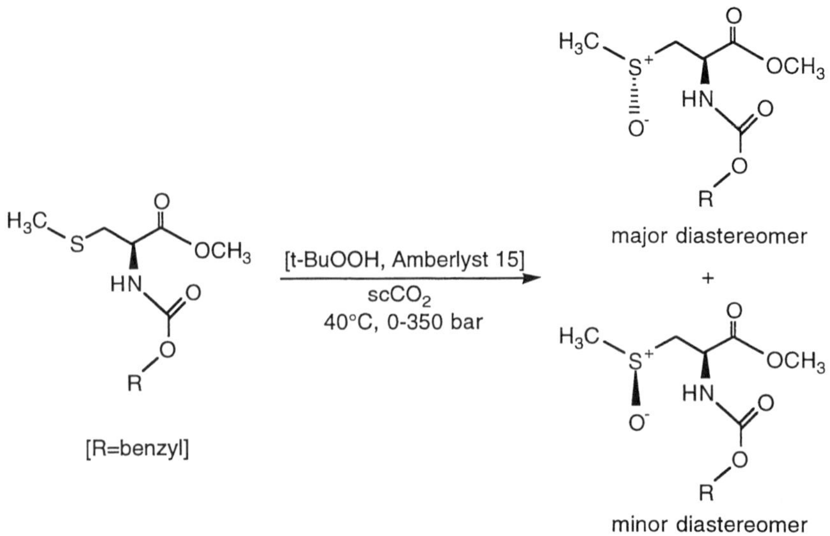

Scheme 20

cocatalyst and reaction conditions of 40°C and 120–130 bar. This produced the desired dimethyl acetal in 97% selectivity at 99% conversion. A strong pressure dependence on the reaction was also reported. Walther and coworkers (266) report the catalytic epoxidation of *cis*-cyclooctene to cyclooctene oxide with $(CH_3)_3OOH$ using $Mo(CO)_6$ catalyst. At 85 bar and the relatively low operating temperature of 45°C, they report 100% selectivity to the epoxide product for a 16-h reaction. The reaction proceeded less selectively under similar conditions using Titan(IV)-isopropylate catalyst ($Ti[(OCH(CH_3)_2]_4$) whereby the epoxide product was formed with 95% selectivity along with 4% 1,2-cyclooctanediol. Jessop (34) reports the epoxidation of 2,3-dimethyl-2-butene by cumene hydroperoxide in $scCO_2$ using 1,1,2,2-tetrachloroethane as cosolvent and $Mo(CO)_6$ as catalyst. He obtained an epoxide yield of about 17% at 85°C and 227 bar, and observed no cyclic carbonates as byproducts.

G. Free-Radical Reactions

In addition to the selective oxidation reactions above, a number of other free-radical reactions are summarized herein. Tanko and Blackert (267,268) report the free-radical side-chain bromination of toluene and ethylbenzene in $scCO_2$ using bromide radicals initiated photochemically from molecular bromine. They report the production of the corresponding benzylic bromides in high yield with selectivities essentially identical to that observed in a conventional chlorinated organic solvent, carbon tetrachloride. For example, at 40°C and 229 bar, ethylbenzene bromination yielded the brominated product 1-bromo-1-phenylethane in 95% yield. This same study also reports that $scCO_2$ is an effective alternative to carbon tetrachloride for use in the classical Ziegler bromination with *N*-bromosuccinimide.

Tanko et al. (118,136) evaluated the chlorine atom cage effect in the free-radical chlorination of alkanes in a subsequent study of the effect of viscosity and the possible role of solvent clusters on cage lifetimes and reactivity for reactions carried out in SCF solvents. These experiments included chlorination of cyclohexane, neopentane, 2,3-dimethylbutane, and propane, and they were conducted in $scCO_2$ at 40°C and various pressures with parallel experiments in conventional liquid- and gas-phase solvents. The results of these experiments provided no indication of an enhanced cage effect near the critical point in $scCO_2$ that might be attributable to solvent-solute clustering. The magnitude of the cage effect observed in $scCO_2$ at all pressures was within what was anticipated on the basis of extrapolation from conventional solvents. The authors conclude that reported enhancement of cage effects (e.g., 81,86,103,133) is thus unique to the specific reactions studied. Chlorine atom selectivities in $scCO_2$ were observed to vary slightly with pressure (viscosity) and were intermediate between gas- and liquid-phase values. The higher selectivity for monochlorinated products observed in $scCO_2$ was attributed to the lower viscosity of the SCF media relative to conventional liquid solvents.

Metzger and coworkers have shown in a series of papers (269–276) that alkanes can be added to alkenes and alkynes in thermally initiated free-radical chain reactions in the neat reactants at SCF conditions. These reactions have been demonstrated with a wide variety of substrates investigating various effects, including the influence of steric and polar substituents as well as product regioselectivity. The radical chain is initiated by a bimolecular reaction of the alkane with the alkene or alkyne to give two radicals. Addition, rearrangement, and elimination reactions have also been observed. No effect on the reaction rate constant near the critical point was observed on varying the physical state of the reaction mixture from liquid to supercritical to gas-phase conditions (276).

H. Cyclization and Other Coupling Reactions

The Diels–Alder cycloaddition is an important reaction class involving the reaction of unsaturated carbonyl compounds with conjugated dienes. The carbon atoms at the 1- and 4-positions of the conjugated diene system become attached to the doubly bonded carbons of the unsaturated carbonyl compound to form a ring structure. Diels–Alder chemistry is used in the synthesis of a variety of complex organic compounds of commercial interest, including insecticides, fragrances, plasticizers, and dyes (92). These bimolecular reactions have been used for fundamental studies in SCF solvents because they generally proceed cleanly with no side reactions, they are believed to follow a similar mechanism over a wide range of solvent densities spanning from gas-phase to liquid-phase reactions, and they have been well characterized in liquid-phase systems including at elevated pressures (92). Several early studies on Diels–Alder reactions in SCF media (61,92,133,134,277,278) have been reviewed by Savage et al. (31). These will only be cited here for completeness.

In these early studies, three groups have evaluated the Diels–Alder cycloaddition reaction of maleic anhydride with isoprene in scCO_2 (92,133,277). Paulaitis and Alexander (92) investigated the use of pressure to control reaction kinetics in SCF mixtures. They point out that pressure can influence the reaction rate through an effect on the concentrations of reactant species as well as through the pressure dependence of the rate constant. The latter can be substantial in SCF systems in the vicinity of the mixture critical point. Kim and Johnston (277) also studied the effect of pressure on the rate constant and calculated activation volumes at various reaction conditions. Ikushima et al. (133) used the Lewis acid catalyst $AlCl_3$ for the reaction and monitored the progress with in situ FTIR spectroscopy. Kim and Johnston (61) extended their study on the thermodynamic pressure effect on rate constants using the Diels–Alder cycloaddition reaction of methylacrylate and cyclopentadiene. They demonstrated control of the selectivity between the product endo and exo adducts of this parallel reaction mechanism by adjustment of the pressure at constant temperature. Ikushima et al. (134,278) investigated pressure effects on the kinetics of the Diels–Alder addition of isoprene and methyl acrylate at 50°C and 70–200 bar. They report increasing rates with increasing pressure as well as a change in the selectivity between the two cyclohexene product isomers, methyl 4-methyl-3-cyclohexene-1-carboxylate and methyl 3-methyl-3-cyclohexene-1-carboxylate. This pressure-dependent selectivity was attributed to clustering of solvent molecules about the reacting species in the vicinity of the critical point.

In more recent work, Ikushima et al. (279) applied the solubility parameter concept to the isoprene and methyl acrylate reaction to evaluate the solvent properties of scCO_2 as well as the mutual affinity among the various chemical species present in the reaction mixture. They estimated the pressure dependence of the solubility parameter of the activated complex through transition state theory at 50°C over the pressure range of 70–200 bar to study the nature of the complex and the effect of the solvent on the reaction. They observed that the solubility parameter of the activated complex approaches that of the reactants as the pressure approaches the critical point. This suggests that the nature of the activated complex becomes more similar to that of the reactants, hence the energy needed for formation of the complex becomes smaller near the critical point. That is, the reaction rate for formation of the complex is enhanced in the vicinity of the critical point, thus driving the overall reaction to the product.

Knutson et al. (280) measured the kinetics of the Diels–Alder reaction of maleic anhydride (MA) and 2,3-dimethyl-1,3-butadiene (DMB) in SCF propane solutions at 100–140°C and 46–141 bar. Reaction to the product 4,5-dimethyl-*cis*-1,2,3,6-tetrahydrophthalic anhydride (DMTA) was evaluated with excess DMB as a reactive cosolvent and 2,2,2-trifluoroethanol (TFE) as an unreactive cosolvent (Scheme 21). Near-critical effects and cosolvent effects on reaction rates were analyzed from transition state theory. Rate constants increased with increasing pressure at 140°C, but were not significantly affected at 100°C and 120°C at near-critical densities. A similar lack of pressure dependence has been reported by Reaves and Roberts (281) for the Diels–Alder reaction of MA with isoprene in subcritical propane at 80°C. This minimal pressure effect is in contrast to those noted above for Diels–Alder reactions in $scCO_2$ where the reactants were at approximately equal and dilute concentrations. The influence of the unreactive cosolvent, TFE, on reaction rates was found to be minimal. These results suggest that the local reactant composition, as well as pressure, temperature, and cosolvent, can be used to control the reaction rate of such reactions in the near-critical region.

Tester and coworkers (282) have investigated the Diels–Alder reaction of cyclopentadiene and ethyl acrylate in $scCO_2$ from 38°C to 88°C and 80 to 210 bar. They observed an increase in reaction rate with increasing pressure at 38°C and developed a traditional Arrhenius expression to correlate kinetic data at a constant density of 0.5 g/cm^3 over the temperature range 38–88°C. The corresponding activation energy was calculated to be 40 ± 2 kJ/mol. They further demonstrated that the mole fraction–based rate constants showed an approximately linear dependence on density, which they suggest is the true independent property. Hence, the density dependence of the rate is reflected primarily in the preexponential factor of the Arrhenius relationship. Evaluation of rate constant data from previous investigations noted above (92,278) were also found to follow this pattern, except at low density. Thus, the authors suggest that all that is required to predict the global rate constant for this reaction is an empirically determined activation energy and a preexponential term with a linear dependence on the solution density.

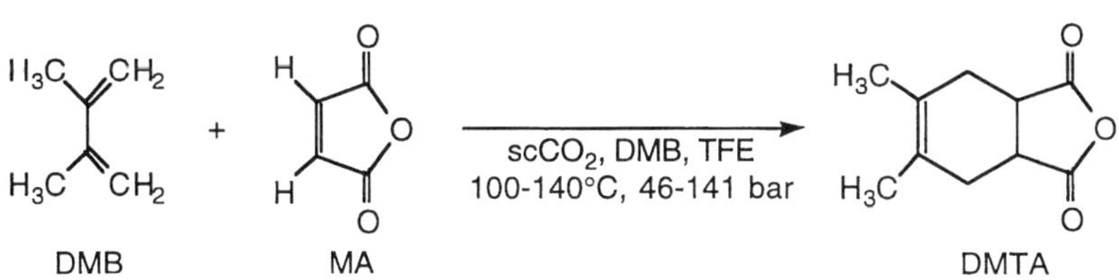

Scheme 21

Tester and coworkers next investigated the effect of changes in $scCO_2$ pressure and density on regioselectivity in several Diels–Alder reactions (135). Initial efforts sought to reproduce the pressure-dependent selectivity effect noted above by Ikushima et al. (134,278) for the cycloaddition of methyl acrylate and isoprene. Using a batch view cell reactor, they observed two-phase behavior under the conditions reported in the previous study and, in contrast to those reported results, saw no dramatic selectivity effects. The authors suggest that the previously reported results may have been based on nonrepresentative sampling due to the presence of two phases in the reaction mixture. Subsequent evaluations were designed to probe the effects of steric and electronic effects of substituent groups on regioselectivity. Reactions evaluated included acrylate with 2-*tert*-butyl-1,3-butadiene and 2-(trimethylsiloxy)butadiene with methyl acrylate. In all cases investigated, the authors observed no significant effect of reaction conditions on Diels–Alder regiochemistry in $scCO_2$-based reactions.

Tester and coworkers (283) report in a recent study the use of amorphous fumed silica (SiO_2) with a high specific surface area ($\approx 400\ m^2/g$) as a promoter for Diels–Alder reactions in $scCO_2$. The addition of silica was shown to increase both yields and selectivities of several such reactions in this media. Acid doping of the silica promoter enhanced the activating effect. Most of the work utilized the cycloaddition of methyl vinyl ketone and penta-1,3-diene at 80°C as a test reaction. The selectivity for this reaction was found to be relatively independent of pressure or fluid density. However, in general, yields were seen to decrease with increasing pressure. This effect was shown to be attributable to adsorption effects on the surfaces of the silica particles. As the pressure is increased, the ratio of the amount of reactants adsorbed on the silica surface to the amount of reactants in the fluid phase decreases, thus causing the yield to decrease.

Clifford et al. (284,285) have studied the selectivity to endo and exo products in the Diels–Alder reaction between cyclopentadiene and methyl acrylate in $scCO_2$. The ratio of the endo to exo product was observed to pass through a maximum on varying the density of the medium. However, the maximum did not occur at the critical density, so the authors suggest that the observed maximum is not associated with long-range critical phenomena as has been previously suggested (134). Instead, they propose that this effect is associated with variation in the average distance between the solute molecules and the transition states as the pressure and density vary. Theoretical arguments and equation of state calculations are presented to support this proposal. The authors suggest that the solvation energy of the solute can thus be controlled by density, which will cause variation in physical or chemical equilibrium. The chemical reaction rate is then indirectly controlled by the transition state equilibrium species. The authors term this concept "potential tuning."

Another type of cyclization reaction, known as the Pauson–Khand reaction, involves the catalytic cocyclization of alkynes with alkenes and carbon monoxide

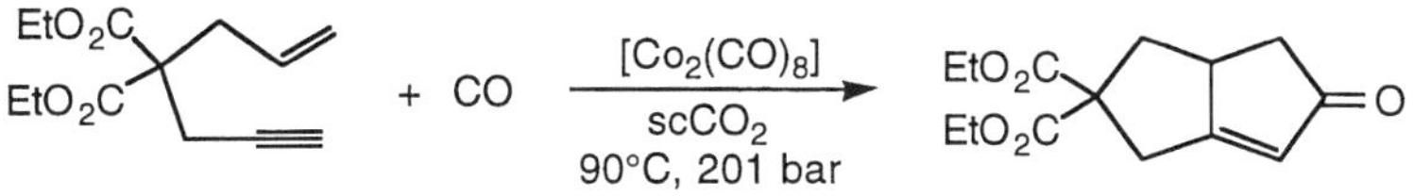

Scheme 22

to yield cyclopentenones. Jeong et al. (286) have demonstrated this chemistry in $scCO_2$ using dicobalt octacarbonyl as catalyst at relatively high concentrations of 2–5 mol %. For the reaction of Scheme 22, they report a product yield of 82% under the conditions shown.

Carroll and Holmes (287) report palladium-catalyzed carbon–carbon bond–forming reactions in $scCO_2$. This investigation included the preparation of novel fluorinated phosphine palladium complexes to successfully enhance the solubility of the metal complex in $scCO_2$ for homogeneous catalysis of the coupling reactions. The soluble palladium complexes were then used to catalyze three classic carbon–carbon bond–forming reactions, namely, the Heck, Suzuki, and Sonogashira couplings (Scheme 23). The indicated yields for the Suzuki and Sonogashira couplings are comparable to those obtained in conventional liquid solvents, but the Heck coupling shows a significant yield enhancement at 91% vs. that in acetonitrile of only 50% (35).

Tumas and coworkers (288) have also examined the palladium-catalyzed Heck coupling reaction with both fluorinated and nonfluorinated phosphine complexes. Coupling of phenyl iodide with both styrene and methyl acrylate at 90°C and 345 bar gave high conversions (>94%) and selectivities (>91%)

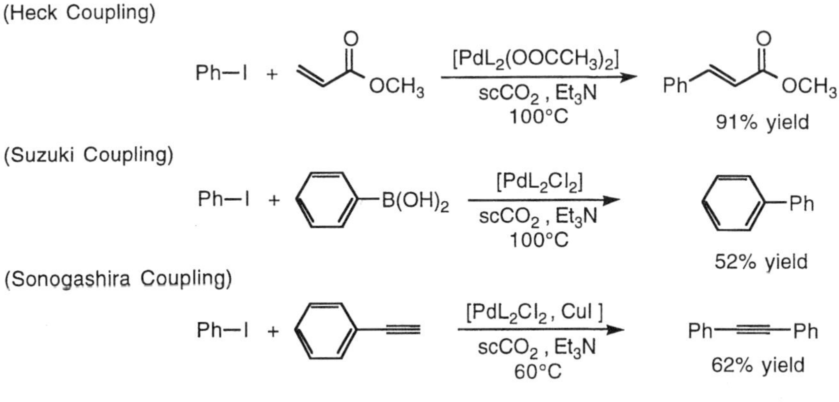

$L = (C_6F_{13}CH_2CH_2)_2PPh$

Scheme 23

with fluorinated phosphine ligands, but significantly lower conversions (20–35%) with nonfluorinated ligands. This suggested an enhanced solubility of the catalyst complex with the fluorinated phosphines, but the authors note the reaction mixture still appeared dark and opaque at their operating conditions. This group also investigated palladium-catalyzed Stille coupling reactions with these same catalyst complexes and the coupling reaction of phenyl iodide and tributyl(vinyl)tin (Scheme 24). Best results were obtained with the indicated tris[3,5-bis(trifluoromethyl)phenyl]phosphine ligand. The indicated conversion of 99% was comparable to that obtained in liquid THF (95%). This work and that of Carroll and Holmes cited above demonstrate that such coupling reactions can proceed in $scCO_2$ using fluorinated phosphine palladium complexes, but overall reaction efficiencies are comparable to that of conventional organic liquid solvents.

Rayner and coworkers (289) have expanded on the work of Carroll and Holmes (287) and Tumas et al. (288) in palladium-catalyzed coupling reactions in $scCO_2$ by using both commercially available fluorinated palladium sources [$Pd(OCOCF_3)_2$ and $Pd(F_6\text{-acac})_2$] and fluorinated phosphine ligands. They reasoned that for efficient palladium-mediated catalysis in $scCO_2$, the solubilities of any dissociated species should be considered, including the solubility of the initial palladium source. The use of commercially available materials rather than specially synthesized alternatives should make this synthesis more widely adopted. In general, the use of the fluorinated palladium sources in $scCO_2$ gave superior results to those previously reported, including low catalyst loadings (≈2%) at moderate temperatures. They investigated the classic Heck coupling reaction of phenyl iodide and methyl acrylate shown in Scheme 23, but at reaction conditions of 75–85°C and 110 bar. They also used the two palladium sources noted above and various ligands, including triphenylphosphine, tricyclohexylphosphine, and tris(2-furyl)phosphine [$P(2\text{-furyl})_3$]. The tris(2-furyl)phosphine ligand was the best of those investigated, giving greater than 95% conversion with high yields in 15 h, but others that are usually regarded as poor ligands for Heck reactions (e.g., tricyclohexylphosphine) gave appreciable conversions (81%) as well. The authors also utilized the [$Pd(OCOCF_3)_2$]/[$P(2\text{-furyl})_3$] system with good efficiency on other Heck

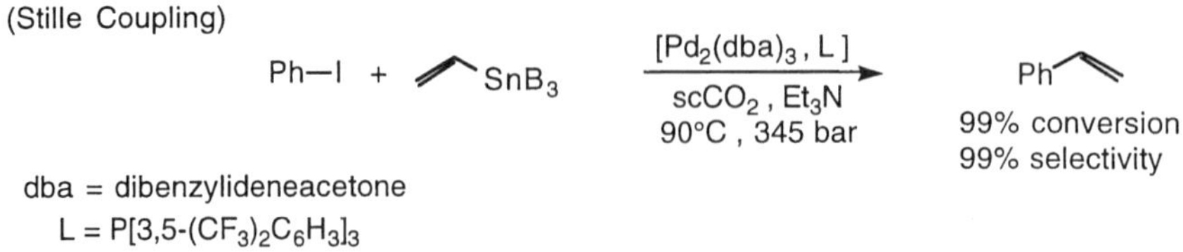

Scheme 24

reactions as well as Stille and Suzuki couplings to demonstrate the generality of these complexes.

Another recent study of palladium-catalyzed coupling reactions in $scCO_2$ is reported by Cacchi et al. (290) for the heterogeneously catalyzed reaction of phenyl iodides or vinyl triflates with a variety of olefins at 80°C and 100 bar in the presence of triethylamine and a carbon-supported Pd catalyst. Product yields were on the order of 40–80% depending on the reactants. This preliminary study demonstrates that the potential advantages of heterogeneously catalyzed reactions can be extended to palladium-catalyzed carbon–carbon bond–forming reactions in $scCO_2$.

I. Isomerization and Catalyst Deactivation

Isomerization has been a long-studied class of reactions in SCF media dating back to initial investigations by Tiltscher and coworkers (291–293). This was an important early study which was perhaps the first to demonstrate the potential utility of SCF media for conducting heterogeneous catalytic reactions (31,40). This group initially investigated 1-hexene isomerization to *cis*/*trans*-2-hexene on γ-Al_2O_3 catalysts in gas, liquid, and SCF phases to compare the effects of operation under such conditions (291,292). The researchers observed a pronounced positive effect of temperature and pressure on the cis isomer selectivity in the SCF phase, whereas there was little temperature effect for low-pressure gas conditions and a declining selectivity with increasing pressure in the liquid phase. This initial study also demonstrated that catalyst deactivation by deposition of hexene oligomers through coking or fouling mechanisms could be minimized by continuous operation at SCF conditions. The authors also indicated that a catalyst that had been deactivated under gas- or liquid-phase conditions by such deposition mechanisms could be reactivated by operating in the SCF phase to extract these materials.

Manos and Hofmann (294) studied coke removal from a stable HY zeolite catalyst using disproportionation of ethyl benzene at SCF conditions as the test reaction. They showed that complete reactivation of a spent catalyst by extraction with the SCF-phase ethyl benzene was not possible due to the decreasing solubility and pore diffusion rates of the polyaromatic coke precursors with increasing molar mass. However, in the study they also demonstrated that by operating at these conditions with fresh catalyst, deposition of the initial low-molecular-weight coke precursors is minimized because they are simultaneously extracted as they form. Thus, maintenance of activity is enhanced. They also suggest that an optimum operating temperature should exist at which the deactivation is significantly reduced but balanced against the corresponding increased reaction rates (including coke precursor formation).

Subramaniam and coworkers (62,182,183,295–307) have conducted an extensive study over the past decade of catalytic 1-hexene isomerization at SCF conditions along with the associated catalyst deactivation issues that arose during this work. The early papers through 1994 have been extensively reviewed by Savage et al. (31) and more recently by Baiker (33), so only a brief synopsis will be included here. Baiker also cites the more recent reports. Isomerization of 1-hexene was selected as a test reaction for several reasons (297): (a) it facilitates direct comparison with previous work by Tiltscher et al. (291–293); (b) olefins such as 1-hexene are direct coke precursors; (c) the critical properties of 1-hexene ($T_c = 230.8°C$, $P_c = 31.7$ bar) are moderate and permit investigations over a wide range of reduced pressures, temperatures, and densities; (d) the physicochemical properties of 1-hexene and corresponding isomers are similar, so that the phase behavior of the reaction mixture is virtually unaffected by conversion; and (e) the high sensitivity of isomerization reactions to catalyst coking lead to measurable drops in catalytic activity within reasonable processing times. Carbon dioxide was used as a diluent in these studies, and the catalyst employed was a commercial microporous Pt/γ-Al_2O_3-reforming catalyst supplied by Engelhard Industries.

Initial studies (295,296) determined phase and reaction equilibria to establish the thermodynamic constraints of phase boundaries and equilibrium conversion in the system. Continuous experiments in a fixed-bed flow reactor showed a decline in catalyst activity at subcritical conditions, whereas no loss of catalyst activity was observed at supercritical pressure and a nearly identical temperature. This catalyst deactivation was attributed to deposition of oligomers within the catalyst pores which were effectively extracted under SCF conditions (296). Subsequent work (297,298) indicated that the addition of CO_2 as a cosolvent afforded milder operating temperatures and pressures as well as lower reactant concentrations while maintaining high solvent densities, all of which were shown to favor maintenance of catalyst activity. A single-pore mathematical model to describe coking and activity characteristics of porous catalysts was developed (299) which was qualitatively consistent with the experimental observations noted above and predicted an optimal density at which the catalyst activity is maximized between the competing effects of coke deposition and pore diffusion limitations. These qualitative predictions were confirmed with subsequent experimental measurements (300) which also demonstrated a twofold increase in isomerization rates at near-critical conditions compared with subcritical operation. Based on these studies, the authors concluded that supercritical reaction mixtures provide an optimum combination of solvent and transport properties for maximizing the isomerization rates and for minimizing catalyst deactivation rates.

Subramaniam and McCoy (302,303) have developed a kinetic scheme and a continuous-mixture kinetic model that describes the formation of coke compounds from olefinic oligomers followed by their reversible adsorption and desorption between the catalyst and the SCF. The model predictions regarding conversion and activity profiles, activity maintenance or decay, and formation of oligomers and coke compounds are consistent with reported experimental observations. The model predicts catalyst activity maintenance under SCF conditions due to enhanced desorption of coke-forming compounds under these conditions.

Subsequent work (62,305) showed that the presence of organic peroxides in the 1-hexene feed stream substantially contributed to the formation of hexene oligomers in the bulk fluid, which contributed to catalyst deactivation. The feed peroxides were essentially eliminated by pretreatment with activated alumina and by deoxygenation of the feed stream. With these enhancements, a several-fold reduction in coke formation and nearly constant isomerization activity were observed in a continuous fixed-bed reactor with about 65% conversion under SCF conditions of 281°C and 70 bar. During an extended 42-h run, catalyst activity was nearly constant with neither measurable coke deposition nor surface area/pore volume losses in the spent catalyst. Subramaniam and Ginosar (304) expanded the previous mathematical model (303) to incorporate the effect of organic peroxides and the addition of inert cosolvents. They present a detailed and comprehensive mathematical model describing the coking and isomerization activity of a Pt/γ-Al_2O_3 catalyst in a continuous fixed-bed reactor. Results from the enhanced model are consistent with the experimental observations of a dramatic decrease in oligomer formation with the elimination of feed peroxides and a corresponding increase in maintenance of catalyst activity.

Following these substantial efforts to minimize the formation of oligomeric coke precursors, Clark and Subramaniam recently reported (306) the measurement of intrinsic kinetic parameters for the Pt/γ-Al_2O_3-catalyzed isomerization of 1-hexene under steady, nondeactivating conditions. Several steps were taken to minimize formation of coking materials including deaerating the hexene feed, pretreating it with activated alumina to minimize peroxide impurities to less than 2 ppm, and passivating the reactor surface with a silicosteel coating. These steps in conjunction with operation under SCF conditions to enhance oligomer desorption from the catalyst surfaces resulted in steady isomerization activity with no measurable coke deposition. Arrhenius plots of the effective rate constants, estimated from the steady isomerization rates assuming plug flow and first-order kinetics, provided an average intrinsic activation energy of 109 kJ/mol in the 235–270°C temperature range. These results confirmed that hexene isomerization to geometrical isomers is the dominant reaction pathway under the conditions studied and that optimum conditions exist in the critical region that maximize the catalyst activity and effectiveness factor.

Amelse and Kutz (308) disclose a patented catalytic process for the isomerization of mixed xylenes to *p*-xylene in the presence of ethyl benzene under SCF conditions. The preferred heterogeneous catalyst is a silicate molecular sieve having 0.1–8 wt % gallium incorporated into the silica matrix. The inventors attribute several process advantages to operating under SCF conditions vs. the commercial gas phase process: (a) catalyst deactivation is minimized in the SCF process without substantial addition of hydrogen, as is practiced in the commercial process; (b) fuel savings are realized due to the higher heat transfer coefficients of the SCF media; and (c) the magnitude of a temperature pinch point in the reactor feed/effluent heat exchangers of the commercial process was substantially reduced by isobarically cooling the reactor effluent to the liquid phase at a pressure exceeding the mixture critical pressure.

Eckert and coworkers (309) report the tuning of reaction rates for the thermal cis-trans isomerization of several azobenzene dye molecules in supercritical ethane and CO_2 using both pressure changes and the addition of cosolvents. The variation in rate constants with density correlates with the solvent density for pure supercritical ethane and CO_2, and a mechanism consistent with the data is proposed. However, the addition of as little as 0.5 mol % cosolvent resulted in a 15-fold increase in the reaction rate. This tremendous rate enhancement was attributed to a combination of local composition enhancements (i.e., clustering), which provided an enriched dielectric environment about the solute, and specific interactions with strong hydrogen-bonding cosolvents, which lowered the activation energy barrier and resulted in a change of mechanism.

J. Alkylation

Yuan and coworkers (310,311) report an investigation of catalyst deactivation and regeneration in the alkylation of benzene with ethylene to produce ethyl benzene in a continuous fixed-bed reactor. Results are compared for operation in both liquid (250°C, 65 bar) and SCF (275°C, 65 bar) phases using a Y-type zeolite catalyst. The authors report improved selectivity to ethyl benzene in SCF-phase operation resulting from a significant decrease in the production of byproduct xylenes relative to liquid phase operation. The catalyst activity decreased after about 12 h of operation at liquid phase conditions but remained constant over a 55-h run under SCF conditions. Operation at higher temperatures of 290°C and 300°C in the SCF phase resulted in increasing deactivation, but the activity was insensitive to pressure. Ethyl, diethyl, triethyl, and tetraethyl benzene were observed in the products of the liquid-phase reaction. The SCF-phase reaction also showed the presence of 2-ethylbiphenyl, 2,2-biethylbiphenyl, and 1,5,6-trimethylanthracene. Based on liquid extraction with benzene of these latter components from the spent catalyst from SCF operations, the authors propose that these multiring compounds are coke precursors that lead to catalyst deactivation by coke formation. The authors unsuccessfully attempted to regenerate a catalyst charge that had been deactivated in liquid-phase operation by operating in situ under SCF conditions. Little change in activity was observed.

Fan et al. (312) report the alkylation on Y-type zeolite catalysts of two olefin/paraffin systems: (a) isobutene with isopentane and (b) isobutene with isobutane. The isopentane and isobutane, respectively, were also used as the solvents in this work, and results were compared for gas-, liquid-, and SCF-phase conditions. Alkylations conducted in the SCF phase were reported to exhibit higher activity and longer catalyst lifetimes in comparison with reactions conducted in the gas and liquid phases. Catalyst deactivation observed in the gas- and liquid-phase results was attributed to deposition of high-molecular-weight olefinic oligomers on the Lewis acidic sites that were determined to be active for the alkylation reactions. Operation under SCF-phase conditions resulted in successful extraction of these oligomers in situ and extended the catalyst life. Additional aspects of this report are discussed in a subsequent commentary (313) and rebuttal (314).

Clark and Subramaniam (315) report the SCF-phase alkylation of 1-butene and isobutane using a molar excess of CO_2 as a diluent to lower the mixture critical temperature (<135°C), thus providing lower operating temperatures in the SCF-phase regime. The solid acid catalysts used in this investigation included both microporous-type H-USY zeolite and mesoporous sulfated zirconia, and the experiments were conducted in a continuous fixed-bed reactor. The effects of adding the CO_2 diluent on alkylate production activity, catalyst coking, and butene conversion are compared with those observed for gas- and liquid-phase reactions without the addition of CO_2. For identical space velocities and feed ratios, the alkylate formation continuously decreased with time on stream when the reaction was conducted without the added diluent. In contrast, operation with added CO_2 resulted in virtually steady alkylate (trimethylpentanes and dimethylhexanes) production for experimental durations of approximately 2 days. At 50°C, 155 bar, a space velocity of 0.25 g 1-butene/gcat/h, and a feed CO_2/isobutane/olefin ratio of 86:8:1, approximately 5% alkylate yield (i.e., alkylates/C_{5+}) and 20% butene conversion were observed at steady-state operating conditions. At the high temperatures (>135°) required for SCF operation without added CO_2, coking and cracking reactions were significant as evidenced by a broad product distribution and substantial surface area and pore volume losses (up to 90%) in the spent catalysts.

Hitzler et al. (316) report the Friedel–Crafts alkylation of mesitylene ($C_6H_3Me_3$) and anisole with propene or 2-propanol using a heterogeneous polysiloxane-supported solid acid catalyst (Degussa's Deloxan) in a small fixed-bed continuous reactor (10-ml volume) using SCF propene or CO_2 as the reaction solvent. For the alkylation of mesitylene with propene at 160–180°C and 200 bar, yield of the monoalkylated product (1-isopropyl-2,4,6-trimethylbenzene) was only approximately 25% due to the formation of the dialkylated product as well as dimers and trimers of propene. Selectivity to the monoalkylated product was significantly higher (40% yield) for alkylation with 2-propanol in $scCO_2$.

The authors demonstrate some success in tuning the product selectivity through control of temperature, pressure, and reactant concentrations, and they report that another potential advantage for conducting the process in an SCF solvent is reduction of catalyst deactivation due to coking. However, as Baiker has noted (33), assessing the relative merits of this selectivity tuning is difficult because no comparable evaluation is reported for continuous alkylation in a conventional solvent. These authors and additional researchers extend this work to include similar acylation reactions in an issued patent (317) and conclude that Friedel–Crafts alkylation or acylation reactions may be effected using a heterogeneous catalyst in a continuous-flow reactor process operating in the SCF regime.

K. Esterification

Mouloungui and coworkers (318,319) evaluated the use of $scCO_2$ for the esterification of oleic acid by methanol using *p*-toluenesulfonic acid monohydrate and the sulfonic acid cationic exchange resins K2411 and K1481 as catalysts. This is a reversible equilibrium reaction that has been widely studied in conventional media, and results are compared with those obtained using liquid *n*-hexane as an example. The effect of temperature, methanol concentration, and catalytic sulfonic site concentration was found to be similar in both mediums, although the reaction rate is faster in $scCO_2$. The authors report that a homogeneous reaction mixture was obtained with $scCO_2$ at 40°C, and the equilibrium was shifted to the ester at high pressure (190 bar), resulting in production of the methyl oleate product in high yield. Reaction with the cationic exchange resins was less successful and gave relatively low product yields, comparable to those obtained in conventional organic solvents of similar hydrophobicity. The authors attribute this to limited swelling of the resin in the presence of the hydrophobic $scCO_2$, so that access of the reactants to catalytic sites within the resin macropores was limited. Consequently, esterification only took place on the external surfaces.

Brennecke and coworkers (90,320) investigated the uncatalyzed esterification of phthalic anhydride with methanol in $scCO_2$ as a probe reaction to show that augmented local densities and cosolvent compositions in the near-critical region represent enhanced reactant concentrations that result in increased reaction rates. The authors report kinetic data for the esterification reaction at both 40°C and 50°C and pressures ranging from 97.5 to 166.5 bar. Based on bulk fluid compositions, a dramatic pressure effect was exhibited for the measured bimolecular rate constants. For example, at 50°C the value of the rate constant decreased 25-fold from 0.0348 L/mol-min at 97.5 bar to 0.00138 L/mol-min at 166.5 bar, which represents one of the largest pressure effects ever reported for a reaction in an SCF. Based on calculations of the thermodynamic pressure effect on the rate constant from transition state theory and estimates of the local concentrations from literature data, the authors conclude that the dramatic pressure effect on the esterification rate is due to both a slight thermodynamic pressure effect on the rate constant and a substantial increase in the local concentration of the methanol around the phthalic anhydride.

L. Amination and Ammonolysis

Felthouse and Mills (321) report the amination of methyl *tert*-butyl ether (MTBE) and isobutene to *tert*-butylamine using alumino- and borosilicate pentasil molecular sieve catalysts. The ether and alkene amination reactions were found to proceed preferentially under SCF conditions at temperatures on the order of 330°C and pressures greater than 193 bar. The study showed that MTBE can be used as a substitute raw material for *tert*-butylamine manufacture, but MTBE decomposition products of isobutene, methanol, and methanol conversion products are produced that require a more complicated product separation process than with isobutene as the only C_4 substrate.

Baiker and coworkers (322,323) investigated the heterogeneously catalyzed amination of aliphatic alcohols in supercritical ammonia ($scNH_3$) in the temperature range of 150–210°C and total pressure range of 50–150 bar. The motivation for this work was to evaluate the potential for a one-step continuous process for the amination of alkanediols, which has a complex reaction network of possible products. In the initial study (322), the authors used a continuous fixed-bed reactor at 195°C to determine the influence of reactor pressure on conversion and selectivity in the amination of 1,3-propandediol on an unsupported Co-Fe catalyst. This showed a pronounced enhancement of selectivity to both the monoaminated product 3-amino-1-propanol and the diamine 1,3-diaminopropane on increasing the pressure from 50 to over 100 bar at 195°C. To decouple the consecutive amination reaction, the intermediate 3-amino-1-propanol was fed to the reactor under similar conditions, and the influence of pressure on this reaction is shown in Figure 10. On increasing pressure through the region of the phase boundary, the conversion decreases slightly, but the diamine selectivity increases dramatically from about 2% to approximately 35% under supercritical conditions. The presence of a homogeneous SCF phase at 130 bar and 200°C for an approximate reaction composition was confirmed by visual observation in a quartz cell. The pronounced pressure effect seems general in that similar enhancement of amination selectivity was also observed with another substrate (2,2-dimethyl-1,3-propanediol) and catalyst (a supported nickel). The second paper (323) reports a detailed evaluation of the role of the catalyst in controlling the product distribution and included screening of a variety of potential metal catalysts and promoters. Cobalt was identified as a leading candidate, and promotion of the cobalt with iron or lanthanum was found to significantly improve both the diamine selectivity and stability to deactivation. The best catalyst identified was a 5 wt % Fe-promoted Co catalyst that showed optimal efficiency and

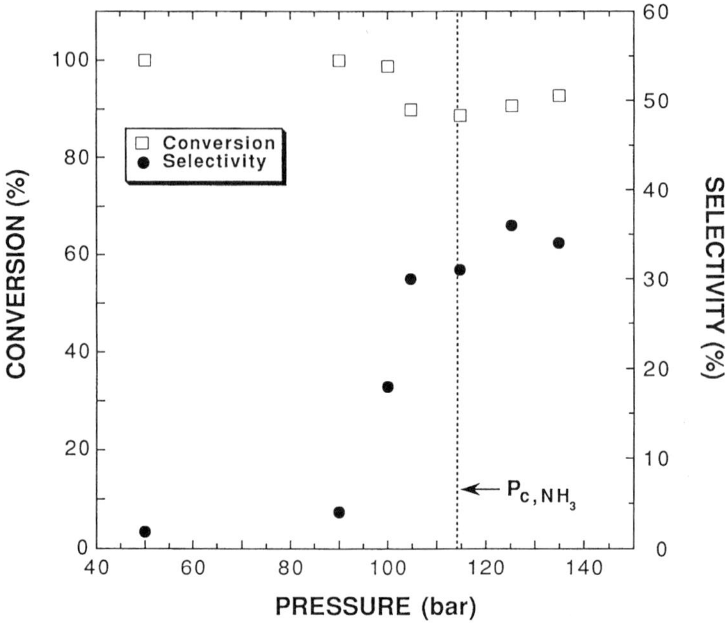

Figure 10 Effect of pressure on the conversion of 3-amino-1-propanol and selectivity to 1,3-diaminopropane at 195°C. (Reprinted with permission from Ref. 322. Copyright 1999 Wiley-VCH.)

stability. No significant deactivation was observed with this catalyst after 10 days of use at 150–210°C. The absence of strong acidic and basic sites on the catalyst was found to be crucial for the selective production of the diamine and for the suppression of various side reactions, including hydrogenolysis, oligomerization, and disproportionation. They also report that the presence of 1–5 mol % hydrogen in the feed minimized undesired dehydrogenation reactions leading to the formation of nitriles and carbonaceous deposits. The authors suggest that the changes in amination selectivity and alkanediol conversion in the vicinity of the mixture critical point can be attributed to reduced mass transfer resistances in the homogeneous SCF phase and to the resulting increased ammonia concentration at the catalyst surface, which favors amination and further suppresses the noted side reactions.

In a separate study (324), the Baiker group reports a similar study on the amination of 1,4-cyclohexanediol in $scNH_3$ using this Co-Fe catalyst. Diaminocyclohexanes are typically manufactured by the catalytic hydrogenation of aromatic amines such as *p*-phenylenediamine (324), so that direct amination of cyclohexanediols provides an attractive process alternative. Operation of a

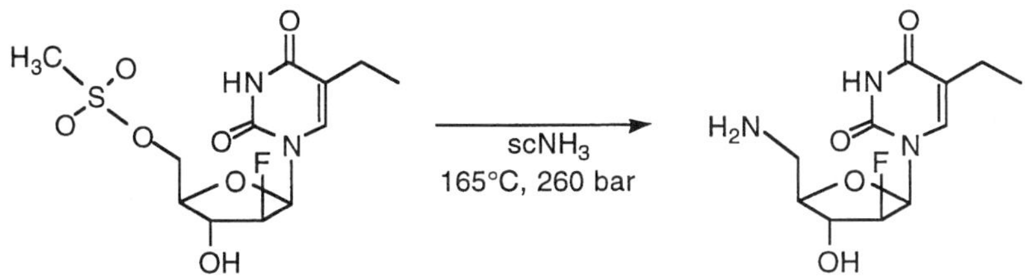

Scheme 25

continuous fixed-bed reactor at 135 bar and 165°C gave amination products of 4-aminocyclohexanol and 1,4-diaminocyclohexane with a cumulative selectivity of 97% at 76% conversion. An excess of ammonia and short contact times favored the desired amination reactions. At low and high conversions, the amination selectivity decreased due to the formation of oligomers and degradation products. Diamine yields could be further boosted by recycling of the unreacted diol and amino alcohol intermediate. These results, combined with the process advantages of continuous operation and relatively simple product isolation from the $scNH_3$ solvent, provide the basis for a potentially attractive alternative process for the synthesis of diaminocyclohexanes.

Wang et al. (325) report the ammonolysis of a mesylate with anhydrous $scNH_3$. Scheme 25 shows the particular mesylate investigated and the corresponding amine product, which is of interest for a pharmaceutical application. This synthesis route was investigated as a potential alternative to the more conventional azide-based routes to avoid the toxicity and explosivity of such reagents. At a reaction temperature of 165°C and total pressure of 60 bar, they report 99.4% conversion with 96.6% selectivity to the desired amine product. This was significantly better than the approximately 80% selectivity obtained at subcritical temperature and pressure. The authors successfully fit the experimental data to a triangular global reaction network involving both a parallel and sequential reaction pathway of the reactant and product and lumping of the byproducts into a single pseudo-byproduct component. This study demonstrated that ammonolysis in $scNH_3$ is a viable route for introduction of an amino functionality by direct substitution of a mesylate. This synthesis alternative avoids the use of azides and removes one step from the more conventional sequence of azide substitution followed by reduction.

M. Disproportionation

Collins et al. (326) investigated the disproportionation of toluene to *p*-xylene as a probe to examine solute-solvent clustering effects in the near-critical region. Benzene and mixed xylenes are the primary products over the unmodified ZSM-

5 zeolite used as the catalyst in this study. The authors reasoned that solvent (toluene) clustering around the desired product solute (*p*-xylene) would inhibit secondary isomerization on the catalyst surfaces and enhance the *p*-xylene yield. On varying the pressure at a constant temperature of 320°C ($T_R = 1.002$), the *p*-xylene selectivity was found to undergo a pronounced maximum near the toluene critical pressure of 41 bar. The selectivity effect was much less pronounced at a reduced temperature of 325°C ($T_R = 1.01$).

Hofmann and coworkers (327–330) have reported a series of studies on the deactivation kinetics for the heterogeneously catalyzed disproportionation of ethyl benzene to benzene and diethyl benzene under SCF conditions. Kinetic studies have been conducted in both a loop reactor using a protonated Y-faujasite (Z-14) catalyst (327) and in a continuous concentration-controlled recycle reactor using an HY-zeolite (HYZ) (329,330) and USY-zeolite, H-ZSM-5, and H-mordenite (328) under supercritical conditions ($T > 373$°C, $P > 45$ bar). Coke extraction by SCFs was found to be strongly dependent on the type of catalyst used, and the Lewis acid centers were determined to play an important role in the coke formation and activity of the catalysts. A simple kinetic model for the catalyst deactivation was proposed (329) for SCF conditions and high ethyl benzene concentration. Based on the relatively high estimated deactivation energy of about 147 kJ/mol and first-order deactivation, the authors concluded that the catalyst deactivates much slower under SCF conditions than under atmospheric pressure.

N. Cracking

Moser and coworkers (331–334) studied the catalytic cracking of *n*-heptane ($T_c = 267$°C, $P_c = 27.4$ bar) over a commercial Y-type zeolite (Promoted Octacat) at conditions significantly above the critical temperature (475°C) and at pressures below (6.7 and 16.7 bar) and above (41.6 bar) the critical pressure. They monitored the reaction in situ using a cylindrical internal reflection infrared technique (CIR-FTIR) developed by the team that permits real-time analysis of SCFs and heterogeneous catalytic processes at temperatures up to 500°C and at 69 bar. They report that the catalyst maintained higher activity during catalytic cracking under SCF conditions, whereas subcritical pressures led to rapid catalyst deactivation. Higher amounts of carbonaceous deposits under subcritical pressures were also observed. The subcritical IR spectra showed dramatic reductions in the concentration of the acid sites and in the interactions of the acid sites with the olefinic and paraffinic reaction products. SCF *n*-heptane showed an unusually dense behavior within the pores of the zeolite, which was significantly denser than estimated by equations of state under identical conditions. This dense phase served to continuously remove coke from the pores and active sites, resulting in the maintenance of the activity of zeolites under high conversion cracking conditions. However, the high density also led to a severe reduction of molecular diffusion within the pore structure, which greatly affected selectivity. Reactivation studies on a deactivated catalyst resulted in partial regeneration of the catalyst under SCF conditions.

O. Miscellaneous Reactions

Jia et al. (335) report the palladium(II)-catalyzed carbonylation of norbornene in $scCO_2$ in the presence of triethylamine and methanol. Leitner and coworkers (336) report the ring-closing metathesis of dienes in $scCO_2$ to form various ring structures using metal-carbene complexes as catalysts. A substantial density effect was observed on the product distribution, with rings being favored at high densities and oligomers the primary product at low densities. Several studies of conducting dimerization reactions in $scCO_2$ have also been reported. Saito and coworkers (337) investigated the dimerization of benzoic acid using an FT-IR spectroscopic technique. Sun and coworkers (338) conducted a systematic study of the photodimerization of anthracene over a range of CO_2 densities and observed an order-of-magnitude increase in product yields relative to conventional liquid solvents. Sakanishi et al. (339) report the selective dimerization of benzothiophene using a solid acid catalyst (aluminum sulfate supported on porous silica gel) to isolate and recover benzothiophene from crude naphthalene.

DeSimone and coworkers (340) investigated the dimerization of α-methylstyrene using DuPont Nafion catalysts. They observed a rate enhancement over conventional liquid solvents such as cumene and *o*-cresol, which they attributed in part to plasticization of the perfluorinated catalyst resin with the $scCO_2$ combined with the enhanced mass transfer characteristics afforded by the SCF solvent. In a subsequent study, DeSimone and coworkers (341) measured the thermal decomposition rates of two perfluoroalkyl diacyl peroxides [bis(trifluoroacetyl) and [bis(perfluoro-2-*n*-propoxyprionyl) peroxides] in liquid and $scCO_2$ and compared rates with similar measurements made in Freon-113. Both peroxides displayed activation energies approximately 5–6 kcal/mol lower than that obtained in Freon-113, which the authors attribute to differences in solvent viscosity.

Wynne and Jessop (68) demonstrated the influence of a pressure-dependent dielectric constant for $scCHF_3$ solvent on an enantioselective reaction. They selected the cyclopropanation of styrene and methylphenyldiazoacetate as a model reaction and used *tert*-butylbenzenesulfonyl-L-prolinate dirhodium as catalyst. By varying the pressure from 52 to 180 bar at 30°C, they showed a corresponding variation in the enantiomeric excess of 77% ee to 40% ee, respectively. The static dielectric constant varied from about 2 to approximately 7 over this range of conditions. Varying the pressure and density in the analogous CO_2-mediated reaction had little effect on enantiomeric excess in the absence of this significant dielectric constant change.

P. Phase Transfer Catalysis

Phase transfer catalysis (PTC) is widely used in industry, with primary applications including oxidations, reductions, polymerizations, transition metal–cocatalyzed reactions, synthesis and subsequent reactions of carbenes, addition reactions, and condensations. These applications are often part of a multistep synthesis process in pharmaceuticals, agricultural chemicals, and other fine-chemicals manufacture. Starks et al. (342) estimated that there are as many as 500 commercial processes containing at least one PTC step, with combined sales of products from such processes exceeding $10 billion a year. Naik and Doraiswamy (343) recently presented an excellent comprehensive review on this topic, with emphasis on combining the predominantly chemistry-intensive literature with insights from engineering analysis of mechanisms and kinetics and through mathematical modeling of process fundamentals.

PTC uses catalytic quantities of phase transfer agents that facilitate the interfacial transfer of reactive components by shuttling reactant and product ions from one phase to the other, thus making possible chemical reactions between reagents in two immiscible phases. For soluble systems, such reactions can be categorized into two main classes: (a) solid–liquid and (b) liquid–fluid, where the fluid can be a liquid, gas, or SCF. Examples of both solid–liquid and liquid–SCF PTC research have appeared in the literature (24,344,345). Use of an SCF phase reduces mass transfer limitations relative to a liquid organic phase in conventional processes and could potentially facilitate isolation of the products and catalysts, which is one of the primary disadvantages of conventional PTC processes (343).

Eckert and coworkers (344) reported the first example of solid–SCF PCT involving a halide exchange reaction between solid potassium bromide and benzyl chloride to yield benzyl bromide (Figure 11). The mechanism of the catalytic cycle is similar to the conventional PTC mechanism and involves shuttling the reactant Br^- anion from the solid phase to the SCF phase using either a quaternary ammonium cation (illustrated in Figure 11) or a macrocyclic multidentate ligand such as a crown ether, both of which were used in this study. Tetraheptylammonium bromide (THAB) was selected as an example of a quaternary ammonium salt, and 18-crown-6 was chosen as a crown ether. The choice of PTC catalyst is limited by the solubility of the catalyst-anion complex in the SCF phase, and the authors found that the addition of a polar or protic cosolvent, such as acetone, was required to provide sufficient solubility for catalytic activity. A cosolvent mixture consisting of 5 mol % acetone in CO_2 was se-

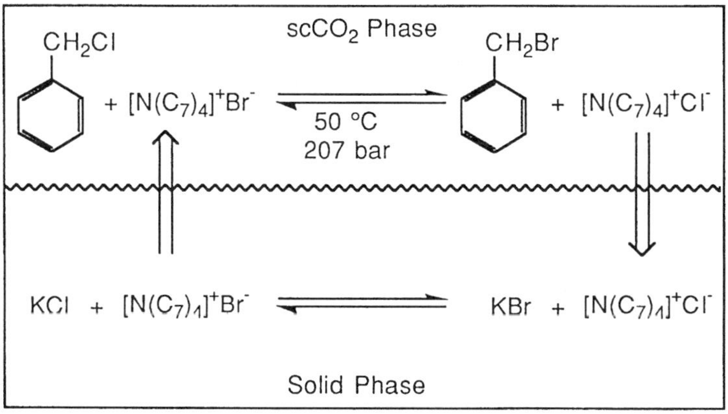

Figure 11 Phase transfer catalytic cycle for the nucleophilic displacement of benzyl chloride with a bromide ion in $scCO_2$. (From Ref. 344.)

lected as the SCF phase, and the THAB solubility in this solution was reported as 1.6×10^{-5} mole fraction under the reaction conditions of 50°C and 207 bar. The 18-crown-6/KBr complex was found to be essentially insoluble at 50°C, but was reported to be catalytically active at 75°C, where the authors estimated its solubility to be on the order of 10^{-7} mole fraction. In a kinetic study with the THAB catalyst and an excess of KBr, the exchange reaction was found to follow reversible pseudo-first-order kinetics and reach equilibrium conversions of approximately 60%. Experiments with the 18-crown-6 catalyst at 75°C also resulted in approximately 60% conversion, but the reactions were found to follow zero-order kinetics. The authors attributed this finding to the rate-limiting step being that of mass transfer of the solid salt to the crown ether, rather than the SCF-phase reaction. This study demonstrated that SCF-phase transfer catalysis reactions may be carried out successfully in a process relatively free of conventional organic solvents.

In a subsequent study (345), this group examined the detailed mechanism of a solid salt/SCF phase transfer–catalyzed reaction. They selected a reaction similar to that depicted in Figure 11, that of the irreversible nucleophilic displacement of benzyl chloride with potassium cyanide to form phenylacetonitrile and potassium chloride. The study primarily used the catalyst THAB as in the previous study. The effects of various factors on the reaction kinetics were investigated, including the amount of catalyst, the amount of KCN, the presence of acetone cosolvent, and temperature. Measured kinetic data were consistent with irreversible pseudo-first-order kinetics in the catalyst concentration. However, the reaction rate was found to be linearly dependent on the catalyst concentration even at two orders of magnitude above the catalyst solubility in the SCF phase. Kinetic measurements with and without added acetone showed that the reaction rate increased when no acetone was present in the system. The measured kinetic data and catalyst solubility measurements suggested that the operating reaction mechanism is a three-phase system consisting of an $scCO_2$ phase, a catalyst-rich phase, and a solid salt phase, and that the reaction actually occurs in the catalyst-rich phase. This mechanism is consistent with conventional phase transfer catalysis in organic/solid systems where the addition of a small amount of water produces a catalyst-rich phase on the salt surface, which is termed the omega phase. Results using two alternative nonionic phase transfer catalysts, 18-crown-6 and poly(ethylene glycol), were consistent with the proposed three-phase mechanism. The authors point out that such a three-phase PTC system with reaction occurring in a catalyst-rich phase could provide important advantages over conventional two-phase systems. For example, CO_2 could be used as the SCF-phase reaction solvent with conventional PTC catalysts but without the addition of organic cosolvents, despite their low solubility in the CO_2 phase. Furthermore, catalyst removal and recovery procedures are often simplified in three-phase PTC systems.

Tumas and coworkers (24) have investigated the two-phase oxidation of cyclohexene to adipic acid in the $scCO_2$/aqueous system using RuO_4 as the phase transfer catalyst, as depicted in Figure 12. Several oxidants were used, including $NaIO_4$, Ce(IV), and peroxyacetic acid. High selectivity (99%) to adipic acid was observed, but extremely low catalyst turnovers of 5 or less occurred. The authors attributed the low TON to the formation of a Ru(IV) complex with bicarbonate, which resulted from reaction between the CO_2 and water. Another

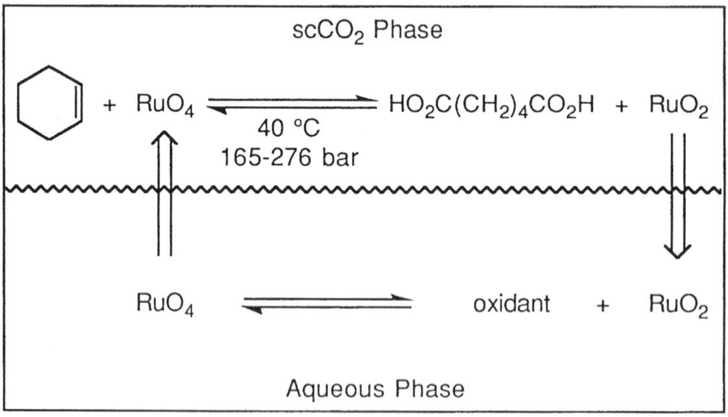

Figure 12 Phase transfer catalytic oxidation of cyclohexene to adipic acid in an aqueous/$scCO_2$ medium. (From Ref. 24.)

concern with this system is that the aqueous phase is forced to approximately pH 3 by the equilibrium between water, CO_2, and carbonic acid (346):

$$CO_2 + H_2O \rightleftharpoons H_2CO_3 \rightleftharpoons H^+ + HCO_3^-$$

Thus, aqueous two-phase reactions requiring basic conditions are not candidates for solvent substitution by CO_2. These two issues suggest significant potential limitations to the transfer of two-phase reactions from organic/aqueous systems to $scCO_2$/aqueous systems in phase transfer catalysis applications, although alternative solvents should be viable for the SCF phase.

IV. CONCLUSION

SCFs offer a number of potential advantages as solvent media for conducting chemical transformations relative to conventional liquid solvents, although they are not a panacea for all reaction applications. Unless one or more of these advantages are realized by operation at SCF conditions, process economics will tend to favor conventional operations, which generally operate at lower pressures and without the complexity of process control near a solution critical point, where density fluctuations may perturb process stability. However, given the possibilities of optimizing a reaction environment by tuning the solvent density and related properties, the potential environmental benefits of waste minimization and operation with more environmentally benign solvents, and the opportunities for process simplification through exploitation of phase behavior effects, SCFs offer an attractive process alternative as a chemical reaction media. The challenge is to identify those specific applications where use of SCF conditions is technically feasible, practical, and economically viable.

The breadth and quantity of papers presented in this chapter illustrate the rapidly increasing level of active research and development in this exciting field of supercritical fluid reactions. The now successful commercialization of SCF reaction processes and the ongoing construction of others will further reduce technical and financial uncertainties about investment in this technology. Thus, SCF media for conducting chemical transformations will likely achieve greater acceptance for manufacturing operations in future applications.

ACKNOWLEDGMENTS

I thank Tom Johns and Steve Reynolds for conducting literature and patent searches in support of this review and Ann Leonardi for acquiring many of the cited documents. Special thanks to Joan Brennecke, Tapan Das, and Mike Harold for reviewing the manuscript and making many helpful suggestions.

REFERENCES

1. M McCoy. Industry intrigued by CO_2 as solvent. Chem Eng News 77(24):11–13, 1999.
2. KP Johnston. Supercritical fluid separation processes. In: RH Perry, DW Green, JO Maloney, eds. Perry's Chemical Engineers' Handbook. 7th ed. New York: McGraw-Hill, 1997, pp 22.14–22.19.
3. KP Johnston, RM Lemert. Supercritical fluid technology: theory and application. In: JJ McKetta, GE Weismantel, eds. Encyclopedia of Chemical Processing and Design. Vol. 56. New York: Marcel Dekker, 1996, pp 1–45.
4. PG Jessop, W Leitner. Supercritical fluids as media for chemical reactions. In: PG Jessop, W Leitner, eds. Chemical Synthesis Using Supercritical Fluids. Weinheim: Wiley-VCH, 1999, pp 1–36.
5. JW Tester, HR Holgate, FJ Armellini, PA Webley, WR Killilea, GT Hong, HE Barner. Supercritical water oxidation technology: process development and fundamental research. In: DW Tedder, FG Pohland, eds. Emerging Technologies in Hazardous Waste Management III. ACS Symposium Series No. 518. Washington, D.C.: American Chemical Society, 1993, pp 35–76.
6. H Schmieder, N Dahmen, J Schön, G Wiegand. Industrial and environmental applications of supercritical fluids. In: R van Eldik, CD Hubbard, eds. Chemistry Under Extreme or Non-Classical Conditions. New York: Wiley-Spektrum Akademischer Verlag, 1997, pp 273–316.
7. KW Hutchenson, NR Foster. Innovations in supercritical fluid science and technology. In: KW Hutchenson, NR Foster, eds. Innovations in Supercritical Fluids: Science and Technology. ACS Symposium Series No. 608. Washington, D.C.: American Chemical Society, 1995, pp 1–31.
8. T Clifford, K Bartle. Chemical reactions in supercritical fluids. Chem Ind (London) 1996(12):449–452, 1996.
9. M Härröd, P Møller. Hydrogenation of substrate and products manufactured according to the process. PCT Patent No. WO 96/01304, 1996.
10. M Härröd, P Møller. Hydrogenation of fats and oils at supercritical conditions. In: P Rudolf von Rohr, C Trepp, eds. Process Technology Proceedings. Vol. 12. High Pressure Chemical Engineering. Amsterdam: Elsevier, 1996, pp 43–48.
11. KH Pickel, K Steiner. Supercritical fluids solvents for reactions. In: G Brunner, M Perrut, eds. Proceedings of the 3rd International Symposium on Supercritical Fluids, Strasbourg, France, Vol. 3, 1994, pp 25–29.
12. M Poliakoff, S Howdle. Supercritical chemistry: synthesis with a spanner. Chem Br 31(2):118–121, 1995.
13. M Jansen, C Rehren. Hydrogenation of organic compounds using amorphous metal catalysts. EP Patent No. 841314, 1998.
14. SC Stinson. High times for fine chemicals. Chem Eng News 75(28):37–55, 1997.
15. M McCoy. DuPont, UNC R&D effort yields results. Chem Eng News 77(17):10, 1999.
16. JM DeSimone, Z Guan, CS Eisbernd. Synthesis of fluoropolymers in supercritical carbon dioxide. Science 257:945–947, 1992.

17. JM DeSimone. Method of making fluoropolymers. U.S. Patent No. 5,496,901, 1996.
18. JM DeSimone, T Romack. Nonaqueous polymerization of fluoromonomers. U.S. Patent No. 5,674,957, 1997.
19. FE DeBrabander, PD Brothers. Polymerization of fluoromonomers in carbon dioxide. PCT Patent No. WO 98/28351, 1998.
20. B Subramaniam, MA McHugh. Reactions in supercritical fluids—a review. Ind Eng Chem Process Des Dev 25:1–12, 1986.
21. Catalytica. Study No. 4189 NM. Catalysis in Novel Media. Vol. II. Supercritical Fluids and Cyclodextrins. Mountain View, CA: Catalytica, Inc., 1990.
22. BC Wu, SC Paspek, MT Klein, C LaMarca. Reactions in and with supercritical fluids—a review. In: TJ Bruno, JF Ely, eds. Supercritical Fluid Technology: Reviews in Modern Theory and Applications. Boca Raton, FL: CRC Press, 1991, pp 511–524.
23. AA Clifford. Reactions in supercritical fluids. In: E Kiran, JMH Levelt Sengers, eds. Supercritical Fluids: Fundamentals for Application. NATO ASI Series. Series E: Applied Sciences, Vol. 273. Dordrecht: Kluwer Academic, 1994, pp 449–479.
24. DA Morgenstern, RM LeLacheur, DK Morita, S Borkowsky, S Feng, GH Brown, L Luan, MF Gross, MJ Burk, W Tumas. Supercritical carbon dioxide as a substitute solvent for chemical synthesis and catalysis. In: PT Anastas, TC Williamson, eds. Designing Chemistry for the Environment. ACS Symposium Series No. 626. Washington, D.C.: American Chemical Society, 1996, pp 132–151.
25. M Poliakoff, SM Howdle, MW George. Clean chemistry in supercritical fluids. In: P Rudolf von Rohr, C Trepp, eds. Process Technology Proceedings, Vol. 12. High Pressure Chemical Engineering. Amsterdam: Elsevier, 1996, pp 67–72.
26. CA Eckert, BL Knutson, PG Debenedetti. Supercritical fluids as solvents for chemical and materials processing. Nature 383:313–318, 1996.
27. E Dinjus, R Fornika, M Scholz. Organic chemistry in supercritical fluids. In: R van Eldik, CD Hubbard, eds. Chemistry Under Extreme or Non-Classical Conditions. New York: Wiley and Spektrum Akademischer Verlag (co-publication), 1997, pp 219–271.
28. CM Wai, F Hunt, M Ji, X Chen. Chemical reactions in supercritical carbon dioxide. J Chem Ed 75(12):1641–1645, 1998.
29. M Poliakoff, MW George. Intermediates in organometallic and organic chemistry: spectroscopy, polymers, hydrogenation and supercritical fluids. J Phys Org Chem 11:589–596, 1998.
30. M Poliakoff, MW George, SM Howdle, VN Bagratashvili, B-X Han, H-K Yan. Supercritical fluids: clean solvents for green chemistry. Chin J Chem 17:212–222, 1999.
31. PE Savage, S Gopalan, TI Mizan, CJ Martino, EE Brock. Reactions at supercritical conditions: applications and fundamentals. AIChE J 41:1723–1778, 1995.
32. PE Savage. Catalysis in supercritical media. In: G Ertl, H Knözinger, J Weitkamp, eds. Handbook of Heterogeneous Catalysis. Vol. 3. Weinheim: VCH, 1997, pp 1339–1347.

33. A Baiker. Supercritical fluids in heterogeneous catalysis. Chem Rev 99:453–473, 1999.
34. PG Jessop. Homogeneously-catalyzed syntheses in supercritical fluids. Top Catal 5:95–103, 1998.
35. PG Jessop, T Ikariya, R Noyori. Homogeneous catalysis in supercritical fluids. Chem Rev 99:475–493, 1999.
36. PG Jessop, W Leitner, eds. Chemical synthesis using supercritical fluids. Weinheim: Wiley-VCH, 1999.
37. W Herrmann, B Cornils. Homogeneous catalysis—quo vadis? In: B Cornils, WA Herrmann, eds. Applied homogeneous catalysis with organometallic compounds. Vol. 2. Weinheim: VCH, 1996, pp 1167–1197.
38. PG Jessop, T Ikariya, R Noyori. Homogeneous catalysis in supercritical fluids. Science 269:1065–1069, 1995.
39. MA McHugh, VJ Krukonis. Supercritical Fluid Extraction: Principles and Practice. Boston: Butterworth, 1986.
40. MA McHugh, VJ Krukonis. Supercritical Fluid Extraction: Principles and Practice. 2nd ed. Boston: Butterworth-Heinemann, 1994.
41. TJ Bruno. A summary of the patent literature of supercritical fluid technology (1982–1989). In: TJ Bruno, JF Ely, eds. Supercritical Fluid Technology. Boca Raton, FL: CRC Press, 1991, pp 525–574.
42. RW Shaw, TB Brill, AA Clifford, CA Eckert, EU Franck. Supercritical water: a medium for chemistry. Chem Eng News 69(51):26–39, 1991.
43. PE Savage. Organic chemical reactions in supercritical water. Chem Rev 99:603–621, 1999.
44. KM Scholsky. Polymerization reactions at high pressure and supercritical conditions. J Supercrit Fluids 6:103–127, 1993.
45. E Kiran. Polymer formation, modifications and processing in or with supercritical fluids. In: E Kiran, JMH Levelt Sengers, eds. Supercritical Fluids: Fundamentals for Application. NATO ASI Series. Series E: Applied Sciences, Vol. 273. Dordrecht: Kluwer Academic, 1994, pp 541–588.
46. J Kendall, DA Canelas, JL Young, JM DeSimone. Polymerizations in supercritical carbon dioxide. Chem Rev 99:543–563, 1999.
47. TW Randolph, HW Blanch, DS Clark. Biocatalysis in supercritical fluids. In: JS Dordick, ed. Biocatalysis in Industry. New York: Plenum Press, 1991, pp 219–237.
48. O Aaltonen, M Rantakylä. Biocatalysis in supercritical CO_2. Chemtech 21(4): 240–248, 1991.
49. AJ Russell, EJ Beckman, AK Chaudhary. Studying enzyme activity in supercritical fluids. Chemtech 24(3):33–37, 1994.
50. K Nakamura. Biological applications of supercritical fluids. In: G Brunner, M Perrut, eds. Proceedings of the 3rd International Symposium on Supercritical Fluids, Strasbourg, France, Vol. 3, 1994, pp 121–130.
51. AJ Mesiano, EJ Beckman, AJ Russell. Supercritical biocatalysis. Chem Rev 99: 623–633, 1999.
52. SK Kumar, KP Johnston. Modelling the solubility of solids in supercritical fluids with density as the independent variable. J Supercrit Fluids 1:15–22, 1988.

53. S Angus, B Armstrong, KM de Reuck, eds. International Thermodynamic Tables of the Fluid State Carbon Dioxide. Oxford, England: Pergamon Press, 1976.
54. AG Aizpiri, A Rey, J Dávila, RG Rubio, JA Zollweg, WB Streett. Experimental and theoretical study of the equation of state of CHF_3 in the near-critical region. J Phys Chem 95:3351–3357, 1991.
55. K Reuter, S Rosenzweig, EU Franck. The static dielectric constant of CH_3F and CHF_3 to 468 K and 2000 bar. Physica A 156:294–302, 1989.
56. TA Rhodes, K O'Shea, G Bennett, KP Johnston, MA Fox. Effect of solvent–solute and solute–solute interactions on the rate of a Michael addition in supercritical fluoroform and ethane. J Phys Chem 99:9903–9908, 1995.
57. DJ Dixon, KP Johnston. Supercritical fluids. In: JI Kroschwitz, M Howe-Grant, eds. Kirk–Othmer Encyclopedia of Chemical Technology. Vol. 23. New York: Wiley, 1997, pp 452–477.
58. RC Reid, JM Prausnitz, BE Poling. The Properties of Gases and Liquids. 4th ed. New York: McGraw-Hill, 1987.
59. KP Johnston. Safer solutions for chemists. Nature 368:187–188, 1994.
60. CA Eckert, K Chandler. Tuning fluid solvents for chemical reactions. J Supercrit Fluids 13:187–195, 1998.
61. S Kim, KP Johnston. Adjustment of the selectivity of a Diels–Alder reaction network using supercritical fluids. Chem Eng Commun 63:49–59, 1988.
62. B Subramaniam, A Jooma. In situ mitigation of coke buildup in porous catalysts with supercritical reaction media: effect of feed peroxides. In: KW Hutchenson, NR Foster, eds. Innovations in Supercritical Fluids: Science and Technology. ACS Symposium Series No. 608. Washington, D.C.: American Chemical Society, 1995, pp 246–256.
63. Y Sun, RN Landau, J Wang, C LeBlond, DG Blackmond. A re-examination of pressure effects on enantioselectivity in asymmetric catalytic hydrogenation. J Am Chem Soc 118:1348–1353, 1996.
64. PG Jessop, T Ikariya, R Noyori. Homogeneous catalytic hydrogenation of supercritical carbon dioxide. Nature 368:231–233, 1994.
65. GB Combes, F Dehghani, FP Lucien, AK Dillow, NR Foster. Asymmetric catalytic hydrogenation in CO_2 expanded methanol—an application of gas anti-solvent reactions (GASR). In: MA Abraham, RP Hesketh, eds. Reaction Engineering for Pollution Prevention. New York: Elsevier Science, 1999.
66. D Kodra, V Balakotaiah. Modeling of supercritical oxidation of aqueous wastes in a deep-well reactor. AIChE J 38:988–1002, 1992.
67. DJ Bochniak, B Subramaniam. Fischer–Tropsch synthesis in near-critical n-hexane: pressure-tuning effects. AIChE J 44:1889–1896, 1998.
68. DC Wynne, PG Jessop. Cyclopropanation enantioselectivity is pressure dependent in supercritical fluoroform. Angew Chem Int Ed Engl 38:1143–1144, 1999.
69. T Aida, TG Squires. Organic chemistry in supercritical fluid solvents: photoisomerization of trans-stilbene. In: TG Squires, ME Paulaitis, eds. Supercritical Fluids: Chemical and Engineering Principles and Applications. ACS Symposium Series No. 329. Washington, D.C.: American Chemical Society, 1987, pp 58–66.
70. JF Brennecke, CA Eckert. Phase equilibria for supercritical fluid process design. AIChE J 35:1409–1427, 1989.

71. MP Ekart, KL Bennett, SM Ekart, GS Gurdial, CL Liotta, CA Eckert. Cosolvent interactions in supercritical fluid solutions. AIChE J 39:235–248, 1993.
72. CJ Peters. Multiphase equilibria in near-critical solvents. In: E Kiran, JMH Levelt Sengers, eds. Supercritical Fluids: Fundamentals for Application. NATO ASI Series. Series E: Applied Sciences, Vol. 273. Dordrecht, The Netherlands: Kluwer Academic, 1994, pp 117–145.
73. GM Schneider. Physico-chemical properties and phase equilibria of pure fluids and fluid mixtures at high pressures. In: E Kiran, JMH Levelt Sengers, eds. Supercritical Fluids: Fundamentals for Application. NATO ASI Series. Series E: Applied Sciences, Vol. 273. Dordrecht, The Netherlands: Kluwer Academic, 1994, pp 91–115.
74. JM Smith, HC Van Ness, MM Abbott. Introduction to Chemical Engineering Thermodynamics. 5th ed. New York: McGraw-Hill, 1996.
75. JM Prausnitz, RN Lichtenthaler, EG de Azevedo. Molecular Thermodynamics of Fluid-Phase Equilibria. 3rd ed. Englewood Cliffs, NJ: Prentice-Hall, 1999.
76. SM Walas. Phase Equilibria in Chemical Engineering. Boston: Butterworth, 1985.
77. FP Lucien, NR Foster. Phase behavior and solubility. In: PG Jessop, W Leitner, eds. Chemical Synthesis Using Supercritical Fluids. Weinheim: Wiley-VCH, 1999, pp 37–53.
78. KP Johnston, C Haynes. Extreme solvent effects on reaction rate constants at supercritical fluid conditions. AIChE J 33:2017–2026, 1987.
79. BC Wu, MT Klein, SI Sandler. Solvent effects on reactions in supercritical fluids. Ind Eng Chem Res 30:822–828, 1991.
80. M Buback. Kinetics and selectivity of chemical processes in fluid phases. In: E Kiran, JMH Levelt Sengers, eds. Supercritical Fluids: Fundamentals for Application. NATO ASI Series. Series E: Applied Sciences, Vol. 273. Dordrecht, The Netherlands: Kluwer Academic, 1994, pp 481–497.
81. JR Combes, KP Johnston, KE O'Shea, MA Fox. Influence of solvent-solute and solute-solute clustering on chemical reactions in supercritical fluids. In: FV Bright, MEP McNally, eds. Supercritical Fluid Technology: Theoretical and Applied Approaches to Analytical Chemistry. ACS Symposium Series No. 488. Washington, D.C.: American Chemical Society, 1992, pp 31–47.
82. AK Dillow, KP Hafner, SLJ Yun, F Deng, SG Kazarian, CL Liotta, CA Eckert. Cosolvent tuning of tautomeric equilibrium in supercritical fluids. AIChE J 43: 515–524, 1997.
83. R van Eldik, T Asano, WJ le Noble. Activation and reaction volumes in solution. 2. Chem Rev 89:549–688, 1989.
84. CA Eckert. High pressure kinetics in solution. Annu Rev Phys Chem 23:239–264, 1972.
85. F-G Klärner, MK Diedrich, AE Wigger. Effect of pressure on organic reactions. In: R van Eldik, CD Hubbard, eds. Chemistry Under Extreme or Non-Classical Conditions. New York: Wiley and Spektrum Akademischer Verlag (co-publication), 1997, pp 103–161.
86. JF Brennecke. Spectroscopic investigations of reactions in supercritical fluids: a review. In: E Kiran, JF Brennecke, eds. Supercritical Fluid Engineering Science.

ACS Symposium Series No. 514. Washington, D.C.: American Chemical Society, 1993, pp 201–219.
87. MHM Caralp, AA Clifford, SE Coleby. Other uses for near-critical solvents: chemical reaction and recrystallisation in near-critical solvents. In: MB King, TR Bott, eds. Extraction of Natural Products Using Near-Critical Solvents. Glassgow, UK: Chapman & Hall, 1993, pp 50–83.
88. MG Evans, M Polanyi. Some applications of the transition state method to the calculation of reaction velocities, especially in solution. Trans Faraday Soc 31: 875–894, 1935.
89. SW Benson. Thermochemical Kinetics. 2nd ed. New York: Wiley-Interscience, 1976.
90. JB Ellington, KM Park, JF Brennecke. Effect of local composition enhancements on the esterification of phthalic anhydride with methanol in supercritical carbon dioxide. Ind Eng Chem Res 33:965–974, 1994.
91. SD Hamann. Chemical kinetics. In: RS Bradley, ed. High Pressure Physics and Chemistry. Vol. 2. London: Academic Press, 1963, pp 163–207.
92. ME Paulaitis, GC Alexander. Reactions in supercritical fluids. A case study of the thermodynamic solvent effects on a Diels–Alder reaction in supercritical carbon dioxide. Pure Appl Chem 59:61–68, 1987.
93. O Reiser. Application of high pressure in catalysis. In: M Beller, C Bolm, eds. Transition Metals for Organic Synthesis: Building Blocks and Fine Chemicals. Vol. 2. Weinheim: Wiley-VCH, 1998, pp 442–449.
94. CB Roberts, JE Chateauneuf, JF Brennecke. Unique pressure effects on the absolute kinetics of triplet benzophenone photoreduction in supercritical CO_2. J Am Chem Soc 114:8455–8463, 1992.
95. D-Y Peng, DB Robinson. A new two constant equation of state. Ind Eng Chem Fundam 15:59–64, 1976.
96. A Drljaca, CD Hubbard, R van Eldik, T Asano, MV Basilevsky, WJ le Noble. Activation and reaction volumes in solution. 3. Chem Rev 98:2167–2289, 1998.
97. Q Ji, EM Eyring, R van Eldik, KP Johnston, SR Goates, ML Lee. Laser flash photolysis studies of metal carbonyls in supercritical CO_2 and ethane. J Phys Chem 99:13461–13466, 1995.
98. Q Ji, CR Lloyd, EM Eyring, R van Eldik. Probing the solvation properties of liquid versus supercritical fluids with laser flash photolysis of $W(CO)_6$ in the presence of 2,2′-bypyridine. J Phys Chem 101:243–247, 1997.
99. CA Eckert, DH Ziger, KP Johnston, S Kim. Solute partial molal volumes in supercritical fluids. J Phys Chem 90:2738–2746, 1986.
100. S Kim, KP Johnston. Clustering in supercritical fluid mixtures. AIChE J 33:1603–1611, 1987.
101. JF Brennecke, PG Debenedetti, CA Eckert, KP Johnston. Letter to the editor. AIChE J 36:1927–1928, 1990.
102. R-S Wu, LL Lee, HD Cochran. Structure of dilute supercritical solutions: clustering of solvent and solute molecules and the thermodynamic effects. Ind Eng Chem Res 29:977–987, 1990.
103. TW Randolph, C Carlier. Free-radical reactions in supercritical ethane: a probe of supercritical fluid structure. J Phys Chem 96:5146–5151, 1992.

104. CB Roberts, J Zhang, JF Brennecke, JE Chateauneuf. Laser flash photolysis investigation of diffusion-controlled reactions in supercritical fluids. J Phys Chem 97:5618–5623, 1993.
105. CB Roberts, J Zhang, JE Chateauneuf, JF Brennecke. Diffusion-controlled reactions in supercritical CHF_3 and CO_2/acetonitrile mixtures. J Am Chem Soc 115: 9576–9582, 1993.
106. CA Eckert, B Knutson. Molecular charisma in supercritical fluids. Fluid Phase Equilib 83:93–100, 1993.
107. C Carlier, TW Randolph. Dense-gas solvent-solute clusters at near-infinite dilution: EPR spectroscopic evidence. AIChE J 39:876–884, 1993.
108. JA O'Brien, TW Randolph, C Carlier, S Ganapathy. Quasicritical behavior of dense-gas solvent-solute clusters at near-infinite dilution. AIChE J 39:1061–1071, 1993.
109. TW Randolph, JA O'Brien, S Ganapathy. Does critical clustering affect reaction rate constants? Molecular dynamics studies in pure supercritical fluids. J Phys Chem 98:4173–4179, 1994.
110. CB Roberts, JF Brennecke, JE Chateauneuf. Solvation effects on reactions of triplet benzophenone in supercritical fluids. AIChE J 41:1306–1318, 1995.
111. D Andrew, BT Des Islet, A Margaritis, AC Weedon. Photo-Fries rearrangement of naphthyl acetate in supercritical carbon dioxide: chemical evidence for solvent-solute clustering. J Am Chem Soc 117:6132–6133, 1995.
112. S Ganapathy, C Carlier, TW Randolph, JA O'Brien. Influence of local structural correlations on free-radical reactions in supercritical fluids: a hierarchical approach. Ind Eng Chem Res 35:19–27, 1996.
113. J Zhang, DP Roek, JE Chateauneuf, JF Brennecke. A steady-state and time-resolved fluorescence study of quenching reactions of anthracene and 1,2-benzanthracene by carbon tetrabromide and bromoethane in supercritical carbon dioxide. J Am Chem Soc 119:9980–9991, 1997.
114. C Bunker, Y-P Sun, JR Gord. Time-resolved studies of fluorescence quenching in supercritical carbon dioxide: system dependence in the enhancement of bimolecular reaction rates at near-critical densities. J Phys Chem A 101:9233–9239, 1997.
115. K Takahashi, CD Jonah. The measurement of an electron transfer reaction in a non-polar supercritical fluid. Chem Phys Lett 264:297–302, 1997.
116. K Takahashi, K Abe, S Sawamura, CD Jonah. Spectroscopic study of 4-aminobenzophenone in supercritical CF_3H and CO_2: local density and Onsager's reaction cavity radius. Chem Phys Lett 282:361–368, 1998.
117. SC Tucker, MW Maddox. The effect of solvent density inhomogeneities on solute dynamics in supercritical fluids: a theoretical perspective. J Phys Chem B 102: 2437–2453, 1998.
118. B Fletcher, NK Suleman, JM Tanko. Free radical chlorination of alkanes in supercritical carbon dioxide: the chlorine atom cage effect as a probe for enhanced cage effects in supercritical fluid solvents. J Am Chem Soc 120:11839–11844, 1998.
119. JL deGrazia, TW Randolph, JA O'Brien. Rotational relaxation in supercritical carbon dioxide revisited: a study of solute-induced local density augmentation. J Phys Chem A 102:1674–1681, 1998.

120. M Khajehpour, JF Kauffman. Chain relaxation kinetics in the intramolecular charge transfer reaction of 1-(9-anthryl)-3-(4-dimethylaniline)propane in supercritical CO_2: bulk solvent properties characterize reactivity. Chem Phys Lett 297: 141–146, 1998.
121. DP Roek, MJ Kremer, C Roberts, JE Chateauneuf, JF Brennecke. Spectroscopic studies of solvent effects on reactions in supercritical fluids. Fluid Phase Equilib 158–160:713–722, 1999.
122. JF Brennecke, JE Chateauneuf. Homogeneous organic reactions as mechanistic probes in supercritical fluids. Chem Rev 99:433–452, 1999.
123. SC Tucker. Solvent density inhomogeneities in supercritical fluids. Chem Rev 99: 391–418, 1999.
124. RA Ferrieri, I Garcia, JS Fowler, AP Wolf. investigations of acetonitrile solvent cluster formation in supercritical carbon dioxide, and its impact on microscale syntheses of carbon-11-labeled radiotracers for PET. Nucl Med Biol 26:443–454, 1999.
125. CA Eckert, DH Ziger, KP Johnston, TK Ellison. The use of partial molal volume data to evaluate equations of state for supercritical fluid mixtures. Fluid Phase Equilib 14:167–175, 1983.
126. AA Chialvo, PG Debenedetti. Molecular dynamics study of solute-solute microstructure in attractive and repulsive supercritical mixtures. Ind Eng Chem Res 31:1391–1397, 1992.
127. JW Tom, PG Debenedetti. Integral equation study of microstructure and solvation in model attractive and repulsive supercritical mixtures. Ind Eng Chem Res 32: 2118–2128, 1993.
128. CB Roberts, J Zhang, JE Chateauneuf, JF Brennecke. Laser flash photolysis and integral equation theory to investigate reactions of dilute solutes with oxygen in supercritical fluids. J Am Chem Soc 117:6553–6560, 1995.
129. S Ganapathy, JA O'Brien, TW Randolph. Effect of solute–solute correlations on rapid reactions in supercritical fluids. AIChE J 41:346–356, 1995.
130. S Ganapathy, JA O'Brien, TW Randolph. Do solute–solute interactions affect activation-limited reactions? A Brownian dynamics study. J Supercrit Fluids 9: 51–55, 1996.
131. CR Yonker, RD Smith. Solvatochromic behavior of binary supercritical fluids: the carbon dioxide/2-propanol system. J Phys Chem 92:2374–2378, 1988.
132. IB Petsche, PG Debenedetti. Solute–solvent interactions in infinitely dilute supercritical mixtures: a molecular dynamics investigation. J Chem Phys 91:7075–7084, 1989.
133. Y Ikushima, N Saito, M Arai. High-pressure Fourier transform infrared spectroscopy study of the Diels–Alder reaction of isoprene and maleic anhydride in supercritical carbon dioxide. Bull Chem Soc Jpn 64:282–284, 1991.
134. Y Ikushima, N Saito, M Arai. Supercritical carbon dioxide as reaction medium: examination of its solvent effects in the near-critical region. J Phys Chem 96: 2293–2297, 1992.
135. A Renslo, RD Weinstein, JW Tester, RL Danheiser. Concerning the regiochemical course of the Diels–Alder reaction in supercritical carbon dioxide. J Org Chem 62:4530–4533, 1997.

136. JM Tanko, NK Suleman, B Fletcher. Viscosity-dependent behavior of geminate caged-pairs in supercritical fluid solvent. J Am Chem Soc 118:11958–11959, 1996.
137. BC Wu, MT Klein, SI Sandler. Influence of supercritical fluid solvent density on benzyl phenyl ether pyrolysis: indications of diffusional limitations. Energy & Fuels 5:453–458, 1991.
138. ET Sundquist. The global carbon dioxide budget. Science 259:934–941, 1993.
139. A Behr. Carbon dioxide activation by metal complexes. Weinheim: VCH, 1988.
140. A Behr. Carbon dioxide as an alternative C_1 synthetic unit: activation by transition-metal complexes. Angew Chem Int Ed Engl 27:661–678, 1988.
141. W Leitner. Carbon dioxide as a raw material: the synthesis of formic acid and its derivatives from CO_2. Angew Chem Int Ed Engl 34:2207–2221, 1995.
142. A Baiker, RA Koeppel. Chemicals from CO_2 by heterogeneous catalytic routes. Proceedings of the Third International Conference on Carbon Dioxide Utilization, Norman, OK, 1995.
143. PG Jessop, T Ikariya, R Noyori. Homogeneous hydrogenation of carbon dioxide. Chem Rev 95:259–272, 1995.
144. E Dinjus, R Fornika. Carbon dioxide as a C_1-building block. In: B Cornils, WA Herrmann, eds. Applied Homogeneous Catalysis with Organometallic Compounds. Vol. 2. Weinheim: VCH, 1996, pp 1048–1072.
145. W Leitner. The coordination chemistry of carbon dioxide and its relevance for catalysis: a critical survey. Coord Chem Rev 153:257–284, 1996.
146. PG Jessop, Y Hsiao, T Ikariya, R Noyori. Homogeneous catalysis in supercritical fluids: hydrogenation of supercritical carbon dioxide to formic acid, alkyl formates, and formamides. J Am Chem Soc 118:344–355, 1996.
147. DH Gibson. The organometallic chemistry of carbon dioxide. Chem Rev 96:2063–2095, 1996.
148. A-AG Shaikh, S Sivaram. Organic carbonates. Chem Rev 96:951–976, 1996.
149. M Aresta, E Quaranta. Carbon dioxide: a substitute for phosgene. Chemtech 27(3): 32–40, 1997.
150. T Sakakura, Y Saito, M Okano, J-C Choi, T Sako. Selective conversion of carbon dioxide to dimethyl carbonate by molecular catalysis. J Org Chem 63:7095–7096, 1998.
151. MT Reetz, W Könen, T Strack. Supercritical carbon dioxide as a reaction medium and reaction partner. Chimia 47:493, 1993.
152. T Ikariya, Y Hsiao, PG Jessop, R Noyori. A method for producing formic acid or its derivatives. EP Patent No. 652202, 1995.
153. T Ikariya, PG Jessop, R Noyori. Method for producing formic acid or its derivatives. US Patent No. 5,639,910, 1997.
154. T Ikariya, PG Jessop, Y Hsiao, R Noyori. Method for producing formic acid or its derivatives. US Patent No. 5,763,662, 1998.
155. F Gassner, W Leitner. Hydrogenation of carbon dioxide to formic acid using water-soluble rhodium catalysts. J Chem Soc, Chem Commun 1993:1465–1466.
156. CS Pomelli, J Tomasi, M Solà. Theoretical study on the thermodynamics of the elimination of formic acid in the last step of the hydrogenation of CO_2 catalyzed by rhodium complexes in the gas phase and supercritical CO_2. Organometallics 17:3164–3168, 1998.

157. PG Jessop, Y Hsiao, T Ikariya, R Noyori. Methyl formate synthesis by hydrogenation of supercritical carbon dioxide in the presence of methanol. J Chem Soc, Chem Commun 1995:707–708.
158. O Kröcher, RA Köppel, M Fröba, A Baiker. Silica hybrid gel catalysts containing group(viii) transition metal complexes: preparation, structural, and catalytic properties in the synthesis of N,N-dimethylformamide and methyl formate from supercritical carbon dioxide. J Catal 178:284–298, 1998.
159. PG Jessop, Y Hsiao, T Ikariya, R Noyori. Catalytic production of dimethylformamide from supercritical carbon dioxide. J Am Chem Soc 116:8851–8852, 1994.
160. O Köcher, RA Köppel, A Baiker. Sol-gel derived hybrid materials as heterogeneous catalysts for the synthesis of N,N-dimethylformamide from supercritical carbon dioxide. Chem Commun 1996:1497–1498.
161. O Kröcher, RA Köppel, A Baiker. Synthesis of N,N-dimethylformamide by heterogeneous catalytic hydrogenation of supercritical carbon dioxide. In: P Rudolf von Rohr, C Trepp, eds. Process Technology Proceedings. Vol. 12. High Pressure Chemical Engineering. Amsterdam: Elsevier, 1996, pp 91–96.
162. O Kröcher, RA Köppel, A Baiker. Highly active ruthenium complexes with bidentate phosphine ligands for the solvent-free catalytic synthesis of N,N-dimethylformamide and methyl formate. Chem Commun 1997:453–454.
163. O Kröcher, RA Köppel, A Baiker. Novel homogeneous and heterogeneous catalysts for the synthesis of formic-acid derivatives from CO_2. Chimia 51:48–51, 1997.
164. A new route to dimethyl carbonate is phosgene-free. Chem Eng 106(6):23, 1999.
165. C Vieville, JW Yoo, S Pelet, Z Mouloungui. Synthesis of glycerol carbonate by direct carbonatation of glycerol in supercritical CO_2 in the presence of zeolites and ion exchange resins. Catal Lett 56:245–247, 1998.
166. T Mizuno, H Tsutsumi, K Ohta, A Saji, H Noda. Photocatalytic reduction of CO_2 with dispersed TiO_2/Cu powder mixtures in supercritical CO_2. Chem Lett 1994: 1533–1536, 1994.
167. S Kaneco, H Kurimoto, Y Shimizu, K Ohta, T Mizuno. Photocatalytic reduction of CO_2 using TiO_2 powders in supercritical fluid CO_2. Energy 24:21–30, 1999.
168. RL Augustine. Heterogeneous Catalysis for the Synthetic Chemist. New York: Marcel Dekker, 1996.
169. Y Sun, C LeBlond, J Wang, DG Blackmond, J Laquidara, JR Sowa Jr. Observation of a $[RuCl_2((S)\text{-}(—)\text{-tol-binap})]_2 \cdot N(C_2H_5)_3$-catalyzed isomerization hydrogenation network. J Am Chem Soc 117:12647–12648, 1995.
170. CY Tsang, WB Streett. Phase equilibria in the H_2/CO_2 system at temperatures from 220 to 290 K and pressures to 172 MPa. Chem Eng Sci 36:993–1000, 1981.
171. T Tacke, S Wieland, P Panster. Hardening of fats and oils in supercritical CO_2. In: P Rudolf von Rohr, C Trepp, eds. Process Technology Proceedings. Vol. 12. High Pressure Chemical Engineering. Amsterdam: Elsevier, 1996, pp 17–21.
172. T Tacke, C Rehren, S Wieland, P Panster, SK Ross, J Toler, MG Hitzler, F Smail, M Poliakoff. Continuous hydrogenation in supercritical fluids. In: FE Herkes, ed. Catalysis of Organic Reactions. New York: Marcel Dekker, 1998, pp 345–356.
173. T Tacke, S Wieland, P Panster, M Bankmann, R Brand, H Mägerlein. Hardening of unsaturated fats, fatty acids or fatty acid esters. U.S. Patent No. 5,734,070, 1998.

174. AG Zwahlen, A Bertucco. CSTR-system for kinetic investigation for hydrogenation reactions. In: P Rudolf von Rohr, C Trepp, eds. Process Technology Proceedings. Vol. 12. High Pressure Chemical Engineering. Amsterdam: Elsevier, 1996, pp 37–42.
175. A Bertucco, P Canu, L Devetta, AG Zwahlen. Catalytic hydrogenation in supercritical CO_2: kinetic measurements in a gradientless internal-recycle reactor. Ind Eng Chem Res 36:2626–2633, 1997.
176. AG Zwahlen, JB Agnew. Isobutane dehydrogenation kinetics determination in a modified berty gradientless reactor. Ind Eng Chem Res 31:2088–2093, 1992.
177. L Devetta, P Canu, A Bertucco, K Steiner. Modelling of a trickle-bed reactor for a catalytic hydrogenation in supercritical CO_2. Chem Eng Sci 52:4163–4169, 1997.
178. L Devetta, A Giovanzana, P Canu, A Bertucco, BJ Minder. Kinetic experiments and modeling of a three-phase catalytic hydrogenation reaction in supercritical CO_2. Catal Today 48:337–345, 1999.
179. MG Hitzler, M Poliakoff. Continuous hydrogenation of organic compounds in supercritical fluids. Chem Commun 1997:1667–1668.
180. MG Hitzler, FR Smail, SK Ross, M Poliakoff. Selective catalytic hydrogenation of organic compounds in supercritical fluids as a continuous process. Org Proc Res Dev 2:137–146, 1998.
181. M Poliakoff, T Tacke, MG Hitzler, SK Ross, S Wieland. Supercritical hydrogenation. PCT Patent No. WO 97/38955, 1997.
182. B Subramaniam, S Saim. In situ mitigation of coke buildup in porous catalysts with supercritical reaction media. PCT Patent No. WO 96/33148, 1996.
183. B Subramaniam, S Saim. In situ mitigation of coke buildup in porous catalysts with supercritical reaction media. U.S. Patent No. 5,725,756, 1998.
184. MJ Burk, S Feng, MF Gross, W Tumas. Asymmetric catalytic hydrogenation reactions in supercritical CO_2. J Am Chem Soc 117:8277–8278, 1995.
185. J Xiao, SCA Nefkens, PG Jessop, T Ikariya, R Noyori. Asymmetric hydrogenation of α,β-unsaturated carboxylic acids in supercritical carbon dioxide. Tetrahedron Lett 37:2813–2816, 1996.
186. T Ohta, H Takaya, M Kitamura, K Nagai, R Noyori. Asymmetric hydrogenation of unsaturated carboxylic acids catalyzed by BINAP–ruthenium(II) complexes. J Org Chem 52:3174–3176, 1987.
187. S Kainz, W Leitner. Catalytic asymmetric hydroformylation in the presence of compressed carbon dioxide. Catal. Lett. 55:223–225, 1998.
188. S Kainz, A Brinkmann, W Leitner, A Pfaltz. Iridium-catalyzed enantioselective hydrogenation of imines in supercritical carbon dioxide. J Am Chem Soc 121: 6421–6429, 1999.
189. B Minder, T Mallat, KH Pickel, K Steiner, A Baiker. Enantioselective hydrogenation of ethyl pyruvate in supercritical fluids. Catal Lett 34:1–9, 1995.
190. B Minder, T Mallat, A Baiker. Enantioselective hydrogenation in supercritical fluids. Limitations of the use of supercritical CO_2. In: P Rudolf von Rohr, C Trepp, eds. Process Technology Proceedings, Vol. 12. High Pressure Chemical Engineering. Amsterdam: Elsevier, 1996, pp 139–144.
191. RB Anderson. The Fischer–Tropsch Synthesis. Orlando, FL: Academic Press, 1984.

192. K Yokota, K Fujimoto. Supercritical phase Fischer–Tropsch synthesis reaction. Fuel 68:255–256, 1989.
193. K Yokota, Y Hanakata, K Fujimoto. Supercritical phase Fischer–Tropsch synthesis. Chem Eng Sci 45:2743–2750, 1990.
194. K Yokota, Y Hanakata, K Fujimoto. Supercritical phase Fischer–Tropsch synthesis reaction. In: A Holmen, K-J Jens, S Kolboe, eds. Studies in Surface Science and Catalysis. Vol. 61. Natural Gas Conversion. Amsterdam: Elsevier, 1991, pp 289–295.
195. L Fan, K Yokota, K Fujimoto. Supercritical phase Fischer–Tropsch synthesis: catalyst pore-size effect. AIChE J 38:1639–1648, 1992.
196. K Yokota, K Fujimoto. Supercritical-phase Fischer–Tropsch synthesis reaction. 2. The effective diffusion of reactant and products in the supercritical-phase reaction. Ind Eng Chem Res 30:95–100, 1991.
197. K Yokota, Y Hanakata, K Fujimoto. Supercritical phase Fischer–Tropsch synthesis reaction. 3. Extraction capability of supercritical fluids. Fuel 70:989–994, 1991.
198. K Fujimoto, L Fan, K Yoshii. New controlling method for product distribution in Fischer–Tropsch synthesis reaction. Top Catal 2:259–266, 1995.
199. K Fujimoto, L Fan, K Yoshii, S Yan, J Zhou. Supercritical phase process for selective synthesis of wax from syngas. Prepr—ACS Div Petrol Chem 41(1):262–265, 1996.
200. L Fan, K Yoshii, S Yan, J Zhou, K Fujimoto. Supercritical-phase process for selective synthesis of wax from syngas: catalyst and process development. Catal Today 36:295–304, 1997.
201. L Fan, K Yokota, K Fujimoto. Characterization of mass transfer in supercritical-phase Fischer–Tropsch synthesis reaction. Top Catal 2:267–283, 1995.
202. L Fan, K Yoshii, K Fujimoto. Selective synthesis of wax from syngas by supercritical phase process: catalyst development. Sekiyu Gakkaishi 40:433–437, 1997.
203. S-R Yan, L Fan, Z-X Zhang, J-L Zhou, K Fujimoto. α-Tetradecene cofeeding effect on the supercritical-phase Fischer–Tropsch synthesis over Co/SiO_2 catalysts. J Nat Gas Chem 7:127–133, 1998.
204. S-R Yan, L Fan, Z-X Zhang, J-L Zhou, K Fujimoto. Supercritical-phase process for selective synthesis of heavier hydrocarbons from syngas over a Ru/Al_2O_3 catalyst. J Nat Gas Chem 7:229–235, 1998.
205. X Lang, A Akgerman, DB Bukur. Steady state Fischer–Tropsch synthesis in supercritical propane. Ind Eng Chem Res 34:72–77, 1995.
206. DB Bukur, X Lang, A Akgerman, Z Feng. Effect of process conditions on olefin selectivity during conventional and supercritical Fischer–Tropsch synthesis. Ind Eng Chem Res 36:2580–2587, 1997.
207. DB Bukur, X Lang, Z Feng. Alpha-olefin selectivity during conventional and supercritical Fischer–Tropsch synthesis. Prepr—ACS Div Fuel Chem 42(2):632–636, 1997.
208. BC Gates. Catalytic Chemistry. New York: Wiley, 1992.
209. GW Parshall, SD Ittel. Homogeneous Catalysis. 2nd ed. New York: Wiley, 1992.
210. CY Tsang, WB Streett. Phase equilibria in the H_2-CO system at temperatures from 70 to 125 K and pressures to 53 MPa. Fluid Phase Equilib 6:261–273, 1981.

211. JW Rathke, RJ Klingler, TR Krause. Propylene hydroformylation in supercritical carbon dioxide. Organometallics 10:1350–1355, 1991.
212. JW Rathke, RJ Klingler, TR Krause. Thermodynamics for the hydrogenation of dicobalt octacarbonyl in supercritical carbon dioxide. Organometallics 11:585–588, 1992.
213. JW Rathke, RJ Klingler. Cobalt carbonyl catalyzed olefin hydroformylation in supercritical carbon dioxide. U.S. Patent No. 5,198,589, 1993.
214. R Fowler, H Connor, RA Baehl. Hydroformylate propylene at low pressure. Hydrocarbon Proc 55(9):247–249, 1976.
215. RJ Klingler, JW Rathke. High-pressure NMR investigation of hydrogen atom transfer and related dynamic processes in oxo catalysis. J Am Chem Soc 116: 4772–4785, 1994.
216. RF Heck, DS Breslow. The reaction of cobalt hydrotetracarbonyl with olefins. J Am Chem Soc 83:4023–4027, 1961.
217. B Cornils. Hydroformylation, oxo synthesis, Roelen reaction. In: J Falbe, ed. New Synthesis with Carbon Monoxide. Berlin: Springer-Verlag, 1980, pp 16–17.
218. KW Kramarz, RJ Klingler, DE Fremgen, JW Rathke. Toroid NMR probes for the in situ examination of homogeneous cobalt hydroformylation catalysts at high pressures and temperatures. Catal Today 49:339–352, 1999.
219. PG Jessop, T Ikariya, R Noyori. Selectivity for hydrogenation or hydroformylation of olefins by hydridopentacarbonylmanganese(I) in supercritical carbon dioxide. Organometallics 14:1510–1513, 1995.
220. TE Nalesnik, JH Freudenberger, M Orchin. Radical hydroformylation and hydrogenation of cyclopropenes with $HCo(CO)_4$ and $HMn(CO)_5$. J Organomet Chem 236:95–100, 1982.
221. Y Matsui, M Orchin. Olefin reactions with $HMn(CO)_5$: product selectivity by micelle sequestering. J Organomet Chem 244:369–373, 1983.
222. TE Nalesnik, M Orchin. Stoichiometric hydroformylation with $HMn(CO)_5$. J Organomet Chem 222:C5–C8, 1981.
223. Y Guo, A Akgerman. Hydroformylation of propylene in supercritical carbon dioxide. Ind Eng Chem Res 36:4581–4585, 1997.
224. Y Guo, A Akgerman. Determination of selectivity for parallel reactions in supercritical fluids. J Supercrit Fluids 15:63–71, 1999.
225. I Wender, P Pino. Organic Syntheses via Metal Carbonyls. New York: Wiley, 1977.
226. S Kainz, D Koch, W Baumann, W Leitner. Perfluoroalkyl-substituted arylphosphanes as ligands for homogeneous catalysis in supercritical carbon dioxide. Angew Chem Int Ed Engl 36:1628–1630, 1997.
227. D Koch, W Leitner. Rhodium-catalyzed hydroformylation in supercritical carbon dioxide. J Am Chem Soc 120:13398–13404, 1998.
228. W Leitner, D Koch. Hydroformylation with unmodified rhodium catalysts in supercritical carbon dioxide. PCT Patent No. WO 99/03810, 1999.
229. W Leitner, S Kainz, D Koch, K Wittmann, C Six. Use of perfluoroalkyl substituted phosphorus compounds as ligands for homogeneous catalysis in supercritical carbon dioxide. PCT Patent No. WO 98/32533, 1998.

230. DR Palo, C Erkey. Homogeneous catalytic hydroformylation of 1-octene in supercritical carbon dioxide using a novel rhodium catalyst with fluorinated arylphosphine ligands. Ind Eng Chem Res 37:4203–4206, 1998.
231. DR Palo, C Erkey. Homogeneous hydroformylation of 1-octene in supercritical carbon dioxide with $[RhH(CO)(P(p\text{-}CF_3C_6H_4)_3)_3]$. Ind Eng Chem Res 38:2163–2165, 1999.
232. B Cornils, WA Herrmann, eds. Applied Homogeneous Catalysis with Organometallic Compounds. Vols. 1&2. Weinheim: VCH, 1996.
233. CK Brown, G Wilkinson. Homogeneous hydroformylation of alkenes with hydridocarbonyltris(triphenylphosphine)rhodium(I) as catalyst. J Chem Soc A 1970: 2753–2764, 1970.
234. DR Palo, C Erkey. Kinetics of the homogeneous catalytic hydroformylation of 1-octene in supercritical carbon dioxide with $HRh(CO)[P(p\text{-}CF_3C_6H_4)_3]_3$. Ind Eng Chem Res 38:3786–3792, 1999.
235. I Bach, DJ Cole-Hamilton. Hydroformylation of hex-1-ene in supercritical carbon dioxide catalysed by rhodium trialkylphosphine complexes. Chem Commun 1998: 1463–1464.
236. I Ojima, M Tzamarioudaki, C-Y Chuang, DM Iula, Z Li. Catalytic carbonylations in supercritical carbon dioxide. In: FE Herkes, ed. Catalysis of Organic Reactions. New York: Marcel Dekker, 1998, pp 333–343.
237. ZY Ding, MA Frisch, L Li, EF Gloyna. Catalytic oxidation in supercritical water. Ind Eng Chem Res 35:3257–3279, 1996.
238. ER Birnbaum, RM Le Lacheur, AC Horton, W Tumas. Metalloporphyrin-catalyzed homogeneous oxidation in supercritical carbon dioxide. J Mol Cat A 139:11–24, 1999.
239. KM Dooley, FC Knopf. Oxidation catalysis in a supercritical fluid medium. Ind Eng Chem Res 26:1910–1916, 1987.
240. FC Knopf, T-H Pang, KM Dooley. Catalyzed reactions of alkylaromatic hydrocarbons dissolved in supercritical fluids. Prepr—ACS Div Fuel Chem 32(3):359–367, 1987.
241. RN Occhiogrosso, MA McHugh. Critical-mixture oxidation of cumene. Chem Eng Sci 42:2478–2481, 1987.
242. GJ Suppes, RN Occhiogrosso, MA McHugh. Oxidation of cumene in supercritical reaction media. Ind Eng Chem Res 28:1152–1156, 1989.
243. AM Gaffney, JA Sofranko. Selective oxidation of propylene to propylene glycol in supercritical media. Prepr—ACS Div Petrol Chem 37(4):1273–1279, 1992.
244. AM Gaffney, JA Sofranko. Oxidation process. EP Patent No. 0,385,631, 1990.
245. AM Gaffney, JA Sofranko. Oxidation of olefin to glycol. US Patent No. 5,210,336, 1993.
246. P Srinivas, M Mukhopadhyay. Oxidation of cyclohexane in supercritical carbon dioxide medium. Ind Eng Chem Res 33:3118–3124, 1994.
247. M Mukhopadhyay, P Srinivas. Multicomponent solubilities of reactants and products of cyclohexane oxidation in supercritical carbon dioxide. Ind Eng Chem Res 35:4713–4717, 1996.

248. P Srinivas, M Mukhopadhyay. Influence of the thermodynamic state on cyclohexane oxidation kinetics in carbon dioxide medium. Ind Eng Chem Res 36: 2066–2074, 1997.
249. X-W Wu, Y Oshima, S Koda. Aerobic oxidation of cyclohexane catalyzed by Fe(III)(5,10,15,20-tetrakis(pentafluorophenyl)porphyrin)Cl in sub- and supercritical CO_2. Chem Lett 1997:1045–1046, 1997.
250. S-I Murahashi, T Naota, N Komiya. Metalloporphyrin-catalyzed oxidation of alkanes with molecular oxygen in the presence of acetaldehyde. Tetrahedron Lett 36: 8059–8062, 1995.
251. DR Pesiri, DK Morita, W Glaze, W Tumas. Selective epoxidation in dense phase carbon dioxide. Chem Commun 1998:1015–1016.
252. L Fan, T Watanabe, K Fujimoto. Reaction phase effect on tertiary-butyl alcohol synthesis by air oxidation of isobutane. Appl Catal A 158:L41–L46, 1997.
253. L Fan, Y Nakayama, K Fujimoto. Air oxidation of supercritical phase isobutane to tert-butyl alcohol. Chem Commun 1997:1179–1180.
254. L Fan, T Watanabe, K Fujimoto. Supercritical-phase oxidation of isobutane to t-butanol by air. In: A Parmaliana, ed. Natural Gas Conversion V. Studies in Surface Science and Catalysis Vol. 119. Amsterdam: Elsevier Science, 1998, pp 581–586.
255. AK Suresh. Isobutane oxidation in the liquid and supercritical phases: comparison of features. J Supercrit Fluids 12:165–176, 1998.
256. EG Foster. Process for oxidation of isobutane. US Patent No. 4,408,081, 1983.
257. HJ Baumgartner. Oxidation of isobutane in the dense phase and at low oxygen concentration. US Patent No. 4,408,082, 1983.
258. EF Lutz, EG Foster. Oxidation of isobutane under super-critical conditions. US Patent No. 4,404,406, 1983.
259. AK Suresh, T Sridhar, OE Potter. Autocatalytic oxidation of cyclohexane—modeling reaction kinetics. AIChE J 34:69–80, 1988.
260. C-T Wang, RJ Willey. Oxidation of methanol over iron oxide based aerogels in supercritical CO_2. J Non-Cryst Solids 225:173–177, 1998.
261. GR Haas, JW Kolis. The diastereoselective epoxidation of olefins in supercritical carbon dioxide. Tetrahedron Lett 39:5923–5926, 1998.
262. GR Haas, JW Kolis. Enantioselective epoxidation of olefins in supercritical carbon dioxide. Proceedings of the 217th ACS National Meeting. Anaheim, CA, 1999.
263. GR Haas, JW Kolis. Oxidation of alkenes in supercritical carbon dioxide catalyzed by molybdenum hexacarbonyl. Organometallics 17:4454–4460, 1998.
264. RS Oakes, AA Clifford, KD Bartle, MT Pett, CM Rayner. Sulfur oxidation in supercritical carbon dioxide: dramatic pressure dependent enhancement of diastereoselectivity for sulfoxidation of cysteine derivatives. Chem Commun 1999:247–248.
265. L Jia, H Jiang, J Li. Palladium(II)-catalyzed oxidation of acrylate esters to acetals in supercritical carbon dioxide. Chem Commun 1999:985–986.
266. U Kreher, S Schebesta, K Walther. Organometallics of transition metals in supercritical carbon dioxide: solubilities, reactions, catalysis. Z Anorg Allg Chem 624: 602–612, 1998.
267. JM Tanko, JF Blackert. Free-radical side-chain bromination of alkylaromatics in supercritical carbon dioxide. Science 263:203–205, 1994.

268. JM Tanko, JF Blackert, M Sadeghipour. Supercritical carbon dioxide as a medium for conducting free-radical reactions. In: PT Anastas, CA Farris, eds. Benign by Design: Alternative Synthetic Design for Pollution Prevention. ACS Symposium Series No. 577. Washington, D.C.: American Chemical Society, 1994, pp 98–113.
269. JO Metzger, J Hartmanns, D Malwitz, P Köll. Thermal organic reactions in supercritical fluids. In: ME Paulaitis, JML Penninger, RD Gray Jr., P Davidson, eds. Chemical Engineering at Supercritical Fluid Conditions. Ann Arbor, MI: Ann Arbor Science, 1983, pp 515–533.
270. JO Metzger. Formation of alkyl radicals by thermal bimolecular reaction of alkanes and alkenes. Angew Chem Int Ed Engl 22:889, 1983.
271. J Hartmanns, K Klenke, JO Metzger. Thermal addition of alkanes to alkenes. II. Addition of cyclohexane to acrylate in a free radical chain reaction. Chem Ber 119:488–499, 1986.
272. J Hartmanns, JO Metzger. Thermal addition of alkanes to alkenes. III. Reactivity of C-H bonds of alkanes in the additon to alkenes. Chem Ber 119:500–507, 1986.
273. JO Metzger, K Klenke, J Hartmanns, D Eisermann. Thermal addition of alkanes to alkenes. IV. Regioselectivity in the addition of cyclohexane to 1,2-disubstituted alkenes. Chem Ber 119:508–513, 1986.
274. JO Metzger, M Blumenstein. Stereoselectivity of the thermally initiated free-radical chain addition of cyclohexane to 1-alkynes. Chem Ber 126:2493–2499, 1993.
275. JO Metzger, F Bangert. Kinetics of the addition of cyclohexane to phenylethyne under supercritical fluid conditions. Chem Ber 127:673–675, 1994.
276. JO Metzger. Thermal organic reactions at supercritical fluid conditions: Functionalization of alkanes by free radical additions to unsaturated compounds. In: G Brunner, M Perrut, eds. Proceedings of the 3rd International Symposium on Supercritical Fluids, Strasbourg, France, Vol. 3, 1994, pp 99–102.
277. S Kim, KP Johnston. Effects of supercritical solvents on the rates of homogeneous chemical reactions. In: TG Squires, ME Paulaitis, eds. Supercritical Fluids: Chemical and Engineering Principles and Applications. ACS Symposium Series No. 329. Washington, D.C.: American Chemical Society, 1987, pp 42–55.
278. Y Ikushima, S Ito, T Asano, T Yokoyama, N Saito, K Hatakeda, T Goto. A Diels–Alder reaction in supercritical carbon dioxide medium. J Chem Eng Jap 23:96–98, 1990.
279. Y Ikushima, N Saito, O Sato, M Arai. Implications of a transition state in a Diels–Alder reaction in supercritical carbon dioxide. Bull Chem Soc Jpn 67:1734–1736, 1994.
280. BL Knutson, AK Dillow, CL Liotta, CA Eckert. Kinetics of a Diels–Alder reaction in supercritical propane. In: KW Hutchenson, NR Foster, eds. Innovations in Supercritical Fluids: Science and Technology. ACS Symposium Series No. 608. Washington, D.C.: American Chemical Society, 1995, pp 166–178.
281. JT Reaves, CB Roberts. Subcritical solvent effects on a parallel Diels–Alder reaction network. Ind Eng Chem Res 38:855–864, 1999.
282. RD Weinstein, A Renslo, RL Danheiser, JG Harris, JW Tester. Kinetic correlation of Diels–Alder reactions in supercritical carbon dioxide. J Phys Chem 100:12337–12341, 1996.

283. RD Weinstein, A Renslo, RL Danheiser, JW Tester. Silica-promoted Diels–Alder reactions in carbon dioxide from gaseous to supercritical conditions. J Phys Chem B 103:2878–2887, 1999.
284. AA Clifford, K Pople, WJ Gaskill, KD Bartle, CM Rayner. Reaction control and potential tuning in a supercritical fluid. Chem Commun 1997:595–596.
285. AA Clifford, K Pople, WJ Gaskill, KD Bartle, CM Rayner. Potential tuning and reaction control in the Diels–Alder reaction between cyclopentadiene and methyl acrylate in supercritical carbon dioxide. J Chem Soc, Faraday Trans 94:1451–1456, 1998.
286. N Jeong, SH Hwang, YW Lee, JS Lim. Catalytic Pauson–Khand reaction in super critical fluids. J Am Chem Soc 119:10549–10550, 1997.
287. MA Carroll, AB Holmes. Palladium-catalysed carbon–carbon bond formation in supercritical carbon dioxide. Chem Commun 1998:1395–1396.
288. DK Morita, DR Pesiri, SA David, WH Glaze, W Tumas. Palladium-catalyzed cross-coupling reactions in supercritical carbon dioxide. Chem Commun 1998: 1397–1398.
289. N Shezad, RS Oakes, AA Clifford, CM Rayner. Use of fluorinated palladium sources for efficient Pd-catalysed coupling reactions in supercritical carbon dioxide. Tetrahedron Lett 40:2221–2224, 1999.
290. S Cacchi, G Fabrizi, F Gasparrini, C Villani. Carbon–carbon bond forming reactions in supercritical carbon dioxide in the presence of a supported palladium catalyst. Synlett 1999:345–347, 1999.
291. H Tiltscher, H Wolf, J Schelchshorn. A mild and effective method for the reactivation or maintenance of the activity of heterogeneous catalysts. Angew Chem Int Ed Engl 20:892–894, 1981.
292. H Tiltscher, H Wolf, J Schelchshorn. Utilization of supercritical fluid solvent-effects in heterogeneous catalysis. Ber Bunsenges Phys Chem 88:897–900, 1984.
293. H Tiltscher, H Hofmann. Trends in high pressure chemical reaction engineering. Chem Eng Sci 42:959–977, 1987.
294. G Manos, H Hofmann. Coke removal from a zeolite catalyst by supercritical fluids. Chem Eng Technol 14:73–78, 1991.
295. S Saim, B Subramaniam. Chemical reaction equilibrium at supercritical conditions. Chem Eng Sci 43:1837–1841, 1988.
296. S Saim, DM Ginosar, B Subramaniam. Phase and reaction equilibria considerations in the evaluation and operation of supercritical fluid reaction processes. In: KP Johnston, JML Penninger, eds. Supercritical Fluid Science and Technology. ACS Symposium Series No. 406. Washington, D.C.: American Chemical Society, 1989, pp 301–316.
297. S Saim, B Subramaniam. Isomerization of 1-hexene on Pt/γ-Al_2O_3 catalyst at subcritical and supercritical reaction conditions: pressure and temperature effects on catalyst activity. J Supercrit Fluids 3:214–221, 1990.
298. S Saim, B Subramaniam. Isomerization of 1-hexene over Pt/γ-Al_2O_3 catalyst: reaction mixture density and temperatue effects on catalyst effectiveness factor, coke laydown, and catalyst micromeritics. J Catal 131:445–456, 1991.
299. S Baptist-Nguyen, B Subramaniam. Coking and activity of porous catalysts in supercritical reaction media. AIChE J 38:1027–1037, 1992.

300. DM Ginosar, B Subramaniam. Coking and activity of a reforming catalyst in near-critical and dense supercritical reaction mixtures. In: B Delmon, GF Froment, eds. Catalyst Deactivation 1994: Studies in Surface Science and Catalysis. Vol. 88. Amsterdam: Elsevier, 1994, pp 327–334.
301. DM Ginosar, B Subramaniam. Olefinic oligomer and cosolvent effects on the coking and activity of a reforming catalyst in supercritical reaction mixtures. J Catal 152:31–41, 1995.
302. B Subramaniam, BJ McCoy. Catalyst activity maintenance or decay: a model for formation and desorption of coke. Ind Eng Chem Res 33:504–508, 1994.
303. BJ McCoy, B Subramaniam. Continuous-mixture kinetics of coke formation from olefinic oligomers. AIChE J 41:317–323, 1995.
304. B Subramaniam, DM Ginosar. Enhancing the activity of solid acid catalysts with supercritical reaction media: experiments and theory. In: P Rudolf von Rohr, C Trepp, eds. Process Technology Proceedings. Vol. 12. High Pressure Chemical Engineering. Amsterdam: Elsevier, 1996, pp 3–9.
305. MC Clark, B Subramaniam. 1-Hexene isomerization on a Pt/γ-Al_2O_3 catalyst: the dramatic effects of feed peroxides on catalyst activity. Chem Eng Sci 51: 2369–2377, 1996.
306. MC Clark, B Subramaniam. Kinetics on a supported catalyst at supercritical nondeactivating conditions. AIChE J 45:1559–1565, 1999.
307. B Subramaniam, S Saim, MC Clark. In situ mitigation of coke buildup in porous catalysts by pretreatment of hydrocarbon feed to reduce peroxides and oxygen impurities. US Patent No. 5,690,809, 1997.
308. JA Amelse, NA Kutz. Catalyzed xylene isomerization under supercritical temperature and pressure conditions. U.S. Patent No. 5,030,788, 1991.
309. AK Dillow, JS Brown, CL Liotta, CA Eckert. Supercritical fluid tuning of reaction rates: the cis–trans isomerization of 4,4′-disubstituted azobenzenes. J Phys Chem A 102:7609–7617, 1998.
310. Y Gao, Y-F Shi, Z-N Zhu, W-K Yuan. Coking mechanism of zeolite for supercritical fluid alkylation of benzene. In: P Rudolf von Rohr, C Trepp, eds. Process Technology Proceedings. Vol. 12. High Pressure Chemical Engineering. Amsterdam: Elsevier, 1996, pp 151–156.
311. Y Gao, Z Zhu, W Yuan. Alkylation of benzene for ethylbenzene under supercritical conditions. Prog Nat Sci 6:625–630, 1996.
312. L Fan, I Nakamura, S Ishida, K Fujimoto. Supercritical-phase alkylation reaction on solid acid catalysts: mechanistic study and catalyst development. Ind Eng Chem Res 36:1458–1463, 1997.
313. LF Albright. Comments on "Supercritical-phase alkylation reaction on solid acid catalysts: mechanistic study and catalyst development." Ind Eng Chem Res 37: 296–297, 1998.
314. L Fan, I Nakamura, S Ishida, K Fujimoto. Rebuttal to the comments of Lyle F. Albright. Ind Eng Chem Res 37:298–299, 1998.
315. MC Clark, B Subramaniam. Extended alkylate production activity during fixed-bed supercritical 1-butene/isobutane alkylation on solid acid catalysts using carbon dioxide as a diluent. Ind Eng Chem Res 37:1243–1250, 1998.

316. MG Hitzler, FR Smail, SK Ross, M Poliakoff. Friedel–Crafts alkylation in supercritical fluids: continuous, selective and clean. Chem Commun 1998:359–360.
317. TM Swan, SK Ross, M Poliakoff, MG Hitzler, FR Smail, T Tacke, S Wieland. Alkylation and acylation reactions. PCT Patent No. WO 98/15509, 1998.
318. C Vieville, Z Mouloungui, A Gaset. Esterification of oleic acid by methanol catalyzed by p-toluenesulfonic acid and the cation-exchange resins K2411 and K1481 in supercritical carbon dioxide. Ind Eng Chem Res 32:2065–2068, 1993.
319. C Vieville, Z Mouloungui, A Gaset. Kinetics of the oleic acid esterification by methanol in the presence of solid acid catalysts in supercritical carbon dioxide. In: G Brunner, M Perrut, eds. Proceedings of the 3rd International Symposium on Supercritical Fluids, Strasbourg, France, Vol. 3, 1994, pp 19–24.
320. JB Ellington, JF Brennecke. Pressure effect on the esterification of phthalic anhydride in supercritical CO_2. J Chem Soc, Chem Commun 1993:1094–1095.
321. TR Felthouse, PL Mills. Catalytic amination of methyl tertiary-butyl ether to tertiary-butylamine over pentasil molecular sieves. Appl Catal A 106:213–237, 1993.
322. A Fischer, T Mallat, A Baiker. Continuous amination of propanediols in supercritical ammonia. Angew Chem Int Ed Engl 38:351–354, 1999.
323. A Fischer, T Mallat, A Baiker. Cobalt-catalyzed amination of 1,3-propanediol: effects of catalyst promotion and use of supercritical ammonia as solvent and reactant. J Catal 183:373–383, 1999.
324. A Fischer, T Mallat, A Baiker. Synthesis of 1,4-diaminocyclohexane in supercritical ammonia. J Catal 182:289–291, 1999.
325. S Wang, M Karpf, F Kienzle. Ammonolysis with supercritical NH_3. J Supercrit Fluids 15:157–164, 1999.
326. NA Collins, PG Debenedetti, S Sundaresan. Disproportionation of toluene over ZSM-5 under near-critical conditions. AIChE J 34:1211–1214, 1988.
327. F Niu, G Kolb, H Hofmann. Deactivation kinetics and modelling of coke removal under supercritical conditions for the example of ethylbenzene disproportionation. Chem Eng Technol 18:278–283, 1995.
328. F Niu, H Hofmann. Investigation of various zeolite catalysts under supercritical conditions. In: P Rudolf von Rohr, C Trepp, eds. Process Technology Proceedings. Vol. 12. High Pressure Chemical Engineering. Amsterdam: Elsevier, 1996, pp 145–150.
329. F Niu, H Hofmann. Studies on deactivation kinetics of a heterogeneous catalyst using a concentration-controlled recycle reactor under supercritical conditions. Appl Catal A 158:273–285, 1997.
330. F Niu, H Hofmann. Investigation of coke extraction from zeolite-HY under supercritical and near-critical conditions. Can J Chem Eng 75:346–352, 1997.
331. Z Dardas, MG Süer, YH Ma, WR Moser. High-temperature, high-pressure in situ reaction monitoring of heterogeneous catalytic processes under supercritical conditions by CIR-FTIR. J Catal 159:204–211, 1996.
332. M Süer, Z Dardas, YH Ma, WR Moser. An in situ CIR-FTIR study of n-heptane cracking over a commercial Y-type zeolite under subcritical and supercritical conditions. J Catal 162:320–326, 1996.

333. Z Dardas, MG Süer, YH Ma, WR Moser. A kinetic study of n-heptane catalytic cracking over a commercial Y-type zeolite under supercritical and subcritical conditions. J Catal 162:327–338, 1996.
334. WR Moser, MG Suer, Z Dardas, YH Ma. Mechanism of the catalytic cracking of heptane under supercritical fluid conditions. Prepr—ACS Div Petrol Chem 43(3): 450–453, 1998.
335. L Jia, H Jiang, J Li. Selective carbonylation of norbornene in $scCO_2$. Green Chem 1(2):91–93, 1999.
336. A Fürstner, D Koch, K Langemann, W Leitner, C Six. Olefin metathesis in compressed carbon dioxide. Angew Chem Int Ed Engl 36:2466–2469, 1997.
337. H Tsugane, Y Yagi, H Inomata, S Saito. Dimerization of benzoic acid in saturated solution of supercritical carbon dioxide. J Chem Eng Jap 25:351–353, 1992.
338. C Bunker, HW Rollins, JR Gord, Y-P Sun. Efficient photodimerization reaction of anthracene in supercritical carbon dioxide. J Org Chem 62:7324–7329, 1997.
339. K Sakanishi, H Obata, I Mochida, T Sakaki, M Shibata. Selective dimerization of benzothiophene using supported aluminum sulfate under supercritical CO_2 conditions. J Supercrit Fluids 13:203–210, 1998.
340. JP DeYoung, BE Kipp, HC Wei, JM DeSimone. A quantitative kinetic study of alpha-methylstyrene dimerization using Nafion solid acid catalyst in supercritical carbon dioxide. Prepr—ACS Div Polym Chem 39(2):833–834, 1998.
341. JF Kadla, JP DeYoung, JM DeSimone. The thermal decomposition of perfluoroalkyl peroxides in carbon dioxide. Prepr—ACS Div Polym Chem 39(2):835–836, 1998.
342. CM Starks, CL Liotta, M Halpern. Phase-Transfer Catalysis: Fundamentals, Applications, and Industrial Perspectives. New York: Chapman & Hall, 1994.
343. S Naik, LK Doraiswamy. Phase transfer catalysis: chemistry and engineering. AIChE J 44:612–646, 1998.
344. AK Dillow, SLJ Yun, D Suleiman, DL Boatright, CL Liotta, CA Eckert. Kinetics of a phase transfer catalysis reaction in supercritical fluid carbon dioxide. Ind Eng Chem Res 35:1801–1806, 1996.
345. K Chandler, CW Culp, DR Lamb, CL Liotta, CA Eckert. Phase-transfer catalysis in supercritical carbon dioxide: kinetic and mechanistic investigation of cyanide displacement on benzyl chloride. Ind Eng Chem Res 37:3252–3259, 1998.
346. KL Toews, RM Shroll, CM Wai. pH-Defining equilibrium between water and supercritical CO_2. Influence on SFE of organics and metal chelates. Anal Chem 67:4040–4043, 1995.

APPENDIX: A Summary of the Patent Literature on SCF Reactions (1990–1999)

Patent no.	Inventor(s)	Title	Assignee	Issue date	Summary
US 5,030,788	Amelse, Jeffrey A. Kutz, Nancy A.	Catalyzed Xylene Isomerization under Supercritical Temperature and Pressure Conditions	Amoco Corp. (US)	9 July 1991	A process is described to catalytically isomerize xylenes and ethylbenzene to *p*-xylene under SCF conditions followed by isobaric cooling to the liquid state.
US 5,142,070	Fullington, Michael C. Pennington, Buford T.	Process for the Direct Oxidation of Propylene to Propylene Oxide	Olin Corp. (US)	25 August 1992	A process is described for the direct oxidation of propylene to propylene oxide with oxygen at supercritical conditions relative to the propylene.
US 5,198,589	Rathke, Jerome W. Klingler, Robert J.	Cobalt Carbonyl Catalyzed Olefin Hydroformylation in Supercritical Carbon Dioxide	U.S. Dept. of Energy (US)	30 March 1993	A process is described for the hydroformylation of olefins with hydrogen and carbon monoxide and with a carbonyl catalyst in the presence of an SCF reaction solvent.
US 5,210,336 EP 385631	Gaffney, Anne M. Sofranko, John A.	Oxidation of Olefin to Glycol	ARCO Chemical Technology (US)	11 May 1993 5 September 1990	This invention describes the oxidation of olefins to the corresponding glycols by reaction over a heterogeneous catalyst in an SCF reaction mixture.
US 5,254,735	Smith, Kim R. Chen, Y.-D. Mark Smith, Rebecca F. Borland, James E. Sauer, Joe D.	Process for Preparing Solid Amine Oxides	Ethyl Corp. (US)	19 October 1993	A process is described for oxidizing a tertiary amine with aqueous hydrogen peroxide (70–90 wt %) to form an amine oxide. Claims include operation with the reaction mixture in the SCF state.
US 5,296,640	Jacobson, Stephen E. Ely, Wayne B.	Process for Preparing Perhaloacyl Chlorides	DuPont (US)	22 March 1994	A process is disclosed for preparing perhaloacyl chlorides such as trifluoroacetyl chloride by oxidizing lower perfluoroalkyl and monochloroperfluoroalkyl dichloromethanes with oxygen within the supercritical region of the compounds and in the absence of water.
US 5,304,698 WO 94/03415	Husain, Altaf	Solid Catalyzed Supercritical Isoparaffin-Olefin Alkylation Process	Mobil Oil Corp. (US)	19 April 1994 17 February 1994	A process is described for improving the octane rating of gasoline by alkylating an olefin with an isoparaffin over a zeolite catalyst above the critical point of the isoparaffin to improve the catalyst longevity.
US 5,321,151	Lange, Barry C.	Process for Preparation of Iodopropargyl Carbamates	Rohm and Haas (US)	14 June 1994	A process is disclosed for preparing iodopropargyl carbamate compounds (used in a fungicide) by reaction of an alkylamine, liq. or $scCO_2$, a propargyl alcohol, and optionally a catalyst to form *N*-alkylpropargyl carbamate, followed by reaction with an iodinating agent. The process avoids the use of phosgene and isocyanates.

US 5,523,420	Lowack, Rainer Meyer, Joachim Eggersdorfer, Manfred Grafen, Paul	Preparation of Alpha-Tocopherol and Alpha-Tocopheryl Acetate in Liquid or Supercritical Carbon Dioxide	BASF (DE)	4 June 1996	A process is described for preparing α-tocopherol or tocopheryl acetate by cyclocondensation of trimethylhydroquinone with phytol or isophytol in the presence of an acid catalyst in liq. or $scCO_2$, optionally followed by acetylation.
US 5,639,910 US 5,763,662 EP 652202	Ikariya, Takao Jessop, Philip G. Hsiao, Yi Noyori, Ryoji	Method for Producing Formic Acid or Its Derivatives	Research Development Corp. of Japan (JP) and NKK Corp. (JP)	17 June 1997 9 June 1998 10 May 1995	A process is described for producing formic acid or derivatives from carbon dioxide in the SCF state by reaction of carbon dioxide with a compound containing an active hydrogen group such as alcohols, amines, or carbamates.
US 5,690,809	Subramaniam, Bala Saim, Said Clark, Michael C.	In Situ Mitigation of Coke Buildup in Porous Catalysts by Pretreatment of Hydrocarbon Feed to Reduce Peroxides and Oxygen Impurities	Center for Research (US)	25 November 1997	A process is described for minimizing coke buildup in porous catalysts used in the processing of hydrocarbon feed stocks by pretreatment of the feed to reduce organic peroxides and dissolved oxygen.
US 5,725,756 WO 96/33148	Subramaniam, Bala Saim, Said	In Situ Mitigation of Coke Buildup in Porous Catalysts with Supercritical Reaction Media	Center for Research (US)	10 March 1998 24 October 1996	A method is described to minimize catalyst deactivation rate and coke deposition and to maximize a desired reaction rate in processing under SCF conditions.
US 5,734,070	Tacke, Thomas Wieland, Stefan Panster, Peter Bankmann, Martin Brand, Reinhold Mägerlein, Hendrik	Hardening of Unsaturated Fats, Fatty Acids or Fatty Acid Esters	Degussa (DE)	31 March 1998	A process is disclosed for continuously hydrogenating unsaturated fats, fatty acids, or fatty acid esters on a shaped catalyst in a solid bed in the presence of an SCF solvent medium.
US 5,792,894 WO 97/03933	Huff, Jr., George A. Mehlberg, Robert L. Train, Peter M.	Conversion of Aromatic and Olefins	Amoco Corp. (US)	11 August 1998 6 February 1997	A process is disclosed for alkylating a volatile aromatic compound with an aliphatic olefin over a solid alkylation catalyst above the critical point of the reactant mixture.
US 5,831,116	Wang, Chien-Tsung Willey, Ronald J.	Catalytic Process for Making Ethers, Aldehydes, Esters, and Acids from Alcohols Using a Supercritical Fluid	Northeastern University (US)	3 November 1998	A process is described for partially oxidizing alcohols to the indicated products with molecular oxygen over solid acid catalysts in an SCF solvent.
US 5,866,733	Gehrer, Eugen Massonne, Klemens Harder, Wolfgang	Preparation of Diarylethanes	BASF (DE)	2 February 1999	A process is described for preparing alkylated diarylethanes by reacting benzene with an alkylated styrene in the liquid or SCF phase and in the presence of a strongly acidic large-pore zeolite.

APPENDIX: (*Continued*)

Patent no.	Inventor(s)	Title	Assignee	Issue date	Summary
US 5,914,031 WO 96/17680	Sentagnes, Dominique Berdeu, Bernard Demazeau, Gérard Garrabos, Yves Largeteau, Alain	Process in a Reducing Medium of Chemically Transforming Complex Chemical Structures in a Supercritical Fluid	L'Electrolyse (FR)	22 June 1999 13 June 1996	A process is described for chemically transforming a complex chemical structure into a final product that involves a reduction reaction in a solvent in the SCF state.
CA 2069373 EP 532153	Bhinde, Manoj V. Hsu, Chao Yang	Isomerization of Hydrocarbons with Solid Superacid Catalyst	Sun Co. (US)	4 December 1992 17 March 1993	A process is described for isomerization of straight chain C_4–C_{24} paraffins with a solid superacid catalyst at supercritical or near-critical conditions. Cited advantages include an optimized product selectivity and mitigation of catalyst fouling.
DE 19529679	Mueller-Markgraf, Wolfgang	Catalytic Direct Oxidation of Propylene to Propylene Oxide	Linde (DE)	13 February 1997	A process is described for catalytically epoxidizing propylene to propylene oxide under SCF conditions.
DE 3836180	Oeste, Franz D.	Production of Polycyclic Aromatic Hydrocarbons	Oeste, Franz D. (DE)	26 April 1990	A process is described for the production of polycyclic aromatic hydrocarbons by chemical or catalytic conversion of aromatic hydrocarbons in the presence of an SCF solvent.
EP 614883	Rescalli, Carlo	Process for Synthesizing Urea from Ammonia and Carbon Dioxide, with Total Carbon Dioxide Conversion	Snamprogetti (IT)	14 September 1994	A process is described for synthesizing urea from ammonia and carbon dioxide in a reactive distillation column where carbamate is formed as an intermediate in an SCF phase reaction in the column.
EP 841314	Jansen, Michael Rehren, Claus	Hydrogenation of Organic Compounds using Amorphous Metal Catalysts	Hoffmann-La Roche (CH)	13 May 1998	A method is described for the catalytic hydrogenation of organic compounds such as fatty acids, aromatics, alkynes, and dehydroisophytol in a near- or supercritical solvent using an amorphous metal alloy catalyst.
EP 882722	Breuninger, Manfred	Manufacture of Alpha-Tocopherol	Hoffmann-La Roche (CH)	9 December 1998	A process is described for the selective preparation of α-tocopherols from other tocopherols by catalytic permethylation with a mixed-oxide catalyst. A reaction mixture in the SCF state and including methanol or the $H_2/CO/CO_2$ equivalent of methanol is claimed.
EP 893451	Harris, Rosemarie Jureller, Sharon H. Kerschner, Judith L. Trzasko, Peter T. Humphreys, Robert W.R.	Polysaccharide Modification in Densified Fluid	National Starch and Chemical Investment Holding Corp. (US)	27 January 1999	A process is described for chemically modifying polysaccharides in an SCF fluid, including esterification and etherification of a starch in $scCO_2$.

EP 916655	Bourne, Stephen W. Oldenhove, Pieter	Process for Preparing Organic Hydroperoxides	Shell International Research (NL)	19 May 1999	This invention describes a process for preparing organic hydroperoxides by oxidation of a hydrocarbon feed with molecular oxygen under SCF conditions in the presence of a separate liquid water film on the reactor walls to inhibit decomposition.
JP 03099072	Takada, Hiroshi Saito, Takashi	Preparation of 5-(Hydroxymethyl)-2-Furancarboxaldehyde	Kao Corp (JP)	24 April 1991	A process is described for preparing 5-hydroxymethyl-2-furan carboxyaldehyde by dehydration of hexoses in $scCO_2$ and oxygen catalyst.
JP 04247045	Yamamoto, Koji Mifuji, Yutaka	Preparation of Dialkylnaphthalenes	Kobe Steel (JP)	3 September 1992	A process is described for preparing dialkylnaphthalenes by alkylation of naphthalene or 2-alkylnaphthalenes in aromatic solvents at SCF conditions in the presence of solid acid catalysts.
JP 06172223	Fujimoto, Kaoru Oodan, Kyoji Yoshii, Kyotaka	Manufacture of Hydrocarbon Waxes by Fischer–Tropsch Synthesis	Ube Industries (JP) and K. Fujimoto (JP)	21 June 1994	A process is described for producing C_{20}–C_{40} hydrocarbons by Fischer–Tropsch synthesis in a C_5–C_8 hydrocarbon solvent in the SCF state. Cited advantages include maintenance of catalyst activity.
JP 07145388	Fujimoto, Kaoru Oodan, Kyoji Toshii, Kyotaka	Manufacture of Waxes	Ube Industries (JP) and K. Fujimoto (JP)	6 June 1995	A process is described for producing wax by Fischer–Tropsch synthesis of a mixture of H_2, CO, and hydrocarbons under SCF conditions using a specified supported cobalt catalyst resulting in high selectivity.
JP 09059205	Oono, Mitsuru	Preparation of Aromatic Acyl Compounds by Friedel–Crafts Acylation	Daicel Chem (JP)	4 March 1997	A process is described for producing aromatic acyl compounds by Friedel–Crafts acylation using supercritical C_1–C_4 hydrocarbons or $scCO_2$ as the reaction media. Examples include the synthesis of 4-methylacetophenone.
JP 09255594	Fujimoto, Kaoru Oodan, Kyoji Yoshii, Kiyotaka	Preparation of Waxes by Fischer–Tropsch Method	Ube Industries (JP) and K. Fujimoto (JP)	30 September 1997	A process is described for the catalytic production of wax by the Fischer–Tropsch synthesis of a mixture of H_2 and CO, C_5–C_{20} unsaturated hydrocarbons containing at least a terminal unsaturated bond, and C_4–C_{20} saturated hydrocarbons under liquid or SCF conditions.
JP 10226679	Kiyoura, Tadamitsu Kato, Kozo	Preparation of Dialkylimidazolidinones as Solvents	Mitsui Chemicals (JP)	25 August 1998	A process is described for producing 1,3-dialkyl-2-imidazolidinones by reaction of ethylene carbonate with monoalkylamines in $scCO_2$.
JP 10251202	Ikariya, Takao Iwasa, Seiji Noyori, Ryoji	Preparation of Cinnamate Esters	Foundation for Scientific Technology Promotion (JP) and Nipppon Kokan (JP)	22 September 1998	A process is described for preparing cinnamate esters by treatment of aromatic halides with α,β-unsaturated carboxylate esters in the presence of amines and group VIII transition metal–based catalysts in $scCO_2$ or $scCHF_3$. Examples include the synthesis of ethyl cinnamate from phenyl iodide and ethyl acrylate.

APPENDIX: (*Continued*)

Patent no.	Inventor(s)	Title	Assignee	Issue date	Summary
JP 11005763	Sakakura, Toshiyasu Sako, Takeshi	Preparation of Aromatic Aldehydes or Aromatic Alcohols	Agency of Industrial Sciences and Technology (JP)	12 January 1999	A process is described for preparing oxygen-containing compounds by reaction of aromatic hydrocarbons with CO in liquid or scCO_2 solvent and photoirradiating in the presence of a transition metal complex containing a phosphine compound. Examples include synthesis of benzaldehyde and benzyl alcohol from benzene.
JP 11071308	Ogawa, Hakaru Murata, Keiji Onoda, Yuko Hori, Michio	Plants and Method for Manufacture of Oxygen-Containing Hydrocarbons	Toshiba (JP)	16 March 1999	A process is described utilizing carbon oxides, hydrogen, and hydrocarbons to synthesize oxygen-containing hydrocarbons in a homogeneous SCF phase. A plant for manufacturing methanol from CO_2, CO, and H_2 using SCF hexane is exemplified. Cited advantages include the high thermal conductivity and thermal diffusion afforded by operation in the SCF phase.
WO 91/09826 EP 509003	Saleh, Ramzi Y. Livingston, Joel R. Mathys, Georges M.K.	Process for the Preparation of Octenes	Exxon Chemical (US)	11 July 1991 21 October 1992	A process is disclosed for the preparation of a mixture of isomeric octenes by dimerizing *n*-butene in the SCF state with a NiO catalyst on a silica-alumina support.
WO 94/20444	Hase, Anneli Alapeijari, Maija Aaltonen, Olli Hae, Tapio	Method for Oxidation	Valtion Teknillinen Tutkimuskeskus (FI)	15 September 1994	A process is described for the direct oxidation of benzene into phenols in a homogeneous SCF phase including benzene, molecular oxygen, and hydrogen in the presence of a solid palladium catalyst.
WO 96/01304	Härröd, Magnus Möller, Poul	Hydrogenation of Substrate and Products Manufactured According to the Process	Härröd (SE) and Möller (DK)	18 January 1996	A process is described for conducting hydrogenation reactions in a homogeneous SCF phase including hydrogenation of C=C bonds in lipids, of COOR to produce fatty alcohols, and of oxygen to hydrogen peroxide.
WO 97/23525	Elsbernd, Cheryl L. Smith, Richard S.	Process for Making (Thio)Urethanes under Superatmospheric Conditions	Minnesota Mining and Manufacturing (US)	3 July 1997	A method is described for producing urethanes and thiourethanes by reaction of isocyanates with compounds containing hydroxyl or thiol groups in a solvent under SCF conditions.
WO 97/30967	Laitinen, Antero	Hydrogenation of Aromatic Nitrocompounds to Aromatic Amines	Valtion Teknillinen Tutkimuskeskus (FI)	28 August 1997	A method is described for the catalytic hydrogenation of aromatic nitrocompounds to aromatic amines in a solvent under SCF conditions.
WO 97/38955	Poliakoff, Martyn Swan, Thomas M. Tacke, Thomas Hitzler, Martin G. Ross, Stephen K. Wieland, Stefan	Supercritical Hydrogenation	Thomas Swan & Co. (GB) and Degussa (DE)	23 October 1997	A process is described for the continuous selective hydrogenation of aliphatic or aromatic substrates using heterogeneous catalysts under supercritical or near-critical conditions.

WO 98/15509	Swan, Thomas M. Ross, Stephen, K. Poliakoff, Martyn Hitzler, Martin G. Smail, Fiona R. Tacke, Thomas Wieland, Stefan	Alkylation and Acylation Reactions	Thomas Swan & Co. (GB) and Degussa (DE)	16 April 1998	A method is described for alkylating or acylating aromatic substrates under SCF conditions. In particular, a method of conducting Friedel–Crafts alkylation or acylation reactions is disclosed using a heterogeneous acid catalyst in a continuous-flow reactor under these conditions. Specifically claimed is the means of controlling product selectivity by varying temperature, pressure, reactant concentrations, and flow rates.
WO 98/32533	Leitner, Walter Kainz, Sabine Koch, Daniel Wittmann, Klaus Six, Christian	Use of Perfluoroalkyl Substituted Phosphorus Compounds as Ligands for Homogeneous Catalysis in Supercritical Carbon Dioxide	Studiengesellschaft Kohle MBH (DE)	30 July 1998	This patent describes the use of catalytic phosphorus compounds containing perfluoroalkyl chains to enhance their solubility in a $scCO_2$ reaction mixture. Furthermore, this invention also describes the use of such reaction mixtures for chemoselective hydrogenation of polyenes, olefin hydroformylations, enantioselective hydrogenation of imines, and for C-C cross-linking reactions.
WO 98/40341	Jeong, Nakcheol Hwang, Sung-Hee	Process for Preparation of Cyclopentenones in Supercritical Fluids	Hanil Synthetic Fiber Co. (KR)	17 September 1998	A process is described for preparation of cyclopentenones (commonly used in pharmaceuticals) by reacting either acetylenes and olefins or enynes with carbon monoxide using an SCF reaction solvent in the presence of a homogeneous cobalt catalyst.
WO 98/47891	Fürstner, Alois Leitner, Walter Koch, Daniel Langemann, Klaus Six, Christian	Selective Olefin Metathesis of Bifunctional or Polyfunctional Substrates in Compressed CO_2 as Reaction Medium	Studiengesellschaft Kohle MBH (DE)	29 October 1998	A method is described for the preparation of cyclic or polymer products by selective olefin metathesis of bi- or polyfunctional substrates in the presence of homogeneous or heterogeneous catalysts in compressed CO_2. A claim includes control of the product distribution by tuning the density of the reaction medium.
WO 98/56739	Subramaniam, Bala Clark, Michael C.	Improved Solid Acid Supercritical Alkylation Reactions Using Carbon Dioxide and/or Other Co-Solvents	University of Kansas (US)	17 December 1998	A process is described for alkylating an isoparaffin and an olefin with a solid alkylation catalyst in an inert solvent where the reaction mixture is under SCF conditions.
WO 99/03810	Leitner, Walter Koch, Daniel	Hydroformylation with Unmodified Rhodium Catalysts in $scCO_2$	Studiengesellschaft Kohle MBH (DE)	28 January 1999	This invention describes the hydroformylation of substrates with C=C double bonds using unmodified rhodium catalysts in an $scCO_2$ reaction mixture to preferentially form the branched isomeric products and the separation of the product and catalyst from the SCF mixture.

4

Homogeneous Catalysis in Supercritical Carbon Dioxide

Can Erkey
University of Connecticut, Storrs, Connecticut

I. HOMOGENEOUS CATALYSIS

Homogeneous catalysis by soluble transition metal complexes offers many advantages over heterogeneous catalysis, such as milder reaction conditions, higher activities and selectivities, and better control of operating conditions. Such superior performance arises from the ability of transition metals to complex with a wide variety of ligands in a number of geometries and to easily change from one oxidation state to another. Even though heterogeneous catalysts dominate the industrial scene today (about 85% of all existing catalytic processes), the importance and market share of homogeneous catalysts are growing at a very fast pace. This growth is being fueled by a combination of factors:

1. Environmental and economic pressures for cleaner processes
2. Growth in specialty and fine chemicals
3. Scientific advances in organometallic chemistry

and is expected to continue in the coming decade. A list of some of the chemicals that are manufactured using homogeneous catalysts is given in Table 1 (1–3).

II. HOMOGENEOUS CATALYSIS IN SUPERCRITICAL CARBON DIOXIDE

A. Introduction

A supercritical fluid (SCF) is a fluid that has been heated and compressed above its critical temperature and pressure. Under these conditions, SCFs have densities

Table 1 Chemicals Produced by Homogeneous Catalysis on Industrial Scale

Hydroformylation	Olefin polymerizations and oligomerizations
C_3–C_{24} aldehydes	α-Olefins
Alcohols	Propylene dimers
Oxidation	Cyclooctadiene
Benzoic acid	Cyclododecatriene
Terephthalic acid	Polyethylene
Adipic acid	Polypropylene
Acetic acid	Ethylene/propylene copolymers
Propylene oxide	Asymmetrical Hydrogenation
Ethanal	Levodopa
Carbonylation	Metolachlor
Acetic acid	Cilazapril
Acetic anhydride	Asymmetrical Isomerization
Ethanal	L-Menthol
Propionic acid	Asymmetrical Oxidation
Ibuprofen	Disparlure
Hydrocyanation	Glycidol
Adiponitrile	
Metathesis	
Norbornene	
Cyclooctene	
Dicyclopentadiene	

that are greater than those of gases but comparable to those of liquids, which enables them to function as solvents. The low viscosities of SCFs and high diffusivities of solutes in SCFs combined with very high buoyant forces (which cause significant density gradients across the interface) may result in superior mass transfer characteristics compared with conventional solvents. As a result of these favorable properties as a solvent, extensive research and development work on SCF use has been conducted in laboratories around the world for a wide variety of applications. Today there are approximately 60 supercritical fluid extraction plants operating around the world (4). Among the SCFs, attention is particularly focused on supercritical carbon dioxide ($scCO_2$) since it is nontoxic, environmentally acceptable, cheap, and has a low critical temperature (31.1°C) and a moderate critical pressure (73.8 bar). An excellent introduction to the field is provided in a monograph by McHugh and Krukonis (5).

While $scCO_2$ has lately been the solvent of choice for many extractions, much less experimentation has been done to explore its uses as a reaction medium. Investigations on using $scCO_2$ as a reaction medium started after Zaks and Klibanov discovered that enzymes can function as catalysts in nearly anhydrous organic solvents (6). It was then correctly hypothesized that $scCO_2$, with its solvent properties similar to those of organic liquids and its other favorable properties, could be exploited as a solvent in biocatalysis (7). Subsequently, a number of studies were published which demonstrated that a wide variety of reactions can proceed in $scCO_2$, sometimes with results better than the corresponding reactions in conventional solvents. These are summarized in various review papers (8–11). Even though $scCO_2$ is currently not used as a reaction medium on an industrial scale, a promising recent development is the planned construction of a pilot plant by DuPont for the manufacture of fluoropolymers in $scCO_2$ (4).

The first study on homogeneous catalytic reactions in $scCO_2$ appeared in the literature as late as 1991. Rathke and Klingler (12) reported the results of their investigation on hydroformylation of propylene in $scCO_2$ catalyzed by $HCo(CO)_4$. Even though $scCO_2$ has many favorable properties as a solvent for homogeneous catalysis, there have been relatively few studies. A summary of all the studies in the literature is given in Tables 2 and 3 in chronological order. The number of works published annually (excluding review articles and non-English journals), shows that the field is attracting a lot of interest, with the number nearly doubling every year since 1997. There has been a comprehensive review article published in the general area of homogeneous catalysis in SCFs which covers the work in the literature until the middle of 1998 (13). Therefore, this chapter is primarily focused on a detailed evaluation of the studies which were published after 1998 and specifically on $scCO_2$.

B. Advantages of Using $scCO_2$ in Homogeneous Catalysis

Supercritical CO_2 as a reaction solvent offers many advantages over conventional organic solvents, especially in reactions involving gaseous substrates. Many gases exhibit higher solubilities in SCFs than in organic solvents. For example, the solubility of hydrogen in hexane at 17°C and 1 atm is around 0.01 mol % (14), whereas at the same temperature hydrogen and carbon dioxide mixtures are miscible in all proportions under supercritical conditions above 209 atm (15). Thus, based simply on increased concentration of reactants, faster reactions are expected to occur within the $scCO_2$ phase provided that the reaction order with respect to the gaseous substrate is positive. Conducting the reaction in a single phase can also eliminate the problems when the reaction rate is controlled by mass transfer across the gas–liquid interface. Furthermore, the design, scale-up, and operation of reactors operating in single phase are much simpler than multiphase reactors. Studies using $scCO_2$ may also improve our understanding of homogeneous catalytic reactions. The exact mechanisms of a large number of homogeneous catalytic reactions, especially those involving gaseous substrates, are not known. One of the problems in studying such reactions in conventional solvents has been the very low concentrations of the gases in solutions, studies of which have been limited to very narrow concentration ranges. With $scCO_2$ as

Table 2 Summary of the Studies in the Literature on Homogeneous Catalysis in $scCO_2$ for 1991–1998

Ref.	Year	Reaction	Catalyst precursor	Conditions
12	1991	Hydroformylation of propylene	$Co_2(CO)_8$	80°C, 210 bar
77	1992	Hydrogenation of $Co_2(CO)_8$	$Co_2(CO)_8$	60–180°C,
78	1993	Reaction of $scCO_2$ with hex-3-yne	$Ni(cod)_2/Ph_2P(CH_2)_4PPh_2$	102°C
22	1994	Production of dimethylformamide	$RuCl_2[P(CH_3)_3]_4$	100°C, 210 bar
94	1994	Hydrogenation of Co and Mn carbonyls	$Co_2(CO)_8$, $Mn_2(CO)_{10}$	80–200°C, up to 370 bar
90	1994	Hydrogenation of $scCO_2$	$RuH_2[P(CH_3)_3]_4$, $RuCl_2[P(CH_3)_3]_4$	50°C, 200–220 bar
79	1995	Hydrogenation/hydroformylation of olefins	$MnH(CO)_5$	35–60°C, 200–230 bar
21	1995	Asymmetrical hydrogenation of enamides	[(R,R)-Et-DuPHOS-Rh](BARF) [(R,R)-Et-DuPHOS-Rh](CF_3SO_3)	40°C, 330 bar
53	1996	Asymmetrical hydrogenation of tiglic acid	(S)-H_8-BINAP-Ru, (R)-BINAP-Ru	20–50°C, 170–180 bar
80	1996	Hydrogenation of $scCO_2$	$RuH_2[PPh_3]_4$, $RuCl_2[P(CH_3)_3]_4$, $RuCl(O_2CCH_3)[P(CH_3)_3]_4$, $RuH_2[P(CH_3)_3]_4$	50–100°C, 200–210 bar
89	1997	Cocylization of alkynes with alkenes, CO	$Co_2(CO)_8$	48–94°C, 96–198 bar
88	1997	Olefin methathesis	$Ru(Cl)_2(PCy_3)_2R$ where R = CH=CPh$_2$, Ph	23–56°C, 56–115 bar
81	1997	Hydroformylation of propylene	$Co_2(CO)_8$	66–108°C, 100–200 bar
82	1997	Oxidation of cyclohexane	Fe(tpfpp)Cl	32–70°C, 60-80 bar
26	1997	Hydroformylation of 1-octene	[(cod)Rh(hfacac)]/PR_3 where R = 4-$(CH_2)_2(CF_2)_6F$-C_6H_4	60°C, 220 bar
83	1998	Hydroformylation of 1-octene	$RhCl(CO)[P(p\text{-}CF_3C_6H_4)_3]_2$	70°C, 270 bar
34	1998	Hydroformylation of olefinic substrates	[(cod)Rh(hfacac)]/PR_3 where R = 3 or 4-$(CH_2)_2(CF_2)_6F$-C_6H_4, 4-$(CH_2)_2(CF_2)_6F$-C_6H_4O	40–65°C, 200–235 bar
48	1998	Oxidation of olefins to diols and epoxides	$Mo(CO)_6$	80–103°C, 510–580 bar
87	1998	Normal and enantioselective epoxidation of allylic and homoallylic alcohols, cyclooctene	$VO(OiPr)_3$, $Mo(CO)_6$, $Ti(OiPr)_3$	0–95°C, 290 bar
84	1998	Coupling reactions of phenyl iodide	$Pd(OOCCH_3)_2$ or Pd(dba)/PR_3 where R = $[3,5\text{-}(CF_3)_2C_6H_3]_3$, PPh_3	75–90°C, 310–345 bar
50	1998	Enantioselective epoxidation of olefinic alcohols	VO(salen*)	40°C, 210 bar
38	1998	Hydroformylation of 1-hexene	$Rh_2(OAc)_4/P(C_2H_5)_3$, $P[(CH_2)_8H]_3$	100°C, up to 250 bar
85	1998	Coupling reactions of phenyl iodide	$PdCl_2L_2$ where L = $PPh_{3\text{-}n}[(CH_2)_2(CF_2)_6F]_n$ $n = 1, 2$	60–100°C

Table 3 Summary of the Studies in the Literature on Homogeneous Catalysis in $scCO_2$ for 1999–January 2000

Ref.	Year	Reaction	Catalyst precursor	Conditions
36	1999	Hydroformylation of 1-octene	$RhH(CO)[P(p\text{-}CF_3C_6H_4)_3]_3$	50°C, 171–273 bar
54	1999	Enantioselective hydrogenation of imines	Ir complex	40°C, 100–130 bar
44	1999	Hydroformylation of propylene	$Co_2(CO)_8$	66–108°C, 91–194 bar
52	1999	Oxidation of methyl acrylate	$PdCl_2(CH_3CN)_2/CuCl$, $PdCl_2/CuCl_2$	27–50°C, 90–130 bar
57	1999	Reactions of acrylates with cyclopentadiene	$Sc(OTf)_3$	50°C, 140–2000 bar
56	1999	Carbonylation of 2-iodobenzyl alcohol	$PdCl_2L_2$ where L = CH_3CN, $P(OC_2H_5)_3$, $P(OCH_3)_3$, $P(OPh)_3$, $PPh(OCH_3)_2$, $PPh_2(OCH_3)$, PPh_3, $P(CH_3)_3$	130°C, 200 bar
51	1999	Oxidation of cyclohexene	Fe(tfpp)Cl, $Fe(tfppBr_8)Cl$	40–80°C, 340 bar
39	1999	Hydroformylation of 1-octene	$RhH(CO)[P(p\text{-}CF_3C_6H_4)_3]_3$	50°C, 273 bar
55	1999	Hydrovinylation of styrenes	Ni complex	1–40°C
91	1999	Carbonylation of norbornene		
86	1999	Heck coupling of iodobenzene with methyl acrylate	$Pd(OCOCH_3)_2$ or $Pd(OCOCF_3)_2/PPh_3$, PCy_3, $P(2\text{-furyl})_3$, PBu_3, $P(o\text{-tolyl})_3$	75–85°C, 110 bar
59	1999	Biphasic hydrogenation of cinnamaldehyde	$RuCl_3/P(3\text{-}SO_3NaC_6H_4)_3$, $Pd(OAc)_2/P(3\text{-}SO_3Na\text{-}C_6H_4)_3$	40°C, 180 bar
37	1999	Hydroformylation of olefinic substrates	$RhH(CO)[P(4\text{-}CF_3C_6H_4)_3]_3$	50°C, 273 bar
45	1999	Hydroformylation	$Co_2(CO)_8$	100°C
60	1999	Heck coupling of iodobenzene with butyl acrylate and styrene	$Pd(OAc)_2/P(3\text{-}SO_3NaC_6H_4)_3$	60°C, 80–140 bar
49	1999	Epoxidation of olefinic alcohols	$VO(OIPr)_3$, $VO(acac)_2$, $VO(tfac)_2$, $VO(hfacac)_2$, $VO(dmac)_2$	25°C, 103–310 bar
61	1999	Biphasic hydrogenation of styrene	$RhCl[P(3,5\text{-}(SO_3Na)_2C_6H_3)_3]_3$	40°C, 270 bar
40	2000	Hydroformylation of 1-octene	$Rh(CO)_2acac/PR_3$ where R = $3,5\text{-}(CF_3)_2C_6H_3$, $4\text{-}CF_3C_6H_4$, $3\text{-}CF_3C_6H_4$, $4\text{-}CF_3OC_6H_4$, $4\text{-}CF_3(CF_2)_3(CH_2)_3C_6H_4$	50°C, 273 bar

a solvent, mechanistic information can be obtained at states with substantially higher gaseous substrate concentrations.

The unusual solvent properties of $scCO_2$ may lead to increased rates or selectivities due to solvent effects in homogeneous catalysis. Solvent effects are poorly understood in both homogeneous reactions and homogeneous catalytic reactions. The monograph by Reichardt (16) provides a comprehensive overview of such effects on homogeneous reactions. Transition metal complexes in catalytic cycles participate in only a few types of reactions, such as ligand exchange, insertion, oxidation, and reduction. The overall rate of the reaction and in some cases the selectivity depends on the equilibria between the reactive intermediates in solution. Such equilibria and distribution of the intermediates are surely affected by the nature of the solvent. The use of $scCO_2$ may shift such equilibria in the desired direction.

Homogeneous catalytic reaction mixtures are multicomponent mixtures wherein the reactant concentrations are on the order of 10 wt %. On the other hand, the catalytic species in the solution exist at very low concentrations; typical [reactant]/[catalyst] values are around 1000. The reaction proceeds through a series of elementary steps that involve transformations of coordinated reactant species in a concerted manner. Such systems are very different from the widely investigated elementary reactions in which the reactants are in very dilute concentrations in a solvent close to its critical point (17). The results of these investigations have shown that rate constants for elementary reactions increase or decrease orders of magnitude with pressure in the vicinity of the critical point of the solvent due to extremely large or negative activation volumes. It has also been observed that if the solvent is sufficiently far from the critical point, the rate constants do not change appreciably with pressure. To find out if such effects were present in a concentrated supercritical solvent mixture, Suppes et al. (18) investigated the oxidation of cumene in $scCO_2$ near the critical point of the initial reacting mixture. No effect of pressure on the rate of oxidation was observed. The homogeneous catalytic reaction system is a hybrid between those two extremes. The reactant concentrations are high, but the solution is dilute in terms of catalytically active species. Therefore, it may be possible to obtain some rate enhancements with pressure if the reaction is conducted in the vicinity of the critical point of the reacting mixture.

Furthermore, in homogeneous catalytic reactions involving transition metal complexes, solvent molecules may coordinate to unsaturated intermediates and transition states, playing an important role in determining the reaction pathway. For example, it is believed that a solvent molecule coordinates to three- and five-coordinated unsaturated intermediates in the catalytic cycle of olefin hydrogenation by $RhCl(PPh_3)_3$ (92). Such effects have recently been quantified by the Ab Inito Molecular Orbital Study of the Full Cycle of Olefin Hydroformylation Catalyzed by $RhH(CO)_2(PH_3)_2$ (93). Coordination of an olefin representative of a solvent molecule into various intermediates was found to have a dramatic effect on the Gibbs free-energy profile of the entire catalytic cycle. For example, for the elementary step that involves oxidative addition of H_2 to the acyl intermediate, coordination of the solvent to the acyl intermediate increased the free-energy barrier dramatically from 15 to 23 kcal/mol. Even though CO_2 is generally a more weekly coordinating solvent than ethene, such theoretical studies can in principle be carried out to investigate the effects of coordination of CO_2 to the unsaturated intermediates in the energy profiles for a wide variety of reactions. Such effects may be more pronounced than those associated with the solvation spheres of the intermediates.

A major drawback for homogeneous catalysis lies in the difficulty of catalyst recovery and recycling. Many of the homogeneous catalysts are complexes of expensive metals, and some of the ligands used can be manufactured only by long and tedious syntheses. Therefore, the cost associated with their recovery is a very important factor in the economics of a process. Only a small minority of organometallic reactions has so far cleared the hurdles for use on an industrial scale due to lack of effective methods for catalyst recovery and recycle. Catalyst recovery is still a very active research area, and the methods currently under investigation in many laboratories around the world are numerous. Some of these include membrane separations, heterogenizing homogeneous catalysts by anchoring them to polymeric supports, supported aqueous- and liquid-phase catalysis, fluorous and aqueous biphasic catalysis. Using $scCO_2$ as a solvent may have great advantages in catalyst recovery. The solubilities of solutes in $scCO_2$ are strong functions of temperature and pressure in the vicinity of the critical point. Therefore, the catalyst, products, and reactants can possibly be separated in an efficient manner through temperature and/or pressure programming.

C. Ligand Modification for Solubility Enhancement

Typical catalyst concentrations employed in homogeneous catalysis are on the order of 1.0 mM in conventional solvents. The development of the field was hampered by the low solubilities of the commonly used homogeneous catalysts in $scCO_2$ having a detrimental effect on activity. For example, Palo and Erkey (19) reported a detailed study on the solubilities of the homogeneous catalyst dichlorobis(triphenylphosphine)nickel(II) in $scCO_2$ up to 300 atm in the temperature range 308–328 K. The maximum solubility measured ($T = 328$ K, $P = 300$ atm, $\rho = 0.83$ g/ml) was a mere 0.01 mM. Likewise, the solubility of $RhCl(PPh_3)_3$ (Wilkinson's catalyst) in $scCO_2$ ($T = 318$ K, $P = 273$ atm, $\rho = 0.88$ g/ml) was found to be no more than 0.02 mM (20). The solubility of a cationic rhodium complex with a highly lipophilic counteranion {tetrakis[3,5-bis(trifluoromethyl)phenyl]} borate in $scCO_2$ was determined as 0.03 mM at 350 atm and 313 K (21). Along the same lines, Jessop et al. (22) had to use $RuH_2[P(CH_3)_3]_4$ for hydrogenation of carbon dioxide to formic acid instead of

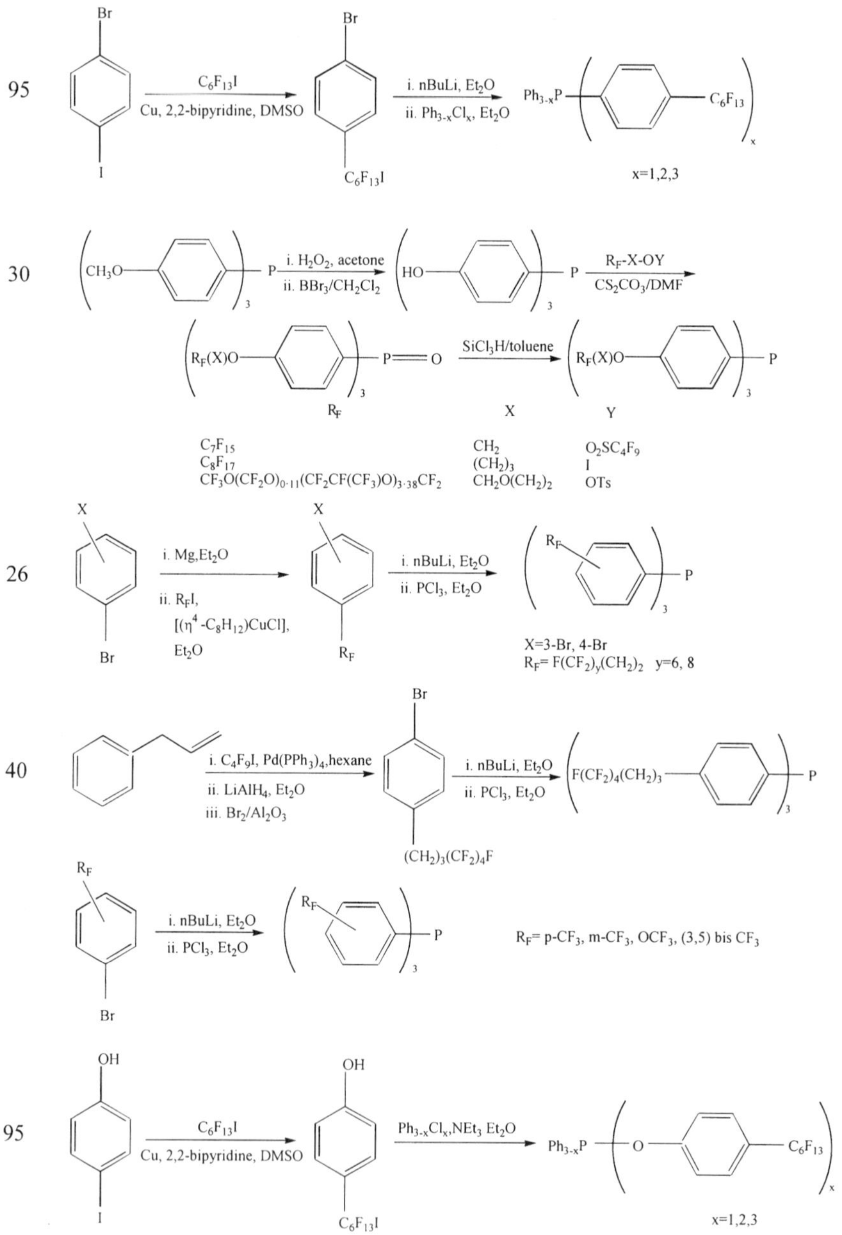

Figure 1 Preparation of arylphosphines substituted with fluorous groups.

the conventional catalyst $RuH_2[P(C_6H_5)_3]_4$ because the conventional catalyst was found to be inactive due to its very low solubility in $scCO_2$.

These studies indicated the necessity of modification or redesign of conventional transition metal catalysts, or some other ways to dissolve catalytic amounts of the complexes in $scCO_2$. One way to increase solubility in $scCO_2$ is to utilize CO_2-philic moieties such as fluoroether, fluoroalkyl, fluoroacrylate, siloxane, or phosphazene. In the pioneering example on developing chelating agents for $scCO_2$ extraction of heavy metals from aqueous solutions, fluorination of the ethyl groups of the diethyldithiocarbamate (DDC) ligand was found to enhance the solubilities of $Cu(DDC)_2$, $Ni(DDC)_2$, and $Co(DDC)_3$ in $scCO_2$ by three orders of magnitude (23). This discovery has been one of the key developments in $scCO_2$ research. Subsequently, a wide variety of reagents functionalized with CO_2-philic groups were developed for $scCO_2$ applications (24,25).

During the past 3 years, significant advances have also been made in the development of synthetic methods for catalysts that exhibit high solubilities in $scCO_2$. These started with the pioneering work of Kainz and Leitner who prepared triarylphosphine ligands with perfluoroalkyl tails (26). The rhodium complexes of these ligands were found to be sufficiently soluble in $scCO_2$ and could catalyze the hydroformylation of olefins. Along the same time period, an interesting coincidence and an important development was the invention of fluorous biphasic systems (FBSs) by Horvath and Rabei (27) as a means to tackle the problems associated with catalyst recovery/recycling in homogeneous catalysis. In an FBS, the catalyst is dissolved in a fluorous solvent and contacted with the organic reactant phase. At high temperatures, the two phases become miscible enabling the reaction to be carried out homogeneously. Once the reaction is complete, the mixture is cooled and separates into a product phase and a fluorous phase that contains the catalyst. The success of this approach depends on development of homogeneous catalysts that will favor partitioning into the fluorous phase. In his pioneering example, Horvath synthesized perfluoroalkyl phosphines. Subsequently, quite a few studies appeared in the literature on synthesis, characterization, and reactive properties of fluorinated ligands, primarily phosphines (28–32). The synthetic schemes for arylphosphines substituted with fluorous groups are summarized in Figure 1. The first route is particularly attractive because the phosphine oxide preparation does not require any special precaution. Furthermore, the phosphine oxides can be stored on the shelf and reduced when needed.

III. STUDIES ON HOMOGENEOUS CATALYSIS IN $scCO_2$

A. Hydroformylation

The majority of the studies in the literature has focused on hydroformylation of olefins, which is one of the largest applications of homogeneous catalysis. More than 6 million tons of aldehydes are produced annually by the homogeneous catalytic hydroformylation of olefins (33). This reaction involves the formation of branched or linear aldehydes by the addition of H_2 and CO to a double bond according to Scheme 1. The linear aldehydes are the preferred products and the selectivity in such reactions is usually expressed as *n*/iso, which is the ratio of the linear aldehyde to the branched aldehyde. The catalysts generally employed are of the form $H_xM_y(CO)_zL_n$; the two transition metals utilized are rhodium and cobalt, and the most commonly utilized ligands are phosphines (PR_3 where R = C_6H_5 or n-C_4H_9). The shares of the various aldehydes are as follows: C_3 (2%), C_4 (73%), C_5–C_{12} (19%), and C_{13}–C_{18} (6%). Production of C_4 aldehydes from hydroformylation of propene is dominated by rhodium-based catalysts whereas higher aldehydes are produced mainly by cobalt catalysts. Since rhodium is about 1000 times more active than cobalt, processes based on Rh catalysts operate at significantly lower temperatures and pressures than processes based on Co catalysts. For example, the Union Carbide Corporation (UCC) liquid recycling process for hydroformylation of propene that uses $HRh(CO)[P(C_6H_5)_3]_3$ operates in the temperature range 85–90°C and at a pressure of 18 bar. In contrast, the BASF process for hydroformylation of 1-octene, which uses $HCo(CO)_4$, operates in the temperature range 160–190°C and in the pressure range 250–300 bar. Therefore, substantial savings in operating and capital costs can be achieved if hydroformylation of higher olefins is conducted using Rh-based catalysts.

One of the major issues in switching to Rh is the difficulty of the separation of products and catalyst. This is illustrated in Figure 2, which shows the two industrial processes for propylene hydroformylation that employ Rh-based catalysts. In the UCC process, propene, H_2/CO mixture, and a high-boiling aldehyde condensation products solution which contains the dissolved catalyst is fed to a reactor. The liquid effluent stream from the reactor is subjected to a complex catalyst recovery scheme consisting of a separator, a pressure let-down valve, a flash evaporator, and two distillation columns in series. The second distillation column operates at subatmospheric pressures around 130°C. The high boiling point of aldehydes beyond C_7 makes such an operation impractical even under reduced pressure due to thermal stability considerations for the catalyst. In the more elegant Ruhrchemie/Rhone-Poulenc (RCH/RP) process, propene and H_2/CO mixture are fed to a continuous stirred tank reactor (CSTR) that contains an aqueous solution with a water-soluble catalyst. The effluent from the reactor is passed through a phase separator. The aqueous solution is recycled back to the

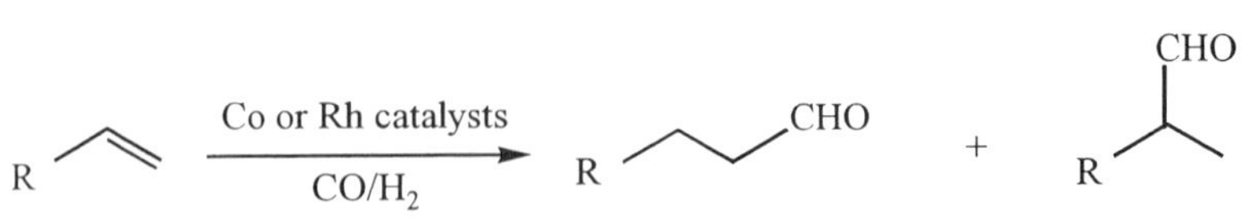

Scheme 1

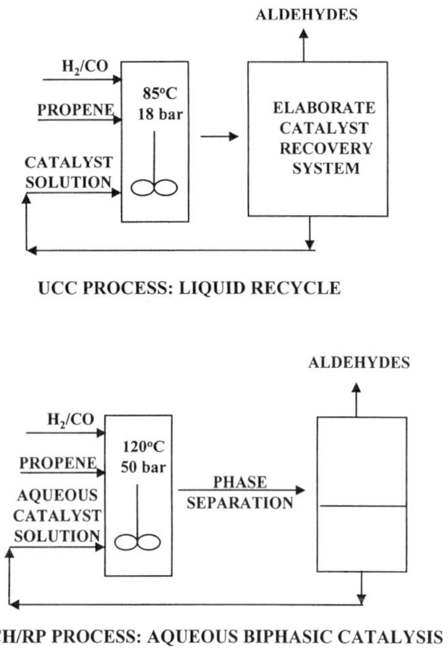

Figure 2 Industrial processes for hydroformylation of propene.

reactor and the crude aldehydes are sent to a distillation column. The process is utilized for production of C_4 and C_5 aldehydes; however, application of this concept to higher-olefin production is highly unlikely due to the extremely low solubilities of higher olefins in water. An alternative may be to utilize $scCO_2$ as the hydroformylation solvent for hydroformylation of higher olefins where the catalyst can be separated from the reaction mixture and recycled by temperature/pressure tuning.

Koch and Leitner (34) reported on the use of perfluoroalkyl-substituted arylphosphines in rhodium-catalyzed hydroformylation of olefins. The catalysts were formed in situ from [(cod)Rh(hfacac)] and PR_3 where R = 4-$(CH_2)_2$-$(CF_2)_6$F-C_6H_4, 3-$(CH_2)_2(CF_2)_6$F-C_6H_4, and 4-$(CH_2)_2(CF_2)_6$F-C_6H_4O. The methylene spacers were used to keep structural and electronic changes at the rhodium center to a minimum due to electron withdrawing effects of fluorine. At 65°C and 200 bar, 99% conversion of 1-octene to C_9 aldehydes was achieved at 1-octene/Rh ratios up to 2650:1. The selectivity obtained with the phosphite ligand was significantly higher than the selectivities obtained with both arylphosphine ligands. Increasing the [P]/Rh ratio from 4 to 10 with P[3-$(CH_2)_2(CF_2)_6$F-$C_6H_4]_3$ increased the selectivity from 3.2 to 5.6. Additional increases in [P]/Rh led to a dramatic reduction of the reaction rate. All three catalysts showed similar reaction profiles in $scCO_2$ and in toluene at comparable conditions and the maximum turnover frequency (TOF) in toluene for the unmodified PPh_3 was slightly lower than the substituted phosphines. Remarkably, the selectivities obtained in $scCO_2$ with substituted phosphines were significantly higher than selectivities obtained in toluene with unsubstituted PPh_3. Furthermore, commonly observed isomerization with phosphite ligands in conventional organic solvents were found to be absent in $scCO_2$ with modified phophites. This was attributed to the absence of mass transfer limitations in the $scCO_2$ system. The reaction studies conducted in the absence of any phosphorous modifiers proceeded efficiently; however, the selectivities were low.

Palo and Erkey (83) described the synthesis of the catalyst *trans*-RhCl(CO)[P(*p*-$CF_3C_6H_4)_3]_2$. The catalyst formed a bright yellow solution in $scCO_2$, exhibiting a solubility of at least 5.5 mM ($T = 343$ K, $P = 273$ atm, $\rho = 0.77$ g/ml) in the reaction mixture, a value comparable to that obtained by Kainz et al. of 4.4 mM for the complex [Rh(hfacac)$(PR_2)_2$, where R = 4-$(CH_2)_2(CF_2)_6$F-C_6H_4]. The high solubility of *trans*-RhCl(CO)[P(*p*-$CF_3C_6H_4)_3]_2$ showed the dramatic enhancement made possible by incorporation of small trifluoromethyl groups into the aryl rings of the phosphine. Nuclear magnetic resonance (NMR) and Fourier transform infrared spectroscopy (FTIR) results indicated only slight alteration in the electronic properties of the catalyst with addition of the CF_3 groups. Not only was the new catalyst highly soluble in $scCO_2$, it was also active for the hydroformylation of 1-octene at 343 K and 273 atm, producing C_9 aldehydes with negligible hydrogenation or isomerization. Complete conversion of 1-octene (0.9 M) was obtained in 27 h using a catalyst concentration of 2.0 mM and a total H_2/CO pressure of 68 atm (at 343 K). A significant induction period was observed at the beginning of the reaction. Selectivity for nonanal was good, with a normal to branched ratio of 2.4 ± 0.1 throughout the reaction, which compared well with the results of Evans et al. (35) for the hydroformylation of 1-pentene (3.9 M) using *trans*-RhCl(CO)$(PPH_3)_2$ (10 mM) in benzene solution at 343 K with 100 atm total pressure of 1:1 H_2/CO. They observed complete conversion in 16 h and a normal to branched ratio of about 2.7. These preliminary results indicated that the modified catalyst behaved in $scCO_2$ similarly to the unmodified catalyst in benzene, despite the addition of highly electron-withdrawing trifluoromethyl groups to the phosphines.

Subsequently, Palo and Erkey (36) described the synthesis of RhH(CO)[P(*p*-$CF_3C_6H_4)_3]_3$. At 273 bar and 323 K, 1-octene hydroformylation ($[CO]_0 = [H_2]_0 = 1.05$ M, $[\text{1-octene}]_0 = 0.95$ M, [catalyst] = 0.63 mM) proceeded with no observable isomerization or hydrogenation of 1-octene and produced exclusively C_9 aldehydes. Nearly complete conversion of 1-octene was achieved in about 3.5 h, with an *n/iso* ratio of 3.0, which was comparable to the selectivity of the conventional catalyst RhH(CO)$(PPh_3)_3$ in benzene. Experiments performed at different total pressures showed no effect on either the reaction rate or the selectivity. Below 160 atm, the reactants, catalyst, and $scCO_2$ were not completely miscible at the concentrations employed. The reaction rate had a nearly first-order dependence on catalyst concentration. Catalyst concentration also had a direct effect on selectivity; the *n/iso* ratio increased from 2.96 to 3.83 as the catalyst concentration increased from 0.63 mM to 7.61 mM. This was similar to the behavior of the conventional catalyst in organic solvents.

Erkey and Palo (37) investigated the performance of RhH(CO)-[P(*p*-$CF_3C_6H_4)_3]_3$ at 273 bar and 323 K for hydroformylation of a wide variety of unsaturated compounds. Reactions involving substrates containing unsubstituted terminal double bonds (1-dodecene, 1-decene, styrene, allylbenzene) had roughly the same initial rate and showed similar behavior throughout the reaction. For each reaction, 80–90% conversion was achieved in around 2 h. Not surprisingly, the reaction rates for compounds with unsubstituted terminal double bonds were more than an order of magnitude higher than for compounds with substituted or internal double bonds such as 2-octene. Furthermore, the reaction rate for cyclohexene was an additional order of magnitude lower than for 2-octene, whereas 1-octyne was not converted at all. The trends in reaction rate and selectivity were quite similar to those obtained by Wilkinson using the standard triphenylphosphine catalyst in benzene. The selectivity behavior of the unsubstituted terminal double bonds was similar to that observed previously for 1-octene with *n/iso* ratios between 2.7 and 3.5. However, hydroformylation of styrene produced an 11:1 ratio in favor of the branched product. In the case of 2-methyl-1-heptene, 3-methyloctanal was formed exclusively, and hydroformylation of *trans*-2-octene produced almost equal amounts of the two isomers, 2-methyloctanal and 2-ethylheptanal.

Bach and Cole-Hamilton (38) described the use of commercial $P(C_2H_5)_3$ ligands for rhodium-catalyzed hydroformylation in $scCO_2$. The catalyst was formed in situ from $Rh_2(OAc)_4/P(C_2H_5)_3$ with [Rh] around 6.5 mmol/L. The reaction mixture was homogeneous at 100°C even at such high catalyst concentrations. The selectivities were around 2.5 and decreased with increasing CO concentration and with increasing free-phosphine concentration. The TOFs obtained in $scCO_2$ were found to be very close to TOF in toluene at identical conditions. Replacing one ethyl group in $P(C_2H_5)_3$ with a fluorinated chain ($-CH_2CH_2C_6F_{13}$) increased the reaction rate slightly, which was attributed to a slight change of the electron density on the rhodium.

Palo and Erkey (39) reported the first kinetic study of the rhodium-catalyzed homogeneous hydroformylation of olefins in $scCO_2$. The kinetics of the hydroformylation of 1-octene using HRh(CO)[P(*p*-$CF_3C_6H_4)_3]_3$ in $scCO_2$ was investigated at 50°C and 273 atm. The expression that best represented the experimental data was

$$r_{(1\text{-}octene)} = \frac{kA^aC^cD^d}{1 + K_BB^b}$$

where $A = [H_2]$, $B = [CO]$, $C = [\mathbf{1}]$, and $D = [\text{1-octene}]$, the rate is in units of mol dm^{-3} min^{-1}, and concentrations are expressed in mol dm^{-3}. The optimized rate parameters were determined to be $k = 6.2 \pm 1.2\ \mathrm{dm}^{2.46}\ \mathrm{min}^{-1}\ \mathrm{mol}^{-0.82}$, $K_B = 0.69 \pm 0.16\ \mathrm{dm}^{6.6}\ \mathrm{mol}^{-2.2}$, $a = 0.48 \pm 0.04$, $b = 2.2 \pm 0.3$, $c = 0.84 \pm 0.03$, $d = 0.50 \pm 0.05$. The observed kinetic behavior was found to differ significantly from behavior of conventional systems employing the nonfluorinated analogue, $HRh(CO)(PPh_3)_3$, in organic solvents. Most notable were the approximately 0.5 order H_2 dependence of the reaction rate, the lack of substrate inhibition, and the absence of a critical catalyst concentration. The altered kinetic behavior relative to conventional systems may be due to $scCO_2$ solvent effects, the modified phosphine ligand, or the high concentrations of reactant gases in the system.

Palo and Erkey (40) described the synthesis and characterization of several fluoroalkyl- or fluoroalkoxy-substituted arylphosphines and investigated the effect of ligand modification for homogeneous hydroformylation of 1-octene in $scCO_2$. The variations of 1-octene concentration with time for all the different phosphines are given in Figure 3. The activity of the rhodium complex [formed in situ from $Rh(CO)_2(acac)$ and L] increased with decreasing basicity of the phosphine according to the series $P[3,5\text{-}(CF_3)_2C_6H_3]_3 > P(4\text{-}CF_3C_6H_4)_3 \approx P(3\text{-}CF_3C_6H_4)_3 > P(4\text{-}CF_3OC_6H_4)_3 > P[4\text{-}F(CF_2)_4(CH_2)_3C_6H_4]_3$. Among the various catalyst systems, the initial rates of reaction differed between the fastest

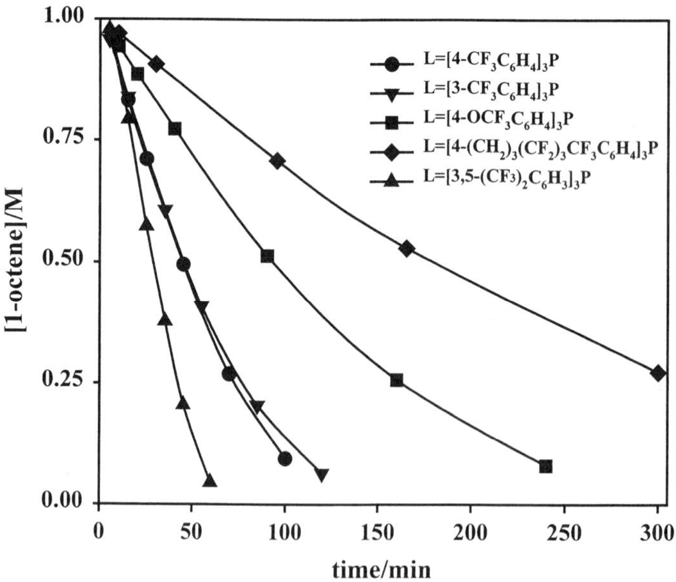

Figure 3 Effect of ligand modification on rhodium-catalyzed hydroformylation in $scCO_2$.

and the slowest by more than fivefold. The rate of hydroformylation correlated very well with the IR stretching frequency of metal carbonyls, ν(CO), for the various catalysts, showing a linear dependence. The ν(CO) is a sensitive indicator of electron density at the metal center, yielding a relative measure of the amount of π-back-bonding from an occupied metal d orbital to the empty π* orbital of the carbonyl (41–43). The trend in ν(CO) corresponded well with the amount and proximity of the electron-withdrawing fluoroalkyl or fluoroalkoxy groups of the phosphines, with $P[3,5\text{-}(CF_3)_2C_6H_3]_3$ having the greatest effect and $P[4\text{-}F(CF_2)_4(CH_2)_3C_6H_4]_3$ having the least effect relative to the value for PPh_3. The phosphines $P(4\ CF_3C_6H_4)_3$ and $P(3\ CF_3C_6H_4)_3$ showed identical electronic behavior and almost identical activity/selectivity behavior, while $P[3,5\text{-}(CF_3)_2C_6H_3]_3$, with a much higher ν(CO) value, also had a significantly higher activity. The results were in agreement with an earlier report that the electronic effect of trifluoromethyl substitution is independent of *ortho*, *meta*, or *para* placement and that the effects of multiple trifluoromethyl substitutions are cumulative (42). The ability of oxygen and methylene groups to insulate against the electron withdrawing effects of the fluoroalkyl moieties could be observed for phosphines $P(4\text{-}CF_3OC_6H_4)_3$ and $P[4\text{-}F(CF_2)_4(CH_2)_3C_6H_4]_3$, where significant decreases in ν(CO) were accompanied by 50% and 70% decreases in activity, respectively. The results confirmed the observation by Leitner that methylene spacers effectively insulated the phosphorus lone pair from the electron withdrawing effect of the fluoroalkyl chain and also showed that oxygen spacers exhibited a similar, though less profound, effect. While these spacers are effective insulators, they actually decrease the catalytic activity relative to "noninsulated" phosphines.

There were also two studies published in 1999 related to propene hydroformylation in $scCO_2$ by $Co_2(CO)_8$. Guo and Akgerman (44) reported on the effect of reaction conditions on selectivity. The experiments were conducted in a 300-ml reactor and initial charge was 3.8 bar propene, 9.3 bar CO, 0.3 bar H_2, and 0.5 g catalyst. At a constant temperature of 88°C, selectivity increased from 1.6 at 90.6 bar to 2.6 at 194.1 bar. At a constant pressure of 166.5 bar, selectivity decreased from 3.5 at 69°C to 2.3 at 88°C. Selectivity was found to be constant throughout the duration of each run.

Kramarz et al. (45) reported the use of toroid NMR probes to examine the behavior of $P(C_4H_9)_3$-substituted and unsubstituted $Co_2(CO)_8$ catalysts in a wide variety of solvents, including $scCO_2$ and $scCO_2$/toluene mixtures. In pure $scCO_2$, difficulties were encountered with low solubilities of the catalytic species. When toluene was used as the cosolvent at 100°C, ^{31}P NMR spectra indicated that the solubility of the dimer $[CoP(C_4H_9)_3(CO)_3]_2$ was enhanced and the dimer was stable for at least 8 h. However, upon addition of CO and hydrogen, the dimer was lost from solution, which was attributed to the formation of a species that is insoluble in the solvent mixture. However, the use of in situ ^{31}P NMR spectroscopy using high-pressure toroid probes will no doubt contribute significantly to our understanding of homogeneous catalytic reactions in $scCO_2$ that utilize phosphorous ligands.

Francio and Leitner (46) investigated the enantioselective asymmetrical hydroformylation in $scCO_2$ (Scheme 2). The problems associated with the low solubility of (*R,S*)-BINAPHOS in $scCO_2$ was handled by incorporating perfluoroalkyl chains to aryl groups. The 98% conversion of styrene could be achieved in less than 16 h at 60°C and 156 bar at a substrate to rhodium ratios of 1000:1 with an enantiomeric excess (ee) of 91%. Furthermore, a very high regioselectivity of 93% was achieved. The reactions with Cl- and Bu-substituted vinyl arenes proceeded as efficiently giving ee values between 90% and 94% and regioselectivities between 93% and 96%. Studies conducted in other organic solvents, such as benzene and hexane, gave very similar results, indicating that the high regioselectivities were mainly due to ligand substitution effects rather than the reaction medium.

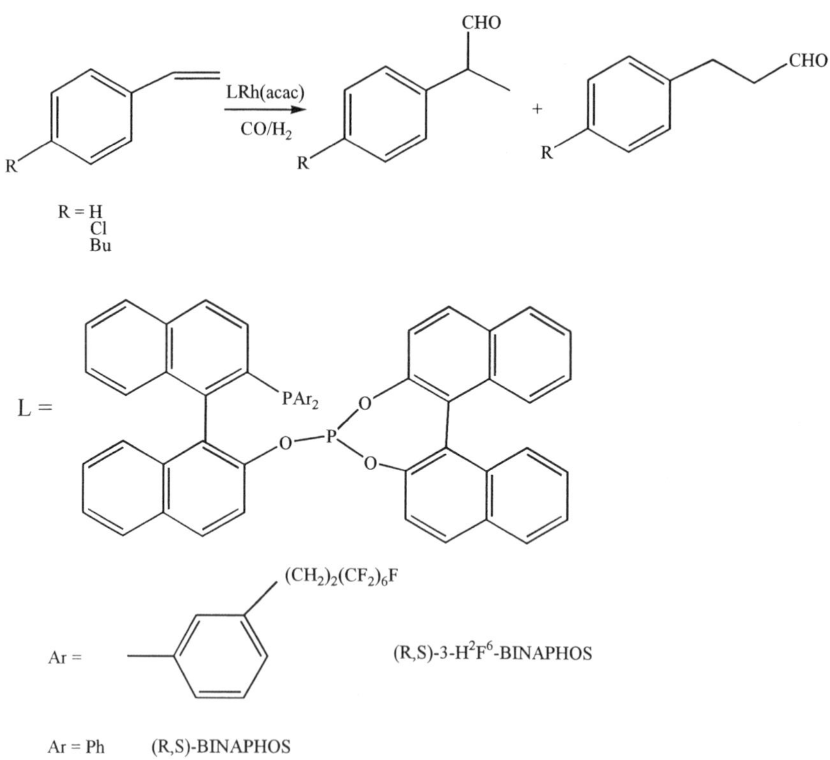

Scheme 2

B. Oxidation

Epoxides, particularly ethylene and propene oxides, are key intermediates used in the production of a wide variety of chemicals and polymers, such as glycols, gylcol ethers, alkanolamines, polyesters, and polyurethanes (47). In the 1960s, Halcon and Arco independently developed processes for the production of epoxides using an alkylhydroperoxide in the presence of homogeneous catalysts based on molybdenum, vanadium, tungsten, titanium, zirconium, and other metals. The two different versions of this process (oxirane technology) are currently used for manufacturing of propene oxide by Arco, and these differ in the hydrocarbon (isobutane or ethylbenzene) that is the precursor to the hydroperoxide [*t*-butyl hydroperoxide (*t*-BuOOH) or ethylbenzene hydroperoxide (EBHP)]. In both of these processes, epoxidation of propene is performed at 100–130°C using 10–300 ppm of Mo. The other commercial application of oxirane technology is the regioselective epoxidation of an allyl alcohol, namely, geraniol. Recent development of the titanium(IV) tartrate catalyst for the asymmetrical epoxidations of allylic alcohols has also resulted in commercial scale production of both (*R*)- and (*S*)-glycidol using *t*-BuOOH. Such epoxidation reactions are usually carried out in hydrocarbon or halogenated hydrocarbon solvents. It may be advantageous to carry out such reactions in $scCO_2$.

Haas and Kolis (48) investigated the oxidation of a variety of olefins to epoxides or diols using $Mo(CO)_6$ as the catalyst precursor and *t*-BuOOH as the oxidant at 200 atm (Scheme 3). Epoxides were obtained in high yields at temperatures below 90°C; however, diols were the predominant products above 90°C. Water had a big effect on selectivity, resulting in hydrolysis of the epoxide at high temperatures. Not surprisingly, attempts to carry out epoxidations with $MoO_2(acac)_2$ or $VO(acac)_2$ were unsuccessful due to the low solubilities of these compounds in $scCO_2$. The lack of reactivity observed at temperatures below 75°C was attributed to the reaction being limited by the dissociation of CO from the $Mo(CO)_6$. Furthermore, cyclohexene was also oxidized in both neat alkene and in benzene in the same reactors under the same conditions.

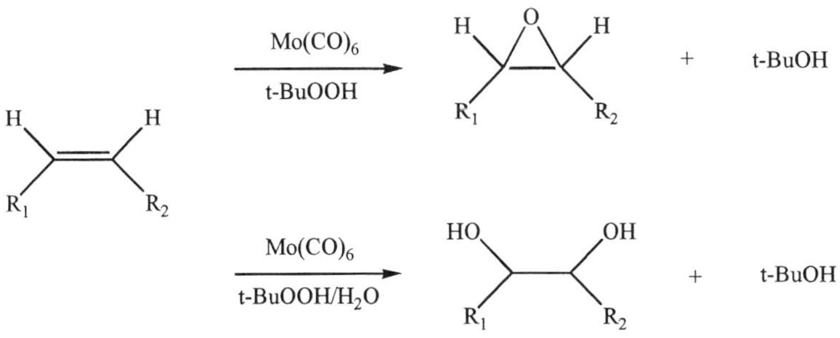

Scheme 3

In both of these cases, the yields were significantly lower than in $scCO_2$ with numerous byproducts. The molybdenum catalyst was found to remain active at the end of the reaction and could oxidize consecutive batches of olefin without any loss of activity.

Pesiri et al. (49) investigated the epoxidations of olefinic alcohols in liquid CO_2 (Scheme 4). The vanadium complex, oxovanadium(V) triisopropoxide [$VO(OiPr)_3$], catalyzed the epoxidation of a wide range of allylic and homoallylic alcohols in liquid CO_2 with *t*-BuOOH as the oxidant. Conversions obtained with the majority of substrates were in excess of 90%. The reaction rate indicated nearly a first-order dependency on concentrations of substrate, catalyst, and the oxidant and a negative 0.4 order dependency on *t*-BuOH, the reduced product of *t*-BuOOH. The rate constants for the $scCO_2$ system were found to be greater than in tetrahydrofuran (THF), *n*-hexane, and CCl_4 and slightly lower than in CH_2Cl_2, acetonitrile, and toluene. Very low conversions were obtained with $VO(acac)_2$, which was attributed to the very low solubility of the catalyst in the reaction mixture. To overcome this problem, complexes of VO with fluorinated β-diketones [1,1,1-trifluoro-2,4-pentanedione (tfac), 6,6,7,7,8,8,8-heptafluoro-3,5-octanedione (hfac)] were synthesized and tested. Significant increases in rates were obtained with $VO(tfac)_2$ and $VO(hfac)_2$, in accordance with a first-order dependency of the reaction rate on concentration of the catalyst in solution.

Haas and Kolis (50) investigated the diastereoselective epoxidation of allylic alcohols in the presence of a vanadylsalen oxo transfer catalyst in $scCO_2$ (Scheme 5). Based on the results obtained with $VO(acac)_2$/*t*-BuOOH catalyst system in their previous study, the authors designed a novel vanadium catalyst with a salen backbone. The rates of epoxidation with V^{IV}(salen*)/*t*-BuOOH were found to be slightly slower than those obtained with many traditional oxygen transfer catalysts in CH_2Cl_2 for epoxidation of a wide variety of allylic alcohols. The olefins without alcohols were found to be epoxidized poorly in yield less than 30%. The results suggested the coordination of the alcohol substrate to the metal center prior to oxygen transfer.

Using *t*-BuOOH as an oxidant results in production of large quantities of *t*-BuOH, which must be handled. Use of air as the oxygen source is more economical. Birnbaum et al. (51) investigated the oxidation of cyclohexene using air catalyzed by halogenated iron porphyrins, namely, Fe(tfpp)Cl and $Fe(tfppBr_8)Cl$ (Scheme 6). The concentrations of these compounds in $scCO_2$ were determined to be approximately 18 and 10 μM. It was found that cyclohexene was oxidized

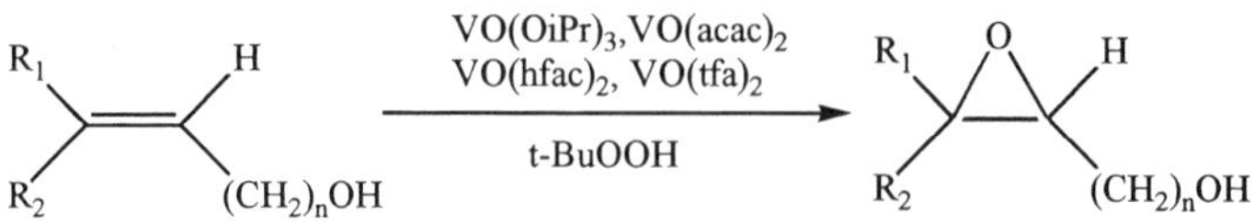

Scheme 4

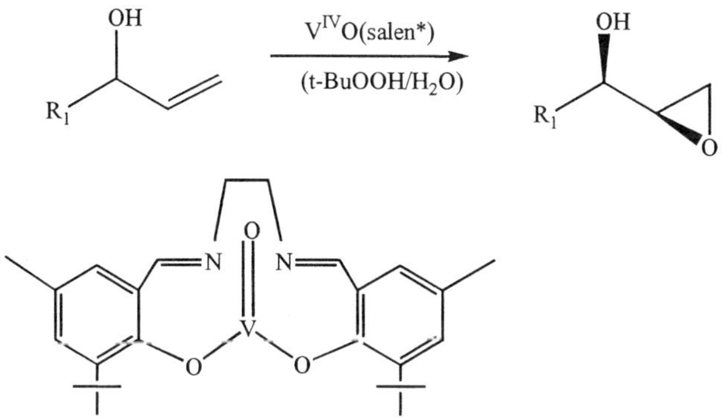

Scheme 5

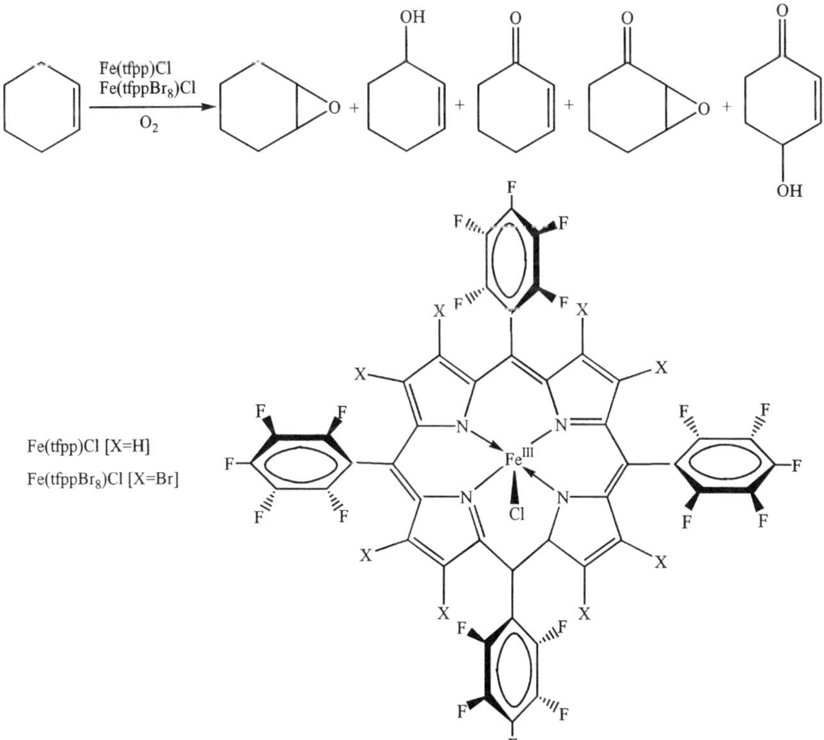

Scheme 6

Scheme 7

to a mixture of compounds. Cyclohexene oxide was derived from epoxidation and 2-cyclohexen-1-ol was derived from allylic oxidation. The other three products were higher oxidation products. The substrate conversions ranged from 0.9% to 12% for Fe(tfpp)Cl and from 3.4% to 22% for Fe(tfppBr8)Cl. Increasing reaction time, temperature, or concentration of oxygen all resulted in higher yields. A UV-visible spectrum of the solution after the reaction indicated significant degradation of the porphyrin. While reactions in methylene chloride produced cyclohexene oxide, 2-cyclohexen-1-ol, and 2-cyclohexen-1-one, reactions in $scCO_2$ also produced the multiply oxidized products. Both epoxides comprised 43–53% of the products using Fe(tfpp)Cl and 45–60% of the products using $Fe(tfppBr_8)Cl$. A series of experiments were also conducted using *t*-BuOOH as the oxidant rather than dioxygen.

Jia et al. (52) reported on the Pd-catalyzed oxidation of methyl acrylate in $scCO_2$ to give the dimethylacetal as major product with an excess of methanol (Scheme 7). The reaction afforded high conversions and selectivities at 40°C and 130 bar. In the presence of cocatalysts $CuCl_2$ and CuCl, the catalytic activity of $PdCl_2$ was found to be higher than that of $PdCl_2(MeCN)_2$. Selectivity for the diacetal ranged from 81% to 96.6%. An increase in the partial pressure of oxygen was found to increase both conversion and selectivity slightly. In the study, an excess amount of methanol was required to promote the partial dissolution of the catalytic species in $scCO_2$. The experiments were conducted in a 25-ml autoclave without visual observation of the reaction mixture. The amount of Pd and Cu salts used for each experiment is around 2 mol %, and it is highly probable that a significant portion of the salts was not solubilized in the reaction mixture.

C. Hydrogenation

Homogeneous hydrogenation is not practiced on an industrial scale due to the existence of efficient processes based on heterogeneous catalysts. However, enantioselective hydrogenation by soluble transition metal complexes that display enzyme-like selectivity is carried out on a large scale in production of compounds such as levodopa and L-phenylalanine. The growing demand for optically pure compounds for use as pharmaceuticals, agrochemicals, and flavors and fragrances is driving the need to develop enantioselective catalyst systems. Carrying out such hydrogenation reactions in $scCO_2$ may result in higher enantioselectivities and faster rates due to high solubility of H_2 in $scCO_2$. In the case of pharmaceuticals, the product purity is a very important issue. The adjustable solvent properties of $scCO_2$ may facilitate development of effective purification procedures. These concepts were first exploited by Burk et al. (21) and Xiao et al. (53) in early studies, and promising results were obtained; however, catalyst solubilities were found to be extremely low. Recently, Kainz et al. (54) investigated the enantioselective hydrogenation of imines in $scCO_2$ (Scheme 8). Cationic iridium(I) complexes with chiral phosphinodihydrooxazoles, modified with perfluoroalkyl groups in the ligand or in the anion, were tested in the hydrogenation of *N*-(1-phenylethylidene)aniline. The iridium complexes containing the anions PF_6 and BPh_4 gave significantly lower enantioselectivities in $scCO_2$ than in CH_2Cl_2 at comparable reaction conditions. However, the iridium complexes containing the tetrakis[3,5-bis(trifluoromethyl)phenyl]borate (BARF) anion led to enantioselectivities comparable with those of CH_2Cl_2. Furthermore, more than 90% conversion of the imine (turnover number = TON > 6800) could be achieved within less than 6 h in $scCO_2$, whereas more than 22 h was required to achieve similar conversions using five times more catalyst (TON = 1400). This large enhancement was attributed to the possible shift of the rate-controlling steps in the established catalytic cycle with changes in the solvent system. Interestingly, the substrate *N*-(1-phenylethylidene)benzylamine could not be hydrogenated efficiently in $scCO_2$, whereas high conversions with high ee values were obtained in CH_2Cl_2.

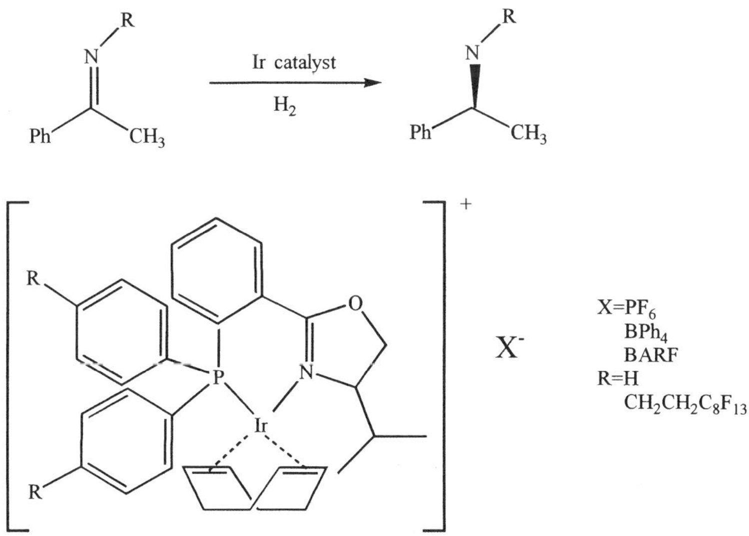

Scheme 8

D. Hydrovinylation

Wegner and Leitner (55) investigated the nickel-catalyzed enantioselective hydrovinylation of styrenes in liquid and supercritical CO_2 (Scheme 9). Three different 4-substituted styrenes were reacted with ethylene in the presence of an (η^3-allyl)-nickel(II) complex containing a chiral 1,2-substituted 1-azaphospholene ligand and a cocatalyst in the temperature range 1–40°C. The use of BARF as the cocatalyst in $scCO_2$ system resulted in activities comparable to the conventional system where the reaction is carried out in CH_2Cl_2 with $Et_3Al_2Cl_3$ as the cocatalyst. Furthermore, the ee values obtained in $scCO_2$ were slightly higher than in CH_2Cl_2 in the temperature range investigated.

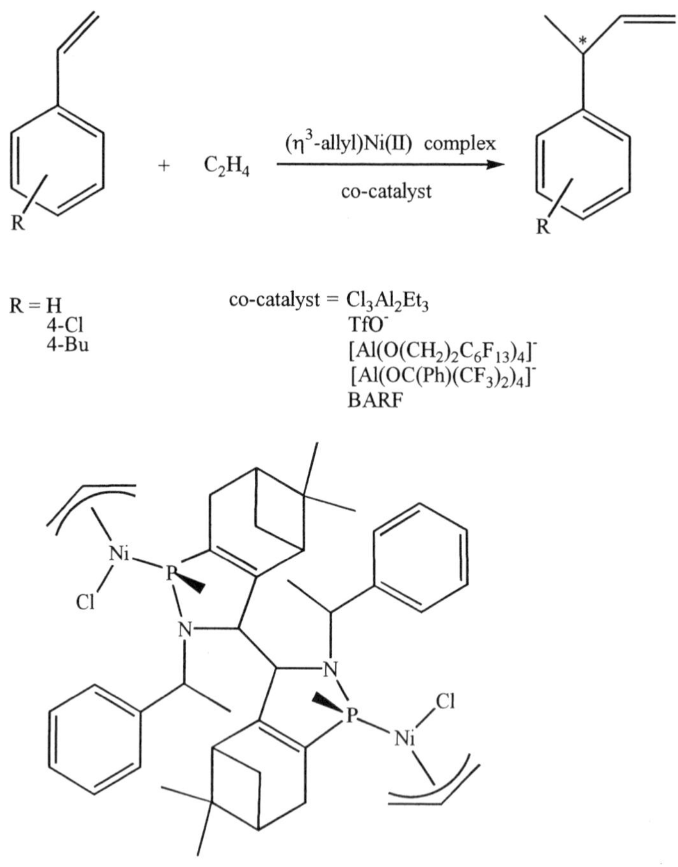

Scheme 9

E. Carbonylation

Kayaki et al. (56) investigated the intramolecular carbonylation of 2-iodobenzyl alcohol in $scCO_2$ using a wide variety of Pd complexes of the form $PdCl_2L_2$ (Scheme 10). The reaction proceeded efficiently in a homogeneous single phase with $P(OEt)_3$ ligands with a TON around 5000. The solubility of the $PdCl_2$-$[P(OEt)_3]_2$ in $scCO_2$ at 130°C and 200 atm was determined to be at least 0.2 mM. On the other hand, the same reaction in toluene at identical conditions resulted in lower TONs ranging from 3100 to 3800. With $PdCl_2(MeCN)_2$, the reaction proceeded to completion, although the complex was observed to be partially soluble at the concentrations employed. $PdCl_2(Me)_2$ was found to be less reactive due to strong coordination of the ligand as observed in conventional organic solvents. $PdCl_2(PPh_3)_2$ was found to be insoluble; however, the reaction proceeded to completion accompanied by metal deposition. The replacement of the phenyl groups in PPh_3 with methoxy groups increased the rate of reaction due to an increase in solubility of the resulting Pd complexes.

F. Diels–Alder Reactions

Oakes et al. (57) investigated the reaction between various acrylates and cyclopentadiene catalyzed by the Lewis acid catalyst $Sc(OTf)_3$ (Scheme 11). The reaction went to completion within 15 h at 50°C. Endo/exo selectivities were found to be a strong function of the density of the $scCO_2$ for all three *n*-butyl, phenyl, and methyl acrylates. As pressure increased, the selectivity rose to a maximum and then began to decrease. For *n*-butyl acrylate, the highest selectivity of 24:1 was observed around a density of 1 g/cm^3. On the other hand, the maximal selectivities that could be obtained at atmospheric conditions in toluene and chloroform were 10:1 and 11:1, respectively. The occurrence of the highest

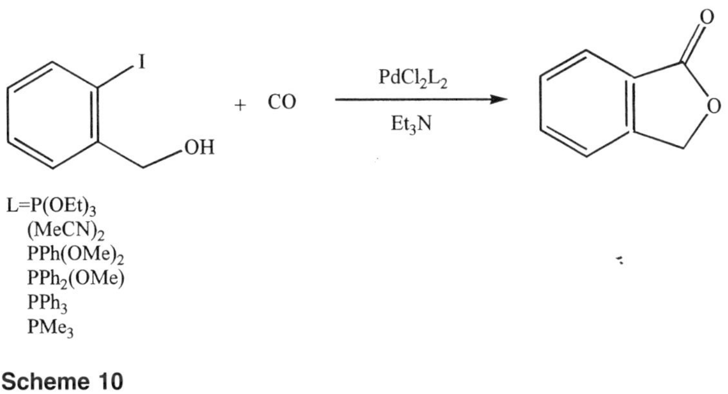

Scheme 10

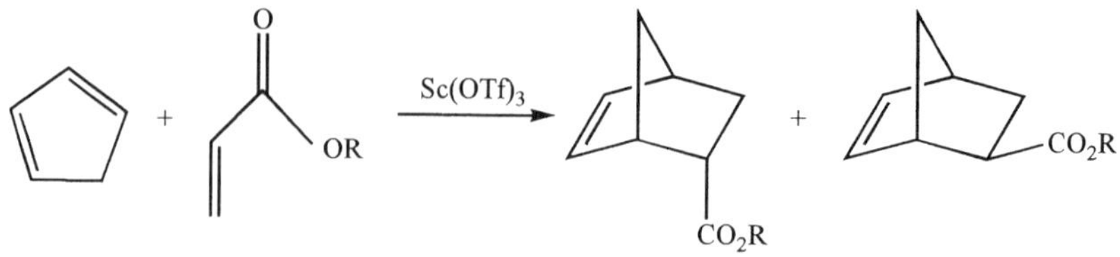

Scheme 11

selectivity significantly above the critical point of the mixture was attributed to the position and number of nearest solvent molecules with respect to a particular transition state.

G. Glaser Coupling Reaction

Li and Jiang (58) reported the first glaser coupling reaction of terminal acetylenes to diacetylenes with $CuCl_2$ in the presence of NaOAc in $scCO_2$ at 40°C and 140 bar (Scheme 12). Addition of methanol was found to be necessary since neither NaOAc and $CuCl_2$ was soluble in $scCO_2$. Visual observations in a cell equipped with a sapphire window showed that addition of methanol (4 vol %) resulted in partial solubilization of these species in $scCO_2$. For all terminal acetylenes, reactions proceeded to completion in 4 h.

H. Biphasic Systems

Biphasic systems were developed to tackle the problems associated with catalyst recovery in homogenoeus catalysis. In such systems, an aqueous phase containing a water-soluble catalyst is contacted with an organic phase containing the reactants. The reaction occurs in the water phase or at the water–organic interface. Upon completion of the reaction, the organic phase, which contains the products, is separated easily from the aqueous phase, which is recycled. For reactions to proceed at appreciable rates, the reactants should have appreciable solubilities in the aqueous phase. The use of $scCO_2$ instead of an organic solvent in aqueous biphasic systems may be advantageous for reactions involving

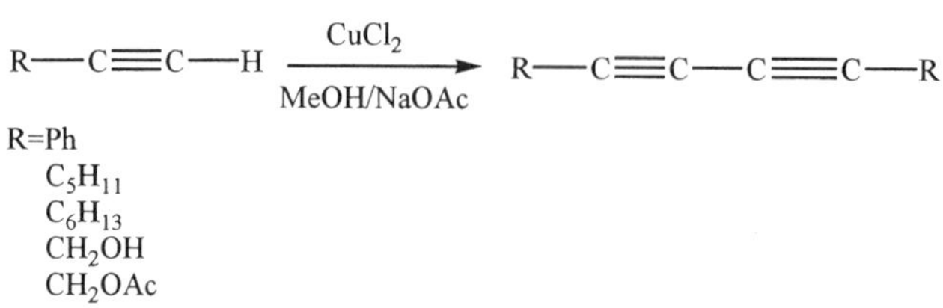

Scheme 12

gaseous substrates if there are some problems associated with mass transfer. Along these lines, Bhanage et al. (59) investigated the hydrogenation of cinnamaldehyde in the water-$scCO_2$ biphasic system using water-soluble catalysts. Using $RuCl_3/P(C_6H_4SO_3Na)_3$ as the catalyst precursor, cinnamaldehyde was converted to unstaurated alcohol with a selectivity of 99%. On the other hand, use of $RhCl_3/P(C_6H_4SO_3Na)_3$ resulted in 100% selectivity toward the saturated aldehyde. In both cases, conversion was around 35% for a reaction time of 2 h. The significant enhancement of the rate of reaction in the water–$scCO_2$ solvent system over toluene–water solvent system [for $RuCl_3/P(C_6H_4SO_3Na)_3$] was attributed to enhancement of the rate of mass transfer of H_2 across the water–solvent boundary due to higher concentration of H_2 in the $scCO_2$ phase. $Pd(OAc)_2/P(C_6H_4SO_3Na)_3$ in the water–$scCO_2$ system gave significantly lower conversions than $RhCl_3/P(C_6H_4SO_3Na)_3$, but selectivities were comparable. An interesting aspect of this study is the high $scCO_2$/water ratio (50:1) used.

In a similar study, Bhanage et al. (60) investigated the Heck vinylation of iodobenzene with butyl acrylate and styrene reaction using the water-soluble complex $Pd(OAc)_2–P(C_6H_4SO_3Na)_3$. Naturally, such catalysts are insoluble in $scCO_2$ and subsequently the rates were found to be very low. Addition of cosolvents, such as water and ethylene glycol, was found to increase the reaction rates considerably, which was attributed to the solubility enhancement of the catalytic species in the fluid phase. However, it is not possible to dissolve 1 ml of water in the $scCO_2$ phase in a 50-ml reactor at 60°C based on solubility data of water in $scCO_2$. Therefore, the reaction most likely took place in the aqueous phase of a biphasic water–$scCO_2$ system.

Jacobson et al. (61) reported the use of surfactants in the biphasic water–$scCO_2$ systems. The surfactants used were capable of forming water/CO_2 emulsions with droplet diameters of 0.5–15 μm and surface areas of up to 10^5 m^2/L. The authors investigated the hydrogenation of styrene to ethylbenzene using the water-soluble catalyst $RhCl(tppds)_3$ at 40°C and 4000 psia with a 50/50 wt % water/CO_2. Emulsions were formed from shear across a 762-μm-diameter capillary tube created by a recirculating pump. The reaction rates for the emulsion system were found to be considerably larger than conventional biphasic water–organic solvent systems and also greater than water–$scCO_2$ systems without any surfactant. The TOFs were comparable to the rates of single-phase hydrogenations using Wilkinson's catalyst.

IV. FUNDAMENTAL STUDIES

Urdahl et al. (62) reported the first vibrational relaxation measurement of a transition metal complex, $W(CO)_6$, in $scCO_2$. The authors determined the lifetime of the T_{1u} asymmetrical CO stretching mode of $W(CO)_6$ at 1990 cm^{-1} as a function of density at 33°C and 50°C. At the lowest densities, the vibrational lifetime became shorter as the density increased. However, in the vicinity of the critical point, the lifetime was independent of density over a wide range. As the density was increased more, the lifetime decreased. On the other hand, the lifetime decreased with density in a monotonous manner at 50°C. The observance of such a plateu at 33°C suggested that the local density experienced by the $W(CO)_6$ solute was independent of the bulk density. In a subsequent study by the authors (63), the same behavior was also observed for $Rh(CO)_2acac$, indicating a universal phenomena. A theory based on density functional methods was used to show that near the critical point, factors that could lead to density-dependent frequencies and lifetimes scaled out of the problem as result of the divergence of the correlation length of the density fluctuations.

Flash photolysis was extended to study of the ring closure reaction of $M(CO)_5$(1,10-phenanthroline) where M = W, Mo in $scCO_2$ (64). The reaction proceeds through the following elementary steps:

$$
\begin{aligned}
M(CO)_6 &\rightarrow M(CO)_5 + CO \\
M(CO)_5 + \text{solv} &\rightarrow M(CO)_5(\text{solv}) \\
M(CO)_5(\text{solv}) + L &\rightarrow M(CO)_5L + \text{solv} \\
M(CO)_5L &\rightarrow M(CO)_4L + CO
\end{aligned}
$$

where the last reaction is rate determining. For $W(CO)_5$(1,10-phenanthroline), the observed rate constants were found to increase dramatically in the vicinity of the critical point. The activation volume for the reaction at slightly higher temperature and pressure than the critical point of CO_2 (35°C and 1145–1160 psi) was determined as 7.2×10^3 cm^3/mol using the logarithmic dependence of the rate constant on pressure. In contrast, the activation volumes ranged from -4.0 to 3.6 cm^3/mol in organic solvents. This large increase was attributed to the large compressibility of the solvent in the vicinity of the critical point. An elegant analysis based on the van der Waals model that separates the activation volume into repulsive and attractive contributions showed a very large repulsive contribution to the activation volume. Interestingly, the same type of rate enhancement in the vicinity of the critical point at 30°C and 1000 psi was not observed with $Mo(CO)_5$(1,10-phenanthroline) even though the compressibility of CO_2 is also very high at these conditions.

Subsequently, the same group investigated the effect of using cosolvents in $scCO_2$ on activation volumes for the ring closure reaction of $W(CO)_5$(2,2′-bipyridene) (65). With benzene as the cosolvent, the activation volumes were found to be significantly smaller than the phenanthroline system in pure $scCO_2$ and decreased with increasing concentration of cosolvent in the mixture ranging from 66.2 to 6.5 cm^3/mol. This was attributed to many factors, such as the smaller compressibility of the solvent mixture with the cosolvent, a lack of preassociation of bipyridene with the metal center as opposed to phenanthroline, and formation of solute-solvent clusters.

Sun et al. (66) investigated the reaction of $W(CO)_5(CO_2)$ with CO in $scCO_2$ using fast time-resolved infrared (TRIR). The reaction is described by the following elementary steps:

$$W(CO)_5(CO_2) \leftrightarrow W(CO)_5 + CO_2$$
$$W(CO)_5 + CO \rightarrow W(CO)_6$$
$$W(CO)_5(CO_2) + CO \rightarrow W(CO)_5 + CO$$

The first two reactions represent the dissociative pathway and the third reaction represents the associative pathway. The authors measured the rate constants for the reaction as function of CO_2 pressure without altering the CO concentration. The decrease in k with pressure provided evidence that the reaction was predominantly a dissociative process. The order of reactivity with different metals was found to be Cr $\approx$ Mo $>$ W. Significant shifts observed in the visible absorbance spectrum recorded after 100 ns for flash photolysis of $W(CO)_6$ in $scCO_2$ at 2500 psi and 40°C was attributed to the possibility of η^1-O coordination of CO_2 to the metal center.

V. EXPERIMENTAL METHODS

The reactions can be conducted in standard high-pressure vessels (67). It is useful to visually observe the contents of the vessel to make sure that the reaction is taking place in a single phase. Furthermore, many homogeneous catalytic reacting mixtures exhibit a color change during the course of the reaction that may provide clues to the formation of various species. A schematic diagram of the experimental apparatus used in our laboratory for homogeneous catalytic reactions in $scCO_2$ is given in Figure 4. The setup consists of a custom-manufactured, 54-ml stainless steel reactor fitted with two sapphire windows (1 in. ID, Sapphire Engineering), polyether ether ketone O rings (Valco Instruments), a T-type thermocouple assembly (Omega Engineering, DP41-TC-MDSS), pressure transducer (Omega Engineering, PX01K1-5KGV), vent line, and rupture disk assembly (Autoclave Engineers). The reactor rests on a magnetic stir plate and is heated to appropriate temperatures by a circulating heater (Haake FJ) via a machined internal heating coil. The reactor is pressurized with CO_2 from a syringe pump (ISCO, 260D) equipped with a cooling jacket to the desired reaction pressure. For kinetic information, periodic samples can be taken through a high-pressure sample loop by filling with the supercritical fluid mixture, depressurizing into a sample vial, and flushing with solvent from a reservoir. Subsequently, the sample loop is dried with compressed air. It is important to make sure that the volume of the sample loop is greater than the volume of the line from the reactor to the sample loop. Furthermore, care should be exercised in designing the sampling system to ensure that the volume of the sample loop is very small compared with the volume of the reactor.

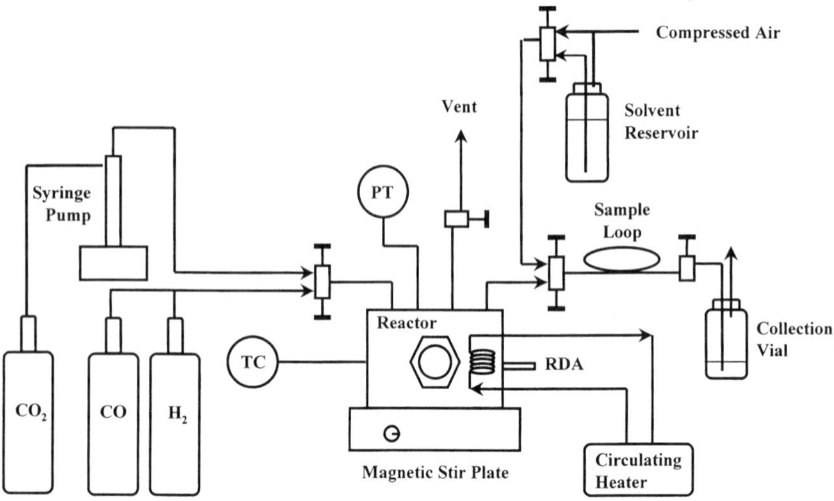

Figure 4 Schematic diagram of reactor setup.

The vessel can easily be used to measure the solubilities of the catalysts in $scCO_2$ by a method recently developed by our group. A detailed schematic diagram of the internal configuration of the pressure vessel for solubility measurements is given in Figure 5. For each experiment, an excess amount of solute and a small magnetic stir bar is placed in a 5-ml glass vial that is then capped with coarse filter paper (Whatman) attached to the vial with Teflon tape. A larger magnetic stir bar is placed inside the vessel; then a perforated stainless steel disk is slid through the thermowell. The sample vial is weighed and placed inside the pressure vessel. The perforated disk isolates the external stir bar from the stir bar in the sample vial and prevents the two stir bars from becoming coupled, which stops both from stirring. The perforations on the disk enable sufficient convection for mixing both sides of the pressure vessel. The vessel is then sealed, connected to the circulating bath, and heated to the desired temperature. Once the desired temperature is reached, stirring is initiated and the vessel is slowly filled with CO_2 until the desired pressure is reached. After allowing sufficient time for equilibration of the $scCO_2$-solute solution, the vessel is depressurized and opened. The vial is removed and reweighed. The solubility of the solute is then readily calculated by the weight difference.

Developing novel homogeneous catalysts for a particular reaction or improving the performance of existing catalysts with minor modifications requires an understanding of the underlying reaction mechanism. Monitoring the concentration of reactants and products is not sufficient to accomplish this goal. Kinetic studies should be coupled with spectroscopic data. Over the past few decades, the utilization of spectroscopic methods in identification of the reac-

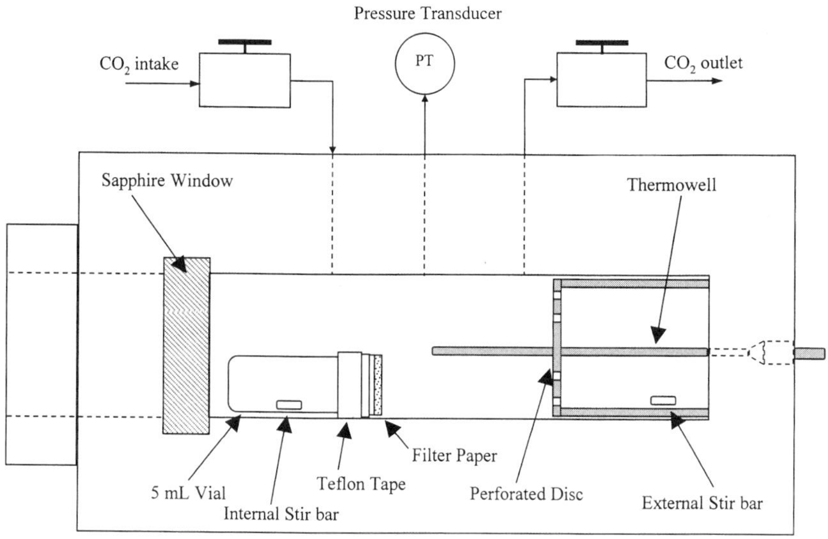

Figure 5 Schematic diagram of vessel internals for solubility measurement.

tive intermediates in solution has contributed significantly to the advances made in transition metal catalysis. Therefore, the utilization of in situ spectroscopic studies should aid development of our current understanding of homogeneous catalytic reactions in $scCO_2$ which, at present, is very limited.

NMR spectroscopy is a particularly powerful tool in homogeneous catalysis. There are three established methods to obtain high-pressure NMR spectra. The oldest method utilizes nonmagnetic metallic high-pressure probes wherein the sample compartment and the radiofrequency coil are in direct contact with the pressure transmitting medium. Among the detectors, toroid detectors first described by Rathke are more efficient than other detectors based on solenoid or Heimholtz coils due to minimization of magnetic coupling with the pressure vessel. The first method has been used extensively by Rathke and coworkers to study $Co(CO)_8$-catalyzed hydroformylation of olefins in $scCO_2$, providing valuable information on the nature of the intermediates. A schematic diagram of the probe is given in Figure 6. Such NMR probes can be operated routinely up to 300 atm and 250°C. The disadvantage of the method is the requirement for specialized equipment, such as wide-bore NMR magnets. The second method, developed by Roe (68), involves using sapphire NMR tubes, which can stand high pressure. The key in the design is a nonmagnetic valve constructed of titanium alloy that provides a means for introducing the fluid and for sealing the tube. The tube is mounted to the titanium alloy flange with an epoxy adhesive, which can withstand pressures up to 1000 bar. Subsequently, Horvath and Ponce (69) reported an improved version with a lighter valve design that is also given

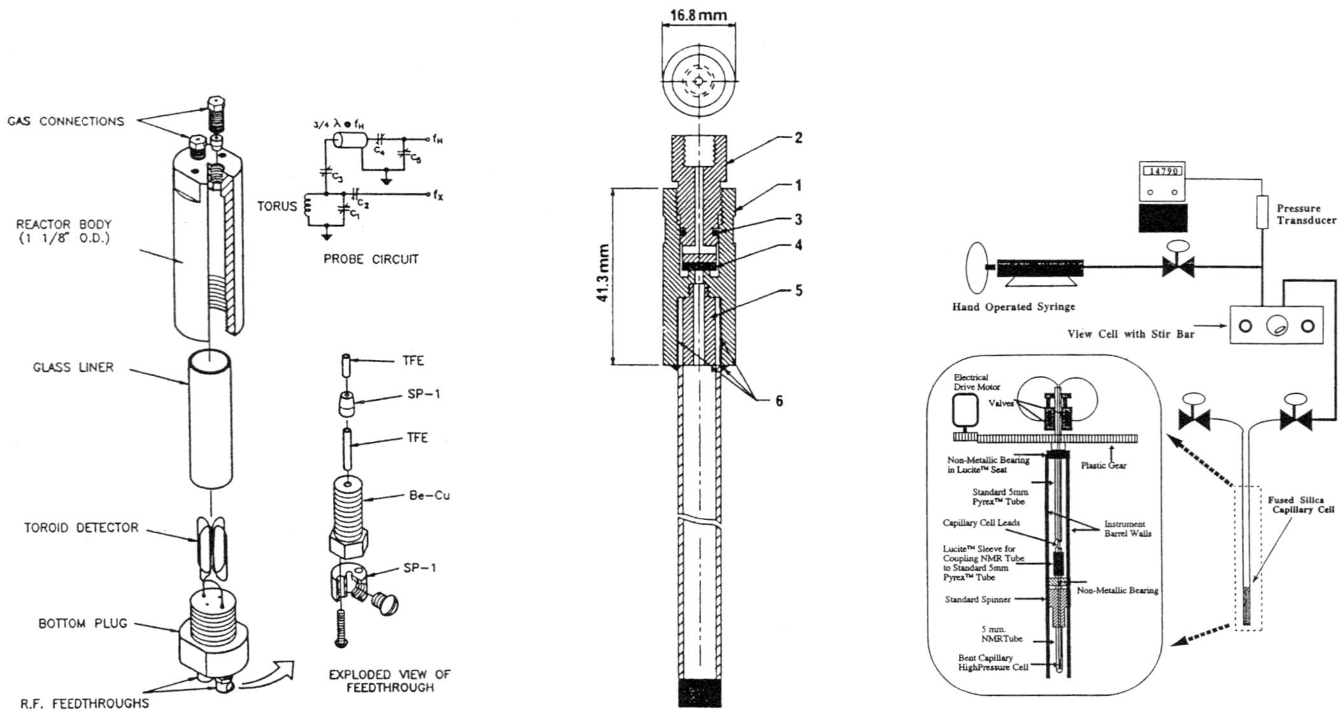

Figure 6 High-pressure NMR methods.

in Figure 6. The third method was developed by Yonker et al. (70) and involves a cell based on fused-silica capillary tubing bent repeatedly to permit multiple passes for use in an existing 5- or 10-mm standard NMR tube. The small internal volume of the tubing eliminates the hazards associated with rupture. The inlet and outlet of the cell are connected to the pressure generation equipment with standard high-pressure fittings. The cell can be used without sample spinning or with spinning with the aid of an external spinner drive assembly, as seen in Figure 6. The last method was recently used to probe and provided evidence for the specific solute–solvent interactions of fluorocarbons dissolved in $scCO_2$ (71). Extension of these studies to probing the interactions of fluorinated homogeneous catalysts with $scCO_2$ and their effects on reactions would be a very valuable contribution to the field. A good review of the literature on use of high-pressure NMR for investigation of reactions involving transition metal complexes is given by Horvath and Millar (72).

In situ IR spectroscopy is another useful method for identification of intermediates, especially in reactions that involve CO due to the strong intensities of carbonyl species. An excellent overview and applications of IR spectroscopy for studying reactions in SCFs is given by Bubback (73). Fyhr (74) described the use of a high-pressure IR flow-through cell that was used to monitor reactions in a high-pressure stirred autoclave. A pump was used to recirculate the fluid from the reactor to the cell and back to the autoclave. The CaF_2 single-crystal windows had dimensions of 40 mm diameter × 15 mm thickness. The cell was situated in a commercial spectrophotometer. Some systems use a high-pressure vessel with a zinc selenide bottom to transmit light back and forth to an FTIR spectrophotometer beneath the lab bench surface. A recent development in the field has been the development of high-pressure probes with diamond tips that can be immersed in reacting mixtures in high-pressure vessels through standard connections. The probes deliver light to the reaction mixture and carry absorption data back to a spectrophotometer. Actually, this system was used by Kainz et al. (54) to monitor the iridium-catalyzed hydrogenation of imines in $scCO_2$.

VI. HIGH-PRESSURE PHASE BEHAVIOR

One of the cited advantages of using $scCO_2$ as a solvent in homogeneous catalytic reactions has been the ease of recovery of the catalyst from the reaction mixture, as well as the ease of separation of reactants and products by temperature and pressure tuning. Unfortunately, the studies reported so far have mainly focused on the kinetic behavior of these reactions in $scCO_2$ without paying much attention to downstream processing aspects of the technology. The primary reason behind this neglect is the enormously complicated nature of the high-pressure phase behavior of multicomponent mixtures in supercritical gases.

A knowledge of the phase behavior of the mixture is necessary for various reasons:

1. Because the chemical reaction changes the chemical composition of the mixture, the critical point of the mixture changes. In some cases, the mixture may phase separate, resulting in an undesirable situation.
2. At the end of the reaction, design of schemes for separation of the products and unconverted reactants from the $scCO_2$ by temperature/pressure tuning requires a knowledge of the phase boundaries for the mixture.

There have been major developments within the past two decades in terms of techniques for determination and prediction of high-pressure equilibria for complex mixtures. A very good introduction to the phase behavior of multicomponent mixtures in supercritical fluids is given by McHugh and Krukonis (5). Some closely related examples are the recent investigations for enzyme-catalyzed reactions in $scCO_2$. Chrisochoo et al. (75) measured and modeled the phase behavior of the composing binaries of a lipase-catalyzed transesterification reaction and predicted the phase behavior of a five-component system, using the Soave–Redlich–Kwong EOS. Stevens et al. (76) measured the vapor-liquid equilibria for the ternary system carbon dioxide + vinyl acetate + 2-butanol. The experimental bubble points were predicted with reasonable success using the Stryjek–Vera modification of the Peng–Robinson equation of state with Wong–Sandler mixing rules. The parameters used were obtained from fits of the binary subsystems.

Another factor that further complicates the problem is the phase behavior of the catalyst, which is in extremely small concentrations in the mixture in comparison with the reactants or products. When the homogeneous product mixture phase separates as a result of a change in temperature and/or pressure, the catalyst distributes itself between the two phases. It is desirable that the catalyst will favorably partition into the CO_2-rich phase. It is the magnitude of this distribution coefficient that governs whether or not the catalyst can be recovered. Naturally, the distribution coefficient is a function of temperature, pressure, and composition. Unfortunately, no distribution coefficient measurements have been reported on relevant systems so far. It is also very difficult to predict such distribution coefficients using standard EOS-based methods since many of the physical parameters required in the calculations are not known. Some of these are vapor pressure, critical temperature and pressure, and eccentricity. On another note, such organometallic complexes are known to decompose before their critical temperatures are reached. It is more appropriate to utilize modified regular solution theories or linear solvation energy relationships for predicting the distribution coefficients for these types of compounds (96). As a first approximation and for screening purposes, the distribution coefficient of the catalyst between the two phases can be approximated by the ratio of the solubilities of the complex in two phases.

VII. CONCLUSIONS AND FUTURE WORK

The field of homogeneous catalysis in $scCO_2$ is still in its early development stages. A large percentage of the literature has been in the form of short communications. These studies have demonstrated that a wide range of homogeneous catalytic reactions can proceed in $scCO_2$ at comparable or better rates than in conventional solvents. The problems associated with extremely low solubilities of conventional homogeneous catalysts in $scCO_2$ have been solved by the incorporation of CO_2-philic groups, specifically fluorous groups, into conventional ligands. In two cases, these modifications have resulted in catalysts, which were found to be superior compared to unmodified catalysts. Detailed investigations of the nature of such effects should lead to development of highly efficient new catalysts for the $scCO_2$ medium. High-pressure spectroscopic methods are just beginning to be used in the field and will contribute to our understanding of the effects of the reaction medium on catalytic cycles.

The catalyst recovery issues must be addressed, and reaction studies should be coupled with thermodynamic studies on the phase behavior of the system under consideration. The recovery and recycling issue should be handled at the ligand design stage. Computational chemistry can also be a valuable tool in elucidating the effects of ligand modification on the catalytic cycle. The field seems to have great potential in terms of development of industrial scale processes that utilize $scCO_2$ as a reaction solvent.

REFERENCES

1. A Mortreux, F Petit, eds. Industrial Applications of Homogeneous Catalysis. Dordrecht: D. Reidel, 1988.
2. GW Parshall, SD Ittel. Homogeneous Catalysis: The Applications and Chemistry of Catalysis by Soluble Transition Metal Complexes. 2nd ed. New York: John Wiley & Sons, 1992.
3. B Cornils, WA Herrmann, eds. Applied Homogeneous Catalysis with Organometallic Compounds. Vol. 1. Weinheim: VCH, 1996.
4. M McCoy. Industry Intrigued by CO_2 as Solvent. C&E News 11–14, 1999.
5. MA McHugh, VJ Krukonis. Supercritical Fluid Extraction: Principles and Practice. 2nd ed. Boston: Butterworths, 1994.
6. A Zaks, A.M. Klibanov. Enzyme-catalyzed processes in organic solvents. Proceedings of National Academy of Sciences 82:3192–3196, 1985.
7. DA Hammond, M Kerel, AM Klibanov. Enzymatic reactions in supercritical gases. Appl Biochem Biotechnol 11:393–400, 1985.

8. B Subramaniam, MA McHugh. Reactions in supercritical fluids: a review. Ind Eng Chem Process Des Dev 25:1–2, 1986.
9. G Kaupp. Reactions in supercritical carbon dioxide. Angew Chem Int Ed Engl 33: 1452–1455, 1994.
10. PE Savage, S Gopalan, TI Mizan, CJ Martino, EE Brock. Reactions at supercritical conditions. AIChEJ 41:1723–1778, 1995.
11. O Kajimoto. Solvation in supercritical fluids: its effects on energy transfer and chemical reactions. Chem Rev 99:355–389, 1999.
12. JW Rathke, RJ Klingler, TR Krause. Propylene hydroformylation in supercritical carbon dioxide. Organometallics 10:1350–1355, 1991.
13. PG Jessop, T Ikariya, R Noyori. Homogeneous catalysis in supercritical fluids. Chem Rev 99:475–493, 1999.
14. C Reid, JM Prausnitz, BE Poling. The Properties of Gases and Liquids. New York: McGraw-Hill, 1985.
15. CY Tsang, WB Street. Phase equilibria in the H_2/CO_2 system at temperatures from 220 to 290 K and pressures to 172 MPa. Chem Eng Sci 36:993–999, 1981.
16. C Reichardt. Solvents and Solvent Effects in Organic Chemistry. Weinheim: VCH, 1988.
17. KP Johnston, C Haynes. Extreme solvent effects on reaction rate constants at supercritical fluid conditions. AIChEJ 33:2017–2026, 1987.
18. GJ Suppes, MA McHugh. Solvent and catalytic effects on the decomposition of cumene hydroperoxide. Ind Eng Chem Res 28:1146–1152, 1989.
19. DR Palo, C Erkey. Solubility of dichlorobis(triphenylphosphine)nickel(II) in supercritical carbon dioxide. J Chem Eng Data 43:47–48, 1998.
20. DR Palo, C Erkey. Unpublished results, 1997.
21. MJ Burk, S Feng, MF Gross, W Tumas. Asymmetric catalytic hydrogenation reactions in supercritical carbon dioxide. J Am Chem Soc 117:8277–8278, 1995.
22. PG Jessop, Y Hsiao, T Ikariya, R Noyori. Catalytic production of dimethylformamide from supercritical carbon dioxide. J Am Chem Soc 116:8851–8852, 1994.
23. KE Laintz, CM Wai, CR Yonker, RD Smith. Solubility of fluorinated metal diethyldithiocarbamates in supercritical carbon dioxide. J Supercrit Fluids 4:194–198, 1991.
24. JM DeSimone, EE Maury, YZ Menceloglu, JR Combes, JB McClain, TJ Romack. Dispersion polymerizations in supercritical carbon dioxide. Science 265:356–359, 1994.
25. AV Yazdi, EJ Beckman. Design, synthesis, and evaluation of novel, highly CO_2-soluble chelating agents for removal of metals. Ind Eng Chem Res 35:3644–3652, 1996.
26. S Kainz, D Koch, W Baumann, W Leitner. Perfluoroalkyl-substituted arylphosphanes as ligands for homogeneous catalysis in supercritical carbon dioxide. Angew Chem Int Ed Engl 36:1628–1629, 1997.
27. IT Horvath, J Rabai. Facile catalyst separation without water: fluorous biphase hydroformylation of olefins. Science 266:72–75, 1994.
28. IT Horvath, G Kiss, RA Cook, JE Bond, PA Stevens, J Rabai, EJ Mozeleski. Molecular engineering in homogeneous catalysis: one-phase catalysis coupled with biphase catalyst separation. The fluorous-soluble $HRh(CO)\{P[CH_2CH_2(CF_2)_5CF_3]_3\}_3$ hydroformylation system. J Am Chem Soc 120:3133–3143, 1998.

29. JJJ Juliette, IT Horvath, JA Gladysz. Transition metal catalysis in fluorous media: practical application of a new immobilization principle to rhodium-catalyzed hydroboration. Angew Chem Int Ed Engl 36:1610–1612, 1997.
30. D Sinou, G Pozzi, EG Hope, AM Stuart. A convenient access to triarylphosphines with fluorous phase affinity. Tetrahedron Lett 40:849–852, 1999.
31. J Fawcett, EG Hope, RDW Kemmitt, DR Paige, DR Russell, AM Stuart, DJ Cole-Hamilton, MJ Payne. Fluorous-phase soluble rhodium complexes: x-ray structure of $[RhCl(CO)(P(C_2H_4C_6F_{13})_3)_2]$. Chem Commun 1127–1128, 1997.
32. J Fawcett, EG Hope, RDW Kemmitt, DR Paige, DR Russell, AM Stuart. Platinum group metal complexes of arylphosphine ligands containing perfluoroalkyl pony-tails; crystal structures of $[RhCl_2(eta\text{-}5\text{-}C_5Me_5)\{P(C_6H_4C_6F_{13}\text{-}4)_3\}]$ and cis- and trans-$[PtCl_2\{P(C_6H_4C_6F_{13}\text{-}4)_3\}_2]$. J Chem Soc, Dalton Trans 3751–3763, 1998.
33. CD Frohning, CW Kohlpaintner, Hydroformylation (oxo synthesis, Roelen reaction), in B Cornils, WA Herrmann, eds. Applied Homogeneous Catalysis with Organometallic Compounds. 1996, VCH: Weinheim, 1996, pp. 29–104.
34. D Koch, W Leitner. Rhodium-catalyzed hydroformylation in supercritical carbon dioxide. J Am Chem Soc 120:13398–13404, 1998.
35. D Evans, JA Osborn, G Wilkinson. Hydroformylation of alkenes by use of rhodium complex catalysts. J Chem Soc A 3133, 1968.
36. DR Palo, C Erkey. Homogeneous hydroformylation of 1-octene in supercritical carbon dioxide with $[RhH(CO)(P(p\text{-}CF_3C_6H_4)_3)_3]$. Ind Eng Chem Res 38:2163–2165, 1999.
37. C Erkey, DR Palo. Rhodium Catalyzed Homogeneous Hydroformylation of Unsaturated Compounds in Supercritical Carbon Dioxide. In: M Abraham, RP Hesketh, eds. Reaction Engineering for Pollution Prevention. Elsevier: Amsterdam, 1999.
38. I Bach, DJ Cole-Hamilton. Hydroformylation of hex-1-ene in supercritical carbon dioxide catalysed by rhodium trialkylphosphine complexes. Chem Commun 1463–1464, 1998.
39. DR Palo, C Erkey. Kinetics of the homogeneous catalytic hydroformylation of 1-octene in supercritical carbon dioxide. Ind Eng Chem Res 38:3786–3792, 1999.
40. DR Palo, C Erkey. Effect of ligand modification on rhodium-catalysed homogeneous hydroformylation in supercritical carbon dioxide. Organometallics 19:81–86, 2000.
41. CA Tolman. Phosphorus ligand exchange equilibria on zerovalent nickel. A dominant role for steric effects. J Am Chem Soc 92:2956, 1970.
42. JAS Howell, JD Lovatt, P McArdle, D Cunningham, E Maimone, HE Gottlieb, Z Goldschmidt. The effect of fluorine, trifluoromethyl and related substitution on the donor properties of triarylphosphines towards $[Fe(CO)_4]$. Inorg Chem Commun 1:118–120, 1998.
43. C Li, SP Nolan, IT Horvath. Solution thermochemical study of a fluorous tertiary phosphine ligand in rhodium and ruthenium systems. Organometallics 17:452, 1998.
44. Y Guo, A Akgerman. Determination of selectivity for parallel reactions in supercritical fluids. J Supercrit Fluids 15:63–71, 1999.
45. KW Kramarz, RJ Klingler, DE Fremgen, JW Rathke. Toroid NMR probes for the in situ examination of homogeneous cobalt hydroformylation catalysts at high pressures and temperatures. Catal Today 49:339–352, 1999.
46. G Francio, W Leitner. Highly regio- and enantio-selective rhodium-catalysed asymmetric hydroformylation without organic solvents. Chem Commun 1663–1664, 1999.

47. RA Sheldon, Synthesis of oxiranes. In: B Cornils, WA Herrmann, eds. Applied Homogeneous Catalysis with Organometallic Compounds. VCH: Weinheim, 1996, pp. 411–423.
48. GR Haas, JW Kolis. Oxidation of alkenes in supercritical carbon dioxide catalyzed by molybdenum hexacarbonyl. Organometallics 17:4454–4460, 1998.
49. DR Pesiri, DK Morita, T Walker, W Tumas. Vanadium-catalyzed epoxidations of olefinic alcohols in liquid carbon dioxide. Organometallics 18:4916–4924, 1999.
50. GR Haas, JW Kolis. The diastereoselective epoxidation of olefins in supercritical carbon dioxide. Tetrahedron Lett. 39:5923–5926, 1998.
51. ER Birnbaum, RML Lacheur, AC Horton, W Tumas. Metalloporphyrin-catalyzed homogeneous oxidation in supercritical carbon dioxide. J Mol Catal A 139:11–24, 1999.
52. L Jia, H Jiang, J Li. Palladium (II)-catalyzed oxidation of acrylate esters to acetals in supercritical carbon dioxide. Chem Commun 985–986, 1999.
53. J Xiao, SCA Nefkens, PG Jessop, T Ikariya, R Noyori. Asymmetric hydrogenation of α,β-unsaturated carboxylic acids in supercritical carbon dioxide. Tetrahedron Lett 37:2813–2816, 1996.
54. S Kainz, A Brinkmann, W Leitner, A Pfaltz. Iridium-catalyzed enantioselective hydrogenation of imines in supercritical carbon dioxide. J Am Chem Soc 121: 6421–6429, 1999.
55. A Wegner, W Leitner. Nickel-catalysed enantioselective hydrovinylation of styrenes in liquid or supercritical carbon dioxide. Chem Commun 1583–1584, 1999.
56. Y Kayaki, Y Noguchi, S Iwasa, T Ikariya, R Noyori. An efficient carbonylation of aryl halides catalysed by palladium complexes with phosphite ligands in supercritical carbon dioxide. Chem Commun 1235–1236, 1999.
57. RS Oakes, TJ Heppenstall, N Shezad, AA Clifford, CM Rayner. Use of scandium tris(trifluoromethanesulfonate) as a Lewis acid catalyst in supercritical carbon dioxide: efficient Diels–Alder reactions and pressure dependent enhancement of endo:exo stereoselectivity. Chem Commun 1459–1460, 1999.
58. J Li, H Jiang. Glaser coupling reaction in supercritical carbon dioxide. Chem Commun 2369–2370, 1999.
59. BM Bhanage, Y Ikushima, M Shirai, M Arai. Multiphase catalysis using water-soluble metal complexes in supercritical carbon dioxide. Chem Commun 1277–1278, 1999.
60. BM Bhanage, Y Ikushima, M Shirai, M Arai. Heck reactions using water-soluble metal complexes in supercritical carbon dioxide. Tetrahedron Lett 40:6427–6430, 1999.
61. GB Jacobson, J C. Ted Lee, KP Johnston, W Tumas. Enhanced catalyst reactivity and separations using water/carbon dioxide emulsions. J Am Chem Soc 121:11902–11903, 1999.
62. RS Urdahl, KD Rector, DJ Myers, PH Davis, MD Fayer. Vibrational relaxation of a polyatomic solute in a polyatomic supercritical fluid near the critical point. J Chem Phys 105:8973–8976, 1996.
63. RS Urdahl, DJ Myers, KD Rector, PH Davis, BJ Cherayil, MD Fayer. Vibrational lifetimes and vibrational line positions in polyatomic supercritical fluids near the critical point. J Chem Phys 107:3747–3757, 1997.

64. Q Ji, EM Eyring, Rv Eldik, KP Johnston, SR Goates, M Lee. Laser flash photolysis studies of metal carbonyls in supercritical CO_2 and ethane. J Phys Chem 99:13461–13466, 1995.
65. Q Ji, CR Lloyd, EM Eyring, Rv Eldik. Probing the solvation properties of liquid versus supercritical fluids with laser flash photolysis of $W(CO)_6$ in the presence of 2,2′-bipyridine. J Phys Chem A 101:243–247, 1997.
66. X-Z Sun, MW George, SG Kazarian, SM Nikiforov, M Poliakoff. Can organometallic noble gas compounds be observed in solution at room temperature? A time-resolved infrared (TRIR) and UV spectroscopic study of the photochemistry of $M(CO)_6$ (M = Cr, Mo, and W) in supercritical noble gas and CO_2 solution. J Am Chem Soc 118:10525–10532, 1996.
67. DA Morgenstern, RM LeLacheur, DK Morita, SL Borkowsky, S Feng, GH Brown, L Luan, MF Gross, MJ Burk, W Tumas. Supercritical carbon dioxide as a substitute solvent for chemical synthesis and catalysis. In: PT Anastas, TC Williamson, eds. Green Chemistry: Designing Chemistry for the Environment. Washington, DC: American Chemical Society, 1996.
68. DC Roe. Sapphire NMR tube for high-resolution studies at elevated pressure. J Magn Reson 63:388–391, 1985.
69. IT Horvath, EC Ponce. New valve design for high-pressure sapphire tubes for NMR measurements. Rev Sci Instr 62:1104–1105, 1991.
70. CR Yonker, TS Zemanian, SL Wallen, JC Linehan, JA Franz. A new apparatus for the convenient measurement of nmr spectra in high-pressure liquids. J Magn Res A 113:102–107, 1995.
71. A Dardin, JM DeSimone, ET Samulski. Fluorocarbons dissolved in supercritical carbon dioxide. NMR evidence for specific solute–solvent interactions. J Phys Chem B 102:1775–1780, 1998.
72. IT Horvath, JM Millar. NMR under high gas pressure. Chem Rev 91:1339–1351, 1991.
73. M Buback. Spectroscopy of fluid phases: the study of chemical reactions and equilibria up to high pressures. Angew Chem Int Ed Engl 30:641–653, 1991.
74. C Fyhr, M Garland. Phenomenological aspects of homogeneous catalysis. The case of equilibrium-controlled precursor conversion. Organometallics 12:1753–1764, 1993.
75. A Chrisochoou, K Schnaber, U Bolz. Phase equilibria for enzyme-catalysed reactions in supercritical carbon dioxide. Fluid Phase Equilibr 108:1–14, 1995.
76. RMM Stevens, A Bakx, PRHvd Neut, TWd Loos, JdS Arons. High-pressure vapor-liquid equilibria of the ternary system carbon dioxide + vinyl acetate + 2-butanol. Experimental and modeling results. J Chem Eng Data 42:1280–1284, 1997.
77. JW Rathke, RJ Klingler, TR Krause. Thermodynamics for the hydrogenation of dicobalt octacarbonyl in supercritical carbon dioxide. Organometallics 11:585–588, 1992.
78. M Reetz, W Koenen, T Strack. Supercritical carbon dioxide as a reaction medium and reaction partner. Chimia 47:493, 1993.
79. PG Jessop, T Ikariya, R Noyori. Selectivity for hydrogenation or hydroformylation of olefins by hydridopentacarbonylmanganese (I) in supercritical carbon dioxide. Organometallics 14:1510–1513, 1995.

80. PG Jessop, Y Hsiao, T Ikariya, R Noyori. Homogeneous catalysis in supercritical fluids: hydrogenation of supercritical carbon dioxide to formic acid, alkyl formates and formamides. J Am Chem Soc 118:344–355, 1996.
81. Y Guo, A Akgerman. Hydroformylation of propylene in supercritical carbon dioxide. Ind Eng Chem Res 36:4581–4585, 1997.
82. X-W Wu, Y Oshima, S Koda. Aerobic oxidation of cyclohexane catalyzed by Fe(III)(5,10,15,20-tetrakis(pentafluorophenyl)porphyrin)Cl in sub- and super-critical CO_2. Chem Lett 10:1045–1046, 1997.
83. DR Palo, C Erkey. Homogeneous catalytic hydroformylation of 1-octene in supercritical carbon dioxide using a novel rhodium catalyst with fluorinated arylphosphine ligands. Ind Eng Chem Res 37:4203–4206, 1998.
84. DK Morita, SA David, W Tumas, DR Pesiri, WH Glaze. Palladium-catalyzed cross-coupling reactions in supercritical carbon dioxide. Chem Commun 1397–1398, 1998.
85. MA Carroll, AB Holmes. Palladium-catalyzed carbon-carbon bond formation in supercritical carbon dioxide. Chem Commun 1395–1396, 1998.
86. N Shezad, RS Oakes, AA Clifford, CM Rayner. Use of fluorinated palladium sources for efficient Pd-catalyzed coupling reactions in supercritical carbon dioxide. Tetrahedron Lett 40:2221–2224, 1999.
87. DR Pesiri, DK Morita, W Glaze, W Tumas. Selective epoxidation in dense phase carbon dioxide. Chem Commun 1015–1016, 1998.
88. A Furstner, D Koch, K Langemann, W Leitner, C Six. Olefin metathesis in compressed carbon dioxide. Angew Chem Int Ed Engl 36:2466–2467, 1997.
89. N Jeong, SH Hwang, YW Lee, JS Lim. Catalytic Pauson-Khand reaction in supercritical fluids. J Am Chem Soc 119:10549–10550, 1997.
90. PG Jessop, T Ikariya, R Noyori. Homogeneous catalytic hydrogenation of supercritical carbon dioxide. Nature 368:231–233, 1994.
91. L Jia, H Jiang, J Li. Selective carbonylation of norbornene in supercritical CO_2. Green Chem 1:91–93, 1999.
92. JA Osborn, FH Jardin, JF Young, G Wilkinson. Preparation and properties of tris(triphenylphosphine)halorhodium(I) and some reactions thereof including catalytic homogeneous hydrogenation of olefins and acetylenes and their derivatives. J Chem Soc A 1711, 1966.
93. M Toshiaki, N Koga, Y Ding, DG Musaev, K Morokuma. Ab inito MO study of the full cycle of olefin hydroformylation catalyzed by a rhodium complex, $RhH(CO)_2(PH_3)_2$. Organometallics 16:1065–1078.
94. RJ Klingler, JW Rathke. High-pressure NMR investigation of hydrogen atom transfer and related dynamic processes in oxo catalysis. J Am Chem Soc 116:4772–4785, 1994.
95. P Bhattacharyya, D Gudmunsen, EG Hope, RDW Kemmitt, DR Paige, AM Stuart. Phosphorus (III) ligands with fluorous ponytails. J Chem Soc, Perkin Trans 1:3609–3612, 1997.
96. AF Lagalante, TJ Bruno. Modeling the water-supercritical CO_2 partition coefficients of organic solutes using a linear solvation energy relationship. J Phys Chem B 102: 907–909.

5
Supercritical Fluid Processing of Polymeric Materials

Mark A. McHugh and Frederick S. Mandel
Virginia Commonwealth University, Richmond, Virginia

J. Don Wang
Consultant, Supercritical Fluid Development, Cleveland, Ohio

I. INTRODUCTION

For more than 100 years, scientists and engineers have been aware that supercritical fluid (SCF) solvents offer the potential of novel processing protocols. However, it is only in the past three decades that SCF solvents have been investigated or applied as solvents for processing foods, nutraceuticals, and polymeric materials; as reaction media for polymerization processes; as environmentally preferable solvents for solution coatings, powder formation, impregnation, encapsulation, cleaning, crystal growth, and antisolvent precipitation; and as mixing/blending aids for crystalline or viscous materials. This broad range of applications could be extended even further if a better understanding of the underlying physics and chemistry of SCF–solute behavior can be established. At present, efficient development of SCF-based processing technology suffers from the limitation that equations of state utilized for process simulation and modeling of SCF–solute mixture behavior are still not facile enough to describe the large changes in solution properties exhibited for an SCF-based process when realistic intermolecular potential functions are used. As a consequence, the approach taken in this chapter is to describe a molecular thermodynamic basis for interpreting SCF–solute phase behavior that relies on a physicochemical interpretation of experimental data. With this approach, the types and the strengths of energetic interactions are related to the chemical nature of the SCF solvent and to the chemical features of the solute.

All of the examples provided in this chapter relate to the application of SCF-based technology for processing polymers. The fundamentals section describes a molecular thermodynamic approach for interpreting SCF–polymer phase behavior. The subsequent section on applications consists of three parts. The first part presents observations on the phase behavior of polymers in SCF solvents with some attention to supercritical CO_2. Examples are given of the impact of polymer architecture and solvent quality on solubility. The second part describes the industrial-scale design of Ferro Corporation's SCF-based technology with emphasis on engineering details such as mixing, heat transfer, and process control schemes. The final part provides a brief description of several SCF-based applications. Although special attention is paid to supercritical CO_2 in this chapter, many of the results can be applied to processing with other SCF solvents.

II. THERMODYNAMIC FUNDAMENTALS OF SCF-SOLUTE MIXTURES

The principles of molecular thermodynamics provide the vehicle for connecting classical thermodynamics with the physicochemical properties of the components in solution. A molecular thermodynamics approach coupled with an experimental protocol in which solute and solvent properties are varied systematically elucidates the underlying chemical features of the components that fix the conditions needed to dissolve the solute in an SCF solvent. This type of molecularly directed experimental approach provides the insight needed to design processes that use SCF solvents at the lowest possible operating temperatures and/or pressures. In addition, it provides a rational methodology for choosing cosolvents to reduce pressure and temperature thresholds for obtaining a single phase.

To form a stable SCF–solute solution at a given temperature and pressure, the Gibbs free energy, shown in Eq. (1), must be negative and at a minimum:

$$\Delta G_{\text{mix}} = \Delta H_{\text{mix}} - T\Delta S_{\text{mix}} \tag{1}$$

where ΔH_{mix} and ΔS_{mix} are the change of enthalpy and entropy, respectively, on mixing (1). Enthalpic interactions depend predominately on solution density and on polymer segment–segment, solvent–solvent, and polymer segment–solvent interaction energies. ΔS_{mix} depends on both the combinatorial entropy of mixing and the noncombinatorial contribution associated with the volume change on mixing, a so-called equation of state effect (2). While the combinatorial entropy always promotes the mixing of a polymer with a solvent, the noncombinatorial contribution can have a negative impact on mixing if segment–segment interactions are very strong (the so-called excluded-volume effect).

For a dense SCF solution, ΔH_{mix} is expected to be approximately equal to the change in internal energy on mixing, ΔU_{mix}, which is shown in Eq. (2):

$$\Delta U_{\text{mix}} \approx \frac{2\pi\rho(P,T)}{kT} \sum_{i,j} x_i x_j \int \Gamma_{ij}(r,T) g_{ij}(r,\rho,T) r^2 dr \tag{2}$$

where x_i and x_j are mole fractions of components i and j, respectively, $\Gamma_{ij}(r,T)$ is the intermolecular pair-potential energy of the solvent and the polymer segments, $g(r,\rho,T)$ is the radial distribution function, r is the distance between molecules, $\rho(P,T)$ is the solution density, and k is the Boltzmann constant (3). The radial distribution function describes the spatial positioning of molecules or segments of molecules with respect to one another, which has embedded in it information on the positioning of the segments of the polymer chain. Since the repeat units of a given chain are connected to one another, the SCF–polymer solution cannot be considered a random mixture of repeat units and SCF molecules. Nevertheless, important generalities can be gleaned from an interpretation of Eq. (2). For example, given that the internal energy of the mixture is roughly proportional to density, the solubility of a polymer is expected to be improved by increasing the system pressure, by using a denser SCF solvent, or by adding a dense liquid cosolvent to an SCF solvent. However, the polymer dissolves only if the energetics of segment–solvent interactions outweigh segment–segment and solvent–solvent interactions. In other words, for certain SCF–polymer solvent mixtures, hydrostatic pressure alone will not overcome a mismatch in energetics between the components in solution. The balance of such interactions in solution is described by the interchange energy, ω, defined as in Eq. (3):

$$\omega = z\left\{\Gamma_{ij}(r,T) - \frac{1}{2}[\Gamma_{ii}(r,T) + \Gamma_{jj}(r,T)]\right\} \tag{3}$$

where z is the coordination number, or number of different pairs in solution (1). Equation (4) presents an approximate form of the electrostatic attractive part of the intermolecular potential energy, $\Gamma_{ij}(r,T)$, for *small, freely tumbling* molecules:

$$\Gamma_{ij}(r,T) \approx -\left[C_1 \frac{\alpha_i \alpha_j}{r^6} + C_2 \frac{\mu_i{}^2 \mu_j{}^2}{r^6\, kT} + C_3 \frac{\mu_i{}^2 Q_i{}^2}{r^8\, kT} + C_4 \frac{\mu_j{}^2 Q_i{}^2}{r^8\, kT} + C_5 \frac{Q_i{}^2 Q_j^2}{r^{10}\, kT} + \text{complex formation}\right] \tag{4}$$

where α is the polarizability, μ is the dipole moment, Q is the quadrupole moment, and C_1 through C_5 are constants. Induction interactions are not shown in Eq. (4) since their contribution tends to be much smaller than dispersion and polar interactions. It is important to note that Eq. (4) is not expected to describe rigorously the interaction of a polymer segment with another segment or with the solvent since segmental motion is constrained by chain connectivity and this architectural feature is not taken into account.

It is informative to utilize the physicochemical property of a few SCF fluids found in Table 1 to demonstrate the importance of Eqs. (1)–(4). The first observation is that nonpolar dispersion interactions, the first term in Eq. (4), depend only on the polarizability of the components in solution and not on temperature. Polarizability scales with molar mass for components in the same chemical family, such as the alkenes shown in Table 1. The critical pressure does not provide any useful information when comparing SCF solvents since critical pressure does not change significantly from one solvent to another. However, critical temperature does change substantially, and within the alkene family the increase in critical temperature is directly proportional to the increase in polarizability. Notice that the polarizability of CO_2 is close in value to that of methane, which is not a good solvent unless the system pressure is exceptionally high or, stated differently, unless the density of methane is increased considerably. By analogy, CO_2 is not expected to be a good solvent at high operating temperatures where dispersion interactions are dominant. It is a tenet of chemistry that CO_2 must have a polar moment because oxygen and carbon have different electron affinities. Due to structural symmetry, CO_2 does not have a dipole moment, but it does have a substantial quadrupole moment (-4.3×10^{-26} erg$^{1/2}$ cm$^{5/2}$) that operates over a much shorter distance than dipolar interactions. CO_2 is a dense solvent at modest temperatures and pressures that magnifies quadrupolar interactions that scale with molar density to the 5/6 power as indicated in Eq. (4) if the distance between the center of mass of two CO_2 molecules is twice the radius of an effective spherical volume occuppied by CO_2 (1). Note that the dipolar and quadrupolar interaction terms in Eq. (4) are inversely proportional to tem-

Table 1 Select Properties of Candidate Supercritical Fluid Solvents

Solvent	T_c (°C)	P_c (bar)	α (Å)	μ (Debye)	Acceptor–donor complex
CO_2	31.0	73.8	26.5	0.0	Both
Methane	−82.6	46.0	26.0	0.0	None
Ethylene	9.2	50.4	42.3	0.0	Weak acceptor
Propylene	91.9	46.2	62.6	0.4	Weak acceptor
Butene	146.5	39.7	82.4	0.3	Weak acceptor
Dimethyl ether	126.9	52.4	51.6	1.3	Acceptor

Data from Ref. 92.

perature, which means that at elevated temperatures polar molecules behave as if they are nonpolar. Hence, it may be possible to dissolve a nonpolar polymer in CO_2 if the temperature is high enough to diminish CO_2–CO_2 quadrupolar interactions relative to CO_2–polymer segment nonpolar dispersion interactions. As will be shown, CO_2 is a weak solvent for polar polymers since the effect of dipole interactions outweighs that of quadrupole interactions, especially at low temperatures where polar interactions are more significant. The challenge that remains is to predict the level of polarity needed in the polymer to make it soluble in CO_2 at modest pressures and temperatures. The reader is encouraged to compare the physicochemical properties of SCF solvents with the properties of common liquid solvents. It should be intuitively obvious that normal liquids have stronger intermolecular interactions or, conversely, stronger "cohesive energy densities"; otherwise the liquid would readily vaporize with the application of a modest thermal input. However, the strong attractive potential of a liquid compared to that of a "gaseous" SCF solvent also precludes the possibility of tuning solvent properties with pressure. In certain applications, in fact, the preferred solvent is the one with "poorer" strength as will be demonstrated subsequently.

An important type of intermolecular interaction not often considered with SCF solvents is temperature-sensitive specific interactions, such as complex or hydrogen bond formation. A good indication of whether a complex or hydrogen bond can form is provided by the value of the dielectric constant of the solvent. If the SCF solvent has a dielectric constant of 4–6, then there is a good possibility for forming a hydrogen bond complex. Unfortunately, the dielectric constant of CO_2 is near 2 and does not change significantly with increasing CO_2 density. However, CO_2 can still form weak acceptor–donor complexes. For example, certain polymers that possess electron-donating groups, such as carbonyls, have been shown to exhibit specific interactions with CO_2 where the carbon atom of CO_2 acts as an electron acceptor and the carbonyl oxygen in the polymer acts as an electron donor (4). The strength of the CO_2–segment complex is generally less than 1 kcal/mol, which makes it only slightly stronger than dispersion interactions, however, in a dense CO_2–polymer solution, this increase in interaction strength can be significant. It will be shown that replacing hydrogen with fluorine increases polymer solubility in CO_2 due to the specific interactions between the basic carbon atom of CO_2 and the fluorine in a C-F bond since fluorine has a higher electron affinity than hydrogen. High-pressure nuclear magnetic resonance (NMR) has been used to elucidate specific CO_2–fluorocarbon interactions in low-molecular-weight fluorocarbon solvents (5) and fluorocarbon repeat units in different fluorinated polymers and copolymers (6).

Polymer solubility in a given solvent also depends on the free-volume, or thermal expansivity, difference between the solvent and the polymer (2). To dissolve a polymer it is necessary for the solvent molecules to solvate the repeat units tethered to a given chain, which reduces the number of conformations available to the pure solvent resulting in a negative entropy of mixing. If the free volume of a polymer is increased, it becomes easier to dissolve it in a given solvent. Polymer free volume increases as chain branches are added to the backbone or as the attractive potential between repeat segments is reduced. Hence, it is difficult to uncouple energetic and entropic contributions to polymer solubility, although several examples are offered in the following section that reveal the impact of polymer architecture on solubility. A more complete compilation of CO_2–polymer studies can be found elsewhere (7).

Before leaving this section it is worthwhile to explicitly offer two other observations concerning SCF–polymer phase behavior. The first observation is that the impact of polymer molecular weight falls off dramatically once the molecular weight increases above 100,000. The second observation is that polymer solubility drops to near zero at temperatures less than the solidification (melting) temperature of the polymer. Both of these observations are also found with liquid solvents.

III. OBSERVATIONS ON THE PHASE BEHAVIOR OF SCF–POLYMER SOLVENT MIXTURES

It is more appropriate to describe the impact of SCF solvent quality and polymer architecture on solution behavior with a select number of examples rather than presenting a large number of phase behavior studies. The reader is directed to a comprehensive review of SCF–polymer phase behavior available in the literature (7).

A. Impact of Solvent Quality

The impact of solvent quality on the conditions needed to obtain a single phase is considered here using examples with low-density, semicrystalline polyethylene (LDPE, $M_w = 108{,}900$; $M_w/M_n = 3.0$; $T_{melt} = 113°C$) in normal alkenes, shown in Figure 1. The pressures needed to dissolve LDPE in ethylene, propylene, and butene (8,9) decrease significantly with increasing size of the alkene, which directly follows the increase in intermolecular interaction with increase in polarizability [see Eq. (4) and Table 1]. However, the reduction in cloud point pressure from one solvent to the next decreases as the size of the solvent increases, or, stated differently, as the solvent quality increases. This diminishing returns of solvent quality with increasing solvent size is a recurring theme with SCF–polymer mixtures that suggests that the largest changes in phase behavior will be observed with much smaller hydrocarbon solvents. It is harder to dissolve nonpolar LDPE in alkene solvents than in alkane solvents since the alkenes

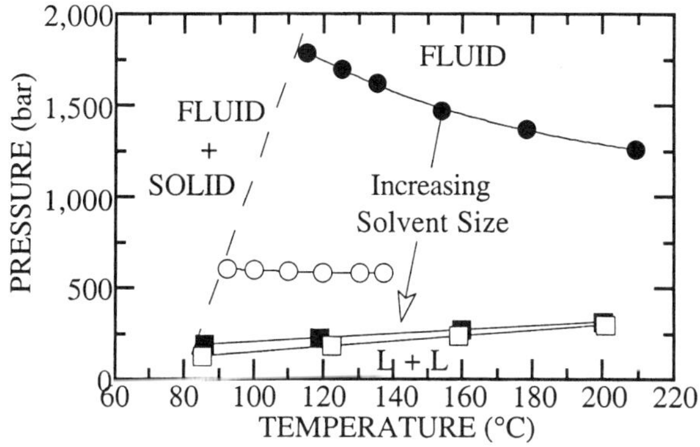

Figure 1 Impact of solvent size, which is related to solvent quality, on the solubility of polyethylene in normal alkenes (closed circles, ethylene; open circles, propylene; closed squares, 1-butene; open squares, 2-butene) (8,9).

possess a quadrupole moment due to the double bond. Hence, the interchange energy [Eq. (3)] is weighted more toward alkene–alkene interactions relative to alkene–PE and PE–PE interactions. The quadrupolar effect is attenuated in propylene and butene because the quadrupole moment is distributed over a larger molar volume, which reduces its effectiveness by a factor of molar volume to the −5/6 power. The impact of an interchange energy that favors solvent–solvent interactions becomes more apparent as the polarity of the solvent increases. Figure 2 shows that the cloud point curve of LDPE in polar dimethyl ether (DMF; dipole moment of 1.3 D) (10) changes slope and increases in pressure dramatically as the temperature is lowered to the region where DME–DME polar

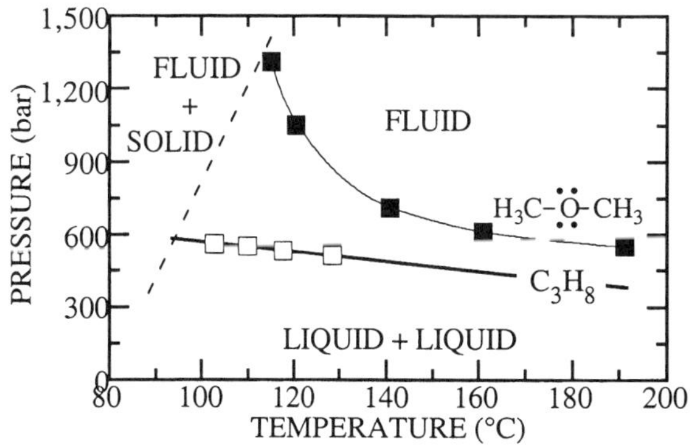

Figure 2 Phase behavior of polyethylene in propane and in dimethyl ether (7,10).

interactions are favored. As a general rule, the cloud point curve will eventually exhibit a negative slope with decreasing temperature for mixtures containing one component that is polar and the other component that is nonpolar unless the solute solidifies.

B. SCF-Induced Changes in Polymer Properties

A very important consideration when processing with SCF solvents is how the presence of the SCF solvent changes the properties of the polymer. For example, the normal melting point of semicrystalline LDPE increases from 113°C to 115°C at 1600 bar in ethylene, decreases to 90°C at 600 bar in propylene, and decreases further to about 80°C at 400 bar in butene. Hydrostatic pressure raises the melting point of PE at a rate of 0.01°C/bar, but the solubility of the SCF solvent in the LDPE-rich liquid phase depresses the melting point at a rate proportional to the SCF solvent concentration. In ethylene, LDPE should melt at 129°C at 1600 bar, but the melting point actually occurs at 115°C due to the solubility of ethylene in the LDPE-rich liquid phase.

Several SCF-based processes take full advantage of the change in melting point of crystalline polymeric resins when forming coated particles, as will be described in a subsequent section of this chapter. Conway and coworkers (11) show that the melting point of a low-molecular-weight, semicrystalline polyester can be reduced from 105°C to 75°C in pure CO_2, to 65°C in CO_2 with 11 wt % acetone, and to 35°C in CO_2 with 11 wt % ethanol. Even though it is not possible to dissolve this polyester in neat CO_2, it is possible to suppress the crystallization of the polyester and to obtain the wetting characteristics needed to coat (nonsoluble) particles at low operating temperatures and pressures. When the pressure is released the polyester coating crystallizes, ensuring that the particles do not agglomerate as long as they remain at temperatures below 105°C (the normal melting point of the polyester). As a polymer imbibes SCF solvent its glass transition temperature (T_g) is also lowered, which provides an opportunity to mix the polymer with other components in an efficient manner as described in subsequent sections.

C. Impact of Polymer Architecture

The pressures and temperatures needed to dissolve a given polymer in an SCF solvent depend intimately on the polymer architecture, which fixes both the strength and type of intermolecular interactions and the free volume of the polymer. Branching increases the free volume of the polymer, which makes it easier to dissolve in an SCF solvent, and branching reduces the intermolecular interactions between polymer segments that would arise due to short-range molecular orientation offered by a high content of linear segments without pendant groups (12). The impact of chain branching is most noticeable if the molecular weight polydispersity is minimized (12–19). For example, the cloud point curves for linear LDPE in ethane increase in pressure with increasing crystallinity or, conversely, with decreasing number of branches off of the main chain. In propane the difference in cloud point pressures is approximately half that found with SCF ethane. Once again it is seen that the weaker of the two solvents magnifies the effect of the polymer properties on the phase behavior.

The phase behavior of vinyl polymers in hydrocarbon SCF solvents is very sensitive to backbone chemical architecture. For example, the pressure needed to dissolve poly(acrylates) in ethylene varies nonlinearly as the length of the acrylate alkyl tail is increased (20). Poly(methyl acrylate) (PMA) does not dissolve in ethylene to pressures of 2500 bar and temperatures to 250°C due to methyl acrylate segment–segment polar interactions that are much stronger than acrylate segment–ethylene interactions. Poly(ethyl acrylate) (PEA) dissolves in ethylene at pressures near 1200 bar and temperatures in excess of 150°C. As the length of the acrylate alkyl tail increases from methyl to ethyl the impact of the polar acrylate interactions decreases since dipolar interactions scale inversely with the square root of the molar volume and quadrupolar interactions scale inversely with the volume to the 5/6 power. It takes progressively less pressure to dissolve the propyl, butyl, and ethylhexyl poly(acrylates) compared to PEA. At high temperatures where configurational polar interactions are reduced, similar cloud point pressures are observed for each of the poly(acrylates). At temperatures near 60°C there is much greater difference in the location of each of the cloud point curves. There is an optimum alkyl tail length that balances the acrylate–acrylate, ethylene–ethylene, and acrylate–ethylene energies of this system. In addition, the free volume of the poly(acrylate) increases as the tail length of the acrylate group increases, which makes it easier to dissolve the poly(acrylate) in highly expanded, supercritical ethylene. However, the polar character of the poly(acrylate) decreases as the alkyl tail length increases since the polarity of the ester is spread over much larger volumes, which reduces its impact. Hence, the reader is cautioned to consider both energetic and entropic effects when interpreting the effect of chemical architecture on SCF–polymer phase behavior.

D. Impact of Chain End Groups

Typically the effect of polymer chain end groups on phase behavior is ignored for polymers with molecular weights in excess of about 100,000. However, as with branching impact on phase behavior, end groups can have a significant effect on the phase behavior in certain situations, especially when the chemical structure of the end group differs considerably from the groups in the main chain. Krukonis and coworkers demonstrated the impact of hydroxyl end groups on the solubility of hydroxy-terminated poly(butadiene) (HTPB) with an M_n of 2960, an M_w of 6250, and a hydroxyl equivalent weight of 1256 (0.796 mEq/g) (21). Carbon dioxide only dissolves about 12% of the HTPB at pressures to 600 bar. Propane at 130°C dissolves virtually all of the parent material at pressures below 600 bar. Krukonis used propane to fractionate the HTPB and obtained fractions with M_n ranging from 780 to 11,750 and M_w from 970 to 21,525. Most of the fractions had polydispersities of about 1.2. However, the hydroxy equivalent weight was nonlinear with molecular weight, especially in the higher molecular weight fractions.

E. Cosolvent–Antisolvent Effects

A cosolvent can greatly enhance polymer solubility in a given solvent due to several factors. If the solvent is highly expanded, the addition of a dense, liquid cosolvent increases the solution density; it also reduces the free-volume difference between the polymer and the solvent, and less pressure is needed to obtain a single phase (22). If the cosolvent provides favorable physical interactions, such as polar interactions, the region of miscibility expands more than that expected from just a density effect as suggested by Equations (1–4) (23). If a polar cosolvent is used with a polar polymer, cloud points monotonically decrease in pressure and temperature (23,24). The cloud point will decrease much more dramatically if the polar cosolvent can form a complex with the polymer since the interaction energy of complex formation, such as hydrogen bonding, is typically an order of magnitude greater than that expected from dispersion or polar interactions. Decoupling the effect of a cosolvent from that of hydrostatic pressure can be complicated since increasing the system pressure also reduces the free-volume difference between the solvent and the polymer and increases the probability of interaction between solvent, cosolvent, and polymer segments in solution (24).

Unreacted monomer can act as a very effective cosolvent when performing polymerization reactions in an SCF solvent. Consider the effect of unreacted butyl acrylate (BA) monomer as a cosolvent for the poly(butyl acrylate) (PBA)-CO_2-BA system (25). It is not possible to dissolve PBA in CO_2 at temperatures less than 50°C since the cloud point curve exhibits a very steep slope at that point, as shown in Figure 3A. However, the addition of about 7 wt % BA to a PBA-CO_2 mixture lowers the cloud point pressure and temperature to a very large extent. Figure 3B shows that adding 32 wt % BA to the PBA-CO_2 solution significantly changes the phase behavior. It is now possible to conceive of polymerizing BA in CO_2 homogeneously at very moderate pressures as long as the conversion is kept below 70 wt %.

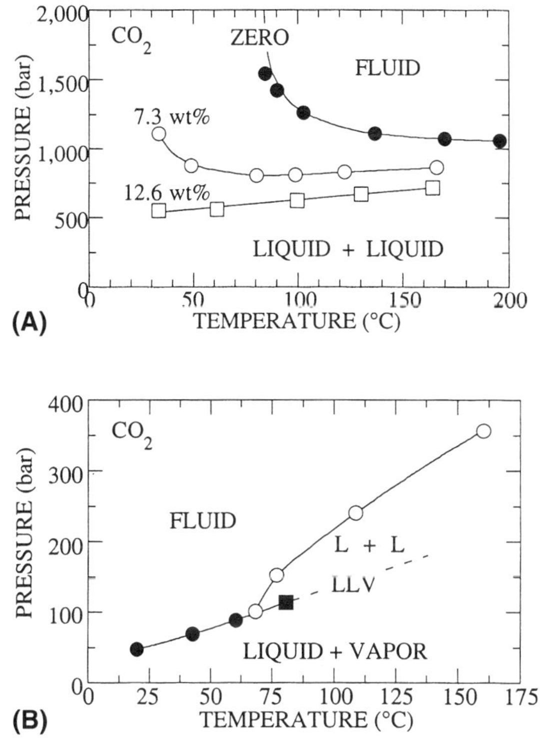

Figure 3 **A**: Impact of unreacted butyl acrylate monomer (wt % on a polymer-free basis) on the phase behavior of the poly(butyl acrylate)-CO_2 system (25). **B**: Impact of 32.0 wt % butyl acrylate monomer (wt % on a polymer-free basis).

There are many cosolvent studies where an SCF solvent is added to a solution for the purpose of decreasing the quality of the liquid solvent to precipitate the polymer from solution. In this instance the supercritical fluid is an antisolvent. The SCF dilates the solvent without raising its temperature, thus avoiding thermal degradation of the polymer (26–31). Many other studies have been reported on the large effect supercritical CO_2 has as an antisolvent (32–36). It should be noted that the antisolvent effect is exacerbated when the operating temperatures are well above the critical point of the supercritical solvent where it is very expanded (37).

F. Supercritical CO_2

Carbon dioxide has been touted as the solvent of choice for many industrial applications because it is non-hazardous and inexpensive. CO_2 has a critical temperature near room temperature, a modest critical pressure, and a density higher than most supercritical fluids. A large body of work has been generated by DeSimone and coworkers demonstrating that CO_2 dissolves polymers containing fluorinated groups [see, for example, (38–40)]. The recurring theme in these studies is that fluorinated groups play a significant role in polymer solubility and that polarity in the polymer also plays a role in fixing solubility levels. It has been suggested that CO_2 may either form a weak complex as it preferentially clusters near the highly negative fluorine atom of the C-F bonds that are more polar than C-H bonds (4). The polar fluorinated groups on a side chain can shield the hydrocarbon main chain from interacting with CO_2. However, fluorination alone does not insure that the polymer will be soluble in CO_2 at temperatures below 100°C (41).

CO_2 at or near room temperature and at pressures typically below 600 bar can dissolve many poly(dimethyl) and poly(phenylmethyl) siloxane, perfluoroalkylpolyethers, chloro- and bromo-trifluoroethylene polymers, and poly(perfluoropropylene oxide) (21,42–47). The polymers reported to have solubility in CO_2 all possess some degree of polarity due to oxygen or other electronegative groups, such as chlorine or bromine, incorporated into the backbone of the polymer. In addition, the high solubility of the silicones in CO_2 is likely due to the very flexible nature of these polymers that endows them with much larger free volumes than other polymers.

CO_2 does not dissolve polyolefins to any great extent unless the molecular weight is very low (48,49). CO_2 can dissolve octane, hexadecane, and squalane, but these nonpolar solutes have molecular weights in the range of 100–500 (50). The interchange energy, given in Eq. (3), is dominated by CO_2–CO_2 quadrupolar self-interactions rather than CO_2–polymer dispersion or induction cross-interactions for mixtures of CO_2 with alkanes or polyolefins. CO_2 can dissolve very low-molecular-weight, slightly polar polymers, but solubility levels quickly drop when the molecular weight surpasses 10,000 (21,43). In contrast, very polar polymers or polymers that are water soluble, such as poly(acrylic acid), do not dissolve in CO_2 even at temperatures in excess of 270°C and pressures of several kilobar.

Interpreting the phase behavior of polymers in CO_2 or any SCF solvent is a challenge since the connectivity of the monomer units and the conformation of the polymer in solution affects both the energy and entropy of mixing. For example, PMA dissolves in CO_2 at high pressures, although PMMA remains insoluble even though the strength of PMA–CO_2 intermolecular interactions are expected to be similar to PMMA–CO_2 interactions. However, the T_g of PMMA is about 100°C higher than that of PMA, suggesting that the rotation of a methacrylate chain segment is more hindered than the rotation of an acrylate chain segment, which likely leads to enhanced PMMA segment–segment interactions. CO_2 does dissolve poly(butyl methacrylate) (PBMA), although in this case the T_g for PBMA is only 70°C higher than that of PBA.

As further evidence of the challenge inherent in predicting polymer solubility in CO_2, consider a comparison between cloud point curves for PMA and poly(vinyl acetate) (PVAc) (49). At 30°C the PMA cloud point curve is more than 1500 bar higher than the PVAc curve even though the molecular weight of PVAc is four times greater than that of PMA. Both polymers are polar, but the T_g for PVAc is approximately 21°C higher than the T_g of PMA, which is likely due to stronger polar interactions between vinyl acetate groups compared to methyl acrylate groups. CO_2 can more easily access the carbonyl group in PVAc than in PMA, which facilitates the formation of a weak CO_2–acetate complex, especially at moderate temperatures.

The characteristics needed to make a fluoropolymer soluble in CO_2 can be ascertained from Figure 4, which shows the difference in cloud point curves for poly(vinylidene fluoride) (PDVF) (51), a statistically random copolymer of 78 mol % vinylidene fluoride and 22 mol % hexafluoropropylene (VDF-HFP$_{22}$), and a nonpolar copolymer with 81 mol % tetrafluoroethylene and 19 mol % hexafluoropropylene (TFE-HFP$_{19}$). The PVDF–CO_2 curve is at 1500–1700 bar

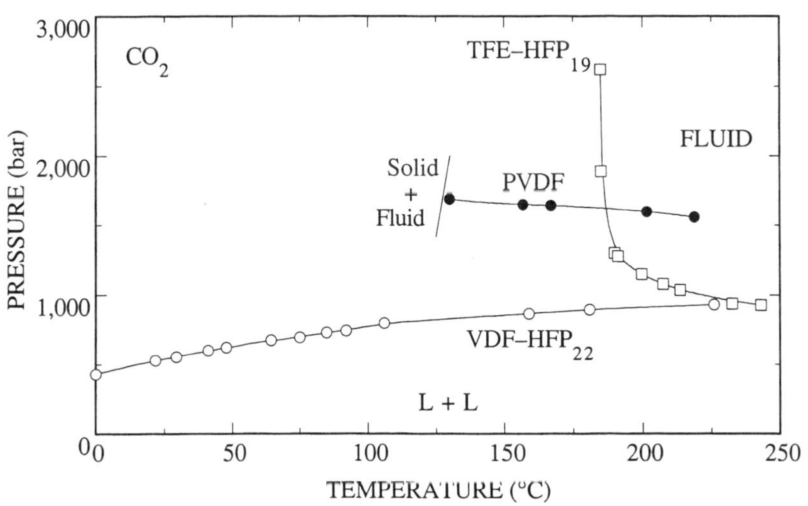

Figure 4 Influence of fluoropolymer architecture on the cloud-point behavior of poly(vinylidene fluoride) (PVDF) (51), poly(vinylidene fluoride-*co*-22 mol % hexafluoropropylene) (VDF-HFP$_{22}$), and poly(tetrafluoroethylene-*co*-19 mol % hexafluoropropylene) (TFE-HFP$_{19}$) in CO_2. The polymer and copolymer concentrations are about 5 wt % in each case.

and terminates at a crystallization boundary at 130°C. In contrast, the VDF–HFP_{22}–CO_2 curve is at pressures less than 1000 bar and extends to 0°C and 400 bar. VDF-HFP_{22} is less polar than PVDF and the copolymer has a higher free volume. Notice that the TFE–HFP_{19} curve exhibits a very steep increase in pressure at about 185°C, which shows that nonpolar TFE-HFP_{19} does not dissolve in CO_2 until very high temperatures are obtained to reduce CO_2–CO_2 interactions. It should also be noted, however, that TFE-HFP_{19} dissolves in CO_2 whereas polyolefins do not, which lends credence to the observation that fluorinating a polyolefin makes it soluble in CO_2. However, it is obvious by comparison of the VDF–HFP_{22} and TFE–HFP_{22} curves that just fluorinating a polyolefin does not guarantee its dissolution at modest pressures and temperatures.

The phase behavior examples presented here demonstrate that SCF solvents are, in fact, quite weak solvents for polymers with only a few exceptions. Because these solvents are so weak it is possible to tune them by increasing the pressure, which increases the solvent density. In the following section, an engineering description is provided on the processes developed by Mandel (52–58) that capitalizes on the observation that the polymer is processible even though it does not dissolve in supercritical CO_2, which attests to the flexibility inherent in the use of SCF solvents for polymer processing.

IV. MATERIALS PROCESSING PRINCIPLES

In this section a commercial-sized process is described for mixing and reaction using supercritical CO_2 as a media to enhance the wetting of refractory materials with thermally labile additives that are subsequently reacted to form the final product structure (52–58). The ultimate range of materials produced from SCF-based processes is remarkably wide and has been extended to the biomaterials arena. A few SCF-based material processing schemes and the products produced from these processes will be described to illustrate the capability of this evolving technology. Since the specific subject matter in this section is that of industrial research and development, emphasis is given to issues such as scale-up, commercial viability, and market specificity. In the applications described here, SCF CO_2 is the solvent of choice because it is readily available, inexpensive, and environmentally benign; therefore, attention is paid to accurately describing the properties of this fluid. As will be shown subsequently, it is not always necessary to dissolve a solute in CO_2 to take advantage of this unique solvent. In fact, a project team lead by Mandel (52–58) showed how to produce powder coatings with the use of CO_2 as a plasticization aid for amorphous thermoset polymers. In this instance, the T_g of the polymer is reduced in the presence of CO_2, which makes it possible to intimately mix the softened polymer with the other components of the formulation. More information is provided on this application in the following paragraphs.

A. Equations of State, Density Tuning, and Process Operation

Figure 5 shows a schematic diagram of a process that is used for supercritical CO_2 processing. For the manufacture of powder particles with this process it is necessary to know the ratio of the mass of CO_2 utilized to the mass of "loaded" polymer product that is processed. In this instance, CO_2 dissolves in the polymer and depresses the T_g of the polymer so that, with appropriate mixing, intimate contact can be obtained between the polymer and the insoluble additives added to the vessel. It is assumed that virtually no polymer dissolves in the CO_2-rich phase. Several equations of state for carbon dioxide (59–63) have been evaluated since accurate CO_2 densities are needed in several stages of the processes. An accurate prediction of solution densities aids in ascertaining the proper configuration and energy requirements for mixing in the multiple scale-up configurations evaluated by Mandel. An additional requirement is that accurate control systems (56) are needed for the laboratory, pilot, demonstration, and commercial processing facilities that utilize CO_2, which means that reliable algorithms are needed for calculating the fluid densities of neat CO_2 via mass of the fluid delivered to a fixed reactor volume. The effect of the reactor internals

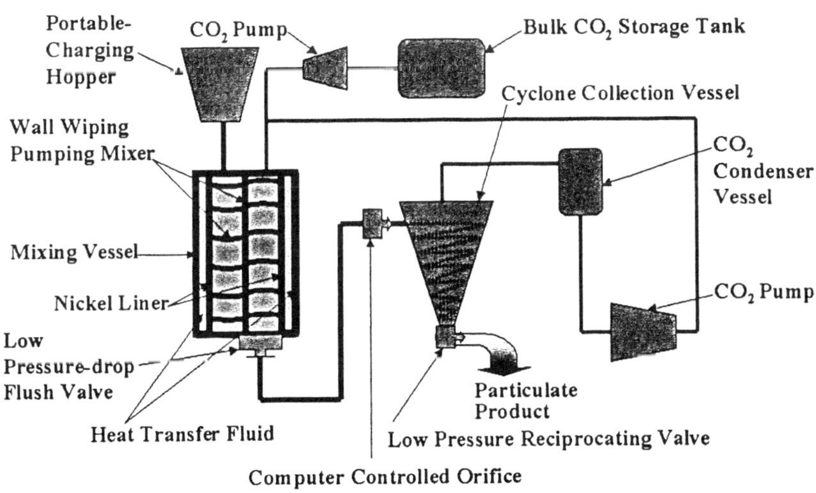

Figure 5 Schematic diagram of a supercritical CO_2 process.

on the processing conditions is also determined via careful evaluation of the experimentally observed pressure and temperature profile with CO_2 in the vessels as compared with the simulated profile using an appropriate equation of state. The goal is to operate at a pressure and temperature where equal densities are obtained for pure CO_2 and for the CO_2-swollen polymer. The operating experience to date with laminar mixing configurations has indicated consistent, predictable, and stable results for the CO_2-based processing facilities developed by Mandel and coworkers (52–58).

The properties of pure CO_2 adhere closely to the modified Benedict–Webb–Rubin (BWR) equation of state when the volume of reactor internals is taken into account. Details on this equation of state, which is currently utilized by NIST (64,65), can be found elsewhere (66–68), and a very useful vapor pressure equation for CO_2 is also available in the literature (69). The multitermed BWR equation for carbon dioxide has facilitated accurate prediction of density, compressibility, entropy, enthalpy, and other salient thermodynamic properties. Development of scale-up mass balances, energy balances, and equipment specifications is carried out utilizing the BWR equation of state. Accurate mass delivery of CO_2 to the pilot mixing vessel shown in Figure 5 is accomplished by storing CO_2 in a set of heated pressure cylinders mounted on precision load cells. The combined CO_2 capacity of the system is 2.3 metric tons with a maximum pressure capability of 306 bar at 60°C. CO_2 delivery is controlled by a control valve supplied by Kammerer valves, and the pressure and temperature are monitored during delivery to ensure accurate metering of the fluid.

B. Supercritical Advanced Manufacturing Process

Many of the polymers and resins that are extruded are amorphous and have a characteristic T_g. These materials swell in the presence of an SCF (52,70–75) with a concomitant decrease in T_g, which means that these materials begin to soften to the point that it is possible to mix them at low temperatures in the presence of an SCF solvent. The development of SCF-based processing of polymers by Mandel and coworkers focused on several potential scale-up issues. The principal area of concern involved the amount of heat rejection required by the process due to the high power requirements of the agitation system and the nature of viscous mixing (76–78). Without sufficient heat rejection, it is not possible to process thermally labile reactants together with the polymers. The concern about heat transfer in the scaling of a process from bench to commercial levels results from the physical reality that the area of the heat transfer surface increases as a square of the vessel radius while the material volume increases as a cube of the radius. Due to the highly viscous character of the products envisioned to be produced by the SCF process, overall heat transfer coefficients were assigned values similar to those of viscous film-forming oils. On the other hand, it is estimated that the overall heat transfer coefficient should be 2 Btu/h·ft^2·°F, a low value suggesting that there is a significant buildup of polymer on the walls of the vessel. However, this "engineering" estimate is incorrect since careful experiments showed that the temperature rise in a vessel of known dimensions resulting from the power input from mixing was less than that estimated with the assigned heat transfer coefficient. This example highlights the pitfall in using estimates of physical properties rather than using experimental values obtained from careful SCF experimentation.

C. Mixing

The primary concern when bringing a pilot- or laboratory-scale mixing process to commercial scale is the prediction of the time required to mix at the commercial scale (79,80). The three regimes of mixing are *turbulent*, *transitional*, and *laminar*. The turbulent is a regime described by the dimensionless Reynolds number (Re) greater than 10,000. This is the regime in which most conventional, dispersion-type mixing processes occur and in which the mixing is considered to be density driven. The transitional regime occurs when Re resides between 10,000 and 100, and the mathematical description of this regime is complex. The program utilized by Mandel avoids the turbulent and transitional mixing regimes and primarily focuses on the laminar region where Re is less than 100. The laminar regime is viscosity driven and is the regime where the highly loaded polymer products have been found to be amenable to SCF-aided processing.

A usual consideration in the design of agitation systems is to minimize the power utilized in the process (79,80). The practical design limitations include the size and nature of the gear box, the pressure–velocity relationship for the mechanical seals, and the diameter of the shaft (81). Many manufacturers of agitation systems limit the power output of the system to ensure ease of design and fabrication of mixers. For systems at elevated pressures, the rotational speed is limited to 240 rpm for conventional agitation systems to avoid shaft deflection and consequent potential damage to the mechanical seal faces and the gear box. Mandel considered many design concepts to facilitate the task of dispersing highly loaded pigment–polymer systems. Since work is necessary to disperse the components of the mixture and the laminar region is the practical regime to accomplish mixing, a helical ribbon agitator equipped with an anchor is chosen, as shown schematically in Figure 5. Traditionally this agitator configuration is avoided due to high energy and power requirements. The helical ribbon forces material to the inside wall of the vessel where the anchor aids in dispersion via a smearing action. Although little or no particle comminution results from the action of the helical ribbon, the power required to rotate this ensemble is greater than that associated with pumping or dispersive impellers. The force required to drive the helical ribbon is observed as a heat input to the vessel. Therefore, management of mixing with energy-intensive impellers becomes management of heat rejection arising from the power input.

D. Mechanical Seals

In order to mix highly loaded polymer systems in the presence of an SCF solvent, the employment of mechanical seals is necessary. A mechanical seal consists of two plane-parallel, highly polished smooth faces, being sufficiently lubricated, gliding on one another in such a way that only a very low leakage rate results at low friction (81).

Loads on the seal rings by secondary seals (mostly O rings), drive pins, eccentricities, and temperature gradients can lead to distortion of the seal faces and consequently to increased leakage or higher wear. In SCF service, care must be taken in the selection of secondary seal rings. Ideally the materials chosen will not be susceptible to diffusion of the SCF and consequent swelling. Seal rings work optimally only if there is minimal distortion of the seal faces from mechanical and thermal loads or if the formation of the lubrication film is supported with increasing load with no increase in leakage. The seal rings are pressed together by springs, so that the gap is already reliably closed before pressurizing. The pressure of the medium to be sealed presses the seal faces closer together, so that the medium to be sealed cannot escape via the seal gap. Materials of construction for seal faces are tungsten carbide, silicon/silicon carbide, and graphite. Applications of these sealing face materials range from hard–hard to soft–soft surface interactions.

E. Agitation System Drives

The polymer charge to the vessel swells as it imbibes high-pressure gas that is being heated to the supercritical regime. The initial swelling produces a tacky mixture that is difficult to stir with conventional equipment. Since all mixing devices have power load protection, the tackiness often results in increased power requirements that can exceed the protection limits of the drives. The result of exceeding this limit is a shut-down of the agitation system. Restarting the system in laboratory and pilot scale is possible several times an hour but would be limited to once in an hour for a large electric motor. The starting and stopping of this motor also produces an electric load spike that adversely impacts the facility electric costs.

The initial pilot plant agitation system was powered by a 40-kW AC motor with a standard gearbox. A problem with AC drives is that they do not deliver full power below 10 rpm. In the situation where the processing conditions produce a tacky material the rotational speed falls to below 10 rpm and the system does not deliver the power needed to maintain rotation. On the other hand, DC motors can develop full power at zero rpm, which motivated the conversion of the agitation motor from AC to DC. This change results in continuous nonstalling operation of the pilot plant agitation system. Since stalls were eliminated it is no longer necessary to design for the possibility of interrupted and time-dependent restarting of the agitation system.

F. Heat Transfer

The second time-dependent issue is related to heat transfer. The most succinct way to discuss this topic is by comparing three process scale-ups employing vessels of similar geometry with height/diameter (H/D) ratios of 2.0. The examples shown in Table 2 are for vessels of 4, 120, 4000, and 14,400 L. The rise in temperature, obtained from Eq. (5), provides a measure of the amount of heat that must be dissipated for each vessel with a given available heat transfer surface.

$$Q = U \cdot A \cdot \Delta T \tag{5}$$

Preliminary experiments indicated that the overall heat transfer coefficient is approximately 48 Btu/h·ft^2·°F, although this is obviously a conservative number as heat input to the system is always required in practice. It should be noted that carbon dioxide aids in heat transfer in these systems (82–85) due to its high mobility and density. The pilot plant designed by Mandel monitors the rate of heat transfer by a continuous method and alerts any process anomalies such as wall build up of polymer.

High-nickel alloys are used for the *internal* heating/cooling jacket to provide a very high rate of heat transfer. The search for manufacturers of internally

Table 2 Heat Transfer Data Used for Process Scale-Up[a]

Vessel volume	4 L	120 L	4000 L	14,400 L
Area (ft^2)	1.310	14.73	107	240
Ratio of scale-up	1	11.2	7.3	2.2
U, overall heat transfer coefficient	48	48	48	48
Kilowatts	0.6	4.2	196	704
Horsepower	0.8	5.6	263	944
Q (Btu/h), heat transfer rate	2040	14,300	669,600	2.4 million
ΔT (°F), temp. rise	32	20	130	209

[a]Assuming a constant heat transfer rate although the process only employs maximum horsepower for brief periods of time.

nickel-jacketed vessels within the ASME Division 8 Section 2 code demanded a worldwide effort. The engineering goals are based on a system designed for maximum rejection of the heat developed from the large impeller system needed for the 14,000-L vessel.

Examination of the heat-up times for varying agitation power inputs identified a potential problem not in heating rate but in rate of heat rejection. Since the initial materials processed were thermoset resin recipes, excessive heat buildup during processing is undesirable as it leads to unwanted chemical reactions in this stage of the process. The sources of heat input are from the heating system itself, from the CO_2 loaded into the vessel, from the agitation input, and from the viscous heating of the batch material. In the pilot plant, the thermal lag is noted between the external heating system and the equilibrium temperature within the reaction/mixing vessel. At this stage the vessels have modest wall thickness, whereas the commercial process vessels would have thickness from four to six times greater than those of the pilot vessels. Although numerous attempts were made to adjust the proportional, integral, and derivative (PID) loop to gain tight process control, product-formulation-dependent PID loops were required to maintain the highest degree of flexibility with the pilot plant vessels. As an alternative design strategy, the concept of internal heating passages was employed to eliminate the effect of thermal lag from external heating devices. Five candidate fabrication and engineering groups were identified that indicated that they had the resources to design and fabricate an internal-walled vessel. Since this is a breech lock vessel with a moving body, the evaluation of the vessels and manufacturers became a major part of the project.

Heat transfer profiles were developed for each candidate vessel design accounting for designed fluid flow rates via analysis of passages, filming heat transfer coefficients, and optimizing the internal jacket wall as a function of thickness and strength. The thermal conductivity of the internal jacket for most steels and alloys averages 115 $Btu/h{\cdot}ft^2{\cdot}°F$. Nickel is chosen because it possesses a high thermal conductivity of 400 $Btu/h{\cdot}ft^2{\cdot}°F$ and has sufficient strength to withstand the pressure environment. Relative to stainless steel, nickel provides a 25% increase in effective wall heat exchange. The passage analysis and required flow rates as a function of simulated heat rejection were calculated for each design variation to determine the optimal design that allowed for the most rapid heat transfer.

The division of the passages into three or more manifold zones was necessary to keep the flow in the turbulent heat transfer regime. A large pressure drop was calculated for one-zone heating, which would decrease the flow rate to below the turbulent for both organic and inorganic heat transfer fluids. Passages were analyzed for each design and iterations presented to potential manufacturers. The pilot vessels currently in operation were retrofit with a scaled replica of the three-zone internal nickel jacket to ascertain if the design assumptions were correct and to develop better heat transfer data. The resultant data demonstrated that some of the viscous, highly swollen polymer–additives–CO_2 mixtures had heat transfer greater than 250 Btu/h·ft^2·°F.

G. Isobaric Transfer of Product, Particle Size Reduction, and Atomization

The preferred method for recovery of products is atomization, which is a form of particle generation from supercritical solutions or suspensions (PGSS) (86). Several operating models have been developed to study the direct atomization of polymer-loaded material. High velocity and shear thinning within the transfer channels are necessary to achieve this goal. Minimization of the pressure drop from the opening of the flush valve to the entrance of the cyclone separator shown in Figure 5 is key to the attainment of direct production of fine particles. With an isobaric control system the real-time processing profiles reveal no diminishment of the pressure during the entire transfer operation.

Assessment of the performance of several types of flush valves was accomplished at the pilot plant, with particular attention paid to the minimization of the pressure drop through the control device. The flush valve initially was viewed as a containment and transfer device but rapidly evolved to be a control device. Fluid mechanics calculations demonstrated the precise condition required to achieve atomization. The atomization was modeled as an effervescent atomizer due to the influence of the orifice diameter and the pressure drop on the size of the collected particles (87–90).

The milling rates observed to date demonstrate that it is possible to obtain increased production rates for SCF-produced products. The milling rates for SCF-produced products are three to four times faster with 35% less energy requirements than that observed with the same materials produced from conventional methods. SCF-produced products have a foam-like character that, when milled, yield a finer particle size distribution for the finished product than that achieved via flake milling of conventional, melt-mixed material.

H. Vessels for Processing

Care was taken to maintain the height-to-diameter dimensions of the vessels for all stages of scale-up to minimize any potential problems. Table 3 gives the dimensions of each vessel utilized at each stage of the SCF development project.

The pilot vessel is designed with a hydraulically driven breech lock for rapid opening and closing. In addition, the vessel body is hydraulically raised or lowered and translated away from the centerpoint, which means that the vessel is expeditiously cleaned by exposing the reactor internals and allowing for access to the helical ribbon agitator. Since both the pilot facility and the

Table 3 Scale-Up Dimensions for Vessels Used in the Development of the SCF Process

Vessel volume	4 L	40 L	120 L	4800 L	14,400 L
Tank diameter, in.	5	12.5	14.5	54	78
Tank height, in.	12	29	38	120	168
Viscosity, poise	29.2	29.2	29.3	29.2	29.2
Blade-width diameter, in.	4.9	12.2	14.2	52	76
Tank height/blade-width diameter	2.44	2.37	2.67	2.37	2.21
Impeller rpm	100	100	100	100	100

commercial facility handle a large number of different product recipes, it is vital to thoroughly clean the vessel internals between each run. In the pilot facility the weight of the mobile section of the vessel is 800 kg, whereas for the commercial vessel this weight could range from 130,000 kg for division 2 vessels to 400,000 kg for division 1 vessels. The lifting and movement of a division 1 vessel would necessitate the design and development of specialized hydraulic systems while it is possible to utilize standard hydraulic systems with the lighter weight division 2 vessel with the appropriate forged material. The designed wall thickness and dimensions of the vessel suggested that several materials could meet the requirements of SCF processing, but few forges have the capability of producing these materials at the scale needed by the project.

The expense of replaceable O rings for sealing the vessel was determined to be a potential problem. Various materials and mechanical configurations were reviewed to find a multirun material of construction. Carbon dioxide diffuses into many of the sealing materials, which potentially shortens their cycle life. These materials were subjected to stress cycling in CO_2 and analyzed both dimensionally and via dynamic mechanical analysis as well as thermomechanical analysis.

V. SELECT APPLICATIONS

The SCF batch process has proven time and again to be capable of producing numerous types of materials. This section describes strategic applications of SCF-based technologies.

A. Organic Synthesis and Polymer Modification

One of the major advantages of using CO_2 for processing is the absence of conventional organic solvents (91). For example, Figure 6 shows a polymer

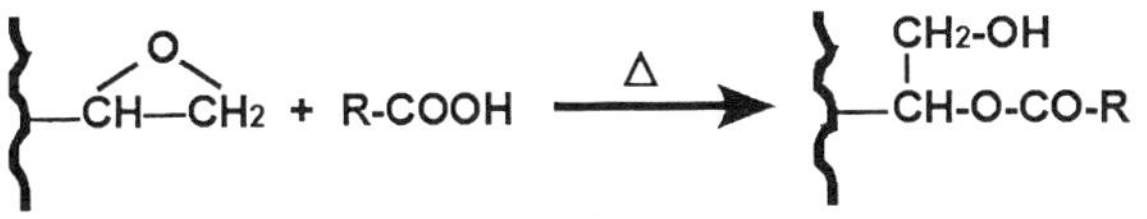

Figure 6 Polymer modification reaction of an epoxy functional group that can be run in supercritical CO_2.

modification reaction that may require refluxing in toluene for many hours, after which considerable work must be done to remove the solvent. Higher yields are found for this same reaction carried out in supercritical CO_2 at 200 bar in place of toluene at similar temperatures and reaction times. Because CO_2 is vented as a gas when the pressure is released, there is no solvent residue in the product, which reduces production cost and VOC emissions are also reduced. CO_2-aided processing was employed to modify many polymers to produce improved, reactive polymers for coatings and other end uses. Other syntheses suitable for SCF-based processing include polymer grafting, synthesis of esters, metathetical neutralization, pigments, and limited types of polymerizations and oligomerizations.

B. Biomaterial Processing

In contrast to melt extrusion and other conventional mixing or compounding processing methods, the process developed by the Mandel team has the very important advantage that it can be used to process polymers at relatively low temperatures. The effect of contacting a polymer with supercritical CO_2 can lead to a lowering of T_g (91), which means that the supercritical CO_2 process provides a facile way to perform "melt-mixing" at close to room temperatures. This extremely mild processing technique has produced several series of novel biological or medicinal materials, two of which are described here.

C. Bone Replacement Materials

Hydroxyapatite and other calcium salts have been incorporated into an array of polymers or polymer mixtures using supercritical CO_2 processing. As much as 50–75 wt % of inorganic material is distributed homogeneously throughout the polymer, which is rendered as a powder or monolithic solid in its final form. The loaded polymer powder can be pressed or otherwise shaped whereas the solid loaded polymer can be cut into bone-like parts. When a biodegradable polymer, e.g., polylactide-*co*-glycolide (PLGA), is employed, the products represent a unique class of bone replacement materials in that the entire composition is bioresorbable. When this type of material is implanted in the body, two processes take place in a parallel fashion. The first process is that of calcification centered around the hydroxyapatite particles, making healthy bone. The second process is degradation of the polymer matrix, converting the polymer to monomeric species, e.g., lactic acid and glycolic acid. The biocompatibility and the kinetics of the two resorption processes depend, in part, on the porosity of the composite material. The resultant material produced from SCF processing possesses controlled porosity from controlled release of the fugitive solvent. The pore sizes and pore types have been tuned to mimic those in natural bone structure so as to allow adequate growth of blood vessels (91).

D. Drug Delivery Matrices

Biologically active or medicinal ingredients can be incorporated into polymer substrates similar to those used as bone replacement materials (91). The activity of the active drug molecules is preserved because SCF processing operates at a low temperature. For example, catalase enzyme can be incorporated into the PLGA matrix at slightly above ambient temperature. Little or no loss of enzymatic activity is observed after SCF processing. In addition, precise control is exercised over the loading level, particle size, porosity, and surface structure of the product, all of which are essential factors in designing and constructing a drug delivery system. The rate of release is a function of particle size, pore structure, drug concentration, and rate of polymer degradation, whereas the delay time for the release depends on the surface structure and the shape of the particles (91). In addition to the PLGA-catalase combination, several model systems have been tested proving that the process is capable of manufacturing a wide variety of drug release matrices using homopolymers, copolymers, and polymer alloys.

E. Powder Coatings

Powder coatings are solvent-free, organic polymer–based, solid, thermoset coatings. The conventional manufacturing process involves multiple steps of preblending, extrusion, flaking, and milling. The SCF-based process developed by Mandel and coworkers (52–58) enables production of the same or improved products in one or at most two steps. Figure 5 shows a typical production scheme in which raw materials are loaded into the processing vessel without preblending except when a liquid ingredient requires master batching. The vessel is then closed, filled with CO_2, and heated with agitation. The entire content of the vessel is delivered to the cyclone separator and atomized. The atomization is controllable to yield a powder that meets the particle size distribution requirement. If the powder is not sufficiently fine, very gentle milling is used to supplement the production. Thus, a single-step SCF process replaces preblending, extrusion, and flaking, and in most cases milling is also replaced.

This shortened production cycle increases automation, cuts production cost, and improves product consistency. Numerous powder coating types were tested and found suitable technologically to be produced with SCF-based technology.

VI. CONCLUSIONS

A versatile and flexible supercritical CO_2 manufacturing process has been developed and deployed in the manufacturing of many high performance materials. The proven industrial products include powder coatings, polymers, polymer additives, and pigments. The process also has lead to the creation and industrial-scale production of a variety of novel biomaterials. This SCF-aided processing capitalizes on the change in polymer properties that occurs when the polymer makes contact with an SCF solvent. The success of the SCF-based technology shows that it is often not necessary to dissolve the polymer to be processed. This is a very positive finding since the high pressures and temperatures needed to dissolve polymers in SCFs have a severe negative impact on process economics. Systematic SCF–polymer solubility studies interpreted based on the principles of molecular thermodynamics provide guidelines for the type of repeat groups that lead to polymer solubility in SCF solvents at low temperatures and pressures. As researchers develop SCF-soluble polymers in the near future, the engineering experience derived from processing refractory polymers in supercritical CO_2 will prove invaluable for the development of new materials.

VII. ACKNOWLEDGMENTS

J. D. Wang and F. S. Mandel express gratitude to Gary Tatterson. M. A. McHugh acknowledges the National Science Foundation for partial support of the work under Grant CTS-9729720.

REFERENCES

1. JM Prausnitz, RN Lichtenthaler, EG de Azevedo. Molecular Thermodynamics of Fluid-Phase Equilibria. 2nd ed. Englewood Cliffs, NJ: Prentice-Hall, 1986.
2. D Patterson. J Polym Sci Eng 1982;22:64.
3. LL Lee. Molecular Thermodynamics of Nonideal Fluids. Stoneham, MA: Butterworth, 1988.
4. SG Kazarian, MF Vincent, FV Bright, CL Liotta, CA Eckert. J Am Chem Soc 1996;118:1729.
5. A Dardin, JM DeSimone, ET Samulski. J Phys Chem B 1998;102:1775.

6. A Dardin, JB Cain, JM DeSimone, JCS Johnson, ET Samulski. Macromolecules 1997;30:3593.
7. CF Kirby, MA McHugh. Chem Rev 1999;99:565.
8. BM Hasch, MA Meilchen, S-H Lee, MA McHugh. J Polym Sci Polym Phys Ed 1992;30:1365.
9. S-H Lee, MA McHugh. Polymer 1997;38:1317.
10. S-H Lee, MA LoStracco, BM Hasch, MA McHugh. J Phys Chem 1994;98:4055.
11. SE Conway, JS Lim, MA McHugh, JD Wang, FS Mandel. J Appl Polym Sci 2001; 81:2642.
12. G Charlet, R Ducasse, G Delmas. Polymer 1981;22:1190.
13. LA Kleintjens, R Koningsveld, M Gordon. Macromolecules 1980;13:103.
14. BM Hasch, S-H Lee, MA McHugh, JJ Watkins, VJ Krukonis. Polymer 1993;34: 2554.
15. S-J Chen, M Banaszak, M Radosz. Macromolecules 1995;28:1812.
16. TW de Loos, W Poot, RN Lichtenthaler. J Supercrit Fluids 1995;8:282.
17. R Spahl, G Luft. Ber Buns Phys Chem 1982;86:621.
18. PD Whaley, HH Winter, P Ehrlich. Macromolecules 1997;30:4887.
19. SJ Han, DJ Lohse, M Radosz, LH Sperling. Reprints, Polym Mater Sci Eng 1996; 75:279.
20. M Lora, F Rindfleisch, MA McHugh. J Appl Polym Sci 1999;73:1979.
21. MA McHugh, VJ Krukonis. Supercritical Fluid Extraction: Principles and Practice. 2nd ed. Stoneham, MA: Butterworth, 1994.
22. JMG Cowie, IJ McEwen. J Chem Soc Faraday Trans 1974;70:171.
23. BA Wolf, G Blaum. J Polym Sci Polym Phys Ed 1975;13:1115.
24. MA LoStracco, S-H Lee, MA McHugh. Polymer 1994;35:3272.
25. MA McHugh, F Rindfleisch, PT Kuntz, C Schmaltz, M Buback. Polymer 1998; 39:6049.
26. CA Irani, C Cozewith, SS Kasegrande. US Patent 4,319,021, 1982.
27. CA Irani, C Cozewith. J Appl Polym Sci 1986;31:1879.
28. TL Guckes, MA McHugh, C Cozewith, RL Hazelton. US Patent 4,946,940, 1990.
29. AK McClellan, MA McHugh. J Polym Sci Eng 1985;25:1088.
30. MA McHugh, TL Guckes. Macromolecules 1985;18:674.
31. AJ Seckner, AK McClellan, MA McHugh. AIChE J 1988;34:9.
32. E Kiran, W Zhuang, YL Sen. J Appl Polym Sci 1993;47:895.
33. Y Xiang, E Kiran. J Appl Polym Sci 1994;53:1179.
34. AA Kiamos, MD Donohue. Macromolecules 1994;27:357.
35. Y Xiang, E Kiran. Polymer 1997;38:5185.
36. E Kiran, Y Xiong. J Supercrit Fluids 1998;11:173.
37. HAJ Kennis, TW de Loos, J de Swann Arons, R van der Haegan, LA Kleintjens. Chem Eng Sci 1990;45:1875.
38. DA Canelas, JM DeSimone. Chem Rev 1999;99:543.
39. DA Canelas, DE Betts, JM DeSimone. Macromolecules 1996;29:2818.
40. G Luna-Barcenas, S Mawson, S Takishima, JM DeSimone, IC Sanchez, KP Johnston. Fluid Phase Equil 1998;146:325.
41. CA Mertdogan, H-S Byun, MA McHugh, WH Tuminello. Macromolecules 1996; 29:6548.

42. I Yilgor, JE McGrath, VJ Krukonis. J Polym Bull 1984;12:499.
43. VJ Krukonis. Polym News 1985;11:7.
44. TA Hoefling, RM Enick, EJ Beckman. J Phys Chem 1991;95:7127.
45. TA Hoefling, D Stofesky, M Reid, EJ Beckman, RM Enick. J Supercrit Fluids 1992; 5:237.
46. TA Hoefling, DA Newman, RM Enick, EJ Beckman. J Supercrit Fluids 1993;6:165.
47. Y Xiang, E Kiran. Polymer 1995;36:4817.
48. CJ Gregg, FP Stein, M Radosz. J Phys Chem 1994;98:10634.
49. F Rindfleisch, TP DiNoia, MA McHugh. J Phys Chem 1996;100:15581.
50. KG Liphard, GM Schneider. J Chem Thermodyn 1975;7:805.
51. TP DiNoia, SE Conway, S-J Lim, MA McHugh. J Polym Sci Polym Phys Ed 2000; 38:2832.
52. AD McMaster, JD Wang, FS Mandel. Paper A-24: 18th International Symposium on Supercritical Fluid Chromatography and Extraction, 1998, St. Louis, MO.
53. FS Mandel, CD Green, AS Scheibelhoffer. US Patent 5,399,597, 1995.
54. FS Mandel, CD Green, AS Scheibelhoffer. US Patent 5,548,004, 1996.
55. FS Mandel. Paper 10. Fluorine in Coatings II, 1996, Munich, Germany.
56. FS Mandel. US Patent 5,698,516, 1997.
57. FS Mandel. Fourth International Symposium on Supercritical Fluids, 1997, Sendai, Japan, 493.
58. FS Mandel. Proceedings of the 5th Meeting on Supercritical Fluids, 1998, Nice, France, 69.
59. A Michels, C Michels. Proc R Soc 1935;A153:201.
60. A Michels, C Michels. Proc R Soc 1935;A153:214.
61. KS Pitzer. J Am Chem Soc 1955;77:3427.
62. F-H Huang, M-H Li, LL Lee, KE Starling, FTH Chung. J Chem Eng Jpn 1985; 18:490.
63. JF Ely, WM Haynes, BC Bain. J Chem Thermodyn 1989;21:879.
64. DC Friend, ML Huber. Int J Thermophys 1991;15:1279.
65. NIST Thermophysical Properties of Pure Fluids Database, NIST12; 3.0 ed.; NIST Thermophysical Properties of Pure Fluids Database, NIST12. Version 3.0, National Institute of Standards and Technology, Gaithersburg, MD, 1992.
66. RT Jacobsen, RB Stewart. J Phys Chem Ref Data 1972;2:757.
67. JF Ely, JW Magee. "Experimental measurement and prediction of thermophysical property data of carbon dioxide rich mixtures. Proceedings of the 68th GPA Convention, 1989; 89.
68. JM Ely, WM Haynes, BC Bain. J Chem Thermodyn 1989;24:879.
69. RD Goodwin. J Res Nat Bureau Stand (US) 1961;231.
70. PB Smith, DJ Moll. Macromolecules 1990;23:3250.
71. RG Wissinger, ME Paulitis. Ind Eng Chem Res 1990;30:842.
72. NK Kalospiros, ME Paulitis. Chem Eng Sci 1994;49:659.
73. W-Y Wen. Chem Soc Rev 1993;117.
74. RG Wissinger, ME Paulitis. J Polym Sci Polym Phys Ed 1987;25:2497.
75. PA Hamley, CN Field, B Han, M Poliakoff. Proceedings of the 5th Meeting on Supercritical Fluids, 1998, Nice, France, 895.

76. DS Dickey. In: Chopey NP, ed. Handbook of Engineering Calculations. New York: McGraw-Hill, 1983, chap. 12.
77. SJ Haam, RS Brodkey, JB Fasano. Process Mixing: Chemical and Biological Applications: AIChE Symposium, 1992;77.
78. JY Oldshue. Fluid Mixing Technology in Chemical Engineering. New York: McGraw-Hill, 1983.
79. GB Tatterson. In: Scaleup and Design of Industrial Mixing Processes. New York: McGraw-Hill, 1994;10.
80. GB Tatterson. In Fluid Mixing and Dispersion in Agitated Tanks. New York: McGraw-Hill, 1983, chaps. 3 and 5.
81. EKATO-Handbook of Mixing Technology. Schopfhein, Germany: EKATO Ruht und Mischtechnik GmbH, 1991.
82. AJ Ghajar, A Asadi. Am Inst Aeronaut Astronaut J 1986;24:2020.
83. PJ Rourke, DJ Pulling, LE Gill, WH Denton. Int J Heat Mass Trans 1970;13:1339.
84. RH Sabersky, EG Hauptmann. Int J Heat Mass Trans 1967;10:1499.
85. EA Krasnoshchekov, VS Protoponov. Teplofizika Vysokikh Temp 1966;4:389.
86. M Perrut. 4th Italian Conference on Supercritical Fluids and Their Applications, 1997, Capri, Italy.
87. AH Lefebvre. International Gas Turbine and Aeroengine Congress and Exposition, 1990, Brussels, Belgium.
88. MT Lund, PE Sojka, AH Lefebvre, PG Gosselin. Atomiz Sprays 1993;3:77.
89. JD Whitlow, AH Lefebvre. Atomiz Sprays 1993;3:137.
90. H Zhen, S Yiming, S Shiga, H Nakamura, T Karasawa. Atomiz Sprays 1994;4:123.
91. SM Howdle, MS Watson, MJ Whitaker, VK Popov, MC Davies, FS Mandel, JD Wang, KM Shakesheff. Chem Comm 2001:109.
92. RC Reid, JM Prausnitz, BE Polling. The Properties of Gases and Liquids. 4th ed. New York: McGraw-Hill, 1987.

6
Surfactants in Supercritical Fluids

Janice L. Panza and Eric J. Beckman
University of Pittsburgh, Pittsburgh, Pennsylvania

I. INTRODUCTION

A surfactant (*surf*ace-*act*ive ag*ent*) is an amphiphilic molecule containing both hydrophilic and lipophilic segments. Surfactants reduce interfacial tension and aid in the solubilization of hydrophilic compounds into hydrophobic solvents, or vice versa, due to their amphiphilic nature. Surfactants are capable of forming micelles, spherical aggregates arranged so that the hydrophilic segment interacts with the aqueous phase and the lipophilic segment is oriented to interact with the organic phase. The structure of a typical normal micelle is shown in Figure 1. The opposite structures, called reverse micelles (Fig. 1), are also formed whereby the lipophilic segment interacts with the continuous organic phase and the hydrophilic heads are directed to the core of the micelle, thus interacting with the aqueous phase.

Winsor developed a classification scheme for oil–water–amphiphile emulsions in 1948, dividing behavior into four general types (1). Winsor I and II are two-phase systems. Winsor I involves micelles in equilibrium with an excess oil phase, whereas Winsor II comprises reverse micelles in equilibrium with excess water. Winsor III includes three phases wherein most of the surfactant is found in a middle phase in equilibrium with both excess oil and water. Finally, Winsor IV is a single-phase system.

The solvent characteristics of supercritical fluids have been extensively investigated over the past two decades (2). Supercritical fluids have increased solvent strength versus gases due to their liquid-like densities. The pressure and temperature within the supercritical region can be adjusted to regulate the density and therefore the solvent strength of a supercritical fluid. In addition to the liquid-like density, supercritical fluids exhibit gas-like diffusivity and viscosity.

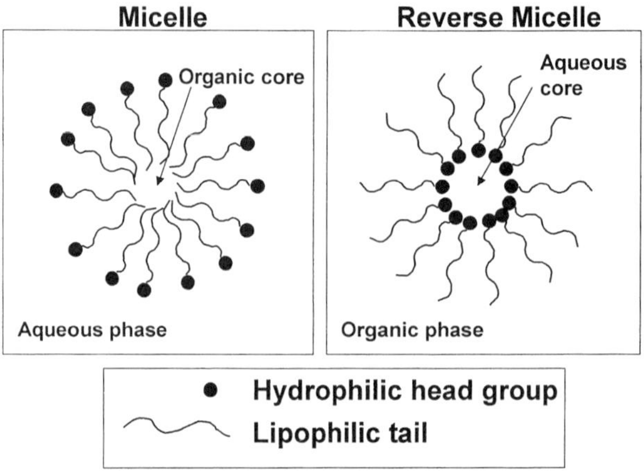

Figure 1 Representation of a micelle and a reverse micelle.

Typically, the critical temperature increases as the polarity increases, rendering the use of polar supercritical fluids, such as water or ammonia, somewhat impractical for most applications. Furthermore, nonpolar supercritical fluids may be limited in some applications due to the insolubility of hydrophilic compounds, necessitating that surfactants be used to assist in solvation. The first studies of surfactants in supercritical fluids were of Aerosol-OT (AOT) in compressed ethane and propane (3), creating the field of microemulsions in supercritical fluids. A review article by Bartscherer et al. discusses microemulsions in supercritical fluids and their applications (4).

Of all the compounds capable of becoming supercritical fluids under relatively moderate temperatures and pressures, supercritical carbon dioxide (CO_2) is unique in that, unlike supercritical alkanes (ethane, propane, and *n*-butane), it is nonflammable and environmentally friendly. Supercritical noble gases, such as krypton and xenon, are benign but very expensive. Although water is environmentally friendly, the critical temperature of 374°C and pressure of 212 atm are much higher that those for CO_2.

There is one major drawback to supercritical CO_2 as a solvent. CO_2 has an extremely low polarizability/volume ratio and hence is a somewhat feeble solvent. Many lipophilic and hydrophilic compounds are not soluble in CO_2. In order to render CO_2 a better solvent for these compounds, surfactants are necessary; yet an extensive study by Consani and Smith showed that commercially available surfactants exhibit low solubility in CO_2 (5). However, Consani did find that some fluorinated surfactants would appreciably dissolve in CO_2. Indeed, hybrid fluorocarbon/hydrocarbon surfactants were shown to form water-in-CO_2 microemulsions (6,7). This chapter will focus on new developments in the design and synthesis of CO_2-soluble surfactants, as well as applications utilizing CO_2-soluble surfactants.

II. DESIGN OF CO_2-SOLUBLE SURFACTANTS

To devise surfactants to be used in CO_2, CO_2-soluble functional groups must be identified. For the most part, organic compounds have been classified as either hydrophilic or lipophilic. However, in the case of CO_2, neither hydrophilic nor lipophilic molecules exhibit appreciable solubility at pressures less than 500 bar. The term "CO_2-philic" (8) has been coined to describe molecules that exhibit high solubility in CO_2 at moderate pressures.

The structure of a surfactant to be used in CO_2 would be amphipathic, like a traditional surfactant, but instead of hydrophilic and lipophilic segments, it would contain CO_2-philic and CO_2-phobic segments. Once the CO_2-philic portion of the surfactant has been identified, the CO_2-phobic segment can be chosen from either hydrophilic or lipophilic molecules, based on the application of the surfactant. The following section will discuss the properties of CO_2 and specific CO_2-philic molecules.

A. Properties of Carbon Dioxide

CO_2 has many properties that make it an interesting solvent; it is abundant, inexpensive, nontoxic, and nonflammable. It has been proposed as a "green" alternative to traditional organic solvents because it is not regulated as a volatile organic chemical (VOC) or restricted in food or pharmaceutical applications. CO_2 attains the supercritical state at near-ambient temperature ($T_c = 31°C$) and a relatively moderate pressure ($P_c = 73$ bar). Supercritical CO_2, like all supercritical fluids, offers many mass transfer advantages over conventional organic solvents due to its gas-like diffusivity, low viscosity, and surface tension.

The major drawback of CO_2, insofar as solvent behavior is concerned, is that it is a poor solvent for many polar and nonpolar compounds, although it can dissolve many small molecules (9). It was once believed that CO_2 had solvent properties similar to those of hexane based on solubility parameter calculations; but in fact, McFann et al. have shown that the quadrapole moment of CO_2 serves to inflate the calculated solubility parameter by 20% (10,11). Consequently, using the solubility parameter as a sole determination of solubility could be misleading. The polarizability/volume has been suggested as a better parameter on which to estimate the solvent power of CO_2, but the polarizability/volume of CO_2 indicates that it is a weak solvent (11). Other properties of CO_2 include low polarizability and electron accepting capacity, since CO_2 is a Lewis acid.

Specific intermolecular interactions between CO_2 and CO_2-soluble polymers have been investigated in molecules that exhibit solubility in CO_2. CO_2 is a weak Lewis acid and can therefore participate in Lewis acid–base interactions. Fourier transform infrared (FTIR) spectroscopy has been used to show that CO_2 interacts with polymers containing electron-donating functional groups (12) and Lewis bases (13). Specific interactions resulting from quadrapole–dipole interactions between CO_2 and certain polymers are also believed to influence solubility (14,15).

O'Neill et al. suggest that cohesive energy density (CED), reflected by surface tension, of a polymer determines solubility in CO_2 (16). Their studies demonstrated that a decrease in the CED of a polymer, closer to that of CO_2, increased the solubility in CO_2. O'Neill suggested that solubility is mainly controlled not by polymer–CO_2 specific interactions but by polymer–polymer interactions.

Although exactly what governs solubility in CO_2 is not entirely clear, several classes of compounds have been identified as having high solubility in CO_2, such as fluoroethers, fluoroacrylates, and silicones. These molecules have been used in the design and synthesis of surfactants for applications in CO_2.

B. Fluoroether-Based Surfactants

Hoefling et al. proposed that incorporation of poly(hexafluoropropylene oxide) (PFPE) (Fig. 2) into a surfactant would lead to high solubility in CO_2 for two reasons: (a) perfluorinated alkanes have low dipolarity/polarizability parameters; (b) fluoroethers have low solubility parameters (17). CO_2 solubility studies of PFPE showed that a MW of 13,000 was soluble up to 10 wt % at a pressure of 17 MPa (18).

Given its high solubility, PFPE was incorporated into surfactants. A fluoroether carboxylic acid (MW 2500) and hydroxyaluminum bis[poly(hexafluoropropylene oxide)] carboxylate (containing two tails, MW 5000) exhibited complete miscibility with CO_2 at 313 K and 11 MPa. An anionic surfactant, sodium PFPE, also showed complete miscibility in CO_2 at 313 K and above 16 MPa.

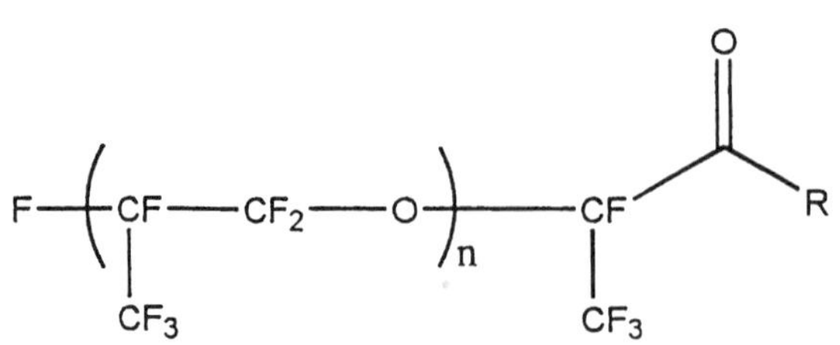

Figure 2 General structure of a PFPE molecule.

The hydroxyaluminum bis[poly(hexafluoropropylene oxide)] carboxylate surfactant was capable of extracting the dye thymol blue from water into CO_2 from CO_2/water/thymol blue mixtures at 23°C (18). Increasing the temperature to 40°C increased the threshold pressure for extraction (due to a decrease in solubility at higher temperatures). Salts of the fluoroether carboxylates were also capable of extracting thymol blue at room temperature, but increasing the temperature to 40°C resulted in the surfactants' preferentially dissolving in the aqueous phase.

Further studies by Hoefling et al. showed that there is a strong correlation between cloud point pressures and the polarity of the hydrophilic head group (17,19). Higher pressures were required to achieve a single phase with the perfluoroether surfactant when the head group was more hydrophilic. The similar cloud point pressures of hydroxyaluminum bis[poly(hexafluoropropylene oxide)] carboxylate (two tails of MW 2500 each) and the fluoroether carboxylic acid (MW 2500) are due to the increased CO_2-philic groups of the two-tailed surfactant and the higher polarity of the carboxylate group.

Newman et al. studied the effects of molecular weight of fluoroether amphiphiles, branching of the fluoroether tails, and the polarity of the head group on the solubility of these amphiphiles in CO_2 (20). In agreement with Hoefling's work, increasing the polarity of the head group shifted the cloud points to higher pressures. Newman also showed that the fluoroether carboxylic salts were capable of extracting the dye thymol blue from an aqueous phase above the vapor pressure of CO_2 (20). An increase in pressure caused the CO_2 phase to grow darker, indicating that more dye was being extracted into the CO_2.

Further studies have demonstrated that PFPE-based surfactants can form microemulsions (with water cores) in supercritical CO_2 (21). At higher water loadings, the CO_2 was saturated with water and micelles began to solubilize water, which demonstrated bulk-like properties using spectroscopic probes. Although the PFPE-ammonium carboxylate surfactant was able to aggregate in CO_2 at low water concentrations, a double-tailed surfactant, $Mn(PFPE)_2$, was not soluble in CO_2 without water. However, in the presence of water, $Mn(PFPE)_2$-based micelles formed and the water core was able to ionize the manganese.

PFPE-based surfactants with sorbitol ester, sulfate, and sulfonate head groups were designed and synthesized, and their phase behavior and emulsion formation in CO_2 were investigated (22). Employing multiple PFPE tails decreased the cloud point pressure of model surfactants. However, a point of diminishing returns was eventually reached where further increases to molecular weight (through the addition of CO_2-philic tails or lengthening of the tails) tended to increase the cloud point pressures. This behavior likely reflects the balance between enthalpy and entropy of mixing, where increasing the level of fluoroether in the molecule renders it more CO_2-philic but increasing the size of the molecule lowers the entropy of mixing. All of the surfactants in this study were capable of forming Winsor I, II, and III emulsions in CO_2, where changes of pressure prompted Winsor I to Winsor III to Winsor II phase transitions. Thus increasing pressure here is analogous to raising the salt concentration in traditional emulsions.

Additional studies using PFPE showed that dendrimers modified with PFPE were able to act as unimolecular micelles in CO_2 (23). Dendrimers are well-defined, highly branched, spherical polymers that are capable of containing small molecules in their loosely packed cores. The dendrimers modified with PFPE were soluble in CO_2 (the precurser dendrimer was insoluble in CO_2) and able to extract hydrophilic compounds from an aqueous solution into CO_2.

Spectroscopic studies were performed on water in supercritical CO_2 microemulsions using an ammonium carboxylate PFPE surfactant (24). FTIR spectoscopy was used to identify a bulk water phase within the microemulsion capable of solubilizing ionic species and supporting inorganic reactions. In addition, the UV-visible spectrum of the solvatochromic probe methyl orange indicated three microenvironments within the microemulsions: (a) a polar microenvironment like that found in dry PFPE reverse micelles; (b) bulk water microenvironment; and (c) an acidic microenvironment due to CO_2 dissolved in water.

Organic synthesis can be conducted in water/CO_2 microemulsions formed with an ammonium carboxylate PFPE surfactant (25). Nucleophilic substitution reactions occurred between hydrophilic nucleophiles and CO_2-soluble reactants. The reaction yields and rate constants were an order of magnitude greater than traditional water-in-oil microemulsions under similar conditions (except pressure), likely due to lower microviscosity of the water/CO_2 microemulsions. Water/CO_2 emulsions also exhibited higher yields than water-in-oil emulsions, which is attributed to lower interfacial tension and viscosity of the water–CO_2 interface than the water–oil interface and higher diffusivity in CO_2 (26). Greater yields were obtained from organic synthesis in water/CO_2 emulsions than in water/CO_2 microemulsions due to the larger amount of water in an emulsion, which allowed for greater excess of the hydrophilic nucleophile in these reactions (26).

C. Silicone-Based Surfactants

Silicone polymers, such as poly(dimethylsiloxane) (PDMS), also show high solubility in CO_2, though not to the extent of poly(hexafluoropropylene oxide) (17). The interest in using silicone-based polymers as surfactants is that they require less expensive raw materials than their fluorinated counterparts; can be synthesized anionically, leading to narrow molecular weight distributions; and are soluble in variety of organic solvents, facilitating their characterization (27).

Hoefling et al. investigated the relationship between structure and solubility of silicone-based amphiphiles in CO_2 (28). Here a silicone copolymer backbone was reacted with allyl glycidyl ether (AGE), a CO_2-phobic group, and allyltris(trimethylsiloxy)silane (ATSS), a CO_2-philic group, to generate copolymers with 1 AGE to 5 ATSS (1:6 AGE) groups to 6 out of 6 AGE branches (6:6 AGE) (Fig. 3). The cloud point curve of the 6:6 AGE was higher than a linear AGE (a linear silicone copolymer with 1 AGE and 1 ATSS end group), likely due to an increase in the molecular weight. The cloud point curves decreased for the 1:6 AGE vs. the 6:6 AGE, due to an increase in the number of CO_2-philic ATSS branches. Functionalization of the AGE branches with amine and ammonium chloride increased the cloud point pressures significantly, showing that the polarity of the head group has a strong effect on solubility. The solubility of the silicone-based surfactants was affected by molecular weight, CO_2–silicone interactions, and branching, but the effects of head group polarity on solubility dominated.

Another study by Hoefling et al. concurred with the above results (19). The cloud point curves for a linear AGE silicone copolymer functionalized with 2-dimethylsiloxane (instead of ATSS) were lower than its amino derivative. Upon further transformation of the amine to ammonium chloride, the surfactant was no longer soluble in CO_2 up to pressures of 50 MPa. The two dimethylsiloxane branches were not enough to balance the addition of the hydrophilic ammonium head group.

D. Fluoroacrylates

The polymer that shows the most favorable mixing thermodynamics with CO_2 is poly(1,1-dihydroperfluorooctylacrylate) [poly(FOA)] (Fig. 4) (29). Poly(FOA)

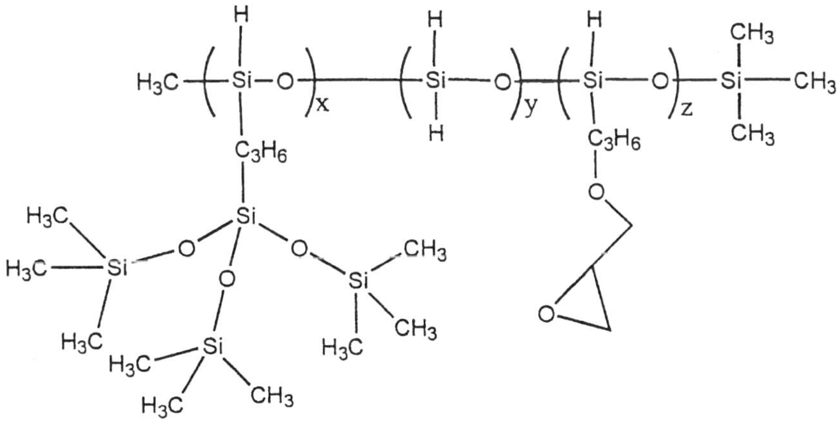

Figure 3 AGE-ATSS-functional silicone where $x + z = 6$.

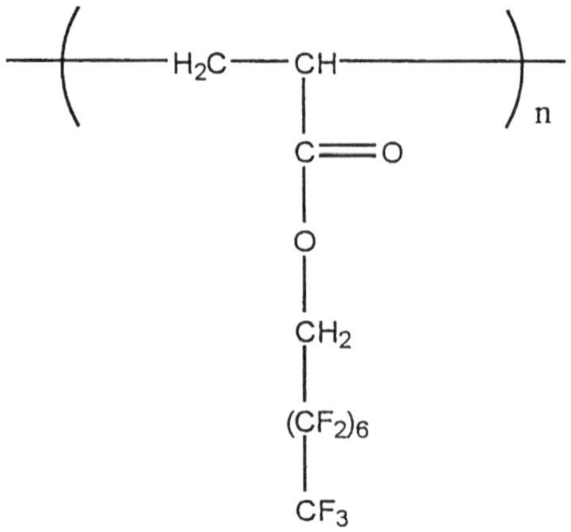

Figure 4 Structure of poly(FOA).

contains a lipophilic, acrylic backbone and a CO_2-philic segment, rendering it amphiphilic. Poly(FOA) can thus be used as a surfactant without further modification.

Traditionally, fluorinated polymers have been synthesized in chlorofluorocarbons (CFCs) as they are insoluble in common organic solvents. Due to the environmental concerns associated with CFCs, CO_2 was investigated as a solvent for the synthesis of poly(FOA) by DeSimone et al. (30). Poly(FOA) was synthesized under homogeneous conditions to high molecular weight (about 2.7×10^5 g/mol), demonstrating the potential of CO_2 as a solvent for fluoropolymer modification (30).

Small-angle neutron scattering (SANS) was used to study the solution properties of poly(FOA) in CO_2 (31). SANS data showed that the second virial coefficient (which describes the interactions between polymer segments and solvent) of poly(FOA) in CO_2 over the range of densities of CO_2 used in the experiment ($0.842 < \rho < 0.943$) is positive, indicating that CO_2 is a thermodynamically "good" solvent for poly(FOA).

McClain et al. used poly(FOA) as the CO_2-philic segment of a nonionic surfactant, where poly(FOA) was copolymerized with a CO_2-insoluble polystyrene (PS) segment to form a block copolymer (Fig. 5) (32). Because of the solubility differences of the two segments, block copolymer molecules assemble into a micelle, where the CO_2-phobic PS segments are found in the micelle core, and are surrounded by the CO_2-philic poly(FOA) segments. These micelles were used to solubilize CO_2-phobic hydrocarbon oligomers, where 99% of the hydrocarbon oligomers localized in the core of the micelle.

Copolymers of poly(FOA) and a poly(ethylene oxide) (PEO) were also able to form aggregates in CO_2, with a PEO-rich core that stabilized small

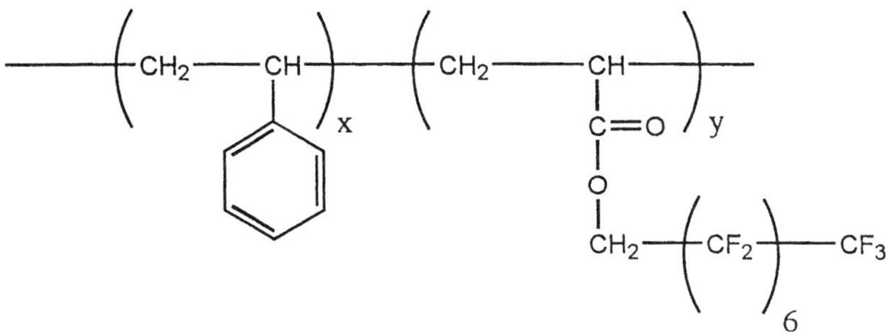

Figure 5 Structure of a PS-poly(FOA) block copolymer.

amounts of water (33). In addition, poly(FOA) and block copolymers made with poly(FOA) were able to stabilize emulsions of poly(2-ethylhexyl acrylate) (PEHA) in liquid and supercritical CO_2 (34,35).

III. APPLICATIONS USING SURFACTANTS IN CARBON DIOXIDE

The discovery of the CO_2-philic molecules and the design and synthesis of CO_2-soluble surfactants with these molecules has opened the door to replacing traditional VOCs with the environmentally friendly CO_2 in several applications. Many applications exist where CO_2 could play an important role provided that CO_2-soluble molecules could be devised and synthesized to function in that particular application. The rest of this chapter deals with specific applications where CO_2 can be used to replace traditional solvents and how the CO_2-philic surfactants were specially designed to operate in the application.

A. Protein Extraction

With the growth in the production of biochemicals, new technologies are needed for bioseparations in order to recover and concentrate biological molecules such as proteins from the media in which they are produced. Liquid–liquid extraction is a common method for separation, but it falls short when applied to bioseparations due to the lack of appropriate solvents. A variety of liquid–liquid extraction techniques are currently being investigating for bioseparations. Liquid–liquid extraction of proteins using reverse micelles has been studied where organic solvents are used as the continuous phase (36–38). Use of CO_2 as the solvent in extraction of proteins with reverse micelles instead of other organic solvents would alleviate the problems of organic waste generation and aqueous stream contamination.

Johnston et al. were the first to show that a protein can be solubilized in a water-in-CO_2 microemulsion, where an ammonium carboxylate PFPE surfactant (MW 740) was used to form the microemulsions (39). Fluorescence was used to monitor the solubilization of bovine serum albumin (MW 67,000) labeled with acrylodan (BSA-Ac) in a stable aqueous environment in CO_2. The fluorescence of BSA-Ac in this water-in-CO_2 microemulsion using the PFPE surfactant (1.4 wt %) was similar to that of native BSA in buffer, pH 7.0.

Ghenciu and Beckman designed an affinity surfactant containing the ligand biotin for the extraction of avidin (Fig. 6) (40). The surfactants were prepared both with and without a polyethylene glycol (PEG) spacer, and the CO_2-philic tail was composed of PFPE. The phase behavior of the surfactant was a function of both the overall molecular weight and the ratio of the number of CO_2-philic to hydrophilic groups. Increasing the length of the PEG spacer (MW 300–600) at constant PFPE chain length (MW 7500) increased the cloud point pressures at 35°C. Increasing the length of the PFPE tail (MW 2500–7500) also increased the cloud point, which suggests that the entropy of mixing dominates in both cases. The surfactant that contained a PEG spacer (PFPE 7500, PEG 600) was capable of extracting more avidin than the surfactant without the spacer (PFPE 7500), probably due to better surface activity of the material with the PEG spacer. An inverse emulsion (20:80 liquid CO_2 to avidin solution) and three-phase emulsion (40:60 liquid CO_2 to avidin solution), both using the PFPE 7500/PEG 600 biotin surfactant, were compared on their abilities to extract avidin. The three-phase emulsion extracted more than double the amount of

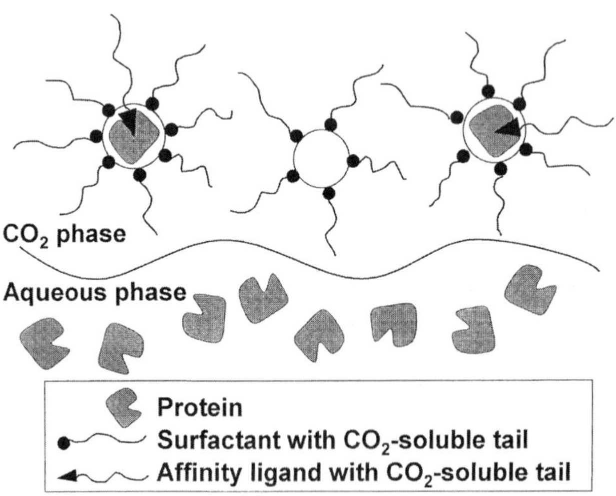

Figure 6 Protein extraction using a CO_2-philic affinity ligand.

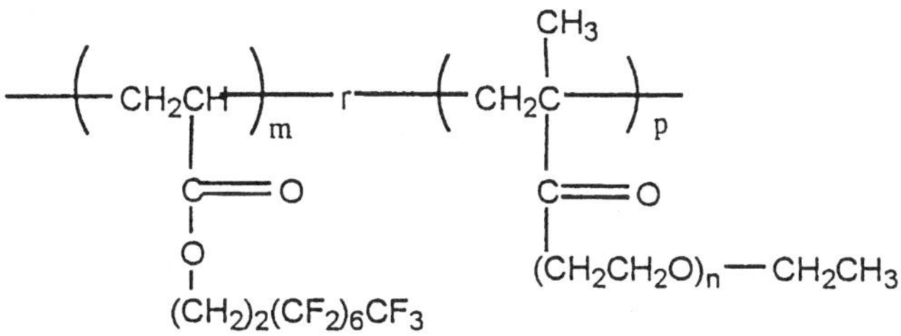

Figure 7 Structure of poly(TAN-*co*-PEG).

protein obtained from the inverse emulsion, possibly because it is better able to partition the surfactant–protein aggregates. To recover the proteins, the three-phase emulsion was stripped with liquid CO_2 until the emulsion broke due to the continuous removal of the surfactant–protein complex.

The biotin-functional surfactant used to extract avidin was a special case, owing to the strong affinity of the avidin–biotin complex. In order to extract other proteins, surfactants must be designed that can interact with proteins yet allow for the easy release of the protein following the extraction. Ghenciu et al. investigated the use of PFPE surfactants with nonionic (PEG groups) and an ionic (sodium sulfate) head groups in the extraction of subtilisin Carlsberg into CO_2 (41). The ionic surfactants and the nonionic surfactants with a PEG molecular weight of 1900 were able to form inverse emulsions in CO_2 (depending on the CO_2 to water ratio), whereas nonionic surfactants with shorter hydrophilic PEG spacers (MW 600 or 900) formed inverse emulsions only in the presence of protein. Larger amounts of subtilisin were solubilized with the ionic surfactant owing to the specific interactions of the negatively charged ionic surfactant with the positively charged protein. As with avidin, subtilisin was effectively stripped from the three-phase emulsion with pure CO_2.

DeSimone et al. designed and synthesized amphiphilic copolymers composed of a perfluoroacrylate polymer, poly(1,1,2,2-tetrahydroperfluorodecyl acrylate) [poly(TAN)] and PEG to form a poly(TAN-*co*-PEG) copolymer (Fig. 7) to be used in bioextractions (42). Poly(TAN-*co*-PEG) was capable of extracting BSA from an aqueous solution into CO_2.

B. Dispersion Polymerization

CO_2 has shown great utility in dispersion polymerizations. A dispersion polymerization begins as a homogeneous solution whereby both the monomers and initiator are soluble in the reaction medium. As the reaction proceeds, oligomers are produced through solution-phase polymerization. Once the oligomers reach a critical size, they begin to precipitate from solution. At this point, specifically designed surfactants are needed to stabilize the precipitated particles in order to prevent flocculation and aggregation. Polymerizations continue in the bulk phase in the stabilized polymer colloid, as shown in Figure 8.

Stabilizers for dispersion polymerizations are specifically designed surfactants that contain a CO_2-phobic region and a CO_2-philic region. The CO_2-phobic region acts as anchor to the growing polymer, either by physical adsorption or by chemical grafting. The CO_2-philic region sterically stabilizes the growing polymer particles, preventing flocculation and precipitation. Extensive research has been done on the design and function of CO_2-soluble stabilizers in dispersion polymerizations.

Methyl methacrylate (MMA) has been polymerized by dispersion polymerization in supercritical CO_2 (204 bar and 65°C) using the polymer poly(FOA) as the surfactant stabilizer (8). Two polymeric stabilizers, a low-molecular-weight (LMW) ($M_n = 1.1 \times 10^4$ g/mol) and a high-molecular-weight (HMW) ($M_n = 2.0 \times 10^4$ g/mol) poly(FOA), and two different initiators [2,2′-azobis(isobutyronitrile) (AIBN) and a fluorinated AIBN (F-AIBN)], were utilized. While product differences owing to changing the initiator were small, the presence of stabilizer had a pronounced effect. Without added stabilizer, the PMMA precipitated on the vessel walls and the reaction proceeded to low conversions only ($<$40%). When either LMW or HWM poly(FOA) was added as the stabilizer, the yield ($>$90%) and molar mass ($> 3.0 \times 10^3$ g/mol) of the product increased dramatically. By increasing the concentration of the stabilizer, smaller and more uniform particles were created. At constant stabilizer concentration, the LMW stabilizer resulted in the formation of more particles with a smaller diameter than with the HMW poly(FOA).

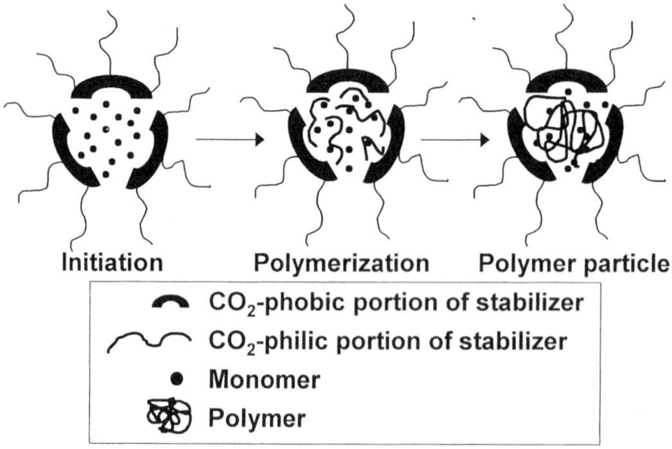

Figure 8 Dispersion polymerization in CO_2 using a CO_2-philic stabilizer.

Additional dispersion polymerizations of MMA were performed using a higher molecular weight poly(FOA) ($M_n = 1.0 \times 10^6$ g/mol) (43). Low concentration of poly(FOA) (0.24 wt %) was sufficient to stabilize the poly(methyl methacrylate) (PMMA) particles to a substantially higher molecular weights ($M_n = 2.55 \times 10^5$ g/mol) when compared with no added stabilizer ($M_n = 8.5 \times 10^4$ g/mol). The particle size decreased from 2.86 to 1.55 μm as the stabilizer concentration increased from 0.24 to 16 wt %. It was speculated that the oligomeric PMMA radicals absorb the higher concentration of stabilizer before aggregation with other particles could occur, resulting in a greater number of smaller particles with higher stabilizer content. In addition, higher concentrations of stabilizer led to the formation of a second population of smaller particles that gave a higher polydispersity index (PDI) at the higher poly(FOA) concentration. Finally, the mode of anchoring of the poly(FOA) stabilizer to the PMMA particle—either adsorption or absorption—may have affected the particle size.

A silicone macromonomer has also been used to stabilize the dispersion polymerization of PMMA (44). The PDMS macromonomer used in this study was a commercially available methacryloxy functional PDMS macromonomer (Fig. 9). The polymerizations were carried out at 340 bar and 65°C for 4 h. When no PDMS macromonomer was added to the polymerization, only low conversions could be achieved and the polymer precipitated. Addition of a small amount of PDMS macromonomer (0.05 wt %) to the polymerizations increased the molecular weight and the yield of the polymer, but at least 3.5 wt % was necessary to obtain monodispersed polymer particles at high yield. PDMS homopolymer, which lacks the reactive MMA functional group, was not effective in stabilizing the polymerization in that a much greater concentration of the

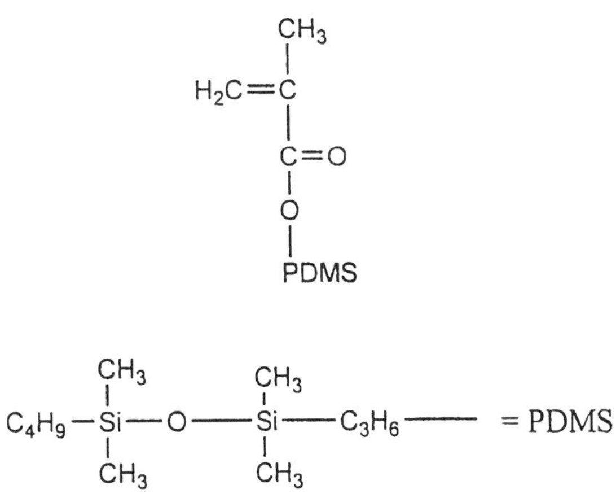

Figure 9 Structure of a PDMS macromonomer.

homopolymer (6.8 wt %) was needed to achieve the same results as 0.05 wt % of the macromonomer. Purification of the PMMA showed that only a fraction of the macromonomer was incorporated into the polymer.

Studies of the particle growth (45) and particle formation regime (46) during MMA polymerization stabilized by the PDMS macromonomer revealed that a pressure of about 3000 psia (about 207 bar) and a stabilizer concentration of about 2 wt % (stabilizer to monomer) was required. Below this threshold pressure, PDMS was poorly solvated by the continuous phase and was unable to stabilize the growing polymer particles. Below the threshold stabilizer concentration, coagulation and precipitation of the polymer occurred due to insufficient steric stabilization. During the particle formation regime, both coagulative nucleation and controlled coagulation processes were identified. In coagulative nucleation, polymerization nuclei coalesced until the surface area was reduced enough to be covered by the stabilizer. This continued until the particle size plateaued and the particle number density and surface area rose to a maximum. At the maximum, the surface area of the particles exceeded that covered by the stabilizer. At this point, the controlled coagulation began, with the particles coalescing to reduce the surface area and the particle number density.

Another fluorinated copolymer stabilizer, poly(methyl methacrylate-*co*-hydroxyethyl methacrylate)-*g*-poly(perfluoropropylene oxide) (PMMA-HEMA-PFPO), was used in the dispersion polymerization of MMA to PMMA (47). This stabilizer was a graft copolymer wherein the PMMA-HEMA acted as the anchor and the graft chain of PFPO was the soluble component (Fig. 10). In this study, the effects of the molecular architecture of the stabilizer, in particular the PMMA-HEMA (backbone) length, the PFPO graft chain density, graft chain length, graft chain distribution, and stabilizer concentration, were investigated. Length of the backbone was found to be the most important factor in determining the polymerization rate, particle size, and particle size distribution. By increasing the length of the backbone, the dispersion polymerization was better stabilized due to better anchorage and surface coverage of the growing polymer particles. The extent of the soluble component was also important; there must be enough CO_2-philic groups to ensure solubilization in the continuous phase. In general, as the number of grafts per backbone increased (graft chain density), the rate of polymerization increased although there was a limit to this effect. The molecular weight, particle size, and size distribution of the polymer did not appear to be affected by the graft density as long as the backbone was long enough to stabilize the growing polymer. The distribution of the grafts along the backbone affected the polymerization, where shorter, more numerous grafts increased the rate of polymerization and produced smaller particles with a narrower size distribution. Finally, the rate of polymerization increased as the concentration of the stabilizer increased until it reached a maximum, at which point the rate of diffusion of the monomers to growing polymer was impeded, decreasing the

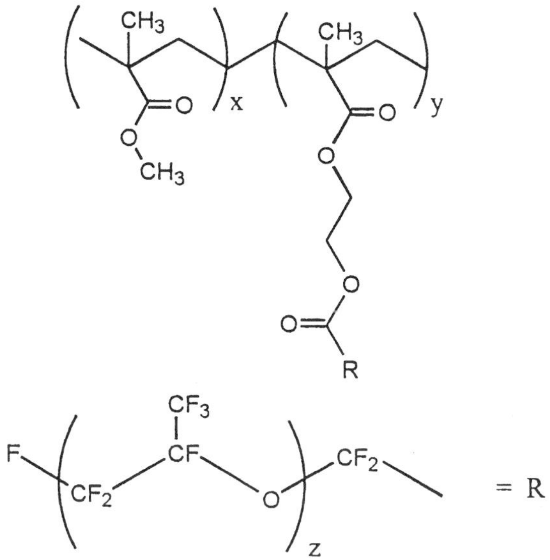

Figure 10 Structure of PMMA-HEMA-PFPO graft copolymer.

rate of polymerization. The particle size and particle size distribution decreased as the stabilizer concentration increased.

Styrene has also been polymerized under dispersion conditions in CO_2. However, the poly(FOA) homopolymer and the PDMS macromonomer were not the best stabilizers for this monomer. Polystyrene (PS) was polymerized efficiently under dispersion conditions using a PS/poly(FOA) diblock stabilizer (48). The PS segment anchored to the growing PS particle, while the poly(FOA) block provided steric stabilization in CO_2. Indeed, it has been shown that the block copolymer reduces the interfacial tension at the PS–CO_2 interface (49). As was shown previously, added stabilizer increased both the yield and molecular weight of the PS when compared with polymerizations without stabilizer. The mean particle diameter and the particle size dispersity decreased as the length of both the PS and the poly(FOA) blocks increased. Poly(FOA) homopolymer did offer some stabilization to the dispersion polymerization of PS when compared with no added stabilizer, but the presence of the PS block greatly enhanced the stabilization of the PS particles.

Block copolymer dispersants were also synthesized using PDMS as the stabilizer and PS as the anchor, as shown in Figure 11 (27). These block copolymers were not soluble in pure CO_2 under supercritical conditions (65°C, 340 bar) but required the presence of styrene as a cosolvent. Similar to the PS/poly(FOA) block copolymer stabilizer, polymerizations with added stabilizers resulted in

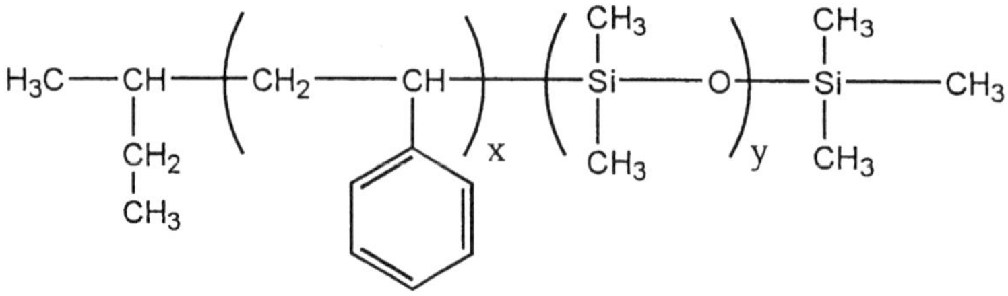

Figure 11 Structure of PS-PDMS block copolymer.

higher conversions and higher molecular weight polymers, whereas the polymerizations conducted without added stabilizer or in the presence of the PDMS homopolymer resulted in precipitated polymers with low conversions and low molecular weight. High yields of PS were obtained when the stabilizer contained a PDMS segment with an M_n of 2.5×10^4 g/mol. The effect of the anchor-soluble balance (ASB), or ratio of the two block lengths (PS to PDMS), was explored. By increasing the size of the PS segment in the stabilizer, the particles became larger and more monodisperse. When a larger PDMS segment was used, low-molecular-weight PS with broad molecular weight distributions were obtained. The authors theorized that the lower ASB (less PS) or the limited solubility of the longer PDMS segments in the CO_2 accounts for the results. Other factors investigated using the PS-PDMS stabilizer were time, temperature, CO_2 pressure, and stabilizer concentration. Finally, as the stabilizer concentration increased, the molecular weight and molecular weight distribution decreased most likely due to surface area arguments.

Although it was originally believed that poly(FOA) was not an effective stabilizer for the dispersion polymerization of PS (48), Shiho and DeSimone demonstrated that poly(FOA) and another fluorinated polymer, poly(1,1-dihydroperfluorooctyl methacrylate) [poly(FOMA)], were actually good stabilizers for PS but required higher pressures than what had been used previously (65°C, 370 bar) (50).

Cationic polymerizations of styrene in CO_2 under dispersion conditions have also been demonstrated (51). The stabilizer was a block copolymer of 2-(*N*-propylperfluorooctanesulfonamido)ethyl vinyl ether (FVE) and methyl vinyl ether (MVE), which formed poly(FVE-*b*-MVE). The poly(FVE) segments served as the soluble block and the poly(MVE) segments served as the anchoring units of PS. These cationic dispersion polymerizations were sensitive to temperature, with the optimum temperature being 15°C, which seemed to prevent chain transfer to monomer. The polymerizations at 15°C and 4 wt % stabilizer resulted in approximately 95% yield and a molecular weight of 1.8×10^4 g/mol.

Poly(vinyl acetate) (PVAc) and ethylene/vinyl acetate copolymers were prepared by dispersion polymerization in supercritical CO_2 using both fluorinated and siloxane-based stabilizers (52). PVAc-poly(FOA) block copolymers with higher molecular weight fluorinated blocks produced smaller particle sizes and higher particle densities due to their ability to stabilize more surface area. However, a sufficient anchor-to-soluble balance was required to achieve adequate adsorption of the stabilizer to the particle. When compared to the poly(FOA)-based stabilizers, the PDMS-based stabilizers gave rise to larger particles and lower particle number densities due to the lower solubility of PDMS in CO_2. The PVAc particles produced by the dispersion polymerizations were larger than PMMA or PS produced under similar conditions, probably due to the higher solubility of PVAc in CO_2.

Recently, novel surfactants (PDMS-*b*-poly(methacrylic acid) [PDMS-*b*-PMA)] were used in the dispersion polymerization of PMMA in supercritical CO_2 to produce latexes that could be transferred to water to form an aqueous latex (53). Surfactants that can stabilize polymer particles in both CO_2 and water were labeled "ambidextrous." The PDMS acted as the soluble block and the PMA block as the anchor. The molecular weight of both the anchor block and the stabilizing block of the PDMS-PMA was less than those used in previously mentioned dispersion polymerizations using PDMS-based stabilizers (27), owing to the harsh synthesis conditions that degraded the PDMS. There are nine ionizable groups per chain on the PDMS-PMA surfactant such that when transferred to an aqueous solution the surfactant on the surface of the polymer ionizes, thereby becoming hydrophilic and producing an electostatically stable latex. The PDMS group collapses to the surface of the particle in the aqueous solution. A commercially available surfactant, PDMS-*g*-pyrrolidonecarboxylic acid (PDMS-*g*-PCA), was also used, but it contains only two ionizable groups per chain. Both surfactants were used to produce PMMA in supercritical CO_2 (345 bar, 65°C) by dispersion polymerization. PDMS-PMA produced polymer particles that precipitated as the reaction proceeded, owing to the lower molecular weight of the PDMS stabilizing block. PDMS-PCA produced smaller, more uniform particles than the PDMS-PMA. However, the PDMS-PMA particles formed electrostatically stable latexes in up to 10 wt % when dispersed in phosphate-buffered solutions of pH 8.17 and 11.36, whereas the PDMS-PCA particles rapidly flocculated in the buffer solutions due to insufficient electrostatic stabilization. Polymer particles produced with a 1:1 mixture of both surfactants were slightly larger than those produced with PDMS-PCA alone and showed improved water dispersibility, but were partially flocculated in the aqueous solutions.

Finally, poly(divinylbenzene) (PDVB) was synthesized by dispersion polymerizations in CO_2 (54). The stabilizer was a block copolymer of MMA and 1H,1H,2H,2H-perfluorooctyl methacrylate. The absence of added stabilizer to the reaction (310 bar, 65°C) resulted in a precipitated polymer; however, increasing the stabilizer concentration to 3 wt % resulted in 95% yield of uniform microspheres of PDVB with a narrow size distribution, suggesting that effective stabilization occurred under these conditions.

As can be seen from the bulk of the work reviewed, CO_2 could play an important role in dispersion polymerizations. Various polymers, including PMMA, PS, PVAc, ethylene/vinyl acetate copolymers, and PDVB, have been synthesized by dispersion polymerizations, with the surfactant stabilizers ranging from homopolymers to copolymers, each one specifically designed for the polymerization. As our knowledge of dispersion polymerization increases, stabilizers can continue to be refined so as to be precisely designed to accommodate each polymer. In addition, today's stabilizers are expensive; therefore, future work should focus on the design of less costly alternatives.

C. Emulsion Polymerizations

Emulsion polymerizations have also been performed in supercritical CO_2. Here the monomer is CO_2 insoluble (or only slightly soluble) but the initiator is CO_2 soluble. Most of the monomer is dispersed as droplets in the CO_2 that are stabilized by surfactant molecules adsorbed to the surface. Micelles are also present in emulsion polymerizations. The initiator is soluble in the CO_2 phase but not in the monomer droplet, and thus the micelles act as the meeting place for the monomer and initiator. The systems contains three types of particles: micelles where polymerization is not occurring, micelles where polymerization is occurring (called the polymer particle), and monomer droplets (Fig. 12).

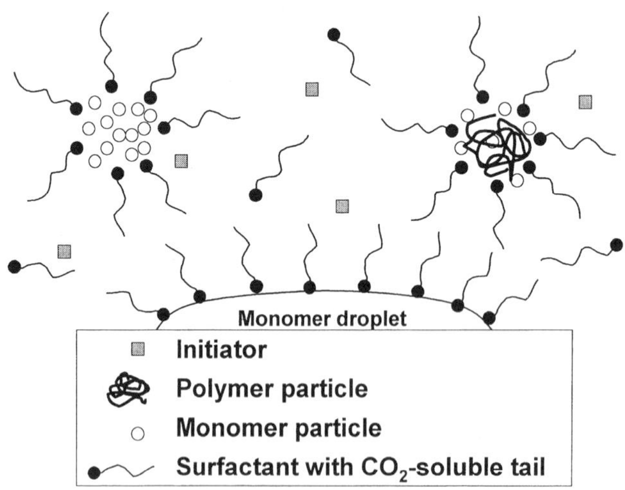

Figure 12 Emulsion polymerization in CO_2 using a CO_2-philic surfactant.

Adamsky and Beckman polymerized acrylamide in inverse emulsions in supercritical CO_2 (345 bar, 60°C) (55). The surfactant used in this study contained a polar head group composed of an amide group, to accomplish micelle formation, and a CO_2-philic tail composed of hexafluoropropylene. The absence of surfactant resulted in precipitation of a single solid mass of polymer in the reaction vessel. When surfactant was present, the solution had a milky-white appearance of an emulsion polymerization with yields higher than when no surfactant was used. The polymers produced exhibited a higher degree of linearity when compared with conventional emulsion polymerization of acrylamide.

D. Metal Extraction

CO_2 as a solvent offers rapid extraction of heavy metals from solids or in applications where clean-up or recovery of metals is necessary. Direct extraction of metal ions into CO_2 is highly inefficient; however, when metal ions are bound by organic chelating agents, they may exhibit solubility in CO_2. Laintz et al. have demonstrated that bis(trifluoroethyl) dithiocarbamate (FDDC), the fluorinated counterpart of the common chelating agent diethyl dithiocarbonate (DDC), can be used for the extraction of copper ions into CO_2 from liquid and solid materials (56). FDDC exhibited two to three orders of magnitude higher solubility in CO_2 than its nonfluorinated counterpart, DDC (57,58). In addition, metal ions (arsenic and antimony) complexed to FDDC have been analyzed by supercritical fluid chromatography using CO_2 as the mobile phase (59). Fluorinated β-diketones have been used in the extraction of different metal ions, such as lanthanide ions (60), thorium and uranium ions (61), copper ions (62), and cadmium, lead, and mercury ions (63). Organophosphorous chelating agents have also been used in supercritical CO_2 extraction of metal ions (60,61,64–66). Solubility studies of metal–chelate complexes in supercritical CO_2 have also been performed (67,68). Supercritical fluid extraction of metal ions has been reviewed recently (69–71).

The aforementioned chelating agents displayed solubility in CO_2; however, they do not necessarily resemble surfactants with amphiphilic character. A distinction will be drawn here between chelating agents that show solubility in CO_2, like those mentioned above, and chelating agents that exhibit amphiphilic properties. CO_2-philic chelating agents have been designed and synthesized by attaching CO_2-philic tails to traditional chelating agents, resulting in an amphipathic molecule with surfactant-like characteristics. The rest of this section focuses on chelating agents that were specifically designed to act like surfactants.

Traditional chelating agents, such as picolylamine, bis(picolylamine), dithiol, or dithiocarbamate, have been derivatized with CO_2-philic PFPE and PDMS (72,73). As with the CO_2-philic surfactants, the polarity of the chelating head group had a pronounced effect on the solubility in CO_2. The PFPE-chelating agents with attached metal have different cloud point curves than the agents alone due to aggregation, charge neutralization, and micellation (73).

Metal extractions from sand in CO_2 were performed using the CO_2-soluble chelating agents. The PDMS-picolylamine and bis(picolylamine) chelating agents showed a significantly higher efficiency in the extraction of lead than the PFPE counterparts (74). A possible explanation is that the higher electron withdrawing capability of the PFPE tail vs. the PDMS tail deactivates the chelating functional group. A propyl spacer was added between the bis(picolylamine) head and the PFPE tail, which served to increase the efficiency of the extraction of lead from CO_2 by 100%. Addition of alkyl spacers (from 0 to 6 carbons) increases the cloud point of the chelating agent, but extraction efficiencies did not increase significantly for spacers longer than two carbons (74).

Metal extraction from water has shown to be difficult due to the low pH of water when contacted with carbon dioxide (carbon dioxide dissolves into water and forms H_2CO_3) (75). PFPE (MW 7500) was attached to piperazine dithiocarbamate, a chelating agent that exhibits high extraction abilities of metals at low pH. The cloud point curve of the fluoroether piperazine dithiocarbamate was found to be very dependent on concentration, unlike the CO_2-soluble chelating agents with picolylamine, bis(picolylamine), dithiol, or dithiocarbamate as the head group (75). However, the fluoroether piperazine dithiocarbamate chelating agent was successful at extracting many metals at pH as low at 1.2.

IV. OTHER APPLICATIONS

CO_2 has been shown to be an ideal solvent for applications such as protein extraction, heterogeneous polymerizations, and metal extraction due to the development of CO_2-soluble surfactants specific for each application. CO_2 could become a beneficial replacement solvent in many other applications, provided that specific surfactants for those functions could also be designed. Such applications include dry cleaning, particle formation, and biocatalysis. The design and formation of specific surfactants for such applications are currently underway.

A. Dry Cleaning

The dry cleaning industry is an excellent example of the use of CO_2 to replace a conventional solvent. The traditional solvent for dry cleaning is perchloroethylene, referred to as PERC. PERC is designated a hazardous air pollutant and regulated under the Clean Air Act. Dry cleaner employees are not the only people affected by PERC in that dry cleaned clothes release PERC into the air of consumers' homes (76). More than 30 billion pounds per year of organic and halogenated solvents are currently estimated to be used in the dry cleaning (77).

Although CO_2 dissolves sweat, oils, and dirt (78), CO_2-soluble surfactants are necessary in the dry cleaning process to form micelles that can capture dirt and grease. After the process is complete, the CO_2 can be returned to the gaseous state, allowing the trapped solutes to precipitate. The CO_2 can then be repressurized and reused. Dry cleaning processes using specially designed CO_2-soluble surfactants are currently being commercialized (79,80).

An early concern of CO_2 in dry cleaning was that CO_2 will cause clothes to shrink. However, preliminary tests indicate that shrinkage is within industry standards (81). Another concern is that CO_2 might swell fabrics, particularly acrylics. Still another concern is how CO_2 will affect some of the dyes used to color fabrics (81).

B. Nanoparticle Formation

There is a growing interest in the preparation of nanometer-size metal materials for use as advanced catalyst materials, pharmaceuticals, pesticides, optical barriers, semiconductor crystallites, lubricants, and others. Current techniques for producing nanoparticles involve harsh process conditions and do not provide adequate control over particle characteristics. The nanometer-size water cores of reverse micelles formed in CO_2 using expressly designed surfactants are proposed to be an ideal environment to produce nanoparticles of uniform size. Changes to CO_2-solvent properties through manipulation of the pressure can affect the growth rate of nanoparticles, their final size, and their size distribution, allowing fine control over the nanoparticular products.

In particle formation (Fig. 13), a metal ion, such as copper (Cu^{2+}), is introduced into a reverse micelle, either as an Cu^{2+} ion-surfactant conjugate or as a copper salt. A reducing agent within the CO_2 continuous phase diffuses into the micelle, reducing the Cu^{2+} ions to form very small copper particles. Intermicellular exchange of the metal particles solubilized within the core of the micelles allows for the growth of the particle by the aggregation and coalescence of the very small particles. Particle growth stops due to the limitation of the particle size that the micelle can support.

Metallic nanoparticles have been formed by reduction of copper ions (82–84) and cobalt ions (83) in reverse micelles using isooctane. Iron-copper alloys have also been produced by this procedure (83). In addition, copper nanoparticles have been synthesized in reverse micelles formed in supercritical fluids (85). CO_2 is believed to be an excellent choice as a solvent replacement for this application due to its environmentally friendly nature and the ability to fine-tune the solvent through pressure changes to manipulate the nanoparticle growth process.

Ji et al. were the first to use CO_2 in the formation of silver nanoparticles (86). $AgNO_3$ was reduced in microemulsion droplets consisting of the

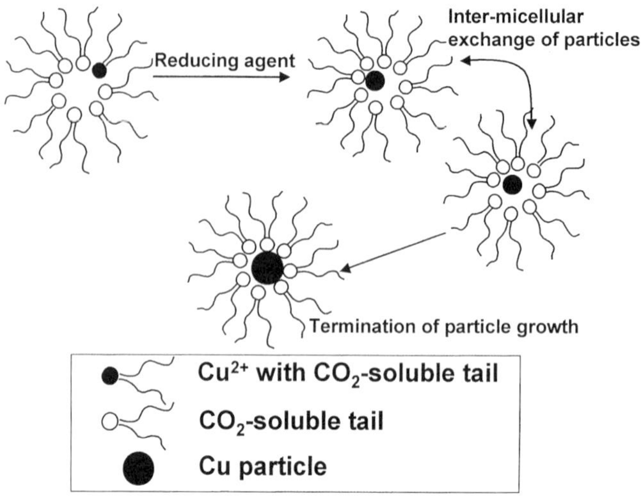

Figure 13 Copper nanoparticle formation in micelles.

surfactants AOT and a perfluoropolyether-phosphate ether (PFPE-PO_4) (average MW 870), where the PFPE-PO_4 acted as a cosurfactant for the AOT micelles. The microemulsions remained optically clear during the entire reaction, even in the presence of Ag particles. The average size of the silver particles was approximately 5–15 nm.

C. Biocatalysis

In a previous section, the extraction of proteins into CO_2 using CO_2-soluble surfactants was discussed. Since it is possible to extract proteins into CO_2, it is proposed that it is possible to solubilize an enzyme in CO_2 using CO_2-soluble surfactants and have the enzyme retain its activity. This hypothesis is currently under investigation.

Enzymes, i.e., protein catalysts, are hydrophilic because they derive from living organisms. However, much research into the use of enzymes in nonaqueous environments, such as organic solvents (87,88) and supercritical CO_2 (89), has been done. Such investigations have shown that enzymes, although not soluble in organic solvents and CO_2, do retain some activity and stability and can catalyze reactions. Taken a step further, enzymes have been modified to allow solubility in organic solvents (90). Solubility occurs either through direct covalent modification of the enzyme with amphiphilic polymers such as PEG or by reverse micelle formation using traditional surfactants. Either case has allowed for organic-soluble, active enzymes.

Holmes et al. reported the first enzyme-catalyzed reactions in water-in-CO_2 microemulsions (91). Two reactions, a lipase-catalyzed hydrolysis and a lipoxygenase-catalyzed peroxidation, were demonstrated in water-in-CO_2 microemulsions using the surfactant di(1H,1H,5H-octafluoro-*n*-pentyl) sodium sulfosuccinate (di-HCF4). A major concern of enzymatic reactions in CO_2 is the pH of the aqueous phase, which is approximately 3 when in contact with CO_2 at elevated pressures. Holmes et al. examined the ability of various buffers to maintain the pH of the aqueous solution in contact with CO_2. The biological buffer 2-(*N*-morpholino)ethanesulfonic acid sodium salt was the most effective, able to maintain a pH of 5, depending on the pressure, temperature, and buffer concentration. The activity of the enzymes in the water-in-CO_2 microemulsions was comparable to that in a water-in-heptane microemulsion stabilized by the surfactant AOT, which contains the same head group as di-HCF4.

Surfactants are currently being formulated to interact specifically with enzymes. These surfactants contain a double tail composed of one of the aforementioned CO_2-philic molecules, such as PDMS, fluoroalkyls, fluoroethers, or others. The head group of these surfactants is a hydrophilic sugar group. It is believed that the hydrophilic sugar group will interact with the enzyme through hydrogen bonding and the CO_2-philic tails will stabilize the enzyme in the CO_2 continuous phase to allow solubility. Figure 14 shows a representation of the surfactant-coated enzyme. Surfactants with this design were used to solubilize a lipase enzyme, which retained its activity, in many organic solvents (92).

Some enzymes require additional compounds to interact with the enzyme and substrate for catalysis, called cofactors. The cofactor may be one or more in-

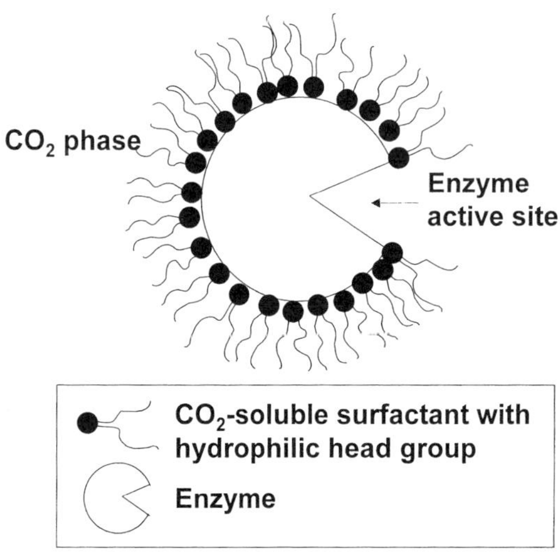

Figure 14 Diagram of a surfactant-coated enzyme.

organic ions or complex organic or metallo-organic molecules called coenzymes. Coenzymes function as transient carriers of specific functional groups. The coenzyme nicatinamide adenine dinucleotide (NAD) participates in oxidation–reduction reactions. NAD accepts a hydride ion from the substrate, allowing the substrate to be oxidized and the NAD to be reduced to NADH. As expected, these molecules are also hydrophilic, but they have been covalently modified to attach them to PEG, solid support, or directly to enzymes (93–95). NAD has been covalently modified with a PFPE molecule, and investigation into its solubility in CO_2 is currently in progress.

V. CONCLUSIONS

CO_2 is an advantageous process solvent for many applications, provided that CO_2-soluble surfactants can be specially designed and synthesized for each application. Such surfactants contain a CO_2-philic segment such as a fluoroether-, fluoroacrylate-, or silicone-based compound with a CO_2-phobic segment made up of a hydrophilic or lipophilic molecule, depending on the application. Applications in which such CO_2-philic surfactants have been formulated and utilized include protein extraction, heterogeneous polymerizations (both dispersion and emulsion), and metal extraction. Due to the success of these applications, investigations into the use of CO_2 as a process solvent in other applications, such as dry cleaning, nanoparticle formation, and biocatalyis, are currently underway.

REFERENCES

1. PA Winsor. Hydrotrophy, solubilisation, and related emulsification processes. Part 1. Trans Faraday Soc 44:376–398, 1948.
2. MA McHugh, VJ Krukonis. Supercritical Fluid Extraction Principles and Practice. 2nd ed. Boston: Butterworth-Heinemann, 1994.
3. RW Gale, JL Fulton, RD Smith. Organized molecular assemblies in the gas phase: reverse micelles and microemulsions in supercritical fluids. J Am Chem Soc 109: 920–921, 1987.
4. KA Bartscherer, M Minier, H Renon. Microemulsions in compressible fluids—a review. Fluid Phase Equilibria 107:93–150, 1995.
5. KA Consani, RD Smith. Observations on the solubility of surfactants and related molecules in carbon dioxide at 50°C. J Supercrit Fluids 3:51–65, 1990.
6. KL Harrison, J Goveas, KP Johnston, EA O'Rear. Water-in-carbon dioxide microemulsions with a fluorocarbon-hydrocarbon hydrid surfactant. Langmuir 10: 3536–3541, 1994.
7. J Eastoe, Z Bayazit, S Martel, DC Steytler, RK Heenan. Droplet structure in a water-in-CO_2 microemulsion. Langmuir 12:1423–1424, 1996.

8. JM DeSimone, EE Maury, YZ Menceloglu, JB McClain, TJ Romack, JR Combes. Dispersion polymerizations in supercritical carbon dioxide. Science 265:356–359, 1994.
9. JA Hyatt. Liquid and supercritical carbon dioxide as organic solvents. J Org Chem 49:5097–5101, 1984.
10. GJ McFann, KP Johnston, PN Hurter, TA Hatton. Carbon dioxide regeneration of block copolymer micelles used for extraction and concentration of trace organics. Ind Eng Chem Res 32:2336–2344, 1993.
11. GJ McFann, SM Howdle, KP Johnston. Solubilization in nonionic reverse micelles in carbon dioxide. AICHE 40:543–555, 1994.
12. SG Kazarian, MF Vincent, FV Bright, CL Liotta, CA Eckert. Specific intermolecular interaction of carbon dioxide with polymers. J Am Chem Soc 118:1729–1736, 1996.
13. JC Merdith, KP Johnston, JM Seminario, SG Kazarian, CA Eckert. Quantitative equilibrium constants between CO_2 and Lewis bases from FTIR spectroscopy. J Phys Chem 100:10837–10848, 1996.
14. F Rindfleisch, TP DiNoia, MA McHugh. Solubility of polymers and copolymers in supercritical CO_2. J Phys Chem 100:15581–15587, 1996.
15. CA Mertdogan, TP DiNoia, MA McHugh. Impact of backbone architecture on the solubility of fluorocopolymers in supercritical CO_2 and halogenated supercritical solvents: comparison of poly(vinylidene fluoride-*co*-22 mol% hexafluoropropylene) and poly(tetrafluoroethylene-*co*-19 mol% hexafluoropropylene). Macromolecules 30:7511–7515, 1997.
16. ML O'Neill, Q Cao, M Fang, KP Johnson, SP Wilkinson, CD Smith, JL Kerschner, SH Jureller. Solubility of homopolymers and copolymers in carbon dioxide. Ind Eng Chem Res 37:3067–3079, 1998.
17. TA Hoefling, D Stofesky, M Reid, EJ Beckman, RM Enick. The incorporation of fluorinated ether functionality into a polymer or surfactant to enhance CO_2-solubility. J Supercrit Fluids 5:237–241, 1992.
18. TA Hoefling, RM Enick, EJ Beckman. Microemulsions in near-critical and supercritical CO_2. J Phys Chem 95:7127–7129, 1991.
19. TA Hoefling, RR Beitle, RM Enick, EJ Beckman. Design and synthesis of highly CO_2-soluble surfactants and chelating agents. Fluid Phase Equilibria 83:203–212, 1993.
20. DA Newman, TA Hoefling, RR Beitle, EJ Beckman, RM Enick. Phase behavior of fluoroether-functional amphiphiles in supercritical carbon dioxide. J Supercrit Fluids 6:205–210, 1993.
21. MP Heitz, C Carlier, J deGrazia, KL Harrison, KP Johnston, TW Randolph, FW Bright. Water core within perfluoropolyether-based microemulsions formed in supercritical carbon dioxide. J Phys Chem B 101:6707–6714, 1997.
22. EJ Singley, W Liu, EJ Beckman. Phase behavior and emulsion formation of novel fluoroether amphiphiles in carbon dioxide. Fluid Phase Equilibria 128:199–219, 1997.
23. AI Cooper, JD Londono, G Wignall, JB McClain, ET Samulski, JS Lin, A Dobrynin, M Rubenstein, ALC Burke, JMJ Frechet, JM DeSimone. Extraction of a hydrophilic compound from water into liquid CO_2 using dendritic surfactants. Nature 389:368–371, 1997.

24. MJ Clarke, KL Harrison, KP Johnston, SM Howdle. Water in supercritical carbon dioxide microemulsions: spectroscopic investigation of a new environment for aqueous inorganic chemistry. J Am Chem Soc 119:6399–6406, 1997.
25. GB Jacobsen, CT Lee Jr, KP Johnston. Organic synthesis in water/carbon dioxide microemulsions. J Org Chem 64:1201–1206, 1999.
26. GB Jacobsen, CT Lee Jr, SRP daRocha, KP Johnston. Organic synthesis in water/carbon dioxide emulsions. J Org Chem 64:1207–1210, 1999.
27. DA Canelas, JM DeSimone. Dispersion polymerizations of styrene in carbon dioxide stabilized with poly(styrene-*b*-dimethylsiloxane). Macromolecules 30:5673–5682, 1997.
28. TA Hoefling, DA Newman, RM Enick, EJ Beckman. Effect of structure on the cloudpoint curves of silicone-based amphiphiles in supercritical carbon dioxide. J Supercrit Fluids 1993, 6:165–171, 1993.
29. G Luna, S Mawson, S Takishima, JM DeSimone, IC Sanchez, KP Johnston. Phase behavior of poly(1,1-dihydroperfluorooctylacrylate) in supercritical carbon dioxide. Fluid Phase Equilibria 146:325–337, 1998.
30. JM DeSimone, Z Guan, CS Elsbernd. Synthesis of fluoropolymers in supercritical carbon dioxide. Science 257:945–947, 1992.
31. JB McClain, D Londono, JR Combes, TJ Romack, DA Canelas, DE Betts, GD Wignall, ET Samulski, JM DeSimone. Solution properties of a CO_2-soluble fluoropolymer via small neutron scattering. J Am Chem Soc 118:917–918, 1996.
32. JB McClain, DE Betts, DA Canelas, ET Samulski, JM DeSimone, JD Londono, HD Cochran, GD Wignall, D Chillura-Martino, R Triolo. Design of nonionic surfactants for supercritical carbon dioxide. Science 274:2049–2052, 1996.
33. JL Fulton, DM Pfund, JB McClain, TJ Romack, EE Maury, M Combes, ET Samulski, JM DeSimone, M Capel. Aggregation of amphiphilic molecules in supercritical carbon dioxide: a small angle x-ray scattering study. Langmuir 11:4241–4249, 1995.
34. ML O'Neill, MZ Yates, KL Harrison, KP Johnston, DA Canelas, DE Betts, JM DeSimone, SP Wilkinson. Emulsion stabilization and flocculation in CO_2: 1. Turbidimetry and tensiometry. Macromolecules 30:5050–5059, 1997.
35. MZ Yates, ML O'Neill, KP Johnston, S Webber, DA Canelas, DE Betts, JM DeSimone. Emulsion stabilization and flocculation in CO_2: 2. Dynamic light scattering. Macromolecules 30:5060–5067, 1997.
36. KE Göklen, TA Hatton. Protein extraction using reverse micelles. Biotechnol Prog 1:69–74, 1985.
37. RS Rahaman, JY Chee, JMS Cabral, TA Hatton. Recovery of an extracellular alkaline protease from whole fermentation broth using reversed micelles. Biotechnol Prog 4:218–224, 1988.
38. MJ Pires, MR Aires-Barros, JMS Cabral. Liquid–liquid extraction of proteins with reversed micelles. Biotechnol Prog 12:290–301, 1996.
39. KP Johnston, KL Harrison, MJ Clarke, SM Howdle, W Heitz, FV Bright, C Carlier, TW Randolph. Water-in-carbon dioxide microemulsions: an environment for hydrophiles including proteins. Science 271:624–626, 1996.
40. EG Ghenciu, EJ Beckman. Affinity extraction into carbon dioxide. 1. Extraction of avidin using a biotin-functional fluoroether surfactant. Ind Eng Chem Res 36: 5366–5370, 1997.

41. EG Ghenciu, AJ Russell, EJ Beckman, L Steele, NT Becker. Solubilization of subtilisin in CO_2 using fluoroether-functional amphipiles. Biotechnol Bioeng 58: 572–580, 1998.
42. JM DeSimone, SA Crette, JM LeClerc, JL Kendall, RG Carbonell. Bioextractions with carbon dioxide. Proc 5th Meeting on Supercritical Fluids 813–819, 1998.
43. Y Hsiao, EE Maury, JM DeSimone, S Mawson, KP Johnston. Dispersion polymerizations of methyl methacrylate stabilized with poly(1,1-dihydoperfluorooctyl acrylate) in supercritical carbon dioxide. Macromolecules 28:8159–8166, 1995.
44. KA Shaffer, TA Jones, DA Canelas, JM DeSimone. Dispersion polymerizations in carbon dioxide using siloxane-based stabilizers. Macromolecules 29:2704–2706, 1996.
45. ML O'Neill, MZ Yates, KP Johnston, CD Smith, SP Wilkinson. Dispersion polymerization in supercritical CO_2 with a siloxane-based macromonomer: 1. The particle growth regime. Macromolecules 31:2838–2847, 1998.
46. ML O'Neill, MZ Yates, KP Johnston, CD Smith, SP Wilkinson. Dispersion polymerization in supercritical CO_2 with a siloxane-based macromonomer: 2. The particle formation regime. Macromolecules 31:2848–2856, 1998.
47. C Lepilleur, EJ Beckman. Dispersion polymerization of methyl methacrylate in supercritical CO_2. Macromolecules 30:745–756, 1998.
48. DA Canelas, DE Betts, JM DeSimone. Dispersion polymerizations of styrene in supercritical carbon dioxide: importance of effective surfactants. Macromolecules 29:2818–2821, 1996.
49. KL Harrison, SRP daRocha, MZ Yates, KP Johnston, DA Canelas, JM DeSimone. Interfacial activity of polymeric surfactants at the polystyrene–carbon dioxide interface. Langmuir 14:6855–6863, 1998.
50. H Shiho, JM DeSimone. Preparation of micron-size polystyrene particles in supercritical carbon dioxide. J Polym Sci A Polym Chem 37:2429–2437, 1999.
51. MR Clark, JL Kendall, JM DeSimone. Cationic dispersion polymerizations in liquid carbon dioxide. Macromolecules 30:6011–6014, 1997.
52. DA Canelas, DE Betts, JM DeSimone, MZ Yates, KP Johnston. Poly(vinyl acetate) and poly(vinyl acetate-*co*-ethylene) latexes via dispersion polymerizations in carbon dioxide. Macromolecules 31:6794–6805, 1998.
53. YM Zates, G Li, JJ Shim, S Maniar, KP Johnston, KT Lim, S Webber. Ambidextrous surfactants for water-dispersible polymer powders from dispersion polymerizations in supercritical CO_2. Macromolecules 32:1018–1026, 1999.
54. AI Cooper, WP Hems, AB Holmes. Synthesis of highly cross-linked polymers in supercritical carbon dioxide by heterogeneous polymerization. Macromolecules 32: 2156–2166, 1999.
55. FA Adamsky, EJ Beckman. Inverse emulsion polymerization of acrylamide in supercritical carbon dioxide. Macromolecules 27:312–314, 1994.
56. KE Laintz, CM Wai, CR Yonker, RD Smith. Extraction of metal ions from liquid and solid materials by supercritical carbon dioxide. Anal Chem 64:2875–2878, 1992.
57. KE Laintz, CM Wai, CR Yonker, RD Smith. Solubility of fluorinated metal diethyldithiocarbamates in supercritical carbon dioxide. J Supercrit Fluids 4:194–198, 1991.

58. KE Laintz, JJ Yu, CM Wai. Separation of metal ions with sodium bis(trifluoroethyl) dithiocarbamate chelation and supercritical fluid chromatography. Anal Chem 64: 311–315, 1992.
59. KE Laintz, GM Shieh, CM Wai. Simultaneous determination of arsenic and antimony species in environmental samples using bis(trifluoroethyl)dithiocarbamate chelation and supercritical fluid chromatography. J Chromatogr Sci 30:120–123, 1992.
60. Y Lin, CM Wai. Supercritical fluid extraction of lanthanides with fluorinated β-diketones and tributyl phosphate. Anal Chem 66:1971–1975, 1994.
61. Y Lin, CM Wai, FM Jean, RD Brauer. Supercritical fluid extraction of thorium and uranium ions from solid and liquid materials with fluorinated β-diketones and tributyl phosphate. Environ Sci Technol 28:1190–1193, 1994.
62. JM Murphy, C Erkey. Thermodynamics of extraction of copper(II) from aqueous solutions by chelation in supercritical carbon dioxide. Environ Sci Technol 31: 1674–1679, 1997.
63. CM Wai, S Wang, Y Liu, V Lopez-Avila, WF Beckert. Evaluation of dithiocarbamates and β-diketones as chelating agents in supercritical fluid extraction of Cd, Pb, and Hg from solid samples. Talanta 43:2083–2091, 1996.
64. Y Lin, NG Smart, CM Wai. Supercritical fluid extraction of uranium and thorium from nitric acid solutions with organophosphorus reagents. Environ Sci Technol 29: 2706–2708, 1995.
65. F Dehghani, T Wells, NJ Cotton, NR Foster. Extraction and separation of lanthanides using dense gas CO_2 modified with tributyl phosphate and di(2-ethylhexyl)phosphoric acid. J Supercrit Fluids 9:263–272, 1996.
66. NG Smart, TE Carleson, S Elshani, S Wang, CM Wai. Extraction of toxic heavy metals using supercritical fluid carbon dioxide containing organophosphorus reagents. Ind Eng Chem Res 36:1819–1826, 1997.
67. AF Lagalante, BN Hansen, TJ Bruno, RE Sievers. Solubilities of copper(II) and chromium(III) β-diketonates in supercritical carbon dioxide. Inorg Chem 34:5781–5785, 1995.
68. W Cross, A Akgerman, C Erkey. Determination of metal–chelate complex solubilities in supercritical carbon dioxide. Ind Eng Chem Res 35:1765–1770, 1996.
69. NG Smart, TE Carleson, T Kast, AA Clifford, MD Buford, CM Wai. Solubility of chelating agents and metal-containing compounds in supercritical fluid carbon dioxide. Talanta 44:137–150, 1997.
70. CM Wai, S Wang. Supercritical fluid extraction: metals as complexes. J Chromatogr A 785:369–383, 1997.
71. M Ashraf-Khorassani, MT Combs, LT Taylor. Supercritical fluid extraction of metal ions and metal chelates from different environments. J Chromatogr A 774:37–49, 1997.
72. AV Yazdi, EJ Beckman. Design of highly CO_2-soluble chelating agents for carbon dioxide extraction of heavy metals. J Mater Res 3:530–537, 1995.
73. AV Yazdi, EJ Beckman. Design, synthesis, and evaluation of novel, highly CO_2-soluble chelating agents for removal of metals. Ind Eng Chem Res 35:3644–3652, 1996.

74. AV Yazdi, EJ Beckman. Design of highly CO_2-soluble chelating agents. 2. Effect of chelate structure and process parameters on extraction efficiency. Ind Eng Chem Res 36:2368–2374, 1997.
75. J Li, EJ Beckman. Affinity extraction into CO_2 2. Extractions of heavy metals into CO_2 from low-pH aqueous solutions. Ind Eng Chem Res 37:4768–4773, 1998.
76. Anon. Environmentally sound alternative may help dry cleaning industry. J Environ Health 58:37–39, 1995.
77. Anon. Supercritical CO_2 seen as a dry cleaning replacement. Chem Eng Progr 93: 30, 1997.
78. Anon. Liquid carbon dioxide cleans up dry cleaning. MRS Bull. 21:7, 1996.
79. SH Jureller, JL Kerschner, M Bae-Lee, L Del Pizzo, R Harris, C Resch, C Wada. Dry cleaning system using densified carbon dioxide and a surfactant adjunct. US Patent 5,683,977, 1997.
80. JM DeSimone, T Romack, DE Betts, JB McClain. Cleaning process using carbon dioxide as a solvent and employing molecularly engineered surfactants. US Patent 5,783,082, 1998.
81. H Black. Prototype CO_2 dry-cleaning process replaces toxic solvent. Environ Sci Technol 29:A497, 1995.
82. J Tanori, N Duxin, C Petit, I Lisiecki, P Veillet, MP Pileni. Synthesis of nansize metallic and alloyed particles in ordered phases. Colloid Polym Sci 273:886–892, 1995.
83. I Lisiecki, M Björling, L Motte, B Ninham, MP Pileni. Synthesis of copper nanosize particles in anionic reverse micelles: effect of the addition of a cationic surfactant on the size of the crystallites. Langmuir 11:2385–2392, 1995.
84. I Lisiecki, MP Pileni. Copper metallic particles synthesized "in situ" in reverse micelles: influence of various parameters on the size of the particles. J Phys Chem 99:5077–5082, 1995.
85. CB Roberts, JP Cason. Cosolvent effects on AOT reverse micelles and colloidal particle production in supercritical fluid mixtures. Abstr. Pap. Am Chem Soc 215: 254, 1998.
86. M Ji, X Chen, CM Wai, JL Fulton. Synthesizing and dispersing silver nanoparticles in a water-in-supercritical carbon dioxide microemulsion. J Am Chem Soc 121: 2631–2632, 1999.
87. CR Wescott, AM Klibanov. The solvent dependence of enzyme specificity. Biochim Biophys Acta 1206:1–9, 1994.
88. JS Dordick. Designing enzymes for use in organic solvents. Biotechnol Prog 8: 259–267, 1992.
89. AJ Mesiano, EJ Beckman, AJ Russell. Supercritical biocatalysis. Chem Rev 99: 623–633, 1999.
90. G DeSantis, JB Jones. Chemical modification of enzymes for enhanced functionality. Curr Opin Biotechnol 10:324–330, 1999.
91. JD Holmes, DC Steytler, GD Rees, BH Robinson. Bioconversions in a water-in-CO_2 microemulsion. Langmuir 14:6371–6376, 1998.
92. Y Okahata, Y Fujimoto, K Ijiro. A lipid-coated lipase as an enantioselective ester synthesis catalyst in homogeneous organic solvents. J Org Chem 60:2244–2250, 1995.

93. AF Bückmann, MR Kula, R Wichmann, C Wandrey. An efficient synthesis of high-molecular-weight NAD(H) derivatives suitable for continuous operation with coenzyme-dependent enzyme systems. J Appl Biochem 3:301–315, 1981.
94. C Lee, NO Kaplan. Characteristics of 8-substituted adenine nucleotide derivatives utilized in affinity chromatography. Arch Biochem Biophys 168:665–676, 1975.
95. T Eguchi, T Iizuka, T Kagotani, JH Lee, I Urabe, H Okada. Covalent linking of poly(ethyleneglycol)-bound NAD with *Thermus thermophilus* malate dehydrogenase NAD(H)-regeneration unit for a coupled second-enzyme reaction. Eur J Biochem 155:415–421, 1986.

7

In Situ Blending of Electrically Conducting Polymers in Supercritical Carbon Dioxide

Amyn S. Teja and Kimberly F. Webb
Georgia Institute of Technology, Atlanta, Georgia

I. INTRODUCTION

Most organic polymers, such as polyethylene and polystyrene, exhibit negligible or no electrical conductivity and behave as insulators. However, conjugate polymers with alternate double and single bonds, such as polypyrrole and polythiophene, can be "doped" to obtain structures in which the electronic states are delocalized and the polymer becomes conducting. Polymers such as polyacetylene, polyaniline, polypyrrole, and polythiophene exhibit intrinsic conductivities that are in the semiconducting range ($\sim 10^{-8}$ S/cm), and doped forms of these polymers can have conductivities that approach those of highly conducting materials. Doped polyacetylene, for example, can exhibit electrical conductivities as high as that of copper at room temperature ($\sim 10^{3}$ S/cm) (1).

Applications of conducting polymers include batteries (2), antistatic transparent films to protect sensitive microelectronic devices (2), antistatic fabrics (3), and sensors (4). Moreover, conducting polymers that change color when oxidized or reduced are the basis of "smart windows" (3). Blends of electrically conducting polymers have also been developed for applications such as rechargeable batteries (5), chemical and optical sensors (6), nonlinear optical devices (7), and light-emitting diodes (8).

Conducting polymer sensors have been proposed that can detect ppm levels of pollutants (9), as well as chemical warfare agents such as dimethyl methylphosphonate (DMMP) (9). Gases such as CO, NH_3, HCl, and HCN can also be sensed (10) because of the changes in conductivity that occur when the gas is in contact with a conducting polymer. Thus, NH_3 and H_2S can be detected because they neutralize free-electron carriers, reduce the carrier density, and lower the conductivity of the polymer. On the other hand, PCl_3, SO_2, and NO_2 oxidize the reduced form of some conducting polymers, generate free carriers, and increase conductivity (11). Electronic "noses" have been produced based on the gas-phase adsorption of vapors that affect the mobility or availability of free-charge carriers (9). Such noses are capable of detecting components in the head space above beers, thus replacing existing analytical methods such as gas chromatography that are time consuming and expensive. They can also differentiate between standard and artificially tainted beers, and between different commercial brands (12,13).

Several applications of conducting polymers have been investigated in the medical field, particularly for the controlled release of various anions such as glutamate and ferrocyanide (10). Biosensors have been produced in which enzymes, antibodies, and even whole living cells have been incorporated into the polymer structure.

Rechargeable battery electrodes made from conducting polymers have been commercialized by Seiko and Varta (14). In contrast to conventional electrodes, conducting polymer electrodes offer ease of fabrication, low cost, low weight, and processibility. They also have a long life and can produce current densities up to 50 mA/cm^2 and energy densities of 10 W-h/kg. For example, iodine-doped polythiophene stores positive charge in the polymer chain and can serve as a good polymer electrode for battery applications (10).

Electrochromic "smart" windows have been made from films of polythiophenes or polypyrroles because of their ability to change color under the action of sunlight or temperature (15). Moreover, polythiophenes can change color from red to blue when a voltage is applied to the polymer film (10).

Antistatic coatings made from polypyrrole have been produced by Milliken (9) for carpet fibers, fabrics, and packaging materials used in highly sensitive electronic components. Modern electronic components can be damaged by a static discharge of as little as 100 V, whereas static charges generated by walking across a synthetic fiber carpet or sitting on polyurethane cushioning can reach as high as 10,000 V. These charges can be neutralized by fibers coated with conducting materials such as polypyrrole. Packaging materials in the electronics industry must also have good antistatic properties that are independent of atmospheric conditions and have a surface resistivity of less than 10^8 ohms (16). Antistatic fibers have been produced by DuPont, Ormecon, and Philips, which are working together to develop novel uses of polyaniline conductive polymers (17,18). The Philips Company has produced the first all-plastic computer chip that can be used in electronic security devices. Finally, electromagnetic shielding materials have been developed to eliminate spacecraft charging, which requires conductivities on the order of 10^{-12} S/cm (19). These applications require conducting polymers that are stable in the presence of oxygen, moisture, and radiation.

Photoresists, memory devices, nanoswitches, and interconnects represent other potential applications of conducting polymers because of the ability of these polymers to switch readily between nonconducting and highly conducting states (20). This switching of conductive states differentiates polymers from metals and can be achieved by the addition and removal of dopant ions.

Conjugation is a key to the desirable properties of conducting polymers, but it also renders them susceptible to unwanted oxidation. Conducting polymers in doped forms by their nature are difficult to process because the parent systems are rigid and exhibit strong interchain interactions. As a result, products made from such polymers are difficult to fabricate. Moreover, many of these polymers are obtained by solution processing in solvents such as acetonitrile and chloroform (21), which are harmful to the environment.

Blending of conducting polymers with plastics has been proposed to alleviate processibility problems and to optimize the best properties of each polymer in the composite (22–24). Thus, blends of conducting polyaniline with nonconducting poly(bisphenol A carbonate) were developed that exhibit conductivities equal to that of pure polyaniline while retaining mechanical properties similar to those of the host polymer (25). Applications of conducting polymer blends include fuel probes that resist deterioration by galvanic corrosion (26). These probes are more durable than metal probes, allow more flexibility with tank design, and interface more easily with computerized controls. The probes consist of a conductive polymer, a host polymer (polyethylene terephthalate), and some glass reinforcement. Other blends have been shown to exhibit good mechanical as well as efficient blue light–emitting properties (27). Blue light emission has proved difficult to achieve with classical semiconductor materials, and the potential of using these blends in light-emitting diode displays is therefore significant. Oligosilanylene blocks and oligothiophenes have been used to control luminescence wavelength (28), and microwave absorbing materials have been made from blends of polyaniline and an ethylene–vinyl acetate copolymer (EVA) (29). However, many of the blending processes also employ harmful organic solvents, such as chloroform, nitrobenzene, and acetonitrile.

Processing with an environmentally benign supercritical fluid, such as carbon dioxide, is an attractive alternative to conventional processing of electrically conducting polymer blends. The advantages of supercritical carbon dioxide as a solvent for polymerization and blend formation have been outlined by many investigators (30,31). Carbon dioxide is inexpensive, nonflammable, offers high mass transport rates, and allows in situ removal of unreacted monomer and other impurities. It is also known to swell host polymers (32), which facilitates blend formation.

In this chapter, we review processes to obtain blends of conducting polymers and host substrates, with particular emphasis on processes that employ carbon dioxide as the solvent. The effect of the solvent on the synthesis of the conducting polymer, on blend formation, and on doping of the conducting polymer is reviewed. Furthermore, morphologies that lead to high electrical conductivity are identified.

II. ELECTRICAL CONDUCTIVITY IN POLYMERS

Several factors can affect electrical conductivity in polymers. These factors include the extent of conjugation and regioregularity of the polymer, its molecular weight, and the interchain distance (29). Processing variables, such as temperature and pH, are also important (33). These factors will be discussed below using polypyrrole and 3-undecylbithiophene as examples.

Polypyrrole and poly(3-undecylbithiophene) are conjugated polymers with coupling at the 2,5-positions that gives rise to a delocalized π-bond structure shown in Figure 1. The delocalized structure allows mobility of electrons when an electron is removed, e.g., by oxidation. Further oxidization leads to the bipolaron structure shown in Figure 2, which allows additional movement of electrons and causes double-bond shifting that extends over a few monomer units. Oxidants such as ferric chloride, iodine, and fullerenes may be added to remove electrons and hence increase conductivity of the resulting polymer (34). However, it should be added that these models of electrical conductivity are somewhat speculative and the actual structures of doped and undoped conducting polymers are not well understood. Additional mechanisms of conduction, such as interchain hopping of electrons, are also important and are discussed by Cao et al. (35) in the case of doped thiophene oligomers.

Since conjugated polymers, such as polypyrrole, usually yield stiff chains with interchain interactions and little flexibility (36), a long alkyl chain is added at the 3-position to change the characteristics of the polymer. This leads to more regular polymeric structures and higher electrical conductivity (37–39). The mean conjugation length of poly(3-alkylthiophenes) can be increased by

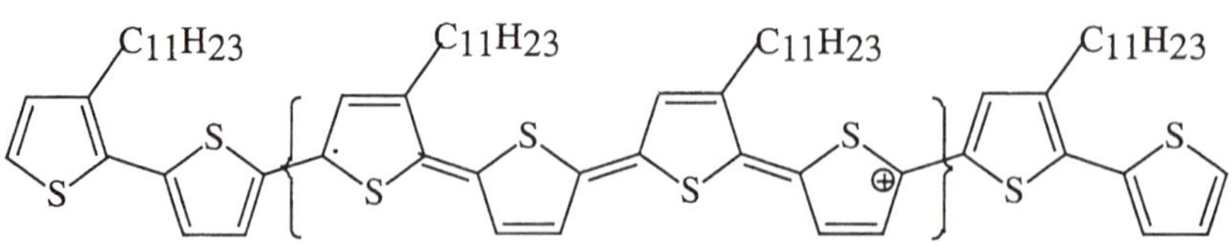

Figure 1 Polaron structure of poly(3-undecylbithiophene).

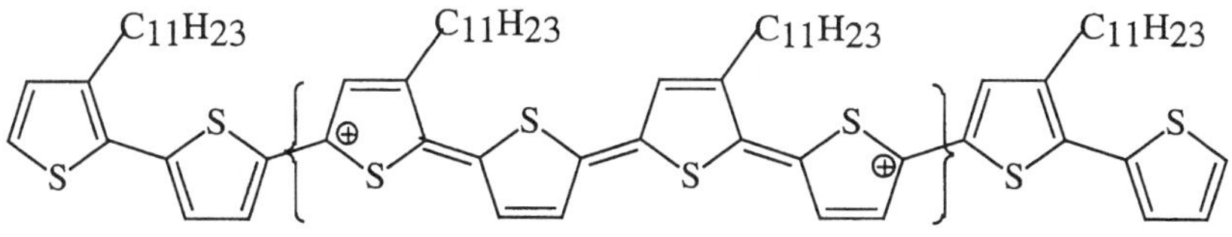

Figure 2 Bipolaron structure of poly(3-undecylbithiophene).

alkyl chains with 8–10 carbon atoms as discussed by Roncali et al. (40). Also, the alkyl groups increase conductivity without altering the π-conjugation structure (41) and increase the solubility of the polymer in solvents such as chloroform, nitromethane, and tetrahydrofuran (42).

Poly(3-undecylbithiophene) displays higher electrical conductivity and greater environmental stability than polypyrrole because the polymerization of 3-undecylbithiophene generally results in regioregular structures with few structural defects or breaks in conjugation (43). Also, both electronic (free spin densities) and steric driving forces produce structures with 2,5 coupling in this polymer rather than 2,4 coupling. Polypyrrole can be linked through the 2- and 4-positions on the ring during bonding, but the 4-position linking can cause irregularity or mislinking in the chain. This leads to lower conductivity due to interruption of the long delocalized π bond of the monomer units already linked at the two position (44). This undesired branching also yields networks that make the polymer insoluble in common solvents and infusible (16). In the case of polyalkylthiophenes, mislinking during bonding could take place on the thiophene ring where no alkyl chain is present at the 4-position, but the alkyl chain on the other ring reduces overall mislinking (45).

Regiochemistry of substitution also leads to polymer chains that do not have chain connectivity, which is necessary for electrical conductivity. In poly(3-alkythiophene), head-to-tail placement during substitution is dominant over head-to-head linkage (45) and leads to high conductivity. In polythiophenes, on the other hand, unfavorable head-to-head coupling causes the thiophene rings to twist out of plane, which results in less conjugation and lower conductivity (46). Biothiophenes exhibit fewer steric interactions because alkyl chains are present on every other ring and are not close enough to interact. Therefore, regioregularity does not play as important a role in the conductivity of the bithiophenes as it does in the case of the thiophenes (43).

III. CONVENTIONAL POLYMERIZATION

A common method for making electrically conducting polymers is by electrochemical polymerization, which is carried out in an electrochemical cell containing the monomer, electrolyte, and a solvent such as acetonitrile. The monomer is oxidized at the anode in the cell and a potential is developed that depends on the electrode materials, solvent, electrolyte, oxygen content, water content, and current density. Growth and film thickness are controlled during the polymerization (14). However, the polymer film that is formed is not very thick or uniform, and large amounts of conducting polymer cannot be synthesized using this approach.

Photochemical methods may also be employed whereby polymerization is carried out in the presence of sunlight and photosensitizers. Other methods include emulsion polymerization and solid-state polymerization. However, these methods are time consuming and involve the use of costly chemicals.

Chemical polymerization is probably the most widely used method for making large amounts of conductive polymer. Both polypyrrole and poly(3-undecylbithiophene) can be made via chemical polymerization with ferric chloride as an initiator as well as the dopant. In the case of polypyrrole, the reaction (Fig. 3) involves coupling and subsequent oxidation of radical cations formed from the one-electron oxidation of the aromatic ring and is carried out in an organic solvent such as nitrobenzene, acetonitrile, or chloroform. The nature of the solvent strongly affects the yield of polypyrrole as well as its conductivity, which ranges from less than 10^{-6} to 45 S/cm as shown in Table 1 (47). Many factors can affect these results, such as the extent of $FeCl_3$ solvation, solvent basicity, solvent dielectric properties, temperature, and residual water. Table 2 shows the yields of polypyrrole from two reactions that employ different reaction media (diethyl ether and tetrahydrofuran) and are carried out at two temperatures (273 K and 295 K) (47). A lower conductivity is obtained at the higher temperature due to increased randomness in the structure of the polymer formed, but this decrease in conductivity is also dependent on the reaction medium. Lower temperatures lead to higher molecular weight products or products of higher conjugation, which is consistent with higher conductivity (47).

The polymerization of 3-undecylbithiophene is similar in many respects to the $FeCl_3$ polymerization of pyrrole. It has been established that chemical synthesis yields more stable poly(3-alkylthiophenes) than electrochemical synthesis (48). The chemical synthesis route uses $FeCl_3$ in nitrobenzene, which

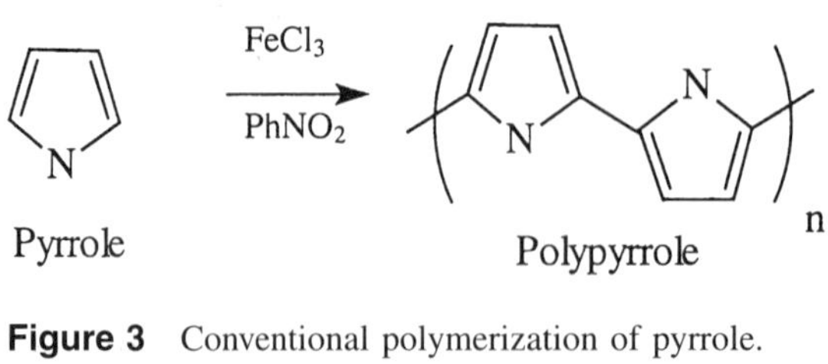

Figure 3 Conventional polymerization of pyrrole.

Table 1 Effect of Reaction Medium on the Polymerization of Pyrrole via the $FeCl_3$ Process

Reaction medium	Polymer yield (g)[a]	Conductivity (S/cm)
Diethyl ether	2.6	45
Tetrahydrofuran	0.6	$<10^{-6}$
Acetonitrile	1.2	3×10^{-3}
Nitromethane	2.8	7×10^{-2}
Dimethyl sulfoxide	0.0	0
Benzene	2.6	2×10^{-2}

[a]Based on a 2-g charge on monomer.
Data from Ref. 47

generally leads to 2,5 coupling of the monomer in the preferred head-to-tail arrangements. Coupling in the 2,4-position is minimized (49), and the method produces standard-quality polymers in good yield (50).

Polymers with alkyl side chains of lengths up to 12 carbons are soluble in conventional solvents and therefore easier to process than the corresponding polymers without side chains. They also exhibit higher conductivity and lower polymerization temperatures (51) as the length of the side chain increases. It is also believed that lower reaction temperatures suppress any competing side reactions. At high temperatures, electron mobility increases and leads to ring distortion and lower electrical conductivity (52). Tashiro et al. (53) found that trans conformations are predominant in poly(3-alkylthiophenes) obtained at temperatures up to 333 K, but gauche forms of the polymer are present at higher temperatures. Reactions in supercritical CO_2 can be carried out at low temperatures, so that more regular conformations would be favored.

Table 2 Effect of Reaction Temperature on the Conductivity of Polypyrrole

Reaction medium	Temp. (K)	Conductivity (S/cm)
Diethyl ether	295	45
Diethyl ether	273	115
Tetrahydrofuran	295	$<10^{-6}$
Tetrahydrofuran	273	3×10^{-2}

Data from Ref. 47.

IV. DOPING

Dopant ions play a major role in achieving electrical conductivity, although the resulting polymers are not always stable. Doping can take place by oxidation (p-type doping) or reduction (n-type doping). During oxidation, the polymers are in contact with electron-attracting substances such as Br_2, I_2, and $HClO_4$, which remove electrons from the polymer chain. Electron-donating dopants, such as Li, Na, K, Cs, and Rb, can also be used to produce radical anions. However, p-type doping is more common with conductive polymers. In the case of polythiophenes, it has been shown (22) that one dopant ion per four monomer units yields a stable polymer that does not have any modifications to its backbone. Also, polymers with long alkyl side chains have been shown to be more stable when doped than polymers with short or no side chains (22).

The nature and concentration of the dopant are important factors in controlling the physiochemical and mechanical properties of the polymer and its conductivity and long-term stability (54). Several dopants have been proposed, including bromine, iodine, lithium perchlorate, ferric chloride, cupric chloride, hydrogen peroxide, lead dioxide, nitrous acid, quinones, ozone (55), as well as 2-naphthalenesulfonate and p-toluenesulfonate (56). Some of these have been studied by Myers (47), who found that ferric chloride under anhydrous conditions yields the most conductive polypyrrole, with electrical conductivity as high as 45 S/cm. Park and Ruckenstein (21) also reported that ferric chloride provides the best performance as an oxidant.

The doping process can be carried out using photochemical, chemical, or electrochemical methods, or by ion implantation with appropriate ion beams (14). Electrochemical and chemical doping are the most common techniques in practice because of convenience and low cost. In electrochemical doping, the polymers are immersed in an organic electrolyte solution or in aqueous electrolytes, and an electric potential is applied that leads to ion migration into the polymer. Photochemical doping involves placing materials such as triarylsulfonium salts in contact with the polymer and exposing the mixture to ultraviolet radiation. Chemical doping can employ several different materials, such as I_2, AsF_5, and H_2SO_4. The polymers can be contacted with the vapor of the dopant or with a liquid solution containing the dopant. The amount of dopant incorporated into the material is dependent on the doping time, the vapor pressure or concentration of the dopant, and the temperature of the doping source and the polymer (14).

V. STABILITY

Environmental stability is an important consideration in many applications. Thus, polypyrroles have found limited application because they exhibit poor stability in oxygen environments (37). On the other hand, polythiophenes have been shown to be stable for more than 8 months in air (57), and polyalkylthiophenes exhibit remarkable stability in oxygen and moisture (58). The bithiophenes are more stable than the thiophenes because the long alkyl side chains are further apart in the bithiophenes and therefore have less chance to interact (59).

Instability can be due to oxygen diffusion and reaction with the conducting polymer, so that the nature of the dopant, chemical reactivity, and oxidation–reduction potentials have a strong affect on the stability (10). Thus, polymethylthiophene (PMeT)—which has an oxidation potential much lower than that of water, and a reduction potential that is much greater than that of oxygen—is stable in an environment containing moisture and oxygen (37). Furthermore, increasing the alkyl chain length causes the oxidation–reduction potentials of polyalkylthiophenes to move further away from those of oxygen and water (51). Polyacetylene (PA) also has oxidation–reduction potentials close to those of PMeT, but the chemical reactivity of oxygen with a double bond leads to lower stability (37).

VI. CONVENTIONAL BLENDING OF ELECTRICALLY CONDUCTING POLYMERS

Electrically conducting polymers, such as polypyrroles and polythiophenes, are often blended with other polymers to yield composites with improved mechanical strength and environmental stability. The method of blending generally involves impregnating the chemically polymerized conducting polymer into an insulating polymer matrix. This can be achieved by modifying the surface of the substrate for chemical adhesion to take place or by simply coating the substrate with the conducting polymer (60,61). Conventional blending involves chemical and physical effects that consist of interlocking and interdiffusion, and these effects contribute to adhesion (62). Many blends of polypyrrole and a host substrate have been prepared using these methods. As discussed below, conductivities of blends vary greatly even with the same host polymer, depending on the method used for blending.

A. Melt Processing

Melt processing has been used by many researchers (22,24,49,57,63) to obtain conducting polymer blends, but high temperatures are needed for processing. Steps in the process include (a) making the conducting polymer, (b) melting the host and conducting polymers, (c) mixing the two polymers, (d) compression molding, and (e) doping with ferric chloride solutions. Organic solvents are needed in steps (a) and (e). Degradation of polymers can take place at high temperatures, although the blends with a conductivity as high as 1 S/cm have been obtained (24). On the other hand, the conducting polymer content can be high and result in low mechanical strength.

B. Synthesis in a Host Polymer

Sun and Ruckenstein (44) prepared a composite of polypyrrole and a host polymer by an emulsion pathway that dispersed a host polymer and surfactant in an aqueous ferric chloride solution, followed by dropwise addition of pyrrole until the final product could be precipitated with methanol. The product was washed several times with methanol and the materials were then hot-pressed to form films. This process produced composites with conductivities as high as 1.3 S/cm, but the processing time was very high because of the multiple steps involved. Also, the process uses several toxic organic solvents.

Chan et al. (64) prepared composites of poly(methyl methacrylate) (PMMA) and polypyrrole using a similar method with several oxidants such as ferric chloride, sodium persulfate, and bromine. The conductivities of the resulting composite films ranged from 1.0×10^{-12} to 0.42×10^{2} S/cm. However, these films were not very uniform. Vincent (65) coated PMMA beads with a thin layer of polypyrrole by infusing pyrrole into PMMA beads and then dispersing these beads in water containing ferric chloride. Others have coated the surface of polypyrrole with PMMA (66). Makhoulki et al. (67) used polyvinyl alcohol as a host substrate and obtained blends with polypyrrole. Conductivities as high as 15 S/cm were obtained. Samir et al. (68) also investigated the use of polyvinyl alcohol as a host for polypyrrole, and reported conductivities as high as 2 S/cm. Another investigation by Byun and Im (69) immersed nylon 6 containing pyrrole into aqueous ferric chloride solutions and obtained blends with a conductivity of 2.35×10^{-3} S/cm. Radhakrishnan and Mandale (70) formed blends by first casting films of poly(ethylene oxide) with ferric chloride on a glass plate and then contacting the films with pyrrole vapor. A similar approach was used by first dissolving a triblock copolymer in THF and then adding ferric chloride to the solution. This solution was cast into a film and then contacted with pyrrole vapors to form a blend (71).

C. Solution Casting

Solution casting generally involves the use of many organic solvents but has been used by a number of researchers (1,72). The process used by Hotta et al. (23) involves (a) formation of the conducting polymer, (b) dissolution of the polymer and a host in an organic solvent, (c) mixing of the resulting solution, (d) casting on glass plates, (e) doping with a solution of dopant in acetonitrile, and (f) evaporation of the solvent. Khedkar and Radhakrishnan (73) used polystyrene as a host for blending with polypyrrole, and these two polymers were then dissolved in tetrahydrofuran. Conductivity of the resulting films was about 0.05 S/cm, although the films were stable in the atmosphere for more than 650 days. Chen et al. (74) produced a blend of polypyrrole and low-density polyethylene with the melting technique as well as solution casting and found that the processing method greatly influenced morphology and electrical conductivity of the blends. Solution casting produced blends with higher conductivities, which was also confirmed by Omastova et al. (75).

D. Absorption Method

In the absorption method, a host polymer is first contacted with a monomer solution and then with an oxidant solution or vice versa. Adequate time is allowed for the monomer and oxidant to impregnate the host substrate, where in situ polymerization to the conducting polymer and simultaneous blend formation occur. Ruckenstein and Park (76), Wu and Chen (60), Mano et al. (77), and Pigois-Landureau (78) have shown that thick conducting composites can be formed using this method. Films with conductivity as high as 0.65 S/cm were obtained using porous, cross-linked polystyrene as the host, polypyrrole as the conducting polymer, and ferric chloride as the oxidant (76). Impregnating the host polymer with ferric chloride first and then with the pyrrole solution yielded higher conductivities than the reverse process. High conductivities were also achieved when using nonaqueous solvents. Morsli et al. (79) also used porous, cross-linked polystyrene as a host polymer in their studies in order to increase the penetration of the conducting polymer into the host.

Thieblemont et al. (56) employed several host substrates in forming polypyrrole blends including woven glass, polyvinyl chloride, and polyester. Their process consisted of dipping these host polymers into a solution of pyrrole in ethanol followed by dipping in an aqueous solution of ferric chloride. Electrical conductivity of the blends was found to be as high as 120 S/cm. However, they also found that water and oxygen caused a loss of electrical conductivity over time. Polyvinyl chloride was employed as a host substrate for blending with polypyrrole by Mano et al. (77) The blend was investigated under a transmission electron microscope, which showed that the polypyrrole layer was only 0.1 μm thick. The blend also exhibited conductivity as high as 10^{-1} S/cm. Balci et al. (80) polymerized polypyrrole in a copolymer of polyvinyl chloride and polyvinyl acetate. Films were prepared by first introducing pyrrole into the polymer matrix, followed by treatment with ferric chloride. Tetrahydrofuran was used as the solvent in this case, and conductivities were in the range of 10^{-4}–10^{-3} S/cm. The blend was not very stable in air and lost its electrical conductivity in one week.

E. Other Methods

Faguy et al. (81) employed spectroscopy to study the polymerization of pyrrole in montmorillonite. An aqueous solution of pyrrole was added to a slurry of montmorillonite clay particles impregnated with ferric chloride, and polypyrrole was formed on the accessible surfaces of the clay particles. The resulting conductivity was on the order of 10^{-4}–10^{-6} S/cm. Polypyrrole was blended into several host substrates, such as polypropylene, polyethylene, polyvinyl chloride, and terpoly(acrylonitrile-butadiene-styrene) by Lenz et al. (1). These blends were synthesized by placing host substrates in a monomer and oxidant solution, and they exhibited conductivities in the range of 10^{-3}–10^{-4} S/cm. The unusual aspect of this research is that the conducting blends were coated with copper in order to prepare metallized films. Polypropylene pellets were used in the formation of a blend with polypyrrole by Omastova et al. (63). A coating of polypyrrole on the surface of these pellets was obtained. The pellets were then compression-molded into films that were 0.2 mm thick. Electrical conductivity attained values from 4×10^{-10} to 10×10^{-3} S/cm, which was much higher than values reached by mechanical mixing of the two polymers. Yin et al. (82) used polyethylene spheres and nylon fibers to form blends with polypyrrole. The host particles were dispersed in a pyrrole solution and then in an aqueous ferric chloride solution, resulting in a coating of polypyrrole. The particles were then hot-pressed into composite films. Conductivity reached as high as 10^{-3} S/cm, but the polypyrrole content was high (20%), resulting in low mechanical strength. Lascelles et al. (61) investigated the coating of polypyrrole onto polystyrene latexes. The latex particles were formed with ferric chloride, and then a solution containing the pyrrole monomer was added. The dimensions of the resulting particles were in the submicrometer range, and the coating thickness of polypyrrole was controlled to vary the electrical conductivity. The highest conductivities were found to be 10 S/cm, but this was only demonstrated using particles with a polystyrene latex center and not films.

Blending of polythiophene with polymer hosts has been studied by a number of research groups (22,44,49,79,83), who have used a variety of hosts as well as different thiophenes. Ruckenstein and Park (83) employed porous, cross-linked polystyrene as the host substrate and formed polythiophene blends using several different oxidants, including copper perchlorate hydrate, iron perchlorate hydrate, ferric chloride, and ferric chloride hexahydrate. All of these oxidants yielded composites with good conductivities. The host polymer was impregnated with an oxidant solution and then with a monomer solution. The resulting blends exhibited electrical conductivity as high as 4.8 S/cm and were stable for more than 52 days. The best results were obtained with acetonitrile as the solvent. Samir et al. (68) also obtained composites of polythiophenes with porous, cross-linked polystyrene that exhibited conductivities as high as 2 S/cm.

Blends of poly(3-alkylthiophenes) have been formed by Pomerantz et al. (49). Chloroform was used as the solvent, and the blends were prepared by melting and by fiber spinning. The melt blends were nonconducting, whereas the fiber blends had conductivities of 10^{-1} S/cm. Copolymers and blends of poly(3-decylthiophene) with polyethylene were also formed with conductivities up to 5 S/cm. Sun and Ruckenstein (44) formed composites of poly(3-methylthiophene) and rubber. An inverted emulsion of rubber was mixed with ferric chloride, and 3-methylthiophene was introduced to the emulsion. Conductivities as high as 1.3 S/cm were obtained. This method for producing composites is preferable to direct mixing and also showed that 3-methylthiophene was a more suitable monomer for obtaining conductive composites than thiophene or 2,2′-bithiophene. Ho et al. (72) produced blends of poly(3-dodecylthiophene) with EVA using a solution casting method. However, the electrical conductivity was not reported.

An electrochemical method was employed by Vatansever et al. (84) to polymerize thiophene on a polyamide-coated electrode at a constant potential. A uniform blend was obtained, but the method is difficult to scale up.

VII. BLENDING IN SUPERCRITICAL CARBON DIOXIDE

Since the solvent has a major effect on electrical conductivity (Tables 1 and 2), the use of supercritical CO_2 as both a solvent and a reaction medium in blending is of considerable interest. However, CO_2 has not been employed in polymer blending until recently, although its use in polymerization reactions (85–89) is quite common. A new route to polymer blends was proposed by Watkins and McCarthy (32), who carried out the polymerization of styrene in supercritical fluid–swollen host polymers. The monomer was allowed to diffuse into several host substrates and then polymerized in situ in the host polymer impregnated with an initiator. The resulting polymers were not soluble in carbon dioxide and not miscible with the host matrices, poly(chlorotrifluoroethylene) or poly(4-methyl-1-pentene), so that discrete phases of the two polymers were obtained in the composite. The use of supercritical fluids in this process offers several advantages because of the increased diffusion rates of penetrants dissolved in the supercritical fluid. Swelling of the polymer substrate, as well as partitioning of reagents between the swollen polymer phase and the supercritical phase, can be adjusted by manipulating the temperature and pressure. Also, the most common supercritical fluid (carbon dioxide) is a gas at atmospheric pressure and can be rapidly dissipated upon release of pressure. Carbon dioxide also causes plasticization, changes in surface properties, and nucleation of voids in the host polymer, which facilitates formation of the blend.

Watkins and McCarthy (32) employed their method of blending to diffuse styrene into several substrates and to subsequently polymerize the styrene in situ in the host polymer. The polymerization involved a free-radical reaction and the amount of polystyrene incorporated in the host was controlled via the reaction time and monomer concentration. Watkins and McCarthy (32) concluded that the blend composition is not limited by the solubility of the monomer in the host polymer but by the solubility of carbon dioxide in the formed polystyrene. They later repeated this work with styrene in poly(chlorotrifluoroethylene) (PCTFE) (90). Diffusion rates in swollen PCTFE were sufficiently high to produce high-molecular-weight polystyrene. These studies emphasize the role of solubility and diffusion of carbon dioxide in the polymers.

Supercritical carbon dioxide is generally a poor solvent for polymers (91). However, it does have the capacity to swell many polymers (92,93), and this can be of considerable advantage in blend formation. Watkins and McCarthy (32) found that the solubility of carbon dioxide in PCTFE reaches a maximum at a temperature of 313 K and a pressure of 10.4 MPa. This maximum represents a mass gain of 4.5% carbon dioxide in the host substrate PCTFE. Wissinger and Paulaitis (94) found that PMMA swells by about 20 vol % after reaching equilibrium in carbon dioxide at a temperature of 305.7 K. However, it is important to note that PMMA undergoes foaming with supercritical carbon dioxide treatment, as observed by Shieh et al. (95,96).

Supercritical carbon dioxide tends to lower the glass transition temperature of glassy polymers, by as much as 60°C in some cases. This makes the polymer more rubbery and flexible and allows carbon dioxide to diffuse into these polymers. Shieh et al. (95) reported that amorphous polymers absorb carbon dioxide to a greater extent than glassy polymers and are therefore subject to increased plasticization. Carbon dioxide diffusivities have been measured to be on the order of 10^{-6}–10^{-7} cm^2/s in rubbery polymers (97), which is of considerable advantage when solutes must be impregnated into host polymers.

The diffusivity of penetrants into host substrates can be enhanced by several orders of magnitude in the presence of supercritical carbon dioxide (98). Nealey et al. (99) found that the diffusivity of a polybutadiene oligomer in polystyrene increased by a factor of 10 at a temperature slightly above the glass transition temperature of the host polystyrene, which had been plasticized with compressed argon. Also, Raleigh scattering measurements of tracer diffusion coefficients showed that the diffusion of azobenzene in glassy polystyrene was enhanced near its glass transition temperature (100).

VIII. RECENT STUDIES INVOLVING IN SITU BLEND FORMATION USING SUPERCRITICAL CARBON DIOXIDE

We have studied the polymerization of pyrrole and 3-undecylbithiophene and blend formation with several host polymers using supercritical carbon dioxide

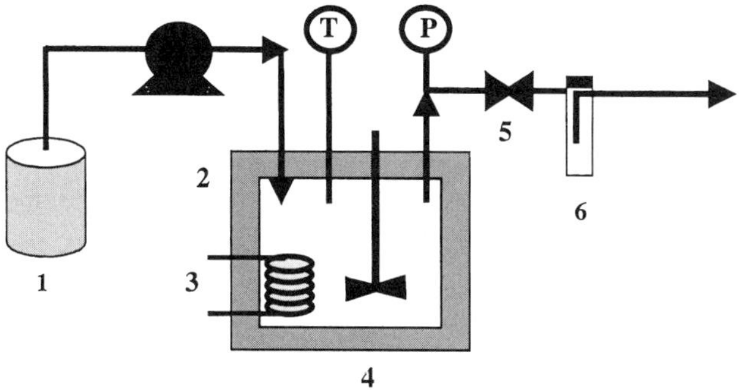

Figure 4 Experimental apparatus for polymerization and blending experiments: (1) CO_2 tank; (2) heating jacket; (3) cooling coil; (4) reactor; (5) depressurization valve; (6) trap.

(102). Ferric triflate was employed as an oxidant in both cases, as it is soluble in supercritical carbon dioxide. The apparatus used in the experiments (Fig. 4) and consists of a stirred high-temperature and high-pressure reactor, a high-pressure syringe pump for pressurizing carbon dioxide, and associated valves and tubing. A cooler with a circulating system was used to circulate coolant to the pump, the cooling coil, and the magnetic stirrer. The temperature in the reactor was controlled by the flow of water through the heating jacket and cooling coil inside the reactor. A calibrated pressure gage was used to measure the system pressure.

A. Polymerization Experiments

The polymerization of pyrrole and of 3-undecylbithiophene in supercritical carbon dioxide was carried out at 10.5 MPa and 313 K. A mole ratio of 4 mol of ferric triflate to 1 mol of monomer was employed in our experiments. The monomer and ferric triflate were placed in the reactor, and the vessel was then pressurized with carbon dioxide until the desired pressure of 10.5 MPa was reached. The polymerization reaction was carried out for 1 h, after which the reactor was depressurized and the polymer removed from the vessel.

The polypyrrole and poly(3-undecylbithiophene) formed in supercritical carbon dioxide were compared with polymers formed by the conventional routes using nitrobenzene as the solvent and ferric chloride as the oxidant. In the case of poly(3-undecylbithiophene), molecular weights were of the order of 20,000 via both routes. UV-visible spectroscopy was used to measure the extent of conjugation of the polymers since higher conjugations can be related to higher electrical conductivity. The maximum λ values obtained were 529.91 nm for the polymer prepared using supercritical carbon dioxide and 527 nm for the polymer prepared in nitrobenzene (43). The extents of conjugation are therefore similar in the two cases. ^{1}H NMR spectra obtained showed that the polymers formed in supercritical carbon dioxide had structures that were very similar to those formed in nitrobenzene. The intrinsic electrical conductivities of the polypyrrole and poly(3-undecylbithiophene) formed in nitrobenzene (43) and carbon dioxide were about 10^{-8} S/cm. This is in the lower end of the semiconducting range, although the conductivity could be increased to 1 S/cm by doping with iodine (102).

B. Blending Experiments

Polymerization of pyrrole and 3-undecylbithiophene was also performed in situ in porous, cross-linked polystyrene (PS). This is a simple two-step batch process, shown in Figure 5, which does not require separation steps like the conventional blending process. The first step involves the impregnation of ferric triflate into the polymer host substrate. This was achieved by placing the host and ferric triflate in a reaction vessel maintained at the desired temperature and then pressurizing the vessel with carbon dioxide until the desired pressure was attained. These conditions were maintained for a length of time necessary for the host substrate to absorb the maximal amount of carbon dioxide [determined in separate experiments according to procedures described elsewhere (101)]. In the second step, the host polymer impregnated with ferric triflate was placed in contact with carbon dioxide loaded with the monomer 3-undecylbithiophene until a maximum was reached in the absorption of carbon dioxide. The polymerization of 3-undecylbithiophene occurred in situ in the polymer host substrate. The vessel was then quickly depressurized and the host substrate weighed to obtain the mass gain due to the poly(3-undecylbithiophene) formed. Blending experiments were performed at 313 K at pressures of 10.5, 20.7, and 34.5 MPa; and at a constant pressure of 10.5 MPa, and temperatures of 313 K, 363 K, and 413 K. Elemental analysis was subsequently performed to determine the actual amount of conducting polymer. Finally, the electrical conductivity of the blends was measured with a four-point probe, and the blends were characterized

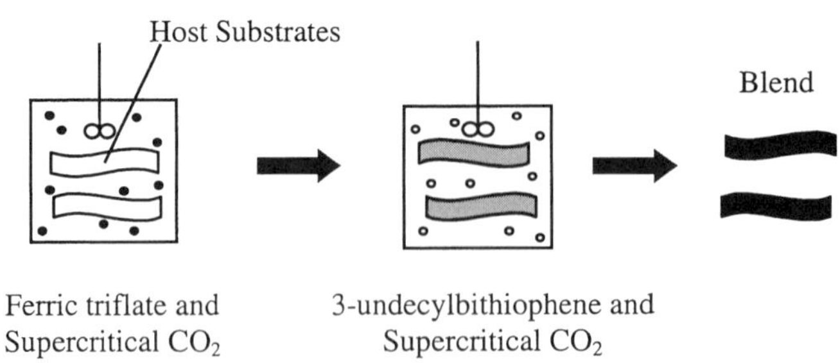

Figure 5 Blending process in supercritical CO_2.

Table 3 Electrical Conductivity of Poly(3-undecylbithiophene) and Porous, Cross-Linked Polystyrene Blends

Temp. (K)	Pressure (MPa)	Conductivity (S/cm)
313	10.5	2×10^{-7}
363	10.5	2×10^{-6}
413	10.5	0
313	20.7	3×10^{-4}
313	34.5	2×10^{-6}

Data from Ref. 102.

using an optical microscope, elemental analysis, and by Differential Scanning Calorimetry (DSC).

As expected, the morphology and electrical conductivity of the blends were greatly influenced by experimental conditions, as summarized in Table 3. Higher electrical conductivities were found at lower temperatures and moderate pressures, although all conductivities fall in the semiconducting range. The only exception is one blend formed at 413 K and 10.5 MPa. Under an optical microscope, the morphology of this blend was found to be different from that of blends obtained at other conditions. This could be due to the high temperature used, which would allow the oxidant to react with the host substrate.

Mixing conditions greatly influenced the amount of conducting polymer formed, as well as the electrical conductivity of the polymer. Optical micrographs of two blends are shown in Figure 6. The blend on the left contained 37 wt % poly(3-undecylbithophene) and had an electrical conductivity of 3×10^{-6} S/cm. The blend on the right contained only 2 wt % poly(3-undecylbithiophene) and no electrical conductivity. Although the temperature and pressure were the same in both experiments (313 K and 10.5 MPa), the blend on the left was formed under high mixing conditions whereas the blend on the right was formed under low mixing conditions. The figure also shows that the morphology that produces electrical conductivity consists of interconnected domains of conducting polymer.

C. Doping Experiments

The blends of porous, cross-linked polystyrene and poly(3-undecylbithiophene) or polypyrrole were doped by contacting them with iodine in a glass vial under atmospheric conditions. The doped blends were removed from the vial at various time intervals and their weight gains monitored as a function of time. Weight gain and time data were recorded until no further weight gain occurred. The electrical

(37 wt % poly 3-undecylbithiophene)

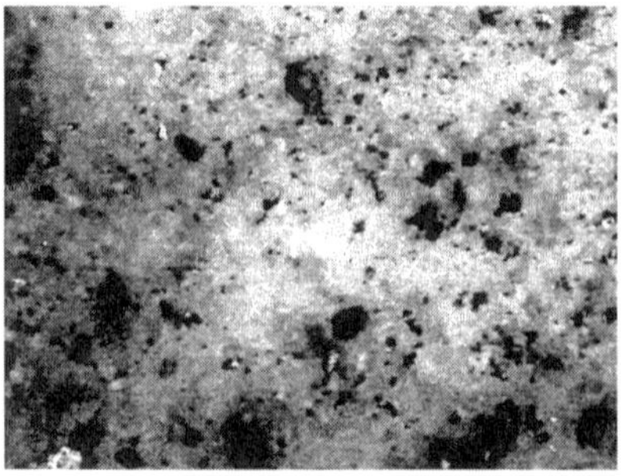

(2 wt % poly 3-undecylbithiophene)

Figure 6 Micrographs of blends of poly(3-undecylbithiophene) and porous, cross-linked polystyrene formed under different mixing conditions (100× magnified and 1000 μm wide).

conductivity before and after doping was measured with a four-point probe. The polystyrene and poly(3-undecylbithiophene) blend was also doped with iodine under supercritical carbon dioxide conditions. The amount of iodine absorbed as a function of time was obtained from a series of absorption experiments.

The results of contacting iodine under atmospheric conditions with a blend of polystyrene and poly(3-undecylbithiophene) are shown in Figure 7. The uptake of iodine after about 8500 min was about 9.3 wt %, and this led to an

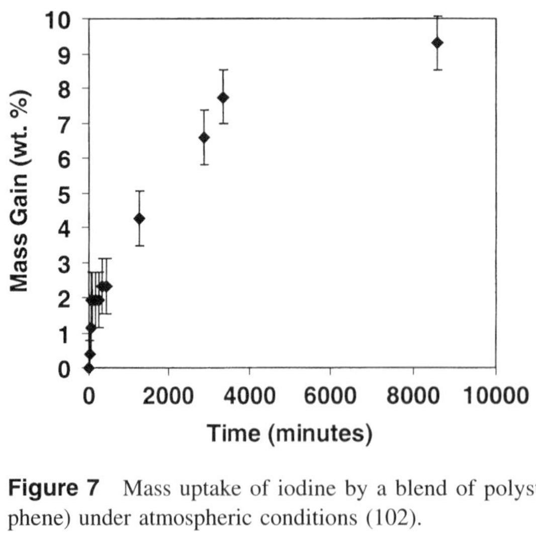

Figure 7 Mass uptake of iodine by a blend of polystyrene and poly(3-undecylbithiophene) under atmospheric conditions (102).

order of magnitude increase in conductivity from 3×10^{-6} to 1×10^{-5} S/cm. The uptake of iodine in the supercritical case was found to be 21.3% mass gain after 60 min. This led to a change in the conductivity of the polymer blend by two orders of magnitude from 2×10^{-6} S/cm to 1×10^{-4} S/cm. Clearly, the supercritical carbon dioxide doping process is superior to the process carried out at ambient conditions.

IX. CONCLUSIONS

The most widely used method for making electrically conducting polymers involves chemical polymerization with ferric chloride as the oxidant and nitrobenzene as the solvent. This chapter reviews the polymerization of pyrrole and 3-undecylbithophene using this method. The extension of this work using supercritical carbon dioxide as the solvent is also described. Since conducting polymers such as polypyrrole and poly(3-undecylbithiophene) are inherently brittle and difficult to process, methods for blend formation to optimize the desirable properties of these polymers are also reviewed. Moreover, a process for the in situ formation of blends of these and other electrically conducting polymers using supercritical carbon dioxide is described, as is a supercritical process for the doping of electrically conducting polymer blends.

Our studies show that supercritical carbon dioxide is a promising solvent for in situ polymerization and the formation of electrically conducting blends (102). The properties of polypyrrole and poly(3-undecylbithiophene) formed in supercritical carbon dioxide were very similar to those of the polymers formed in nitrobenzene in terms of structure, molecular weight, conjugation, and intrinsic electrical conductivity. However, experimental conditions, including the extent of mixing, were found to affect the amount of conducting polymer formed and therefore the electrical conductivity of the blend. Doping of the blends with iodine was found to improve the electrical conductivity by one order of magnitude or more. The larger increases were obtained in the presence of supercritical carbon dioxide and could be due to the ability of carbon dioxide to diffuse into and swell the host polymer. Morphologies containing interconnected domains of conducting polymer were found to exhibit the highest conductivities.

ACKNOWLEDGMENTS

Financial support from the Molecular Design Institute Graduate Student Research Fellowship/Office of Naval Research, the Dow Chemical Minority Mentoring Research Fellowship, and the NSF/GEE fellowship is greatly appreciated.

REFERENCES

1. D Lenz, M Schultz, C Ferreira. Metallization of polymers by using conducting polymers. J Polym Eng 1996;16:295–310.
2. S Roth, W Graupner. Conductive polymers: evaluation of industrial applications. Synth Met 1993:3623–3631.
3. HH Kuhn, AD Child. Electrically conducting textiles. In: TA Skotheim et al., eds. Handbook of Conducting Polymers. New York: Marcel Dekker, 1998:993–1003.
4. CG Koopal, B Eijsma, RJ Nolte. Chronoamperometric detection of glucose by a 3rd-generation biosensor constructed from conducting microtubules of polypyrrole. Synth Met 1993;57:3689–3695.
5. AG MacDiarmid, RB Kaner, K Kaneto, M Maxfield, DP Nairns, PJ Nigrey, AJ Heeger. Lightweight rechargeable batteries using polyacetylene, (CH)//x as the cathode- and/or anode-active material. Energy Technology, Proceedings of the Energy Technology Conference 10th 1983 Government Inst Inc., Rockville, MD, USA, 1983:675–683.
6. J Yue, AJ Epstein. Electronic control of PH at sulphonated polyaniline electrodes. J Chem Soc Chem Commun 1992;21:1540–1542.
7. K Kaneto, K Yoshino, Y Inuishi. Characteristics of electro-optic device using conducting polymers, polythiophene and polypyrrole films. Jpn J Appl Phys 1983; 22:412–414.
8. D Braun, AJ Heeger. Visible light emission from semiconducting polymer diodes. Appl Phys Lett 1991;58:1982–1984.
9. GE Collins, LJ Buckley. Conductive polymer-coated fabrics for chemical sensing. Synth Met 1996;78:93–101.
10. D Kumar, RC Sharma. Advances in conductive polymers. Eur Polym J 1998;34: 1053–1060.
11. A Guiseppi-Elie, GG Wallace, T Matsue. Chemical and biological sensors based on electrically conducting polymers. In: TA Skotheim et al., eds. Handbook of Conducting Polymers. New York: Marcel Dekker, 1998:963–991.
12. TC Pearce, JW Gardner, S Friel, PN Bartett, N Blair. Electronic noses for monitoring the flavour of beers. Analyst 1993;118:371–377.
13. JV Hatfield, P Neaves, PJ Hicks, K Persaud, P Travers. Towards an integrated electronic nose using conducting sensors. Sensors and Actuators B 1994;18–19: 221–228.
14. M Aldissi. Inherently Conducting Polymers: Processing, Fabrication, Applications, Limitations. Park Ridge, NJ: Noyes Data Corporation, 1989.
15. B Wessling. Ger Offen DE 3834526, 1990.
16. F Jonas, G Heywang. Technical applications for conductive polymers. Electrochemica Acta 1994;39:1345–1347.
17. P Fairley. Dupont, Philips test polyaniline. Chem Week 1998;160:73.
18. N Noor-Drugan. Dupont to market Ormecon's conductive polymers. Chem Week 1999;161:14.
19. DB Cotts, Z Reyes. Electrically Conductive Organic Polymers for Advanced Applications. Park Ridge, NJ: Noyes Data Corporation, 1986.

20. TL Rose, SD Antonio, MH Jillson, AB Kon, R Suresh, F Wang. A microwave shutter using conducting polymers. Synth Met 1997;85:1439–1440.
21. JS Park, E Ruckenstein. Conducting polyheterocycle composites based on porous hosts. J Electron Mater 1992;21:205–215.
22. JE Oesterholm, J Laakso, P Nyholm. Melt and solution processable poly(3-alkylthiphenes) and their blends. Synth Met 1989;28:c435–c444.
23. S Hotta, SDDV Rughooputh, AJ Heeger. Conducting polymer composites of soluble polythiophenes in polystyrene. Synth Met 1987;22:79–87.
24. J Laakso, JE Oesterholm, P Nyholm, Conducting polymer blends. Synth Met 1989; 28:c467–c471.
25. S Dogan, U Akbulut, L Toppare. Conducting polymers of aniline I. Electrochemical synthesis of a conducting composite. Synth Met 1992;53:29–35.
26. S Ashley. Fuel probe uses conductive polymers. Mech Eng 1996;118:28.
27. M Belletete, L Mazerolle, N Desrosiers, M Leclerc, G Durocher. Spectroscopy and photophysics of some oligomers and polymers derived from thiophenes. Macromolecules 1995;28:8587–8597.
28. PF van Hutten, RE Gill, JK Herrema, G Hadziioannou. Structure of thiophene-based regioregular polymers and block copolymers and its influence on luminescence spectra. J Phys Chem 1995;99:3218–3224.
29. P Hourquebie, B Blondel, S Dhume. Microwave and optical properties of soluble conducting polymers. Synth Met 1997;85:1437–1438.
30. P Rajagopalan, TJ McCarthy. Two-step modification of chemically resistant polymers: blend formation and subsequent chemistry. Macromolecules 1998;31:4791–4797.
31. E Kung, AJ Lesser, TJ McCarthy. Morphology and mechanical performance of polystyrene/polyethylene composites prepared in supercritical carbon dioxide. Macromolecules 1998;31:4160–4169.
32. JJ Watkins, TJ McCarthy. Polymerization in supercritical fluid–swollen polymers: a new route to polymer blends. Macromolecules 1994;27:4845–4847.
33. GG Wallace, M Smyth, H Zhao. Conducting electroactive polymer-based biosensors. Trends Anal Chem 1999;18:245–251.
34. K Yoshino, K Tada, K Yoshimoto, M Yoshida, T Kawai, H Araki, M Hamaguchi, A Zakhidov. Electrical and optical properties of molecularly doped conducting polymers. Synth Met 1996;78:301–312.
35. Y Cao, D Guo, M Pang, R Qian. Studies on iodine doped thiophene oligomers. Synth Met 1987;18:189–194.
36. VDDS Rughooputh, M Nowak, S Hotta, AJ Heeger, F Wudl. Soluble conducting polymers: the poly(3-alkylthienylenes). Synth Met 1987;21:41–50.
37. G Tourillon, F Garnier. Stability of conducting polythiophene and derivatives. J Electrochem Soc 1983;30:2042–2044.
38. S-A Chen, S-J Lee. Application of molecular-dynamics in determination of conformation change with temperature of poly(3-dodecylthiophene) in crystalline cell. Synth Met 1995;72:253–260.
39. P Garcia, JM Pernaut, V Wintgens, P Valat, F Garnier, D Delabouglise. Effect of end substitution on electrochemical and optical properties of oligothiophenes. J Physique Chemie 1993;97:513–516.

40. J Roncali, M Lemaire, R Garreau, F Garnier. Enhancement of mean conjugation length in conducting polythiophenes. Synth Met 1987;18:139–144.
41. MJ Nowak, SDDV Rughooputh, S Hotta, AJ Heeger. Polarons and bipolarons on a conducting polymer in solution. Macromolecules 1987;20:965–968.
42. K Yoshino, S Nakjima, R Sugimoto. Fusability of polythiophene derivatives with substituted long alkyl chain and their properties. Jpn J Appl Phys 1987;26:11038–11039.
43. J Kowalik, LM Tolbert, S Narayan, AS Abhiraman. Electrically conducting poly(undecylbithiophenels. 1. Regioselective synthesis and primary structure. Macromolecules 2001;34:5471–5479.
44. Y Sun, E Ruckenstein. Poly(3-methylthiophene)-rubber conductive composite prepared via an inverted emulsion pathway. Synth Met 1995;74:145–150.
45. YC Wei, C Chan, J Tian, G Jang, KF Hsueh. Electrochemical polymerization of thiophenes in the presence of bithiophene or terthiophene: kinetics and mechanism of the polymerization. Chem Mater 1991;3:888–897.
46. RD McCullough. The chemistry of conducting polythiophenes. Adv Mater 1998; 10:93–116.
47. RE Myers. Chemical oxidative polymerization as a synthetic route to electrically conducting polypyrroles. J Electron Mater 1986;15:61–69.
48. Y Wang, MF Rubner. Stability studies of the electrical conductivity of various poly(3-alkylthiophenes). Synth Met 1990;39:153–175.
49. M Pomerantz, JJ Tseng, H Zhu, SJ Sproull, JR Reynolds, R Uitz, HJ Arnott. Processable polymers and copolymers of 3-alkyl thiophenes and their blends. Synth Met 1991;41–43:825–830.
50. G Schopf, G Kossmehl. Polythiophenes: Electrically Conductive Polymers. Berlin: Springer-Verlag, 1997.
51. SH Wang, K Takahashi, K Yoshino, T Tanaka, T Yamabe. Dependence of poly(3-alkylthiophene) film properties on electrochemical polymerization conditions and alkyl chain length. Jpn J Appl Phys 1 1990;29:772–775.
52. K-I Iwasaki, H Fujimoto, S Matsuzaki. Conformational changes of poly(3-alkylthiophenes) with temperature and pressure. Synth Met 1994;63:101–108.
53. K Tashiro, K Ono, Y Minagawa, M Kobayashi, T Kawai, K Yoshino. Structure and thermochromic solid-state phase transitions of poly(3-alkylthiophene). J Polym Sci B: Polym Phys 1991;29:1223–1233.
54. MM Chehimi, M Abel, Z Sahraoui. An inverse gas chromatographic study of the PMMA/conducting polypyrrole interface. J Adhes Sci Technol 1996;10:287–303.
55. ET Kang, KG Neoh, TC Tan, YK Ong. Polymerization and oxidation of pyrrole by organic electron acceptors. J Polym Sci A: Polym Chem 1987;25:2143–2153.
56. JC Thieblemont, MF Planche, C Petrescu, JM Bouvier, G Bidan. Stability of chemically synthesized polypyrrole films. Synth Met 1993;59:81–96.
57. H Isotalo, M Ahlskog, H Stubb. Stability of processed poly(3-octylthiophene) and its blends. Synth Met 1993;55–57:3581–3586.
58. JE Oesterholm, P Passiniemi. Synthesis and properties of FeCl4 doped polythiophene. Synth Met 1987;18:213–218.

59. Q Pei, O Inganaes, G Gustafsson, M Granstrom, M Andersson. Routes toward processible and stable conducting poly(thiophene)s. Synth Met 1993;55:1221–1226.
60. CG Wu, CY Chen. Chemical deposition of ordered conducting polypyrrole films on modified inorganic substrates. J Mater Chem 1997;7:1409–1413.
61. SF Lascelles, SP Armes, PA Zhdan, SJ Greaves, AM Brown, JF Watts, SR Leadley, SY Luk. Surface characterization of micrometer-sized, polypyrrole-coated polystyrene latexes: verification of a "core-shell" morphology. J Mater Chem 1997; 7:1349–1355.
62. S Wu. Polymer Interface and Adhesion. New York: Marcel Dekker, 1982.
63. M Omastova, J Pionteck, S Kosina. Preparation and characterization of electrically conductive polypropylene/polypyrrole composites. Eur Polym J 1996;32:681–689.
64. HS Chan, TSA Hor, PKH Ho. Chemical preparation and characterization of conductive polypyrrole composite thin films. J Macromol Sci Chem 1990;A27:1081–1094.
65. B Vincent. Electrically conducting polymer colloids and composites. Polym Adv Technol 1995;6:356–361.
66. ML Abel, JL Camalet, MM Chehimi, JF Watts, PA Zhdan. A solvent effect on the morphology of PMMA-coated polypyrrole surfaces. Synth Met 1996;81:23–31.
67. M Makhlouki, M Morsli, A Bonnet, A Conan, A Pron, S Lefrant. Transport properties in polypyrrole-PVA composites: evidence for hopping conduction. J Polym Sci B: Polym Phys 1992;44:443–446.
68. F Samir, E Benesiddik, B Corraze, C Bernede, M Morsli, A Bonnet, A Conan, S Lefrant. XPS and transport studies of PVA-PPy and PSt-PBTh composites. Synth Met 1995;69:341–342.
69. SW Byun, SS Im. Transparent and conducting nylon 6-based composite films prepared by chemical oxidative polymerization. Synth Met 1993;55–57:3501–3506.
70. S Radhakrishnan, AB Mandale. Polypyrrole growth on crystalline PEO complexes: characterization by WAXD and XPS. Synth Met 1994;62:217–221.
71. S Radhakrishnan, DR Saini. Structure and electrical properties of polypyrrole-thermoplastic elastomer blends. Polym Int 1994;34:111–117.
72. KS Ho, K Levon, J Mao, W Zheng. Phase behavior of undoped poly(3-alkyl thiophene) with ethylene-co-vinylacetates. A solvatochromatic transition in the solid state. Synth Met 1993;55–57:3591–3596.
73. SP Khedkar, S Radhakrishnan. Long-term aging and stability of conductivity in vapour phase deposited polypyrrole films. Polym Degrad Stabil 1997;57:51–58.
74. XB Chen, J Devaux, J-P Issi, D Billaud. The conducting behavior and stability of conducting polymer composites. Polym Eng Sci 1995;35:637–641.
75. M Omastova, S Kosina, J Pionteck, A Janke, J Pavlinec. Electrical properties and stability of polypyrrole containing conducting polymer composites. Synth Met 1996;81:49–57.
76. E Ruckenstein, JS Park. New method for the preparation of thick conducting polymer composites. J Appl Polym Sci 1991;42:925–934.
77. V Mano, MI Felisberti, T Matencio, M-A De Pauli. Thermal, mechanical and electrochemical behavior of poly(vinyl chloride)/polypyrrole blend (PVC/Ppy). Polymer 1996;37:5165–5170.

78. E Pigois-Landureau, YF Nicolau, M Delamar. XPS study of layer-by-layer deposited polypyrrole thin films. Synth Met 1995;72:111–119.
79. M Morsli, A Bonnett, F Samir, S Lefrant. Electrical properties of polybithiophene-polystyrene composites. J Appl Polym Sci 1996;61:213–216.
80. N Balci, E Bayramli, L Toppare. Conducting polymer composites: polypyrrole and poly(vinyl chloride–vinyl acetate) copolymer. J Appl Polym Sci 1997;64:667–671.
81. PW Faguy, RA Lucas, W Ma. An FT-IR-ATR spectroscopic study of the spontaneous polymerization of pyrrole in iron-exchanged montmorillonite. Colloids Surf 1995;105:105–112.
82. XH Yin, K Kobayashi, K Yoshino, H Yamamoto, T Watanuki, I Isa. Percolation conduction in polymer composites containing polypyrrole coated insulating polymer fiber and conducting polymer. Synth Met 1995;69:367–368.
83. E Ruckenstein, JS Park. Polythiophene and polythiophene-based conducting composites. Synth Met 1991;44:293–306.
84. F Vatansever, J Hacaloglu, U Akbulut, L Toppare. A conducting composite of polythiophene: synthesis and characterization. Polym Int 1996;41:237–244.
85. TJ Romack, EE Maury, JM DeSimone. Precipitation polymerization of acrylic acid in supercritical carbon dioxide. Macromolecules 1995;28:912–915.
86. JM DeSimone, EE Maury, JR Combes, YZ Menceloglu, JB McClain, TJ Romack. Dispersion polymerizations in supercritical carbon dioxide. Sci 1994;265:356–359.
87. E Dada, W Lau, RF Merritt, YH Paik, G Swift. Synthesis of poly(acrylic acid)s in supercritical carbon-dioxide. Polymeric Materials Science and Engineering, Proceedings of the ACS Division of Polymeric Materials Science and Engineering, Vol. 74, 1996, p. 427.
88. JM DeSimone, Z Guan, CS Elsbernd. Synthesis of fluoropolymers in supercritical carbon-dioxide. Science 1992;257:945–947.
89. T Pernecker, JP Kennedy. Carbocationic polymerizations in supercritical CO_2. Polym Bull 1994;32:537–543.
90. JJ Watkins, TJ McCarthy. Polymerization of styrene in supercritical CO_2-swollen poly(chlorotrifluoroethylene). Macromolecules 1995;28:4067–4074.
91. MA McHugh, VJ Krukonis. Supercritical Fluid Extraction: Theory and Applications, 2nd ed. Boston: Butterworth-Heineman, 1994.
92. AR Berens, GS Huvard, RW Korsmeyer. Process for incorporating an additive into a polymer and product produced thereby. U.S. Patent 4,820,752, 1986.
93. ML Sand. Method for impregnating a thermoplastic polymer. U.S. Patent 4,598,006, 1986.
94. RG Wissinger, ME Paulaitis. Swelling and sorption in polymer-CO_2 mixtures at elevated pressures. J Polym Sci B: Polym Phys 1987;25:2497–2510.
95. Y-T Shieh, J-H Su, G Manivannan, PHC Lee, SP Sawan, WD Spall. Interaction of supercritical carbon dioxide with polymers. II. Amorphous polymers. J Appl Polym Sci 1996;59:707–717.
96. Y-T Shieh, J-H Su, G Manivannan, PHC Lee, SP Sawan, WD Spall. Interaction of supercritical carbon dioxide with Polymers. I. Crystalline polymers. J Appl Poly Sci 1996;59:695–705.

97. AR Berens, GS Huvard. Interaction of polymers with near-critical carbon dioxide. In: Supercritical Fluid Science and Technology, American Chemical Society, Washington D.C., 1989.
98. NJ Cotton, KS Bartle, AA Clifford, CJ Dowle. Rate and extent of supercritical-fluid extraction of additives from polypropylene: diffusion, solubility, and matrix effects. J Appl Polym Sci 1993;48:1607–1619.
99. PF Nealey, RE Cohen, AS Argon. Effect of gas-pressure on the solubility and diffusion of polybutadiene in polystyrene. Macromolecules 1994;27:4193–4197.
100. BR Chapman. CO_2-enhanced diffusion of azobenzene in glassy polystyrene near the glass transition. Macromolecules 1996;29:5635–5649.
101. KF Webb, AS Teja. Solubility and diffusion of carbon dioxide in polymers. Fluid Phase Equilibria 1999;160:1029–1034.
102. KF Webb, J Kowalik, L Tolbert, AS Teja. The formation of electrically conducting polymer blends in supercritical carbon dioxide (to be published).

8

Hydrothermal Synthesis of Metal Oxide Nanoparticles Under Supercritical Conditions

Tadafumi Adschiri and Kunio Arai
Tohoku University, Sendai, Japan

I. INTRODUCTION

When a metal salt aqueous solution is heated up, metal (hydro) oxides are formed due to the shift of reaction equilibrium. At higher temperatures, the equilibrium shifts further toward the metal oxide formation side.

$$M^{x+} + xOH^- = M(OH)_x \qquad \text{Reaction 1}$$

$$M(OH)_x = MO_{\frac{x}{2}} + \frac{x}{2} H_2O \qquad \text{Reaction 2}$$

This method of producing metal (hydro) oxides from aqueous solutions at elevated temperatures is hydrothermal synthesis.

Hydrothermal synthesis of metal oxides from aqueous solutions is typically utilized at temperatures ranging from 373 K to 473 K. We are developing a new process for hydrothermal synthesis that uses temperatures in the range of 673–723 K, which is just above the critical temperature of water (1–7).

Properties of water, including density, dielectric constant (Figure 1) (8), and ion product, vary greatly around the critical point of water (T_c, 647 K; P_c, 22.1 MPa) and result in a specific hydrothermal reaction atmosphere. The reaction rate, equilibrium, phase behavior, solubility of metal oxides, and distribution of soluble chemical species change greatly at the critical point range. In situ heat treatment of the metal (hydro) oxides formed may occur in a steam-like

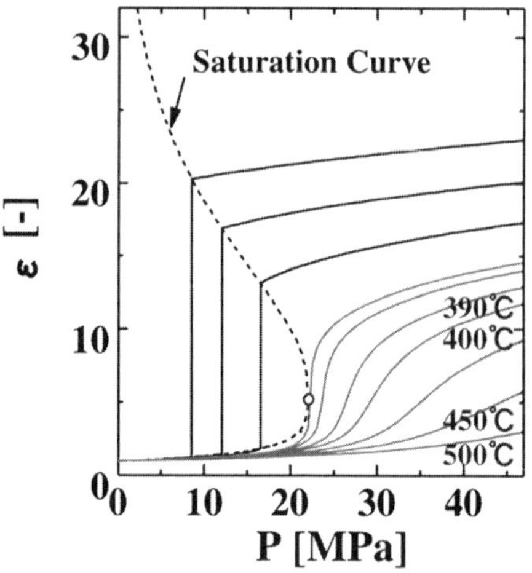

Figure 1 Dielectric constant of water around the critical point.

atmosphere. These supply the specific features of hydrothermal synthesis under supercritical conditions.

In this chapter, first the ionic reaction equilibrium, phase behavior, and solubility of metal oxides in supercritical water are discussed. Next, the specific features of hydrothermal synthesis under supercritical conditions are discussed based on the experimental results. The supercritical hydrothermal crystallization method was applied to the production of functional materials, barium hexaferrite ($BaFe_{12}O_{19}$), metal-doped oxide [$Al_5(Y+Tb)_3O_{12}$, YAG:Tb], and Li ion battery cathode material ($LiCo_2O_4$). The importance of understanding the chemical reaction equilibrium and phase behavior is discussed.

II. SPECIFIC FEATURES OF SUPERCRITICAL WATER FOR HYDROTHERMAL SYNTHESIS

A. Reaction Equilibrium and Reaction Rate

Reaction rate and equilibrium change greatly around the critical point due to the variation in properties of water. Various models have been proposed for describing the variation of reaction rate or equilibrium over a range of supercritical state (9–12). The effect of the dielectric constant on ionic reactions can be expressed by the Born equation, which expresses electrostatic interactions of ions in a dielectric field (13). It is widely known that the Helgeson-Kirkham-Flomers (HKF) model that uses this term can describe successfully the change of ionic reaction equilibrium, although five reference state parameters should be known.

Recently, we demonstrated that even a simple equation is applicable for various ionic reactions in the supercritical region (14), which is given here as:

$$\ln K(T, \rho) = \ln K(T_0, \rho_0) + \frac{\Delta H}{R}\left(\frac{1}{T} - \frac{1}{T_0}\right) - \frac{\omega_j}{RT}\left(\frac{1}{\varepsilon(T, \rho)} - \frac{1}{\varepsilon_0(T_0, \rho_0)}\right) \quad (1)$$

where K is the equilibrium constant, T is the absolute temperature, T_0 is the room temperature, R is the gas constant, ε is the dielectric constant, ω is a parameter determined by the reaction system, ρ is the density, and ρ_0 is the density at the ambient condition. The last term of Eq. (1) is the Born term.

The $(1/\varepsilon - 1\varepsilon_0)$ term is denoted as the dielectric factor. Figure 2 shows the dielectric factor as a function of temperature and pressure. A dielectric constant at ambient temperature and pressure is chosen as the reference state. The heavy dashed line shows the reference density isochore. At pressures from 23 to 50 MPa, where supercritical hydrothermal synthesis are mostly operated, the dielectric factor becomes significant above the critical temperature. This relation suggests that the equilibrium constant change greatly in the region around the critical point.

The reaction rate also changes greatly in the region around the critical point. A similar equation with the Born term is applicable to the reactions in supercritical water (13).

$$\ln k = \ln k_0 - \frac{\omega^{\neq}}{RT}\left(\frac{1}{\varepsilon} - \frac{1}{\varepsilon_0}\right) \quad (2)$$

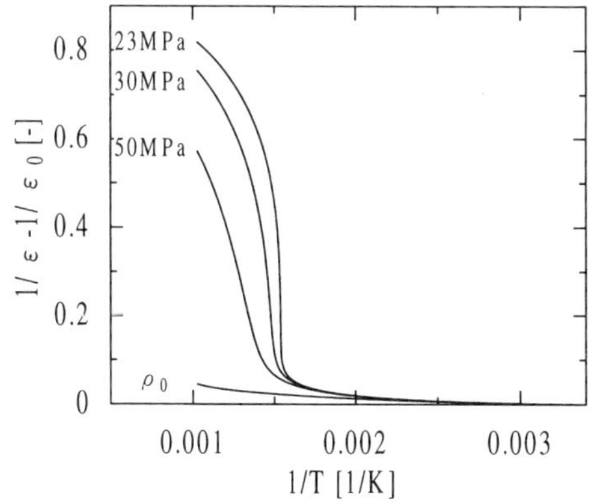

Figure 2 Dielectric factor around the critical point.

where k and k_0 are rate constants corresponding to the dielectric constants ε and ε_0, respectively, $\omega^{\neq}$ is a constant determined by the reaction system, R is the gas constant, and T is the temperature in K.

B. Phase Behavior and Solubility of Metal Oxides

The knowledge on the solubility of the crystalline substances formed and the identification of the ionic species are essential to the understanding of hydrothermal synthesis. As an example, the main equilibrium reactions for AlOOH dissolution in aqueous solution are shown in Table 1. For the estimation of metal oxides, the equilibrium constant K for each reaction is required. Equation (1) can be used for the estimation of the equilibrium constants under supercritical conditions. The data for equilibrium constants or Gibbs free energy changes and enthalpy of reaction for various reactions are available at ambient conditions.

The solubility of copper oxide at 28 MPa in sub- and supercritical water are shown in Figure 3. Curves shown in this figure are the solubility estimated by the model, with the term $\Sigma\omega_i$ being treated as a fitting parameter for each dissociation reaction. However, even by using the value estimated from the ionic radius reported, the trend of the solubility behavior can be estimated fairly well. As illustrated, the experimental results could be correlated successfully by this model to express the increase of solubility with increasing temperature and the dramatic decrease around the critical point (14).

The distribution of ionic species in solution is also important for the understanding of hydrothermal reaction atmosphere. As shown in the figure, from 573 K to 673 K the single charge cationic and anionic copper species sharply decrease whereas the neutral solute increases and dominates the solubility behavior above 673 K at 28 MPa.

Due to the drastic change in properties of water around the critical point, the phase behavior of multicomponents varies greatly. Franck and the colleagues (15,16) have studied the phase change around the critical point for various systems with water. Figure 4 shows critical loci for some light gas–water systems. On the right side of the critical locus, homogeneous phase is formed. This suggests that introducing oxygen or hydrogen gas into the system can control oxidizing or reducing atmospheres of hydrothermal synthesis.

Table 1 Reactions Related to AlOOH Dissolution

$AlOOH + H_2O = Al^{3+} + 3OH^-$	$Al^{3+} + OH^- = Al(OH)^{2+}$	$NaNO_3 = Na^{3+} + NO_3^-$
$AlOOH + H_2O = Al(OH)^{2+} + 2OH^-$	$Al(OH)^{2+} + OH^- = Al(OH)_2^+$	$NaOH = Na^+ + OH^-$
$AlOOH + H_2O = Al(OH)^+ + OH^-$	$Al(OH)_2^+ + OH^- = Al(OH)_3$	$HNO_3 = H^+ + NO_3^-$
$AlOOH + H_2O = Al(OH)_4^- + H^+$	$Al(OH)_3 + OH^- = Al(OH)_4^-$	$H_2O = H^+ + OH^-$
	$Al(NO_3)_3^- = Al^{3+} + 3(NO_3)^-$	

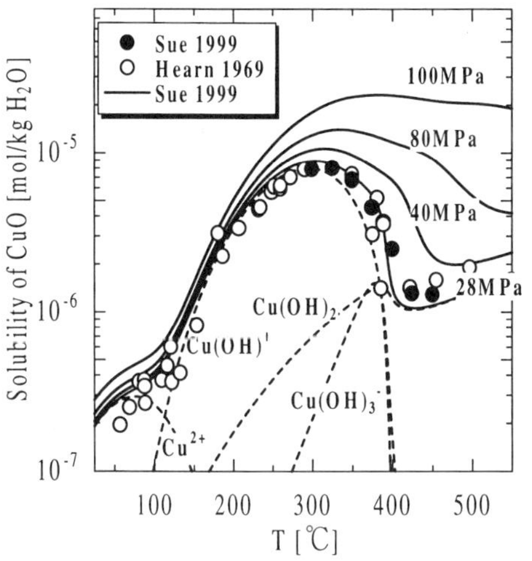

Figure 3 Solubility of lead oxide in high-temperature water. Curves are the correlated curves by Eq. (1).

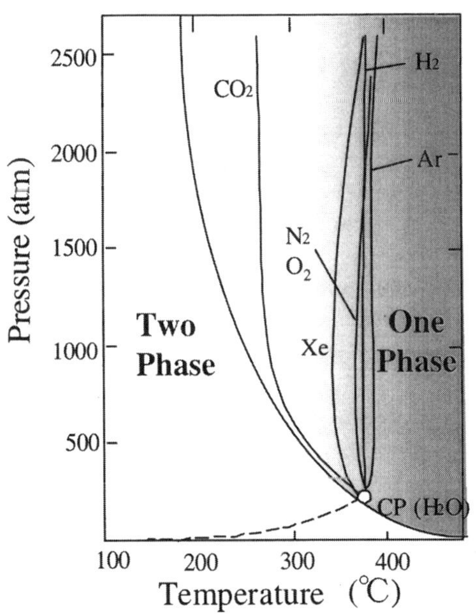

Figure 4 Phase behavior of water–light gas systems.

III. SPECIFIC FEATURES OF HYDROTHERMAL SYNTHESIS UNDER SUPERCRITICAL CONDITIONS

A. Experimental

Figure 5 shows a typical experimental apparatus used. An aqueous metal salt solution is first prepared and fed into the apparatus in one stream. In another stream, distilled water is pressurized and then heated to a temperature above the desired reaction temperature. The pressurized metal salt solution stream and the pure supercritical water stream are then combined at a mixing point, which leads to rapid heating and subsequent reactions in the reactor. After the solution leaves the reactor, it is rapidly quenched. In-line filters are used to remove larger particles. Pressure is controlled with a back-pressure regulator. Fine particles are collected in the effluent. By this rapid heating method, the effect of the heating period on the hydrothermal synthesis is eliminated; thus, specific features of supercritical hydrothermal synthesis can be elucidated.

Table 2 shows the metal salts used in the experiment (1–7), including metals Fe, Al, Ni, Co, Ti, Zr, and anions NO_3^-, Cl^-, SO_4^{2-}, CO_3^-, and ammonium citrate.

B. Specific Features of Supercritical Hydrothermal Synthesis

1. Ultrafine Particles Formation

Table 2 also shows the summary of the obtained products. Particle in the size range 10–1000 nm are produced. In many cases, the particles have well-developed crystal faces, which suggests single crystals.

We have also conducted a series of experiment at subcritical conditions using the same apparatus. However, the particles obtained in subcritical water are larger than those in supercritical water. Figure 6 shows the comparison of the ceria particles obtained at subcritical and supercritical conditions. In the subcritical water experiments there is particle growth with an increase in the residence time, whereas under supercritical conditions such a phenomenon has not been observed.

The mechanism for the fine-particle formation in supercritical water is discussed as follows (Figure 7): The solubility of metal oxides in subcritical water is higher than that at supercritical conditions, as discussed above. Thus, after nucleation, inclusion of precursors (soluble intermediates) takes place to grow crystals. On the other hand, in supercritical water hydrothermal synthesis reaction proceeds faster than that in subcritical water due to the higher temperature and the lower dielectric constant, as expected from Eq. (2). The solubility

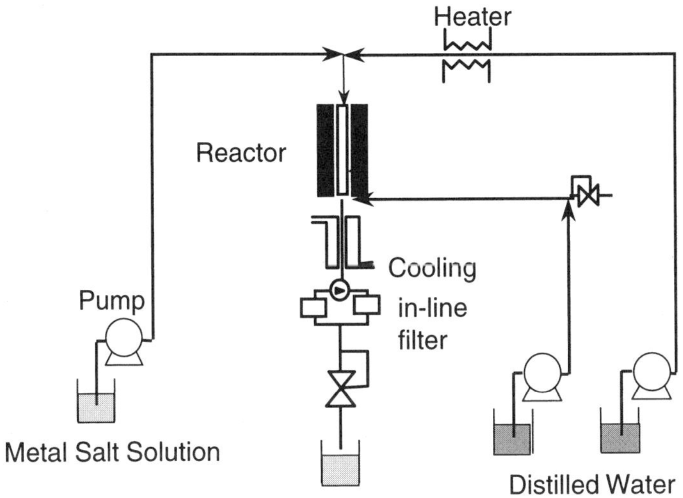

Figure 5 Experimental apparatus.

Table 2 Metal Oxide Particles Produced by Supercritical Water Crystallization

Metal salts used starting material	Products	Particle size (nm)	Morphology	Ref.
$Al(NO_3)_3$	AlOOH	80 ~ 1000	Hexagonal plate Rhombic Needle-like	1,2,4
$Fe(NO_3)_3$	α-Fe_2O_3	~ 50	Spherical	1
$Fe_2(SO_4)_3$	α-Fe_2O_3	~ 50	Spherical	1
$FeCl_2$	α-Fe_2O_3	~ 50	Spherical	1
$Fe(NH_4)2H(C_6H_5O_7)_2$	Fe_3O_4	~ 50	Spherical	1
$Co(NO_3)_2$	Co_3O_4	~ 100	Octahedral	1
$Ni(NO_3)_2$	NiO	~ 200	Octahedral	1
$ZrOCl_2$	ZrO_2	~ 10	Spherical	1
$Ti(SO_4)_2$	TiO_2	~ 20	Spherical	1
$TiCl_4$	TiO_2	~ 20	Spherical	1
$Ce(NO_3)_3$	CeO_2	20 ~ 300	Octahedral	3
$Fe(NO_3)_3$, $Ba(NO_3)_2$	$BaO{\cdot}6Fe_2O_3$	50 ~ 1000	Hexagonal plate	5
$Al(NO_3)_3$, $Y(NO_3)_3$, $TbCl_3$	$Al_5(Y+Tb)_3O_{12}$	20 ~ 600	Dodecahedral	6
$Li(NO_3)$, $Co(NO_3)_2$	$LiCoO_2$	40 ~ 400	Dodecahedral	7

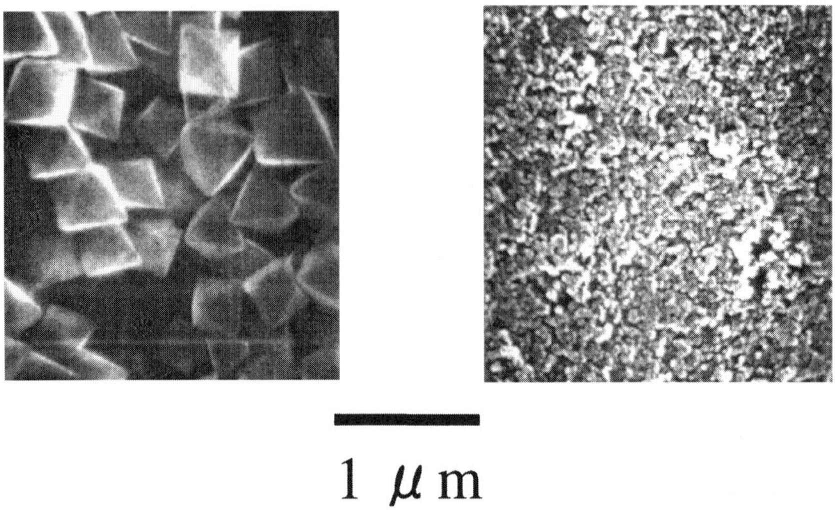

Figure 6 Comparison of ceria particles obtained in subcritical water and supercritical water.

of intermediates is lower than that in subcritical water. Thus, when the metal salt aqueous solution is mixed with high-temperature supercritical water, an extremely high degree of supersaturation is generated and rapid nucleation occurs at the mixing point. Intermediates are consumed at this nucleation stage. This is a reason why ultrafine particles are obtained under supercritical condition (3).

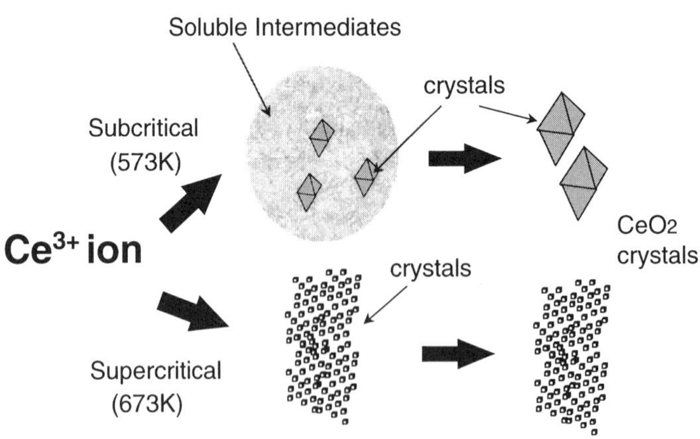

Figure 7 Mechanism of fine-particle production under supercritical conditions.

2. Control of Oxidizing or Reducing Atmosphere with O_2 or H_2

When ferrous ammonium citrate is used as feed, magnetite (Fe_3O_4) particles are formed in a single phase (1). For this case, Fe(III) is reduced to Fe(II), due probably to the reduction of Fe(III) by CO formed in the thermal decomposition of ammonium citrate. Since CO is miscible with supercritical water as shown in Fig. 4, a homogeneous reaction atmosphere is provided, which allows for magnetite particles to form in single phase. This suggests that by introducing O_2 or H_2, the oxidizing and reducing atmosphere can be controlled.

3. Control of Morphology

When aluminum nitrate aqueous solution was employed as feed, various morphologies could be found, as shown in Figure 8 (2). We have studied the relation between crystal habit of boehmite (AlOOH) particles and chemical equilibrium in hydrothermal conditions. Around the critical point, the reaction equilibrium varies greatly with temperature and pressure. This leads to a change in the chemical species distribution, as described earlier. Comparison of the calculation results of the aluminum ion species distribution and the obtained crystal habits suggests that when the positively charged ion is predominant, sword-like particles are obtained, and when the neutral species is predominant, rhombic particles are obtained. It is known that AlOOH crystal has faces with different charges. The results suggest that AlOOH particle morphology is determined by selective adsorption of positively charged species, $Al(OH)^{2+}$ or $Al(OH)_2^+$, on the negatively charged surface of AlOOH crystal (4). Around the critical point, the equilibrium constant of the chemical reaction changes greatly with temperature and pressure, which leads to a change in the chemical species distribution and thus the crystalline habits.

IV. APPLICATIONS OF THE METHOD FOR PRODUCING COMPLEX METAL OXIDES

Perpendicular magnetizing recording method is being used for producing high-density recording media. Barium hexaferrite ($BaO{\cdot}6Fe_2O_3$) is hexagonal and plate-like, and its magnetization axis is perpendicular to the plate plane. However, conventional solid-state methods have difficulty in producing fine hexagonal plates, and they are energy and time consuming. We have employed the supercritical hydrothermal synthesis method for producing barium hexaferrite particles continuously. The experimental apparatus for the production of barium

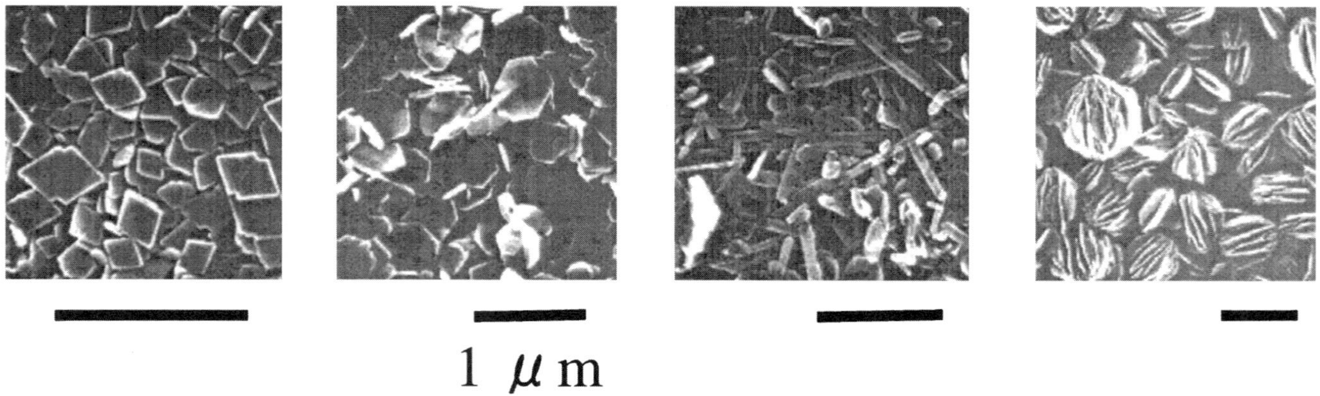

Figure 8 Morphology of AlOOH particles obtained around the critical point.

hexaferrite is shown in Figure 9. A mixture of $Ba(NO_3)_2$ and $Fe(NO_3)_3$ aqueous solutions was mixed with a KOH aqueous solution and, after preheating, the solution was mixed with supercritical water.

By this method, the barium hexaferrite hexagonal plate particles (Figure 10) are produced in a single phase continuously with the reaction time of approximately 1 min (5). It takes more than 1 day by a conventional batchwise solid-state method, and the morphology is more like spherical. The obtained $BaO_6Fe_2O_3$ hexagonal plate particles have magnetic properties comparable with commercially produced barium hexaferrite, whereas particles obtained in subcritical conditions show poorer magnetic properties.

Another example is the YAG/Tb phosphor synthesis (6). The commercial method for synthesizing YAG phosphor is via solid-state reaction at temperatures above 1500 K. Thus, long reaction times are required to transform the two-phase system into single-phase YAG. The doping of Tb(activator ion) onto the $Y_3Al_5O_{12}$(YAG) crystal is made after producing it. In our process, a mixed solution of $Y(NO_3)_3$, $Al(NO_3)_3$, and $TbCl_3$ was used as feed. It was demonstrated that single-phase YAG/Tb particles could be produced at 673K, 30 MPa in 1 min (6). The luminescent properties of the YAG/Tb particles thus obtained were comparable, even without any heat treatment, with that of particles produced by the conventional solid-state reaction at high temperature.

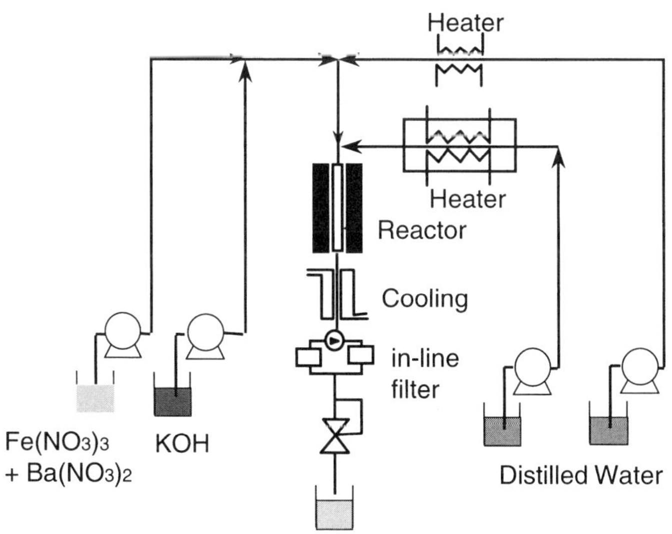

Figure 9 Apparatus for continuous production of barium hexaherrite particles.

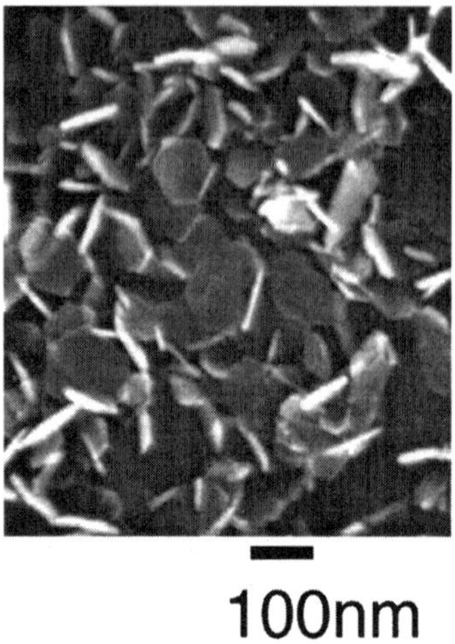

Figure 10 SEM photograph of barium hexaferrite hexagonal plate particles.

The probable cause of barium hexaferrite and phosphor produced at supercritical condition having better properties even without any heat treatment is the in situ treatment of the particle formed at supercritical conditions.

$LiCoO_2$ or $LiMn_2O_4$ is of great importance for lithium ion battery technologies. For hydrothermal synthesis of $LiCoO_2$, where Co^{2+} ion is a starting species, the oxidation conditions for Co^{2+} do not always match that required for the crystallization atmosphere. In most cases oxidants decomposed in high-temperature water.

The previous experimental results show that it might be possible to form $LiCoO_2$ particles under supercritical conditions. Experiments were performed with an apparatus similar to that shown in Fig. 5. $Co(NO_3)_2$ aqueous solution (0.02 M) was first mixed with LiOH aqueous solution (0.4 M) at a mixing point. For oxidizing Co^{2+} ions in the reaction atmosphere, O_2 gas was introduced into the reactor after decomposing H_2O_2 aqueous solution (0.07 M). The mixed solution of LiOH and $Co(NO_3)_2$ was contacted with a high-temperature water–oxygen mixture fed from another line. At this mixing point, solutions were rapidly heated to the reaction temperature (573, 623, or 673 K) for hydrothermal crystallization.

Figure 11 shows x-ray diffraction patterns of the particles produced. The results clearly indicate that $LiCoO_2$ particles (○) were produced in a single phase at 673 K and 30 MPa. The $LiCoO_2$ particles were found to have rock

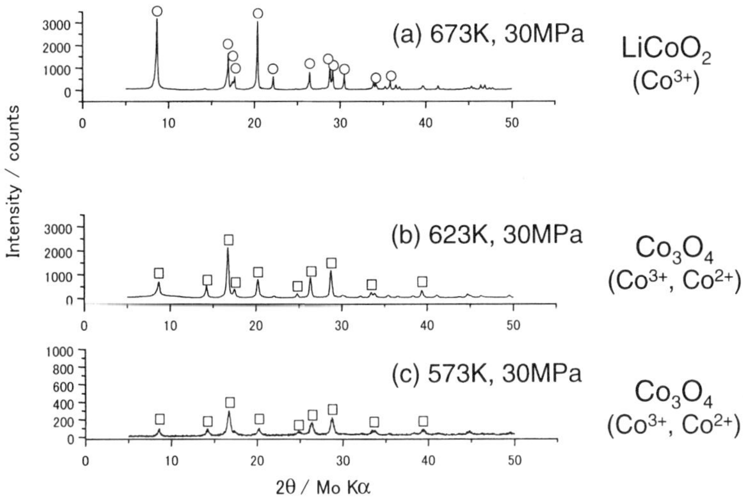

Figure 11 XRD patterns of the products.

salt type of structure belonging to the *Fd3m* space group. However, at 573 K or at 623 K, Co_3O_4 (□), which is composed of Co^{2+} and Co^{3+}, was produced. Thus, at supercritical conditions, more oxidation of Co^{2+} took place more effectively than at 573 K or 623 K. This is probably because the O_2 formed from H_2O_2 is miscible with the supercritical aqueous solution in the reactor; thus, homogeneous reaction atmospheres are generated.

V. CONCLUSIONS

The specific features of the supercritical hydrothermal synthesis method are (a) production of ultrafine particles and (b) control of particle morphology or crystal structure via variation of the pressure and temperature or control of reaction atmosphere (reducing or oxidizing) via introduction of O_2 or H_2 gases. The supercritical hydrothermal synthesis method was applied for the production of barium hexaferrite, YAG/Tb phosphor, and lithium ion battery materials ($LiCoO_2$). By rapid heating of the reactant solutions, continuous production of the metal oxide fine particles could be achieved. For the production of $LiCoO_2$, oxygen gas was introduced into the system for oxidizing Co^{2+} during the hydrothermal synthesis reaction. Under supercritical conditions, $LiCoO_2$ could be formed in single phase, whereas in contrast, under subcritical conditions, Co_3O_4 was the main product formed. This implies effective oxidation could be achieved in supercritical water due to the formation of homogeneous phase for oxygen gas and water.

In conclusion, the general features of the supercritical hydrothermal synthesis method can be expected to open the door to a wide rangc of new applications in material production. Material production with the supercritical hydrothermal synthesis method can be expected to have a bright future.

REFERENCES

1. T Adschiri, K Kanazawa, K Arai. Rapid and continuous hydrothermal crystallization of metal oxide particles in supercritical water. J Am Ceram Soc 75:1019–1023, 1992.
2. T Adschiri, K Kanazawa, K Arai. Rapid and continuous hydrothermal synthesis of boehmite particles in subcritical and supercritical water. J Am Ceram Soc 75: 2615–2620, 1992.
3. Y Hakuta, H Terayama, S Onai, T Adschiri, K Arai. Production of ultra fine ceria particles by hydrothermal synthesis under supercritical conditions. J Mater Sci Lett 17:1211–1213, 1998.
4. Y Hakuta, T Adschiri, H Hirakoso, K Arai. Chemical equilibria and particle morphology of boehmite (AlOOH) in sub and supercritical water. Fluid Phase Equilibria 158–160:733–739, 1999.
5. Y Hakuta, T Adschiri, T Suzuki, K Seino, K Arai. Flow method for rapidly producing single phase barium hexa-ferrite particles in supercritical water. J Am Ceram Soc 81(9):2461–2465, 1998.
6. Y Hakuta, K Seino, H Ura, T Adschri, H Takizawa, K Arai. Production of phosphor (YAG:Tb) fine particles by hydrothermal synthesis in supercritical water. J Mater Chem 9(10):2671–2675, 1999.
7. T Adschiri, Y Hakuta, K Kanamura, K Arai. Continuous production of $LiCoO_2$ fine crystals for lithium batteries by hydrothermal synthesis under supercritical condition. High-Pressure Research 20:373–384, 2001.
8. M Uematsu, EU Franck. Static dielectiric constant of water and steam. J Phys Chem Ref Data 9:1291, 1991.
9. WL Marshall, EU Franck. Ion product of water substance, 0–1000°C, 1–100,000 bars new international formulation and its background. J Phys Chem Ref Data 10: 295–304, 1981.
10. GM Anderson, S Castet, J Schott, RE Mesmer. The density model for estimation of thermodynamic parameters of reactions at high temperatures and pressures. Geochim Cosmochim Acta 55:1769–1779, 1991.
11. HC Helgeson, DH Kirkham, GC Flowers. Theoretical prediction of the thermodynamic behavior of aqueous electrolytes at high pressures and temperatures: IV. Calculation of activity coefficients, osmotic coefficients, and apparent molal and standard and relative partial molal properties to 600°C and 5 kb. Am J Sci 281: 1249–1516, 1981.

12. JC Tanger, HC Helgeson. Calculation of the thermodynamic and transport properties of aqueous species at high pressures and temperatures: revised equations of state for the standard partial molal properties of ions and electrolytes. Am J Sci 288: 19–98, 1988.
13. N Isaacs. Liquid Phase High Pressure Chemistry. New York: John Wiley & Sons, 1987.
14. K Sue, Y Hakuta, RL Smith Jr, T Adschiri, K Arai. Solubility of lead(II) oxide and copper(II) oxide in subcritical and supercritical water. J Chem Eng Data 44: 1422–1426, 1999.
15. TM Seward, EU Franck. The system hydrogen–water up to 400°C and 2500 bar pressure. Ber Bunsenges Phys Chem 85:2, 1981.
16. EU Franck. Special aspects of fluid solutions at high pressures and sub- and supercritical temperatures. Pure Appl Chem 53:1401, 1981.

9

Production of Magnetic Nanoparticles Using Supercritical Fluids

Amyn S. Teja and Linda J. Holm
Georgia Institute of Technology, Atlanta, Georgia

I. INTRODUCTION

Fine magnetic particles have found use in many applications, such as recording media and ferrofluids (1,2). New uses of these particles have also been proposed in high-density information storage (3) and in a number of biomedical applications (4–6). These new applications require excellent control of properties such as particle size, and variables that affect these properties are the focus of an increasing number of studies.

Magnetic particles generally consist of multiple magnetic domains, which are "groups of spins all pointing in the same direction and acting cooperatively" (7). These magnetic domains are separated by domain walls, which have a length scale on the order of 100 nm. Particles of nanometer size are not large enough to support the formation of multiple domains; as a result, they possess unique properties.

An interesting property of single-domain particles is that they exhibit superparamagnetic behavior at certain temperatures. A single particle has a magnetic moment because of alignment of spins within its crystal structure; however, magnetic moments of individual particles in an assembly are not always aligned, and the assembly does not necessarily exhibit a net magnetic moment. An external field must be applied to magnetize such an assembly of particles (or align all magnetic moments). Moreover, there is an energy barrier associated with changing the direction of the magnetic moment, so that moments in an assembly of particles will stay aligned even after the magnetic field is removed (this is termed magnetic remanence and is a property of ferromagnetic materials). The magnetic field strength required to reduce the magnetization to zero is referred to as the coercivity. At high temperatures, nanoparticles of magnetic materials have enough thermal energy to overcome the energy barrier associated with changing the direction of the magnetic moment, and therefore the moment of each individual particle in an assembly returns to its preferred unaligned direction when an external magnetizing field is removed. This behavior is termed superparamagnetism. The transition between superparamagnetic behavior and ferromagnetic behavior occurs at the blocking temperature (Figure 1). Particles are superparamagnetic above the blocking temperature and ferromagnetic below this temperature. Furthermore, the blocking temperature varies from material to material, and decreases with particle size (8).

Magnetic information storage makes use of the coercivity of magnetic particles (the magnetic field strength required to remove remanent magnetization) to store information. By applying magnetic fields to particles embedded in a matrix, information can be written and erased. With recent emphasis on increasing memory and decreasing recording medium size, improvements in magnetic recording in large part rely on decreasing the particle size to increase the density of magnetic particles in a matrix (3). Because of the superparamagnetic effect, the particle size cannot be reduced beyond a certain limit and still be useful in magnetic recording applications. This effect is a significant barrier to increasing the density of longitudinal recording media (3). In addition, it is important that the coercivity be high enough that information is not erased by weak magnetic fields.

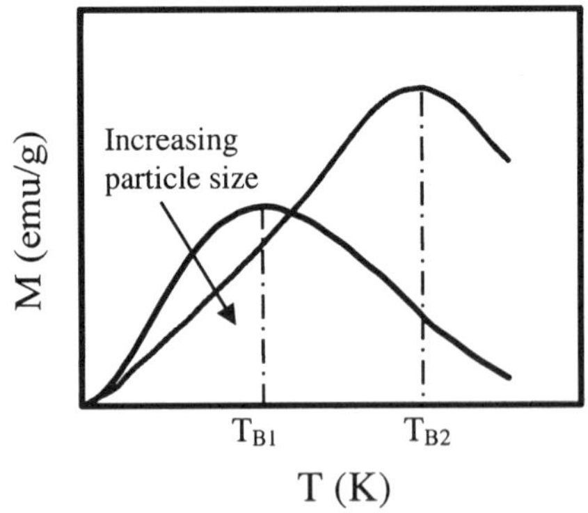

Figure 1 Magnetization as a function of temperature. The blocking temperature T_B is the transition point between ferromagnetic behavior and superparamagnetic behavior.

The size of magnetic particles is also important in ferrofluids, which were originally developed as high-performance seals for space applications (2). Ferrofluids consist of fine magnetic particles coated with a surfactant and suspended in a fluid. External magnets act on the suspended magnetic particles and trap the fluid in a target area. One of the keys to improving ferrofluids is to make the particles smaller and more uniform. Ferrofluids are currently used in sealing computer disk units as well as in vibrating environments in place of conventional seals (9). They have also been suggested for use in eye surgery to repair retinal tears (10).

The use of magnetic particles has also been proposed for targeted drug delivery, immunoassays, magnetic resonance contrast agents, and magnetic hyperthermia. For such applications, small size is an important consideration. However, it is also extremely important that the particles be superparamagnetic. Remnant magnetization could lead to agglomeration of particles due to magnetic attraction, which must obviously be avoided within the body as it could block blood vessels.

In targeted drug delivery (11), a drug is bound to the surface of a coated nanoparticle and an external magnetic field is applied to attract the particle to a specific site in the body. At the site, the chemical unbinds from the nanoparticle due to specific chemical interactions. The external magnetic field is then turned off and the nanoparticle freely circulates throughout the body until it is naturally eliminated. Current biomedical research is concentrated on the biocompatibility of the particles and possible coating materials (4,12–14).

Coated magnetic nanoparticles have also been suggested for use in immunoassays (15). By functionalizing the surface of a coated nanoparticle, molecules such as proteins can be bonded with that surface. These particles can be separated from the rest of the material by applying a magnetic gradient, thus separating the protein from the rest of the matrix. Another novel technique involves measuring the change in the magnetic relaxation of the nanoparticles and relating it to a binding reaction between a vitamin molecule attached to the nanoparticle and a protein present in the fluid around the particle (16). Essentially the technique takes advantage of the change in the magnetic properties of the particle due to a change in the thickness of the particle coating.

In magnetic resonance imaging (MRI), image contrast is improved by introducing fine magnetic particles into the tissue being imaged. Currently, paramagnetic metal ion complexes, such as Gd-EDTA, are used as contrast agents. However, superparamagnetic nanoparticles offer higher molar relaxivities, which would enhance image contrast (5,17,18). As with targeted drug delivery, these particles are coated with biocompatible materials to ensure their natural elimination from the body.

Magnetic nanoparticles have also been proposed for use in hyperthermia. Hyperthermia involves heating certain tissues or organs between 41°C and 46°C for cancer therapy (6). It induces reversible damage to cells and tissues that enhances the efficacy of radiation or chemotherapy. Conventional methods to induce hyperthermia cannot be used easily for shielded organs such as those located near the pelvis, and magnetic hyperthermia is one way to target these organs. Magnetic hyperthermia employs particles dispersed in the tissue which are heated by the application of an external AC magnetic field. One of the keys to this application is the homogeneous uptake of magnetic nanoparticles by the targeted organ or tissue. Jordan and coworkers are currently studying the uptake of dextran-coated magnetite particles by various cancerous tissues (6,19).

II. MAGNETIC MATERIALS

Metals have long been used in magnetic applications because they exhibit the best magnetic saturation and remanent magnetization (1). As a result, the magnetic properties of single transition metals (Fe, Co, Ni) and their alloys (Fe-Co, Fe-Ni) have been studied extensively. These properties can be varied as the composition of the particles is varied. However, metals suffer from corrosion problems, and their use requires further processing to passivate the surface of the particles (20).

Metal oxides, on the other hand, are not prone to corrosion. In particular, ferrites such as γ-Fe_2O_3 and $CoFe_2O_4$ have found use in magnetic recording media, although their coercivity and magnetic moments are not high enough for high-density applications. Decreasing the particle size and hence improving magnetic properties is likely to increase the applicability of these materials in such applications (3).

High coercivity, magnetic moment, and Curie temperature can be obtained with mixed oxides such as $CoFe_2O_4$. Such oxides belong to a class of spinel ferrites, MFe_2O_4, that exhibit ferrimagnetic properties (21). The crystal structure of spinels consists of two intertwined sublattices in which the magnetic moments are oriented in opposite directions (Figure 2) such that the material has a net magnetic moment. The magnetic properties of spinels can be manipulated by altering the composition of the crystal via substitution of other cations, e.g., manganese, zinc, and magnesium, into the lattice. The distribution of the divalent and trivalent cations between the tetrahedral and octahedral interstitial sites (normal, inverse, or random spinel) also changes with different cations and alters the net magnetic moment of the material. As applications that are very sensitive to magnetic properties emerge, the ability to tailor magnetic properties by composition and size offers some interesting opportunities.

A useful crystal structure is exhibited by barium hexaferrite ($BaO \cdot 6Fe_2O_3$). This material can be crystallized as hexagonal platelets in which the magnetic moment is oriented perpendicular to the flat surface. This makes barium hexaferrite useful for perpendicular recording technology. Barium hexaferrite also

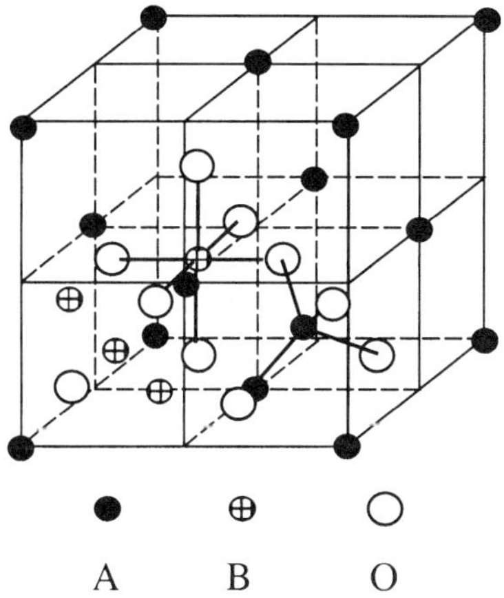

Figure 2 Partially filled unit cell of spinel AB_2O_4. **A** represents tetrahedrally coordinated sites and **B** represents octahedrally coordinated sites.

has a very high coercivity that can be tailored by substituting some of the iron cations with other cations, such as Zn^{2+} and Sn^{4+} (22).

Other materials that also exhibit magnetic properties include perovskite manganates, which have been studied because of their large magnetoresistance behavior (23,24). These materials have potential applications as magnetic sensors and magnetic recording heads. Improvements in magnetic recording heads will also expand the use of ferrites as magnetic recording materials for high-density applications.

III. SYNTHESIS OF FERRITE NANOPARTICLES

In the following section, we review a number of chemical synthesis techniques to produce ferrite nanoparticles. These techniques include coprecipitation, sol-gel, microemulsion, and spray pyrolysis methods. Advantages as well as disadvantages of each technique with respect to size and composition of nanoparticles are examined.

In general, the formation of spinel ferrites (MFe_2O_4) via one of these chemical synthesis techniques occurs by reacting metal salts with a strong base in an aqueous solution to form the spinel ferrite, salt, and water. For example,

$$2Fe(NO_3)_3 + Co(NO_3)_2 + 8NaOH \rightarrow CoFe_2O_4 + 8NaNO_3 + 4H_2O \quad (1)$$

The specific salts and base will, of course, differ in different studies. Reaction (1) proceeds via the formation of metal hydroxides, which are then dehydrated to form the mixed metal oxides. The intermediate hydroxides formed can be quite complex because of the rapid formation of polymeric networks (25). Nevertheless, their precipitation occurs rapidly, and essentially upon contact of the salts with the base. The dehydration rate, on the other hand, is not very rapid at room temperature although it does increase with increasing temperature. The thermodynamically favored product of reaction (1) is the spinel oxide, $CoFe_2O_4$. However, a number of other oxides can also be formed, depending on the relative solubilities and precipitation rates of the intermediate species.

It is very important to control the composition, size, size distribution, and crystallinity of the particles formed by reactions of the type shown above. All of the applications of magnetic particles require small, single-domain, well-crystallized particles. The biomedical applications require particles that are superparamagnetic at body temperature, whereas the information storage applications require that the particles not be superparamagnetic under their normal operating conditions. Each of the methods described below offers some advantages in controlling these properties. In general, water is the solvent of choice in these methods because (a) the reactants are soluble in water and (b) water is a byproduct of the reaction. In addition, the reaction is typically carried out with an excess of base to ensure the complete reaction of the metal salts and to provide a good environment for recrystallization. The strength of the base obviously affects the product composition, as does the relative solubility of the metal ions. Finally, such variables as time needed for growth and recrystallization will also affect the crystallization characteristics. These variables will therefore be emphasized in the description of the techniques described below.

A. Precipitation

Precipitation is the simplest technique for the preparation of spinel iron oxides. The metal salts are dissolved in water and added to a high-pH aqueous solution (26–30), which results in the formation of a dark precipitate consisting of hydroxides and oxides. Because the hydroxides form polymeric networks very rapidly and the dehydration rate is slow, this precipitate must be digested in the mother liquor for a lengthy period to allow for slow recrystallization and ripening. The particles produced are submicrometer in size, have a large size distribution, and tend to be agglomerated. Furthermore, the size and morphology of the product are very sensitive to variables such as pH and temperature. Coprecipitation has been used to synthesize ferrites such as γ-Fe_2O_3 (31,32), substituted barium ferrites (22), and ferrites imbedded in a polymer matrix (33,34).

Although iron oxides precipitate out of water readily, it is difficult to control the precipitation process to form uniform particles of the desired size. The variability in the local environment results in variability in the product. Coprecipitation techniques require strict control of many variables (pH, temperature, mixing times, etc.) and provide no guarantees that a uniform solution environment will be obtained in the precipitation reaction. In addition, a period of digestion is required to induce the recrystallization and formation of iron oxides from the initially formed hydroxides.

B. Sol-Gel Synthesis

Sol-gel synthesis is a traditional method for making fine ceramic particles. It involves forming an aqueous dispersion of oxide particles that is then "gelled" either by concentrating the dispersion by solvent removal or by carrying out a chemical reaction (35). For example, one method of sol-gel synthesis is to start with a metal alkoxide solution and add a small amount of water to control the hydrolysis and condensation of metal hydroxides. As the sol is dried, these metal hydroxides form a polymeric network through the cross-linking of metal oxygen bonds. The method of drying greatly influences the final product morphology. Supercritical drying has been shown to produce soft aggregates that can easily be broken down to a uniform powder (36). The resulting powder is typically subjected to heat treatment to induce the complete dehydration and crystallization of oxide particles. Another method of sol-gel synthesis is to start with a solution of a metal salt and a water-soluble polymer. By adding a base to this solution, the metal salt can be converted to metal hydroxides while the polymer cross-links to form a porous network around these metal hydroxides. Once again, heat treatment is required for dehydration of the hydroxides. In this case, the polymer network serves to prevent significant growth and aggregation of the metal hydroxides, so that it is possible to obtain nanoparticles using this technique. Nanoparticles of a number of ferrites including $CoFe_2O_4$ (37), $Ni_{0.5}Zn_{0.5}Fe_2O_4$-SiO_2 (38), $BaFe_{12}O_{19}$ (39), and $Ge_{0.5}Fe_{2.5}O_y$ (40) have been prepared using sol-gel synthesis.

C. Microemulsion Techniques

Microemulsions may also be used to limit the growth of particles in a solution. The coprecipitation reaction is carried out in a stable microemulsion formed by mixing metal salts, surfactant, organic solvent, and water (41). The metal cations attach to the hydrophilic end of the surfactant molecules and, upon addition of a strong base, the hydroxide ion reacts with the metal cations to precipitate the oxide. As in the case of the coprecipitation method, a period of digestion is required to obtain a crystalline product. However, growth of the particles is limited by the local environment in the individual micelles (42,43). A number of spinel ferrites have been made in microemulsions, including Fe_3O_4 and $CoFe_2O_4$ (41,43–50). Microemulsions have also been used in the formation of magnetic polymeric nanoparticles (50).

The microemulsion method can control the local environment and produce nanoparticles with a very narrow size distribution. On the other hand, agglomeration of particles is still a problem, and the amount of product formed per volume of solution is much smaller than can be obtained by other methods. In addition, separation of the product from the solution is more difficult.

D. Aerosol Spray Pyrolysis

Aerosol spray pyrolysis is another method used to control particle size by limiting the local environment around the particle nucleation sites. In aerosol spray pyrolysis, metal salt solutions are first aerosolized with a carrier gas and then heated in a high-temperature furnace to remove the solvent and form oxides. Heating metal salts to very high temperatures in air results in decomposition of the salts and subsequent formation of metal oxides.

The first particles formed via this method were found to consist of hollow spheres; later modifications of the apparatus include a dryer (to remove the solvent) before the high-temperature furnace (51). Experiments carried out with iron have shown good results in forming crystalline γ-Fe_2O_3, α-Fe_2O_3, and Fe_3O_4 depending on the amount of oxygen present in the system (51). In the case of the mixed system containing barium and iron, the resultant particles required annealing after particle formation (52). Similar results were shown at lower temperatures when barium citrate was used as the barium precursor (53). Another variation of this method incorporates a reacting gas, such as ammonia, after the solution has been aerosolized (54). Smaller, more uniform $NiFe_2O_4$ particles were formed with this approach.

Aerosol spray pyrolysis has the advantage of making particles very quickly. However, the composition and morphology of the particles, as well as their crystallinity, can be difficult to control. Variables such as solution concentration, oxygen concentration, presence of reacting gases, as well as temperature and residence time in the furnace significantly affect the final product.

IV. SYNTHESIS IN SUPERCRITICAL FLUIDS

Supercritical fluids (SCFs) offer the potential for a controlled solution environment because of the tunability of their properties by small changes in temperature and pressure. Indeed, near-critical water and supercritical water are obvious candidates as solvents in nanoparticle formation because water is the most commonly used solvent in conventional synthesis of inorganic particles. However, other solvents, such as carbon dioxide, can also be used. Several methods that take advantage of SCF behavior are described below. Not all have been employed in the production of magnetic nanoparticles. However, they represent a natural bridge between methods that are carried out mainly in the liquid state and those that are carried out in the gaseous state.

A. Hydrothermal Techniques

Hydrothermal synthesis in near-critical water has been used to make a number of metal oxide particles, including ferrites, using the precipitation reaction (1) (55). A variety of spinel ferrites and substituted barium ferrites have been prepared from metal chlorides, sulfates, and nitrates as described by Dias (56), Diamandescu (57), Komarneni (58), and Sugita (59). Other starting materials, such as metal hydroxides, oxides, or metal powders in combination with metal hydroxides, have also been used to prepare barium ferrites and spinel ferrites in a hydrothermal reaction environment (60–64). Although the particles produced are usually highly crystalline and do not require additional heat treatment, the product size and morphology cannot be controlled very well. Most of the particles produced via hydrothermal synthesis are micrometer-sized, in large part due to the long processing time required to form a well-crystallized product. There is also no means of limiting the growth of particles. Furthermore, the hot, compressed water environment favors crystal growth, which leads to larger size, since the solubilities of these materials are slightly higher at these conditions than at ambient conditions.

B. Solvothermal Techniques

The main difference between hydrothermal and solvothermal methods for the synthesis of ferrites is the substitution of some or all of the water with another solvent. The choice of solvent depends on the goal of the experiment. In the case of ferrite formation, the properties of the solvent are manipulated by adding alcohol, such as ethanol or isopropanol, to water (65). The alcohol influences the reaction path by providing a reducing atmosphere. This is advantageous when, for instance, it is desired to make Fe_3O_4 from $Fe(NO_3)_3$. By changing the solvent composition, the reaction as well as the solvent properties can be manipulated. However, once again, particle size is typically closer to a micrometer. While there are a number of other supercritical solvents that can be used, the formation of oxides requires a source of oxygen or hydroxide anions. Therefore, water and alcohols are best suited for this case.

C. Rapid Expansion of Supercritical Solutions

The rapid expansion of supercritical solutions (RESS) has found limited use in the area of metal oxide nanoparticle formation due to the very low solubility of transition metal oxides in supercritical fluids. However, the RESS technique has been used to produce particles of other metal oxides and organometallic complexes. Researchers have been able to produce submicrometer particles of SiO_2 and GeO_2 from supercritical water solutions (66–68) because SiO_2 and GeO_2

have relatively high solubilities in supercritical water. RESS has also been used to make fine particles of various metal complexes, such as iron pentacarbonyl (69), which are soluble in supercritical carbon dioxide. However, these precursor particles require further reaction or heat treatment to form the desired metal oxides.

D. Aerosol Spray Pyrolysis with SCF

A modified aerosol spray pyrolysis technique (70–72) has been reported in which an aqueous solution of metal salts is aerosolized by addition of supercritical carbon dioxide followed by expansion of the solution through a nozzle (Figure 3). To produce nanometer-size particles, the solution is expanded into a high-temperature furnace as in conventional spray pyrolysis. One of the benefits of using carbon dioxide is to reduce volatile organic compound (VOC) emissions by substituting CO_2 for organic solvents. Furthermore, in typical aerosol techniques, the aerosol agent can strip the solvent from the solution, leaving behind a more concentrated solution. With the use of carbon dioxide and aqueous solutions, the solutions are completely nebulized in the mixing tee and nozzle configuration. Because of the experimental design, a uniform solution is delivered throughout the process. Furthermore, carbon dioxide is more soluble in water than other gases used to make aerosols. While this solubility is still rather low, it is sufficient to cause the formation of very small droplets

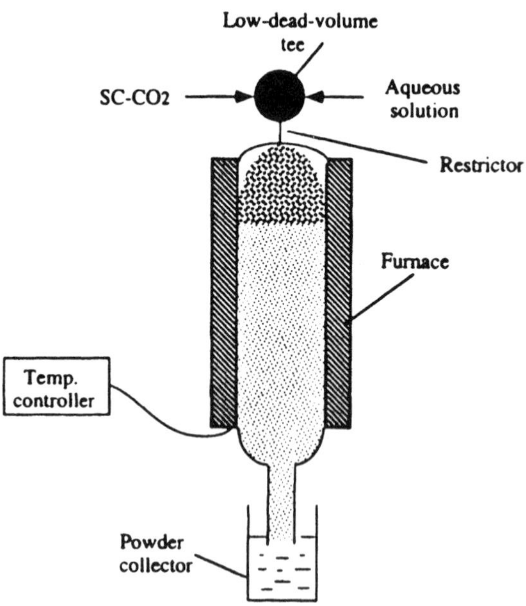

Figure 3 Aerosol spray pyrolysis with SCF. (From Ref. 72.)

upon depressurization of the solution across the nozzle. Particles formed using supercritical carbon dioxide were found to be smaller than those formed using standard aerosol gases (71).

E. Alkoxide Decomposition in SCF

Metal alkoxides can by decomposed in supercritical alcohols to produce various oxides, such as TiO_2 (73–75). Metal alkoxides also react readily with water to form hydroxide species that are easily dehydrated into oxides upon heating. As a solution of the alkoxides and alcohol is heated, some of the alcohol decomposes into water and aldehyde. The metal alkoxides react with the water formed and then undergo dehydration at the elevated temperatures. Compared with samples made via conventional sol-gel synthesis, powders synthesized in supercritical ethanol exhibit a higher degree of crystallinity and contain less hydroxide.

F. SCF Microemulsions

One of the first reactions carried out in a SCF micelle system was the formation of $Al(OH)_3$ from $Al(NO_3)_3$ and ammonia (76). This microemulsion system consisting of $Al(NO_3)_3$ dissolved in water, supercritical propane, and surfactant was contacted with dry ammonia to form $Al(OH)_3$. The ammonia, which is soluble in supercritical propane, diffused into the micelles and reacted with the $Al(NO_3)_3$ in the aqueous phase. Particle size was shown to be a function of concentration.

More recently, titanium dioxide particles have been produced using supercritical CO_2 (77). In this case, carbon dioxide is used as the organic phase of the microemulsion. Two containers were placed inside a pressure vessel with one container partially filled with an alkoxide precursor and the other with a solution of water and fluorinated surfactants. The vessel was then pressurized with carbon dioxide and heated until supercritical conditions were attained. The alkoxide is soluble in carbon dioxide, and the aqueous surfactant solution forms a stable microemulsion with the supercritical carbon dioxide. Over time, the alkoxides mix with the micelles and react with the water to form titanium dioxide. The product characteristics depend significantly on the solubility of the alkoxides in supercritical carbon dioxide.

V. FLOW PROCESS FOR THE PRODUCTION OF SPINEL IRON OXIDE

Methods that require long processing times generally yield a well-crystallized product. However, the particles are generally micrometer-sized rather than nanometer-sized. This can be prevented to some extent by restricting the growth environment, as in a micelle. SCFs allow the solution environment to be "tuned"; furthermore, a flow environment allows the growth processes to be curtailed. Therefore, flow processes that employ SCFs are examined in this section.

A flow technique was developed by Matson and coworkers to produce nanoparticles of catalytic materials using supercritical water (78–80). In this method, pressurized solutions of metal salts are fed through a heated region of tubing (Figure 4) where they react to form hydroxides, which are then rapidly dehydrated to form oxides. The solution is depressurized across a nozzle and the particles are collected in a suspension. The oxides are only sparingly soluble in supercritical water and therefore precipitate out of solution. Single oxides are relatively easy to form in this apparatus. However, certain oxides require the addition of urea to the starting solutions. As urea is heated, it decomposes to form ammonia and carbon dioxide, which react with the metal species. In the case of iron, Fe_2O_3 or iron hydroxides are formed without the addition of urea. With the addition of urea, Fe_3O_4 is formed. Urea was also necessary in the formation of $NiFe_2O_4$.

Another flow technique was developed by Adschiri and Arai, who explored the continuous hydrothermal synthesis of fine metal oxide particles using supercritical water (see Chapter 8). In this process, the aqueous metal salt solutions are rapidly heated by combining with a stream of supercritical water. After the

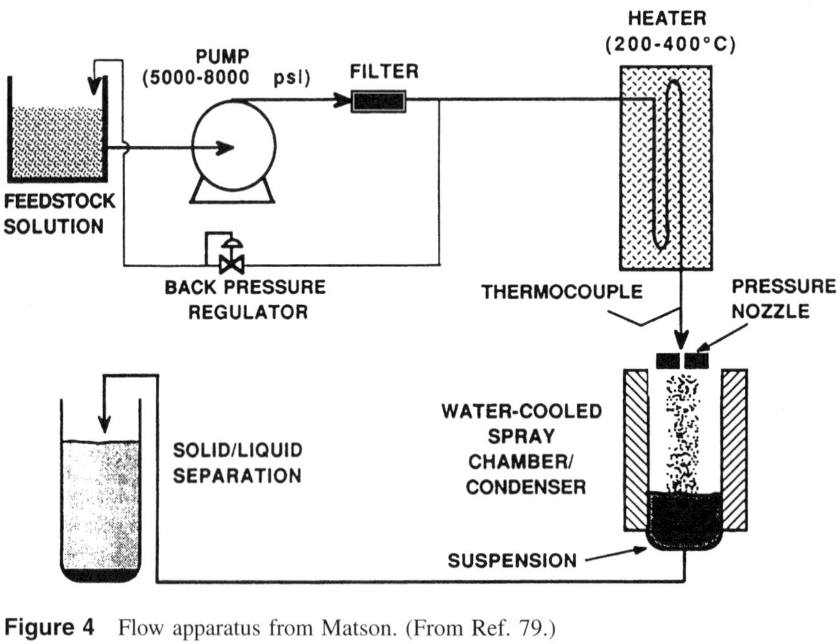

Figure 4 Flow apparatus from Matson. (From Ref. 79.)

heated solution passes through a reactor, the solution is cooled in a condenser. Larger particles are trapped in a filter upstream of the back-pressure regulator used to control the pressure. Finer particles are collected at the outlet of the back-pressure regulator. A number of modifications of this process have been described to control the reaction of the metal salts and form desired products. Nanoparticles of several types of metal oxides have been produced, including Fe_3O_4 and barium hexaferrite.

We have investigated the production of spinel iron oxides in an apparatus shown in Figure 5 that is similar to the one employed by Adschiri (81). Several cation solutions consisting of $Fe(NO_3)_3$ and/or $Co(NO_3)_3$ dissolved in water were reacted with sodium hydroxide solutions, and the conditions for the formation of oxide and mixed oxide nanoparticles were examined. The hydroxide and cation streams were first contacted in a tee and then combined with a hot, compressed water stream before being pumped through a heated tubular reactor. The flow rate of water was varied to maintain a set temperature (200°C, 300°C, or 400°C) in the reactor. The solution was then cooled and filtered using a 0.5-μm stainless steel filter. The filtrate was collected and any solids were washed and dried. Acidic suspensions of the particles were also collected to prepare transmission electron microscope (TEM) samples. Experiments were conducted with and without the addition of sodium hydroxide. In the mixed oxide experiments, the ratio of iron to cobalt concentration was set at the stoichiometric ratio of 2:1. An excess of sodium hydroxide was used to ensure complete reaction of the metal cations.

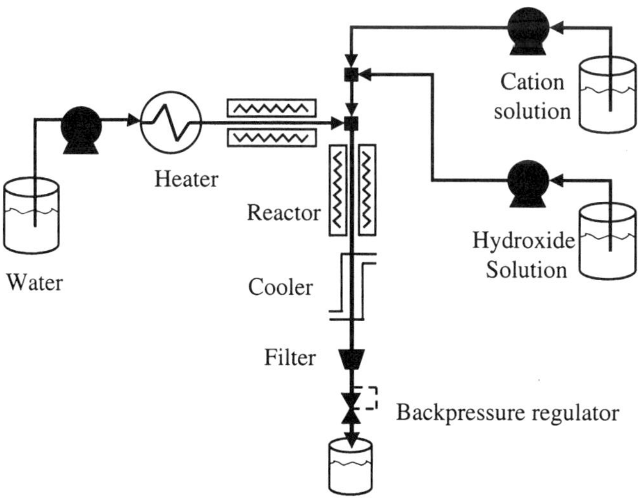

Figure 5 Schematic of experimental apparatus.

A. Single-Oxide Results

Single oxides were studied so as to eliminate the role of relative solubility (of the two precipitating cations) in the experiments. Fe^{3+} solutions were reacted with and without sodium hydroxide, and the resulting particles are shown in Figure 6. The particles were identified via powder x-ray diffraction to be Fe_2O_3 when $Fe(NO_3)_3$ was reacted with or without the addition of hydroxide. Changing the temperature from 200°C to 400°C did not have a discernible effect on the species precipitated. It is important to emphasize that Fe^{3+} precipitates out of these high-temperature aqueous solutions as Fe_2O_3 even without the addition of sodium hydroxide.

In the case of $Co(NO_3)_3$ at 200°C, the powder was shown to consist of a mixture of Co_3O_4 and CoO(OH). At the higher temperatures, 300°C and 400°C, only Co_3O_4 particles were produced (Figure 7). The presence of a cobalt hydroxide in the low-temperature sample indicates that dehydration was incomplete. Thus, Co^{2+} did not precipitate out of solution unless sodium hydroxide was present, and then only at the higher temperatures.

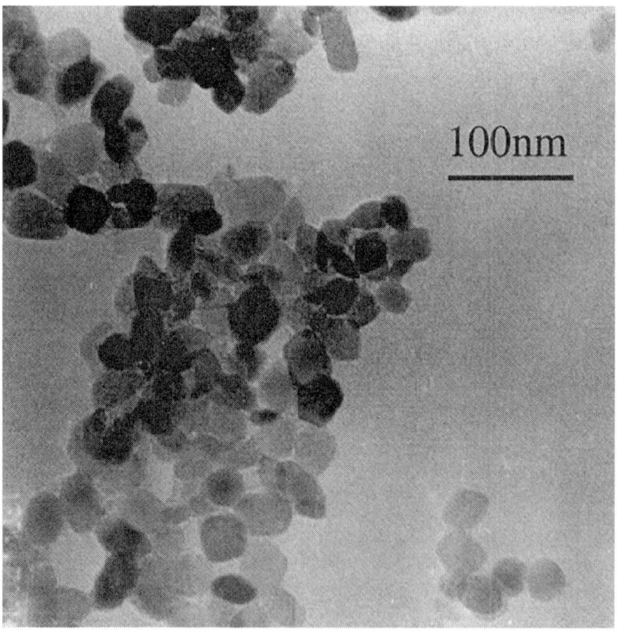

Figure 6 Transmission electron micrograph of Fe_2O_3 particles produced from $Fe(NO_3)_3$.

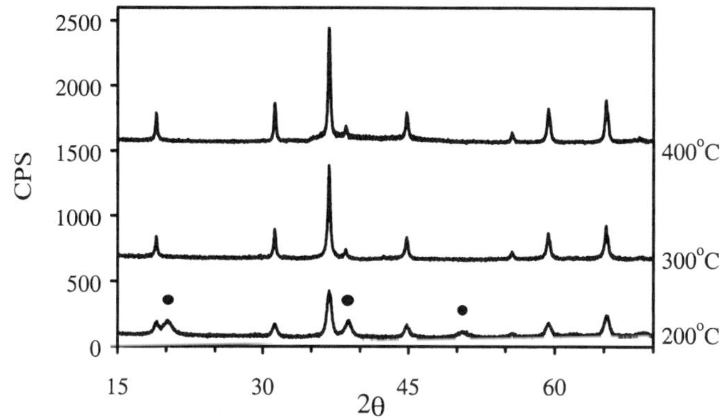

Figure 7 Powder x-ray diffraction patterns of cobalt samples. ●, (CoO)OH peaks.

Particle sizes were calculated from powder x-ray diffraction data and are shown in Table 1. Unfortunately, it is difficult to draw conclusions concerning the affect of temperature on particle size because residence times were different in these experiments. However, it is interesting to note that the particle size of Fe_2O_3 increased significantly with the addition of sodium hydroxide. Sodium

Table 1 Summary of Experimental Results

Cation feed	Hydroxide feed	Reactor temp. (°C)	Residence time (s)	Particle size (nm)
$Fe(NO_3)_3$	—	202	27	44
$Fe(NO_3)_3$	—	300	23	—
$Fe(NO_3)_3$	—	389	7	43
$Fe(NO_3)_3$	Excess	202	20	69
$Fe(NO_3)_3$	Excess	295	20	70
$Fe(NO_3)_3$	Excess	389	7	65
$Co(NO_3)_2$	Excess	200	20	
$Co(NO_3)_2$	Excess	295	18	38
$Co(NO_3)_2$	Excess	383	10	38
$Fe(NO_3)_3 + Co(NO_3)_2$	Excess	200	20	12
$Fe(NO_3)_3 + Co(NO_3)_2$	Excess	300	19	13
$Fe(NO_3)_3 + Co(NO_3)_2$	Excess	383	8	14

hydroxide reacts with the ferric salt to yield iron hydroxides earlier in the flow process. These hydroxide particles then have more time to grow before they are dehydrated.

B. Mixed-Oxide Results

Powder x-ray diffraction patterns showed that the mixed oxides in our experiments were predominantly $CoFe_2O_4$ together with a very small amount of Fe_2O_3 (Figure 8). Also, the amount of Fe_2O_3 decreased with temperature, although the effect of temperature could not be separated from the residence time effect in this set of experiments. Table 1 shows that the particle sizes of the mixed oxides are significantly smaller than those of the single oxides. The magnetic properties of the mixed samples were measured and compared with the properties of $CoFe_2O_4$ nanoparticles prepared using microemulsion techniques and were found to be similar.

C. Discussion

Powder x-ray diffraction patterns and TEM micrographs (Figure 9) clearly show that very fine, uniform particles can be produced in a hot, compressed-water environment. The particles are generally well formed and crystalline. However, their composition depends on the effect of the solution environment on the relative solubility of the precipitating cations. The presence of Fe_2O_3 in the mixed oxide samples indicates that not all of the iron is reacting to form the thermodynamically favored product $CoFe_2O_4$. The single oxide results also show that there is a significant solubility difference between the Fe^{3+} and Co^{2+} species in near-critical water, so that the formation of iron hydroxide, and therefore

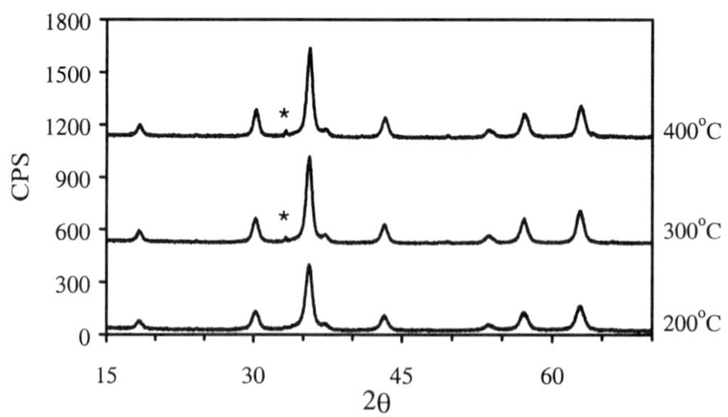

Figure 8 Powder x-ray diffraction patterns of mixed samples. *, Fe_2O_3 peaks.

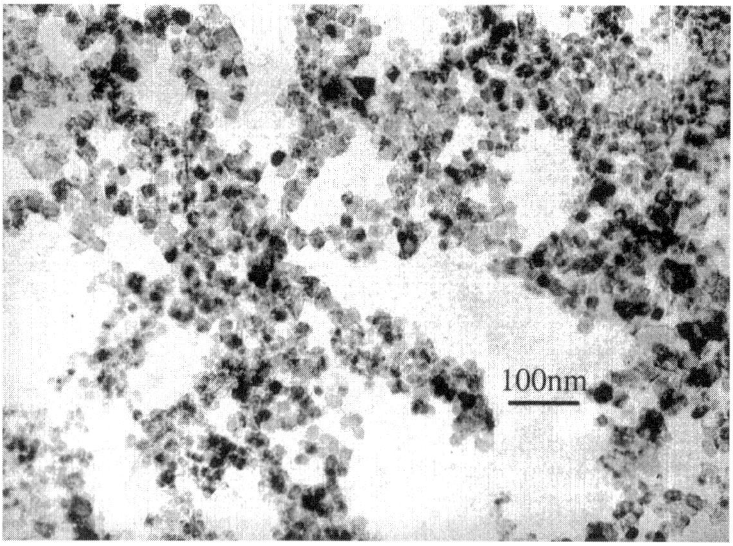

Figure 9 Transmission electron micrograph of particles produced in flow experiment from $Co(NO_3)_2$ and $Fe(NO_3)_3$ reacted with NaOH.

iron oxide, proceeds more easily than in the case of cobalt. The need to always add sodium hydroxide in order to precipitate cobalt indicates that the Co^{2+} species are much more soluble in near-critical water and that it is necessary to shift the equilibrium to the formation of cobalt hydroxide. The predominant product of the cobalt experiments was Co_3O_4, which requires that the cobalt undergo oxidation. In the mixed-cation case, the cobalt is not oxidized; therefore it easily incorporates with the iron to form $CoFe_2O_4$. The solubility of the Fe^{3+} species is so low that even with excess hydroxide some Fe_2O_3 is formed. Therefore, even though $CoFe_2O_4$ is the thermodynamically favored product, the relative solubility of the cations has a major effect on the product composition.

The particles produced in the single-oxide experiments were larger than desired, whereas the mixed-oxide particles were of the appropriate size for attaining superparamagnetic behavior at reasonable temperatures. The size distribution of the particles produced is also very narrow, due to the limited time spent in the reactor and uniform residence time for all of the particles. The hot, compressed-water environment therefore shows promise of providing nanoparticles of the desired size and composition.

One aspect of this process that has not been mentioned is the possible occurrence of reduction–oxidation reactions involving the metal cations. This is of little concern in the systems described above because cobalt is relatively stable in near-critical water. However, other divalent cations, such as Mn^{2+}, would be more susceptible to oxidation in these experiments. Reduction–oxidation potentials are a function of temperature and pH, so that the solution environment must be adjusted via pH and temperature to favor the desired species. In addition, the ability of supercritical water to provide a good oxidation environment in the presence of dissolved oxygen must be overcome.

VI. CONCLUSIONS

Supercritical fluids have already been shown to be useful in the production of particles, including those of metal oxides and other metal species. The near-critical and supercritical fluid environment therefore offers a means for the control of size and composition of mixed oxide nanoparticles. Supercritical fluids also have the potential for providing facile separation of impurities from desired species, such as $CoFe_2O_4$. The role of processing variables such as temperature, pressure, pH, residence time, and relative solubility has been discussed briefly in this work. However, a complete understanding of the reaction and subsequent particle formation will require much further work. Nevertheless, the examples discussed in this work clearly demonstrate the potential of near-critical and supercritical fluids in this area.

REFERENCES

1. G Bate. Magnetic recording since 1975. J Magn Magn Mater 1991;100:413.
2. RE Rosensweig. Magnetic fluids. Sci Am 1982;247:136.
3. DE Speliotis. Magnetic recording beyond the first 100 Years. J Magn Magn Mater 1999;193:29.
4. C Bergemann, D Müller-Schulte, J Oster, L à Brassard, AS Lübbe. Magnetic ion-exchange nano- and microparticles for medical, biochemical and molecular biological applications. J Magn Magn Mater 1999;194:45.
5. JWM Bulte, M de Cuyper, D Despres, JA Frank. Preparation, relaxometry, and biokinetics of PEGylated magnetoliposomes as MR contrast agent. J Magn Magn Mater 1999;194:204.
6. A Jordan, R Scholz, P Wust, H Schirra, T Schiestel, H Schmidt, R Felix. Endocytosis of dextran and silan-coated magnetite nanoparticles and the effect of intracellular hyperthermia on human mammary carcinoma cells in vitro. J Magn Magn Mater 1999;194:185.
7. DL Leslie-Pelecky, RD Rieke. Magnetic properties of nanostructured materials. Chem Mater 1996;8:1770.

8. C Liu, B Zou, AJ Rondinone, ZJ Zhang. Chemical control of superparamagnetic properties of magnesium and cobalt spinel ferrite nanoparticles through atomic level magnetic couplings. J Am Chem Soc 2000;122:6263.
9. M Ozaki. Synthesis of Magnetic Particles. In: Chemical Processing of Ceramics; Lee, B.I., Pope, E.J.A., Eds.; Marcel Dekker, New York, 1994.
10. JP Dailey, JP Phillips, C Li, JS Riffle. Synthesis of silicone magnetic fluid for use in eye surgery. J Magn Magn Mater 1999;194:140.
11. AS Lübbe, C Bergemann, J Brock, DG McClure. Physiological aspects in magnetic drug-targeting. J Magn Magn Mater 1999;194:149.
12. OA Kuznetsov, NA Brusentsov, AA Kuznetsov, NY Yurchenko, NE Osipov, FS Bayburtskiy. Correlation of coagulation rates and toxicity of biocompatible ferromagnetic nanoparticles. J Magn Magn Mater 1999;194:83.
13. ZGM Lacava, RB Azevado, LM Lacava, EV Martins, VAP Garcia, CA Rébula, APC Lemos, MH Sousa, FA Tourinho, PC Morais, MF Da Silva. Toxic effects of ionic magnetic fluids in mice. J Magn Magn Mater 1999;194:90.
14. ZGM Lacava, RB Azevado, EV Martins, LM Lacava, MLL Freitas, VAP Garcia, CA Rébula, APC Lemos, MH Sousa, FA Tourinho, MF Da Silva, PC Morais. Biological effects of magnetic fluids: toxicity studies. J Magn Magn Mater 1999; 201:431.
15. C Grüttner, J Teller. New types of silica-fortified magnetic nanoparticles as tools for molecular biology applications. J Magn Magn Mater 1999;194:8.
16. R Kötitz, W Weitschies, L Trahms, W Brewer, W Semmler. Determination of the biding reaction between avidin and biotin by relaxation measurements of magnetic nanoparticles. J Magn Magn Mater 1999;194:62.
17. JWM Bulte, RA Brooks, BM Moskowitz, LH Bryant, Jr., JA Frank. Relaxometry, magnetometry, and EPR evidence for three magnetic phases in the MR contrast agent MION-46L. J Magn Magn Mater 1999;194:217.
18. CW Jung, JM Rogers, EV Groman. Lymphatic mapping and sentinel node location with magnetite nanoparticles. J Magn Magn Mater 1999;194:210.
19. A Jordan, R Scholz, P Wust, H Fähling, R Felix. Magnetic fluid hyperthermia (MFH): Cancer treatment with AC magnetic field induced excitation of biocompatible superparamagnetic nanoparticles. J Magn Magn Mater 1999;201:413.
20. M Jacoby. Data Storage: New materials push the limits. Chem Eng News June 12, 2000; 37.
21. B Viswanathan. Chemistry of Ferrites In: Ferrite Materials: Science and Technology; Viswanathan, B., Murthy, V.R.K., Eds. Springer-Verlag, New York, 1990.
22. HC Fang, Z Yang, CK Ong, Y Li, CS Wang. Preparation and magnetic properties of (Zn-Sn) substituted barium hexaferrite nanoparticles for magnetic recording. J Magn Magn Mater 1998;187:129.
23. F Rivadulla, LE Hueso, J Rivas, MC Blanco, MA López-Quintela, RD Sánchez. Effects of electrochemical reduction on the magnetotransport properties of $La_{0.67}Ca_{0.33}MnO_{3\pm\delta}$ nanoparticles. J Magn Magn Mater 1999;203:253.
24. C Vázquez-Vázquez, MA López-Quintela, RD Sánchez, D Caeiro, J Rivas, SB Oseroff. Characterization of sol-gel nanoparticles of magnetoresistive $La_{0.67}Ca_{0.33}MnO_{3+\delta}$ Mater Sci For 1998;278–281:606.

25. MA Blesa, E Matijevic. Phase transformations of iron oxides, oxohydroxides, and hydrous oxides in aqueous media. Adv Colloid Interf Sci 1989;29:173.
26. E Auzans, D Zins, E Blums, R Massart. Synthesis and properties of Mn-Zn ferrite ferrofluids. J Mater Sci 1999;34:1253.
27. JP Chen, CM Sorensen, KJ Klabunde, GC Hadjipanayis, E Devlin, A Kostikas. Size-dependent magnetic properties of $MnFe_2O_4$ fine particles synthesized by coprecipitation. Phys Rev B 1996;54:9288.
28. Q Chen, AJ Rondinone, BC Chakoumakos, ZJ Zhang. Synthesis of superparamagnetic $MgFe_2O_4$ nanoparticles by coprecipitation. J Magn Magn Mater 1999; 194:1.
29. E Hasmonay, J Depeyrot, MH Sousa, FA Tourinho, J-C Bacri, R Perzynski. Optical properties of nickel ferrite ferrofluids. J Magn Magn Mater 1999;201:195.
30. S Lefebure, E Dubois, V Cabuil, S Neveu, R Massart. Monodisperse magnetic nanoparticles: Preparation and dispersion in water and oils. J Mater Res 1998;13: 2975.
31. MP Morales, M Andres-Vergés, S Veintemillas-Verdaguer, MI Montero, CJ Serna. Structural effects on the magnetic properties of γ-Fe_2O_3 nanoparticles. J Magn Magn Mater 1999;203:146.
32. D Prodan, C Chanéac, E Tronc, JP Jolivet, R Cherkaour, A Ezzir, M Noguès, JL Dormann. Adsorption phenomena and magnetic properties of γ-Fe_2O_3 nanoparticles. J Magn Magn Mater 1999;203:63.
33. DY Godovsky, AV Varfolomeev, GD Efremova, VM Cherepanov, GA Kapustin, AV Volkov, MA Moskvina. Magnetic properties of polyvinyl alcohol-based composites containing iron oxide nanoparticles. Adv Mater Opt Electron 1999;9:87.
34. M Kryszewski, JK Jeszka. Nanostructured conducting polymer composites—superparamagnetic particles in conducting polymers. Synth Met 1998;94:99.
35. JDF Ramsay. Sol-gel processing. In: Controlled Particle, Droplet and Bubble Formation; Wedlock, D.J., Ed.; Butterworth-Heinemann, Oxford, 1994.
36. ZS Hu, JX Dong, GX Chen. Preparation of nanometer titanium oxide with n-butanol supercritical drying. Powder Technol 1999;101:205.
37. T Sugiomoto, Y Shimotsuma, H Itoh. Synthesis of uniform cobalt ferrite particles from a highly condensed suspension of β-FeOOH and β-$Co(OH)_2$ particles. Powder Technol 1998;96:85.
38. AS Albuquerque, JD Ardisson, WAA Macedo. A study of nanocrystalline NiZn-ferrite-SiO_2 synthesized by sol-gel. J Magn Magn Mater 1999;192:277.
39. W Zhong, W Ding, Y Jiang, N Zhang, J Zhang, Y Du, Q Yan. Preparation and magnetic properties of barium hexaferrite nanoparticles produced by the citrate process. J Am Ceram Soc 1997;80:3258.
40. HH Hamdeh, K Barghout, JC Ho, RJ Willey, MJ O'Shea, J Chaudhuri. Structural and magnetic characterization of aerogel-produced $Ge_{0.5}Fe_{2.5}O_y$ nanoparticles. J Magn Magn Mater 2000;212:112.
41. JA López Pérez, MA López Quintela, J Mira, J Rivas, SW Charles. Advances in the preparation of magnetic nanoparticles by the microemulsion method. J Phys Chem B 1997;101:8045.

42. S Bandow, K Kimura, K Kon-no, A Kitahara. Magnetic properties of magnetite ultrafine particles prepared by W/O microemulsion method. Jpn J Appl Phys 1987; 26:713.
43. V Pillai, DO Shah. Synthesis of high-coercivity cobalt ferrite particles using water-in-oil microemulsions. J Magn Magn Mater 1996;163:243.
44. CT Seip, EE Carpenter, CJ O'Connor, VT John, S Li. Magnetic properties of a series of ferrite nanoparticles synthesized in reverse micelles. IEEE Trans Magn 1998;34:1111.
45. AT Ngo, MP Pileni. Nanoparticles of cobalt ferrite: influence of the applied field on the organization of the nanocrystals on a substrate and on their magnetic properties. Adv Mater 2000;12:276.
46. II Yaacob, AC Nunes, A Bose. Magnetic nanoparticles produced in spontaneous cationic-anionic vesicles: room temperature synthesis and characterization. J Colloid Interf Sci 1995;171:73.
47. GX Cheng, F Shen, LF Yang, LR Ma, Y Tang, KD Yao, PC Sun. On properties and structure of the AOT-water-isooctane reverse micellar microreactor for nanoparticles. Mater Chem Phys 1998;56:97.
48. L Shen, PE Laibinis, TA Hatton. Bilayer surfactant stabilized magnetic fluids: Synthesis and interactions at interfaces. Langmuir 1999;15:447.
49. AJ Rondinone, ACS Samia, ZJ Zhang. Superparamagnetic relaxation and magnetic anisotropy energy distribution in $CoFe_2O_4$ spinel ferrite nanocrystallites. J Phys Chem B 1999;103:6876.
50. PA Dresco, VS Zaitsev, RJ Gambino, B Chu. Preparation and properties of magnetite and polymer magnetite nanoparticles. Langmuir 1999;15:1945.
51. GC Hadjipanayis, S Gangopadhyay, L Yiping, CM Sorensen, KJ Klabunde. Ultrafine magnetic particles. In: Science and Technology of Nanostructured Magnetic Materials; Hadjipanayis, G.C., Prinz, G.A.; Eds.; Plenum Press, New York, 1991.
52. GC Hadjipanayis, ZX Tang, S Gangopadhyay, L Yiping, CM Sorensen, KJ Klabunde, A Kostikas, V Papaefthymiou. Preparation of fine particles. In: Studies of Magnetic Properties of Fine Particles and their Relevance to Materials Science; Dormann, J.L., Fiorani, D., Eds.; Elsevier Science, New York, 1992.
53. T González-Carreño, MP Morales, CJ Serna. Barium ferrite nanoparticles prepared directly by aerosol pyrolysis. Mater Lett 2000;43:97.
54. H-F Yu, AM Gadalla. Preparation of $NiFe_2O_4$ powder by spray pyrolysis of nitrate aerosols in NH_3. J Mater Res 1996;11:663.
55. A Rabenau. The role of hydrothermal synthesis in preparative chemistry. Angew Chem Int Ed Engl 1985;24:1026.
56. A Dias, VTL Buono. Hydrothermal synthesis and sintering of nickel and manganese-zinc ferrites. J Mater Res 1997;12:3278.
57. L Diamandescu, D Mihăilă-Tărăbăşanu, V Teodorescu, N Popescu-Pogrion. Hydrothermal synthesis and structural characterization of some substituted magnetites. Mater Lett 1998;37:340.
58. S Komarneni, E Fregeau, E Breval, R Roy. Hydrothermal preparation of ultrafine ferrites and their sintering. J Am Ceram Soc 1988;71:C26.

59. N Sugita, M Maekawa, Y Ohta, K Okinaka, N Nagai. Advances in fine magnetic particles for high density recording. IEEE Trans Magn 1995;31:2854.
60. X Yi, Q Yitai, L Jing, C Zuyao, Y Li. Hydrothermal preparation and characterization of ultrafine powders of ferrite spinels MFe_2O_4 (M = Fe, Zn and Ni). Mater Sci Eng 1995;B34:L1.
61. H Kumazawa, H-M Cho, E Sada. Hydrothermal synthesis of barium ferrite fine particles from goethite. J Mater Sci 1993;28:5247.
62. M-L Wang, Z-W Shih, C-H Lin. Reaction mechanism producing barium hexaferrites from goethite and barium hydroxide by hydrothermal method. J Crystal Growth 1993;130:153.
63. M-L Wang, Z-W Shih, C-H Lin. Reaction of α-Fe_2O_3 with $Ba(OH)_2$ under hydrothermal conditions. J Crystal Growth 1994;139:47.
64. A Ataie, MR Piramoon, IR Harris, CB Ponton. Effect of hydrothermal synthesis environment on the particle morphology, chemistry and magnetic properties of barium hexaferrite. J Mater Sci 1995;30:5600.
65. T Dubois, G Demazeau. Preparation of Fe_3O_4 fine particles through a solvothermal process. Mater Lett 1994;19:38.
66. DW Matson, RC Petersen, RD Smith. Formation of silica powders from the rapid expansion of supercritical solutions. Adv Ceram Mater 1986;1:242.
67. DW Matson, RC Petersen, RD Smith. Production of powders and films by the rapid expansion of supercritical solutions. J Mater Sci 1987;22:1919.
68. DW Matson, RD Smith. Supercritical fluid technologies for ceramic-processing applications. J Am Ceram Soc 1989;72:871.
69. JR Williams, AA Clifford, KD Bartle, TP Kee. The production of fine particles of metal complexes using supercritical fluids. Powder Technol 1998;96:158.
70. C Xu, A Watkins, RE Sievers, X Jing, P Trowga, CS Gibbons, A Vecht. Submicron-sized spherical yttrium oxide based phosphors prepared by supercritical CO_2-assisted aerosolization and pyrolysis. Appl Phys Lett 1997;12:1643.
71. RE Sievers, U Karst. US Patent No. 5,639,441, June 17, 1997.
72. C Xu, RE Sievers, U Karst, BA Watkins, CM Karbiwnyk, WC Anderson, JD Schaefer, CR Stoldt. Supercritical carbon dioxide assisted aerosolization for thin film deposition, fine powder generation, and drug delivery. In: Green Chemistry; Frontiers in Benign Chemical Syntheses and Processes; Anastas, P.T., Williamson, T.C., Eds.; Oxford University Press, New York, 1998.
73. C Pommier, K Chhor, JF Bocquet, M Barj. Reactions in supercritical fluids: a new route for oxide ceramic powder elaboration. Synthesis of the spinel $MgAl_2O_4$. Mater Res Bull 1990;25:213.
74. V Gourinchas Courtecuisse, JF Bocquet, K Chhor, C Pommier. Modeling of a continuous reactor for TiO_2 powder synthesis in a supercritical fluid—experimental validation. J Supercrit Fluids 1996;9:222.
75. K Chhor, JF Bocquet, C Pommier. Synthesis of submicron TiO_2 powders in vapor, liquid and supercritical phases, a comparative study. Mater Chem Phys 1992;32: 249.
76. DW Matson, JL Fulton, RD Smith. Formation of fine particles in supercritical fluid micelle systems. Mater Lett 1987;6:31.

77. ME Tadros, CLJ Adkins, EM Russick, MP Youngman. Synthesis of titanium dioxide particles in supercritical CO_2. J Supercrit Fluids 1996;9:172.
78. DW Matson, JC Linehan, JG Darab, MF Buehler, MR Phelps. Neuenschwander a flow-through hydrothermal method for the synthesis of active nanocrystalline catalysts. In: Advanced Catalysts and Nanostructured Materials: Modern Synthetic Methods; Moser, W.R., Ed.; Academic Press, San Diego, 1996.
79. DW Matson, JC Linehen, JG Darab, MF Buehler. Nanophase iron-based liquefaction catalysts: synthesis, characterization, and model compound reactivity. Energy and Fuels 1994;8:10.
80. DW Matson, JL Fulton, JC Linehan, RM Bean, TD Brewer, TA Werpy, JG Darab. US Patent No. 5,652,192, July 29, 1997.
81. T Adschiri, K Arai. Chapter 8 in this text.

10

Metal Processing in Supercritical Carbon Dioxide

Chien M. Wai
University of Idaho, Moscow, Idaho

I. INTRODUCTION

In the past two decades, supercritical fluid (SCF) extraction technology has attracted considerable attention from chemists and engineers for its potential applications as an environmentally friendly solvent for chemical processing (1,2). Supercritical fluids exhibit gas-like mass transfer properties yet have liquid-like solvation capabilities. Because of its high diffusivity and low viscosity, an SCF is capable of penetrating into porous solid materials to dissolve organic compounds. Since the density of an SCF can be altered continuously by manipulating pressure and temperature, the solvation strength of the fluid is tunable. Thus, selective dissolution of certain groups of solutes in an SCF may be achieved by optimizing density of the fluid phase. This tunable solvation capability is a unique property that makes SCFs different from conventional liquids. Another major advantage of SCF extraction is rapid separation of solutes that can be easily achieved by reduction of pressure. Carbon dioxide is one of the most widely used gas for SCF applications because of its moderate critical constants ($T_c = 31.1°C$, $P_c = 72.8$ atm, and $\varphi_c = 0.47$ g cm^{-3}), nontoxic nature, low cost, and availability in pure form. Examples of large-scale industrial applications of the SCF CO_2 extraction technology include the preparation of decaffeinated coffee and hop extracts.

Solvent extraction associated with acid or base dissolution is commonly used in industrial operations for separation, processing, and cleaning of metals. These industrial operations often generate large quantities of liquid wastes creating environmental problems for handling and disposal of used solvents.

Development of environmentally friendly dissolution/extraction techniques for metal processing is highly desirable due to changing environmental regulations. Supercritical CO_2 is considered an environmentally friendly solvent for industrial processes. One difficulty of using this green solvent for metal dissolution is that metal ions are not soluble in supercritical CO_2 because of the charge neutralization requirement and the weak interactions between CO_2 and metal ions. However, when metal ions are bound to organic ligands, they may become quite soluble in supercritical CO_2 (3). Systematic studies of metal chelate solubilities in supercritical CO_2 started in the early 1990. Wai and coworkers first reported quantitative measurements of metal dithiocarbamate solubilities in supercritical CO_2 in 1991 (4). In this report, the authors showed that fluorine substitution in a dithiocarbamate chelating agent could enhance the solubility of its metal chelates by two to three orders of magnitude relative to the nonfluorinated analogues. The same authors demonstrated in 1992 that copper ions spiked in liquid and solid samples could be effectively extracted by supercritical CO_2 containing the fluorinated chelating agent bis(trifluoroethyl)dithiocarbamate (5). Since then, more than 80 journal publications and a number of review articles related to SCF extraction of metal species have appeared in the literature (6–11). A variety of chelating agents, including dithiocarbamates, β-diketones, organophosphorous reagents, and macrocyclic ligands, have been tested for extraction of metal species from different matrixes (9). These studies have provided a basis for our understanding of metal dissolution processes in supercritical CO_2.

A new development in dissolution of metal species in supercritical CO_2 is the discovery that electrolytes, such as metal salts, can be stabilized in a water-in-supercritical CO_2 microemulsion (12,13). The water-in-CO_2 microemulsions with diameters typically on the order of several nanometers allow metal species and high polar compounds to be dispersed in nonpolar supercritical CO_2. This type of microemulsion may be regarded as a new solvent for dissolution and transport of metal species in SCFs. It has also been shown that the water-in-CO_2 microemulsion can be used as a reactor for chemical synthesis of nano-sized materials in SCFs. The first report on the synthesis of nano-sized metal particles in a water-in-CO_2 microemulsion appeared in 1999 (13). In this report, Wai and coworkers showed that silver ions in the water core of a water-in-supercritical CO_2 microemulsion could be reduced to nano-sized metallic silver particles by a reducing agent dissolved in the fluid phase. Further reports from this research group indicate that by mixing two microemulsions containing different ions in the water cores, exchange of ions can take place leading to chemical reactions (14). For example, by mixing two water-in-supercritical CO_2 microemulsions, one containing $AgNO_3$ and the other NaI in the water core, formation of AgI nanoparticles was observed. These reports suggest the feasibility of utilizing the water-in-CO_2 microemulsions as nanoreactors for synthesizing a variety of nanoparticles in supercritical CO_2. The water-in-CO_2 microemulsion has also been shown to be an effective medium for conducting voltammetry and electrolysis in supercritical CO_2 (15). The results imply that the microemulsion may even be considered as a medium for various electrosyntheses in SCFs.

This chapter presents some of the new developments in metal dissolution using supercritical CO_2 as a solvent and explores potential applications of the technology for metal processing in the future. Examples given in this chapter include potential applications of the SCF technology for the nuclear industry and for synthesis of nano-sized materials.

II. SOLUBILITY OF METAL COMPLEXES IN SUPERCRITICAL CO_2

Solubility of solutes in supercritical CO_2 is perhaps the most important factor in determining the efficiency of a SCF extraction process. The following three techniques are commonly used for determining solubility of solutes in SCFs: gravimetric, chromatographic, and spectroscopic methods. Spectroscopic methods generally offer rapid determination of solubility, with increased sensitivity, and require small amounts of samples. If a metal complex has characteristic absorption bands in the ultraviolet-visible (UV-Vis) region, spectroscopic method is often a good choice for determining its solubility in supercritical CO_2. Time-resolved laser-induced fluorescence (TRLIF) has also been used for determination of solubility of uranium and lanthanide complexes in supercritical CO_2 (16). The TRLIF method is highly sensitive but is limited to certain metals that have relatively long fluorescence lifetimes. Brief descriptions of these spectroscopic techniques for determination of solubilities of metal complexes in supercritical CO_2 are given in the next section.

A. Spectroscopic Methods for Solubility Measurement

A stainless steel high-pressure view cell with quartz windows was used originally by Laintz et al. for measuring the solubilities of a number of metal dithiocarbamate complexes in supercritical CO_2 (4). The view cell containing CO_2 and a dissolved metal complex was placed in a Cary 1 UV-Vis spectrometer and the absorbance of the metal complex at a selected wavelength was measured. The temperature and pressure of the fluid phase were monitored using a thermal couple and a pressure gauge. The concentration of the metal complex in the supercritical CO_2 phase was determined from the absorbance using the Beer–Lambert law. One problem of using a view cell for solubility measurement is its fixed pathlength (usually in the order of several centimeters) which limits the concentration range of the solubility measurement. For highly soluble metal complexes, the absorbance may be out of the linear range of the

Beer–Lambert law. Recently, a high-pressure fiberoptic system was described in the literature for measurement of solubility of metal complexes in supercritical CO_2 (17). The fiberoptic system, consisting of three fiberoptic cells with pathlength ranging form 38 μm to 1 cm, enabled compounds of high or low solubility to be measured over a concentration range of several orders of magnitude (Figure 1). The fiberoptic cells were housed in a high-performance liquid chromatography (HPLC) oven to allow precise temperature control. To measure absorbance through optical fibers using a conventional Cary 1E UV-Vis spectrometer, a fiberoptic interface is required. The fiberoptic system reported by Carrott and Wai was capable of withstanding pressure in excess of 300 atm, and spectra over the entire UV-Vis range (200–900 nm) could be obtained (17). This type of fiberoptic system is easy to assemble, and the cost for its manufacturing is only about one-tenth of that for a typical high-pressure stainless steel view cell. Using this high-pressure fiberoptic system, the solubility of some uranyl complexes, including $UO_2(NO_3)_2{\cdot}2TBP$ and $UO_2(tta)_2TBP$, in supercritical CO_2 was measured, where tta is thenoyltrifluoroacetylacetone and TBP is tri-*n*-butyl phosphate (17,18). The molar absorptivity for a uranyl complex was determined by introducing standards of the metal complex in hexane, through a syringe, into the 1-cm pathlength cell. Hexane was used as the solvent for calibration because it has a polarity similar to that of CO_2 and has been shown to exhibit similar extinction coefficient and negligible shifts in the position of absorption maxima. The molar absorptivity for the complex was calculated from the slope of a linear calibration curve of absorbance versus concentration. A modified high-pressure fiberoptic system was used later by Waller to measure the solubility of a number of $UO_2(tta)_2X$ complexes in supercritical CO_2, where $X =$ H_2O, an organophosphate, or an organophosphine oxide (19). The solubilities of these uranyl complexes were found to vary by orders of magnitude depending on the nature of the coordinated X. In Waller's system, a view cell with sapphire windows was placed upstream of the fiberoptic system so that the phase behavior of the solute could be observed visually (Figure 1).

To measure the rate of dissolution of solutes and rate of chemical reactions in supercritical CO_2, Hunt et al. designed a high-pressure fiberoptic cell connected to a CCD (charge-coupled device) array UV-Vis spectrometer (20). The CCD array spectrometer allows rapid measurement of the UV-Vis spectra of solutes (e.g., one spectrum per second) in supercritical CO_2. Rate information can be derived from data regarding variation of absorbance with time obtained from the fiberoptic reactor. A schematic diagram of the high-pressure fiberoptic reactor is shown in Figure 2. The fiberoptic reactor was used to measure the rate of formation of metal nanoparticles in a water-in-supercritical CO_2 microemulsion (20). The reactor was also used to measure the rate of dissolution of ferrocene in supercritical CO_2 (20).

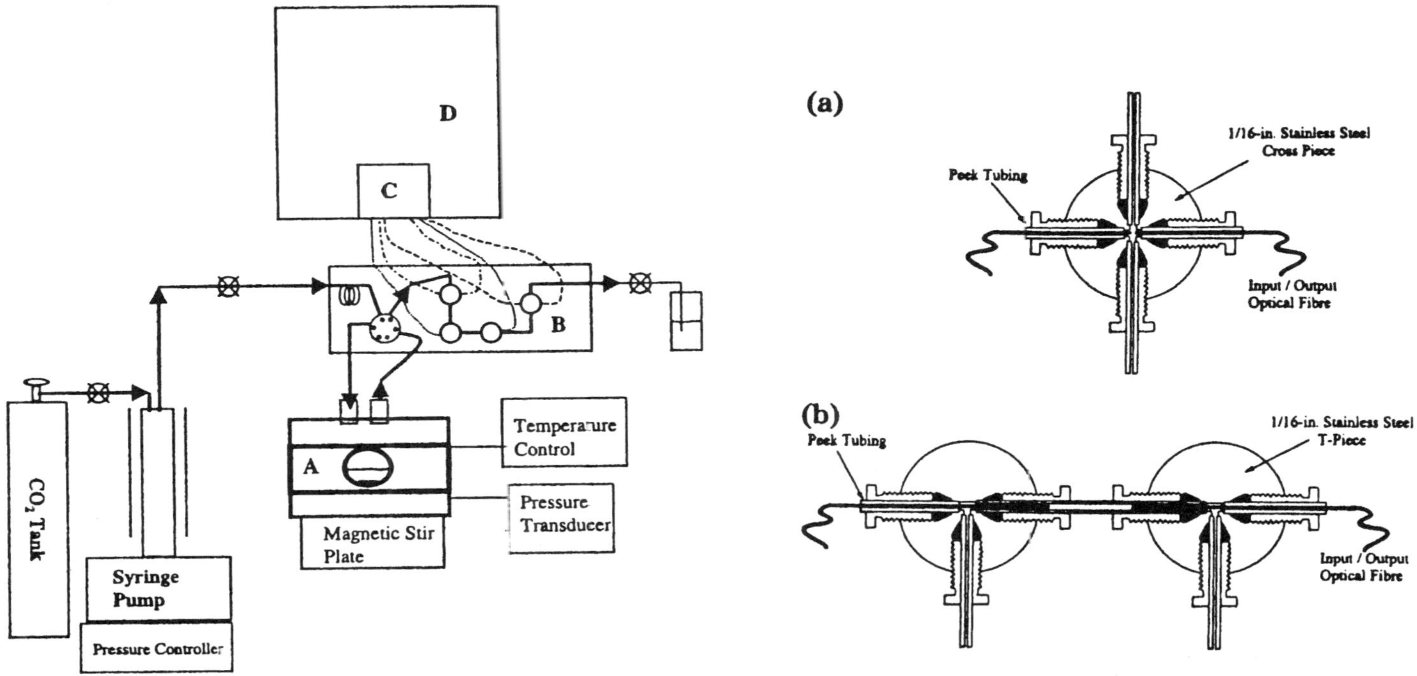

Figure 1 Schematic diagram of a fiberoptic system for sclubility measurements in supercritical fluids. (From Ref. 19). Diagrams on the right (a, b) show the structures of the short pathlength (<1 mm) and the 1-cm fiberoptic cells, respectively. (From Ref. 17.)

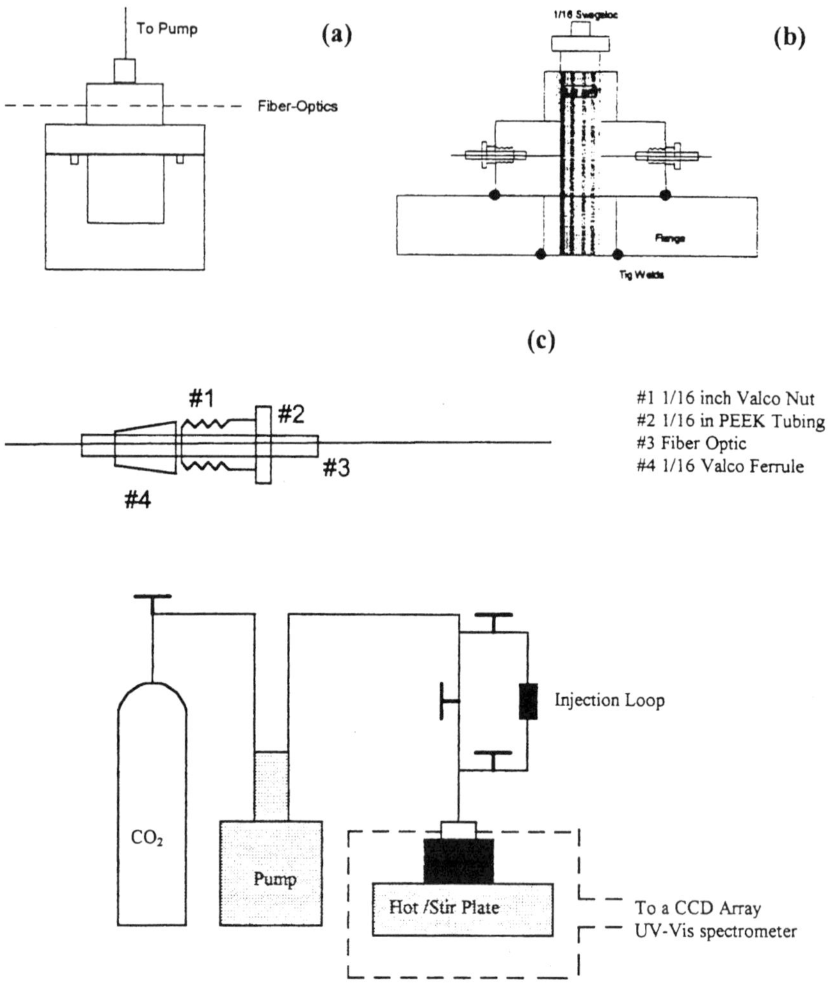

Figure 2 Schematic diagram of the high-pressure fiberoptic reactor system from Ref. 20. (a) Fiberoptic cell, (b) flange and fiberoptic mounting fixture, and (c) fiberoptic sealing system.

Very recently, Addleman et al. described a high-pressure cell for the study of TRLIF of uranyl complexes in supercritical CO_2 (21). A schematic of the optical cell is shown in Figure 3. The cell has two perpendicular optical paths that are both orthogonal to the SCF flow, allowing absorption, fluorescence, and Raman measurements. The cell body was machined from stainless steel with an internal volume of 0.3 ml. The cell windows were made of 2-mm-thick synthetic

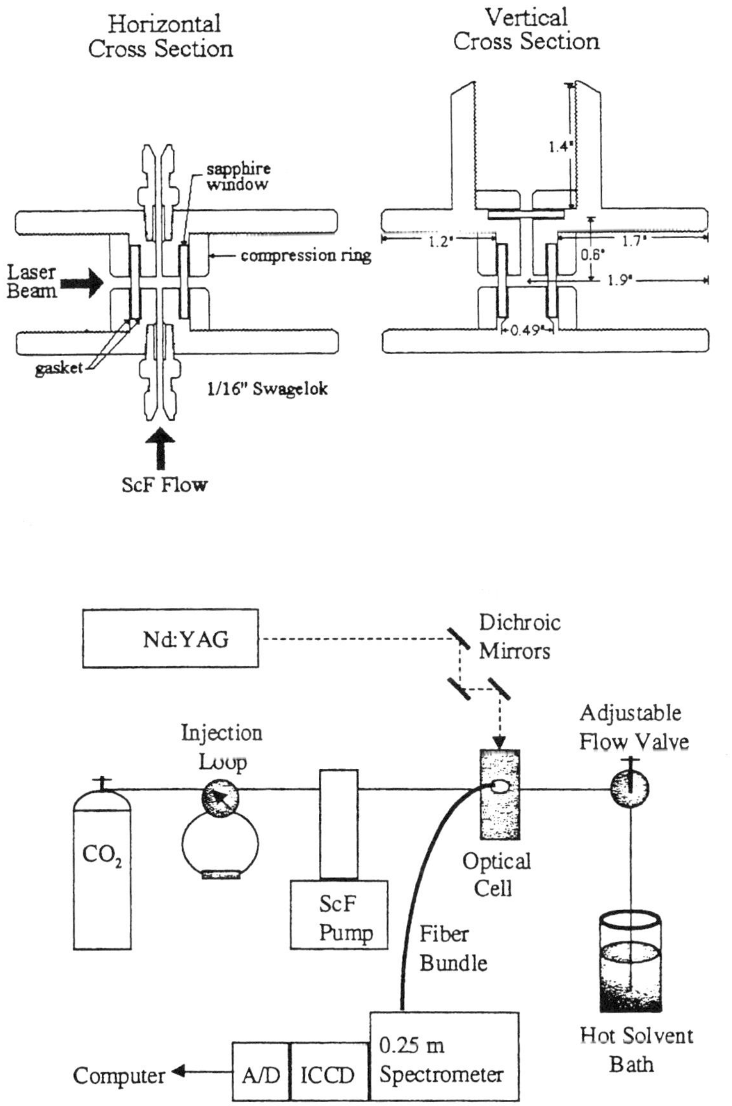

Figure 3 Schematic diagram of the TRLIF system for supercritical fluid experiments. The high-pressure optical flow cell is shown in the top of the figure. (From Ref. 21.)

sapphire. Synthetic sapphire was selected for window material because of its high strength, resistance to corrosion, good thermal conductivity, high-energy damage threshold, and wide transmission range (150–6000 nm). The solubility of $UO_2(NO_3)_2 \cdot 2TBP$ in supercritical CO_2 measured by the TRLIF technique agreed with those obtained by the UV-Vis spectroscopic method (22). The spectroscopic techniques briefly described above are effective tools for studying dissolution and reactions of metal species in SCFs.

B. Solubility Prediction for Metal Complexes

Several methods for predicting solubilities of metal complexes in SCF CO_2 have been mentioned in the literature. Lagalante et al. studied the solubilities of copper(II) β-diketonates and chromium(III) β-diketonates with different β-diketone ligands and showed that the solubility of the metal complexes is correlated with the Hildebrand solubility parameter (δ) of the free ligands calculated using a group contribution method (Figure 4) (23). The chemical group contributions to the solubility parameters of molecules were reported by Fedor (24). The solubility parameter of supercritical CO_2 at different density can be calculated using Gidding's equation (25):

$$\delta = 1.25 P_c^{1/2} (\rho_r / \rho_{rbp}) = 1.25 P_c^{1/2} (\rho_r / 2.66) \tag{1}$$

where ρ_{rbp} is the reduced density at the normal boiling point of the fluid which for conventional liquid solvents is roughly 2.66. Supercritical CO_2 has a low solubility parameter (δ) compared with organic ligands. Fluorine substitution in a ligand tends to lower the δ value of the ligand relative to a nonfluori-

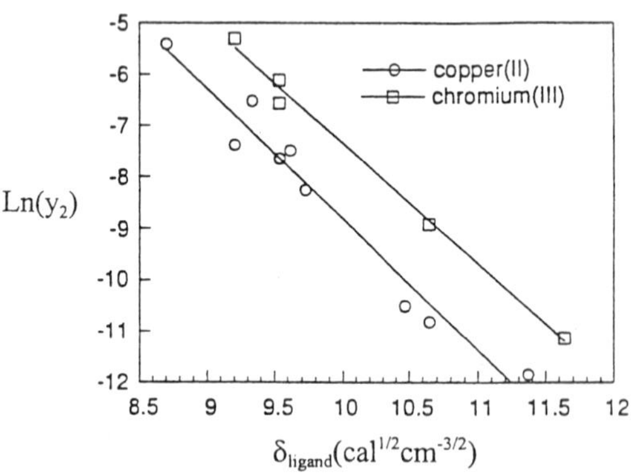

Figure 4 Plot of solubility, $\ln(y_2)$, of various copper(II) and chromium(III) β-diketonates in supercritical CO_2 at a density of 0.874 g/ml vs. Fedor's solubility parameter of the free ligand. (From Ref. 23.)

nated analogue. It is perhaps this "like dissolves like" property that renders fluorinated metal chelates more soluble in supercritical CO_2 relative to the nonfluorinated analogues. Wai and Wang also showed that the solubilities of a series of copper(II), zinc(II), and mercury(II) dithiocarbamates in supercritical CO_2 are correlated to the solubility parameter values of the free ligands (26). Increasing the chain length of an alkyl substitution group in dithiocarbamate tends to lower the solubility parameter of the ligand and results in a higher solubility of the metal dithiocarbamate in supercritical CO_2. For example, if we increase the alkyl chain length of the bis(dialkyl)dithiocarbamate ligand from ethyl to *n*-hexyl, the calculated solubility parameter of the molecule tends to decrease from 10.39 to 9.50, respectively (Table 1). On the other hand, the measured solubility of the metal dithiocarbamate chelates in supercritical CO_2 increases with chain length of the alkyl group and correlates nicely with the solubility parameters of the free ligands. If fluorine replaces the six terminal hydrogen in the bis(diethyl)dithiocarbamate ligand, the resulting bis(trifluoroethyl)dithiocarbamate has a lower δ value (8.75) than bis(dihexyl)-dithiocarbamate ligand. The solubilities of the metal bis(trifluoroethyl)dithiocarbamate chelates in supercritical CO_2 are much higher than the corresponding metal chelates with the alkyl group substitution up to 6-carbon length in the ligand (Table 1). According to the Scatchard–Hildebrand equation, the solubility of a solute in supercritical CO_2 depends on molar volume of the solute and the solubility parameter difference between solute and solvent (27):

$$\ln x_2 = \ln a_2 - (V_2 \phi_1^2 / RT)(\delta_2 - \delta_1)^2 \qquad (2)$$

where a_2 is the activity of solute, x_2 is the mole fraction of solute, V_2 is the molar volume of solute, ϕ_1 is the volume fraction of solvent, and δ_1 and δ_2 are the solubility parameter values of solvent and solute, respectively. Typical

Table 1 Solubility Parameters of Dithiocarbamate Ligands and Solubilities of Their Copper Complexes in Supercritical CO_2 at 60°C and 230 atm

Ligand	δ_2 (cal$^{1/2}$/cm$^{3/2}$)	Solubility (mol/L) Cu-dithiocarbamate	m.p. (°C)
$(C_2H_5)_2NCS_2H$	10.4	1.1×10^{-5}	189–191
$(n\text{-}C_3H_7)_2NCS_2H$	10.0	1.2×10^{-4}	101–102
$(n\text{-}C_4H_9)_2NCS_2H$	9.8	7.2×10^{-4}	73–75
$(n\text{-}C_5H_{11})_2NCS_2H$	9.6	1.8×10^{-3}	63–65
$(n\text{-}C_6H_{13})_2NCS_2H$	9.5	2.8×10^{-3}	44–46
$(CF_3CH_2)_2NCS_2H$	8.8	4.0×10^{-3}	144–146

δ_1 of CO_2 at 60°C and 230 atm is about 6.6.
Data from Ref. 26.

δ_1 values for supercritical CO_2 are in the range 4–6, while those for ligands and metal complexes range from 7 to 14. Equation (2) predicts that increase in V_2 tends to decrease solubility of a solute whereas decrease in δ_2 relative to δ_1 increases a solute's solubility. Increasing the size of the R group attached to the ligand tends to decrease the δ_2 value but increases V_2. Thus, a compromise between molar volume V_2 and solubility parameter δ_2 must be attained. It was also noted by Smart et al. that a *tert*-butyl group tends to lower δ_2 more than an *n*-butyl group in a ligand. Smart et al. showed that a *tert*-butyl-substituted ligand with between six and eight carbon atoms in each R group is perhaps the most favorable alkyl substitution to achieve high solubility in supercritical CO_2 (27). This method of predicting the solubility of metal complexes in supercritical CO_2 depends on the availability of solubility parameter data of ligands and metal complexes. Group contribution data for metal–sulfur and metal–oxygen bonds are generally not available.

A simple solvato complex model has also been used to predict the solubility of metal complexes in supercritical CO_2 (19,28). According to this model, the molecules of a solute and those of a solvent would associate with one another to form a solvato complex as shown in the following equation:

$$A + kB \rightarrow AB_k \tag{3}$$

Equation (3) shows that one molecule of a solute A is surrounded by k molecules of a solvent B to form one molecule of a solvato complex AB_k. The equilibrium constant K is represented by Eq. (4):

$$K = [AB_k]/[A][B]^k \tag{4}$$

Expressing Eq. (4) in logarithmic form, we have

$$\ln[AB_k] = k \ln[B] + \ln[A] + \ln K \tag{5}$$

Table 2 Solubility of Some Uranyl Complexes in Supercritical CO_2 (temp = 40°C)

	Solubility (mol/L)	
Uranyl complex	200 atm	250 atm
$UO_2(NO_3)_2 \cdot 2TBP$	4.2×10^{-1}	
$UO_2(tta)_2 \cdot TBP$	7.5×10^{-3}	1.0×10^{-2}
$UO_2(tta)_2 \cdot TEP$	3.2×10^{-3}	4.6×10^{-3}
$UO_2(tta)_2 \cdot TOPO$	1.5×10^{-3}	3.4×10^{-3}
$UO_2(tta)_2 \cdot TBPO$	2.1×10^{-4}	3.2×10^{-4}
$UO_2(tta)_2 \cdot H_2O$	1.0×10^{-4}	1.7×10^{-4}

Data taken from Refs. 18 and 19.

The equilibrium constant K is related to the enthalpy of solvation and temperature and [B] is a function of density (D). According to Eq. (5), the concentration of the solvato complex or solubility (S) can be related to the density of the fluid phase by the relation

$$\ln S = k \ln D + C \tag{6}$$

where C is a constant related to the enthalpy of solvation.

Solubility data of some uranyl complexes obtained from spectroscopic measurements were used to test the solvato complex model (Table 2). Figure 5a shows the solubility of $UO_2(NO_3)_2 \cdot 2TBP$ at different temperature and pressure and Figure 5b shows a plot of $\ln S$ vs. $\ln D$ of the solubility data. All of the solubility data fall on a straight line when they are plotted in the form of $\ln S$

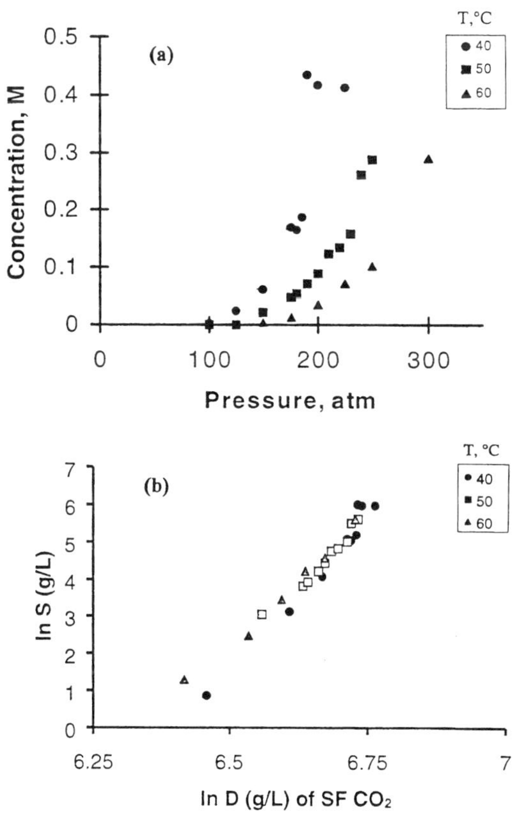

Figure 5 (a) Solubility of $UO_2(NO_3)_2 \cdot 2TBP$ in supercritical CO_2 at 40, 50, and 60°C and different pressure. (b) Plot of ln S vs. ln D for the solubility data given in (a). (Data from Ref. 18.)

vs. ln D. Figure 6a shows the solubility data of five $UO_2(tta)_2 \cdot X$ complexes in supercritical CO_2 at 40°C and different pressure, where $X = H_2O$, TBP, TEP (triethyl phosphate), TBPO (tributylphosphine oxide) and TOPO (trioctylphosphine oxide). When these solubility data are plotted in terms of ln S vs. ln D, all of them exhibit a linear relationship. The solvato complex model appears satisfactory for modeling the urany complexes solubility data. Using the solvato complex model, the solubilities of uranium complexes in supercritical CO_2 may be predicted over a wide density range based on a few experimental measurements. Smart et al. also showed earlier that Eq. (6) could be used to predict the solubilities of a number of transition metal complexes in supercritical CO_2 (10).

According to the model, the k value should be related to the average number of solvent molecules associated with the metal complex. According to Figures 5 and 6, the most soluble complex appears to be most highly solvated, e.g., the k value for $UO_2(NO_3)_2 \cdot 2TBP$ is about 17. For the $UO_2(tta)_2 \cdot X$ complexes,

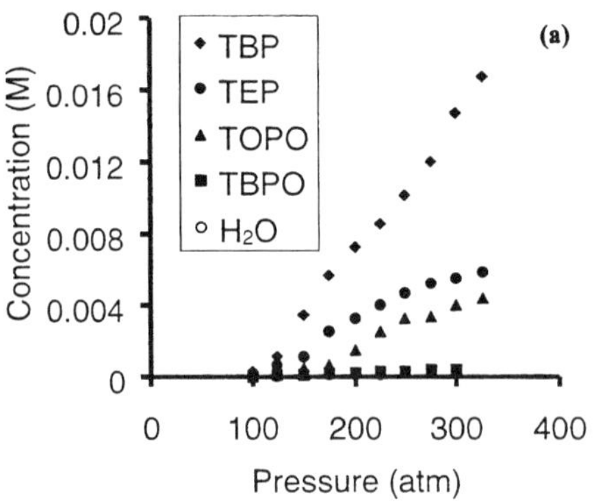

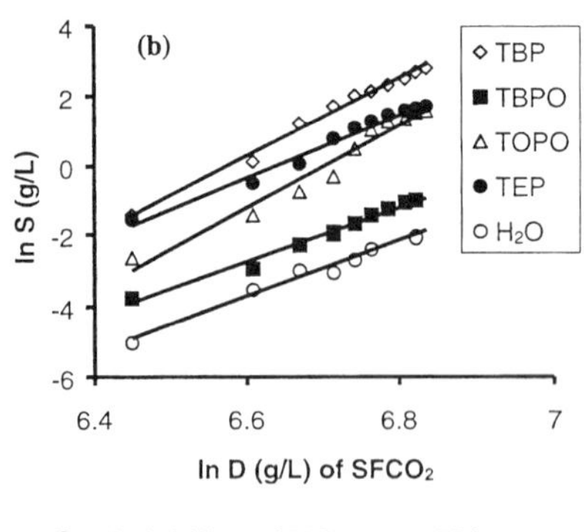

Figure 6 Solubility of $UO_2(tta)_2 \cdot X$ in supercritical CO_2 at 40°C and different pressure. $X =$ TBP, TEP, TOPO, TBPO, and H_2O. (b) Plot of log S vs. ln D for the solubility data shown in (a). (From Ref. 19.)

the phosphate adducts are more soluble than the phosphine oxide adducts. For the phosphate adducts, the one with the higher *k* value [$UO_2(tta)_2 \cdot TBP$, $k = 11$] is more soluble than the one with the lower *k* value [$UO_2(tta)_2 \cdot TEP$, $k = 9$]. A similar trend is also observed for the phosphine oxide adducts. The physical meaning of the *k* value is not well understood; it is probably related to the number of CO_2 molecules solvated the metal complexes in supercritical CO_2.

Use of the Peng–Robinson equation of state for predicting the solubility of metal complexes in supercritical CO_2 was also reported in the literature (29). This approach requires a number of molecular parameters, such as critical constants, molar volume, and van der Waals' interaction parameters, which are not easily found in the literature. Akgerman et al. recently developed a dynamic measurement technique to determine solubilities of chelating agents and metal chelates in supercritical CO_2 (30). The authors measured the solubility of a number of metal-containing complexes for thermodynamic modeling and prediction of solubility in supercritical CO_2 at varying operational conditions. The Peng–Robinson equation of state and van der Waals' mixing rules were employed in their modeling. The solubility of copper diethyldithiocarbamate in supercritical CO_2 was found to agree with the results reported earlier by Laintz et al. (4). The predicted solubilities fitted well with experimentally measured data for Cu(II) acetylacetonate and Cu(II) diethyldithiocarbamate.

The solubility models presented in this section provide general guidelines for predicting solubilities of metal chelates in supercritical CO_2. Based on some experimental measurements, solubilities of the metal chelates of interest may be estimated using these models. However, accurate calculation of metal chelate solubilities in supercritical CO_2 is still difficult to accomplish.

III. FLUID EXTRACTION TECHNOLOGY FOR THE NUCLEAR INDUSTRY

The supercritical fluid technology is of particular interest to the nuclear industry because of the volume of hazardous liquid and solid wastes generated by the industry. The predominant waste production operations in the nuclear industry are mining and milling of uranium ores, reprocessing of used materials, and decommissioning of redundant facilities. In managing such wastes, the main objective is to minimize the production of secondary waste, while converting as much as possible of both the radioactive and inactive wastes to a small-volume solid suitable for long-term storage and ultimate disposal.

Nuclear power reactors operate by converting energy released by the fission of actinides such as ^{235}U and ^{239}Pu into electrical energy. This process produces fission products and reduces the fissile content of the fuel until eventually further fission becomes uneconomical and the spent fuel is removed from the reactor. The spent fuel contains the bulk of the original isotopes with a relatively small amount of fission products and higher atomic number actinides. Reprocessing recovers the residual uranium and plutonium for reuse. The nuclear industry currently treats spent fuel by the Purex (*p*lutonium *u*ranium *e*xtraction) process. Essentially, this process involves dissolving the fuel rod, minus the cladding, in 6 M nitric acid. Uranium and plutonium are subsequently extracted into an organic solvent via the formation of a nonpolar complex with TBP, i.e., $UO_2(NO_3)_2 \cdot 2TBP$. The fission products remaining in the acid solution after the extraction are vitrified and stored as high-level waste (31). The Purex process, though highly efficient, has the limitations of a liquid–liquid separation system. SCF CO_2 with its gas-like property and minimal waste generation capability may overcome some of these limitations.

A. Reprocessing Spent Nuclear Fuels in Supercritical CO_2

The feasibility of replacing supercritical CO_2 for the organic solvent required for the extraction of $UO_2(NO_3)_2 \cdot 2TBP$ was first reported in 1995. Lin et al. showed that uranyl ions and thorium ions in 3–6 M nitric acid solutions could be effectively extracted by supercritical CO_2 containing TBP (Figure 7) (32). Meguro et al. also reported the characteristics of uranyl ion extraction from nitrate acid using TBP-modified supercritical CO_2 (33,34). The efficiencies of the SCF extraction were shown by Lin et al. to be comparable to those used in the conventional Purex process (32). The extracted $UO_2(NO_3)_2 \cdot 2TBP$ complex has a solubility of 0.43 mol/L at 40°C and 230 atm, a concentration comparable to that used in the Purex process (18). A possible scheme using supercritical CO_2 as a solvent in the Purex process is shown in Figure 8 (31). The process is similar to that used in the Purex process with the following advantages:

1. The distribution coefficient, D, can be controlled by varying temperature and pressure (or density) due to the tunable solvation strength of the fluid, thus leading to more efficient separation of uranium and plutonium.
2. The mass transport properties of supercritical CO_2 are much higher than those of a conventional solvent extraction system, resulting in faster extraction.
3. Faster extraction should in turn lead to fewer solvent degradation problems, minimizing the formation of any interfacial crud.
4. The nature of the contact between nitric acid and the extraction fluid will be simpler because of the low-density, gas-like properties of the SCF.

Another approach to reprocessing spent nuclear fuels is by direct dissolution of uranium oxides in supercritical CO_2 (19,31). For example, UO3 can be

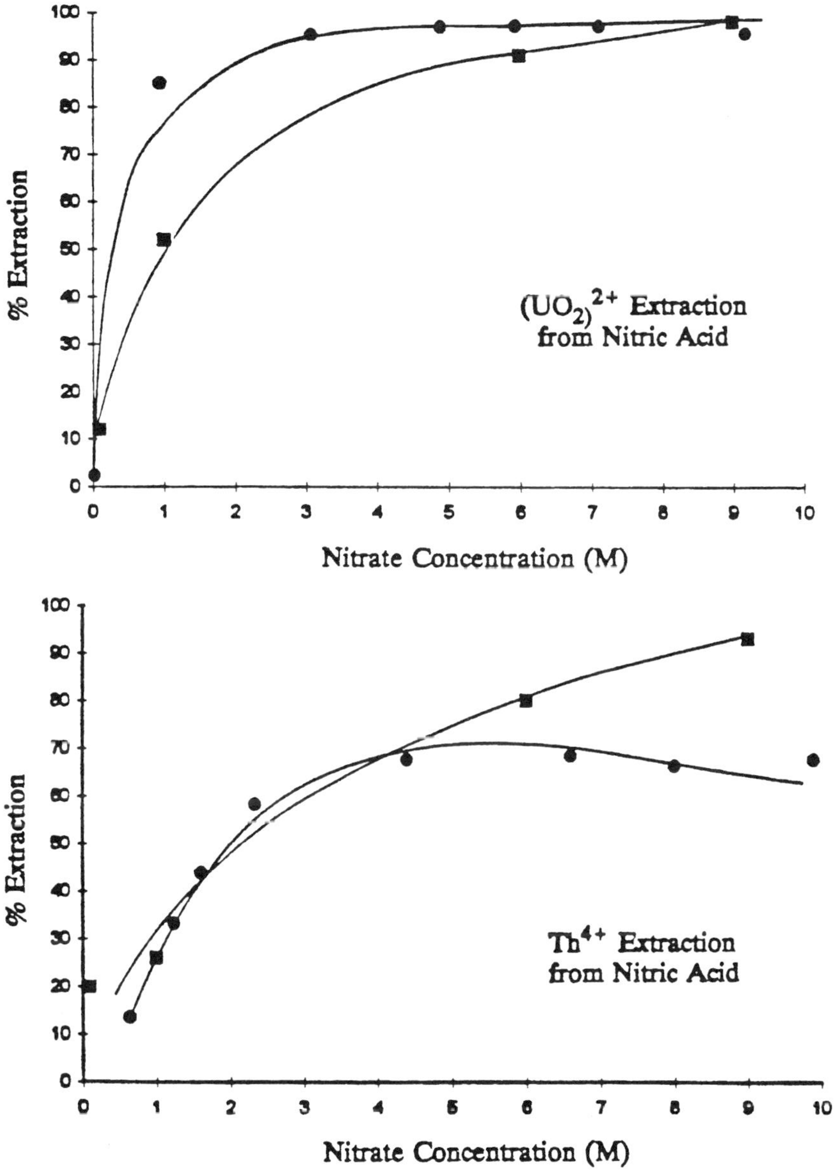

Figure 7 Supercritical fluid extraction of uranyl ions and thorium ions from nitric acid solutions with supercritical CO_2 containing saturated TBP at 60°C and 200 atm. ●, Supercritical CO_2 extraction; ■, solvent extraction with 19% v/v TBP in kerosene. (From Ref. 32.)

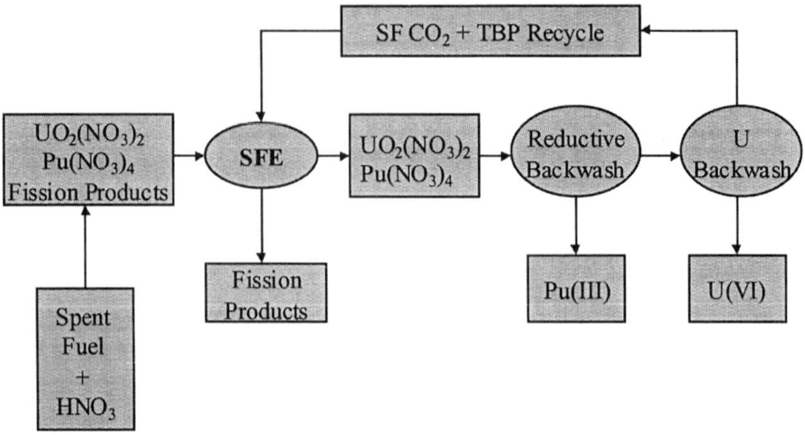

Figure 8 A proposed supercritical fluid Purex process using CO_2 to replace organic solvent. (From Ref. 31.)

dissolved in supercritical CO_2 in the presence of a fluorinated β-diketone such as thenoyltrifluoroacetylacetone (Htta) according to the following equation:

$$UO_3(s) + Htta \rightarrow UO_2(tta)_2{\cdot}H_2O \tag{7}$$

This dry dissolution process proceeds with a much higher efficiency in the presence of TBP in CO_2. Because TBP is a stronger Lewis base, it replaces the coordinated H_2O to form the more soluble adduct complex $UO_2(tta)_2{\cdot}TBP$ in supercritical CO_2 according to Eq. (8):

$$UO_3(s) + 2Htta + TBP \rightarrow UO_2(tta)_2{\cdot}TBP + H_2O \tag{8}$$

The solubility of $UO_2(tta)_2{\cdot}TBP$ in supercritical CO_2, though less than that of $UO_2(NO_3)_2{\cdot}2TBP$, is still sufficiently high to be potentially useful, thus making reprocessing of spent nuclear fuel by this dry process a realistic proposition (20). Figure 9 shows a possible scheme of a dry process utilizing Eq. (8) for dissolution of UO_3 in supercritical CO_2. Regeneration of the ligand may be achieved on-line using hydrogen reduction in the SCF phase according to the following reaction:

$$UO_2(tta)_2{\cdot}TBP + H_2 \rightarrow UO_2(s) + 2Htta + TBP \tag{9}$$

The dry process for reprocessing of nuclear spent fuel should in principle greatly reduce generation of liquid wastes such as those in the current Purex process. Reaction (9) occurs efficiently for UO_3 in supercritical CO_2 but not for UO_2. According to our experiments, in the case of UO_2, an oxidizing agent such as H_2O_2 has been shown to promote the dissolution of the oxide in supercritical CO_2 (35).

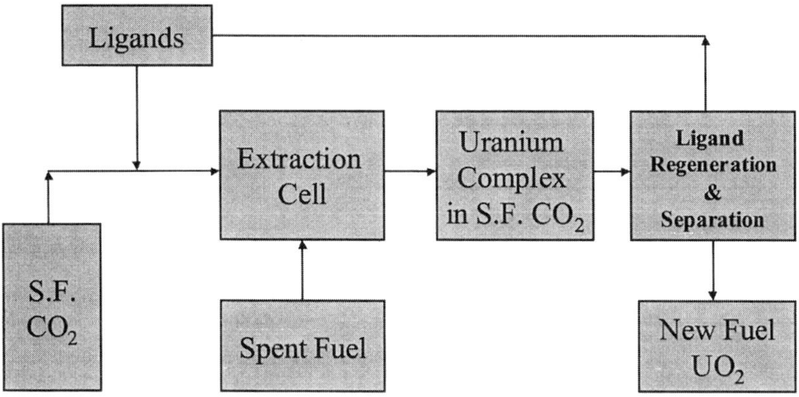

Figure 9 A proposed dry supercritical fluid process for reprocessing spent nuclear fuels. (From Ref. 31.)

B. Supercritical Fluid Treatment of Nuclear Wastes

Another area where the SCF technology may find important applications is the clean-up of low-level radioisotope-contaminated materials generated by nuclear power plants, hospitals, industrial plants, government and defense laboratories, and reactors. Such waste takes a variety of forms, such as medical treatment materials, contaminated wiping papers and towels, used filters, protective clothing, hand tools, equipment, components of decommissioned nuclear power plants, and so forth. The matrixes of such wastes often consist of porous materials. Supercritical CO_2, with its high diffusivity, is an ideal solvent for removing contaminated radioisotopes from such materials. Supercritical CO_2 extraction of lanthanides, actinides (Table 3), and toxic metals from solid and liquid materials using different chelating agents has been reported by several research groups (35–44). Recently, extraction of plutonium from soil using Htta and TBP in supercritical CO_2 has been demonstrated (45). Selective extraction of strontium and cesium using crown ethers and fluorinated counteranions has also been reported (Table 4) (46,47). Crown ethers are known to be selective ligands for solvent extraction of the alkali metals and the alkaline earth metals based on the ionic radius–crown cavity size compatibility concept. For example, dicyclohexano-18-crown-6 (DC18C6) is known to be selective for complexation with Sr^{2+}. To render the Sr-DC18C6 complex soluble in supercritical CO_2, a perfluorinated counteranion [a carboxylate or a sulfonate from pentadecafluoro-*n*-octanoic acid or perfluoro-1-octanesulfonate salt, (PFOSA), respectively] was used by Wai et al. to neutralize the charge and to transport the Sr complex as an ion pair into supercritical CO_2 (46). For supercritical CO_2 extraction of Cs^+, dicyclohexano-21-crown-7 (DC21C7) and perfluoro-1-octanesulfonate (PFOSA)

Table 3 Extraction of Uranium from Mine Water and from Contaminated Soil with Supercritical CO_2 (60°C and 150 atm)

Sample	U conc.	Extractant	% Extraction
Mine water A	9.6 μg/ml	Htta + TBP	81 ± 4
Mine water B	18.0 μg/ml	Htta + TBP	78 ± 5
Soil A	63 μg/g	Htta + TBP	82 ± 5
		Hhfa + TBP	91 ± 4
Soil B	154 μg/g	Htta + TBP	77 ± 4
		Hhfa + TBP	89 ± 5

Mine water: 4-ml sample, 200 μmol each of Htta and TBP; soil sample: 100-mg sample, 200 μmol each of Htta or Hhfa and TBP.
Hhfa, hexafluoroacetylacetone; Htta, thenoyltrifluoroacetylacetone.
Data from Ref. 39.

were found to be effective (47). The studies described above have enhanced our understanding of the factors controlling the efficiency of SCF extraction of radioisotopes from different matrixes.

Mixed wastes usually are referred to as wastes containing both hazardous chemical components, subject to the requirement of the Resource Conservation and Recovery Act (RCRA), and radioactive components, subject to the require-

Table 4 Extraction of Sr^{2+} and Cs^+ from Aqueous Solutions with Supercritical CO_2 Containing a Crown Ether and a Fluorinated Counteranions

Mole ratio	Matrix	% Extraction		
		Sr^{2+}	Ca^{2+}	Mg^{2+}
Sr^{2+}:DC18C6:PFOSA-K				
1 : 10 : 10	Water	97	8	2
1 : 10 : 50	1.3 M HNO_3	60	8	2
		Cs^+	K^+	Na^+
Cs^+:DC21C7:PFOSA-$N(C_2H_5)_4$				
1 : 100 : 100	Water	55	39	0

PFOSA-K = $CF_3(CF_2)_6CF_2SO_3K$; PFOSA-$N(C_2H_5)_4$ = $CF_3(CF_2)_6CF_2SO_3N(C_2H_5)_4$; DC18C6 = dicyclohexano-18-crown-6; concentration of Sr^{2+} or Cs^+ ion = 5.6×10^{-5} M; pH of water in equilibrium with CO_2 under the experimental conditions ≈ 2.9; for water samples, 20 min static followed by 20 dynamic extraction (flow rate 2 ml/min) at 60°C and 100 atm; for the nitric acid sample, 35°C and 200 atm.
Data from Refs. 46 and 47.

Table 5 Sequential Extraction of PCB (BZ#54) and Uranium from Solid Samples

Sample matrix	First extraction (%) neat CO_2 for PCB	Second extraction (%) CO_2 + Htta + TBP for U
Sand[a]	98	94
Idaho soil[a]	97	82
UTS-4 tailings[b]	98	80

[a]Spiked with 200 μg of BZ#54 (2,2′,6,6′-tetrachlorobiphenyl), 200 μg U; 300 mg Htta and 200 μl of TBP; 150°C and 200 atm for the first extraction and 80°C and 200 atm for the second extraction; 30 min static followed by 30 min dynamic extraction at a flow rate of 1.5 ml/min.
[b]A standard uranium tailings obtained from CANMET and contained 1010 μg U/g soil; same extraction conditions as in a.
Data from Ref. 19.

ments of the Atomic Energy Act. One example is the storm sewer sediments existing at the Oak Ridge National Laboratory that contains polychlorinated biphenyls (PCBs), mercury, and uranium (T. Conley, Mercury Working Group under Mixed Waste Forcus Area, Chemical Technology Division, ORNL, 1998, personal communication). This type of mixed waste can be treated using a sequential SCF extraction procedure. For example, the organic pollutants can be first removed using neat supercritical CO_2. This is followed by the addition of a suitable chelating agent to remove metals or radioisotopes. Using the sequential SCF extraction technique, organic pollutants, toxic metals, and radioisotopes can be extracted separately, thus enormously simplifying disposal of such wastes. An example illustrating the sequential extraction of PCB and uranium from soil is shown in Table 5 (19).

Commercialization of the SCF technology, though still years away, should offer many benefits for the 21st century nuclear industry. Such benefits should be realized in terms of improved efficiency of chemical processes, reducing secondary waste generation, and cost saving associated with waste disposal.

IV. WATER-IN-CO_2 MICROEMULSIONS FOR MATERIALS SYNTHESIS

There has been much interest in recent years in exploiting the properties of microemulsion phases in SCFs (48–52). A reverse micelle or microemulsion system of particular interest is one based on CO_2 because of its minimal environmental impact in chemical applications. Since water and CO_2 are the two most abundant, inexpensive, and environmentally compatible solvents, the application of such a system could have tremendous implications for the chemical industries of the 21st century. Reverse micelles and microemulsions formed in supercritical CO_2 allow highly polar or polarizable compounds to be dispersed in this nonpolar fluid. However, most ionic surfactants with long hydrocarbon tails, such as sodium bis(2-ethylhexyl)sulfosuccinate (AOT), are insoluble in supercritical CO_2. Nonionic surfactants have a greater affinity for CO_2. However, they have little tendency to aggregate and take up water due to weak electrostatic interactions of the head groups. The use of ionic surfactants with fluorinated tails provides a layer of a weakly attractive compound covering the highly attractive water droplet cores, thus preventing their short-range interactions that would destabilize the system. However, our experiments indicate that these CO_2-philic ionic surfactants do not form very stable water-in-CO_2 microemulsions, especially when the water cores contain high concentrations of electrolytes. A recent communication shows that with the use of a conventional AOT and a fluorinated cosurfactant, a very stable water-in-CO_2 microemulsion containing a relatively high concentration of silver nitrate can be formed (12).

A high-pressure system used for studying the stability of the water-in-CO_2 microemulsions in our laboratory is illustrated in Figure 10. The system consists of a high-pressure cell (20-ml volume) with sapphire windows. The

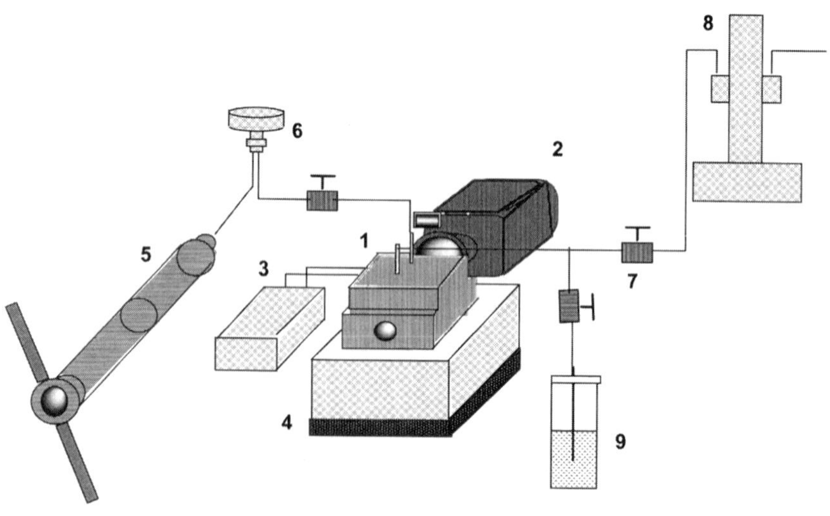

Figure 10 A high-pressure system for evaluating stability of water-in-CO_2 microemulsions in supercritical CO_2 used in the author's laboratory. (1) High-pressure view cell; (2) video camera; (3) temperature controller; (4) stirrer; (5) hand-operated syringe pump; (6) pressure transducer; (7) high pressure valve; (8) ISCO syringe pump; (9) collection vessel.

phase behavior of a surfactant system can be observed through the windows visually or using a video camera. A hand-operated syringe pump (HIP model 62-6-10) and a pressure transducer (Setra Systems model 204) were used to deliver the water to the view cell at a given pressure. The liquid CO_2 was metered with an ISCO syringe pump (model 260D) and pump controller. The liquid CO_2 was introduced to the view cell via stainless steel tubing (1/16 in. o.d. × 0.03 in. i.d.). After loading a surfactant, H_2O, and CO_2 to the high-pressure view cell, the cell was heated with a thermal block and the temperature was kept constant and controlled with a J-type sensor and an Omega digital controller. The fluid was stirred to allow dissolution of the surfactants and formation of an optically transparent water-in-CO_2 microemulsion. After that, water was added carefully (10 μl each time with stirring for 30 min) to the view cell with a hand-operated syringe pump until the stability phase boundary was reached. When the phase boundary was reached, the fluid phase became cloudy and small water droplets precipitated on the bottom of the cell. The amount of water injected was calculated from the vernier on the hand-operated syringe pump using a calibration curve.

A. Synthesizing Metal Nanoparticles in a Water-in-CO_2 Microemulsion

Using a mixture of AOT and a perfluoropolyether (PFPE) surfactant, the resulting water-in-CO_2 microemulsion is stable, with *W* (ratio of water to AOT) values greater than 40 at 40°C and 200 atm, according to experiments conducted in this author's laboratory. The perfluorinated surfactant obtained from Ausimont has the general structure $CF_3O[OCF(CF_3)CF_2]_n(OCF_2)_mOCF_2CH_2OCH_2CH_2O$-$PO(OH)_2$ and an average molecular weight of 870. The initial experiments for synthesizing silver nanoparticles used a mixture of AOT (12.8 mM) and the perfluoropolyether-phosphate ether (PFPE-PO4, 25.3 mM) at a water-to-surfactant ratio of 12. The microemulsion containing $AgNO_3$ (0.33 mM) was optically transparent and stable in supercritical CO_2 for hours. To make metallic silver nanoparticles, a reducing agent, $NaBH(OAc)_3$, was injected into the supercritical fluid microemulsion system to reduce Ag^+ in the water core to Ag (12). The formation of Ag nanoparticle in the microemulsion system was observed within a minute after the introduction of the reducing agent. The formation and stability of the Ag nanopartices was monitored in situ by UV-Vis spectroscopy utilizing the 400-nm band originating from the surface plasmon resonance of nano-sized Ag crystals. The SCF solution remained optically clear with a yellow color due to the absorption of the Ag nanoparticles. Bandwidth analysis indicated that the particles had a size of about 4 nm. The Ag nanoparticle samples were collected via the RESS (rapid expansion of SCF solution) (53) technique using a 50-μm i.d. (1.5-mm o.d.) PEEK restrictor. TEM photographs showed the average size of the Ag particles to be approximately 5–15 nm. Rapid expansion of the microemulsion probably could cause aggregation of the Ag nanoparticles resulting in a slightly larger particles size. Other methods of collecting the synthesized nanoparticles with minimal aggregation are currently under investigation. One possible method of collecting the nanoparticles is by changing pressure or temperature of the SCF to make the microemulsion unstable so that the metal nanoparticles will precipitate rapidly in the fluid phase. This method may be able to deliver metal nanoparticles to small porous areas of a solid substrate placed in the SCF phase.

The reducing agent $NaBH(OAc)_3$ used in the original study by Ji et al. has a limited solubility in supercritical CO_2. Other more soluble reducing agents were tested later by Ohde et al. for making silver nanoparticles in the water-in-supercritical CO_2 microemulsion. Several reducing agents including $NaBH_3CN$ (sodium cyanoborohydride), TMPD (tetramethyl-*p*-phenylenediamine), and ferrocene were tested using the fiberoptic reactor described in Figure 2. The reduction of Ag^+ to Ag by $NaBH_3CN$ and the formation of the Ag nanoparticles in the microemulsion were found to occur very rapidly as shown in Figure 11 (21). The characteristic Ag nanoparticle absorption band around 400 nm increased rapidly and reached a constant value in about 30 s. The spectra shown in Figure 11 were taken at 4-s intervals. The fourth point in Figure 11 represented the time of injection of the reducing agent $NaBH_3CN$. The reduction and formation of Ag nanoparticles occurred almost instantaneously, and after about 20 s the reaction was essentially complete. These results indicate the microemulsion is dynamic in nature. Ionic species in the water core of the microemulsion obviously can interact effectively with molecular species dissolved in the fluid phase. Using the same approach, a number of other metallic nanoparticles, including those of Cu and Pd, have recently been synthesized in supercritical CO_2 in our laboratory.

B. Synthesizing Metal Halide Nanoparticles by Mixing Water-in-CO_2 Microemulsions

The dynamic nature of the microemulsion suggests that, by collision, exchange of contents between two water-in-CO_2 microemulsions probably would take place effectively. This suggests a possibility of synthesizing nanomaterials in supercritical CO_2 by mixing water-in-CO_2 microemulsions containing different ions in the water cores. If this is true, a wide range of nanomaterials can be synthesized in supercritical CO_2 utilizing the microemulsions containing different ionic species as starting materials. The synthesis of silver halide nanoparticles in supercritical CO_2 is used as an example to illustrate this principle. Ohde et al. reported recently that by mixing two microemulsions containing silver ions (Ag^+) and halide ions (X^-) separately, formation of AgX nanoparticles was observed

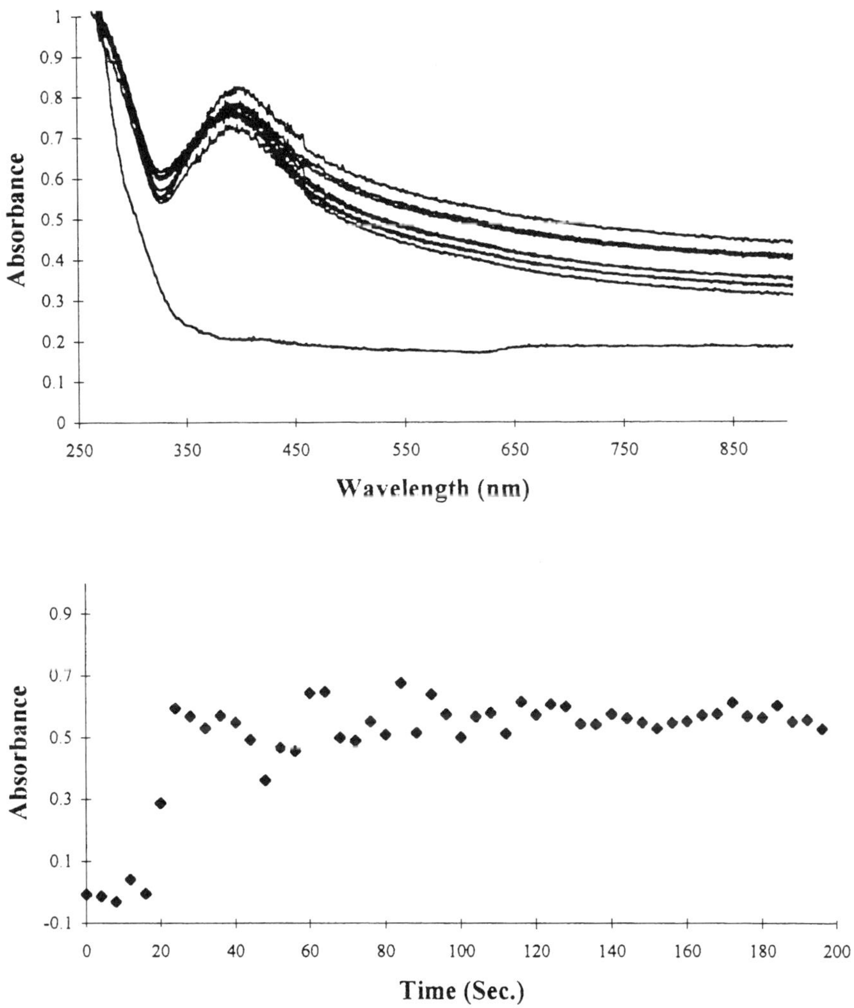

Figure 11 Top: UV-Vis spectra of the silver nanoparticles synthesized in a water-in-CO_2 microemulsion (38°C and 200 atm) obtained by the high-pressure fiberoptic reactor. Reduction of Ag^+ was achieved using sodium cyanoborohydride $NaBH_3CN$. Bottom: Variation of absorbance with time at 400 nm. The absorbance data were collected at 4-s intervals. (From Ref. 20.)

(13). The mechanism of exchanging silver nitrate and sodium halide between two water-in-CO_2 microemulsions is illustrated in Figure 12.

A schematic diagram of the supercritical CO_2 microemulsion reaction system for synthesizing silver halide nanoparticles is shown in Figure 13. Two home-made high-pressure vessels were utilized for the synthesis. One vessel (9.5 ml volume) was equipped with a fiberoptic system (5 mm pathlength) connected to a CCD array UV-Vis spectrometer. The other vessel was a 15-ml high-pressure view cell with sapphire windows. The high-pressure vessels were connected to each other via 1/16-in. stainless steel tubing. Each vessel could be isolated from the other by Valco high-pressure valves. A silver nitrate solution ($AgNO_3$, 3.3 mM) was placed in the 9.5-ml fiberoptic vessel and an NaX solution (X = Cl, Br, or I; 6.6 mM) was placed in the 15-ml view cell. A mixture of AOT and PFPE-PO_4 (12.8 mM and 25.3 mM, respectively) was added to each vessel to make the water-in-CO_2 microemulsions. The view cell allows visual examination of the formation of a homogeneous, optically transparent microemulsion. The vessels were first pressurized to 80 atm. After the formation of the microemulsions was complete, the 15-ml vessel was pressurized to 200 atm; the microemulsion containing the sodium halide solution was pushed

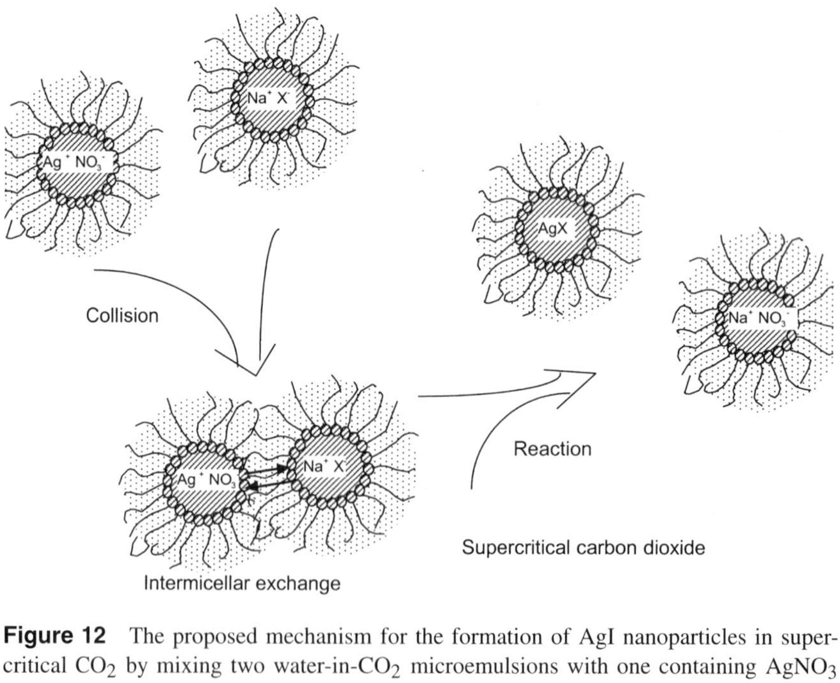

Figure 12 The proposed mechanism for the formation of AgI nanoparticles in supercritical CO_2 by mixing two water-in-CO_2 microemulsions with one containing $AgNO_3$ and the other NaI in the water core.

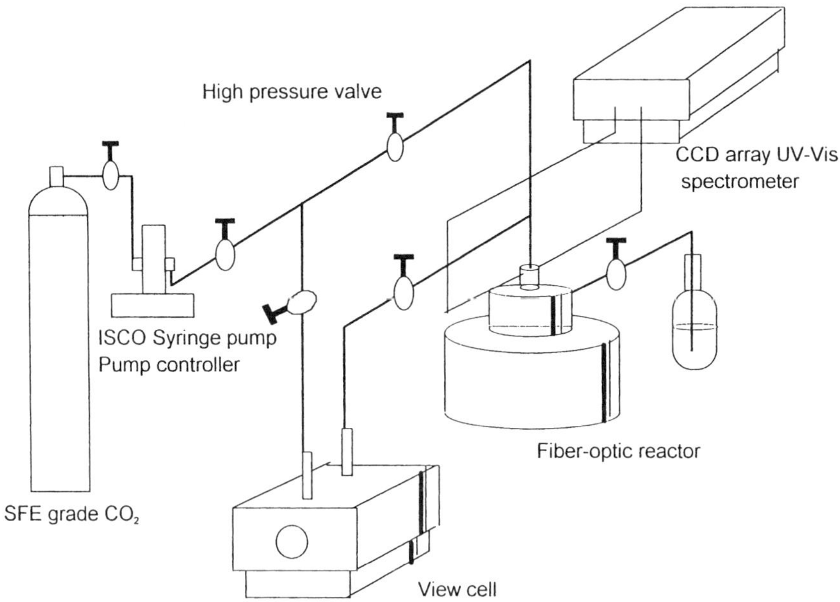

Figure 13 Schematic diagram of the high-pressure system for synthesizing AgI nanoparticles in supercritical CO_2. (From Ref. 14.)

into the fiberoptic cell by the opening of an interconnecting valve and the spectra were recorded. Figure 14 shows the UV-Vis spectra of silver halide nanoparticles observed in the supercritical CO_2 experiments. The spectra for silver iodide showed a broad band in the approximate range of 410–450 nm. The peak wavelength of the absorption band increased with time after the mixing, varying from about 422 to 429 nm. About 30 s after mixing, the peak wavelength reached approximately a constant value at 429 nm and the peak absorption intensity also became constant for the remainder of the experiment. Similar experiments by mixing water/AOT/hexane microemulsions (containing Ag^+ and I^- separately) would require about 3 min to reach a plateau for the formation of AgI nanoparticles. The AgI spectra observed in the SCF microemulsion system are consistent with the UV Vis spectra previously reported by Sato et al. for nano-sized silver iodide particles formed in an AOT/isooctane reverse micelle system (54). The UV-Vis spectra for AgBr and AgCl are also consistent with those reported in the literature for the corresponding silver halide nanoparticles synthesized using conventional water-in-oil microemulsions (54).

Ida et al. showed that the ionic conductivity of silver iodide increased significantly as the particle size decreased to the nanometer range (55). Silver iodide nanoparticle is a direct gap semiconductor. Recently, much attention has

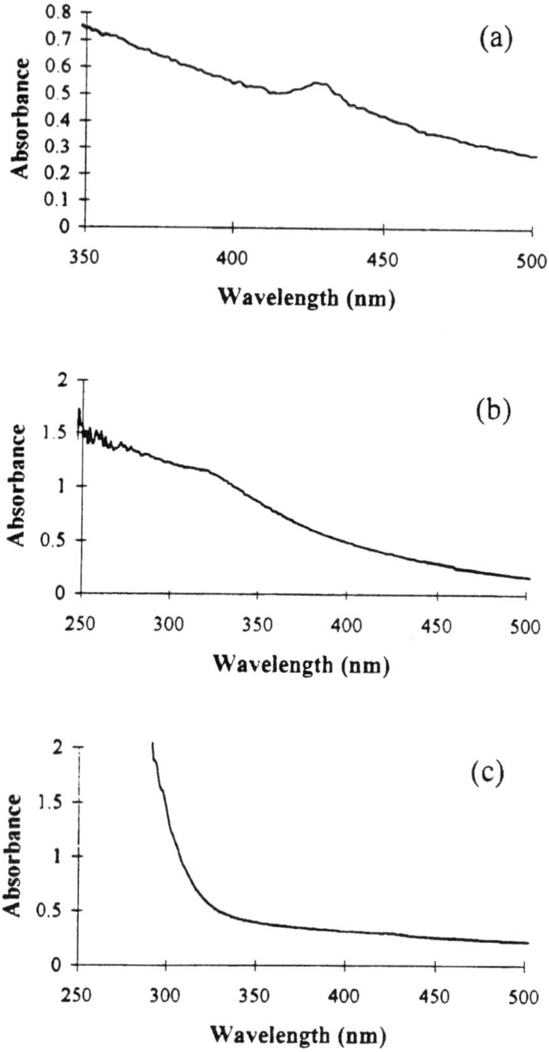

Figure 14 The UV-Vis absorption spectra obtained from the fiberoptic reactor for (a) AgI, (b) AgBr, and (c) AgCl nanoparticles synthesized in supercritical CO_2 by mixing two water-in-CO_2 microemulsions. (From Ref. 14.)

been paid to the changes in band gap energies for small semiconductor particles as the particle radius decreases (56–58). Rosseti et al. explained the particle size dependence of the excited state electronic properties of silver iodide as a consequence of electron and hole localization in the small crystallites (59). The size of the silver iodide nanoparticles can be estimated from the band gap energies of the semiconductor based on the following equation derived by Ravich et al. (60):

$$\sigma h\nu = K(h\nu - E_g)^{1/2} \tag{10}$$

where σ and $h\nu$ are the absorption coefficient and the photon energy, respectively, K is a proportionality constant and E_g is the band gap energy. The average diameter of the nanoparticle, d_p, can be estimated from the band gap energy, E_g, by the same procedures as those described by Hirai et al. (58) for CdS and ZnS and by Sato et al. for silver halide ultrafine particles (54):

$$E_g = E_{g,bulk} + (h^2/2d_p^2)[(1/m_e) + (1/m_h)] - [(3.6e^2)/(4\pi\varepsilon d_p)] \tag{11}$$

where m_e and m_h are the effective mass of an electron and a hole in a semiconductor, respectively. In Eq. (11), $E_{g,bulk}$ is the energy of the band gap of the bulk semiconductor, ε is the dielectric constant of the semiconductor, and e is the charge of an electron. The following parameters (54) were used for the calculations: for AgI, $E_{g,bulk} = 2.83$ eV, $\varepsilon/\varepsilon_o = 4.91$, and $m^*/m_o = 0.2$. For the other symbols, ε_o is the dielectric constant of vacuum, m_o is the rest mass of an electron and m^* is represented by the equation $m^* = (1/m_e + 1/m_h)^{-1}$.

The variation of AgI nanoparticle diameters with respect to time during synthesis in supercritical CO_2 can be calculated using Eq. (11) and the UV-Vis spectroscopic data as shown in Figure 15. The results for silver iodide reflected a sharp increase in particle size in the first few seconds until a stable size of around 3.4 nm in supercritical CO_2 was reached in less than 30 s. For silver

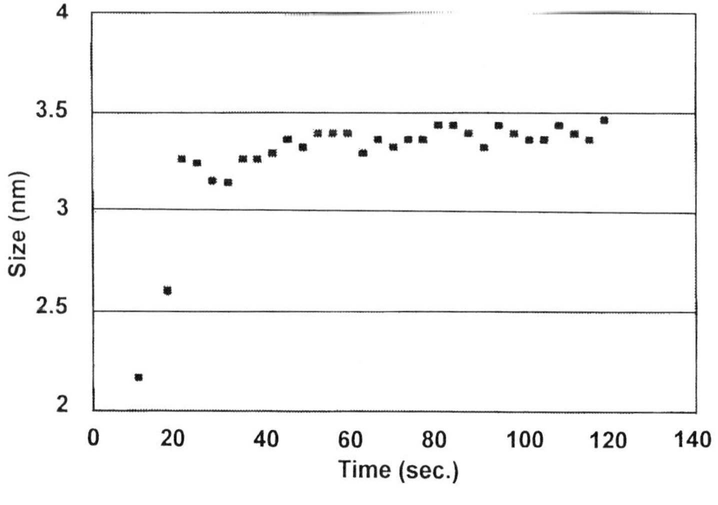

Figure 15 Variation of the size of the synthesized AgI nanoparticles with time calculated from Eq. (11) based on the spectroscopic data reported in Ref. 14.

bromide, a particle size of about 3.0 nm was calculated. The size of the silver chloride particles could not be calculated; since, the effective mass of hole was not available.

The RESS method (53) described in the previous section was used to collect the synthesized silver halide particles. The size distribution of AgI particles collected on copper grit by the RESS method showed that the particle sizes (~3–15 nm) were larger than the calculated AgI size based on the spectroscopic method. To obtain more information on the mechanism of silver halide nanoparticle formation in supercritical CO_2, the absorbance vs. time for the product AgI, AgBr, and AgCl were investigated at 426, 325, and 300 nm, respectively (13). The time required for the absorbance–time curve to reach a plateau was used as a measure of the completion of the reaction. The reaction time followed the order AgI (16 s) > AgBr (28 s) > AgCl (40 s). Formation of silver halide nanoparticles in the water-in-CO_2 microemulsion system involves several processes including the collision, the intermicellar exchange, and the reaction between silver ion and halide ion as shown in Figure 12. The intermicellar exchange process involves the distribution of silver nitrate and sodium halide from water droplets to surfactant phase. This process should depend on the hydrophobicity of sodium halide, i.e., NaI > NaBr > NaCl. The observed reaction speed follows the order of hydrophobicity of sodium halide, suggesting that the most hydrophobic species probably reside close to the interface and exchange faster. The reaction between Ag^+ and I^- is expected to be fast because of the large K_{sp} values for all AgX. The rate-determining step for the formation of silver halide nanoparticles in the water-in-supercritical CO_2 probably is the intermicellar exchange process.

The water-in-CO_2 microemulsion described in this section behaves like a nanoreactor, allowing ionic reactions to take place in the SCF phase. In principle, this system can be used to study chemical reactions and to synthesize nanoparticles involving any aqueous ionic species that are normally not soluble in supercritical CO_2.

C. Electrochemistry in Supercritical CO_2 Utilizing Water-in-CO_2 Microemulsion

Elucidation of solvation characteristics of supercritical fluids is indispensable to their utilization as media for separation or reaction. One powerful method for elucidating chemical equilibrium and solvation in SCFs is voltammetry. However, voltammetric measurement in pure supercritical CO_2 is extremely difficult because CO_2 is nonpolar. Electrochemical processes in several polar SCFs including acetonitrile, ammonia, and sulfur dioxide, were investigated by Bard and coworkers in the late 1980s (61–65). Dombro et al. reported the electrochemical synthesis of dimethyl carbonate from carbon monoxide and methanol in supercritical CO_2-CH_3OH mixtures with a mole fraction of $CH_3OH > 0.35$ (66). Niehaus et al. showed that a well-defined voltammetric wave due to the redox reaction of ferrocene in supercritical CO_2 could be obtained by the addition of small amounts of water and tetrahexylammonium hexafluorophosphate, $(THA)PF_6$ (67). In this case, $(THA)PF_6$ provided a diffusion layer region with ionic conductivity by the formation of a layer of molten salt on the electrode surface.

The water-in-CO_2 microemulsion mentioned previously in this section may provide an effective medium for generating electrical conductivity in supercritical CO_2. In 2000, Ohde et al. first reported the results of voltammetric measurements for the redox reactions of ferrocene (FC) and *N,N,N′N′*tetramethyl-*p*-phenylenediamine (TMPD) in supercritical CO_2 in the presence of a water-in-CO_2 microemulsion (14). The design of their high-pressure electrochemical cell is shown in Figure 16. The same AOT/PFPE-PO_4 water-in-CO_2 microemulsion described in Section IV.A was used in their voltammetric experiments. Well-defined voltammetric waves were obtained for FC and for TMPD in the microemulsion system as shown in Figure 17. An obvious diffusion current for the redox reaction of FC or TMPD was observed. An electrolysis experiment was also performed with TMPD. After the electrolysis at +0.3 V, the UV-Vis absorption spectrum of the sample collected in hexane was measured. The absorption peak wavelength and the shape of the peak were identical to that for $TMPD\cdot^{+}$ in water. The result suggests that $TMPD\cdot^{+}$ produced at the electrode surface was in the water core of the water-in-CO_2 microemulsion, as shown in Eq. (12):

$$TMPD_{(CO2)} + microemulsion_{(CO2)} \rightarrow TMPD\cdot^{+}_{(microemulsion)} + e^{-} \qquad (12)$$

The observation that the electrolysis product that is insoluble in CO_2 can be stabilized in the water core of the microemulsion is significant. It implies a wide range of electrochemical syntheses involving ionic or radical species, as starting materials for synthesis in supercritical CO_2 may be possible utilizing the microemulsion as a medium. Another interesting observation is that the voltammogram obtained from ethanol-modified CO_2 is different from that obtained using the water-in-CO_2 microemulsion as the conductive medium. In the ethanol-modified CO_2 system, a diffusion current for the oxidation of FC was not observed (Figure 17C). This suggests that only adsorbed FC is electroactive and the oxidation product FC^{+} is also adsorbed on the electrode surface. Electrosynthesis in ethanol-modified supercritical CO_2 apparently cannot be done. Using water-in-CO_2 microemulsion as a medium, electrosynthesis in supercritical CO_2 is possible, and this appears to be another promising new technique for preparation of nanomaterials in supercritical CO_2.

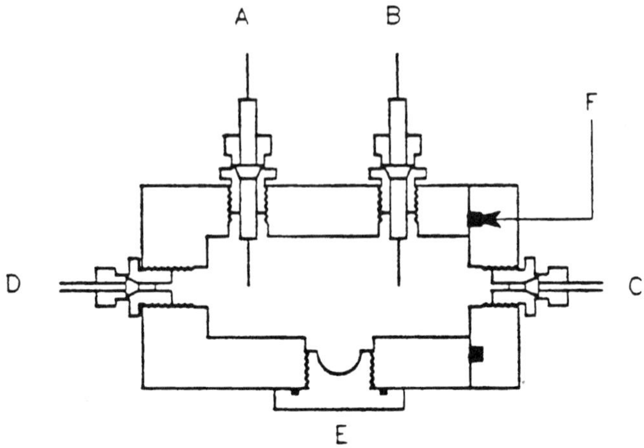

A: Control electrode; B: Reference electrode; C: Supercritical fluid inlet; D: Supercritical fluid outlet; E: Sample port; F: Teflon O-ring.

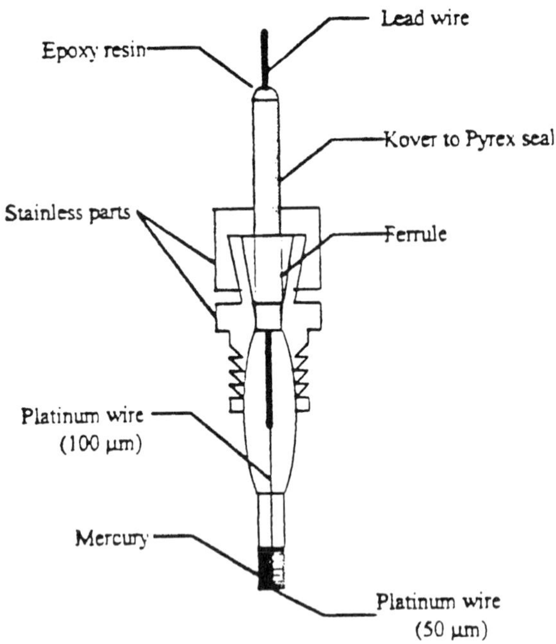

Figure 16 Structures of the high-pressure electrochemical cell and the microelectrode used for the voltammetry experiments in Ref. 15.

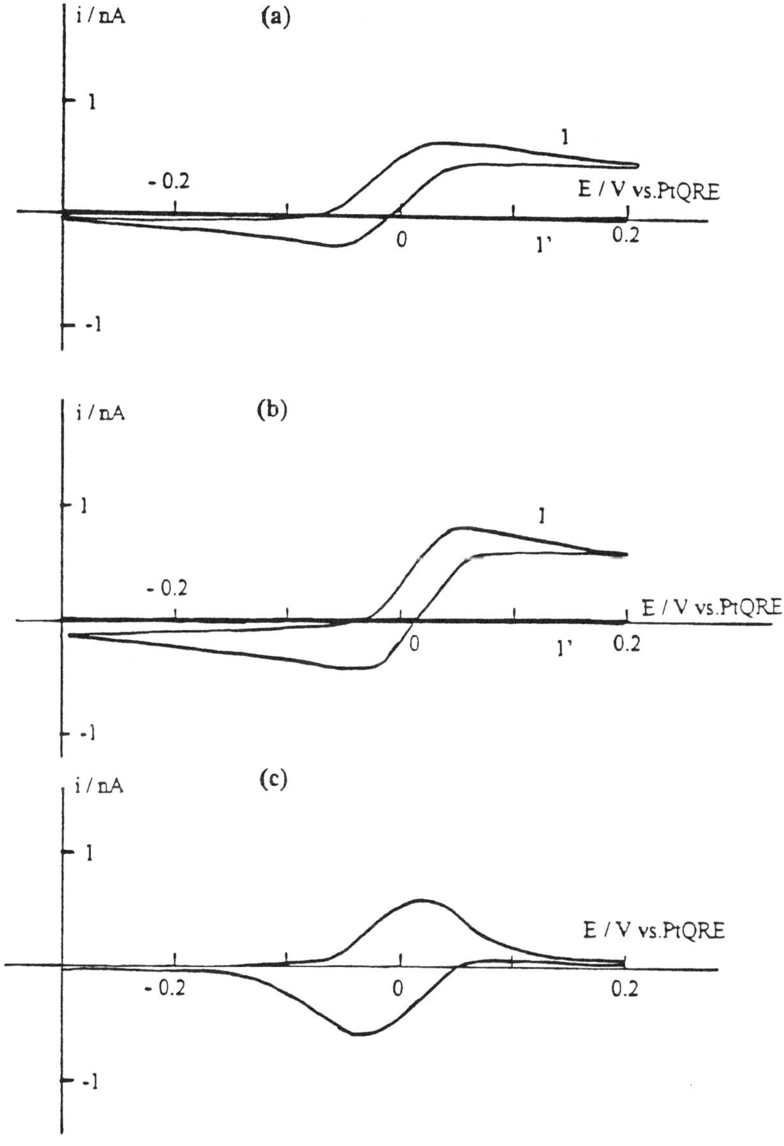

Figure 17 (a) Voltammogram for the redox reaction of ferrocene in the water-in-supercritical CO_2 microemulsion system. (b) Voltammogram for the redox reaction of TMPD in the water-in-supercritical CO_2 microemulsion system. (c) Voltammogram for the redox reaction of ferrocene in the supercritical CO_2–ethanol mixture system. Scan rate: 200 mV s^{-1}. (From Ref. 15.)

V. SUMMARY

Metal species in liquid and in solid materials can be dissolved in supercritical CO_2 containing proper chelating agents. Fluorine- and phosphorus-containing ligands usually form metal chelates that are quite soluble in supercritical CO_2. This in situ chelation/SCF extraction method may be used for treatment of radioisotope- or toxic metals–contaminated waste materials or mixed wastes. Direct dissolution of uranium oxides has also been demonstrated using the ligand-assisted SCF extraction technique. The advantages of using this new extraction technology for metal processing include selectivity, fast kinetics, rapid separation of solutes from solvent, and reduction of waste generation. The ability of this technology to dissolve uranium oxides directly suggests the possibility of developing an SCF-based dry process for reprocessing of spent nuclear fuels.

An important factor in determining the efficiency of metal extraction using the ligand-assisted technique is the solubility of the metal complex in supercritical CO_2. Spectroscopic methods using high-pressure fiberoptic systems are rapid and effective ways of determining solubility and monitoring reaction speed in supercritical CO_2. Several models for predicting solubility of metal complexes in supercritical CO_2 have been reported in the literature. They include solubility parameter approach, solvato complex model, and equation of state calculations. These prediction methods are generally empirical or semiempirical in nature. They provide general guidelines for solubility prediction but not refined enough for accurate calculation of solubility of metal complexes in supercritical CO_2.

The feasibility of using water-in-CO_2 microemulsions as a medium to dissolve and disperse metal species provides new opportunities for metal processing in supercritical CO_2. Metal species that are not soluble in CO_2 can now be dispersed and transported in SCF CO_2 utilizing the microemulsions. Recent reports demonstrating that nanomaterials can be synthesized in supercritical CO_2 utilizing the microemulsions as nanoreactors are significant. A wide range of nanomaterials can now be synthesized in SCFs using ionic species or water-soluble compounds as starting materials in the water cores of the microemulsions. The water-in-CO_2 microemulsions can also be used as a medium for conducting electrochemistry in SCFs. The possibility of performing electrosynthesis in supercritical CO_2 utilizing the microemulsions may provide a new approach for nanomaterials synthesis in SCFs. These new techniques may have tremendous implications for the chemical industries of the 21st century.

REFERENCES

1. MA McHugh, VJ Krukonis. Supercritical Fluid Extraction: Principles and Practice. 2nd ed. Stoneham: Butterworth-Heineman, 1993.

2. CL Phelps, NG Smart, CM Wai. Past, present, and possible future applications of supercritical fluid extraction technology. J Chem Edu 73(12):1163–1168, 1996.
3. CM Wai, KE Laintz. Supercritical fluid extraction. US patent 5,356,538. 1994.
4. KE Laintz, CM Wai, CR Yonker, RD Smith. Solubility of fluorinated metal dithiocarbamates in supercritical carbon dioxide. J Supercrit Fluids 4:194–198, 1991.
5. KE Lainta, CM Wai, CR Yonker, RD Smith. Extraction of metal ions from liquid and solid materials by supercritical carbon dioxide. Anal Chem 64:2875–2878, 1992.
6. Y Lin, NG Smart, CM Wai. Supercritical fluid extraction and chromatography of metal chelates and organometallic compounds. Trends Anal Chem 14:123–133, 1995.
7. J. Darr, M Poliakoff. New directions in inorganic and metal-organic coordination chemistry in supercritical fluids. Chem Rev 99:495–541, 1999.
8. CM Wai. Metal extraction with supercritical fluids. In: R.G. Bautista, ed. Emerging Separation Technologies for Metals II. Warrendale, PA: TMS, the Minerals, Metals & Materials Society, 1996, pp. 233–248.
9. CM Wai, S Wang. Supercritical fluid extraction: metals as complexes. J Chromatogr A 785:369–383, 1997.
10. NG Smart, TE Carleson, T Kast, AA Clifford, MD Burford, CM Wai. Solubility of chelating agents and metal containing compounds in supercritical fluid carbon dioxide—a review. Talanta 44:137–150, 1997.
11. CM Wai, Y Lin, M Ji, KL Toews, NG Smart. Extraction and separation of uranium and lanthanides with supercritical fluids. In: AH Bond, ML Dietz, RD Rogers, eds. ACS Symposium Series 716: Progress in Metal Ion Separation and Preconcentration. Washington, DC: Am Chem Soc, 1999, Chapter 21, pp. 390–400.
12. KP Johnston, KL Harrison, MJ Clarke, SM Howdle, MP Heitz, FV Bright, C Carlier, TW Randolph. Water-in-carbondioxide microemulsions: an environment of hydrophiles including proteins. Science 271:624–626, 1996.
13. M Ji, X Chen, CM Wai, JL Fulton. Synthesizing nd dispersing silver nanoparticles in a water-in-supercritical carbon dioxide microemulsion. J Am Chem Soc 121: 2631–2632, 1999.
14. H Ohde, JM Rodriguez, X Ye, CM Wai. Synthesizing silver halide nanoparticles in supercritical carbon dioxide utilizing a water-in-CO_2 microemulsion. Chem Commun 2353–2354, 2000.
15. H Ohde, F Hunt, S Kihara, CM Wai. Voltammetric measurement in supercritical CO_2 utilizing a water-in-CO_2 microemulsion. Anal Chem 72:4738–4741, 2000.
16. RS Addleman, CM Wai. On-Line time-resolved laser-induced fluorescence of $UO_2(NO_3)_2{\cdot}2TBP$ in supercritical fluid CO_2. Anal Chem 72:2109–2116, 2000.
17. M Carrott, CM Wai. UV-Vis spectroscopic measurement of solubilities in supercritical CO_2 using high pressure fiber optic cells. Anal Chem 70:2421–2425, 1998.
18. Mike J. Carrott, Brenda E. Waller, Neil G. Smart, Chien M. Wai, High solubility of $UO_2(NO_3)_2{\cdot}2TBP$ complex in supercritical CO_2. Chem Commun 373–374, 1998.
19. CM. Wai, B Waller. Dissolution of metal species in supercritical fluids—principles and applications. Ind Eng Chem Res 39:3837–3841, 2000.
20. F Hunt, H Ohde, CM Wai. a high pressure fiber-optic reactor with CCD array UV-Vis spectrometer for monitoring chemical processes in supercritical fluids. Rev Sci Instrum 70(12):4661–4667, 1999.

21. RS Addleman, JW Hills, CM Wai. A high pressure flow cell for on-line absorption, Raman, and time resolved laser induced fluorescence spectroscopy in supercritical fluids. Rev Sci Instrum 69(9):3127–3131, 1998.
22. RS Addleman, MJ Carrot, CM Wai. Determination of solubilities of uranium complexes in supercritical CO_2 by on-line laser induced fluorescence. Anal Chem 72: 4015–4021, 2000.
23. AF Lagalante, BN Hansen, TJ Bruno, RE Sievers. Solubilities of copper(II) and chromium(III) β-diketonates in supercritical carbon dioxide. Inorg Chem 34:5781–5785, 1995.
24. RF Fedor. A method for estimating both the solubility parameters and molar volumes of liquids. Polym Eng Sci 14(2):147, 1974.
25. JC Giddings. High pressure gas chromatography of nonvolatile species. Science 67–73, 1968.
26. CM Wai, S Wang, JJ Yu. Solubility parameter and solubility of metal dithiocarbamates in supercritical carbon dioxide. Anal Chem 68:3516–3519 (1996).
27. NG Smart, TE Carleson, S Elshani, S Wang, CM Wai. Extraction of toxic heavy metals using supercritical fluid carbon dioxide containing organophosphorus reagents. Ind Eng Chem Res 36:1819–1826, 1997.
28. J Chrastil. Solubility of solids and liquids in supercritical gases. J Phys Chem 86: 3016–3021, 1982.
29. TE Carleson, S Chandra, CM Wai, LL Wai, SS Huang. Group contribution method for estimating the solubility of selected hydrocarbon solutes in supercritical carbon dioxide. In: E. Kiran, J.F. Brennecke, eds. Supercritical Fluid Engineering Science, ACS Symposium Series 514, Washington, DC: Am Chem Soc, 1992, Chapter 6, pp. 66–73.
30. W Cross Jr., A Akgerman, C Erkey. Determination of metal-chelate complex solubilities in supercritical carbon dioxide. Ind Eng Chem Res 35:1765–1770, 1996.
31. NG Smart, CM Wai. Supercritical solutions: potential applications in the nuclear industry. Chem Br 34(8):34–36, 1998.
32. Y Lin, NG Smart, CM Wai. Supercritical fluid extraction of uranium and thorium from nitric acid solutions with organophorsphorus reagents. Environ Sci Technol 29:2706–2708, 1995.
33. Y Meguro, S Iso, T Sasaki, Z Yoshida. Solubility of organophosphorus metal extractants in supercritical carbon dioxide. Anal Chem 70:774–779, 1998.
34. Y Meguro, S Iso, Z Yoshida. Correlation between extraction equilibrium of uranium(VI) and density of CO_2 midium in a HNO_3/supercritical CO_2-tributyl phosphate system. Anal Chem 70:1262–1267, 1998.
35. TI Trofimov, MD Samsonov, SC Lee, NG Smart, CM Wai. Ultrasound enhancement of dissolution kinetics of uranium dioxides in supercritical carbon dioxide. J Chem Technol Biotechnol 76:1223–1226, 2001.
36. MD Burford, MZ Ozel, AA Clifford, KD Bartle, Y Lin, CM Wai, NG Smart. Extraction and recovery of metals using a supercritical fluid with chelating agents. Analyst 124:609–614, 1999.
37. S Wang, CM Wai. Supercritical fluid extraction of bioaccumulated mercury from aquatic plants. Environ Sci Technol 30:3111–3114, 1996.

38. CM Wai, S Wang, Y Liu, V Lopez-Avila, WF Beckert. Evaluation of dithiocarbamates and β-diketones as chelating agents in supercritical fluid extraction of Cd, Pb, and Hg from Solid Samples. Talanta 43:2083–2091, 1996.
39. Y Lin, CM Wai, FM Jean, RD Brauer. Supercritical fluid extraction of thorium and uranium ions from solid and liquid materials with fluorinated β-diketones and tributyl phosphate. Environ Sci Technol 28:1190–1193, 1994.
40. KG Furton. L Chen, R. Jaffe. Rapid determination of uranium on solid matrixes by synergestic in situ chelation supercritical fluid extraction and UV absorption spectroscopy. Anal Chim Acta 304:203–208, 1996.
41. KE Laintz, E Tachikawa. Extraction of lanthanide from acidic solution using tributyl phosphate modified supercritical carbon dioxide. Anal Chem 66:2190 2193, 1994.
42. M Ashraf-Khorassani, MT Combs, LT Taylor, Solubility of metal chelates and their extraction from an aqueous environment via supercritical CO_2. Talanta 44:755–763, 1997.
43. M Ashraf-Khorassani, MT Combs, LT Taylor. Supercritical fluid extraction of metal ions and metal chelates from different environments. J Chromatogr A 744:37–49, 1997.
44. CM Wai. Supercritical fluid extraction of trace metals from solid and liquid materials for analytical applications. Anal Sci 11:165–167, 1995.
45. RV Fox, BJ Mincher, RGG Holmes. Extraction of plutonium from spiked INEEL soil samples using the ligand assisted supercritical fluid extraction technique. Idaho National Environmental and Engineering Lab Report, INEEL-EXT-99-00870, p. 13.
46. CM Wai, Y Kulyako, HK Yak, X Chen, SJ Lee. Selective extraction of strontium with supercritical fluid carbon dioxide. Chem Commun 2533–2535, 1999.
47. CM Wai, YM Kulyako, BF Myasoedov. Supercritical carbon dioxide extraction of cesium from aqueous solutions in the presence of macrocyclic and fluorinated compounds. Mendeleev Commun 180–181, 1999.
48. JL Fulton, DM Pfund, JB McClain, TJ Romack, EE Maury, JR Combes, ET Samulski, JM DeSimone, M Capel. Langmuir 11.4241–4249, 1995.
49. TA Hoefling, RM Enick, EJ Beckman. Microemulsions in near-critical and supercritical CO_2. J Phys Chem 95:7127–7129, 1991.
50. RW Gale, JL Fulton, RD Smith. Organized molecular assemblies in the gas phase: reverse micelles and microemulsions in supercritical fluids. J Am Chem Soc 109: 920–921, 1987.
51. K Harrison, J. Goveas, KP Johnston, EA O'Rear. Water-in-carbon dioxide microemulsions with a fluorocarbon-hydrocarbon hybrid surfactant. Langmuir 10:3536–3541, 1994.
52. RG Zielinski, SR Kline, EW Kaler, N Rosov. A small-angle neutron scattering study of water in carbon dioxide microemulsions. Langmuir 13:3934–3937, 1997.
53. RD Smith, DW Matson, JL Fulton, RC Peterson. Rapid expansion of supercritical fluid solutions: solute formation of powders, thin films, and fibers. Ind Eng Chem Res 26:2298–2306, 1987.
54. H Sato, T. Hirai, I Komasawa. Mechanism of formation of silver halide ultrafine particles in reverse micellar systems. J Chem Eng Jpn 29:501–507, 1996.
55. T Ida, H Saeki, K Kimura. The preparation and properties of polycrystals of solid electrolyte ultrafine particles. Surf Rev Lett 3:41–44, 1996.

56. RP Bagwe, KC Khilar. Effects of the intermicellar exchange rate and cations on the size of silver chloride nanoparticles formed in reverse ficelles of AOT. Langmuir 13:6432–6438, 1997.
57. S Chen, T Ida, K Kimura. A novel method for large-scale synthesis of AgI nanoparticles. Chem Commun 2301–2302, 1997.
58. S Chen, T Ida, K Kimura. Thiol-derivatized AgI nanoparticles:synthesis, characterization, and optical properties. J Phys Chem B 102:6169–6176, 1998.
59. R Rossetti, R Hul, JM Gibson, LE Brus. Hybrid electronic properties between the molecular and solid state limits: lead sulfide and silver halide crystallites. J Chem Phys 83:1406–1410, 1985.
60. IY Ravich, BA Efimova, IA Smimov. Semiconducting Lead Chalcogenides. New York: Plenum Press, 1970, pp. 51–53.
61. RM Crooks, FF Fan, AJ Bard. Electrochemistry in near-critical and supercritical fluids. 1. Ammonia. J Am Chem Soc 106:6851–6852, 1984.
62. AC McDonald, FF Fan, AJ Bard. Electrochemistry in near-critical and supercritical fluids. 2. Water. Experimental techniques and the copper(II) system. J Phys Chem 90:196–202, 1986.
63. RM Crooks, AJ Bard. Electrochemistry in near-critical and supercritical fluids. 4. Nitrogen heterocycles, nitrobenzene, and solvated electrons in ammonia at temperatures to 150°C. J Phys Chem 91:1274–1284, 1987.
64. RM Crooks, AJ Bard. Electrochemistry in near-critical and supercritical fluids. 6. The electrochemistry of ferrocene and phenazine in acetonitrile between 25 and 300°C. J Electroanal Chem 243:117–141, 1988.
65. CR Cabrera, AJ Bard. Electrochemistry in near-critical and supercritical fluids. 8. Meethyl viologen, decamethylferrocene, $As(bpy)_3^{2+}$ and ferrocene in acetonitrile and the effect of pressure on diffusion coefficients under supercritical conditions. J Electroanal Chem 273:147–160, 1989.
66. RA Dombro, GA Prentice, MA McHugh. Electro-organic synthesis in supercritical organic mixtures. J Electrochem Soc 135:2219–2223, 1988.
67. D Niehaus, M Philips, A Michael, RM Wightman. Voltammetry of ferrocene in supercritical CO_2 containing water and tetrahexylammoniium heafluoropphosphate. J Phys Chem 93:6232–6236, 1989.

11
Understanding the RESS Process

Markus Weber and Mark C. Thies
Clemson University, Clemson, South Carolina

I. INTRODUCTION

When a supercritical solution that contains a dissolved solute is expanded across a micro-orifice, the solvent density decreases dramatically and the solute is rejected from solution. Petersen et al. (1) were the first to call this process the *r*apid *e*xpansion of *s*upercritical *s*olutions, or RESS for short. Because the characteristic speed of the expansion is the speed of sound, the process is quite rapid, with residence times in the orifice on the order of 1 μs. The rapid pressure reduction across the expansion nozzle leads to both uniform conditions and very high supersaturation ratios in the postexpansion environment. These two characteristics are a key feature of RESS and favor the formation of small particles with narrow size distributions (2). When materials such as polymers are used, other product morphologies are possible. For example, RESS solutions can be sprayed to form thin films (3). In other cases, the very high extensional rates in the expansion nozzle can be used to make microfibers (4–6).

There have been several reviews of RESS over the past decade, with the most comprehensive being the 1991 work of Tom and Debenedetti (7), as it discusses both theory and experimental work in detail. An updated review of their modeling work was presented 2 years later (8). In more recent years, reviews have become more general, discussing RESS as one of several alternatives for processing materials with supercritical fluids (9–11). Such a development is, of course, not surprising, as many of the other techniques (such as supercritical antisolvent (SAS) and precipitation with compressed antisolvent (PCA) processes) have been developed to overcome one of the disadvantages of RESS, namely, the limited solubility of many materials in supercritical carbon dioxide.

In the classic sense, this chapter is not a review of the literature, although many sources that are relevant to understanding RESS were consulted in the preparation of this work. Instead, as implied by the title, we have written this chapter to talk about some of the issues surrounding both the experimental investigation and theoretical modeling of RESS, with an emphasis on those that we believe are important but that are at best only briefly mentioned in the current RESS literature.

This chapter is arranged as follows. In Section II, experimental apparatuses for carrying out RESS are described. In Section III, stagnation conditions and the importance of their proper definition for both RESS experiments and modeling are discussed. In Section IV, the issue of adequate heat transfer to the expansion nozzle is analyzed; afterward, a one-dimensional model for the flow of pure supercritical solvent inside the RESS nozzle is presented. In Section V, we move outside the RESS nozzle and into the free jet region, and use classic compressible flow theory to reveal interesting characteristics of the transsonic and supersonic regions. Section VI presents one-dimensional modeling results for selected examples of the plain orifice and long capillary. Finally, in Section VII we derive a model for RESS that considers both nucleation and condensation (but not coagulation) and discuss the implication of the results for producing nanoparticles.

II. DESCRIPTION OF THE RESS PROCESS

As shown in Figure 1, the RESS process can be considered to consist of four steps:

1. Preparation of the supercritical solution, typically by dissolution of the solute in the supercritical solvent.
2. Setting of the preexpansion conditions just upstream of the expansion device.
3. Rapid expansion of the supercritical solution from the pre-expansion conditions to ambient or near-ambient conditions through a device such as a micro-orifice or capillary.
4. Recovery of the product in the expansion chamber.

The method for preparing the supercritical solution depends on the properties of the solvent and solute. If the supercritical solvent is a liquid at ambient conditions, then dissolution of the solute can be carried out with any kind of conventional mixing device, such as a vessel with stirrer. Typically, however, the solvent is gaseous at ambient conditions and is available as a compressed liquid or a fluid of liquid-like density (e.g., carbon dioxide). In this case, the supercritical solution is prepared by pumping the carbon dioxide (CO_2) through

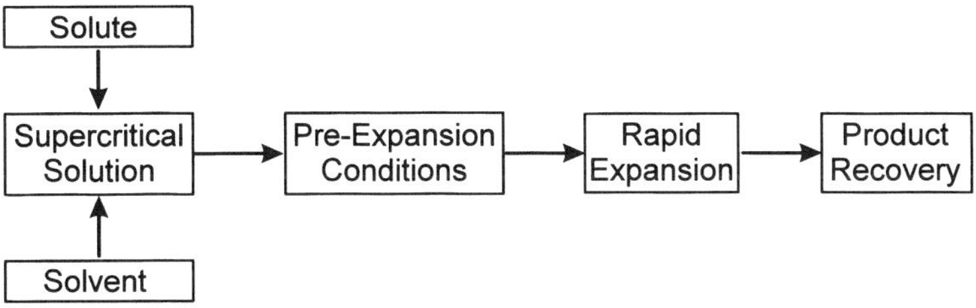

Figure 1 Simplified block diagram of the RESS process.

a packed column (also called the extraction column) consisting of the solid solute of interest mixed with an inert substance, such as glass beads or glass wool (12,13). The so-called "flow" RESS apparatus shown in Figure 2 includes such a setup, as liquefied CO_2 is delivered from a cylinder and through the extraction column at a defined extraction temperature (T_{ext}) and pressure (P_{ext}). If the mean residence time of the solvent in the extraction column is sufficient, the solute attains its equilibrium value at the column outlet. Sometimes multiple columns are required for the solute mole fraction to attain the equilibrium concentration, especially for sparingly soluble systems (14,15). In any case, the operator should always ensure that equilibrium conditions have been established at the extraction column exit; otherwise, the solute concentration will be unknown. If a liquid solute is used, care must be taken to avoid liquid entrainment in the solution exiting the column. As shown in Figure 2, most flow RESS apparatuses

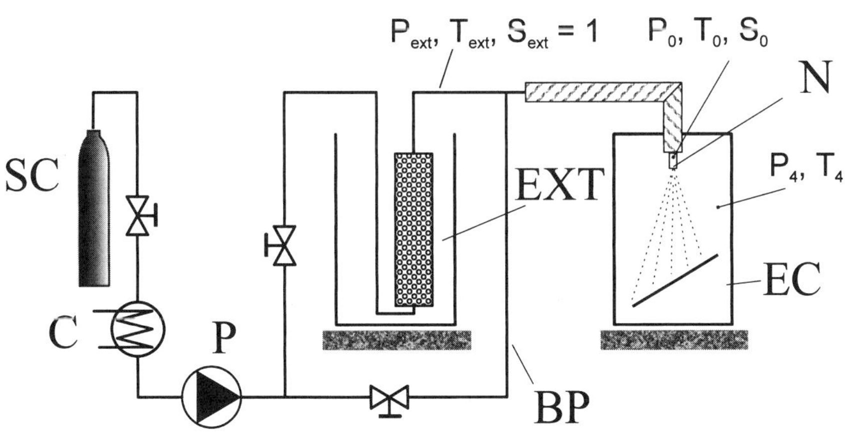

Figure 2 Schematic of a "flow" RESS apparatus: SC, solvent cylinder; C, solvent cooler; P, solvent pump; EXT, extraction column; BP, bypass line for solvent; N, nozzle; EC, expansion chamber.

include a solvent bypass line to establish steady state in the expansion device before charging the solution of interest. This line can also be used to mix the extraction column effluent with pure solvent, thereby adjusting the solute concentration to any desired value lower than the equilibrium concentration under extraction conditions (13,15).

At this point in our discussion, it is useful to define the degree of saturation, S:

$$S = \frac{y}{y^{\mathrm{eq}}} \tag{1}$$

where y is the mole fraction of solute at the temperature (T) and pressure (P) of interest and y^{eq} is the equilibrium mole fraction at the same location in the process, and thus at the same T and P. y is defined assuming that a single phase exists at a given T and P (it may or may not), and thus is an overall mole fraction. For example, referring to Figure 2, the solution leaving the extraction column is saturated, so that $S_{\mathrm{ext}} = 1$.

Another technique for preparing the supercritical solution is to use a batch setup to prepare the supercritical solution; this method has been used primarily for polymers (5,16). A solid polymer is loaded into a variable-volume view cell, and the cell is sealed. Solvent is then charged to the cell via tubing, and polymer and solvent are mixed together at the desired temperature and pressure with a stirring bar to form a homogeneous phase. In comparison with the flow RESS apparatus, the "batch" RESS apparatus has the advantage that the concentration of the solute can be precisely fixed to any value up to the equilibrium concentration, i.e., $S \leq 1$. On the other hand, a relatively small batch of solution is usually expanded, significantly shortening the time available for a RESS experiment.

After a supercritical solution of the desired concentration has been prepared, the conditions just upstream of the expansion device (i.e., the pre-expansion conditions) are established. Whether a flow (Figure 2) or a batch apparatus is used, the pre-expansion pressure (P_0) is determined by the solvent pump and is held constant. The pre-expansion temperature (T_0) is then established by heating the expansion device (i.e., nozzle) and frequently also the tubing that leads to the nozzle. Several issues are involved in setting this temperature. First, T_0 should be set high enough to prevent condensation of the supercritical solvent upon expansion (12). Second, the phase behavior of the solvent–solute system must be considered before deciding on the desired T_0. Thus, for systems that are operated isobarically at pressures above the retrograde region, heating the nozzle increases the solubility of the solute in the solvent (and reduces S_0 to <1) and helps ensure that no precipitation of the solute occurs before the actual rapid expansion process. However, for a system being operated in the retrograde region of pressure, heating the nozzle decreases the solute solubility and causes a saturated solution to precipitate prior to the expansion ($S_0 > 1$). Polymer–solvent systems typically exhibit retrograde behavior, as they can be easily heated to the point at which a liquid–liquid phase split occurs ($S_0 > 1$) (4,5,16). The third issue that needs to be considered when setting T_0 is the thermal stability of the solute. For thermally labile materials, such as pharmaceuticals, an upper limit of pre-expansion temperature must be observed (17,18). The issues surrounding heat transfer and the temperatures used in RESS are analyzed in detail in Section IV.A of this chapter.

After the pre-expansion conditions are established, the supercritical solution is rapidly expanded across a nozzle (Figure 2). During expansion, the density of the supercritical solvent decreases dramatically to gas-like values, resulting in very high supersaturations and the rejection of the solute from solution. Another characteristic of RESS is that the nucleation process is initiated by pressure reduction, which travels at the speed of sound and thus favors the rapid attainment of uniform conditions within the expanding fluid. Typically, the nozzle is a micro-orifice; inner diameters (i.d.'s) from 25 to 100 μm and L/D ratios of 3 to 20 have been used. Capillaries with similar i.d.'s have also been used, but with L/Ds ranging from about 150 to as much as 6000. Of course, the term "rapid expansion" is somewhat a misnomer in the case of the long capillary, as expansion times approach 0.1 s (vs. 10^{-6} s for a micro-orifice). An illustration of the differences in RESS conditions (e.g., outlet temperature, pressure, and velocity) induced by a micro-orifice vs. a long capillary is given in Section VI.

The final step in RESS is the recovery of product in the expansion chamber. Figure 2 shows a common technique used for the RESS of organic crystals: collection onto a glass slide. Paper filters are also frequently employed as the substrate (19). For polymers, scanning electron microscopy (SEM) stages covered with sticky carbon tape are commonly used (5,16). Although the expansion chamber is usually operated at ambient temperature and pressure, it should still be completely contained so that the gaseous solvent can be vented to an exhaust hood (or even condensed and recycled if CO_2 is not being used). Some workers also measure the solvent flow rate as it leaves the expansion chamber (15,18,19). If the expansion chamber is being used to crystallize pharmaceuticals, both temperature control and above-ambient pressure control of the chamber can be used to control postexpansion crystal growth. In this case, an inert gas or CO_2 is used to control the chamber pressure (15).

For particle size characterization, optical microscopy and SEM are commonly employed. Recently, a three-wavelength extinction technique has been used to determine the median diameter of RESS particles on-line (20). Other characterization techniques, such as differential scanning colorimetry (DSC) and x-ray diffraction (XRD), are used to determine crystallinity.

III. DEFINITION AND MEASUREMENT OF STAGNATION CONDITIONS

One of the challenges in modeling the RESS process is to describe the changes in the properties of the supercritical fluid as it flows through the expansion device. Such changes are most conveniently handled if we define some kind of a consistent reference state. A reasonable choice, and the one that is chosen here, is that of a fluid with zero velocity, known as the stagnation state. Two stagnation state quantities that are of particular interest to us are the stagnation temperature T^0 and the stagnation pressure P^0. T^0 is defined as the temperature that a fluid in a given location would have if brought to zero velocity adiabatically, while P^0 is the pressure of a fluid that is brought to zero velocity isentropically (21). The conditions well upstream of the nozzle (i.e., T_0 and P_0 in Figure 2) are typically at stagnation conditions because of the low flow speed at this location; thus $T_0 = T_0^0$) and $P_0 = P_0^0$. Finally, it should be noted that the static temperature T and static pressure P of a fluid in steady motion are always lower than their corresponding stagnation values.

Now consider the rapid expansion of a supercritical fluid through the expansion nozzle depicted in Figure 3, which is in use in our laboratory and is similar to those typically reported in the literature (19,20). A tapered inlet (usually 120° angle) is followed by a cylindrical capillary section in which L/D typically ranges from 3 to 6000. Given the comparatively low viscosity of a supercritical fluid, the effects of acceleration and friction on the pressure are weak at the low flow speeds that exist upstream of the nozzle in the process tubing. Consequently, the pressure drop up to this point is small. However, when the fluid passes into the tapered inlet section of a typical RESS nozzle, the slow expansion gradually turns rapid. If, to simplify analysis, we subdivide the process into an isobaric part followed by a rapid expansion, the question of where the rapid expansion truly starts must be addressed.

A safe guess for the initiation of rapid expansion would be that cross section of the tapered inlet in which the fluid has attained an average velocity 1/100 of that at the exit of the capillary. Neglecting the effects of compressibility and boundary layers, and assuming there is no flow contraction, this cross section would have a diameter D_{ini} 10 times greater than that of the capillary, where $D_1 = D_2$. The axial distance z_{ini} from this cross section to the entrance of the capillary can be readily calculated as follows:

$$\frac{z_{\text{ini}}}{D_2} = \frac{(10-1)}{2 \cdot \tan(\gamma/2)} = \frac{4.5}{\tan(\gamma/2)} \tag{2}$$

For $\gamma = 120°$, Eq. (2) yields $z_{\text{ini}} \approx 2.6 \cdot D_2$ (see Figure 3b). Accounting for the compressibility and the flow contraction leads to even shorter distances for

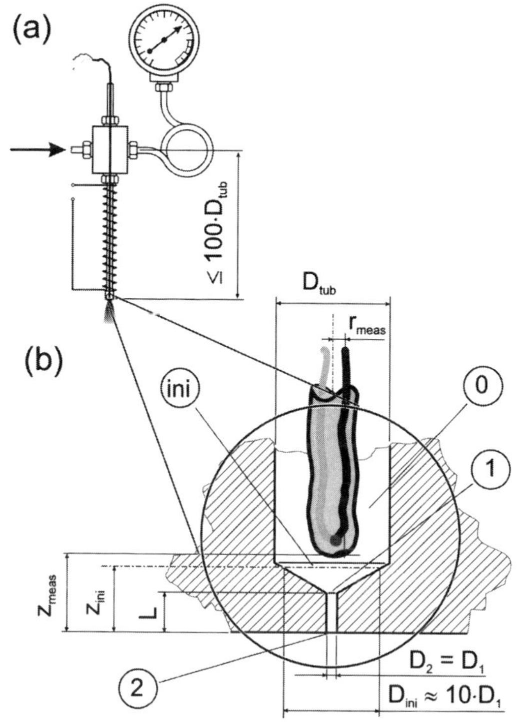

Figure 3 (a) RESS nozzle and associated components. (b) Longitudinal cross section of the inlet and capillary sections of a RESS nozzle with measuring thermocouple: 0, pre-expansion conditions; ini, initiation of rapid expansion; 1, capillary entrance; 2, capillary exit.

z_{ini}, while the effect of a boundary layer (which can be approximated by the choice of a smaller angle γ than that given by the nozzle inlet geometry) leads to greater distances for z_{ini}. Thus, the estimate for z_{ini} is a reasonable one for supercritical systems.

Now consider the pressure drops that occur during rapid expansion. Applying the Bernoulli equation for an inviscid, incompressible fluid, we can readily show that the pressure drop from pre-expansion conditions to the initiation of rapid expansion (i.e., from P_0 to P_{ini}) is smaller by a factor of 10^4 than the pressure drop from pre-expansion conditions to the nozzle outlet (i.e., from P_0 to P_2). Now for a typical RESS application, the pre-expansion pressure P_0 is about 200 bar. Because of the choked-flow conditions that exist at the nozzle exit, the pressure drop is about 100 bar. Accordingly, the pressure at z_{ini} (i.e., P_{ini}) is only about 0.01 bar less than P_0 for incompressible flow, and the pressure drop can be shown to be even less for compressible flow. With such a small pressure drop, we can safely set the static and stagnation pressures at the beginning of rapid expansion equal to each other (i.e., $P_{\text{ini}} = P^0_{\text{ini}}$). Analogous arguments to those made above for pressure can be used to show that the static temperature at the beginning of rapid expansion is essentially equal to the stagnation temperature at this same location (i.e., $T_{\text{ini}} = T^0_{\text{ini}}$). In summary, when discussing the RESS process, we will always define the stagnation pressure and temperature at the beginning of rapid expansion as the static values at P_{ini} and T_{ini}.

Now that we have required that the stagnation pressure P^0_{ini} be at the location z_{ini}, the question of just how far upstream of the nozzle entrance the pressure can be measured without loss of accuracy needs to be addressed. Using the definition of the Fanning friction factor to estimate the pressure drop for fully developed, turbulent flow (22), one can show that P_0 will increase by less than 0.10 bar as much as 100 inner tubing diameters upstream of the nozzle entrance as long as the flow speed inside the tubing does not markedly exceed 3 m/s (Figure 3a).

Unfortunately, the accurate measurement of the stagnation temperature at the beginning of rapid expansion (i.e., at T_{ini}) is less straightforward. Typically, the measuring thermocouple is placed at a distance z_{meas} upstream of the nozzle inlet, measuring the temperature T_{meas} at a radial position r_{meas} in the pre-expansion section (Figure 3b). However, ensuring that T_{meas} is essentially equal to T_{ini} is a nontrivial matter. As shown by Eq. (3) below, any kind of heat flux $\dot{q}(z)$ between the inner wall of the tubing and the fluid over the interval from z_{meas} to z_{ini} produces a temperature difference:

$$T_{\text{ini}} - T_{\text{avg}} = \frac{\pi}{\dot{m}} \cdot \int_{z_{\text{meas}}}^{z_{\text{ini}}} \frac{\dot{q}(z)}{c_{\text{p}}(z)} D_{\text{tub}}(z)dz \tag{3}$$

where the average temperature T_{avg} over the tubing cross section at z_{meas} is given by

$$T_{\text{avg}} = \frac{\displaystyle\int_0^{D_{\text{tub}}/2} T(z_{\text{meas}}, r) \cdot u(z_{\text{meas}}, r) \cdot c_{\text{p}}(z_{\text{meas}}, r) \cdot \rho(z_{\text{meas}}, r) 2\pi r dr}{\displaystyle\int_0^{D_{\text{tub}}/2} u(z_{\text{meas}}, r) \cdot c_{\text{p}}(z_{\text{meas}}, r) \cdot \rho(z_{\text{meas}}, r) 2\pi r dr} \tag{4}$$

Here c_{p} is the heat capacity of the fluid, ρ is the fluid density, D_{tub} is the tubing diameter, and u is the axial fluid velocity. In Eq. (3), $\dot{q}$, c_{p}, and D_{tub} are evaluated over the interval dz between z_{meas} and z_{ini}, while in Eq. (4), T, u, c_{p}, and ρ are evaluated at the cross section z_{meas} over the differential element $2\pi r dr$ from the tubing center to the wall. Equation (4) reflects the fact that there is a temperature profile across the tubing cross section at z_{meas}, such that

$T_{meas}(z_{meas}, r_{meas}) \neq T_{avg}$. Both of these factors are minimized by the use of a well-insulated nozzle (i.e., no heat flux from the fluid toward the wall or vice versa) in which the relative distance between z_{ini} and z_{meas} is very small. Figure 3 shows a nozzle that was especially designed in our laboratory (23) to minimize this distance, as a thin thermocouple (i.d. = 0.5 mm) is placed close to the point in the tapered section where rapid expansion essentially begins. It should be noted that for a conventional RESS setup described in the literature, the distance between z_{meas} and z_{ini} is relatively large, i.e, from several millimeters up to 1 cm. Furthermore, many nozzles are heated all the way to the nozzle tip, increasing the uncertainty in the stagnation temperature T_{ini}.

The unwanted deviation between T_{meas} and T_{avg} can also be reduced if turbulent flow conditions are maintained upstream of the nozzle. [In fact, Halverson (24) reported that, under laminar flow conditions, the temperature measured with a thermocouple that touches the inner tubing wall can be closer to T_{meas} than the one measured along the central axis of the flow.] To estimate the maximum tubing diameter that can be used upstream of the expansion device (i.e., while still maintaining turbulent flow), we proceed as follows: The Reynolds number of the fluid at pre-expansion conditions under turbulent flow is given by

$$\mathrm{Re}_0 = \frac{D_0 \cdot u_0 \cdot \rho_0}{\eta_0} = \frac{4}{\pi} \cdot \frac{\dot{m}_0}{D_0 \eta_0} \geq 2350 \tag{5}$$

where pre-expansion conditions are denoted by the subscript "0" (Figure 3b). Now the mass flow rate at pre-expansion conditions must equal that at the nozzle exit conditions of choked flow (where Ma = 1), so that

$$\dot{m}_0 - \dot{m}_2 = \rho_2 \ u_2 \cdot \frac{\pi}{4} D_2^2 \tag{6}$$

where, as shown in Figure 3b, the subscript "2" refers to conditions at the nozzle exit. Inserting Eq. (6) into Eq. (5) yields an expression representing the turbulent flow conditions, which can be translated into a criterion for the choice of an appropriate tubing size if the diameter of the capillary section is known:

$$\mathrm{Re}_0 = \frac{D_2}{D_0} \cdot \frac{D_2 \cdot u_2 \cdot \rho_2}{\eta_0} \tag{7}$$

Thus

$$\frac{D_0}{D_2} \leq \frac{D_2 \cdot u_2 \cdot \rho_2}{\eta_0 \cdot \mathrm{Re}_0} \tag{8}$$

Inserting typical values for the majority of most experimental setups (e.g., $u_2 =$ 250 m/s, $\rho_2 = 400$ kg/m^3, $\eta_0 = 50$ μPa/s, and $D_2 = 50$ μm), we find that turbulent flow in the pre-expansion section is achieved if $D_0/D_2 < 43$. In other words, for an experimental setup with a typical orifice or capillary diameter of 50 μm, the pre-expansion tubing diameter should be no greater than 2 mm.

Most of the thermocouples and RTDs suitable for high-pressure experiments have outer diameters ranging from 0.25 mm to 1/16 in. (1.58 mm), which is one obvious lower limit for the tubing diameter. Moreover, if the thermocouple inserted into the high-pressure tubing occupies a large fraction of the available cross-sectional area, the Reynolds number can decrease such that flow becomes laminar; in addition, the probability that the thermocouple will touch the tubing wall increases. Another criterion for the maximum pre-expansion tubing diameter is given by our earlier assumption that rapid expansion starts where the cross section of the channel is about 10 times greater than the capillary diameter [see Eq. (2)]. In summary, we find relatively restrictive guidelines for the recommended ratio of pre-expansion tubing diameter to nozzle diameter in a RESS apparatus:

$$10 \leq \frac{D_0}{D_2} \leq 100 \tag{9}$$

The upper limit in Eq. (9) [which is from Eq. (8)] is for a nozzle diameter of about 100 μm. Even less flexibility is available for smaller (e.g., 50 μm) nozzles, whereas the designer has greater freedom in the case of larger capillary diameters ($D_2 > 100$ μm).

IV. SUBSONIC FLOW THROUGH THE NOZZLE DURING RESS

A. Heat Transfer Considerations: A Zero-Dimensional Analysis

Under typical RESS operating conditions, supercritical fluids such as CO_2 must be heated to prevent a significant temperature drop during the expansion process. In general, the actual heating methods used during RESS fall between two idealized limiting cases: (a) the fluid in the pre-expansion section upstream of z_{ini} is brought up to the desired pre-expansion temperature and is then expanded adiabatically through the nozzle, or (b) both the pre-expansion section and the nozzle are heated so that the expansion is isothermal. In the zero-dimensional analysis that follows, we explore the applicability of assuming that the heat transferred from the inner wall of the nozzle to the fluid during expansion can be neglected. In other words, how valid is this assumption of adiabatic expansion that is frequently made in the literature?

We begin our considerations with a duct of length L through which a compressible fluid is expanded adiabatically from a state at the entrance of the duct, where the Mach number $\mathrm{Ma}_{ini} \approx 0$, to sonic conditions at the outlet, where $Ma_2 \equiv 1$. For illustrative purposes, Figure 4a shows an enlargement of the duct given in Figure 3b, but at this point in the analysis the true shape of the duct is undefined. Precluding any specific information about the shape of the duct, we somewhat arbitrarily assume that the temperature of the fluid can be represented by a linear function, dropping from $T_{\mathrm{fl,ini}}$ to $T_{\mathrm{fl,2ad}}$ as z varies from $z_{\mathrm{ini}} = 0$ to $z_2 = L$. (As it is adiabatically expanded, the typical supercritical fluid will always experience a temperature drop, with the limits being between that which occurs during isentropic and isenthalpic expansions.) This is expressed by Eq. (10):

$$\left(\frac{dT_{\mathrm{fl}}}{dz}\right)_{\mathrm{ad}} = -\frac{(T_{\mathrm{fl,ini}} - T_{\mathrm{fl,2ad}})}{L} \tag{10}$$

Now for a real expansion process, some degree of heating or cooling occurs because of heat transfer through the walls of the duct. For a differential length

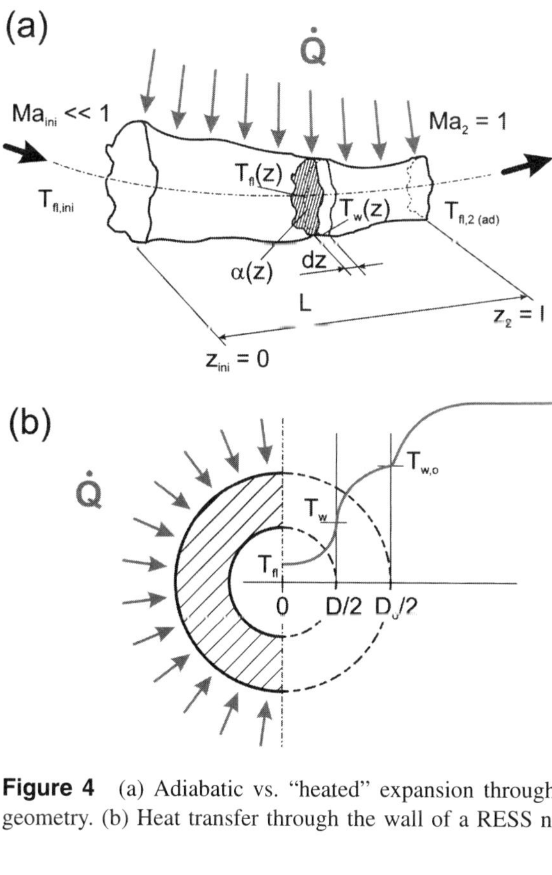

Figure 4 (a) Adiabatic vs. "heated" expansion through a RESS nozzle of undefined geometry. (b) Heat transfer through the wall of a RESS nozzle.

of duct dz we describe this heating process by the following simple expression for the change in fluid temperature with duct length:

$$\left(\frac{dT_{\mathrm{fl}}}{dz}\right)_{\mathrm{ht}} = \frac{c_{\mathrm{h}} \cdot \varsigma}{\dot{m} \cdot c_{\mathrm{p}}} \cdot [T_{\mathrm{w}}(z) - T_{\mathrm{fl}}(z)] \tag{11}$$

Here c_{h} denotes the heat transfer coefficient in W/m^2 K, ς the circumference of the channel cross section, $\dot{m}$ the mass flow rate, and c_{p} the heat capacity of the fluid, respectively. The wall temperature and fluid temperature in dz are given by $T_{\mathrm{w}}(z)$ and $T_{\mathrm{fl}}(z)$, respectively (Figure 4a). Of course, the above expression is valid only for the heating of an incompressible fluid but nonetheless is useful for the arguments that we wish to make below. Introducing the continuity equation $\dot{m} = \rho \cdot u \cdot \alpha$ (where α is the cross-sectional area of the duct) and the dimensionless Stanton number (St) (25,26) into Eq. (11), we obtain

$$\left(\frac{dT_{\mathrm{fl}}}{dz}\right)_{\mathrm{ht}} = \frac{\varsigma}{\alpha} \cdot \mathrm{St} \cdot [T_{\mathrm{w}}(z) - T_{\mathrm{fl}}(z)] \tag{12}$$

where $\mathrm{St} \equiv c_{\mathrm{h}}/u \cdot \rho \cdot c_{\mathrm{p}}$.

Now the actual temperature gradient that occurs during the rapid expansion of a supercritical fluid can be approximated by summing the "purely adiabatic" case of Eq. (10) with the "pure heating" case of Eq. (12):

$$\left(\frac{dT_{\mathrm{fl}}}{dz}\right) = \left(\frac{dT_{\mathrm{fl}}}{dz}\right)_{\mathrm{ad}} + \left(\frac{dT_{\mathrm{fl}}}{dz}\right)_{\mathrm{ht}} \tag{13}$$

If we assume that the wall temperature, T_{w}, can always be maintained constant at the fluid inlet temperature, $T_{\mathrm{fl,ini}}$, then the actual fluid outlet temperature, $T_{\mathrm{fl,2}}$, is obtained by integrating the resulting ordinary differential equation to obtain

$$\Rightarrow \left(\frac{(T_{\mathrm{fl,2}} - T_{\mathrm{fl,2ad}})}{(T_{\mathrm{fl,ini}} - T_{\mathrm{fl,2ad}})} = 1 - \frac{\left[1 - \exp\left(-\overline{\frac{\varsigma}{\alpha} \cdot \mathrm{St}} \cdot L\right)\right]}{\overline{\frac{\varsigma}{\alpha} \cdot \mathrm{St}} \cdot L} \tag{14}$$

Here we have assumed that an average value of the grouping $\varsigma/\alpha \cdot \mathrm{St}$ can be used for the entire duct. Next, we incorporate the Reynolds analogy (25,26), which relates the heat transfer rate to the frictional loss for turbulent flow through a tube of diameter D:

$$\mathrm{St} \equiv \frac{\mathrm{Nu}}{\mathrm{Re\ Pr}} = \frac{f}{2} \tag{15}$$

where f is the Fanning friction factor, which correlates the pressure drop per unit length with the product of the velocity head and density of the bulk flow:

$$-\frac{dP}{dz} = f \cdot \left(\frac{\varsigma}{\alpha} \cdot \frac{\rho}{2} u^2\right) \tag{16}$$

The Reynolds analogy is derived assuming that heat and momentum are transported at the same rate, i.e., that Pr = 1. [It is interesting to note that Pr $\approx$ 1–2 for dense supercritical gases (22).] The Reynolds analogy is substituted into Eq. (14) to obtain the final, desired expression for tubing of length L and diameter D:

$$\frac{(T_{\text{fl},2} - T_{\text{fl},2\text{ad}})}{(T_{\text{fl,ini}} - T_{\text{fl},2\text{ad}})} = 1 - \frac{\left[1 - \exp\left(-2f \cdot \dfrac{L}{D}\right)\right]}{2f \cdot \dfrac{L}{D}} \tag{17}$$

The temperature ratio on the LHS of the above equation indicates how close an expansion is to being perfectly adiabatic (LHS = 0) vs. being perfectly isothermal (LHS = 1.0). Inserting typical values of the friction factor and L/D for the archetypical cases of the plain orifice and the long capillary we obtain the following:

$$\text{Plain orifice:} \quad f = 0.0125, \quad L/D = 5 \rightarrow \frac{(T_{\text{fl},2} - T_{\text{fl},2\text{ad}})}{(T_{\text{fl,ini}} - T_{\text{fl},2\text{ad}})} = 0.06 \tag{18}$$

$$\text{Long capillary:} \quad f = 0.005, \quad L/D = 5000 \rightarrow \frac{(T_{\text{fl},2} - T_{\text{fl},2\text{ad}})}{(T_{\text{fl,ini}} - T_{\text{fl},2\text{ad}})} = 0.98 \tag{19}$$

Thus, if one maintains the inner tubing wall of a long capillary at the temperature of the entering fluid, the expansion will be close to an isothermal process. For a plain orifice, however, such heating will be decidedly inadequate, and the expansion will be approximately adiabatic. Furthermore, the resulting temperature ratio for the plain orifice will be even lower if we take into account the fact that the expansion is primarily a result of acceleration in the tapered inlet section rather than one of friction inside of the cylindrical part. Although simplifying assumptions were used to obtain the above relationships, the assumptions were applied uniformly to both cases, so our conclusions are qualitatively correct. Finally, we note that the effect of heating the duct depends only on a dimensionless length scale and is therefore virtually independent of the absolute value for the residence time, $\tau = L/\bar{u}$, of the fluid inside the duct.

In the above analysis, we assumed that T_w could be heated no hotter than $T_{\text{fl,ini}}$. Ideally, one would always want to operate a RESS process in this manner so as to avoid problems such as overheating. But, as we have seen above, this is not possible for the case of the plain orifice. However, operating the orifice as an essentially adiabatic expansion is not a viable option either, as freezing of the solution at the nozzle can occur. In fact, cooling can even be more pronounced than for the adiabatic case, as the highly accelerated gas stream exiting at the tip of the nozzle can cause further cooling of the front surface of the expansion nozzle. Thus, in the analysis below, we relax the restriction that $T_w = T_{fl,ini}$ and consider the extent to which the outer wall temperature in the expansion duct, $T_{w,o}$, has to be heated so that the inner wall temperature, T_w, is greater than the bulk fluid temperature, T_{fl}.

With RESS, the heating power $\dot{Q}$ is supplied by a heat source that is separated from the inner wall of the expansion device by a solid wall of a thickness suitable to contain the applied static pressure. We assume that the heat is uniformly supplied on the outer surface of tubing of length L and outer diameter D_o (Figure 4b). The heat flow through the tubing wall by conduction must be equal to the heat flow into the fluid from the inner wall:

$$\dot{Q} = k_w \frac{2\pi \cdot L}{\ln(D_o/D)} \cdot (T_{w,o} - T_w) = \bar{c}_h \cdot \pi DL \cdot (T_w - T_{fl})_{avg} \tag{20}$$

where k_w is the thermal conductivity of the wall material, $\bar{c}_h$ is the average heat transfer coefficient, and $(T_w - T_{fl})_{avg}$ is the average temperature difference between the inner wall and bulk fluid temperature. The resulting ratio of the temperature differences becomes a Biot number multiplied by the logarithm of the ratio of the outer and inner tubing diameters:

$$\frac{(T_{w,o} - T_w)}{(T_w - T_{fl})_{avg}} = \mathrm{Bi} \cdot \ln(D_o/D) \tag{21}$$

where $\mathrm{Bi} \equiv (\bar{c}_h/k_w)(D/2)$.

An alternative form of Eq. (21) is obtained by substituting the definition of the Nusselt number ($\mathrm{Nu} \equiv \bar{c}_h D/k_{fl}$) into Eq. (15), rearranging to obtain an expression for $\bar{c}_h$, and using the result in Eq. (21) to obtain

$$\frac{(T_{w,o} - T_w)}{(T_w - T_{fl})_{avg}} = \frac{f}{4} \cdot \mathrm{Re}\ \mathrm{Pr} \cdot \frac{k_{fl}}{k_w} \cdot \ln(D_o/D) \tag{22}$$

Although Eq. (22) does not directly depend on the aspect ratio (L/D) of the expansion device, several terms do. For a plain orifice, the velocity is higher on average than for the capillary because the expansion is driven by acceleration rather than by friction; thus, Re is higher. Furthermore, laser-drilled or drilled orifices are rougher than drawn tubing, so f is higher. Finally, the outer diameter D_o is usually larger for a plain orifice than for a capillary of a given D. Thus, the LHS of Eq. (22) can be up to 100 times higher for a plain orifice than for a capillary of the same inner diameter.

This leads to the following conclusion that is consistent with our earlier analysis: heating a supercritical solution during its rapid expansion is much more difficult in a plain orifice than in a long capillary. As shown by the Biot number, wall conduction resistance is significantly higher than convective fluid resistance. Thus, heating elements on the outer walls of a plain orifice must be maintained at much higher temperatures than for a capillary in order to maintain the same bulk fluid temperature. As a result, the supercritical solution may be heated not only during the expansion but also (inadvertently) prior to the expansion, so that the true pre-expansion temperature is significantly higher than what is actually being measured. Clearly, the careful design of experimental equipment so that the desired temperatures are achieved is a more demanding task when a plain orifice is to be used.

B. One-Dimensional Analysis of Subsonic Expansion Inside the RESS Nozzle

In this section we show how to calculate the pressure, temperature, and velocity distributions inside a RESS nozzle. We apply a one-dimensional model to compressible single-phase flow in a duct with a circular cross section, such as shown in Figure 3. However, unlike several of the one-dimensional models that have been previously presented in the literature, ours accounts for heat exchange between the tubing wall and the fluid, and also allows the cross-sectional area, α, to change with position z along the axis of the duct. The most important underlying assumption with such a one-dimensional model is that neither the velocity of the fluid nor its physical properties change for a given duct cross section of radius r (or at least they can be represented by a value that is averaged over the radius).

We start with the governing equations for the adiabatic expansion of a supercritical fluid through a capillary of constant cross section, as presented by Lele and Shine (27). However, in our approach we allow heat transfer from the duct wall to the fluid, and we allow the cross-sectional area of the expansion device to vary:

Continuity: $$\frac{1}{\rho}\cdot\frac{d\rho}{dz}+\frac{1}{u}\cdot\frac{du}{dz}+\frac{1}{\alpha}\cdot\frac{d\alpha}{dz}=0 \tag{23a}$$

Momentum: $$\rho u\cdot\frac{du}{dz}+\frac{dP}{dz}=-\frac{\rho}{2}\cdot u^2\cdot\frac{f\varsigma}{\alpha} \tag{23b}$$

Energy: $$\rho u^2\cdot\frac{du}{dz}+\rho u\cdot\frac{dh}{dz}=\dot{q}\cdot\frac{\varsigma}{\alpha} \tag{23c}$$

All variables in Eq. (23) have been previously defined except for h, the specific enthalpy of the fluid. The term du/dz can be eliminated from Eqs. (23b) and (23c) by inserting Eq. (23a); Eqs. (24a) and (24b) are then obtained:

$$\frac{dP}{dz}-u^2\cdot\frac{d\rho}{dz}=-\frac{\rho}{2}u^2\cdot f\cdot\frac{\varsigma}{\alpha}+\rho u^2\cdot\frac{1}{\alpha}\cdot\frac{d\alpha}{dz} \tag{24a}$$

$$u\rho\frac{dh}{dz}-u^3\cdot\frac{d\rho}{dz}=\rho u^3\cdot\frac{1}{\alpha}\cdot\frac{d\alpha}{dz}+\dot{q}\cdot\frac{\varsigma}{\alpha} \tag{24b}$$

For a circular cross section (but whose diameter can change with the duct length), dividing the circumference ς by the cross-sectional area α gives

$$\frac{\varsigma}{\alpha} = \frac{\pi \cdot D(z)}{\frac{\pi}{4} \cdot D^2(z)} = \frac{4}{D(z)} \tag{25a}$$

The variation of the cross-sectional area with z is given by

$$\frac{1}{\alpha} \cdot \frac{d\alpha}{dz} = \frac{d}{dz}(\ln \alpha) = 2 \cdot \frac{d}{dz}(\ln[D(z)]) = \frac{2}{D(z)} \cdot \frac{dD(z)}{dz} \equiv \frac{2D_z}{D(z)} \tag{25b}$$

Next, we use the definition of the total differential to obtain expressions for dP/dz and dh/dz as functions of T and ρ:

$$\frac{dP}{dz} = \left.\frac{\partial P}{\partial T}\right|_\rho \cdot \frac{dT}{dz} + \left.\frac{\partial P}{\partial \rho}\right|_T \cdot \frac{d\rho}{dz}, \qquad \frac{dh}{dz} = \left.\frac{\partial h}{\partial T}\right|_\rho \cdot \frac{dT}{dz} + \left.\frac{\partial h}{\partial \rho}\right|_T \cdot \frac{d\rho}{dz} \tag{25c,d}$$

By substituting in Eqs. (25a)–(25d), Eqs. (24a) and (24b) can be rewritten as Eqs. (26a) and (26b):

$$\left.\frac{\partial P}{\partial T}\right|_\rho \cdot \frac{dT}{dz} + \left(\left.\frac{\partial P}{\partial \rho}\right|_T - u^2\right) \cdot \frac{d\rho}{dz} = \frac{-2\rho u^2 f}{D(z)} + \rho u^2 \frac{2D_z}{D(z)} \tag{26a}$$

$$\rho \left.\frac{\partial h}{\partial T}\right|_\rho \cdot \frac{dT}{dz} + \left(\rho \left.\frac{\partial h}{\partial \rho}\right|_T - u^2\right) \cdot \frac{d\rho}{dz} = \rho u^2 \frac{2D_z}{D(z)} + \frac{4 \cdot \dot{q}(z)}{u \cdot D(z)} \tag{26b}$$

Multiplying Eq. (26a) by $\rho \frac{\partial h}{\partial T}|_\rho$ and (26b) by $\frac{\partial P}{\partial T}|_\rho$ (to eliminate dT/dz) and rearranging, we obtain

$$\frac{d\rho}{dz} = \frac{-\rho \left.\frac{\partial h}{\partial T}\right|_\rho \cdot \frac{2\rho u^2 f}{D(z)} + \rho \cdot \left.\frac{\partial h}{\partial T}\right|_\rho \cdot \frac{2D_z \rho u^2}{D(z)} - \left.\frac{\partial P}{\partial T}\right|_\rho \cdot \left(\frac{2D_z \rho u^2}{D(z)} + \frac{4 \cdot \dot{q}}{u \cdot D(z)}\right)}{\left(\rho \cdot \left.\frac{\partial h}{\partial T}\right|_\rho \cdot \left.\frac{\partial P}{\partial \rho}\right|_T - \rho \cdot \left.\frac{\partial h}{\partial T}\right|_\rho \cdot u^2 - \left.\frac{\partial P}{\partial T}\right|_\rho \cdot \rho \cdot \left.\frac{\partial h}{\partial \rho}\right|_T + \left.\frac{\partial P}{\partial T}\right|_\rho \cdot u^2\right)} \tag{27a}$$

After changing the basis of density and enthalpy from kilograms to molecules in Eq. (27) and dividing the numerator and denominator by ρu^2, we obtain

$$\frac{d\rho_m}{dz} = \frac{-\frac{2\rho_m f}{D(z)} \cdot \left.\frac{\partial h_m}{\partial T}\right|_\rho - \left(\left.\frac{\partial P}{\partial T}\right|_\rho - \rho_m \cdot \left.\frac{\partial h_m}{\partial T}\right|_\rho\right) \cdot \frac{2D_z}{D(z)} - \left.\frac{\partial P}{\partial T}\right|_\rho \cdot \frac{N_A \cdot 4 \cdot \dot{q}}{\rho_m u^3 \cdot D(z) \cdot mw}}{\left[\left(\left.\frac{\partial h_m}{\partial T}\right|_\rho \cdot \left.\frac{\partial P}{\partial \rho_m}\right|_T - \left.\frac{\partial P}{\partial T}\right|_\rho \cdot \left.\frac{\partial h_m}{\partial \rho_m}\right|_T\right) \cdot \frac{N_A}{u^2 \cdot mw} + \frac{1}{\rho_m} \left.\frac{\partial P}{\partial T}\right|_\rho - \left.\frac{\partial h_m}{\partial T}\right|_\rho\right]} \tag{27b}$$

where $\rho_m = \rho \cdot N_A/mw$ and $h_m = h \cdot mw/N_A$, where N_A is Avogadro's number and mw is the molecular weight in kg/mol.

From classical thermodynamics, derivatives of the enthalpy with respect to temperature and density can be related to those of the pressure, where $c_{v,m}$ is the constant–volume molecular heat capacity of the pure fluid:

$$\left.\frac{\partial h_m}{\partial T}\right|_\rho = c_{v,m} + \frac{1}{\rho_m} \cdot \left.\frac{\partial P}{\partial T}\right|_\rho, \qquad \left.\frac{\partial h_m}{\partial \rho_m}\right|_T = \frac{1}{\rho_m} \cdot \left.\frac{\partial P}{\partial \rho_m}\right|_T - \frac{T}{\rho_m^2} \cdot \left.\frac{\partial P}{\partial T}\right|_\rho \tag{28}$$

In addition, we introduce the Mach number, which is the ratio of the actual fluid velocity to the speed of sound that the fluid exhibits in the same state ($\mathrm{Ma} = u/u_{snd}$).

$$u_{snd} = \sqrt{\left.\frac{\partial P}{\partial \rho}\right|_S} = \left(\frac{N_A}{mw}\right)^{1/2} \cdot \sqrt{\left.\frac{\partial P}{\partial \rho_m}\right|_T + \frac{T}{c_{v,m}\rho_m^2} \cdot \left(\left.\frac{\partial P}{\partial T}\right|_\rho\right)^2} \tag{29}$$

This gives us the final, desired equation for the change in fluid density along the length of the nozzle:

$$\frac{d\rho}{dz} = -\frac{\left[\left(1 + \frac{1}{c_{v,m}\rho_m} \cdot \left.\frac{\partial P}{\partial T}\right|_\rho\right) \cdot 2f - 2D_z + \frac{1}{c_{v,m}\rho_m} \cdot \left.\frac{\partial P}{\partial T}\right|_\rho \cdot \frac{4\dot{q}}{\rho u^3}\right]}{(\mathrm{Ma}^{-2} - 1)} \cdot \frac{\rho}{D(z)} \tag{30}$$

Incidentally, if we drop the friction and heat transfer term in the numerator of Eq. (30), we obtain an expression for the expansion in a friction-free channel under adiabatic conditions:

$$\frac{d\rho}{dz} = \frac{\mathrm{Ma}^2}{(1 - \mathrm{Ma}^2)} \cdot \frac{2\rho \cdot D_z}{D(z)} \tag{31}$$

For the fluid to expand, the channel must be convergent ($D_z < 0$) under subsonic conditions ($\mathrm{Ma} < 1$) and divergent ($D_z > 0$) under supersonic conditions ($\mathrm{Ma} > 1$). Note that Eq. (31) has a singularity at $\mathrm{Ma} = 1$ and changes its sign when the flow speed $u(z)$ crosses the speed of sound u_{snd}. Because $D_z = 0$ in a cylindrical channel, an adiabatic expansion must be driven by friction (or else the trivial solution $d\rho/dz = 0$ is obtained). The friction factor f in Eq. (30) is always positive; thus, if one starts at subsonic conditions, flow can never be accelerated beyond the speed of sound inside a cylindrical duct. Flow that attains exactly the speed of sound at the exit of the duct is called choked flow. As we have seen above, acceleration of compressible flow beyond the speed of sound requires a conduit of convergent-divergent geometry (21).

By combining Eq. (30) with Eqs. (24a) and (26), the derivatives of velocity and temperature with respect to z are obtained:

$$\frac{du}{dz} = u \cdot \left(\frac{-1}{\rho} \cdot \frac{d\rho}{dz} - \frac{2D_z}{D(z)} \right) \tag{32}$$

$$\frac{dT}{dz} = \left[\frac{2\rho u^2 f}{D(z)} + T \cdot \left. \frac{\partial P}{\partial T} \right|_{\rho} \cdot \frac{1}{\rho} \frac{d\rho}{dz} + \frac{4 \cdot \dot{q}(z)}{u \cdot D(z)} \right] \cdot (c_V \rho)^{-1} \tag{33}$$

Finally, the pressure is given by the equation of state and values for ρ and T at each location z. Calculation of the pressure is a straightforward operation if the equation of state used is available in its pressure-explicit form.

$$P = P(\rho_m, T) \qquad \textit{pressure-explicit equation of state} \tag{34}$$

The equation of state, Eq. (34), is also used to calculate several derivatives of the pressure with respect to temperature and density. These are required for several of the previous equations for describing the compressible flow inside and outside the expansion device. The constant-pressure and the constant-volume heat capacity at a given temperature and density are found in standard textbooks of thermodynamics [e.g., Sandler (28)] and read as follows after minor transformations:

$$c_{p,m}(P, T) = c_{p,m}^{id}(T) - T \cdot \int_{P=0,T}^{P,T} \left[2 \cdot \left. \frac{\partial T}{\partial \rho_m} \right|_P + \rho_m \cdot \left. \frac{\partial^2 T}{\partial \rho_m^2} \right|_P \right] \cdot \left(\rho_m \cdot \left. \frac{\partial T}{\partial \rho_m} \right|_P \right)^{-3} dP \tag{35a}$$

$$c_{v,m}(\rho_m, T) = c_{v,m}^{id}(T) + T \cdot \int_{\rho_m, T}^{\rho_m=0,T} \left. \frac{\partial^2 P}{\partial T^2} \right|_{\rho} \cdot \rho_m^{-2} \cdot d\rho_m, \quad \text{where } c_{v,m}^{id} = c_{p,m}^{id} - k_B \tag{35b}$$

Recall that the ideal-gas heat capacity, $c_{p,m}^{id}$, is a function of the temperature only and yields $c_{v,m}^{id}$ by subtraction of the Boltzmann constant k_B.

In the above analysis, we use the convention of properties with respect to single molecules; such a basis is needed for our model of the RESS process, as the results developed here are to be combined with the equations that describe the formation and growth of particles during RESS (see Section VII).

To calculate axial profiles of temperature, pressure, density, and velocity inside the RESS nozzle, the following procedure is used:

1. Values of P_{ini} and T_{ini} are fixed. Using these values, the equation of state [Eq. (34)] is used to calculate ρ_{ini}. (Recall that Figure 3 shows the location of "ini.") A value of the inlet velocity u_{ini} is assumed such that $Ma_{ini} \ll 1$.
2. The changes in density, velocity, and temperature for an increment of z are obtained by simultaneously integrating Eqs. (30), (32), and (33), respectively. The new values of density and temperature are then used in Eq. (34) to calculate a new value of P at $z_{ini} + \Delta z$.
3. Step 1 is repeated with the new values of P and T. If, during the integration process, the velocity becomes sonic at a point z upstream of z_2, the integration is terminated, the results are rejected, the assumed value for u_{ini} is decreased, and the entire process is repeated.
4. The integration process continues along the nozzle until $z = z_2$. If $Ma_2 = 1$ within $\pm 1\%$, the results for the whole profile are accepted as the final set of values. For the results shown in Section VI, guesses of u_{ini} between 1 and 2 m/s yielded the desired values.

C. Estimating the Friction Factor and Heat Flux Inside the RESS Nozzle

To perform the above calculations, the local friction factor, $f(z)$, and the local specific heat flux from the wall to the fluid, $\dot{q}(z)$, must be known. In the following discussion we show how several characteristics of the RESS process have a significant impact on the calculation of these parameters.

The classic method for determining the friction factor uses the diagram of Moody (22), where the Fanning friction factor f is given as a function of the Reynolds number and the wall roughness ε. An explicit formula for this relationship was developed by Haaland (29):

$$\frac{1}{2\sqrt{f}} = -\frac{1.8}{n} \cdot \log\left[\left(\frac{6.9}{\mathrm{Re}}\right)^n + \left(\frac{\varepsilon}{3.75D}\right)^{1.11n}\right] \tag{36}$$

The exponent n in Eq. (36) has a default value of unity and can be increased to improve accuracy in case of abrupt transitions from turbulent flow in smooth pipes to turbulent flow in rough pipes.

To calculate the local specific heat flux $\dot{q}$, there are two possibilities. In a few cases, $\dot{q}$ can be obtained by subtracting the local heat loss from the known flux generated by an electrical resistance heater:

$$\dot{q}(z) = \dot{q}_{el}(z) - \dot{q}_{loss}(z) \tag{37}$$

Typically, however, we calculate the heat flux as the product of the heat transfer coefficient and the difference between the temperature of the wall and the bulk temperature of the fluid:

$$\dot{q}(z) = c_{\mathrm{h}}(z) \cdot [T_{\mathrm{w}}(z) - T_{\mathrm{fl}}(z)], \qquad \text{where } c_{\mathrm{h}}(z) = \mathrm{Nu}(z) \cdot \frac{k_{\mathrm{fl}}(z)}{D(z)} \tag{38}$$

The solution of Eq. (38) for the RESS process is more difficult than for more conventional heat transfer scenarios for several reasons. For example, we know that both thermodynamic properties (such as the heat capacity) and transport properties of the fluid (such as viscosity and thermal conductivity) undergo significant changes during expansion. As a result, the Prandtl number of the fluid can decrease from liquid-like values of about 2–3 to about 0.7 (typical for gases). Thus, the Reynolds analogy [Eq. (15)] that correlates heat and momentum transfer must be replaced by a correlation that allows the Prandtl number to vary. For example, after incorporating f, the correlation of Gnielinski (30) is

$$\mathrm{Nu} = \frac{\frac{f}{2} \cdot (\mathrm{Re} - 1000)\mathrm{Pr}}{\left[1 + 12.7 \cdot \sqrt{\frac{f}{2}} \cdot \left(\mathrm{Pr}^{2/3} - 1\right)\right]} \tag{39}$$

where $\mathrm{Pr} = \eta c_{\mathrm{p}}/k_{\mathrm{fl}}$.

The compressible-flow nature of RESS also affects the solution of Eq. (38). In particular, the wall temperature can be considerably higher than the bulk fluid temperature because as the high-speed fluid is brought to rest, the kinetic energy of the fluid is converted into internal energy. It should be emphasized that this temperature increase occurs even for a perfectly insulated channel wall with no external heating. For the compressible-flow case, Eckert (25) has shown that Eq. (38) can generally be used with the same heat transfer relations used for incompressible flow if the bulk fluid temperature is replaced with the adiabatic wall temperature $T_{\mathrm{ad,w}}$, so that

$$\dot{q}(z) = c_{\mathrm{h}}(z) \cdot [T_{\mathrm{w}}(z) - T_{\mathrm{ad,w}}(z)] \tag{40}$$

Now $T_{\mathrm{ad,w}}$ is not as high as the stagnation temperature of the bulk, free-stream fluid $T^0_{\mathrm{fl}}(z)$, as some of the heat is transferred by viscous dissipation to the surrounding fluid. This fact is reflected in the so-called recovery factor, which can be estimated from theory or measured experimentally and is used to calculate $T_{\mathrm{ad,w}}$ (25):

$$T_{\mathrm{ad,w}}(z) = \mathrm{rf} \cdot [T^0_{\mathrm{fl}}(z) - T_{\mathrm{fl}}(z)] + T_{\mathrm{fl}}(z) \tag{41}$$

As with any heat transfer problem, the actual wall temperature T_w must also be estimated. The only relatively simple case in RESS would be that of the long capillary, with the outside wall being heated by a constant-temperature bath (typically, the capillary is immersed in the bath). Because heat conduction in the axial direction can be neglected, we can write

$$\dot{q}(z) = \bar{c}_h(z) \cdot [T_o - T_{ad,w}(z)],$$

$$\text{where } \bar{c}_h(z) = \left[\frac{D}{\text{Nu}(z) \cdot k_{fl}(z)} + \frac{D \cdot \ln(D_o/D)}{2k_w} + \frac{D}{c_{h,o}(z) \cdot D_o}\right]^{-1} \tag{42}$$

where $\bar{c}_h$ is the local overall heat transfer coefficient, T_o is the temperature of the immersion bath, and $c_{h,o}(z)$ is the local heat transfer coefficient for the bath fluid.

Up to this point, we have presented strategies and formulas necessary to calculate complete pressure, temperature, and velocity profiles for subsonic compressible flow under conditions typical for RESS. The last issue we address is the one of reliability of the various correlations that have been presented. Here again, the facts that the flow is compressible and that the expansion device is a comparatively narrow channel play an important role.

Guo and Wu (31) have pointed out that the effects of acceleration are more important than the effects of friction at high Mach numbers. They conclude that the assumption of a fully developed or locally fully developed velocity profile does not apply to near-sonic flow in a microtube. Because the velocity profile becomes flatter, higher velocity gradients are expected near the wall. The results of their two-dimensional flow model and their experimental findings indicate that the Reynolds number multiplied by the Darcy friction factor can be as high as 80 as compared with 64 from the standard literature (22). Accordingly, they also found a Nusselt number that was about 25% higher than that predicted by conventional correlations.

The arguments that Adams (32) makes against the straightforward applicability of the classical momentum and heat transfer correlations are concerned with the absolute dimensions of the flow channels used in RESS. He points out that the theory built on measurements carried out for flat plates is assumed to be applicable to flow inside circular channels. However, a dimensionless analysis indicates that the validity of this assumption is questionable. Comparing experimental results for flow in channels with inner diameters of 1.09, 0.76, and 0.102 mm to predictions based on Eq. (39), Adams concluded that heat transfer can be significantly enhanced in microchannels. He defined this enhancement in terms of a factor Φ:

$$\Phi \equiv \frac{\text{Nu}_{exp}}{\text{Nu}} = 1 + F(\text{Re}, D) \tag{43}$$

where

$$F = 7.6 \cdot 10^{-5} \cdot \mathrm{Re} \cdot \left[1 - \left(\frac{D}{1.164\ \mathrm{mm}}\right)^2\right] \tag{44}$$

In the Adams experiments, Φ was found to range from 1.0 to 2.5. Equation (44) is valid in the range $2000 \leq \mathrm{Re} \leq 23{,}000$, $1.53 \leq \mathrm{Pr} \leq 6.43$, and $0.102\ \mathrm{mm} \leq D \leq 1.109\ \mathrm{mm}$.

V. SUPERSONIC FLOW IN THE FREELY EXPANDING JET: A ZERO-DIMENSIONAL ANALYSIS

In the previous section, we discussed how one-dimensional flow models can be used to calculate pressure and temperature profiles in cylindrical tubing under subsonic conditions; in general, reasonable results are obtained (see Section VI). At first glance, such an approach might seem to be worth pursuing for idealized free jets of circular cross section, with the transition from one regime to the other (where $\mathrm{Ma} \equiv 1$) being the only unresolved singularity. However, a freely expanding supersonic jet always creates a transsonic flow field in addition to its supersonic parts by mixing with its surroundings. In such a flow field, Mach numbers ranging from values below unity up to values well above unity are spatially distributed, and the flow field contains loci, called shock waves, in which the velocity of the flow undergoes a sudden transition. Bier and Schmidt (33) were the first to observe the shock wave pattern for a freely expanding jet; a schlieren apparatus was used with expanding nitrogen gas. As shown in Figure 5a, the core streamlines of the expanding jet pass through a normal shock wave, the Mach disk, whereas the outer streamlines cross a zone of oblique shocks, the so-called barrel shock zone. As flow passes through the Mach disk, which is essentially of zero thickness, supersonic flow is transformed into subsonic flow. Oblique shocks, on the other hand, generally reduce only the magnitude of the flow velocity and change its direction.

The presence of shock waves during rapid expansion implies two seemingly insurmountable difficulties for any one-dimensional flow model. First, the assumption that the flow through a given cross section can be represented by one particular state has been shown to be highly flawed. Second, even if the flow pattern could be subdivided into zones of purely subsonic and purely supersonic flow, one-dimensional models offer no help in determining the location and shape of the various shock waves.

Two-dimensional models can overcome these difficulties. Generalized procedures are available to simultaneously solve for the location of the shock waves and the spatial distribution of pressure, temperature, and velocity (34). Unfortu-

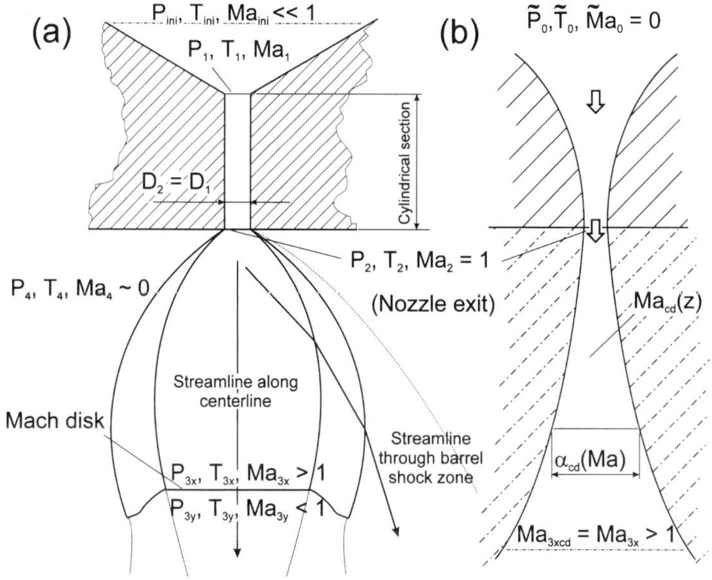

Figure 5 Freely expanding supersonic jet through (a) a plain orifice and (b) a convergent-divergent duct.

nately, accounting for the effects of both turbulence and fluid nonideality renders these models very computer time costly, even if (as is commonly done) the fluid dynamic problem is only solved for a pure solvent. Nevertheless, work is underway to solve the two-dimensional problem for the rapid expansion of pure nonideal CO_2 through a RESS nozzle (35).

More qualitative descriptions of the flow pattern of a freely expanding supersonic jet have been available since the early 1960s. Although this work is, strictly speaking, applicable only to ideal gases, it is useful for helping us understand the behavior of a rapidly expanding supercritical fluid in the free-jet region. In this section, we combine these classic, zero-dimensional models for the isentropic expansion of an ideal gas in both the supersonic and transsonic regions with the experimental and theoretical results of Ashkenas and Sherman (36), which describe the flow properties of a supersonic fluid between the nozzle exit and the Mach disk. Evidence will be presented that the pressure, temperature, and velocity profiles along the centerline are not the best choice to represent the entire flow regime.

We perform all of our analyses for an ideal gas in which the ratio of the heat capacities, $\gamma = c_p/c_v$, is constant. Expansions are assumed to be isentropic, which is a reasonable assumption for a plain orifice (where L/D is small) and even more so for the subsequent expansion of the fluid on its way to the Mach disk. For expansions through capillaries with high L/D, the analysis given below can also be used for conditions from the capillary outlet outward. However, a method that accounts for the effect of friction must be used to calculate the conditions of temperature and pressure that would exist at the capillary inlet.

Assuming uniform conditions across the entire Mach disk, the mass flow rate of the fluid passing through the disk can be calculated from the density and velocity that it has attained immediately before (ρ_{3x} and u_{3x}) passing through the disk, and from the area of the Mach disk, α_M (see Figure 5a). The mass flow rates entering and leaving (ρ_{3y} and u_{3y}) the Mach disk are identical because the Mach disk is of virtually zero thickness, so fluid can be neither added nor withdrawn in the cross-flow direction.

$$\dot{m}_M = \rho_{3x} \cdot u_{3x} \cdot \alpha_M = \rho_M \cdot u_M \cdot \alpha_M = \rho_{3y} \cdot u_{3y} \cdot \alpha_M \tag{45}$$

In Eq. (45), ρ_{3x} and u_{3x} are functions of the initial stagnation conditions P^0 and T^0 and of the Mach number of the entering fluid, Ma_{3x}, while the density ρ_{3y} and the velocity u_{3y} are calculated using the Mach number of the fluid that leaves the disk, Ma_{3y}. The supersonic and subsonic Mach numbers, Ma_{3x} and Ma_{3y}, respectively, are related by normal-shock relations. If the Mach number on one side of the disk is known, all properties of the flow on the other side can be derived (21). Finally, we note that the mass flow rate of the entire jet (i.e., not just the portion that passes through the Mach disk) is given by the flow under choked conditions ($\mathrm{Ma} \equiv 1$) at the nozzle exit (see Figure 5a):

$$\dot{m}_2 = \rho_2 \cdot u_2 \cdot \alpha_2 \tag{46}$$

where u_2 is at the speed of sound.

Now consider a duct with variable cross section of the convergent-divergent (cd) type (see Figure 5b). The fluid can attain any given Mach number through isentropic expansion under ideal conditions. When the pressure ratio P_4/P_0 is lower than the critical pressure ratio, cross sections greater than the critical cross section produce subsonic flow in the convergent part and supersonic flow in the divergent part, with the fluid passing through the critical cross section α_{2cd} at exactly the speed of sound. The cross-sectional area for which the fluid attains a given Mach number greater than unity is given by the following equation from Hansen (21):

$$\frac{\alpha_{cd}}{\alpha_{2cd}} = \frac{1}{\mathrm{Ma}} \cdot \left(\frac{1 + \dfrac{\gamma - 1}{2} \cdot \mathrm{Ma}^2}{\dfrac{(\gamma + 1)}{2}} \right)^{\frac{\gamma+1}{2(\gamma-1)}} \tag{47}$$

Next, we calculate the mass flow rate at a particular cross section within the convergent-divergent duct, $\dot{m}_{3xcd}$, that would give one the same Mach number, Ma_{3x}, that a freely expanding jet reaches just before entering the Mach disk. Of course $\dot{m}_{3xcd}$ must be identical to the mass flow rate at the critical cross section, $\dot{m}_{2cd}$. Then we define this flow rate as being equal to that passing through the nozzle exit of the plain orifice, $\dot{m}_2$:

$$\dot{m}_2 = \dot{m}_{2cd} = \dot{m}_{3xcd} = \rho_{3xcd} \cdot u_{3xcd} \cdot \alpha_{3xcd} \tag{48}$$

Thus, the ratio of the mass flow rate through the Mach disk to that of the entire free jet exiting the plain orifice can be estimated from the following equation:

$$\frac{\dot{m}_M}{\dot{m}_{jet}} = \frac{\dot{m}_M}{\dot{m}_2} = \frac{\rho_{3x} u_{3x} \alpha_M}{\rho_{3xcd} u_{3xcd} \alpha_{3xcd}} = \frac{\alpha_M}{\alpha_{3xcd}} = \frac{\alpha_M/\alpha_2}{\alpha_{3xcd}/\alpha_{2cd}} \tag{49}$$

Note that both the plain orifice and the convergent-divergent nozzle are assumed to undergo an isentropic expansion from the same initial conditions; thus, the velocities and densities in each device are identical. The denominator $\alpha_{3xcd}/\alpha_{2cd}$ in Eq. (49) can be calculated with Eq. (47) if the Mach number that the flow attains when entering the Mach disk is known. To calculate α_M/α_2, one needs to know only the diameter of the Mach disk D_M and of the nozzle exit D_2:

$$\frac{\alpha_M}{\alpha_2} = \left(\frac{D_M}{D_2}\right)^2 \tag{50}$$

Based on an extensive experimental study, Bier and Schmidt (36) developed a general correlation between the dimensionless diameter of the Mach disk and its dimensionless distance from the nozzle exit for ideal gases:

$$\frac{D_M}{D_2} = 0.42 \cdot (1 + K_M) \cdot \frac{z_M}{D_2} \quad \text{at} \quad \frac{\tilde{P}_0}{P_4} = 20 \tag{51a}$$

$$\frac{D_M}{D_2} = 0.48 \cdot (1 + K_M) \cdot \frac{z_M}{D_2} \quad \text{at} \quad \frac{\tilde{P}_0}{P_4} = 1000 \tag{51b}$$

Here z_M is the distance from the nozzle exit to the Mach disk, K_M is a constant that depends on γ, and $\tilde{P}_0$ is the stagnation pressure at the nozzle inlet conditions for a perfectly isentropic expansion. The dimensionless distance, z_M/D_2, was found by Ashkenas and Sherman (36) to be a function of the isentropic expansion pressure ratio:

$$\frac{z_M}{D_2} = 0.67 \cdot \sqrt{\frac{\tilde{P}_0}{P_4}} \tag{52}$$

Combining Eq. (52) with a log-linear interpolation between Eqs. (51a) and (51b), we obtain a generalized expression for the size of the Mach disk. The resulting expression indicates that the dimensionless diameter of the Mach disk generated by a freely expanding supersonic jet is a function of only the expansion pressure ratio and the thermodynamic properties of the gas:

$$\frac{D_M}{D_2} = 0.67 \cdot \left\{ 0.42 + \frac{(0.48 - 0.42)}{\ln(1000/20)} \cdot \ln\left(\frac{1}{20} \cdot \frac{\tilde{P}_0}{P_4}\right)\right\} \cdot (1 + K_M) \cdot \sqrt{\frac{\tilde{P}_0}{P_4}} \tag{53}$$

Ashkenas and Sherman (36) were also able to obtain an expression for the Mach number of the flow at any position z between the nozzle exit and the Mach disk:

$$\mathrm{Ma}(z) = A_M \left(\frac{z}{D_2} - \frac{z_{M0}}{D_2}\right)^{\gamma-1} - \frac{1}{2} \cdot \frac{(\gamma+1)}{(\gamma-1)} \cdot A_M \left(\frac{z}{D_2} - \frac{z_{M0}}{D_2}\right)^{1-\gamma} \tag{54}$$

For Eq. (54) to be valid, the distance from the nozzle exit, z, must be at least 2.5 capillary diameters. Like K_M, the constants A_M and z_{M0} are functions of γ. The variation of these constants with γ is given in Table 1. Finally, the above equations can be used to obtain the desired expression for the fraction of the free-jet flow that passes through the Mach disk. Equation (53) is substituted into Eq. (50); then this result and Eq. (47) are substituted into Eq. (49) to obtain the desired result:

$$\frac{\dot{m}_M}{\dot{m}_2} = \frac{0.1399 \cdot \left\{\left[1 + 0.041 \cdot \ln\left(\frac{\tilde{P}_0}{P_4}\right)\right] \cdot (1 + K_M\right\}^2 \cdot 0.67^2 \cdot \frac{\tilde{P}_0}{P_4}}{\frac{1}{\mathrm{Ma}_{3x}} \cdot \left[\frac{2 + (\gamma-1) \cdot \mathrm{Ma}_{3x}^2}{(\gamma+1)}\right]^{\frac{(\gamma+1)}{2(\gamma-1)}}} \tag{55}$$

Table 1 Constants for Calculating D_M/D_2 [Eq. (53)] and Ma(z) [Eq. (54)] as a Function of γ

$\gamma = c_p/c_v$	K_M	A_M	z_{M0}/D_2
1.667	−0.20	3.26	0.075
1.400	0.0	3.65	0.40
1.286	0.25	3.96	0.85

By combining Eqs. (52) and (54), an expression for the Mach number immediately before the Mach disk, Ma_{3x}, is obtained:

$$Ma_{3x} = A_M \left(0.67 \sqrt{\frac{\tilde{P}_0}{P_4}} - \frac{z_{M0}}{D_2} \right)^{\gamma - 1} - \frac{1}{2} \frac{(\gamma + 1)}{(\gamma - 1)} \cdot \frac{1}{A_M} \left(0.67 \sqrt{\frac{\tilde{P}_0}{P_4}} - \frac{z_{M0}}{D_2} \right)^{1 - \gamma} \quad (56)$$

Substituting Eq. (56) into Eq. (55), we can calculate the mass fraction of flow through the Mach disk as a function of the expansion pressure ratio. The results of these calculations are given in Figure 6 and are somewhat surprising because they indicate that the percentage of the total mass flow that passes through the Mach disk never exceeds 20% at pressure ratios typical of those used in RESS processing (i.e., $\tilde{P}_0/P_4 = 100$–200). Of course, these results are strictly valid only for ideal gases of constant γ, but nevertheless they suggest that profile analyses of the RESS process should not focus on streamlines that

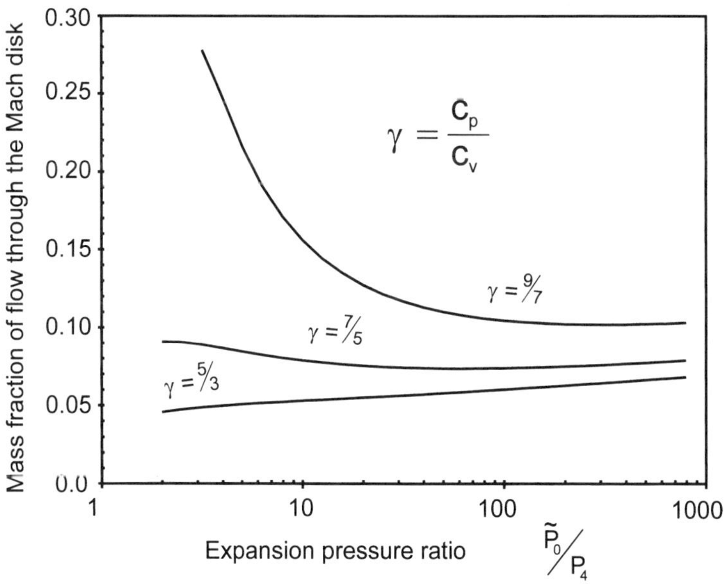

Figure 6 The mass fraction of total free jet flow that passes through the Mach disk is shown for an ideal gas as a function of the expansion pressure ratio.

pass through the Mach disk, as those that circumvent the disk may actually be more representative of the overall jet behavior.

Zero-dimensional, compressible-fluid theory can also be used to analyze the pressure change that occurs across the Mach disk. Momentum, energy, and continuity equations are written for both the upstream (3x) and downstream (3y) sides of the Mach disk, with the ideal gas law being used as the equation of state. In addition, the second-law requirement that the entropy must either increase or remain constant across the disk is used. Simultaneous solution of these equations (the classical solution technique is graphical) followed by considerable algebraic manipulation yields the downstream Mach number M_{3y} as a function of γ and the upstream Mach number M_{3x} (21):

$$\mathrm{Ma}_{3y}^2 = \frac{\mathrm{Ma}_{3x}^2 + \dfrac{2}{(\gamma - 1)}}{\dfrac{2\gamma}{\gamma - 1}\mathrm{Ma}_{3x}^2 - 1} \tag{57}$$

The static pressures of the fluids entering and leaving the Mach disk are derived from the application of the momentum and continuity equations, along with the ideal gas law:

$$\frac{P_{3y}}{P_{3x}} = \frac{1 + \gamma \mathrm{Ma}_{3x}^2}{1 + \gamma \mathrm{Ma}_{3y}^2} \tag{58}$$

Now it is easy to show that the stagnation and static pressures for an ideal gas are related by

$$\frac{P^0}{P} = \left(1 + \frac{\gamma - 1}{2}\mathrm{Ma}^2\right)^{\gamma/\gamma - 1} \tag{59}$$

Thus, the stagnation pressures on each side of the Mach disk are related by

$$\frac{P_{3y}^0}{P_{3x}^0} = \frac{P_{3y}}{P_{3x}} \cdot \frac{\left(1 + \dfrac{(\gamma - 1)}{2}\mathrm{Ma}_{3y}^2\right)^{\frac{\gamma}{\gamma - 1}}}{\left(1 + \dfrac{(\gamma - 1)}{2}\mathrm{Ma}_{3x}^2\right)^{\frac{\gamma}{\gamma - 1}}} \tag{60}$$

Finally, we realize that for an isentropic expansion, P_{3x}^0 must equal $\tilde{P}_0$, the inlet pressure to the nozzle, which is essentially at stagnation. Applying this equality to Eq. (60), substituting Eq. (58) into Eq. (60), and dividing Eq. (60) by the stagnation pressure in the expansion chamber, P_4, we obtain an expression for the dimensionless stagnation pressure downstream of the Mach disk:

$$\frac{P_{3y}^0}{P_4} = \frac{\tilde{P}_0}{P_4} \cdot \frac{(1 + \gamma \mathrm{Ma}_{3x}^2)}{(1 + \gamma \mathrm{Ma}_{3y}^2)} \cdot \frac{\left(1 + \frac{(\gamma - 1)}{2} \mathrm{Ma}_{3y}^2\right)^{\frac{\gamma}{\gamma-1}}}{\left(1 + \frac{(\gamma - 1)}{2} \mathrm{Ma}_{3x}^2\right)^{\frac{\gamma}{\gamma-1}}} \tag{61}$$

A plot of P_{3y}^0/P_4 vs. the expansion pressure ratio, $\tilde{P}_0/P_4$, is given as Figure 7 and shows that, for typical RESS conditions, the stagnation pressure on the downstream side of the Mach disk never exceeds 1 or 2.

From the above analysis of the Mach disk, we can draw several conclusions. First, the normal shock that occurs at the Mach disk is an effective dissipator that takes the fluid from supersonic to subsonic Mach numbers. In fact, Eq. (57) shows that the higher the upstream Mach number, Ma_{3x}, the lower the downstream Mach number, Ma_{3y}. In RESS, the entropy of the fluid always increases, and the largest fraction of this entropy increase (particularly

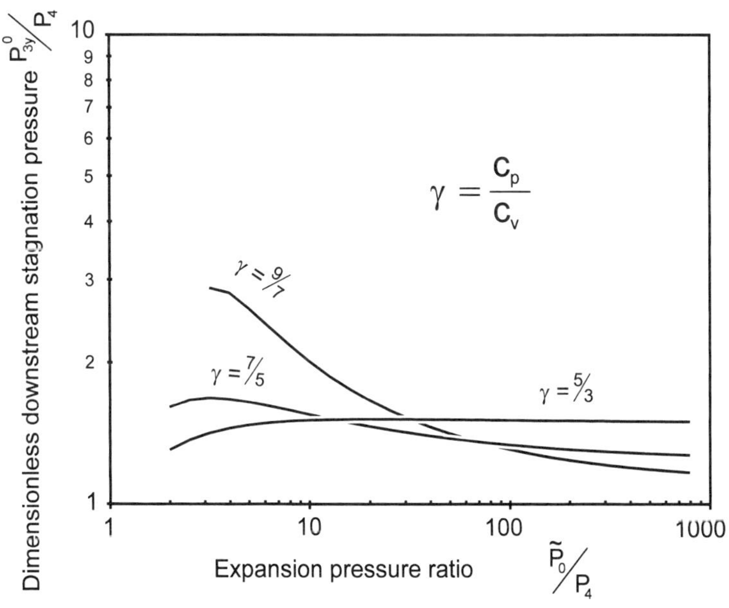

Figure 7 The dimensionless stagnation pressure on the downstream side of the Mach disk is shown for an ideal gas as a function of the expansion pressure ratio.

for the case of plain orifices with low L/D values) occurs as the fluid passes through the normal shock of the Mach disk. Thus, we can also consider the Mach disk as a device that converts kinetic energy to heat. Second, the low stagnation pressures given in Figure 7 illustrate that the fluid in the core of the jet itself has little potential to reaccelerate and expand as its stagnation pressure decreases to equal that in the chamber. Instead, the core flow is reaccelerated by the annular flow that circumvents the Mach disk. This major portion of the jet does not pass through the Mach disk but through a series of oblique shocks that are weaker than the normal shock in the Mach disk.

VI. FLUID DYNAMIC CALCULATIONS FOR RESS: THE PLAIN ORIFICE VS. THE LONG CAPILLARY

In the preceding sections we have discussed the basic principles of RESS and shown how to carry out simplified calculations of the fluid dynamics of a rapid expansion. We now apply these calculations to the two limiting types of expansion devices: the plain orifice and the long capillary. For the plain orifice, L/D for the cylindrical section is typically of the order of unity. We picked $L/D = 5$ as a value typical of many investigations, with the holes being short enough to be drilled either mechanically (23) or with the use of a laser (e.g., 12,20). For the capillary, we chose $L/D = 5000$ as a typical aspect ratio, which can be either an individual tube or part of a microchannel that leads through a porous frit plate (37).

Consider the rapid expansion of pure CO_2 from a reservoir at the stagnation conditions of $P_0 = 200$ bar and $T_0 = 403$ K into an expansion chamber that is set to a stagnation pressure of $P_4 = 1$ bar. (The subscript nomenclature is identical to that used in Figures 3 and 5.) A temperature–entropy diagram (Figure 8) can be used to depict expansion processes in general. In the analysis below, we limit our calculations to adiabatic expansions, so the temperature T_4 in the expansion chamber is a function of the two pressures P_0 and P_4, of the temperature T_0, and of the PvT properties of carbon dioxide.

First, consider the two limiting cases of an adiabatic expansion. The isenthalpic expansion from P_0 and T_0 along a constant h line to P_4 yields $(T_4)_h$ (Figure 8). Here the specific kinetic energy of the fluid never changes. On the other hand, for the isentropic expansion (a vertical line from P_0 and T_0 downward) the maximal possible amount of enthalpy is transformed into kinetic energy because there are no losses due to friction. Actual RESS processes follow paths that lie between the two limiting cases, with certain segments of the process essentially coinciding with the ideal paths. For example, with the nozzle shown in Figure 3 (a plain orifice) the fluid undergoes an almost per-

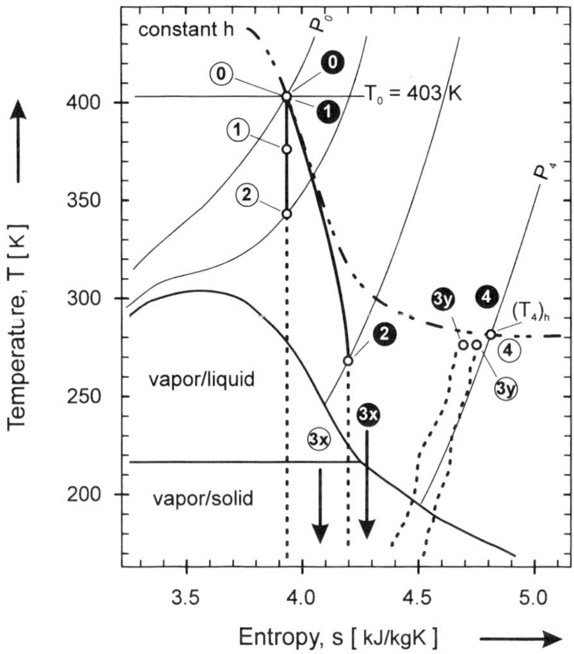

Figure 8 Adiabatic expansion of pure CO_2 from 200 bar and 403 K to 1 bar through a plain orifice ($L/D = 5$; *light circles*) vs. a long capillary ($L/D = 5000$; *dark circles*).

fectly isentropic expansion in the convergent inlet section, with frictional losses occurring during the subsequent expansion inside the cylindrical duct. The fluid then undergoes another isentropic expansion after exiting the nozzle under sonic conditions (choked flow) until it traverses a shock zone (which includes the Mach disk) in which its entropy increases stepwise.

Using the formula for one-dimensional compressible flow presented in Section IV.B, we calculated the pressure, temperature, and velocity profiles describing the subsonic adiabatic expansion of pure carbon dioxide inside the orifice and the capillary up to the nozzle exit (i.e., point 2 in Figures 3 and 5). Both the Bender (38) and Carnahan–Starling–van der Waals (39) equations of state were used to calculate the necessary PvT properties for CO_2, and results using either of the two equations were essentially identical. Downstream of the nozzle exit, we calculated the pressures and temperatures on the upstream and downstream sides of the Mach disk by using the formulas of Ashkenas and Sherman (36) (see Section V). These formulas assume an ideal gas with $\gamma = 1.286$, close to the value of CO_2 at ambient conditions. We should remember, however, that our earlier calculations indicate that the pressure profile through the Mach disk represents no more than about 20% of the mass flow rates in most freely expanding supersonic jets.

The calculated expansion paths are given in Figure 8. It can be seen that the subsonic expansion (i.e., from 0 to 2) is very close to an isentropic expansion in a plain orifice, while the expansion is significantly affected by friction in the long capillary. According to ideal gas calculations (21), the temperature and pressure then drop to very low values when the flow reaches the Mach disk at 3x (such that CO_2 can either freeze or liquefy), but they increase stepwise (along with the entropy) on the downstream side of the Mach disk at 3y, ending up at conditions close to those corresponding to an isenthalpic expansion (Figure 8).

Although the results of the above thermodynamic calculations are independent of the choice of scalable variables, such as nozzle length and diameter, residence time, and mass flow rate (as long as the initial conditions of P_0 and T_0 are the same), the aerosol dynamic processes (discussed in the next section) are substantially affected by these choices. It is therefore of interest to compare the plain orifice to the long capillary on the basis of these scalable variables (Table 2). For the capillary, L, D, τ_{1-2} (residence time in the cylindrical portion of the nozzle between points 1 and 2), and m (mass flow rate) were held constant (column 2), while for the plain orifice each of these variables were alternately set to the corresponding value for the capillary. For all cases, P_0 and T_0, as well as L/D for both the orifice and the capillary, were kept constant; thus, the calculated thermodynamic properties were independent of the scalable variables (Table 2) and are also applicable to the results shown in Figure 8.

The effects of setting the diameters (i.e., $D_5 = D_{5000}$), nozzle lengths ($L_5 = L_{5000}$), residence times ($\tau_5 = \tau_{5000}$), and mass flow rates ($\dot{m}_5 = \dot{m}_{5000}$) equal are shown in columns 3–6, respectively. Column 3 shows the effect of holding the diameter constant; as would be expected, very large differences in length L and residence time $\tau_{1\text{-}2}$ are observed. Column 4 shows that setting the length of the cylindrical section equal for both expansion devices results in mass flow rates that are more than a million times higher for the plain orifice than for the capillary. In column 5, the effect of setting the residence times equal in each of the cylindrical sections results in even larger discrepancies in the mass flow rates and diameters that are 1000 times larger. The comparisons in columns 4 and 5 would be impractical to make in the laboratory because completely different solvent supply systems would be required. Finally, the comparison in column 6, that of equal mass flow rates, gives similar results to the constant-diameter comparison.

As was discussed above, it is difficult to compare rapid expansion in the orifice and the capillary on the basis of similar residence times. Thus, we consider an alternative analysis by calculating the residence time in each section

Table 2 Rapid Expansion of Pure CO_2 Through a Plain Orifice vs. a Long Capillary: Comparison of Scalable Parameters

$P_0 = 200$ bar $T_0 = 403$ K	Long capillary $L/D = 5000$	Plain orifice $L/D = 5$ $D_5 = D_{5000}$	$L_5 = L_{5000}$	$\tau_5 = \tau_{5000}$	$\dot{m}_5 = \dot{m}_{5000}$
D	20 μm	**20 μm**	20 mm	121 mm	7.9 μm
L	100 mm	100 μm	**100 mm**	606 mm	39.5 μm
τ_{1-2}	3.06 ms	0.51 μs	0.505 ms	**3.06 ms**	0.2 μs
$\dot{m}$	9.8 g/h	62.6 g/h	17.4 kg/s	640 kg/s	**9.8 g/h**
P_2	16 bar	95 bar →			
T_2	267 K	343 K →			
$v_{sound} = v_2$	237 m/s	244 m/s →			
$\overline{P}_0$[a]	32 bar	197 bar →			
z_M/D_2	7.65	9.65 →			
Ma_{3x}	5.67	6.89 →			
v_{3x} (ideal gas)	660 m/s	770 m/s →			
v_{3y} (ideal gas)	108 m/s	110 m/s →			
$\dot{m}_{Ma}/\dot{m}$	12%	15% →			
P^0_{3y} (ideal gas)	1.6 bar	1.2 bar →			

[a]The initial reservoir pressure if isentropic expansion of an ideal gas results in above-calculated values of P_2 and T_2.

of the RESS process relative to the time spent between the entrance and exit of the cylindrical section. We call the time spent in the cylindrical section between points 1 and 2 the "capillary residence time" and denote it by $\tau_{1\text{-}2}$; for the other times, an analogous nomenclature is used. Calculations were performed using the methods described above (i.e., the same methods used to generate the results for Figure 8 and Table 2); results are shown in Figure 9. Assuming (as we did) that each nozzle has an entrance region with an angle of 120° (Figure 3), the time needed for the fluid to travel from z_{ini} to z_1 is negligible for the case of the long capillary (i.e., $\tau_{ini\text{-}1}/\tau_{1\text{-}2} \approx 0.1$), but it translates to an additional 10–20 capillary residence times for the case of a plain orifice. By definition, the time in the capillary section for each expansion device is $\tau_{1\text{-}2}/\tau_{1\text{-}2} = 1$ (Figure 9). After expansion, the free jet attains essentially stagnation conditions after a maximal flow length of 100 nozzle exit diameters. Again, this is equivalent to only a small fraction of the capillary residence time in the case of the long capillary ($\tau_{2-4}/\tau_{1\text{-}2} \approx 0.1$), but it adds 2–20 capillary residence times for the case of the plain orifice. Based on our estimates, then, the residence time relevant for the RESS process is about equal to or even less than the capillary residence time

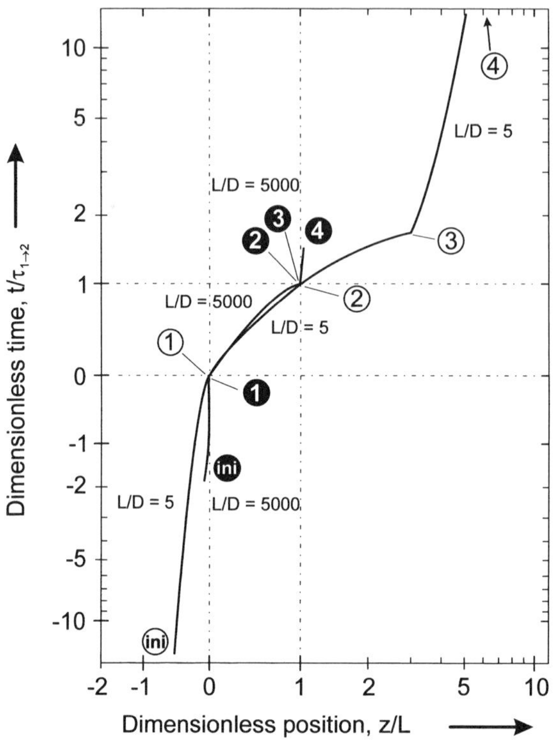

Figure 9 The time (in units of capillary residence times) that the fluid spends in each part of the rapid expansion process for a plain orifice (*light circles*) vs. a long capillary (*dark circles*). The scale is condensed by plotting units in terms of $\sinh^{-1}$.

for those cases where $L/D = 5000$, whereas for the plain orifice with $L/D = 5$ it can be as much as 40 times longer than the capillary residence time.

VII. PRECIPITATION AND GROWTH PROCESSES: NUCLEATION, CONDENSATION, AND COAGULATION

A. Background

Assume that a supersaturated homogeneous solution is flowing through a smooth duct containing no solid impurities either in the fluid or on the inner walls, precluding the possibility of heterogeneous nucleation. Whenever the equilibrium mole fraction of a solute, y_j^{eq}, is lower than its mole fraction, the difference between the actual state of the solution (which is usually metastable) and the equilibrium state provides a driving force for the processes of precipitation, namely, homogeneous nucleation and condensation. The process of homogeneous nucleation is mathematically described by two variables: the number of molecules forming a critical nucleus, n_{crit}, and the rate of formation of these nuclei per unit volume, $\overset{*}{J}_{\mathrm{ncl}}$ $[\$/\mathrm{m}^3\ \mathrm{s}]$. The expressions for n_{crit} and $\overset{*}{J}_{\mathrm{ncl}}$ (given below) are based on the assumption that, in a metastable solution, solute clusters of random size can form and decay according to the laws of statistical mechanics, and that only clusters above a certain critical size will exhibit stable growth, i.e., growth accompanied by a net release of energy.

In his recent book, Debenedetti (40) presents a comprehensive discussion of classical nucleation theory. Thus, only a brief summary is given here. Classical nucleation theory uses two parameters to describe the size of a critical nucleus n_{crit} and its rate of formation $\overset{*}{J}_{\mathrm{ncl}}$. The first parameter, the interfacial tension σ, is a macroscopic measure of the work per unit surface area that is added when a droplet or a particle grows. The second parameter, $\Delta\mu$, is the difference in the chemical potential between the solute molecule in the metastable solution and in the equilibrium phase. By defining a dimensionless interfacial tension $\overset{*}{\sigma}$ and a dimensionless difference in the chemical potential δ in Eqs. (62a) and (62b), we convert the well-known formulas for n_{crit} and $\overset{*}{J}_{\mathrm{ncl}}$, Eqs. (63a) and (64a), to more convenient forms, which are given as Eqs. (63b) and (64b), respectively:

$$\text{Definitions:}\quad \overset{*}{\sigma} = \frac{\sigma \cdot v_{\mathrm{m,j}}^{2/3}}{k_{\mathrm{B}}T} \qquad \delta = \frac{-\Delta\mu}{2k_{\mathrm{B}}T} \tag{62a,b}$$

$$n_{\mathrm{crit}} = \frac{32\pi \cdot \sigma^3 \cdot v_{\mathrm{m,j}}^2}{-3 \cdot \Delta\mu^3} \Rightarrow n_{\mathrm{crit}} = \frac{4\pi}{3} \cdot \left(\frac{\overset{*}{\sigma}}{\delta}\right)^3 \tag{63a,b}$$

$$\overset{*}{J}_{\mathrm{ncl}} = 2\beta_{\mathrm{ncl}}\rho_{\mathrm{m}}y_{\mathrm{j}} \cdot \sqrt{\frac{\overset{*}{\sigma} v_{\mathrm{m,j}}^2}{k_{\mathrm{B}}T}} \cdot \exp\left(-\frac{16\pi\overset{*}{\sigma}^3}{3k_{\mathrm{B}}T}\left(\frac{v_{\mathrm{m,j}}}{\Delta\mu}\right)^2\right) \Rightarrow \overset{*}{J}_{\mathrm{ncl}}$$

$$= 2\beta_{\mathrm{ncl}}\rho_{\mathrm{m}}y_{\mathrm{j}} \cdot v_{\mathrm{m,j}}^{2/3} \cdot \sqrt{\overset{*}{\sigma}} \cdot \exp(-n_{\mathrm{crit}} \cdot \delta) \tag{64a,b}$$

In the above equations, $v_{\mathrm{m,j}}$ denotes the volume of a molecule of species j, ρ_{m} is the density of the metastable phase in molecules per cubic meter, y_{j} is the mole fraction of the solute in the metastable phase, and β_{ncl} is a transport coefficient associated with the process of nucleation.

Once a cluster has attained the critical size, the precipitation of any additional solute molecules on its outer surface is described by the laws of condensation. The number of solute molecules that precipitate per unit time and particle on the outer surface of an existing particle of spherical shape with diameter D_{p} is given by (41)

$$\overset{*}{J}_{\mathrm{cnd}} = 2\pi D_{\mathrm{p}} \cdot \beta_{\mathrm{cnd}} \cdot \mathcal{D}_{\mathrm{g}} \cdot N_{\mathrm{A}}(c_{\mathrm{j}} - c_{\mathrm{j}}^{\mathrm{eq}}) \Rightarrow$$

$$\overset{*}{J}_{\mathrm{cnd}} = 2\pi D_{\mathrm{p}} \cdot \beta_{\mathrm{cnd}} \cdot \mathcal{D}_{\mathrm{g}} \cdot \rho_{\mathrm{m}}(y_{\mathrm{j}} - y_{\mathrm{j}}^{\mathrm{eq}}) \tag{65a,b}$$

The correction factor β_{cnd} in Eq. (65) is dimensionless and accounts for non-continuum effects. The mobility of solute molecules in the metastable gas phase is expressed by the gas-phase diffusion coefficient $\mathcal{D}_g$, and the driving force for condensation is the difference in mole fraction between the metastable phase (y_j) and a gas phase that is in equilibrium with the precipitated phase (y_j^{eq}). Nucleation and condensation are mechanisms of precipitation that compete with each other. Condensation depends on the availability of the outer surface area of particles that have been generated by nucleation. Both processes lower the actual solute mole fraction y_j and therefore diminish the driving force for either process.

Once precipitation is complete, the formation of new particles will not continue. However, condensation is not the only growth mechanism; particles can continue to grow by the process of coagulation, in which two particles collide and combine to form a larger particle. The number of collisions, C_{12}, that particles of species 1 undergo with particles of species 2 per unit volume and time is an expression that is linear in the number concentration of either species. This is valid under the assumption of a low volume fraction occupied by the droplets or particles. For this scenario, Seinfeld and Pandis (41) give

$$C_{12} = K_{12} \cdot N_1 \cdot N_2 \tag{66}$$

The collision coefficient K_{12} in Eq. (66) is independent of the number densities N_1 and N_2, which have units of particle number per cubic meter. Instead, this coefficient is a function of several parameters describing the size of the particles and their ordered and random motion. For a quiescent fluid, the only driving force leading to collisions between particles is Brownian motion. Expressions for K_{12} describing this phenomenon are linear in the sum of both particle diameters, D_{P1} and D_{P2}. However, for a flow field that exhibits significant velocity gradients, different expressions for K_{12}, both for laminar and for turbulent flow fields, are required (42,43). Both formulas contain the sum of the particle diameters raised to the third power, which has two important consequences. First, collisions between two particles with large size differences are favored over those between particles of about the same size. Second, C_{12} increases dramatically with the particle diameter. This reflects the fact that velocity gradients of the flow field do not affect the frequency of collisions between particles that are smaller than a certain threshold. The higher the shear rates, the smaller this threshold.

Once the collision rate C_{12} has been obtained from Eq. (66), the coagulation rate can be obtained if particle break-up is excluded and if the percentage of "successful" collisions between two particles to form a bigger one is known. Von Smoluchowski (42) calculated the effect of introducing a so-called sticking factor (i.e., the fraction of collisions that are successful) on the coagulation rate.

In practice, sticking factors of less than 1 can be induced, for example, by the addition of a surfactant to the solution.

As was pointed out in the discussion above, the coagulation rate increases with increasing mean particle diameter and with increasing number density of the particles. Therefore, coagulation is initiated and accelerated by the precipitation processes of nucleation and condensation. On the other hand, coagulation influences these precipitation processes only indirectly and unevenly. Particle growth by coagulation reduces the overall available particle surface area, which would be expected to hinder condensation. It might be concluded, then, that coagulation tends to favor nucleation. On the other hand, bigger particles are more likely to exhibit a considerable slip velocity (i.e., a velocity difference between a particle and the surrounding fluid) than smaller ones. Slip velocity is a prerequisite for convective mass transport toward the outer particle surface, which favors condensation.

In Figure 10, we show the interrelationship between the three microphysical processes of nucleation, condensation, and coagulation, which result in the precipitation and growth of particles during RESS. As described below, arrows 1–6 illustrate how a given process (at the head of the arrow) affects a second process (at its tail). Arrow 1: Only when nucleation has produced a sufficient number of particles does condensation become an effective precipitation process.

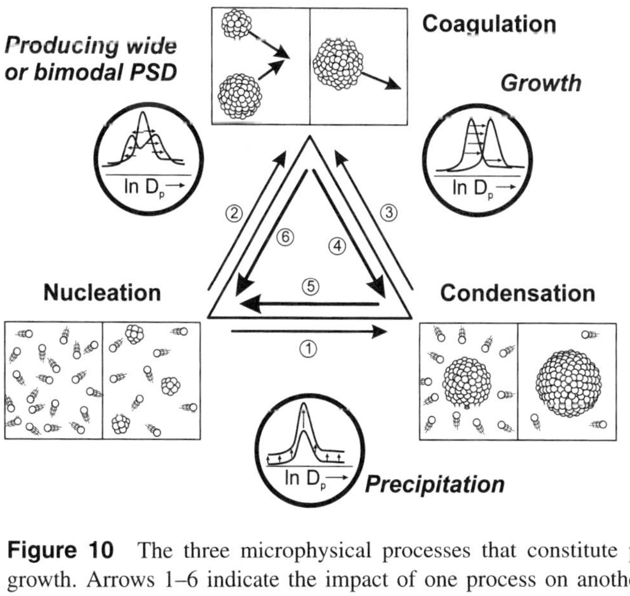

Figure 10 The three microphysical processes that constitute particle formation and growth. Arrows 1–6 indicate the impact of one process on another. Changes in particle size distribution caused by a given pair of processes are also shown.

Arrows 2 and 3: As the number density and size of particles increase (via nucleation and condensation, respectively), coagulation increases. Arrow 4: Coagulation reduces the number of particles and, consequently, the particle surface area available for condensation. Arrow 5: Condensation competes with nucleation and weakens it by lowering its driving forces. Arrow 6: Therefore, coagulation could be looked at as a process indirectly favoring nucleation. In fact, complete precipitation and particle growth without significant condensation is a possible scenario.

Also shown in the figure is the effect of a given pair of microphysical processes on particle size distributions (PSDs), schematics of which are located in the circles. Beginning with the bottom circle, nucleation and condensation are competing precipitation processes and as such increase the zeroth moment of the mass-based PSD. In contrast, coagulation does not change the total mass that is precipitated. Moving to the right circle, condensation and coagulation are growth processes and therefore increase the first moment of any kind of PSD. By definition, the particles generated by homogeneous nucleation are the smallest to persist (as any smaller cluster randomly decays in a very short time). Thus, nucleation reduces the mean particle diameter. As shown in the left circle, a scenario that only allows for nucleation and coagulation would most certainly result in a wide or even bimodal PSD. Condensation can either enhance or detract from this effect. Diffusion-driven condensation can narrow PSDs, as it provides faster growth (on a log scale) for smaller particles than for larger particles. However, as discussed earlier, condensation driven by convective mass transfer favors larger over smaller particles, and thus can also lead to a wide or bimodal PSD.

B. Controlling Particle Size in RESS: A Simplified Model for Nucleation and Condensation

A primary motivation for investigating RESS is the potential it offers for the control of particle size and distribution. Furthermore, with the recent national push toward nanotechnology, the production of particles significantly smaller than 1 μm is of particular interest. However, if particle growth is to be controlled the interrelationship between nucleation and growth processes needs to be more clearly understood.

Consider, as an example, the precipitation of an organic solute with a molecular weight of 200 and a density of 1000 kg/m^3. From classical nucleation theory, a typical critical nucleus (n_{crit}) produced during RESS would contain about 10 molecules and thus would have a diameter ($D_{\mathrm{p,crit}}$) of about 2 nm. Even if an unreasonable estimate of critical nucleus size (e.g., 10,000 molecules) were made, or if the nucleus were assumed to consist of polymeric molecules with a molecular weight of 2×10^5, the corresponding nucleus would still have a diameter of only about 20 nm. Thus, nanometer-sized particles can be produced by RESS if their growth is suppressed after their formation by homogeneous nucleation. Although, as was discussed above, growth processes are generally promoted by higher solute concentrations, operating RESS under even more dilute conditions than is already imposed by the solubility limits may not be economically viable. Clearly, it is important to consider how we might otherwise impede growth.

In Section VII.A, we discussed how particle growth by coagulation can occur because of either Brownian motion or shear. Coagulation due to Brownian motion is a comparatively slow growth process under most conditions, but particle growth due to shear becomes significant beyond a certain threshold size. (The threshold size is defined as the particle diameter for which the coagulation rate due to shear increases to the rate due to Brownian motion.) Consequently, "fighting" growth due to coagulation becomes more difficult as the particles grow bigger.

According to Seinfeld and Pandis (41), the third power of the threshold diameter ($D_{p,thr}$) scales inversely with the shear rate. This is shown by Eq. (67a) for a laminar flow field, where Γ is the shear rate, and by Eq. (67b) for a turbulent flow field, where the amount of kinetic energy ε_k dissipated per unit volume and time is taken as a measure of the shear rate:

$$\text{Laminar: } \frac{1}{D_{p,thr}^3} \propto \Gamma \qquad \text{turbulent: } \frac{1}{D_{p,thr}^3} \propto \sqrt{\frac{\varepsilon_k}{\nu}} \tag{67a,b}$$

While a typical value for the threshold diameter of an aerosol particle in atmospheric flows is between 2 and 5 μm, our estimates show that threshold diameters in a typical RESS flow field would be 10–100 times smaller. In summary, then, (a) the process of nucleation produces nanoparticles of a highly desirable size and (b) coagulation does not become a significant factor in particle growth until particles achieve diameters of about 100 nm. Thus, the key to making nanoparticles by RESS is to prevent the growth of particles to their threshold size ($D_{p,thr}$) by condensation.

In order to elucidate the factors leading to particle growth by condensation, we propose a simplified two-step model for particle formation and growth that only considers nucleation and condensation (Figure 11). The model assumes that particles precipitate from a metastable solution containing the solute j at a mole fraction $y_{j,ini}$ in two distinct steps. First, homogeneous nucleation produces particles that all incorporate the same number of molecules, n_{crit}. Second, all solute molecules precipitate by condensation uniformly on the outer surface of the existing particles, and no new ones are generated.

The rate per unit volume at which solute molecules precipitate by the mechanism of homogeneous nucleation (i.e., the molecular nucleation rate $\overset{*}{n}_{ncl}$)

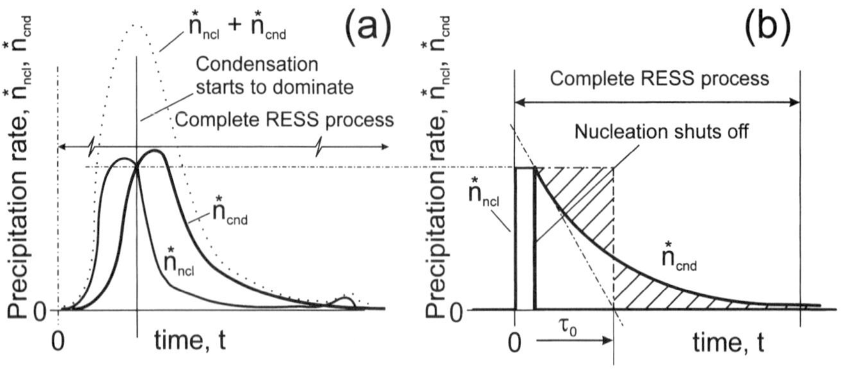

Figure 11 (a) The rates at which molecules precipitate during RESS processing are complex functions of time. (b) The simplifying assumptions of our two-step model are shown.

is given by the nucleation rate [Eq. (64b)] multiplied by the number of solute molecules contained in each nucleus [Eq. (63b)]:

$$\overset{*}{n}_{\mathrm{ncl}} = \overset{*}{J}_{\mathrm{ncl}} \cdot n_{\mathrm{crit}} = \frac{8\pi\beta_{\mathrm{ncl}}\upsilon_{\mathrm{m,j}}^{2/3}}{3} \cdot \rho_{\mathrm{m}} y_{\mathrm{j,ini}} \cdot \overset{*}{\sigma}^{7/2} \cdot \delta^{-3} \cdot \exp\left(-\frac{4\pi \cdot \overset{*}{\sigma}^{3}}{3\delta^{2}}\right) \tag{68}$$

The particles generated are all of the same size and provide the basis for the next step, condensation, to take place.

Next, we calculate the initial rate per unit volume at which solute molecules precipitate by condensation (i.e., the molecular condensation rate $\overset{*}{n}_{\mathrm{cnd,ini}}$), which is obtained by multiplying the initial condensation rate $\overset{*}{J}_{\mathrm{cnd}}$ [see Eq. (65b)] by the number density of particles, N_{2SM}, that have been generated in the first step (i.e., nucleation) of the process (2SM refers to two-step model):

$$\overset{*}{n}_{\mathrm{cnd,ini}} = \overset{*}{J}_{\mathrm{cnd}} \cdot N_{\mathrm{2SM}} = 2\pi \cdot D_{\mathrm{p}} \cdot \beta_{\mathrm{cnd}} \cdot \mathcal{D}_{\mathrm{g}} \cdot \rho_{\mathrm{m}} \cdot y_{\mathrm{j,ini}} \cdot N_{\mathrm{2SM}} \tag{69}$$

Note in the above equation that we have assumed that the nucleation step does not alter the solute mole fraction $y_{\mathrm{j,ini}}$ significantly, as the equilibrium mole fraction $y_{\mathrm{j}}^{\mathrm{eq}}$ is assumed to be negligibly small compared with $y_{\mathrm{j,ini}}$. Thus, the above equation gives us $\overset{*}{n}_{\mathrm{cnd}}$ at that instant in time when nucleation terminates and condensation begins, i.e., when $\overset{*}{n}_{\mathrm{ncl}} = \overset{*}{n}_{\mathrm{cnd}}$.

Because nucleation has just ended and condensation has just begun, the diameter of a particle in Eq. (69) is given by

$$D_{\mathrm{p,crit}} = \left(\frac{6}{\pi} \cdot n_{\mathrm{crit}} \cdot V_{\mathrm{m,j}}\right)^{1/3} \tag{70}$$

Substituting Eqs. (70) and (63b) into Eq. (69), we obtain

$$\overset{*}{n}_{\mathrm{cnd,ini}} = 4\pi \cdot v_{\mathrm{m,j}}^{1/3} \cdot \frac{\overset{*}{\sigma}}{\delta} \cdot \beta_{\mathrm{cnd}} \cdot \mathcal{D}_{\mathrm{g}} \cdot \rho_{\mathrm{m}} \cdot y_{\mathrm{j,ini}} \cdot N_{\mathrm{2SM}} \tag{71}$$

The number density N_{2SM} of the particles formed when nucleation terminates and condensation begins (i.e., when $\overset{*}{n}_{\mathrm{ncl}} = \overset{*}{n}_{\mathrm{cnd}}$) is obtained by equating Eqs. (68) and (71), and then rearranging to obtain

$$N_{\mathrm{2SM}} = \frac{2}{3} \cdot \frac{\beta_{\mathrm{ncl}} \cdot v_{\mathrm{m,j}}^{1/3}}{\beta_{\mathrm{cnd}} \cdot \mathcal{D}_{\mathrm{g}}} \cdot \overset{*}{\sigma}^{5/2} \cdot \delta^{\ 2} \cdot \exp\left(-\frac{4\pi \cdot \overset{*}{\sigma}^{3}}{3\delta^{2}}\right) \tag{72}$$

Note that the number (and thus, the number density) of particles does not change during the condensation step because we assume that the onset of condensation "shuts off" nucleation (and we neglect the effects of coagulation).

The final particle diameter when condensation is complete and all solute molecules have precipitated, $D_{\mathrm{p,f}}$, can be calculated by equating the mass of solute in the metastable phase at initial time to the mass of the particles when condensation is complete:

$$\rho_{\mathrm{m}} \cdot y_{\mathrm{j,ini}} \cdot v_{\mathrm{m,j}} \cdot \rho_{\mathrm{j}} = \frac{\pi}{6} \cdot D_{\mathrm{p,f}}^{3} \cdot \rho_{\mathrm{j}} \cdot N_{\mathrm{2SM}} \tag{73}$$

where ρ_{j} is the mass density (kg/m^3) of a solute molecule j. Substituting Eq. (72) into Eq. (73) and rearranging, the resulting expression for $D_{\mathrm{p,f}}$ is obtained:

$$D_{\mathrm{p,f}} = \Omega \cdot D_{\mathrm{p,2SM}} \tag{74}$$

where

$$\Omega = \left[\overset{*}{\sigma}^{5/2} \cdot \delta^{-2} \cdot \exp\left(-\frac{4\pi \cdot \overset{*}{\sigma}^{3}}{3\delta^{2}}\right)\right]^{-1/3} = \overset{*}{\sigma}^{-5/6} \cdot \delta^{2/3} \cdot \exp\left(\frac{4\pi \cdot \overset{*}{\sigma}^{3}}{9\delta^{2}}\right) \tag{75}$$

and

$$D_{\mathrm{p,2SM}} = \left(\frac{9\beta_{\mathrm{cnd}} \cdot \rho_{\mathrm{m}} \cdot y_{\mathrm{j,ini}} \cdot \mathcal{D}_{\mathrm{g}} \cdot v_{\mathrm{m,j}}^{2/3}}{\pi \cdot \beta_{\mathrm{ncl}}}\right)^{1/3} \tag{76}$$

The groups Ω and $D_{\mathrm{p,2SM}}$ have been defined because Ω is solely a function of the dimensionless interfacial tension $\overset{*}{\sigma}$ and chemical potential difference δ, while $D_{\mathrm{p,2SM}}$, a reference particle diameter, indicates the relative strength of condensation vs. nucleation.

Equation (76) can be further simplified by substituting in Eqs. (77a)–(77c) (41):

$$P = Z \cdot \rho_m \cdot k_B T \quad \beta_{ncl} = \frac{y_{j,ini} \cdot P}{\sqrt{\frac{2\pi \cdot mw_j \cdot k_B T}{N_A}}} \quad \bar{u} = \sqrt{\frac{8 \cdot N_A k_B T}{\pi \cdot mw_j}} \tag{77a,b,c}$$

to obtain

$$D_{p,2SM} = \left(\frac{36}{\pi \cdot Z} \cdot \frac{\beta_{cnd} \cdot \mathcal{D}_g \cdot v_{m,j}^{2/3}}{\bar{u}} \right)^{1/3} \tag{78}$$

Note that $D_{p,2SM}$ is essentially independent of the initial solute mole fraction $y_{j,ini}$, suggesting that the size of a primary particle (i.e., a particle that forms by nucleation, not coagulation) generated during RESS is not directly influenced by solute concentration.

In addition to calculating $D_{p,f}$, our two-step model can also be used to calculate the time required for the nucleation and condensation processes to be completed (see Figure 11). If, as we have assumed, essentially all solute molecules precipitate by condensation, the solute mole fraction will decrease from its initial value $y_{j,ini}$ to zero during the second step of our model. Consequently, the driving force for condensation, i.e., the difference between actual and equilibrium mole fractions, will also approach zero. (To some extent, condensation will be enhanced as the particles grow in diameter and surface area, but this effect is believed to be relatively small and was neglected.) It was also assumed that nucleation never becomes significant again despite the decrease in $\overset{*}{n}_{cnd}$. The condensation rate, $\overset{*}{n}_{cnd}$, was assumed to be proportional to the number of solute modules present in the metastable phase (i.e., exponential-decay behavior):

$$\overset{*}{n}_{cnd} = \frac{-d(y_j \cdot \rho_m)}{dt} = \frac{1}{\tau_0}(y_j \cdot \rho_m) \tag{79}$$

The solution to Eq. (79) is

$$y_j = y_{j,ini} \exp(-t/\tau_0) \tag{80}$$

where τ_0 is a time constant. If we assume that the process is complete when 99.5% of the molecules have precipitated (i.e., when $y_j/y_{j,ini} = 0.005$), then the final time, t_f, is given by $5.3\tau_0$. Thus, at the onset of condensation Eq. (79) can be written as

$$\overset{*}{n}_{cnd,ini} \cdot \tau_0 = y_{j,ini} \cdot \rho_m \tag{81}$$

which is also equivalent to the total number of solute molecules initially present in the metastable phase (and which precipitate). Finally, we can rewrite Eq. (81) in terms of the time required for 99.5% of the precipitation to occur:

$$t_f = (5.3) \cdot \frac{(y_{j,ini} \cdot \rho_m)}{\overset{*}{n}_{cnd,ini}} \tag{82}$$

By substituting Eqs. (77a–c) into Eq. (68), an expression for $\overset{*}{n}_{ncl}(= \overset{*}{n}_{cnd,ini})$ is obtained that can be inserted into Eq. (82) to yield the desired expression for t_f:

$$t_f = (5.3) \cdot \tau_{2SM} \cdot \frac{\delta}{\overset{*}{\sigma}} \cdot \Omega^3 \tag{83}$$

where

$$\tau_{2SM} = \frac{3}{2\pi \cdot Z \cdot \rho_m \cdot y_{j,ini} \cdot v_{m,j}^{2/3} \cdot \overline{u}} \tag{84}$$

It is interesting to note that the grouping defined as the reference time, τ_{2SM}, is inversely proportional to the density of the fluid, the initial solute mole fraction, and the volume that the particles finally attain.

Under conditions typical for the RESS processing of organics, such as pharmaceuticals with a micro-orifice, values on the order of 1 ns and 1 nm are obtained for τ_{2SM} and $D_{p,2SM}$, respectively. Thus, the potential for the most significant variation in particle diameter with respect to processing time would seem to lie in the behavior of Ω, $\overset{*}{\sigma}$, and δ [see Eqs. (74) and (83)].

To help us understand the effect of chemical potential difference and interfacial tension on final particle size and processing times, a graph was constructed (see Figure 12). The construction of contour lines attributable to constant particle diameter and constant processing times, respectively, were facilitated by the introduction of two dimensionless parameters:

$$\zeta = \sqrt{\frac{3\delta^2}{4\pi\overset{*}{\sigma}^3}} \tag{85}$$

and

$$\Xi = \frac{\delta}{\overset{*}{\sigma}} \cdot \Omega^3 \tag{86}$$

With these new definitions, we see that $D_{p,f} = D_{2SM} \cdot \Omega$ and $t_f = (5.3) \cdot \tau_{2SM} \cdot \Xi$. Using Eqs. (75), (85), and (86), the equations necessary for generating

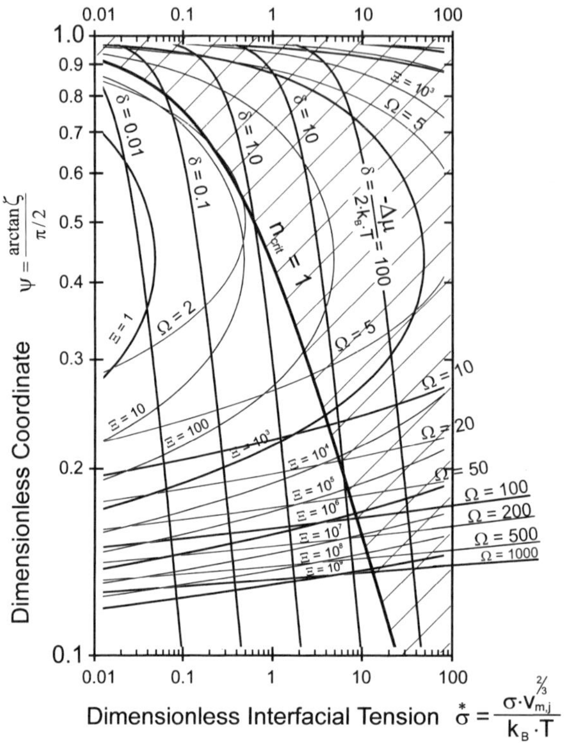

Figure 12 Our two-step model for nucleation and condensation can be used to examine the effects of chemical potential difference and interfacial tension on final particle diameter and processing times.

the contour lines for Ω and Ξ as functions of $\overset{*}{\sigma}$ and ζ (and thus of δ) are obtained:

$$\overset{*}{\sigma} = \left(\frac{3 \cdot \exp\left(\frac{-1}{\zeta^2}\right)}{4\pi \cdot \zeta^2} \right)^2 \cdot \Omega^6 \quad \text{and} \quad \overset{*}{\sigma} = \left(\frac{3 \cdot \exp\left(\frac{-2}{3\zeta^2}\right)}{4\pi \cdot \zeta^2} \right)^{3/2} \cdot \Xi \tag{87a,b}$$

Although contour lines of constant Ω or Ξ can be found in a σ-δ plane by varying ζ in the above equations, most of the contour lines in the resulting plot are so close together that reading becomes awkward and difficult. Therefore, for the sake of legibility a transform variable, ψ, was defined:

$$\psi = \frac{\arctan\zeta}{\pi/2} \tag{88}$$

As shown in Figure 12, contour lines of constant δ, Ω, and Ξ were then plotted in a ψ vs. $\overset{*}{\sigma}$ plane. The reader should note that not all areas of the ψ-$\overset{*}{\sigma}$ plane shown in Figure 12 are accessible. Obviously, the smallest size that a critical nucleus can have during homogeneous nucleation is that of one solute molecule. By rearranging Eq. (63b) and setting $n_{\text{crit}} = 1$, the following expression is obtained, which, along with Eq. (85), was used to generate the $n_{\text{crit}} = 1$ line shown in Figure 12.

$$\delta = \sqrt[3]{\frac{4\pi}{3}} \cdot \overset{*}{\sigma} \quad (n_{\text{crit}} = 1) \tag{89}$$

This line separates the regime of nucleation and growth (to the left) from the region of spinodal decomposition. A physical limit also exists on the left side of Figure 12 because the final diameter of a particle formed by nucleation and condensation cannot be smaller than the critical nucleus formed by homogeneous nucleation. Furthermore, $D_{\text{p,f}}$ must be considerably larger than n_{crit}, so that our initial assumption is not violated (i.e., that only an insignificant fraction of the solute molecules are precipitated by nucleation). The expression of interest is obtained by dividing Eq. (74) by Eq. (70):

$$\frac{D_{\text{p,f}}}{D_{\text{p,crit}}} = \frac{\Omega \cdot D_{\text{p,2SM}}}{2v_{\text{m,j}}^{1/3} \cdot \dfrac{\overset{*}{\sigma}}{\delta}} \gg 1, \qquad \text{or} \qquad \frac{\Omega \cdot \delta}{\overset{*}{\sigma}} \gg \frac{2v_{\text{m,j}}^{1/3}}{D_{\text{p,2SM}}} \tag{90a,b}$$

Figure 12 can be used to help us understand the relationship between final particle size and processing conditions both for RESS and for materials processing with supercritical fluids in general. For example, consider the production of nanoparticles by RESS. Starting at large values of Ω and holding $\overset{*}{\sigma}$ constant at 1, we see that δ increases from about 0.5 to 1.0 when we reach the $\Omega = 5$ line. (Recall that $D_{\text{p,2SM}}$ is on the order of a nanometer, so that $\Omega = 5$ corresponds to $D_{\text{p,f}} \approx 5$ nm.) This confirms what is intuitively known and has also been found by some experimentalists, i.e., that the final particle diameter decreases with increasing supersaturation ratio if coagulation is not possible or not accounted for. Now let us assume that particles of this size are desirable but that high supersaturation ratios (and, thus, high δ values) cannot be attained. How would one operate the RESS process to obtain nanoparticles? The answer, according to Figure 12, is to move left along the $\Omega = 5$ line until you reach a value of δ that can be attained in your RESS process. For example, at the location where the $\Omega = 5$ line crosses the $\delta = 0.01$ line we observe that (a) the dimensionless interfacial tension $\overset{*}{\sigma}$ is now only about 0.05 and (b) the time frame, Ξ, within which the particle formation process is completed is about one-third of that for when $\overset{*}{\sigma} = 1$. From this exercise, we conclude that particles of virtually any size can be produced by RESS if the interfacial tension in the system is sufficiently low.

As another example, consider how one would produce micrometer-sized particles by RESS. Assuming, as before, that $D_{p,2SM}$ is on the order of a nanometer, particles with diameters in the μm range will correspond to the $\Omega = 1000$ contour line. Within the range of our diagram, this implies time scales of $10^8 < \Xi < 10^9$, which translates into absolute times in the range of 1 s. This is much longer than the time scale attributable to any typical RESS setup, be it the residence time in the cylindrical section of a micro-orifice (about a microsecond), or the time scale of the whole spray process, which still is a maximum of only a few milliseconds. The most obvious explanation for the discrepancy between experiment and prediction is that particles attain micrometer-sized diameters because of coagulation. Another possibility is that the basic molecular data used to calculate $D_{p,2SM}$ and τ_{2SM} are off by more than an order of magnitude, but this is rather unlikely, as reasonable estimates of the terms in Eqs. (78) and (84) can be made with confidence. For example, even an uncertainty in σ of one order of magnitude does not explain the observed discrepancy between the experimentally observed and predicted time.

In closing, the reader should be reminded that even comprehensive numerical simulations, e.g., simulations that include coagulation and use a two-dimensional flow field, have similar difficulties in predicting the formation of the micrometer-sized particles that are found in RESS experiments with micro-orifices. Our two-step model can therefore be seen as a practical tool that predicts these difficulties and aids us in the search for solutions.

NOMENCLATURE

Universal Constants

π	3.1415926..	[-]
e	2.7182818..	Euler's constant
N_A	$6.02214 \cdot 10^{23}$	Avogadro's constant [#/mol]
k_B	$1.3807 \cdot 10^{-23}$	Boltzmann's constant [J/#K]

General Dimensionless Symbols

#	[-]	Molecule (or atom)
$	[-]	Particle (or nucleus)
€	[-]	Event

Dimensionless Numbers

Bi	[-]	Biot number
Kn	[-]	Knudsen number
Ma	[-]	Mach number
Nu	[-]	Nusselt number
Pr	[-]	Prandtl number

Re	[-]	Reynolds number
St	[-]	Stanton number

Symbols

A_M	[-]	Constant in "Mach analysis"
C_{12}	[€/m^3 s]	Collision frequency
c_h	[W/m^2 K]	Heat transfer coefficient
c_j	[mol/m^3]	Concentration of substance j
c_p	[J/kg K]	Heat capacity at constant p
c_v	[J/kg K]	Heat capacity at constant ρ
D	[m]	Diameter
D_z	[-]	Derivative of D with resp. to z
F	[-]	Excess function for heat transfer
f	[-]	Fanning friction factor
h	[J/kg]	Specific enthalpy
$\overset{*}{J}$	[$/m^3 s]	Nucleation rate
$\overset{*}{J}_{cnd}$	[#/$ s]	Condensation rate
K_{12}	[€ m^3/$^{2} s]	Collision coefficient
K_M	[-]	Constant in Mach disk analysis
k	[W/m K]	Thermal conductivity
L	[m]	Length of the cylindrical channel
$\dot{m}$	[kg/s]	Mass flow rate
mw	[kg/mol]	Molecular mass
N	[$/m^3]	Number distribution of the part
$\overset{*}{n}$	[#/m^3 s]	Phase transition frequency
n_{crit}	[#/$]	Number of molecules in a critical nucleus
P	[Pa]	Pressure
$\dot{Q}$	[W]	Heating power
$\dot{q}$	[W/m^2]	Heat flux per unit area
r	[m]	Position in radial direction
rf	[-]	Recovery factor
S	[-]	Supersaturation ratio
s	[J/kg K]	Specific entropy
T	[K]	Temperature
t	[s]	Time
u	[m/s]	Velocity in axial direction
v_m	[m^3/#]	Molecular volume
x_M	[-]	Constant for Mach disk
y	[#/#]	Mole fraction
Z	[-]	Compressibility
z	[m]	Position on the symmetry axis
z_{M0}	[m]	Offset distance in Mach disk analysis

Script Symbols

$\mathcal{D}$	[m^2/s]	Diffusion coefficient

Greek Symbols

α	[m^2]	Cross-sectional area
β_{ncl}	[#/m^2 s]	Mass transfer coefficient for nucleation
β_{cnd}	[-]	Mass transfer coefficient for condensation
γ	[-]	Cone angle of a tapered section
γ	[-]	Ratio of the heat capacities
δ	[-]	Dimensionless difference in chemical potential
ε	[mm]	Surface roughness
ε_k	[m^2/s^3]	Energy dissipation rate
ζ	[-]	Dimensionless parameter in two-step model
η	[kg/ms]	Dynamic viscosity
θ	[-]	Dimensionless time t/τ
λ	[m]	Free mean path length
μ	[J/kg]	Chemical potential
ν	[m^2/s]	Kinematic viscosity
Ξ	[-]	Time function in the two-step model
ρ	[kg/m^3]	Density
σ	[N/m]	Interfacial tension
$\overset{*}{\sigma}$	[-]	Dimensionless interfacial tension
ς	[m]	Circumference
τ	[s]	Residence time inside channel
Φ	[-]	Microchannel correction
ψ	[°]	Transform variable in Figure 12
Ω	[-]	Diameter function in the two-step model

Subscripts

avg	Average
ad	Adiabatically
cnd	Due to condensation
cd	Convergent-divergent
ext	Extraction conditions
el	Electrical
f	Final
fl	Fluid
ht	Heating
i	Denoting particle species *i*
ini	Immediately prior to experiment
j	Denoting chemical species *j*
loss	Lost to the environment
jet	Referring to the jet

m	Molecular
M	At the Mach disk
meas	Measurement
ncl	Due to nucleation
o	Outside/outer surface
p	Referring to a particle/particles
snd	Referring to the sound
tub	In the tubing
thr	Threshold
w	At the wall
x	Upstream of the Mach disk
y	Downstream of the Mach disk
0	Far upstream of the nozzle
1	Entrance of cylindrical section
2	Exit of cylindrical section, Ma = 1
3	Location of the Mach disk
4	Stagnant region in the expansion chamber
2SM	Referring to two-step model

Superscripts

eq	Referring to equilibrium
id	Ideal gas
0	At stagnation

ACKNOWLEDGMENTS

This work was supported in part by the ERC program of the National Science Foundation under Award Number EEC-9731680. M. Weber thanks D. Keith Walters for helpful discussions.

REFERENCES

1. RC Petersen, DW Matson, RD Smith. Rapid precipitation of low vapor pressure solids from supercritical fluid solutions: the formation of thin films and powders. J Am Chem Soc 108:2102–2103, 1986.
2. PG Debenedetti. Homogeneous nucleation in supercritical fluids. AIChE J 36:5507–5515, 1990.
3. J-J Shim, MZ Yates, KP Johnston. Polymer coatings by rapid expansions of suspensions in supercritical carbon dioxide. Ind Eng Chem Res 38:3655–3662, 1999.
4. AK Lele, AD Shine. Effects of RESS dynamics on polymer morphology. Ind Eng Chem Res 33:1476–1485, 1994.

5. S Mawson, KP Johnston, JR Combes, JM DeSimone. Formation of poly(1,1,2,2-tetrahydroperfluorodecyl acrylate) submicron fibers and particles from supercritical carbon dioxide solutions. Macromolecules 28:3182–3191, 1995.
6. NE Aniedobe, MC Thies. Formation of cellulose acetate fibers by the rapid expansion of supercritical carbon dioxide solutions. Macromolecules 30(9):2792–2794, 1997.
7. JW Tom, PG Debenedetti. Particle formation with supercritical fluids—a review. J Aerosol Sci 22(5):555–584, 1991.
8. PG Debenedetti, JW Tom, X Kwauk, S-D Yeo. Rapid expansion of supercritical solutions: fundamentals and applications. Fluid Phase Equilib 82:311–321, 1993.
9. B Subramaniam, RA Rajewski, K Snavely. Pharmaceutical processing with supercritical carbon dioxide. Journal Pharm Sci 86(8):885–890, 1997.
10. P Subra, P Jestin. Powders elaboration in supercritical media: comparison with conventional routes. Powder Technol 103:2–9, 1999.
11. I Kikic, P Sist. Applications of supercritical fluids to pharmaceuticals: controlled drug release systems. In: E Kiran, PG Debenedetti, CJ Peters, eds. Supercritical Fluids: Fundamentals and Applications. NATO Science Series E 366. Dordrecht: Kluwer Academic, 2000, pp. 291–306.
12. RS Mohamed, PG Debenedetti, RK Prud'homme. Effects of process conditions on crystals obtained from supercritical mixtures. AIChE J 35(2):325–328, 1989.
13. E Reverchon, G Donsi, D Gorgoglione. Salicylic acid solubilization in supercritical CO_2 and its micronization by RESS. J Supercritic Fluids 6:241–248, 1993.
14. GJ Griscik, AS Teja. Crystallization of n-octacosane by the rapid expansion of supercritical solutions. J Cryst Growth 155:112–119, 1995.
15. P Alessi, A Cortesi, I Kikic, NR Foster, SJ Macnaughton, I Colombo. Particle production of steroid drugs using supercritical fluid processing. Ind Eng Chem Res 35:4718–4726, 1996.
16. A Blasig, MC Thies. Effect of concentration and degree of saturation on RESS of a CO_2-soluble fluoropolymer. Ind Eng Chem Res (submitted).
17. KA Larson, ML King. Evaluation of supercritical fluid extraction in the pharmaceutical industry. Biotechnol Prog 2(2):73–82, 1986.
18. E Reverchon, G Della Porta, R Taddeo, P Pallado, A Stassi. Solubility and micronization of griseofulvin in supercritical CHF_3. Ind Eng Chem Res 34:4087–4091, 1995.
19. EM Berends, OSL Bruinsma, GM van Rosmalen. Nucleation and growth of fine crystals from supercritical carbon dioxide. Journal Cryst Growth 128:50–56, 1993.
20. M Türk. Formation of small organic particles by RESS: experimental and theoretical investigations. J Supercrit Fluids 15(1):79–89, 1999.
21. AG Hansen. Fluid Mechanics. New York: John Wiley and Sons, 1967, Chapter 7.
22. JN Tilton. Fluid and particle dynamics. In: RH Perry, DW Green, eds. Perry's Chemical Engineers' Handbook. 7th ed. New York: McGraw-Hill, 1997, pp. 6–10.
23. M Weber, MC Thies. Improving the prediction of RESS particle sizes. Presented at the 5th International Symposium on Supercritical Fluids, Atlanta, GA, 2000.
24. DS Halverson. Precipitation from supercritical fluids: effects of process conditions on the morphology and particle size of precipitation products. MS thesis, Princeton University, Princeton, NJ, 1989.

25. JP Holman. Heat Transfer. 5th ed. New York: McGraw-Hill, 1981, pp. 209–222.
26. WL McCabe, JC Smith, P Harriott. Unit Operations of Chemical Engineering. 5th ed. New York: McGraw-Hill, 1993, pp. 86–87.
27. AK Lele, AD Shine. Morphology of polymers precipitated from a supercritical solvent. AIChE J 38:742–752, 1992.
28. SI Sandler. Chemical and Engineering Thermodynamics, 2nd ed. New York: John Wiley and Sons, 1989.
29. SE Haaland. Simple and explicit formulas for the friction factor in turbulent pipe flow. Transactions of the ASME, J Fluids Eng 105:89–90, 1983.
30. V Gnielinski. New equations for heat and mass transfer in turbulent pipe and channel flow. International Chemical Engineering 16(2):359–368, 1976.
31. ZY Guo, XB Wu. Further study on compressibility effects on the gas flow and heat transfer in a microtube. Microscale Thermophys Eng 2:111–120, 1998.
32. TM Adams. Turbulent convection in microchannels. PhD dissertation, Georgia Institute of Technology, Atlanta, GA, 1998, pp. 71–105.
33. K Bier, E Schmidt. Zur Form der Verdichtungsstöße in frei expandierenden Gasstrahlen. Zeitschrift für Angewandte Physik 13(11):493–500, 1961.
34. E Dick. Steady transonic flow. In: NP Cheremisinoff, ed. Encyclopedia of Fluid Mechanics, Vol. 1: Flow Phenomena and Measurement. Houston: Gulf Publishing, 1986, pp. 510–532.
35. RK Franklin, JR Edwards, Y Chernyak, RD Gould, RG Carbonell. Formation of perfluoropolyether coatings by the rapid expansion of supercritical solutions (RESS) process. Part 2: numerical modeling. Ind Eng Chem Res (in press).
36. H Ashkenas, FS Sherman. The Structure and utilization of supersonic free jets in low density wind tunnels. In: JH de Leeuw, ed. Supplement 3: Rarefied Gas Dynamics. In: HL Dryden, T von Kármán. Advances in Applied Mechanics, vol. 2. New York: Academic Press, 1965, pp. 84–105.
37. C Domingo, EM Berends, GM van Rosmalen. Precipitation of ultrafine organic crystals from the rapid expansion of supercritical solutions over a capillary and a frit nozzle. J Supercrit Fluids 10:39–55, 1997.
38. G Brunner. Gas Extraction: An Introduction to Fundamentals of Supercritical Fluids and the Application to Separation Processes. Darmstadt: Steinkopff, 1994, pp. 10–19.
39. KP Johnston, CA Eckert. An analytical Carnahan-Starling-van der Waals model for solubility of hydrocarbon solids in supercritical fluids. AIChE J 27(5):773–779, 1981.
40. PG Debenedetti. Metastable Liquids: Concepts and Principles. Princeton, NJ: Princeton University Press, 1996.
41. JH Seinfeld, SN Pandis. Atmospheric Chemistry and Physics: From Air Pollution to Climate Change. New York: John Wiley & Sons, 1998, pp. 664–665.
42. M von Smoluchowski. Versuch einer mathematischen Theorie der Koagulationskinetik kolloider Lösungen. Zeitschrift für physikalische Chemie 92:129–160, 1917.
43. PG Saffmann, JS Turner. On the collision of drops in turbulent clouds. J Fluid Mech 1:16–30, 1956.

12

Pharmaceutical and Biological Materials Processing with Supercritical Fluids

Srinivas Palakodaty, Raymond Sloan, and Andreas Kordikowski
Bradford Particle Design plc, Bradford, West Yorkshire, England

Peter York
University of Bradford, Bradford, West Yorkshire, England

I. INTRODUCTION

A. Background

Materials processing, often the last stage in a series of unit operations in the pharmaceutical industry, is perhaps the most critical step in drug synthesis. A suitable formulation and an appropriate form of drug delivery to the patient should effectively complement all the multistep operations of drug synthesis. A range of drug delivery systems are available for the administration of medicines, with virtually all requiring drug substances and/or functional formulation constituents in powder form at some stage in their preparation. The tablet, prepared in a variety of types but principally for oral delivery, remains the most popular and widely used form of drug delivery, with the drug mixed and processed with excipient(s) prior to compression into a tablet. Compact strength and hence the drug dissolution rate strongly depend on particle properties. An alternative route of administration for drug delivery that is increasing in interest and use is the respiratory tract (1). For deep lung deposition, the drug is inhaled directly from a metered dose inhaler (MDI) or a dry powder inhaler (DPI) whereupon it acts locally and/or is absorbed into the bloodstream in the lungs. This form of drug delivery generally requires the particle size of the drug in the narrow size range of 1–5 μm to be effective (2). Another class of delivery systems is for modified drug release. In one approach, the drug is trapped in a matrix of a suitable soluble carrier/excipient, and fine tuning the morphological structures of the drug carriers is important in determining and controlling the modified release profile. In another approach, the drug is coformulated along with a polymer to form a solid dispersion. Polymorphic purity is frequently another important issue in pharmacy due to different physicochemical properties and therapeutic effects of the alternative forms (3). Carefully controlled crystallization procedures linked to expensive separation methods are currently employed to obtain pure stable polymorphic forms. The degree of crystallinity of drug substances in powder form is also a critical factor not only in drug stability but also in achieving longevity of final products.

In many cases, drugs are required as micrometer-sized particles for targeting purposes and to enhance dissolution. The current methods of particle formation, size reduction, and materials processing most prevalent in the industry are (a) precipitation, (b) jet milling, (c) spray-drying, (d) freeze-drying, and (e) crystallization from solution. Solvent-based crystallization and precipitation processes generally cannot directly provide these fine particles for pharmaceutical formulations. Thus, additional downstream operations are required, such as drying, milling, and sieving. The energic milling operations used give rise to highly charged and cohesive materials that often exhibit poor secondary processing characteristics in mixing and flow operations. More specifically for DPI formulations, such materials give poor lung depositions. The spray- and freeze-drying operations are multivariable, complex, and expensive processes, which have found application especially for particulate products of biomolecules, although many challenges remain (1,4). Hence, the search for new and efficient processes for particle formation is ever present in industry, particularly when coupled with the increasing demands of novel drug formulation and delivery systems.

B. Supercritical Fluid Processing

Over the past decade supercritical fluid (SCF) processing has shown great promise in addressing many of the challenges faced by the industry in particle formation highlighted above. Techniques such as rapid expansion of supercritical solution (RESS) (5), precipitation from gas saturated solutions (PGSS) (6), and the gas antisolvent (GAS) process (5) have all been deployed in pharmaceuticals processing in a range of applications. The principles of these processes are described below.

1. Rapid Expansion of Supercritical Solutions (RESS)

The process, as shown in Figure 1, is a simple and efficient technique wherein a supercritical fluid (SCF), such as carbon dioxide with or without a cosolvent,

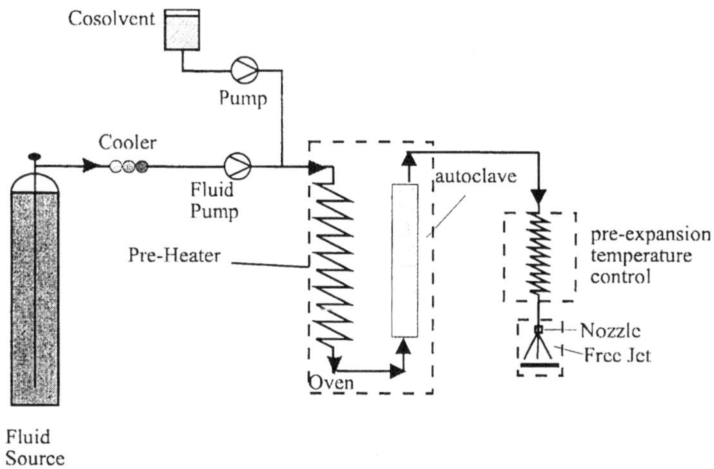

Figure 1 Schematic flow diagram of the rapid expansion of supercritical solution (RESS) process.

is passed over a fixed bed of material. The supercritical solution now containing the solute is then subjected to a rapid pressure decrease by expanding through a fine nozzle. The solvent capacity of the SCF is thus reduced, causing high supersaturation resulting in the formation of very fine particles. One of the major drawbacks with this process is the isenthalpic expansion that results in large temperature drops. Thus the nozzle is usually heated to avoid freezing of the solid and the carbon dioxide. The preexpansion and postexpansion temperatures determine the phase changes taking place, thereby affecting product characteristics. Furthermore, nozzle geometry plays an important role in determining the final particle size (7). For pharmaceutical processing, the RESS process has limited applications due to very low/negligible solubility of most drug substances in supercritical carbon dioxide, the most commonly used SCF. In addition, there is limited control over the process itself due to the effect of the microscopic variables listed above.

2. Gas Antisolvent (GAS) Process

The principle of this technique is similar to conventional liquid antisolvent crystallization. The high solubility of SCFs, which is an antisolvent for the solute, in most common organic solvents causes a volume expansion and a subsequent decrease in the solvent density by nearly twofold (8). Such a reduction in the solvent capacity causes phase changes wherein the solute molecules nucleate and particles separate from the solution. The process is typically carried out in a batch operation. The process variables include the solvent to antisolvent ratio and the rate of antisolvent addition. The latter also determines the rate of attaining supersaturation, thereby affecting the rate of crystallization and particle formation. A problem often encountered with the addition of a compressed gas to a solvent or solution is the apparent increase in the temperature of the system (9). Thus, the process requires increased control and, since particles are formed in solution, suffers from the need of additional downstream processing operations such as filtration and drying, negating the potential benefits of SCF processing. Hence, the process had been modified by several groups of investigators with the primary objective of making the process continuous and avoiding secondary processing of product. In almost all of the modified processes, an appropriate nozzle arrangement sprays the solution into a supercritical carbon dioxide medium. The modified processes are briefly discussed below.

3. Supercritical Antisolvent (SAS) Process

A schematic diagram of this process is shown in Figure 2. A solution of drug dissolved in a suitable solvent is sprayed into a continuum of supercritical carbon dioxide (10). The nozzle is typically a fine capillary tube, which generates droplets in the supercritical medium. The technique has been used to prepare polymers and proteins from both organic and organic–water solvents (11).

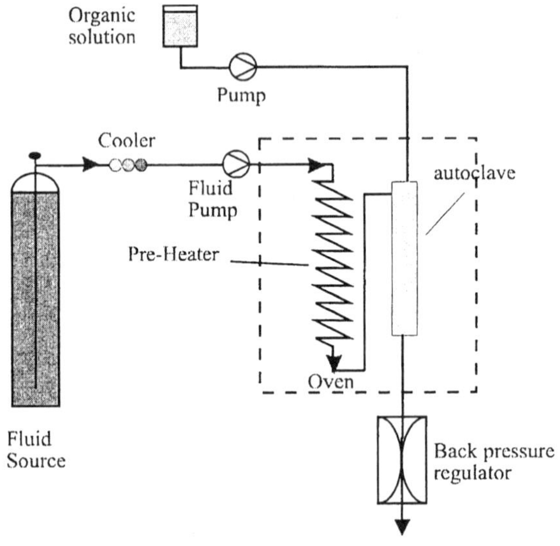

Figure 2 Schematic flow diagram of the supercritical antisolvent (SAS) process.

4. Solution-Enhanced Dispersion by Supercritical Fluid (SEDS) Process

The schematic diagram of the process is illustrated in Figure 3. The process is based on the concept of coupling the use of an SCF as a dispersing agent, by means of a coaxial nozzle, in addition to its primary role as an antisolvent and a vehicle to extract the solvent from a solution of drug (12). The technique has also been extended to process water-soluble compounds, including biologicals, by introducing an organic antisolvent together with the SCF and aqueous drug solution as separate streams via a three-concentric-opening nozzle, as shown in Figure 3 (13). As an alternative and more flexible approach for such processes involving aqueous media, the SEDS process has been further modified by mixing the liquid antisolvent with supercritical carbon dioxide to form a homogeneous mixture and subsequently introduced into the two-fluid nozzle along with the aqueous drug solution. A schematic of such operation is also shown in Figure 3 (14,15). The SEDS process has been widely used for a large number of applications in the pharmaceutical industry compared to the other SCF processes, and some details of results are presented in this chapter.

5. Precipitation with Compressed Antisolvent (PCA) Process

While the original work (16) reported used a single-capillary nozzle to disperse the solution, the technique was modified by introduction of both the SCF carbon dioxide and the solution through a coaxial nozzle. In principle, the modified process is similar to the SEDS process described above except for the nozzle design. While the SEDS nozzle has the provision for a mixing chamber, the nozzle design reported in the PCA process (17) is a simple arrangement of concentric tubes as shown in Figure 3. Although the process has been demonstrated for processing small polymer particles, it has a relatively low level of control in achieving the high degree of mixing often necessary for efficient and rapid mass transfer of solvent into the supercritical medium.

6. Aerosol Solvent Extraction System

With the principle of this process being antisolvent addition, the process does not differ significantly from the SAS technique. As in SAS process, the drug or polymer solution is sprayed into a bulk of SCF, typically carbon dioxide, for a fixed period of time. This step is then followed by passing supercritical carbon dioxide to extract and remove the solvent and dry the precipitated product (18). Reports on the application of this technique focus on particle size reduction (18–22).

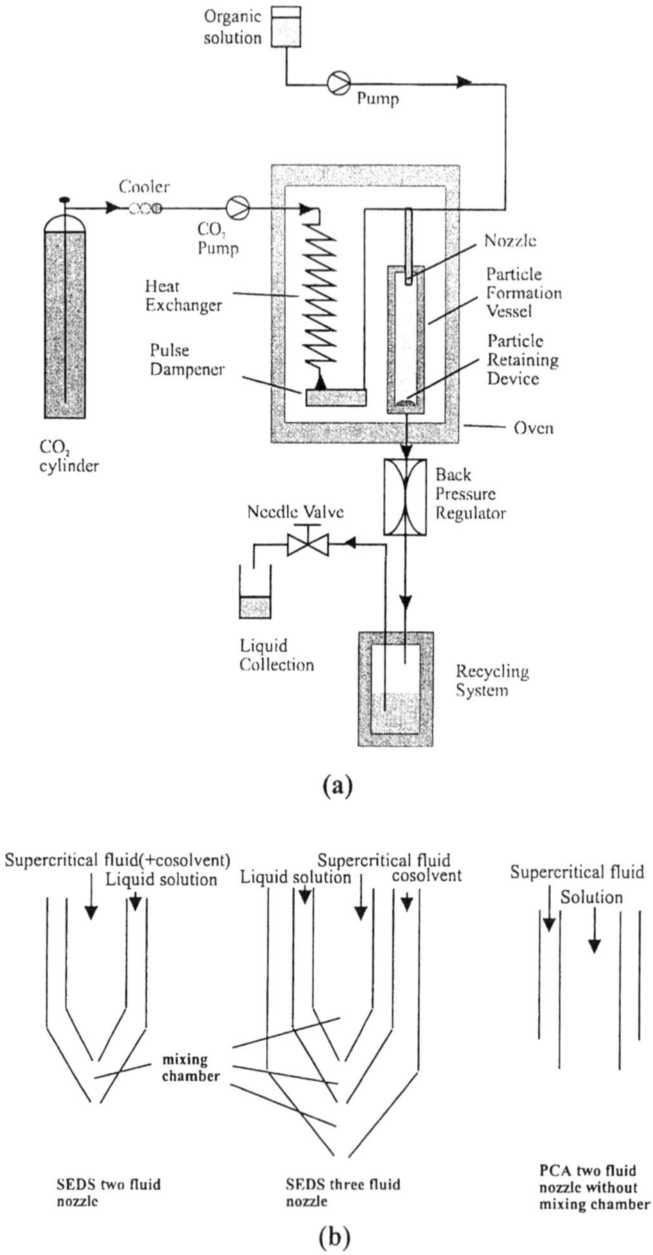

Figure 3 (a) Schematic flow diagram of the solution-enhanced dispersion by supercritical fluids (SEDS) process. (b) Nozzle configurations of the SEDS and the precipitation by compressed antisolvent (PCA) processes.

7. Precipitation from Gas-Saturated Solutions (PGSS)

This technique involves the concept of melting the material to be processed, which then dissolves an SCF under pressure (6). The saturated solution is then expanded across a nozzle where the SCF, which is more volatile, escapes, leaving dry fine particles. The process has been demonstrated to coprecipitate a calcium antagonist drug, nifedipine, along with polyethylene glycol (PEG) 4000 (23) to enhance dissolution rates in water. However, the process has limited applications in the pharmaceutical industry due to the fundamental issue of avoiding high-temperature processing necessary to melt the material.

The droplet size generated in all of the spray processes described above is dependent on the nozzle diameter and the relative flow rates of solution to antisolvent. High flows of solution also determine the pressure drop in the nozzle. The droplet generation is often realised by the Weber number, N_{we}, which is defined as (24):

$$N_{\mathrm{we}} = \rho_{\mathrm{A}} \nu^2 D / \sigma \tag{1}$$

where ρ_{A} is the antisolvent density, ν is the relative velocity, D is the jet diameter, and σ is the interfacial tension. The higher the Weber number, the smaller the droplet size and vice versa. Under supercritical conditions where the solvent and the SCF are completely miscible, the interfacial tension is zero. The Weber number therefore becomes infinite. Nevertheless, the actual increase in the numerical values when changing the system from partially to completely miscible conditions, such as increasing the pressure from below to above binary mixture critical pressure, indicates that the droplet size and hence the particle size of the product decreases.

Although the SCF and solvent are completely miscible in all proportions in the supercritical region, the rate of transfer of solvent into the CO_2 medium determines the physical characteristics of the resulting powder. This step dictates not only the particle size but also the degree of interparticle aggregation and the residual solvent levels. This step is not a limiting factor when working in the supercritical region but becomes detrimental when operating at conditions below the binary mixture critical point. The rate of mutual transport of SCF into the solvent-rich phase and vice versa is a function of the mass transfer coefficient and is given by the relation (25):

$$N_i = k_{\mathrm{L,SCF}} a (C_{\mathrm{i,e}} - C) \tag{2}$$

where N is the number of moles of component i, a is the mass transfer area, $C_{\mathrm{i,e}}$ is the equilibrium solubility in mole fraction units of the component i in the respective SCF or liquid (L) phase, C is the concentration in mole fraction units at a particular instant of time. $k_{\mathrm{L,SCF}}$ (mol cm^{-2} min^{-1} mol fraction^{-1}) is the overall mass transfer coefficient either in the liquid or SCF phase, which is a function of the system hydrodynamics and derived from the following correlation for flows in circular pipes (25) between the Sherwood number (Sh), Reynolds number (Re), and Schmidt number (Sc):

$$Sh_{\text{L,SCF}} = c\ [Re^{d}_{\text{L,SCF}} Sc^{e}_{\text{L,SCF}}] \tag{3}$$

where c, d, and e are system-dependent constants. Two ways of improving the mass transfer rates under such circumstances are (a) maintaining a significantly high ratio of the relative velocities of SCF to the solvent and (b) increasing the SF to solvent ratio which increases the concentration gradient in Eq. (2). Such high ratios when coupled with high velocities of SCF would also generate smaller droplet sizes, giving an added advantage. This concept was developed by Hanna and York (12) who incorporated a coaxial nozzle design with a mixing chamber to increase the relative velocity of the dispersing SCF.

II. APPLICATIONS

A. Polymorphic Purity

Polymorphism is the ability of a compound to exist in different crystalline forms with different packing of the molecules in the crystal lattice. This phenomenon is important in the pharmaceutical industry due to differences in properties of the polymorphic forms such as chemical and physical stability, dissolution rate, and bioavailability (3).

A single compound system exhibiting two solid states is univariant according to the Gibbs phase rule. Thus, at constant pressure, temperature is sufficient to define the solid state. The temperature at which both solid states (polymorphs) exist in equilibrium is called the transition temperature, T_t. Essentially, there are two types of polymorphic transformations that can occur: (a) reversible enantiotropic and (b) irreversible monotropic. The thermodynamic state of polymorphic transitions can be explained in terms of Gibbs free energy, G, which is given by the relation:

$$G = H - TS = U + PV - TS \tag{4}$$

where, U is the internal energy, S is the entropy, and V is the volume at pressure P and absolute temperature T. In differential form, Eq. (4) can be written as

$$dG = dU - TdS - SdT + PdV + VdP \tag{5}$$

Since

$$dU = TdS - PdV \tag{6}$$

Equation (5) becomes

$$dG = VdP - SdT \tag{7}$$

Hence, at constant temperature

$$\left(\frac{\partial G}{\partial P}\right)_T = V \tag{8}$$

The volume change between two solid states is always positive with pressure. Hence, it follows from Eq. (8) that the curves shown in Figure 4a for an enantiotropic system and in Figure 4b for a monotropic system must have a positive slope. As can be seen, the transition pressure for the enantiotropic system is above the melting pressures while that for a monotropic system is below. Under isobaric conditions:

$$\left(\frac{\partial G}{\partial T}\right)_P = -S \tag{9}$$

Entropy changes are always positive; hence, the plots of isobars exhibit a negative slope with increasing temperature as shown in Figure 5a and b for enantiotropic and monotropic systems, respectively. As seen, the transition temperature for the enantiotropic system is below the melting temperature of the individual forms whereas that for monotropic forms is above. Thus, pressure and temperature are the main variables for polymorphic transformations. Since these two variables are the process parameters for an SCF process like the SEDS, such technique facilitates polymorphic screening.

The most stable polymorphic form is the one with the lowest Gibbs energy. At the transition point where the two polymorphic forms are in equilibrium, the Gibbs energies of the individual forms are identical. Thus,

$$\Delta G = 0 = \Delta(U - TS) + P\Delta V \tag{10}$$

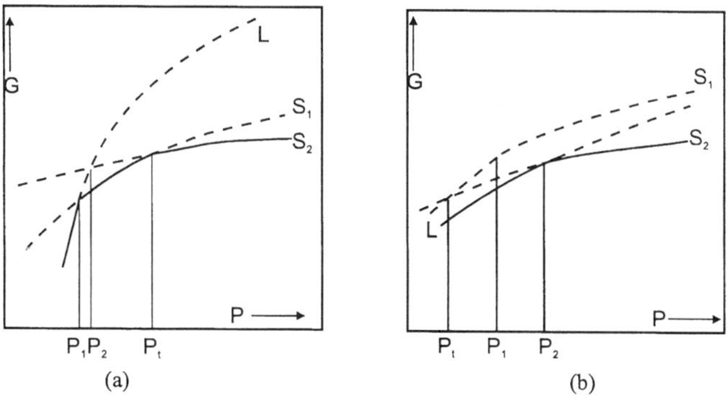

Figure 4 Free-energy isotherms of the liquid and two polymorphic forms S_1 and S_2 exhibiting (a) enantiotropic and (b) monotropic transformation.

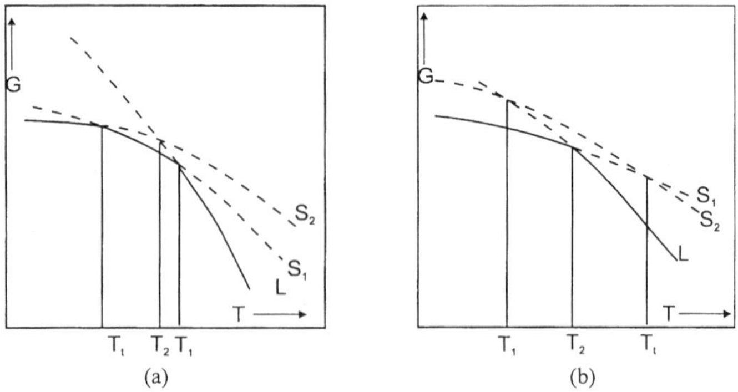

Figure 5 Free-energy isobars of the liquid and two polymorphic forms S_1 and S_2 exhibiting (a) enantiotropic and (b) monotropic transformation.

where Δ represents the difference between the two polymorphic forms. At low pressures, the volume change is negligible for transformations taking place at constant temperature, and the above equation at the transition temperature reduces to

$$\Delta U - T\Delta S = \Delta F = 0 \tag{11}$$

where F is the Helmoltz free energy. At absolute zero temperature, the entropy term vanishes and the Helmoltz energy is equal to the internal energy. The internal energy, which is a function of the lattice energy, determines the stability of the polymorphic form. The lattice structure with the higher lattice energy has the lowest Gibbs energy and hence is the most stable form. On the other hand, at higher temperatures, the stability between the two polymorphic forms depends on the entropy (S), which is a function of the number of ways (W) in which the molecules can be packed in the crystal lattice and is given by the relation

$$S = R \ln W \tag{12}$$

where R is the universal gas constant. Hence, the structural form, which has lower internal energy and higher entropy, is more stable. From Eqs. (4) and (10), at the transition temperature, the entropy term $T_t\ \Delta S$ is equal to the enthalpy difference, which is the enthalpy of transition between the two polymorphic forms. As the melting temperatures of the different polymorphic forms are significantly different, a differential scanning calorimetry (DSC) tracing that reflects enthalpy changes with temperature, shown in Figure 6, can be used as an identification tool. When the transition temperature T_t is below the melting temperatures of the two polymorphs, as in Figure 6a, a crystalline solid with polymorph I when heated slowly transforms into polymorph II at T_t and the second form melts at T_{II}^{M}. Alternatively, when the same sample is heated at a faster

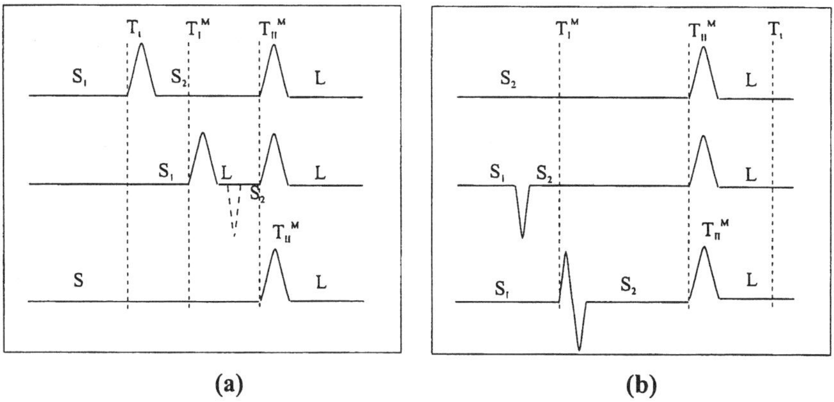

Figure 6 Schematic representation of DSC traces of (a) enantiotropic and (b) monotropic transformation.

rate, the transition from form I to form II may not occur due to insufficient time for equilibrium. But the melt of form I may recrystallize into form II after $T_{\mathrm{I}}^{\mathrm{M}}$ which then is followed by its melt at $T_{\mathrm{II}}^{\mathrm{M}}$ on further heating. Thus, both forms are thermodynamically stable at temperatures below their melting points, and the transformation is reversible and enantiotropic. In contrast, when the transition temperature is above the melting temperatures of the two individual forms, form I always exists as a metastable form relative to form II. Thus, polymorph I transforms irreversibly into form II at temperatures below the melting point of form I, as shown in Figure 6b. If the rate of transformation is slow, the change takes place at the melting point followed by recrystallization and the final melt. This behavior is a monotropic transition.

Salmeterol xinafoate, an antiasthmatic drug delivered via the respiratory route, exhibits enantiotropic polymorphism, as shown in Figure 7. It has been crystallized from an organic solution in pure polymorphic form I and II (26) using the SEDS process in a respirable particle size of less than 5 μm. When prepared by conventional organic solvent crystallization, salmeterol xinafoate is primarily polymorph I (shown in Figure 7a), which melts and recrystallizes into form II prior to final melt. SEDS-prepared material at temperatures below 353 K is pure form I, as can be seen in Figure 7b. Similarly, this process was successful in preparing a stable form II at temperatures above 353 K, as shown in Figure 7c. The absence of the form II peak in Figure 7b could be due to the relatively high rate of heating to which the sample is subjected under DSC. It is therefore presumed that the transformation of form I to form II is quite slow. However, when the conventionally crystallized and SEDS-processed form I of salmeterol xinafoate is subjected to the same rate of heating by DSC, the

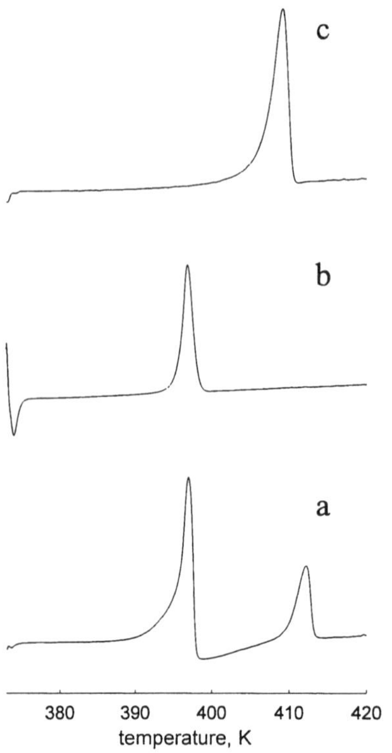

Figure 7 DSC profiles of (a) conventionally crystallized and (b) form I and (c) form II of SEDS processed salmeterol xinafoate. (From Ref. 26.)

transformation of form I to form II in the conventionally crystallized sample is due to the presence of trace amounts of form II, which triggers the polymorph change. The transition pressure is believed to be far below the range of pressures used in the current study and hence pressure is unlikely to have any significant effect on the transformation.

Another example in the literature on the preparation of pure polymorphic forms of drug compounds by SCF processing is that of fluticasone propionate, which is also a drug delivered by the respiratory route (27). As the compound decomposes before melting, DSC traces do not give definitive temperature profiles for the transformation between polymorphs I and II. X-ray diffraction patterns, however, have been used to confirm the presence of different polymorphic forms. The drug recrystallizes as form II from solution by the SEDS process in the presence of supercritical carbon dioxide at all temperatures, while form I is obtained by conventional crystallization methods. The rapid nucleation by SCF antisolvent aids in the formation of the metastable form II. A further advantage of SCF processing is the direct preparation of product with a particle size less than 5 μm for respiratory drug delivery, thereby avoiding the high energy and damaging micronization process required for conventionally crystallized material.

B. Processing Excipients from Aqueous Media

Several compounds are readily soluble in water and virtually insoluble in organic solvents. Processing such materials for pharmaceutical applications, such as size reduction and desired morpohological forms, is extremely difficult with supercritical carbon dioxide. This is due to very low solubility of CO_2 in water, typically less than 3 mol % between 313 K–373 K and pressures between 50–700 bar (28). Examples for such classes of compounds include pharmaceutical excipients, especially carbohydrates and sugars, fillers, and biologicals. It is therefore necessary to use an organic solvent, which performs as an antisolvent in addition to CO_2. In early studies, Tom and coworkers (11) used a mixed ethanol/water (90:10) solvent to precipitate insulin and catalase using supercritical carbon dioxide. The authors reported difficulties in product collection due to the presence of an aqueous-rich phase in the vessel because of the distribution of ethanol into the CO_2 phase. Palakodaty and coworkers (14) have shown that lactose can be rapidly recrystallized in a long, thin band morphology from a 95:5 methanol/water solution with supercritical carbon dioxide using the SEDS process.

While a mixed-solvent approach shows promise for processing water-soluble compounds, not all compounds are sufficiently soluble in organic solvents. Furthermore, biologicals denature in the presence of organic solvents; hence, it would be particularly beneficial to process such materials from an aqueous medium. When selecting operating conditions for such a process, knowledge of the phase behavior of the ternary system organic solvent/water/carbon dioxide is essential for preparing a particulate product with desired properties. Figure 8 illustrates the binary phase behavior of methanol/water/carbon dioxide system at 323 K and 363 K. At a pressure of 150 bar, the vapor and liquid phases are enveloped by a binodal curve at 323 K, whereas at a higher temperature of 363 K, the saturated vapor and liquid curves do not converge until the pressure is raised above the critical pressure of the binary methanol/carbon dioxide system. Nevertheless, due to limited solubility of methanol in CO_2 phase, the ratio of methanol to water in the liquid phase increases with decreasing concentration of carbon dioxide in the system. The behavior with ethanol does not significantly vary from that with methanol but is quite different with higher alcohols such as isopropanol (29) or acetone (30).

Palakodaty and coworkers (15) used such cosolvent distribution in defining the process variables for crystallizing lactose monohydrate from aqueous media in a quasi-continuous operation and reported the effect of methanol and ethanol as cosolvents on the crystallized product. The flow conditions of different

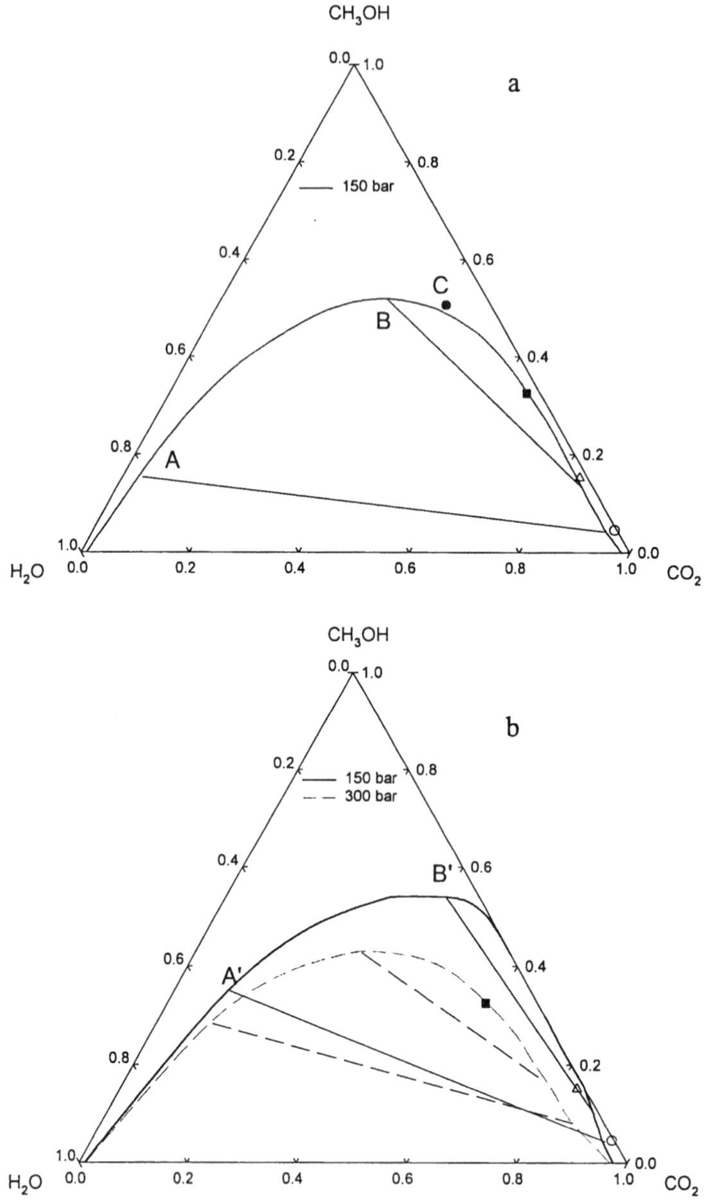

Figure 8 Estimated phase behavior of H_2O-CH_3OH-CO_2 system at (a) 323 K and (b) 363 K. ■, Critical point. Process flow conditions: Aqueous lactose solution = 0.035 ml/min; methanol solution = 0.665 ml/min; liquid carbon dioxide: △ = 5.0 ml/min; ○ = 19.0 ml/min. Condition C (●): aqueous solution flow = 0.32 ml/min, flow rate of ethanol = 6.4 ml/min, flow rate of liquid CO_2 = 4.0 ml/min. (From Ref. 15.)

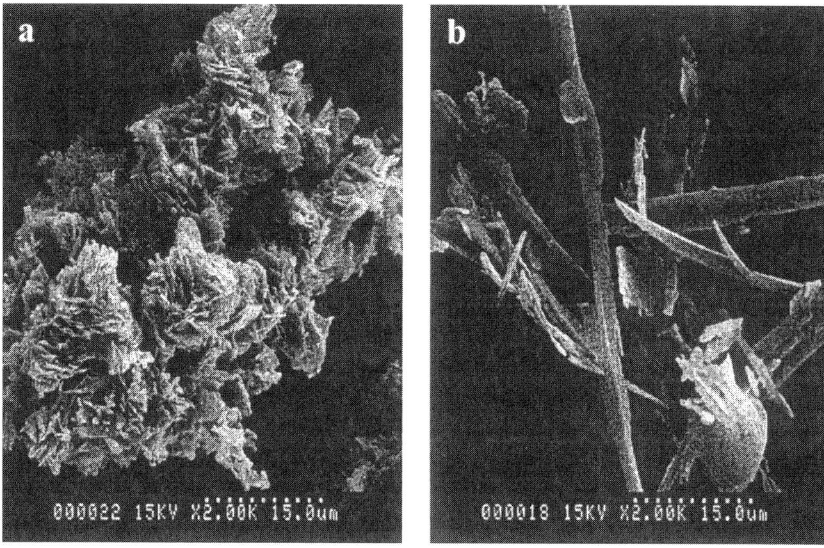

Figure 9 Scanning electron micrographs of the SEDS-processed lactose samples. Process conditions: temperature = 323 K, pressure = 150 bar; lactose concentration = 10% w/v in water; flow rate of solution = 0.035 ml/min; flow rate of ethanol = 0.665 ml/min; flow rate of liquid CO_2 (a) 19.0 ml/min; (b) 5.0 ml/min. (From Ref. 15.)

streams are shown in Figure 8. The process path for a single homogeneous aqueous solution to go into a two-phase solid-SCF phase must follow the closest tie line. Hence, to describe the particle formation process, the corresponding tie lines are also shown in Figure 8. Figure 9 shows the scanning electron microscope (SEM) photomicrographs of lactose samples prepared at 323 K at two different flow rates of carbon dioxide using ethanol as a cosolvent. At higher flow rates of carbon dioxide, lactose is crystallized as fused particles, compared with the more distinctive nonagglomerated particles obtained with lower flow rates. Such morphological changes are typical of those obtained when lactose is rapidly crystallized from saturated aqueous solutions (31). Owing to high affinity of ethanol for the CO_2 phase, the particles nucleate and grow in an essentially aqueous-rich phase (with composition such as those under condition A in Figure 8) prior to drying while working at a higher flow rate of supercritical CO_2. On the other hand, at a lower flow rate of supercritical carbon dioxide and with constant cosolvent and solution flow rates, the particles nucleate rapidly from an organic-rich phase, with composition such as those under condition B in Figure 8. For lactose, which is insoluble in ethanol, there is little tendency for any agglomeration. As a result of these differences in the rate of particle formation, the level of water content in products is as expected as listed in Table 1. The significantly lower amount of water in lactose as compared with a stoichiometric monohydrate form (i.e., 5% w/w) emphasizes the fact that the rate of particle formation in SCF media is relatively fast, though working in different regions of phase behavior. Dry methanol is known to dehydrate lactose, and a stable α-anhydrous form of lactose can be prepared by methanol reflux over a bed of hydrous lactose (32). Thus, by choosing an alternative cosolvent (methanol), it is possible to further reduce the water content, as shown in Table 1.

Table 1 Experimental Conditions and Particle Analysis Results of Lactose Prepared Using the SEDS Process

Concentration of α-lactose monohydrate in H_2O = 10% w/v
Flow rate of aqueous lactose solution = 0.035 ml/min
Flow rate of alcohol = 0.665 ml/min

	Lactose[c]		Liquid carbon dioxide flow rate, ml/min	
	α[d]	β[e]	5.0	19.0
Size[a]	31.17	16.99		
Water[b]	5.27	0.31		
Cosolvent—Methanol				
Set 1: Pressure = 150 bar; temperature = 323 K				
Size[a]			4.20	7.16
Water[b]			1.59	1.47
Set 2: Pressure = 150 bar; temperature = 363 K				
Size[a]			10.68	5.78
Water[b]			1.23	1.16
Set 3: Pressure = 300 bar; temperature = 363 K				
Size[a]			5.89	9.13
Water[b]			1.03	2.59
Co-solvent—Ethanol				
Set 4: Pressure = 150 bar; temperature = 323 K				
Size[a]			12.69	11.12
Water[b]			2.53	3.06

[a] Average value of the geometric diameter measured from volume distribution (micrometers).
[b] Karl Fisher analysis (% w/w).
[c] Unprocessed lactose (Sheffield products).
[d] α-Lactose monohydrate.
[e] Anhydrous β-lactose.
From Ref. 15

The ability to define the process parameters for achieving a desired product with defined morphology and water content is further exemplified by the results obtained at the higher temperature. By working again with higher flow of carbon dioxide, the composition of the crystallizing medium (of condition A′ at 363 K) is different from the composition at condition A at 323 K with a higher ratio of methanol to water. Although operating with the same flow rates of carbon dioxide, cosolvent, and solution, the particle morphology is similar to that obtained by rapid crystallization both at 150 bar and 300 bar as shown in Figure 10. Nevertheless, aggregation is observed at 150 bar, due to fact that the conditions lay in the phase split region with respect to cosolvent/water/CO_2 at 150 bar as opposed to those at 300 bar. Another feature of the SEDS process is the ability to regain the crystalline structure of lactose even when processing at high temperatures, such as 363 K. This is evidenced by the powder x-ray diffractogram in Figure 11. At similar temperatures a conventional particle formation process, such as spray drying, usually gives amorphous product. As can be seen from Figure 11, the SEDS-prepared material is a mixture of both the hydrous and β-anhydrous forms of lactose.

Similar changes in crystal habits of other excipients, such as glycine and D-mannitol, have also been observed when processed by SEDS. Under condi-

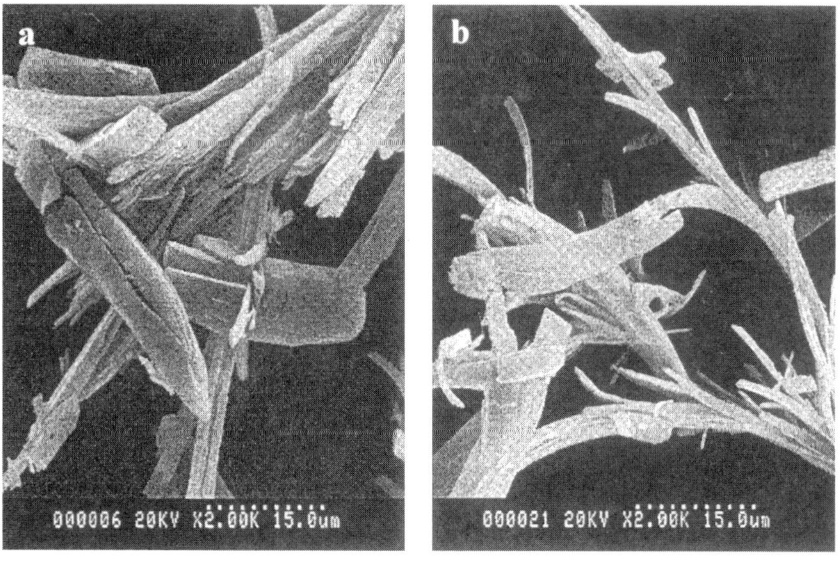

Figure 10 Scanning electron micrographs of the SEDS-processed lactose samples. Process conditions: temperature = 363 K, lactose concentration = 10% w/v in water; flow rate of solution = 0.035 ml/min; flow rate of methanol = 0.665 ml/min; flow rate of liquid CO_2 = 5.0 ml/min. pressure (a) 150 bar (b) 300 bar. (From Ref. 15.)

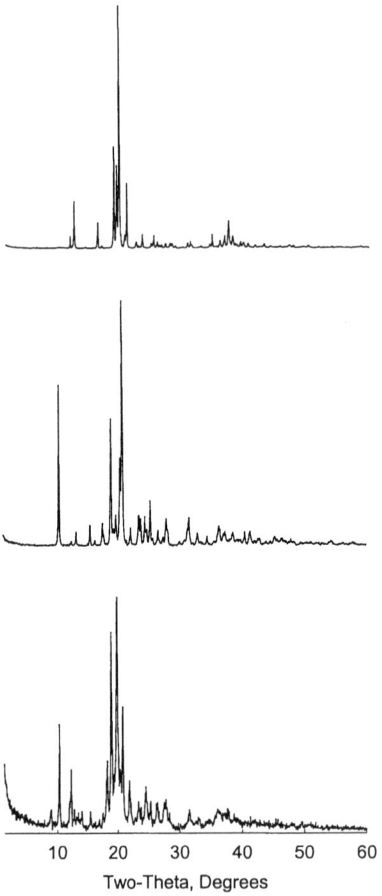

Figure 11 XRPD patterns of unprocessed and SEDS-processed lactose at 363 K and 300 bar pressure. (a) α-Lactose monohydrate. (b) Anhydrous β-lactose. (c) SEDS-processed lactose with methanol as a cosolvent. Experimental conditions: lactose concentration in water = 10% w/v; flow rate of lactose solution = 0.035 ml/min; flow rate of cosolvent = 0.665 ml/min; flow rate of liquid CO_2 = 19.0 ml/min. (From Ref. 15.)

tions far beyond the critical point in the liquid-phase region, such as condition C in Figure 8, glycine is rapidly crystallized from aqueous solution. This is due to the high ratio of antisolvent mixture of ethanol + CO_2 to water. The morphology changes from well-faceted crystals, typical of slow crystallization at high CO_2 ratio, to thin platelets with a substantial decrease in size (Figure 12). Another procedure for changing particle morphology is by processing with an alternate cosolvent with different transport properties. Though the organic cosolvent, water, and CO_2 are completely miscible under conditions beyond the

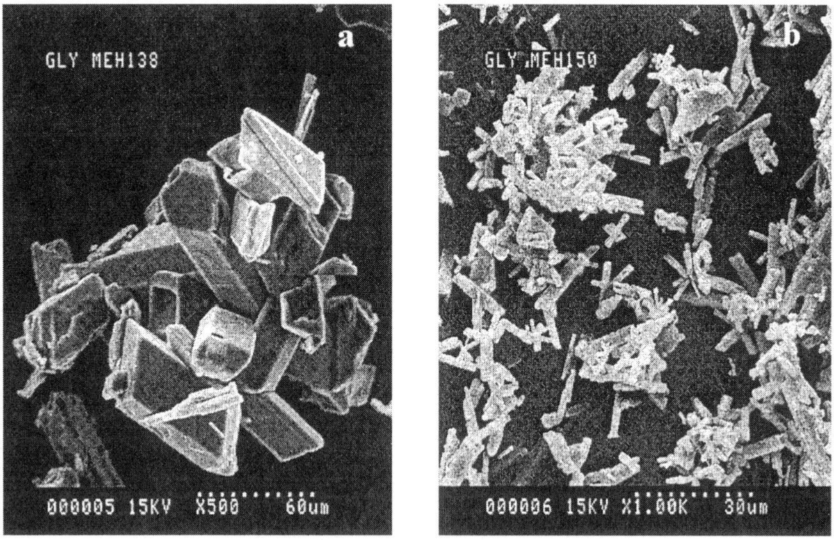

Figure 12 Changes in morphology of glycine achieved by working in different regions of phase diagram. Experimental conditions: glycine concentration in water = 10% w/v; temperature = 313 K; pressure = 100 bar; flow rate of aqueous glycine solution (a) 0.05 ml/min; (b) 0.32 ml/min; flow rate of ethanol (a) 0.5 ml/min; (b) 6.4 ml/min; and flow rate of liquid CO_2 (a) 18.0 ml/min; (b) 4.0 ml/min.

critical point, the mass transfer of CO_2 in the organic solvent and the subsequent transfer of the organic/water solution into the CO_2 phase is strongly dependent on the density and viscosity of the cosolvent. D-Mannitol crystallizes as long needles under rapid conditions, as illustrated in Figure 13, with such products obtained using methanol as cosolvent even while working with high CO_2 flows. The habit changes with increased density and viscosity of the cosolvent, such as with isopropanol, to flat platelets due to delay in cosolvent removal.

These remarkable features of SCF processing is important because the process variables can be defined from knowledge of phase behavior rather than in terms of absolute pressure, temperature, or flow rates of the individual streams. Such flexibility in the choice of process conditions makes the technology particularly suitable for processing labile biologicals, as discussed in the following paragraphs.

C. Morphological Changes

The scientific background and technological strategies of obtaining different morphological forms of water-soluble compounds by SCF processing has been explained. Similar change can also be achieved when materials are recrystallized

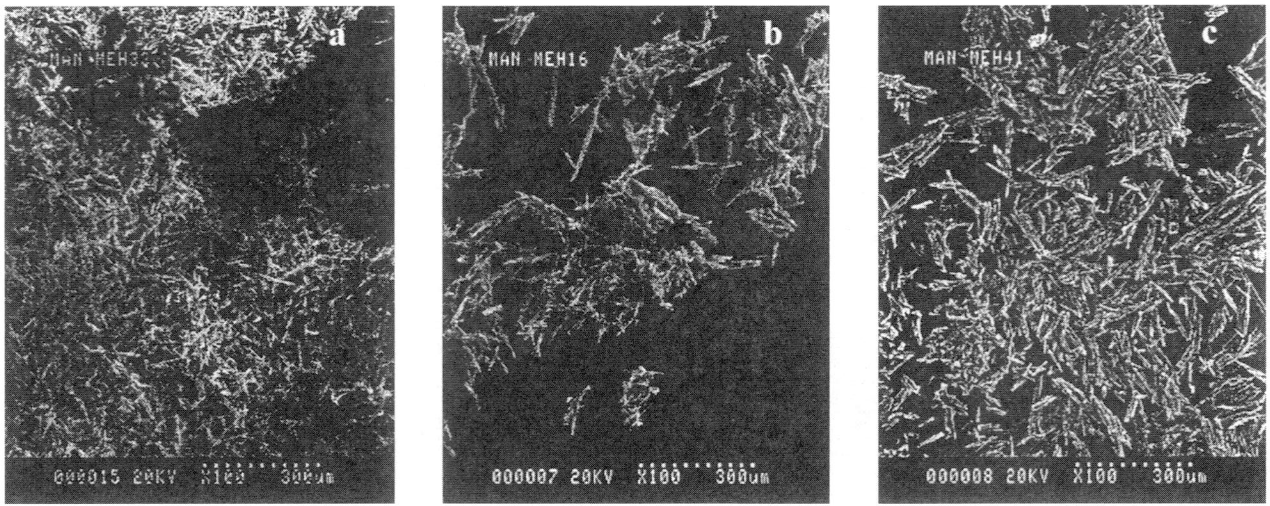

Figure 13 Changes in morphology of D-mannitol achieved by working with different solvents. Experimental conditions: mannitol concentration in water = 10% w/v; temperature = 313 K; pressure = 200 bar; flow rate of aqueous mannitol solution = 0.03 ml/min; flow rate of organic solvent = 0.6 ml/min; and flow rate of liquid CO_2 = 18.0 ml/min; (a) methanol (b) ethanol, and (c) *i*-propanol.

from organic solvents using SCF antisolvent techniques. While materials processing is often carried out in the miscible region of a binary organic solvent–SCF phase diagram, the process can also be run at pressures below the binary mixture critical point. This is important when probing various process conditions for preparing different polymorphic forms, as discussed earlier. However, a major limitation of the process operating in such partially miscible conditions is that of solvent extraction into the SCF medium. The mass transfer is mainly dependent on the mixing and hydrodynamics of the individual streams, as given by Eq. (3). Hence, the mass ratio of solvent to CO_2 is usually kept very low to allow sufficient concentration gradient for the solubility of the solvent into CO_2. Such a high ratio of CO_2 to solvent when passed through a coaxial nozzle into a particle formation vessel would also assist in dispersing the liquid medium and forming smaller particles. Figure 14 shows SEM photomicrographs of a model compound, acetaminophen (paracetamol), processed by the SEDS technique from ethanol solution. As can be seen from Figure 14a, the particles fuse to form agglomerates of smaller spherical particles. The high level of residual solvent in the material is also indicative of a slower rate of mass transfer into the CO_2 medium. By raising the pressure to 200 bar, where ethanol and CO_2 are

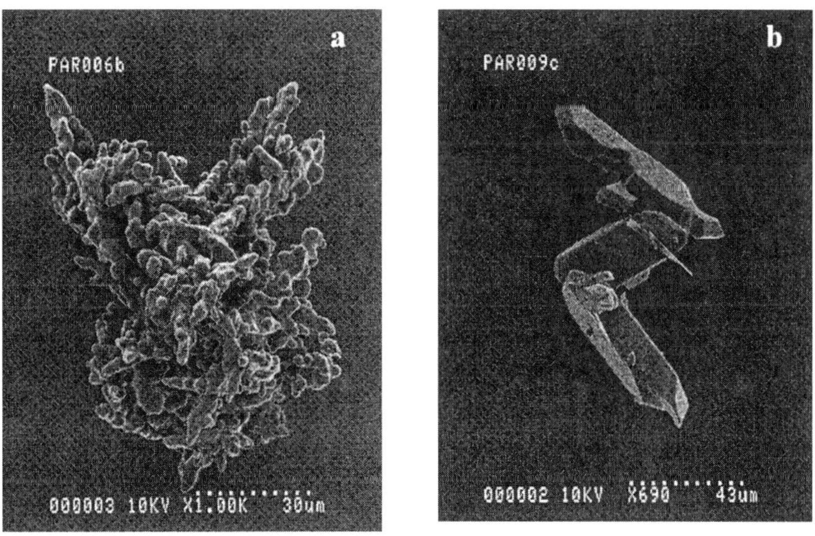

Figure 14 Morphological changes of acetaminophen when processed at different pressures. Experimental conditions: Temperature = 358 K; concentration of acetaminophen in ethanol = 1% w/v; flow rate of solution = 2.2 ml/min; flow rate of carbon dioxide = 200 ml/min; pressure (a) = 80 bar, residual solvent = 5885 ppm and (b) 200 bar, residual solvent = 150 ppm.

completely miscible, the particles are well faceted and agglomeration is absent, as shown in Figure 14b. In parallel, the residual solvent levels are reduced by an order of magnitude.

It should be emphasized that the solubility of carbon dioxide in the solvent decreases with increasing temperature thus reducing the degree of supersaturation for solute separation compared to that of processing at lower temperatures. Figure 15 shows the binary phase behavior of ethanol-CO_2 system at 308 K and 333 K. While data are not available at 358 K, those at 333 K have been used to qualitatively explain the results obtained from SEDS experiments. The composition of the liquid phase at condition A at 308 K is 90 mol % carbon dioxide and at condition B at 333 K is 65 mol % carbon dioxide. It is expected to be even lower at 358 K. Hence, when a particle formation process is operated in the partially miscible regions as mentioned above, the degree of supersaturation is lower than that achievable in the miscible region. Since Weber numbers are large due to high relative velocities, the particle size is expected to be smaller. However, since the solubility of the solvent is much lower in the carbon dioxide phase, solvent removal becomes an issue. As a result, agglomeration would be expected.

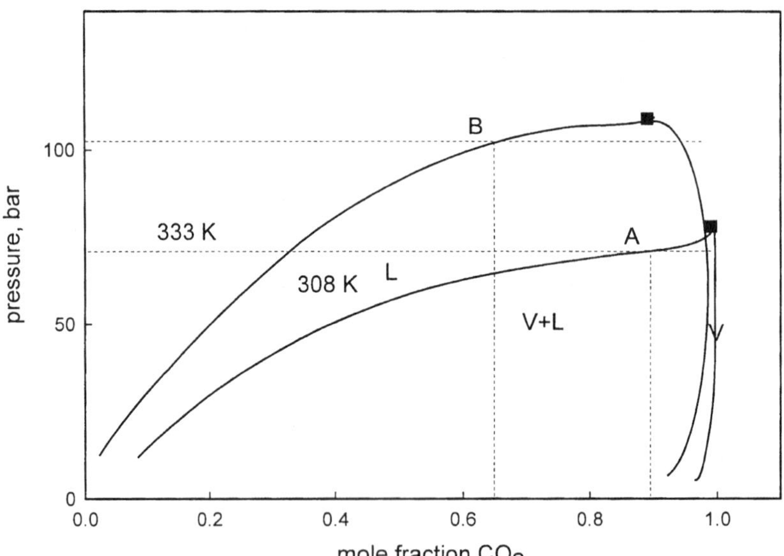

Figure 15 Binary phase diagram of ethanol–carbon dioxide system at 308 K and 333 K. (Data from Refs. 33 and 34.)

D. Crystallinity

Crystallinity is a very important issue in the pharmaceutical industry not only in determining the drug stability but also in affecting its dissolution rate. An amorphous form shows faster dissolution than the crystalline form of a given material (3). It is therefore desirable to control crystallinity during material processing. Recrystallization by SCF technique, either by RESS or by any of the GAS methods, is rapid due to the high supersaturation levels that are developed in a very short period of time compared with the conventional methods of crystallization. It is therefore not surprising that on occasions amorphous forms are obtained after SCF processing. Nevertheless, the propensity for formation of amorphous or partially amorphous products is likely to be dependent on the crystal lattice energies of the materials as well as molecular mobility and packing behavior (3). Jaarmo and coworkers (35) were able to control the crystallinity of an antiallergic drug, sodium cromoglycate, by subjecting it to SEDS processing. They reported an amorphous form when processed from methanol solution. Small amounts of water added to the methanol solution prior to SEDS processing reduced the rate of crystallization, thereby leading to a more crystalline form. Though examples in the literature are scarce, it is evident that the rate of crystallization in relative terms can be modified thereby facilitating control of the crystallinity of the separated phase.

E. Chiral and Isomeric Separation

Impurity removal from a product is an important issue in chemical and pharmaceutical production. Impurities range from unreacted starting material to stereo- and regioisomers to chiral enantiomers (36–39). The final product is usually recrystallized several times in an appropriate solvent to remove any of these possible impurities and unwanted byproducts. Although recrystallization is an everyday tool for chemists, the process can be tedious (e.g., choice of solvent, concentration, temperature gradient) and costly on industrial scale, and frequently the product after recrystallization is not significantly purer than before. To overcome these problems, impurities can also be removed from drug substances using SCFs during particle formation. For studies to date, only carbon dioxide has been used because of its favorable solvation properties for organic molecules and its critical parameters (5). Recently, supercritical carbon dioxide has been successfully used for separating chemical entities from different matrices as diverse as monomers from polymers (40), metal ions from soil (41), isomers of aromatic hydrocarbons (42–47), and enantiomeric (39,48–50) separation of chiral molecules. The following discussion will concentrate on isomeric and chiral separation.

1. Solubility

Any impurity removal or isomeric separation will be successful only if the solubility of the impurity in the solvent or fluid is significantly higher than that in the parent molecule. As pure solvents might not exhibit a sufficiently large solubility difference, mixed solvents or, in the case of SCFs, modifiers can be used (51) that will alter the chemical properties of the solvent. As a result this can enhance the solubility of the unwanted product and therefore facilitate its removal. Solubility in liquid solvents can be expressed as (52):

$$\ln x_1 f_1 = \frac{\Delta H^{\text{fus}}}{R}\left(\frac{1}{T_{\text{m}}} - \frac{1}{T}\right) \tag{13}$$

From Eq. (13), solubility expressed in mole fraction of the solid x_1, is dependent on the heat of fusion ΔH^{fus}, which can be related to the sublimation pressure P^{sub} of the solid, and the melting temperature T_{m} (exactly: triple-point temperature T_{Tr}) of the solid. Assuming an ideal solution, with an activity coefficient f_1 of unity, the solubility of a solid in a liquid can be calculated. In the present case, the magnitude of separation of two species will depend principally on the difference in their melting temperatures ($\Delta T_{\text{m}} = T_{\text{m1}} - T_{\text{m2}}$). Modifying the solvent will produce a nonideal solution with activity coefficients different from unity. In such a case, separation is also dependent on the difference in activity coefficient of both species ($\Delta f = f_1 - f_2$).

By contrast, solubility of solids in SCFs is described as (5):

$$x_i \phi_i P = P_i^{\text{sub}} \exp\left(\frac{(P - P_i^{\text{sub}})V^{\text{s}}}{RT}\right) \tag{14}$$

As can be seen from Eq. (14), the solubility of a solid in an SCF depends not only on solid-state parameters, such as sublimation pressure P^{sub} and molar volume V^{S}, but additionally on the fugacity coefficient ϕ_i. The fugacity coefficient is the "supercritical analogue" to the activity coefficient (5). The fugacity coefficient varies not only with the type of fluid but with temperature and pressure (53). Therefore, solubility of solids can be significantly influenced by changing the density of SCFs on alteration of temperature and/or pressure. The fugacity coefficient is the key variable that explains the different solubility of solids in SCFs compared with ordinary liquids.

2. Separation of Isomers

Although almost any isomeric mixture could be used as a model for investigation, isomers of derivatized aromatic hydrocarbons are frequently used as test substances (42–47). As an example, hydroxybenzoic acid (HBA) demonstrates the different solubility of the ortho and para isomer in supercritical CO_2. From

Figure 16, the solubility of the ortho isomer is seen to be in the region of two orders of magnitude higher than that of the para isomer. Even at 373 K, the para isomer has a lower solubility than the ortho isomer at 308 K. In general, the differing solubility of isomers is explained by the different chemical substitution pattern and the ability to form hydrogen bonds for the ortho isomer (54). In contrast, the behavior for HBA isomers can also be explained by comparing their melting points. The melting point of the ortho isomer is more than 50 K lower than the para isomer. A lower melting point implies a higher sublimation pressure at any given temperature. From Eq. (14), and assuming that all other solid-state parameters are identical, a higher solubility for the ortho isomer is predicted.

Solubility also increases with increasing fluid density and temperature. The increase in solubility with rising temperature can thus be explained using Eq. (14). Higher temperatures increase the sublimation pressure of the solid, resulting in an enhanced solid solubility. The increase with fluid density, ρ, stems from changes of the fugacity coefficient with pressure (5,53). As can be seen from Figure 16, a density change of one order of magnitude at constant temperature enhances the solubility by almost two orders, confirming that ϕ is the most influential parameter in the solubility of solids in SCFs.

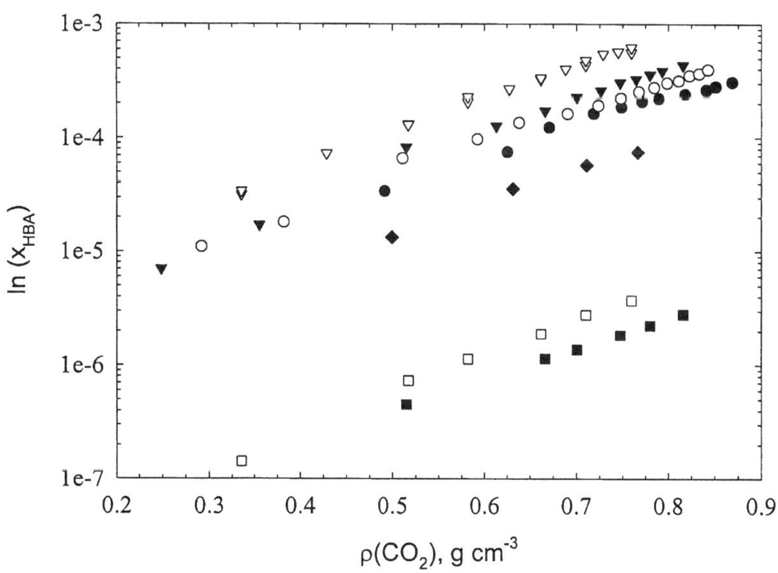

Figure 16 Solubility increase with density of *ortho*-hydroxybenzoic acid at 308 K (●), 313 K (○), 318 K (▼), and 328 K (▽), and *para*-hydroxybenzoic acid at 318 K (■), 328 K (□), and 373 K (◆). (Data from Refs. 36, 42, and 47.)

3. Chiral Separation

As discussed in the previous section, isomers or impurities can be successfully removed from a host material, if solubilities are sufficiently different. Therefore, a worst-case scenario for any impurity removal or isomeric separation is the resolution of a racemate of enantiomers, since enantiomers have equivalent physical and chemical properties in an achiral medium (ordinary solvent) and hence the same solubility (39,55). Furthermore, 50% of the sample can be regarded as impurity.

Many possibilities for the resolution of racemates are known and excellent reviews have been given in the literature (55,56). A particularly attractive way to resolve a racemate is by forming a diastereomeric salt with a suitable chiral agent. In contrast to enantiomers, diastereomers have different physical and chemical properties and therefore exhibit alternative solubilities (39,55).

In order to achieve resolution, both the racemate and chiral agent must be soluble in the supercritical fluid (57). As mentioned above, the diastereomers exhibit different melting points and hence different solubilities in the SCF. Resolution is therefore mainly dependent on the melting point difference, ΔT_{m}, between the diastereomeric salts. As expected the resolution of racemates can be described as an isomeric separation. By reference to standard thermodynamic expressions, resolution can be related to the Gibbs energy difference, $\Delta\Delta G$, between the diastereomers (58). It has been shown that resolution varies with increasing pressure or temperature (39,57). These changes can be explained by changes in molar volume, $\Delta\Delta V$, and entropy, $\Delta\Delta S$, between the diastereomers which can also be derived from the Gibbs energy. Furthermore, differences in isobaric expansion, $\Delta\Delta\alpha_{\mathrm{p}}$, and isothermal compressibility, $\Delta\Delta\kappa_{\mathrm{T}}$, are also accessible via more complicated differential expressions (57).

Many chiral substrates have been partially or completely resolved in common solvents. By contrast only few resolutions have been carried out in supercritical CO_2 (37,39,49). Partial resolution has been achieved for a variety of substrates, but complete resolution is still a challenge. As an example, the enantiomeric resolution of 2′2-binaphthyl-1,1′-diamine (BINAP-DA) as diastereomeric salt with camphorsulfonic acid (CSA) in supercritical CO_2 (57) is shown in Figure 17. As can be seen, resolution decreases with increasing density of the fluid phase, which implies a loss in resolution with an increase in pressure. This decrease is linked to changes in molar volume between the diastereomers. It is understandable that with increasing pressure the difference in volume between the diastereomers becomes smaller. In addition, a decrease in resolution with increasing temperature is observed (see Figure 17). Although this can be explained thermodynamically by a change in entropy difference, one can also argue that the higher temperature enhances the overall solubility, thereby inhibiting precipitation.

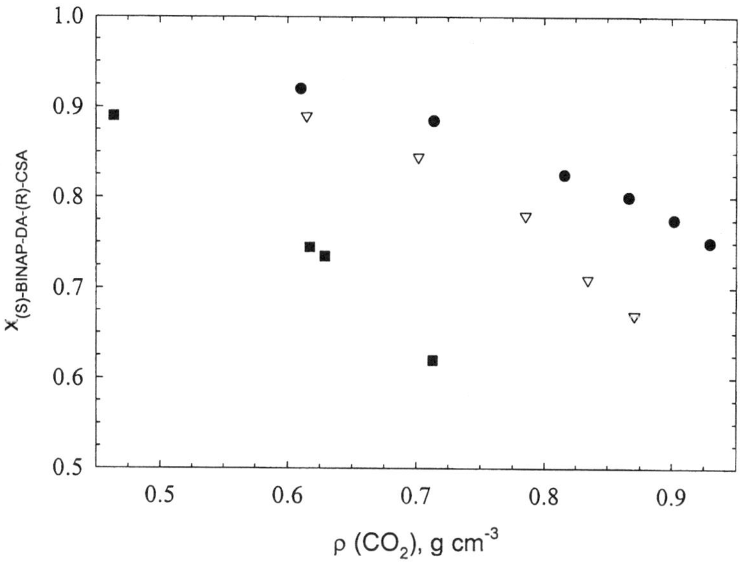

Figure 17 Enantiomeric resolution of 1,1′-binaphthyl-2,2′-diaminocamphorsulfonate with density at 318 K (●), 323 K (▽), and 348 K (■). (Data from Ref. 57.)

Thus, chiral resolution, isomeric separation, and impurity removal can all be treated in a similar manner. While the aim for each process is completely different, the same thermodynamic principles can be applied. Although operating with SCFs is sometimes thought to be completely different from routine conventional chemical manipulation, no fundamental differences are apparent. However, working with SCFs provides the opportunity to achieve goals that might be remote or even impossible with common liquid solvents, due to the unique possibility of tuning the characteristics of the working fluid.

F. Proteins and Biologicals

The largest current growth area of pharmaceuticals in research and development is biotechnology-derived human therapeutic agents, and sales of approved biotechnology products are expected to undergo rapid growth (59). Many of the biotechnology products are therapeutically important proteins and peptides, but there is considerable scientific effort in the development of gene therapy products, requiring the production and delivery of plasmid DNA. The first regulatory approval was expected in the year 2000 (60), but according to the FDA website none have been approved to date. The advances in production and availability of such therapeutic agents has focused research on noninvasive delivery methods, including pulmonary delivery using dry powder inhalers (1). This has required consideration of the physical properties of powdered bioactives, in addition to macromolecular stability during formulation and delivery.

Proteins and peptides are polymeric chains comprising a defined sequence of amino acids. Unlike small drug molecules, the folding of the peptide chain is very important in addition to the primary amino acid structure and sequence (61). The primary amino acid sequence folds into a secondary structure, which can consist of ordered structures including α helices, β sheets, and β turns. This secondary structure then folds further to form the overall shape of the protein, termed ternary structure. In some proteins, association of a number of these folded structures (or subunits) occurs, resulting in formation of a multimeric protein, which is then said to have quaternary structure. The three-dimensional conformation thus formed is required for biological activity, whether that be enzyme activity or receptor binding. As a consequence, it is important that the three-dimensional conformation be retained following any processing of the protein. Furthermore, the common protein degradation paths of aggregation, deamination, and oxidation that can lead to changes in biological activity, metabolic half-life, and immunogenicity must be minimized during the preparation and formulation of protein particles (62).

The production of microparticles of therapeutic proteins and peptides is required for the developing methods and new strategies for drug delivery, including modified-release systems and pulmonary delivery. Conventional approaches for producing particulate proteins include spray drying and lyophilization together with the associated postprocessing steps of milling and sieving. These approaches can often result in unsuitably sized particles, requiring secondary processing, or can lead to protein denaturation. Spray drying can produce particles of suitable dimensions (<5 μm diameter for respiratory delivery), but exposure to heat and air during the process can denature biological compounds (1). Lyophilization is not always readily applicable to many proteins, and the protein materials formed by the process often require downstream secondary processing, such as milling (4). While milling generates particles of suitable sizes, these highly energetic processes can result in denaturation of proteins. Furthermore, the relatively large quantities of material required and the loss of material associated with these size reduction methods can make them impractical for highly potent and expensive therapeutic proteins.

The problems associated with conventional size reduction methods, with respect to biologicals, make particle formation processes using SCF technology an attractive single-step alternative. In particular, the application of the GAS-type processes has demonstrated significant promise for the preparation of microparticulate protein material (10). The low solubility of water in supercritical CO_2 has required many workers to use nonaqueous media, such as dimethylsulfoxide (DMSO) and dimethylformamide (DMF), as solvents for protein material. These are not ideal choices due to the relatively low solubility of proteins in such solvents and the evidence for the loss of protein secondary and tertiary structures when dissolved in these solvents (63,64). The generation of microparticulates of several proteins (bovine insulin, lysozyme, and trypsin) from such nonaqueous solutions has been demonstrated by several workers (10,65). In general, spherical or spheroidal particles were observed with particle sizes in the range 1–5 μm. In each case, essentially complete biological activity was recovered on reconstitution in aqueous media. Spectroscopic analysis (Raman and FTIR) of the process steps suggested that changes in the secondary structure of the proteins occurred due to their dissolution in the nonaqueous solvents (66). The SF-induced precipitation partially removed the structural perturbation caused by exposure to the nonaqueous solvent, which was reversed on reconstitution in an aqueous medium (65).

Current thinking on the storage of dried proteins suggests that the protein native conformation state is desirable for long-term stability (67). However Winters and coworkers have demonstrated the long-term and ambient-temperature stability of SAS-precipitated insulin, lysozyme, and trypsin powders, despite the previously non-native structural conformations (68).

Tom and coworkers obtained microparticulate protein powders of bovine insulin and catalase from solutions in ethanol/water mixtures by application of the SAS process (11). The experimental design required the proteins to be dissolved in ethanol/water 90:10 mixtures, which permitted only low concentrations of protein in the process flow. Particles of uniform size, but variable morphology, were obtained, but difficulties in product recovery were reported.

Exploitation of the differing solubilities of several proteins in CO_2-expanded DMSO, has enabled the partial purification or separation of binary mixtures of proteins from DMSO solutions (69). In most, but not all, cases, substantial levels of biological activity were recovered from the precipitates.

Particle formation of biological material from purely aqueous solutions is highly desirable because it permits the processing of macromolecules, including proteins, peptides, and nucleic acids, which often have limited solubility in potentially denaturing organic solvents. The development of the SEDS process, used for the processing of small water soluble molecules, has been applied successfully to a range of water soluble biological materials. The design of the SEDS-process allows a potentially labile macromolecule to be maintained in a favorable aqueous environment until rapid processing with modified SF CO_2 (13). The aqueous feed is only contacted with potentially denaturing organic solvent and supercritical fluid at the specially designed co-axial nozzle. In this way the contact time between the aqueous feed and the organic solvent-modified supercritical CO_2 is minimized. The more prolonged contact of labile macromolecules with denaturing organic solvents, such as DMSO required in the other GAS-type methods, can thus be avoided.

Sloan and coworkers have demonstrated with the SEDS process the preparation of one micrometer-sized particles of lysozyme and trypsin from aqueous solutions (70). Scanning electron microscopy confirmed the size of particles which exhibited a spherical morphology with minimal surface defects. Recovery of enzymatic activity of lysozyme was >95% after processing, whereas only 40% of trypsin activity was retained using similar processing conditions of 200 bar and 328 K. Improved recovery of trypsin enzyme activity was achieved by changing the processing conditions, with 85% activity recovered at a lower operating temperature of 308 K (see Table 2) (71). The differences between the stabilities of the two proteins under SF processing reflect their relative thermal stabilities, with lysozyme and trypsin having unfolding transition temperatures (T-m) of 347 K and 330 K respectively (72,73).

Zagrobelny and Bright demonstrated conformational changes of trypsin in supercritical CO_2 caused by a fluid compression step, with decompression having little effect on conformation (74). SEDS processing of both lysozyme and trypsin was found to be independent of the operating pressure with regard to recovery of enzyme activity (Table 2). This finding may reflect the constant pressure conditions during the SEDS particle formation or that the reported compression-induced conformation changes reported are not associated with the active site of the enzyme or are reversed on reconstitution of the dried powder.

A comparison of freeze-drying, spray-drying, and SEDS, with respect to the physical and chemical properties of processed lysozyme powders, has been reported (75). Spray-drying and SEDS processing gave particles with narrow size distribution (see Table 3), whereas freeze-drying gave no such control and material manufactured in this way for pulmonary delivery would require secondary processing such as micronization. SEDS processing generated spherical particles that were smaller than those produced by spray-drying, without evidence of the surface deformation often observed in spray-dried proteins (Figure 18). Retention of enzymatic activity was highest after SEDS processing and freeze-drying

Table 2 Effect of SEDS Processing Operating Temperature and Pressure on the Recovery of Trypsin Enzyme Activity[a]

Pressure (bar)	Temp. (°C)	Activity recovered (%)
Unprocessed	Unprocessed	100
100	35	84
100	55	47
200	35	78
200	55	54

[a]Trypsin was processed from aqueous solution (25 mg/ml in 10 mM HCl) at a flow rate of 0.03 ml/min and contacted with 1.2 ml/min ethanol and 5.0 ml/min CO_2. Reported percent activities are relative to the as-received commercial protein.

Table 3 Aerodynamic Volume Diameter for Particulate Lysozyme Prepared by Freeze Drying, Spray Drying, and SEDS Processing[a]

	Aerodynamic volume diameter (μm)		
Process	d_{10}	d_{50}	d_{90}
Unprocessed	7.51	16.0	25.5
Spray-dried	1.70	3.14	4.79
Freeze-dried	3.44	6.68	10.9
SEDS	0.77	1.34	2.67

[a]Particle size was determined by time-of-flight measurements using the Aerosizer.

with substantial loss detected after spray-drying (see Table 4). High sensitivity differential scanning calorimetry of reconstituted protein samples showed that perturbation had occurred during the spray-drying process, with the thermal unfolding event occurring over a broad temperature range with a maximum of 338 K (Table 4 and Figure 19). In contrast, the unfolding event for unprocessed, freeze-dried and SEDS-processed samples occurred over a narrower range and at a common maximum of 347 K. The relatively high processing temperatures employed in spray-drying, in contrast to freeze-drying and SEDS-processing, are higher than the unfolding T-m of the protein and must result in perturbation of the protein conformation including the active site resulting in a loss of biological activity.

In addition to perturbation of the structural conformation of a protein during processing, many proteins are prone to aggregation of the monomeric units, leading to formation of soluble or insoluble aggregates. Insoluble aggregates are unacceptable for pharmaceutical applications and soluble aggregates in a protein formulation can have a significant deleterious effect on the pharmacokinetics and immunogenicity of the protein. Aggregation can occur as a result of interaction or adsorption at hydrophobic surfaces, which can be additionally influenced by the thermal and solvent environment. The SEDS product prepared from an aqueous solution of a therapeutic protein, known to be prone to aggregation, has shown limited perturbation (Figure 20), suggesting that the environment encountered during the SEDS process is not detrimental to this protein in respect to aggregation.

The moisture content of dried protein material is important in ensuring recovery of the biological activity on rehydration by reconstitution (76,77). SEDS processing allows the retention of bound water in the dried protein particles, with residual levels similar to those found with spray- and freeze-drying techniques. Lysozyme and trypsin have been found to retain 10(±0.5)% (w/w) water

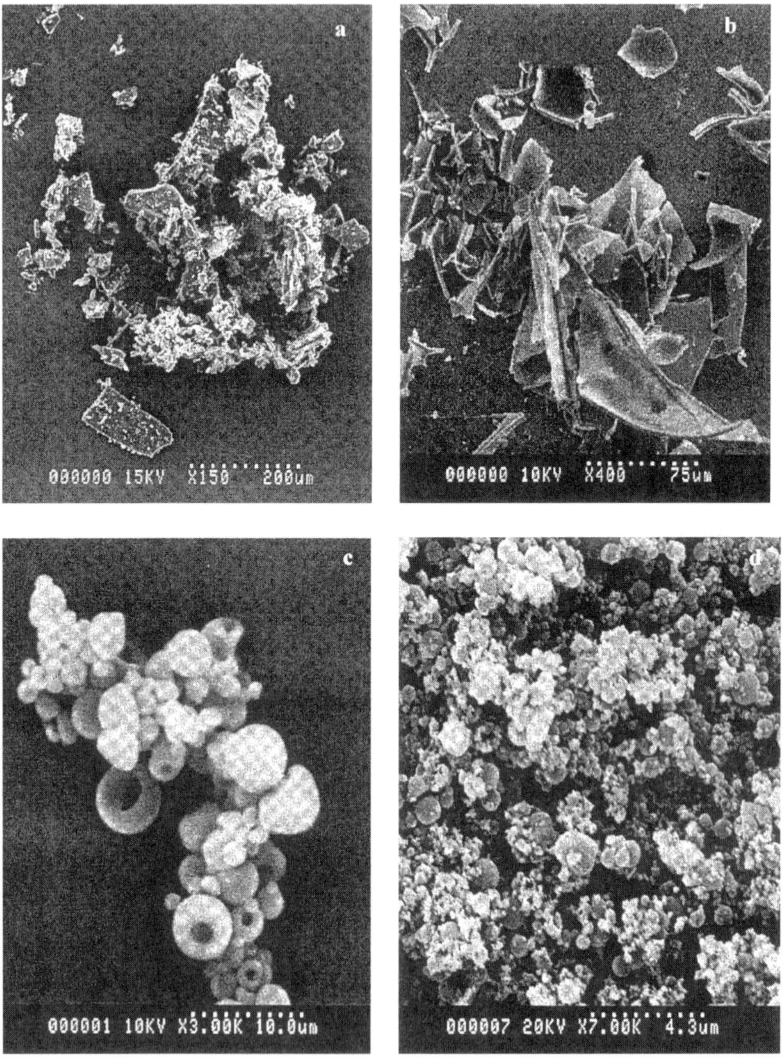

Figure 18 Scanning electron micrographs of (a) commercially supplied lysozyme particles and material produced by (b) freeze-drying, (c) spray-drying, and (d) SEDS processing. (From Ref. 75.)

Table 4 Enzymatic Activity Retention and Unfolding Transition Temperature (*T-m*) of Reconstituted Lysozyme Following Various Particle Formation Methods[a]

Process	% activity relative to unprocessed material	Transition temp., *T-m* (°C)
Unprocessed	100	73.3
Spray-dried	85.3	64.6
Freeze-dried	89.3	73.7
SEDS lab scale	95.0	73.7
SEDS pilot scale	95.0	76.0

[a]The activity is expressed relative to the unprocessed protein. The transition temperatures were determined by high-sensitivity differential scanning calorimetry.
From Ref. 75.

following SEDS processing (70). These results contrast with the water content of proteins prepared from DMSO solutions, which have been reported as low as 2% (w/w) (68). This reflects the exchange of water associated with dissolution of the protein in DMSO.

The SEDS-processing of plasmid DNA from an aqueous solution has been reported (78,79). The preferred conformation of DNA used for transfection is the closed circle supercoil (60). In this form the polynucleotide strands of the double-stranded DNA are circular. If there is one single-strand interruption, the DNA is termed open circular with further interruptions leading to the formation of linear DNA and a loss of ability to cause transfection. Any processing of plasmid DNA must aim to maximize the recovery of the supercoiled form. The SEDS processing of an aqueous solution of the pSVβ plasmid was performed with isopropanol as an antisolvent and modifier for SF CO_2. In an unbuffered system, low recovery of the supercoiled DNA form was observed (Figure 21). Supercritical CO_2 and water mixtures have been shown to have an acidic pH (80), and DNA is susceptible to depurination and apuritic site-strand cleavage at acidic pH (81). Inclusion of a buffer (sodium acetate) in the aqueous solution to reduce depurination of the DNA, resulted in a substantial recovery of the preferred supercoiled DNA (Figure 21).

Particle formation, using supercritical fluid processing, offers an exciting alternative to the conventional approaches of lyophilization and spray-drying for the preparation of particulate biological material for emerging drug delivery methods. Enhancement of the GAS-type procedures by allowing the processing of aqueous solutions, such as that achievable by the SEDS process, widens the potential further. Tailoring the aqueous feed, based on the aqueous biochemistry of the drug macromolecule, can be used both to control the particle formation and

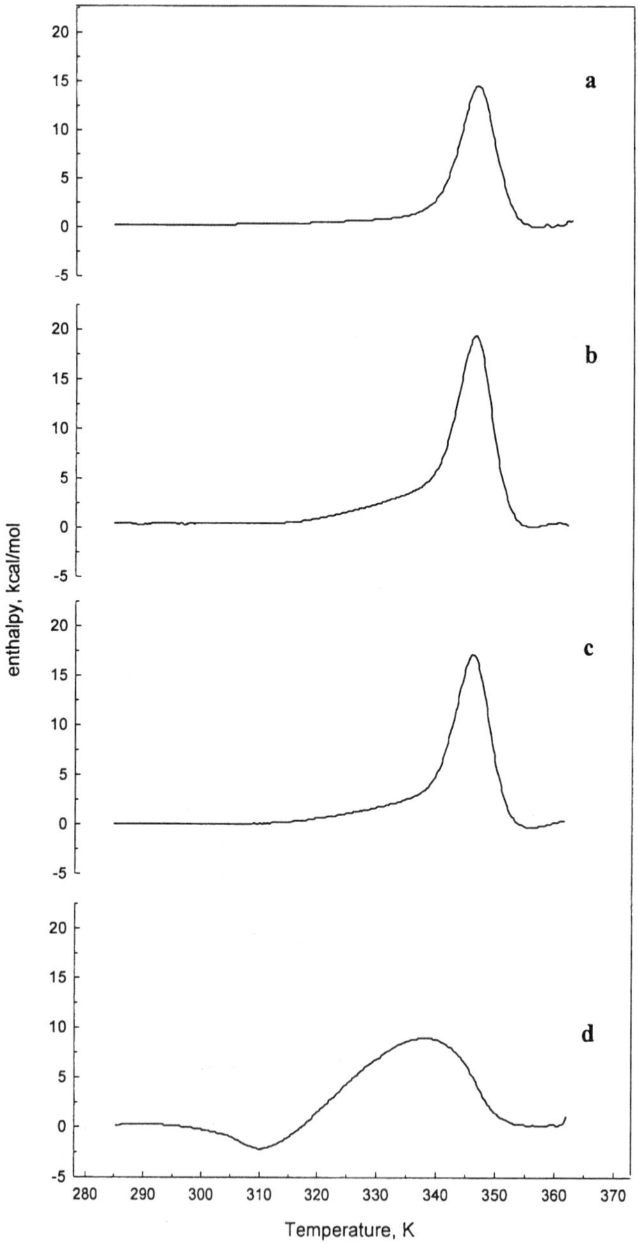

Figure 19 High-sensitivity DSC scans of reconstituted lysozyme samples subject to different process. (a) Unprocessed, (b) freeze-dried, (c) SEDS-processed, (d) spray-dried. (From Ref. 75.)

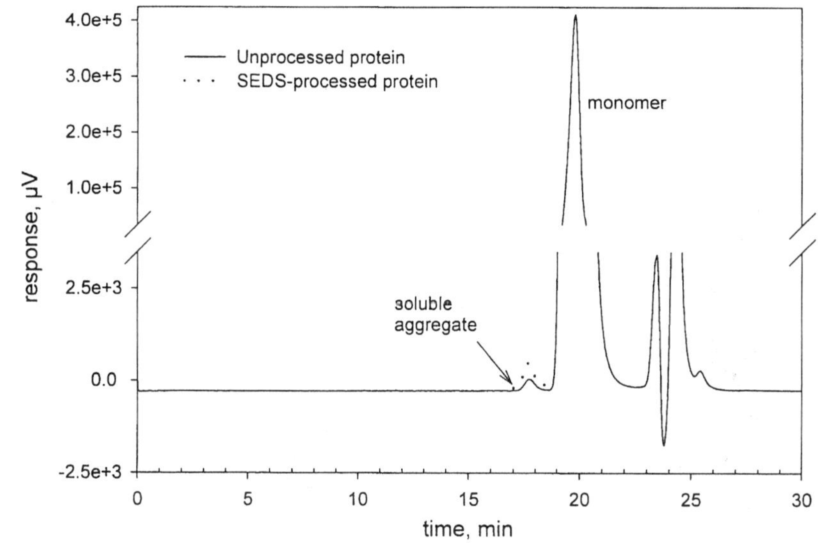

Figure 20 Size exclusion HPLC of unprocessed and SEDS-processed protein, demonstrating the limited aggregation resulting from the processing environment.

to protect and stabilize labile molecules during processing to generate particles with desirable and targeted properties.

G. Formulations

Drug–drug and polymer–drug systems provide an alternative formulation strategy for specific therapeutic objectives. Drug–drug formulations find special applications in combining short and long-acting drugs. Blending powdered forms of two drugs does not generally give a sufficiently homogeneous mixture, leading to out of specification products. An attractive option is to recrystallize both components from a single solution with the correct proportions of individual drugs in the final particulate products. Drug–polymer coformulations are primarily designed for modifying drug dissolution, either to increase or to slow down the dissolution rate. When slowing down dissolution, it is also important to sustain the release of the drug from the polymer matrix in a controlled manner over a period of time.

Several investigations are reported in the literature on coformulating a drug with a biodegradable polymer either by the RESS or by the SCF antisolvent technique. Tom and coworkers (11) reported a coprecipitation of lovastatin with poly(l-lactic acid) (L-PLA) by RESS and observed that the final drug loading is a function of its solubility in supercritical CO_2. They found that with

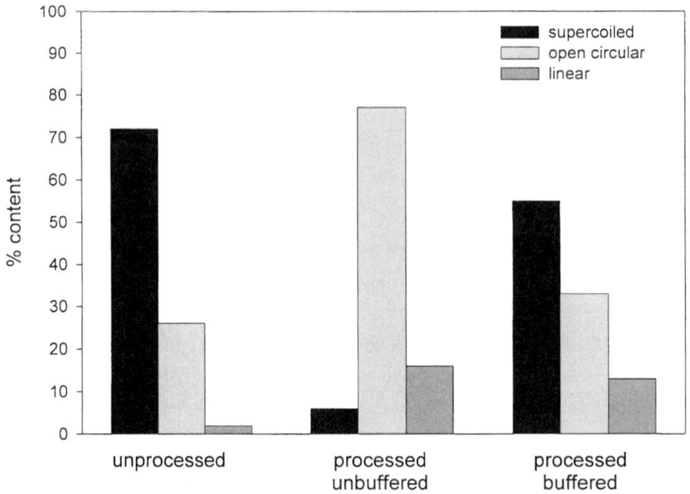

Figure 21 Retention of supercoiled conformation of plasmid DNA pSVβ following SEDS processing. Particulate material was prepared from unbuffered and buffered (sodium acetate) aqueous solutions of plasmid DNA (5 μg/ml) and mannitol (50 μg/ml) using SEDS processing with 0.03 ml/ml aqueous, and 0.9 ml/min propan-2-ol, and 10 ml/min carbon dioxide at a pressure of 200 bar and temperature of 323 K.

increasing solubility of lovastatin in SCF CO_2 a drug loading of up to 36 wt % in polymer could be achieved. In a later study, Kim and coworkers (82) coprecipitated 3.75 wt % naproxin in L-PLA and confirmed the encapsulation with reflected-light confocal laser scanning microscopy. Bleich and Muller (21) coprecipitated four individual drugs with variable polarity with L-PLA using the ASES process. They report an incorporation of up to 20% of indomethacin and hyoscine-butylbromide, 7% of piroxicam and 5% of thymopentin peptide into L-PLA.

Wilkins and coworkers (83) have studied the coprecipitation of a model drug compound, indomethacin, with three different polymers of varying chemical structure and polarity. They found that the specific molecular and macromolecular interactions between the polymer and the drug play an important role in the amount of drug that can be dispersed in the polymer in an amorphous phase. Thus, hydroxypropyl methylcellulose (HPMC), a more polar polymer compared to ethylcellulose, has the propensity to establish hydrogen bonds with the carboxyl group in indomethacin. It is therefore not surprising to find an increase in the amount of drug, which can be maintained in amorphous form to increase from 25% w/w in ethylcellulose to 35% w/w in HPMC. The effect is more pronounced with poly-(vinylpyrrolidone) (PVP), which has a free carbonyl group and therefore potential for additional specific interactions with the hydroxyl group of indomethacin with up to 60% w/w drug incorporated into the amorphous state. Figure 22 shows the degree of crystallinity, assessed using DSC, against drug loading. Above the limiting level of amorphous content (i.e., 0% crystallinity) and higher drug loading, the product becomes partially crystalline, suggesting that phase separation is taking place.

III. IN VITRO STUDIES

The formulations developed for drug delivery systems are routinely examined by in vitro techniques as a guide for, and indication of, in vivo performance. With one of the initial targets of SF processing being micron-sized particles, several reports of in vitro studies in the literature have been in the field of respiratory drug delivery. For respiratory drug delivery of micron-sized drug particles, the performance of dry powder inhalers (DPIs) is tested using a cascade impactor. A typical impactor consists of a series of stages, which progressively retain smaller particles. The powder sample is introduced at the top of the stack and is drawn through by vacuum in an airstream. Depending on aerodynamic size, the particles carried by the airstream are retained on a metal plate at individual stages. Thus, smaller particles are carried farther and hence deposited at the lower stages compared to large particles. The fractions thus sequentially separated are grouped typically into eight sizes for the Anderson cascade impactor in terms of aerodynamic diameter. For most of the respiratory drugs, particles with an aerodynamic diameter between 1–5 μm are deposited deep in the lungs. While size of particles plays a critical role, it is increasingly recognized that the surface energetics of the particulate material are also important in the efficiency of such drug delivery systems. While small particle size is critical for delivery in the deep airways of the lungs, a high surface area is available and high cohesive energy of such surfaces would tend to cause particles to aggregate and adhere to the walls and components of drug delivery devices such as DPIs and propellant borne MDIs. Feeley (84) has measured the surface properties of drugs for respiratory drug delivery by inverse gas chromatography (IGC). Table 5 shows comparative results for unprocessed, micronized, and SEDS-prepared sample of salbutamol sulfate (albuterol sulfate), a widely used asthma drug. The γ_s^d value represents the dispersive component of the surface free energy and K_A and K_B are the specific acid and base parameters. As can be seen, the dispersive component of the micronized sample is higher than the unprocessed material, implying more energetic surfaces for nonpolar and dispersive interactions compared to the unprocessed material. The SEDS-processed sample, on the other hand, has a lower surface dispersive energy compared with the micronized sample. Though the micronized and SEDS-prepared samples are both amphoteric in

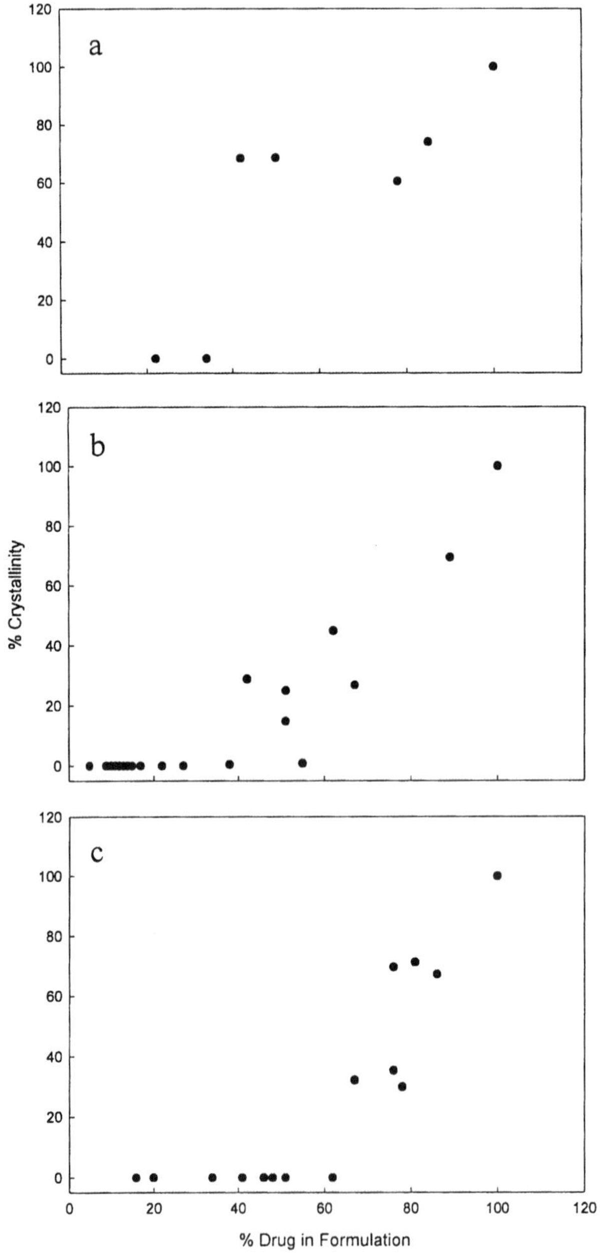

Figure 22 Indomethacin loading in (a) ethylcellulose (EC), (b) hydroxypropylmethylcellulose (HPMC), and (c) polyvinylpyrrolidone (PVP) in an amorphous state. (From Ref. 83.)

Table 5 Surface Characteristics of Unprocessed, Micronized, and SEDS-Processed Samples of Salbutamol Sulfate

Sample	γ_s^d (mN m^{-1})	K_a	K_b
Unprocessed	49.07	0.46	0.61
Micronized	58.27	0.31	0.67
SEDS (S1)	38.45	0.24	0.45

From Ref. 84.

nature, the presence of higher values in the basic parameter of the micronized sample as compared to the unprocessed materials indicates stronger basic or electron donor interactions at the surface. In contrast, the SEDS-prepared material is less energetic and thus is likely to perform better in drug delivery from DPIs.

Figure 23 shows representative cascade impactor results of salbutamol sulfate (84). Three samples, including one prepared by micronization (mass median

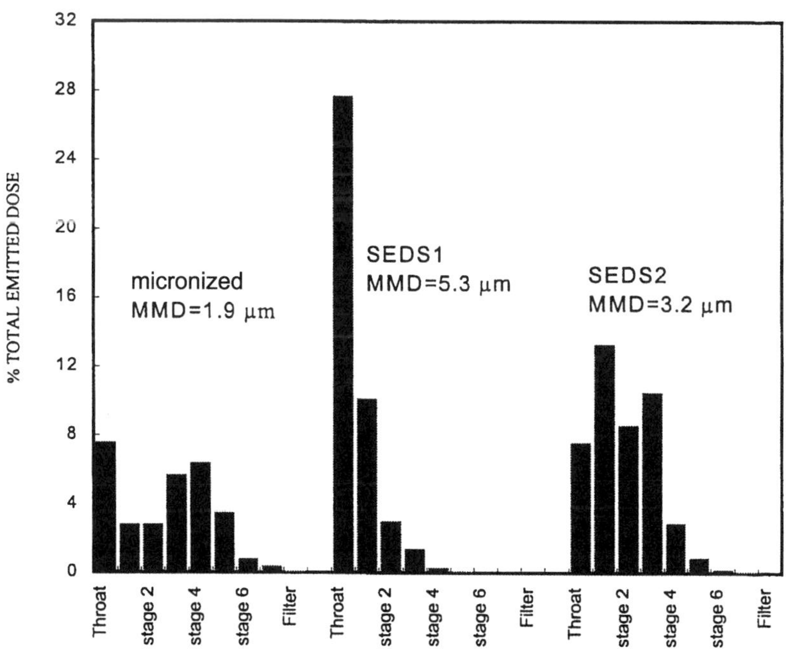

Figure 23 Deposition patterns of micronized and SEDS-prepared salbutamol sulfate. (From Ref. 84.)

diameter 1.9 μm) and two SEDS-prepared sample with mean mass diameters of 5.3 μm (SEDS1) and 3.2 μm (SEDS2), respectively, were tested. The drug was blended with α-lactose monohydrate before filling into blister-pack units for operation by the DPI device (Rotahaler). As can be seen, the deposition of the micronized sample in the metal throat attachment prior to the impactor apparatus is low compared with the SEDS1 material, although the distribution is wide throughout the cascade impactor. The deposition of SEDS1 sample is negligible from stages 5 downward. However, by reducing the particle size such as sample SEDS2, the throat deposition was dramatically reduced and the total emitted dosage on stages 1–5, representing particle size of 1–4 μm, of the cascade impactor increased by a factor of 2 compared with the micronized drug. A similar increase in in vitro drug delivery, simulating transport to the deep lungs, was reported from a twin impinger study of SEDS-processed salmeterol xinafoate (85). A twofold increase in drug deposition in the lower stage representing peripheral lung delivery was observed with the SEDS-processed salmeterol xinafoate compared with micronized material. Another report showing improved in vitro drug delivery performance of the SCF-processed material is that for fluticasone propionate (19). In this study on formulation of MDIs with propellants other than CFCs, the authors report in vitro twin impinger test results and compared the results between two formulations with a micronized and ASES-processed drug. Of the total emitted dose, 47% was detected as fine-particle deposition (size <6.5 μm) for the micronized product, dispersed with an additive, compared with 23% for the ASES product. Nevertheless, an ASES product prepared with 5% lecithin during processing gave 60% deposition of the total emitted dose. This study reflects the added advantage of a single-stage process in tailoring the particle characteristics by SCF processing.

York and Hanna (85) studied the cohesive energy and agglomeration tendency of particles of salmeterol xinafoate prepared by micronization and SEDS process by measuring the mean aerodynamic particle size by dispersing the samples at various shear forces in an Aerodisperser/Aerosizer, a particle-sizing instrument. The results are shown in Figure 24. As can be seen, a high-shear force is needed for the micronized sample to disperse the sample into primary particles as compared with the SEDS-prepared sample. This finding is consistent with the practical observation that the majority of SEDS-processed materials exhibit low charge and are easily flowing materials even at 5-μm particle size. This finding is of major importance in designing and optimizing powder formulations for DPIs, since any device dependency on drug particle fluidization and aerosolization can be dramatically reduced or even eliminated.

Figure 25 shows the dissolution data of coformulates of indomethacin with hydroxypropylmethylcellulose (HPMC) and theophylline with ethylcellulose (EC) in water. Data are also presented for a physical mix of 20 wt % drug and polymer. Figure 25a shows that the SEDS-prepared formulation of 15 wt %

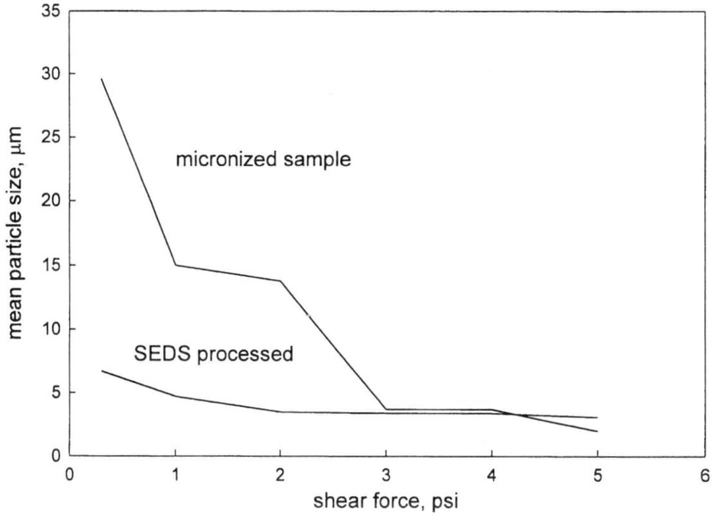

Figure 24 Effect of shear force on dispersion of a micronized and SEDS-prepared sample of salmeterol xinafoate. (From Ref. 85.)

indomethacin in HPMC dissolves much faster than a physical mix. On the other hand, the dissolution of theophylline is retarded considerably by coformulating with ethylcellulose, as shown in Figure 25b. These results demonstrate important changes that can be achieved in particle engineering for drug formulations by SCF processing in a single-stage operation.

IV. SCALE-UP

Successful scale-up of SCF particle formation processes is crucially important for industrial viability. The SEDS process has been scaled up from a 50-ml vessel (Keystone, USA) with a CO_2 pump capacity of 1.8 kg/h (Milton Roy, Pont Saint Pierre, France) to a vessel of 2 L (HIP, Erie, PA) with a CO_2 pump capable of delivering 18 kg/h (Thar Designs, Pittsburgh, PA). The plant has all of the salient safety features normally required for processing powder pharmaceuticals. The plant is housed in a walk-in downward laminar airflow cabinet to protect the operator from any powder spillage during handling. The pressure is maintained within ±1 bar by a computer-controlled back-pressure regulator (Tescom, Coatbridge, UK) and the temperature of the vessel is maintained and controlled within ±1°C by means of an electrical heating jacket. An important feature of the plant is the 70-L surge tank into which the contents of the particle formation vessel can be expanded in case of emergency pressure release. The

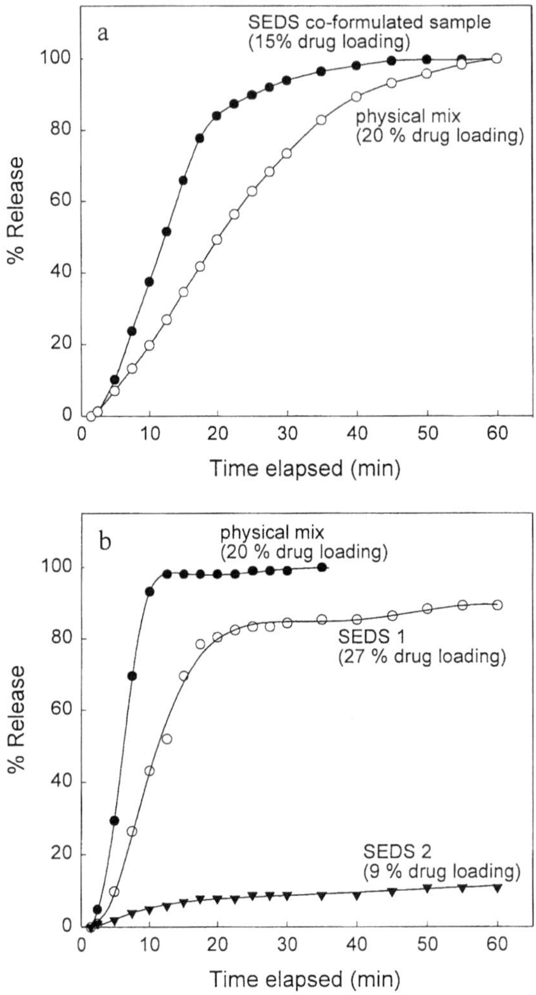

Figure 25 Dissolution profiles of (a) indomethacin-HPMC and (b) theophylline-EC conformulates in water.

solvent and any dissolved solute are separated using cyclones (Thar Designs, Pittsburgh, PA) before venting CO_2.

The SEDS technique, with wide applications of pharmaceutical processing as discussed above and successfully operating on the bench scale, has now been transferred and tested on the pilot plant for its robustness, consistency, and scalability. The results of studies with two model compounds—paracetamol crystallized from ethanol and a protein, lysozyme, precipitated from aqueous solution—are presented. Figure 26 shows comparative SEM photomicrographs of paracetamol prepared on a laboratory scale and the pilot plant scaled up 10 times in respective flows of carbon dioxide and solution. As can be seen, the morphology and particle size of products is similar, a result that emphasizes the scalability of the process in terms of achieving particles with similar physicochemical characteristics. Another critical issue when testing the robustness of the process is to study the effect of the process variables temperature, pressure, and flow rates of the individual streams. Figure 27 shows the effect of solution flow on the crystallization of paracetamol. By increasing the solvent-to-CO_2 ratio and working in a region close to the critical point in the miscible region of the phase diagram, the particles crystallize at a much slower rate resulting in crystal growth, which is typically seen in conventional solvent-based crystallization. Also, as paracetamol is soluble in ethanol-modified SCF CO_2 (86), the particles are ripened in a saturated medium of supercritical solution, which may be detrimental in extended production runs. Nevertheless, the effect is minimal when the flow rate is increased from 2.2 ml/min to 10 ml/min and these results

Lab scale Pilot Plant

Figure 26 Comparison of results from laboratory scale and pilot plant of SEDS-processed paracetamol compound. Experimental conditions: temperature = 328 K; pressure = 150 bar; solute concentration in ethanol = 1% w/v; CO_2 flow rate: lab scale = 18 ml/min; pilot plant = 200 ml/min; soln. flow rate: lab scale = 0.2 ml/min; pilot plant = 2.2 ml/min.

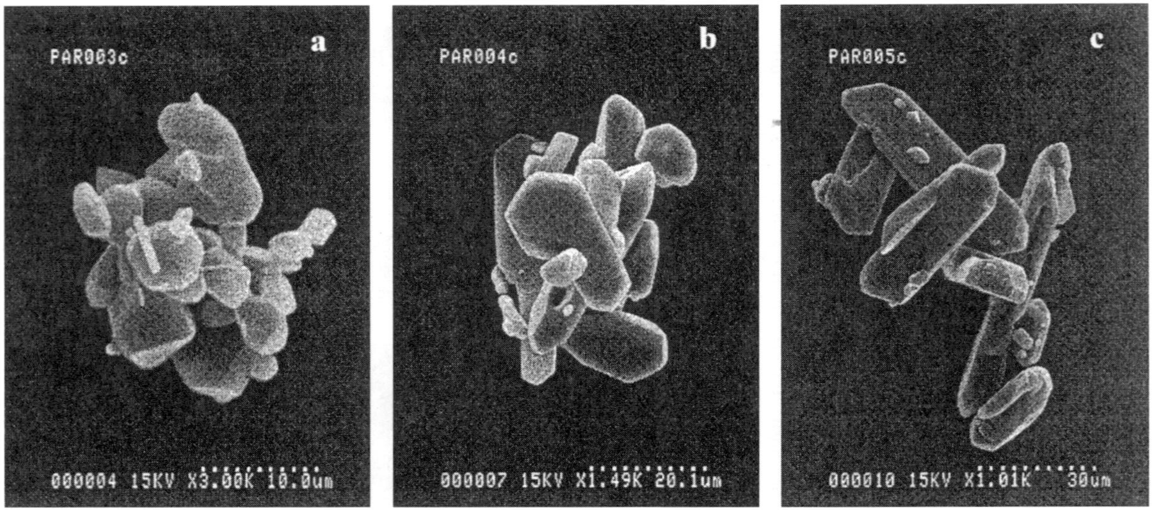

Figure 27 Morphological changes of acetaminophen when processed at different solution flow rates. Experimental conditions: temperature = 328 K; pressure 150 bar; concentration of acetaminophen in ethanol = 1% w/v; flow rate of carbon dioxide = 200 ml/min; flow rate of solution (a) 2.2 ml/min; (b) 10 ml/min; and (c) 20 ml/min.

emphasize the potential of production capacities that can be achieved. Temperature and pressure effects are pronounced when working at conditions where carbon dioxide and solvent are only partially miscible where solvent removal is an issue. The effects would be minimal when operating in the miscible region as can be seen from similar crystal behavior as shown in Figures 14b and 27c.

The successful scale-up, by a factor 30, of the generation of protein particles from aqueous solution has been achieved using the SEDS process. Using the model protein lysozyme, similar particulate morphology was observed by SEM analysis, and importantly the enzyme activity was completely recovered as reported at laboratory scale (Table 4). High-sensitivity DSC analysis confirmed the retention of the macromolecular conformation on reconstitution of the powder (Table 4).

A. Market Demand (Example)

Large commercial high-pressure supercritical extraction plants are currently employed for decaffeination of coffee beans. Therefore, the technology of high-pressure engineering for food grade applications is readily available. It is thus not an altogether new task of designing a plant that is compliant with current Good Manufacturing Practice (cGMP). Further, production of highly potent drugs does not necessitate a large commercial plant. Table 6 lists a typical market requirement of an antiasthmatic drug, salbutamol sulfate, in the United States where the annual requirement is in the region of 500–1000 kg. This is an attractive scenario, as a relatively small plant does not require huge capital costs and provides sufficient material. Where multipurpose processing is required, a larger and more expensive plant will be needed.

Apart from the capital cost, other issues must be considered. cGMP requires the industry to follow strict guidelines as laid out by the relevant regulatory authorities (87). Broadly this may be classified into two categories: (a) properly designed, installed, and commissioned equipment and (b) a reliable and consistent process. Validation is crucial for both process and equipment and must be successfully completed and demonstrated prior to the manufacture of any drug for clinical trial or for sale. The robustness of the process can be

Table 6 Market Requirements for Salbutamol Sulfate in the United States

Dosage levels	100–200 μg
Doses required	500/yr
Total asthmatics in USA	10×10^6 (5% of population)
Total doses required	5000×10^6
Total quantity of albuterol sulfate required	500–1000 kg/yr

tested in the laboratory and pilot plant by a systematic study of the influence of process variables, as discussed above in the scale-up section. This stage of process validation is documented in a Process Qualification (PQ) sheet. Equipment manufacture plays a major role in cGMP. There are certain restrictions regarding the use of materials. Stainless steel 316L and PTFE are generally accepted as safe, least contaminating, and most easily cleaned surfaces. Since most of the high-pressure equipment is made of stainless steel this does not impose a major cost burden. Nevertheless, all the seals need to be constructed from PTFE or similar material and the valves specially designed to minimize dead volumes and to provide easy-to-clean facilities without dismantling. All the steel surfaces, both internal and external, should be highly polished or electroplated to give a smooth surface, with a typical surface roughness of less than 0.5 μm, to aid in efficient and effective cleaning. Furthermore, a complete history of all the materials used in the equipment manufacture must be available. All these procedures require design qualification (DQ) documentation of the complete process equipment, including all individual components such as pipe unions and fittings, valves, pressure and temperature indicators, etc. After the plant construction is completed, the next stage of plant installation is checked and verified with the specifications listed in the DQ and the results are documented in the installation qualification (IQ) report. The equipment performance is then tested against the limits of each component item such as pressure and temperature indicators, pressure control, temperature control, and so forth, and the results are documented in an operation qualification (OQ) sheet. The process is then validated in the plant by processing a model compound, with the resultant exhibiting those characteristics as described in the PQ. Thus, cGMP compliance requires extensive documentation and must be adhered to in terms of the strict guidelines laid out by the regulatory authorities. Another important aspect of cGMP is the recording of all the main process parameters, such as pressure, temperature, flow rates, etc., during a run. Hence, the process benefits from computer control using validated software.

Apart from the equipment point of view, cGMP also has strict procedures for maintaining the highest standards of safety because of potential adverse effects of the drug on the operator. This becomes a major issue during handling of particulate solids. Such particle formation and processing operations typically have to be housed in a controlled-quality air atmosphere. The specified air quality in an operational area for handling particulate solids is termed class 10,000 (which means a maximum number of particulate solids of size less than 0.5 μm to be less than 10,000 per cubic foot) for nonsterile manufacturing. This is not a major issue with SCF processing, such as the SEDS process. Since the process is completely enclosed, the entire process equipment does not need to be in a clean room. Also, as the particulate material is collected in the pressure vessel, it is relatively straightforward to house the vessel in a clean room with controlled temperature, humidity, and class 10,000 air. The process components and main plant comprising the pumps, CO_2-solvent separation lines, and CO_2 recycling lines can be housed outside the clean room.

A complementary and important factor in cGMP is batch-to-batch consistency and traceability of individual batches, especially, for example, where purity or particle size is a target characteristic of the product. Though most drugs do not need a sterile environment for processing, contamination is another major issue. The problem is compounded if the plant is designed to process multiple products. Contamination during processing can be of two types: (a) outside contamination during processing and (b) cross-contamination from previous product when the same equipment is used for processing multiple products. While the SEDS process is completely enclosed, there is little scope for external contamination during processing; the latter issue calls for validated cleaning procedures.

In general, SCF processing can be adapted to cGMP compliance in a straightforward manner without major difficulty. This feature is especially attractive for scaling up, and it will not be surprising to find an SCF plant in the near future manufacturing high-value, low-volume pharmaceutical particulate products.

V. CONCLUSIONS

From the discussion and examples given, it can be seen that SCF processing in general and the SEDS technique in particular is a promising emerging technology that has potential in addressing some of the most prevalent problems in particle formation and particle design for drug delivery in the pharmaceutical industry. The process, being controlled by macroscopic variables such as pressure, temperature, and flow rates of the respective fluids, is exemplified by the ease of scalability. The technique is easily adapted to process a wide range of materials including biologicals from different solvent media. The high-pressure engineering available in the market complements the ease with which this technology can be adapted to comply with cGMP standards, especially with regard to hardware in the form of equipment design. Though the cost of infrastructure is high in terms of high-pressure apparatus, it can be outweighed by the potential benefits the technology has to offer. In particular, the process is best suited for low-volume high-value products.

REFERENCES

1. R Krishnamurty. Protein stability in pulmonary delivery formulations: a review. Pharm Technol 23(3):48–57, 1999.

2. AJ Hackay. Lung deposition and clearance of pharmaceutical aerosols: what can be learned from inhalation toxicology and industrial hygiene? Aersol Sci Technol 18:290–304, 1993.
3. HG Brittain, EF Fiese. Effects of polymorphism and solid-state solvation on solubility and dissolution rate. In: HG Brittain, ed. Polymorphism in Pharmaceutical Solids. New York: Marcel Dekker, 1999.
4. LA Gatlin, SL Nail. Freeze drying: a practical overview. In: RG Harrison, ed. Protein Purification Process Engineering. New York: Marcel Dekker, 1993, pp. 317–367.
5. MA McHugh, VJ Krukonis. Supercritical Fluid Extraction: Principles and Practice. 2nd ed. Boston: Butterworth-Heinemann, 1994.
6. E Weidner, R Steiner, Z Knez. Powder generation from polyethyleneglycols with compressible fluids. In: PR Von Rohr, C Trepp, eds. Process Technology Proceedings 12, High Pressure Chemical Engineering. Netherlands: Elsevier, 1996, pp. 223–228.
7. S Palakodaty, P York. Phase behavioural effects on particle formation processes using supercritical fluids. Pharm Res 16:976–984, 1999.
8. A Kordikowski, AP Schenk, RM Van Nielen, CJ Peters. Volume expansions and vapour-liquid equilibria of binary mixtures of a variety of polar solvents and certain near-critical solvents. J Supercrit Fluids 8:205–216, 1995.
9. FE Wubbolts, C Kersch, V Rosmallen. Semi-batch precipitation of acetaminophen from ethanol with liquid carbon dioxide at a constant pressure. In: M Perrut, P Subra, eds. Proceedings of the 5th Meeting on Supercritical Fluids, Nice, France, 1998, Vol. 1. International Society for the Advancement of Supercritical Fluids, pp. 249–256.
10. SD Yeo, GB Lim, PG Debenedetti, H Bernstein. Formation of microparticulate protein powders using a supercritical fluid antisolvent. Biotechnol Bioeng 41:341–346, 1993.
11. JW Tom, GB Lim, PG Debenedetti, KR Prudhomme. Applications of supercritical fluids in the controlled release of drugs. ACS Symp Series 514:238–257, 1993.
12. MH Hanna, P York. Method and apparatus for the formation of particles. Patent application, published as WO 95/01221, 1995.
13. MH Hanna, P York. Method and apparatus for the formulation of particles. Patent application, published as WO 96/00 610, 1996.
14. S Palakodaty, J Pritchard, P York, M Hanna. Crystallisation of lactose using solution enhanced dispersion by supercritical fluid (SEDS) technique. In: M Perrut, P Subra, eds. Proceedings of the 5th Meeting on Supercritical Fluids, Nice, France, 1998, Vol. 1. International Society for the Advancement of Supercritical Fluids, pp. 275–280.
15. S Palakodaty, P York, J Pritchard. Supercritical fluid processing of materials from aqueous solutions: the application of SEDS to lactose as a model substance. Pharm Res 15:1835–1843, 1998.
16. DJ Dixon, KP Johnston, RA Bodmeier. Polymeric materials formed by precipitation with a compressed fluid antisolvent. AIChE J 39:127–139, 1993.
17. S Mawson, S Kanakia, KP Johnston. Coaxial nozzle for control of particle morphology in precipitation with a compressed fluid antisolvent. J Appl Polym Sci 64:2105–2118, 1997.

18. J Bleich, BW Müller, W Wassmus. Aerosol solvent extraction system—a new microparticle production technique. Int J Pharm 97:111–117, 1993.
19. H Steckel, J Thies, BW Müller. Micronizing of steroids for pulmonary delivery by supercritical carbon dioxide. Int J Pharm 152:99–110, 1997.
20. J Thies, BW Müller. Size controlled production of biodegradable microparticles with supercritical gases. Eur J Pharm Biopharm 45:67–74, 1998.
21. J Bleich, BW Müller. Production of drug loaded microparticles by the use of supercritical gases with aerosol solvent extraction system (ASES) process. J Microencap 13:131–139, 1996.
22. F Ruchatz, P Kleinebudde, BW Müller. Residual solvents in biodegradable microparticles. Influence of process parameters on the residual solvent in microparticles produced by the aerosol solvent extraction system (ASES) process. J Pharm Sci 86:101–105, 1997.
23. S Srcic, P Sencar-Bozic, Z Knez, J Kerc. Improvement of nifedipine dissolution characteristics using supercritical CO_2. In: Proceedings of the 16th Pharmaceutical Technology Conference and Exhibition, Athens, Greece, 1997, pp. 59–69.
24. AH Lefebvre. Atomisation and Sprays. New York: Hemisphere Publishing, 1989.
25. RE Treybal. Mass transfer operations. New York: McGraw-Hill, 1980.
26. S Palakodaty, BY Shekunov, P York. Crystallisation process in modified supercritical fluids. In: Proceedings of the IChemE World Congress on Particle Technology 3, Brighton, UK, 1998. Rugby, UK: Institution of Chemical Engineers, p. 383.
27. SM Cooper. A novel polymorphic crystalline form of fluticasone propionate. A method for its production and pharmaceutical compositions thereof. Patent application, published as WO 98/17676, 1997.
28. R Wiebe. The binary system carbon dioxide-water under pressure. Chem Rev 29:475–481, 1941.
29. M Wendland, H Hasse, G Maurer. Multiphase high-pressure equilibria of carbon dioxide-water-isopropanol. J Supercrit Fluids 6:211–222, 1993.
30. P Traub, K Stephan. High-pressure phase equilibria of the system CO_2-water-acetone measured with a new apparatus. Chem Eng Sci 45:751–758, 1990.
31. BL Herrington. Some physico-chemical properties of lactose. I. The spontaneous crystallisation of supersaturated solutions of lactose. J Diary Sci 17:501–534, 1934.
32. CF Lerk, AC Andreae, AH de Boer, P de Hoog, K Kussendrager, J van Leverink. Alterations of α-lactose during differential scanning calorimetry. J Pharm Sci 7: 856–857, 1984.
33. K Suzuki, H Sue, M Itou, RL Smith, H Inomata, K Arai, S Saito. Isothermal vapour-liquid equilibrium data for binary systems at high pressures: carbon dioxide-ethanol: carbon dioxide-1-propanol, methane-ethanol, methane-1-propanol, ethane-ethanol, and ethane-1-preopanol systems. J Chem Eng Data 35:63–66, 1990.
34. H Tanaka, M Kato. Vapour-liquid equilibrium properties of carbon dioxide+ethanol mixture at high pressures. J Chem Eng Jpn 28:263–266, 1995.
35. S Jaarmo, M Rantakyla, O Aaltonen. Particle tailoring with supercritical fluids: Production of amorphous pharmaceutical particles. In: Proceedings of the 4th International Symposium on Supercritical Fluids, Sendai, Japan, 1997. International Society for the Advancement of Supercritical Fluids, pp. 263–266.

36. FP Lucien, NR Foster. Solubilities of hydroxybenzoic acid isomers in supercritical carbon dioxide. J Chem Eng Data 43:726–731, 1998.
37. B Simandi, S Keszei, E Fogassy, S Kemeny, J Sawinsky. Separation of Enantiomers by supercritical fluid extraction. J Supercrit Fluids 13:331–336, 1998.
38. KP Johnston, SE Barry, NK Nolan, TR Holcomb. Separation of isomers using retrograde crystallization from supercritical fluids. Ind Eng Chem Res 26:2372–2377, 1987.
39. A Kordikowski, P York, D Latham. Resolution of ephedrine in supercritical carbon dioxide: a novel technique for the separation of chiral drugs. J Pharm Sci 88:786–791, 1999.
40. S Alsoy, JL Duda. Supercritical devolatilization of polymers. AIChE J 44:582–590, 1998.
41. CM Wai, KE Laintz. Extraction of metals using carbon dioxide as supercritical fluid with diketone as chelate-forming agent. US patent US5730874A, 1998.
42. FP Lucien, NR Foster. Influence of matrix composition on the solubility of hydroxybenzoic acid isomers in supercritical carbon dioxide. Ind Eng Chem Res 35:4686–4699, 1996.
43. JW Chen, FN Tsai. Solubilities of methoxybenzoic acid isomers in supercritical carbon dioxide. Fluid Phase Equil 107:189–200, 1995.
44. KL Tsai, FN Tsai. Solubilities of methoxybenzoic acid isomers in supercritical carbon dioxide. J Chem Eng Data 40:264–266, 1995.
45. Y Iwai, H Uchido, Y Mori, H Higashi, T Matsuki, T Furuya, Y Arai, K Yamamoto, Y Mito. Separation of isomeric dimethylnaphthalene mixture in supercritical carbon dioxide by using zeolite. Ind Eng Chem Res 33:2157–2160, 1994.
46. HK Lee, CH Kim, S Kim, CS Choi. Solid solubilities of methoxyphenylacetic acid isomer compounds in supercritical carbon dioxide. J Chem Eng Data 39:163–165, 1994.
47. VJ Krukonis, RT Kurnik. Solubility of the solid aromatic isomers in carbon dioxide. J Chem Eng Data 30:247, 1985.
48. B Simandi, S Keszei, E Fogassy, J Sawinsky. Supercritical fluid extraction, a novel method for the production of enantiomers. J Org Chem 62:4390–4394, 1997.
49. S Keszei, B Simandi, E Fogassy, J Sawinsky, CS Niklos, R Lovas. Resolution of ibuprofen and cis-chrysanthemic acid by supercritical carbon dioxide. Process Technol Proc 12:393–398, 1996.
50. E Fogassy, M Acs, T Szili, B Simandi, J Sawinsky. Molecular chiral recognition in supercritical solvents. Tetrahedron Lett 35:257–260, 1994.
51. JG Van Alsten, CA Eckert. Effect of entrainers and of solute size and polarity in supercritical fluid solutions. J Chem Eng Data 38:605–610, 1993.
52. JD Hildebrand, RL Scott. The Solubility of Nonelectrolytes. 3rd ed. New York: Dover Publications, 1964.
53. CT Lira. Physical chemistry of supercritical fluids. A tutorial. ACS Symp Series 366:1–25, 1988.
54. RT Morrison, RN Boyd. Organic Chemistry. 3rd ed. Boston: Allyn and Bacon, 1983.
55. J Jacques, A Collet, SH Wilen. Enantiomers, Racemates and Resolutions. New York: John Wiley and Sons, 1981.

56. SH Wilen. Resolving agents and resolution in organic chemistry. Topics Stereochem 6:107–176, 1971.
57. A Kordikowski, P York. Chiral separation using supercritical carbon dioxide. In: M Poliakoff, MW George, SM Howdle, eds. Proceedings of the 6th Meeting on Supercritical Fluids—Chemistry and Materials, Nottingham, 1999. International Society for the Advancement of Supercritical Fluids, pp. 163–167.
58. ZJ Li, DW Grant. Relationship between physical properties and crystal structures of chiral drugs. J Pharm Sci 86:1073–1078, 1997.
59. DJ Burgess. Drug delivery aspects of biotechnology products. In: JM Pezzuto, ME Johnson, HR Manasse, eds. Biotechnology and Pharmacy. New York: Chapman and Hall, 1993, pp. 116–151.
60. DMF Prazeres, GNM Ferreira, GA Monterio, CL Cooney, JMS Cabral. Large-scale production of pharmaceutical grade plasmid DNA for gene therapy: problems and bottlenecks. Trends Biotechnol 17:169–174, 1999.
61. A Fersht. Enzyme Structure and Mechanism. 2nd ed. New York: Freeman and Company, 1985, pp. 1–17.
62. JL Cleland, MF Powell, SJ Shire. The development of stable protein formulations: a close look at protein aggregation, deamination, and oxidation. Crit Rev Ther Drug Carrier Syst 10:307–377, 1993.
63. M Jackson, HH Mantsch. Beware of proteins in DMSO. Biochim Biophys Acta 1078(2):231–235, 1991.
64. T Knubovets, JJ Osterhout, AM Klibanov. Structure of lysozyme dissolved in neat organic solvents as assessed by NMR and CD spectroscopies. Biotechnol Bioeng 63:242–248, 1999.
65. MA Winters, BL Knutson, PG Debenedetti, HG Sparks, TM Przybycien, CL Stevenson, SJ Prestrelski. Precipitation of proteins in supercritical carbon dioxide. J Pharm Sci 85:586–594, 1996.
66. SD Yeo, PG Debenedetti, Y Patro, TM Przybycien. Secondary structure characterisation of microparticulate insulin powders. J Pharm Sci 83(12):1651–1656, 1994.
67. A Dong, SJ Prestrelski, SD Allison, JF Carpenter. Infrared spectroscopic studies of lyophilization- and temperature-induced protein aggregation. J Pharm Sci 84: 415–424, 1994.
68. MA Winters, PG Debenedetti, J Carey, HG Sparks, SU Sane, TM Przybycien. Long-term and high-temperature storage of supercritically-processed microparticulate protein powders. Pharm Res 14(10):1370–1378, 1997.
69. MA Winters, DZ Frankel, PG Debenedetti, J Carey, M Devaney, TM Przybycien. Protein purification with vapour-phase carbon dioxide. Biotech Bioeng 62:247–258, 1999.
70. R Sloan, ME Hollowood, W Ashraf, GO Humphreys, P York. Supercritical fluid processing: preparation of stable protein particles. In: M Perrut, P Subra, eds. Proceedings of the 5th Meeting on Supercritical Fluids, Nice, France, 1998. Vol. 1. International Society for the Advancement of Supercritical Fluids, pp. 301–306.
71. R Sloan, ME Hollowood, W Ashraf, GO Humphreys, P York. American Association of Pharmaceutical Scientists, Annual Meeting 1999, New Orleans, LA.

72. T Knubovets, JJ Osterhou, PJ Connolly, AM Klibanov. Structure, thermostability, and conformational flexibility of hen egg-white lysozyme dissolved in glycerol. Proc Natl Acad Sci USA 96:1262–1267, 1999.
73. MA Amiza, RKO Apenten. Urea and heat unfolding of cold-adapted Atlantic cod (Gadus morhua) trypsin and bovine trypsin. J Sci Food Agric 70:1–10, 1996.
74. J Zagrobelny, F Bright. In situ studies of protein conformation in supercritical fluids: trypsin in carbon dioxide. Biotechnol Progr 8:421–423, 1992.
75. RT Forbes, R Sloan, I Kibria, ME Hollowood, GO Humphreys, P York. Production of stable protein particles: A comparison of freeze, spray and supercritical drying. In: Proceedings of the IChemE World Congress on Particle Technology 3, Brighton, UK, 1998. Rugby, UK: Institution of Chemical Engineers, p. 180.
76. MJ Hageman. The role of moisture in protein stability. Drug Dev Ind Pharm 14:2047–2070, 1988.
77. PL Poole. The role of hydration in lysozyme structure and activity—relevance in protein engineering and design. J Food Eng 22:349–365, 1994.
78. R Sloan, M Tservistas, ME Hollowood, L Sarup, GO Humphreys, P York, W Ashraf, M Hoare. Controlled particle formation of biological material using supercritical fluids. In: M Poliakoff, MW George, SM Howdle, eds. Proceedings of the 6th Meeting on Supercritical Fluids, Nottingham, UK, 1999. International Society for the Advancement of Supercritical Fluids, pp. 169–174.
79. M Tservistas, MS Levy, MYA Lo-Lim, RD O'Kennedy, P York, GO Humphreys, M Hoare. The formation of plasmid DNA loaded pharmaceutical powders using supercritical fluid technology. Biotech Bioeng 72:12–18, 2001.
80. KL Toews, RM Shroll, CM Wai, NG Smar. pH-Defining equilibrium between water and supercritical CO_2. Influence on SFE of organics and metal chelates. Anal Chem 67:4040–4043, 1995.
81. T Suzuki, S Ohsumi, K Makino. Mechanistic studies on depurination and apurinic site chain breakage in oligodeoxyribonucleotides. Nucleic Acid Res 22:4997–5003, 1994.
82. JH Kim, TE Paxton, DL Tomasko. Microencapsulation of naproxen using rapid expansion of supercritical solutions. Biotechnol Progr 12:650–661, 1996.
83. SA Wilkins, P York, RJ Roberts, RC Rowe, IF McConvey. The formation of indomethacin: polymer coprecipitates by the solution enhanced dispersion by supercritical fluids (SEDS) process. Presented at 136th British Pharmaceutical Conference, Cardiff, UK, 1999.
84. JC Feeley. The design and characterisation of powder particles for respiratory drug delivery. PhD dissertation, University of Bradford, Bradford, UK, 1999.
85. M Hanna, P York. Particle engineering by supercritical fluid technologies for powder inhalation drug delivery. In: RN Dalby, PR Byron, SJ Farr, eds. Proceedings of Respiratory Drug Delivery V, Phoenix, AZ, 1996. Buffalo Grove, IL: Interpharm Press, Inc., pp. 231–239.
86. SC Bristow, BY Shekunov, P York. Determination of solubility of paracetamol in supercritical carbon dioxide. J Pharm Pharmacol 50:53, 1998.
87. Medicines Control Agency. Rules and Guidance for Pharmaceutical Manufacturers and Distributors. London: The Stationary Office, 1997.

13

Preparation and Processing of Nanoscale Materials by Supercritical Fluid Technology

Ya-Ping Sun, Jayasundera Bandara, and Jaouad M. Meziani
Clemson University, Clemson, South Carolina

Harry W. Rollins
Idaho National Engineering and Environmental Laboratory, Idaho Falls, Idaho

Christopher E. Bunker
Wright-Patterson Air Force Base, Ohio

Supercritical fluid (SCF) technology for the preparation and processing of materials, especially nanomaterials, is an emerging research area that has attracted considerable attention recently. This chapter contains an overview of the links between the two seemingly separate topics of SCF technology and materials chemistry and engineering; summarizes exciting applications and recent developments; and discusses the advantages, opportunities, and challenges associated with the use of SCF technology for advanced materials.

I. SCF METHODS AND MATERIALS-RELATED APPLICATIONS

A. Supercritical Fluids

An SCF is loosely defined as a solvent at a temperature above the critical temperature, where the fluid remains a single phase regardless of pressure. However, for practical purposes, such as high density for solubility considerations, fluids of interest are typically at near-critical temperatures. Among the most important properties of an SCF are the low and tunable densities, which can be easily varied from gas-like to liquid-like via a simple change in pressure at constant temperature, and the unusual solvation effects at densities near the critical density (often discussed in terms of solute-solvent and solute-solute clusterings). Numerous investigations have been conducted on SCFs [see, for example, (1–3)]. The results have demonstrated the unique advantages of using SCFs as alternative media for both chemical reactions and materials processing. Examples of the applications and related advantages are the use of SCFs (CO_2 and water, in particular) as environmentally benign solvents (2), the ability to tune selectively chemical reactions or processes (4–10), the enhancement of reaction rates due to the low viscosities or high diffusivities in the fluids (9,10), the ability to solvate or precipitate solutes selectively (11), and the production of fibers and powders via rapidly expanding the SCF solutions (12,13). The parameters for commonly used supercritical fluids are summarized in Table 1.

An SCF may be considered as being macroscopically homogeneous but microscopically inhomogeneous, consisting of clusters of solvent molecules and free volumes. With the macroscopic homogeneity, extremely wide variations in solvent properties may be achieved in a single SCF via changes in the fluid density. In fact, many investigations have focused on the dependence of solvation properties on density in supercritical fluids. For example, spectroscopic techniques, coupled with well-characterized molecular probes, have been employed to examine the local polarities at different densities in an SCF. Among the commonly used molecular probes were those for the Kamlet-Taft π^* polarity scale (14,15), pyrene (16,17), and molecules known to form a twisted-

Table 1 Physical Parameters of Commonly Used Supercritical Solvents

Compound	Boiling point (°C)	Crit. temp. (°C)	Crit. pressure (atm)	Crit. density (g/cm^3)
Ethylene	−10.8	9.2	49.7	0.218
Carbon dioxide	−78.5	31.0	72.8	0.468
Ethane	−88.7	32.2	48.2	0.203
Nitrous oxide	−88.5	36.4	71.5	0.452
Butane	−17.8	91.8	45.6	0.232
Propane	−42.1	96.6	41.9	0.217
Ammonia	−33.5	132.5	111.3	0.235
Acetone	56.0	235.0	46.3	0.277
Methanol	64.6	239.4	79.9	0.272
Ethanol	78.3	243.0	63.0	0.276
THF	65.0	267.0	51.2	0.321
Toluene	110.6	318.6	40.5	0.291
Water	100.0	374.1	217.6	0.322

intramolecular charge transfer (TICT) state (18). Results from these investigations suggest that the solvent strength of an SCF increases with increasing fluid density in a nonlinear fashion, deviating significantly from the prediction based on the classical dielectric-continuum theory (19–21). Interestingly, however, the pattern for the density dependence of solvation properties (determined using molecular probes based on drastically different mechanisms) is nearly universal among SCFs in different categories (from nonpolar to polar and from ambient to high temperature). The results prompted Sun and coworkers (19) to propose a three-density-region solvation model for solute–solvent interactions in SCFs (Figure 1). Experimentally, the solvent effects are strong (increasing significantly with density) in the gas-like region (approximately reduced density $\rho_r < 0.5$), nearly plateau-like in the near-critical region (approximately $0.5 < \rho_r < 1.5$), and moderately density dependent in the liquid-like region ($\rho_r > 1.5$) (19–21). According to this model, the density dependence of solvation in SCF solutions is governed by the intrinsic properties of the neat fluid over the three-density regions. The behavior in the gas-like region at low densities is probably strongly

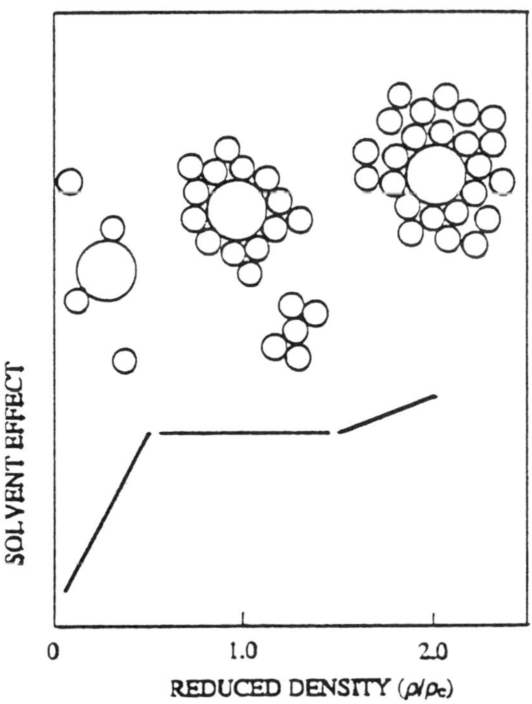

Figure 1 A cartoon illustration of the three-density-region solvation model. (From Ref. 19.)

influenced by short-range interactions in the inner solvation shell of the probe molecule. The strong density dependence of the spectroscopic and other responses is probably associated with a process of saturation of the inner solvation shell. Before saturation of the inner shell, the consequence of increasing the fluid density is microscopically the addition of solvent molecules to the inner solvation shell of the probe, which produces large incremental effects (Figure 1). In the near-critical region where the responses are nearly independent of density changes, the microscopic solvation environment of the solute probe undergoes only minor changes. Such behavior is probably due to the microscopic inhomogeneity of the near-critical fluid—a property that all SCFs share. Despite the dynamics, the fluid in the near-critical region can on average be viewed as consisting of solvent clusters and free volumes that possess liquid-like and gas-like properties, respectively. Changes in bulk density through compression primarily correspond to decreases in the free volumes, with solute–solvent interactions in the solvent clusters being largely unaffected. At the boundary of this region, the free volumes become less significant (consumed), and further increases in bulk density in the liquid-like region alter the microscopic solvation environment of the probe in a manner similar to that in normal liquid solvents, as predicted by the classical dielectric-continuum theory.

In addition to solute–solvent interactions, the effect of solvent local-density augmentation on solute–solute interactions in an SCF solution has been the subject of extensive investigations (13), with the focus being on whether the supercritical solvent environment facilitates solute-solute clustering, which may be loosely defined as the local solute concentration being higher than the bulk solute concentration. An important consequence of solute-solute clustering is the enhancement of bimolecular reactions in SCF solutions, which provides potentially significant opportunities for manipulating chemical reactions and processes under SCF conditions. The investigations employed well-established probes that are sensitive to bimolecular processes. The bimolecular processes and reactions examined in SCF solutions included the entrainer effect in SCF mixtures (22,23), excimer and exciplex formation and dynamics (24,25), photodimerization reactions (10,26), fluorescence quenching due to bimolecular diffusion (27,28), and various energy transfer reactions (29). The results seem to suggest that the solute-solute clustering is system dependent, which makes it difficult to confirm experimentally the existence of local concentration augmentation or solute-solute clustering in an unambiguous fashion. Thus, the effect of supercritical solvent environment on solute–solute interactions remains a somewhat controversial topic.

The highly compressible nature of SCFs (especially in the near-critical density region) has made them uniquely applicable in materials processing—in particular, the production of particles, fibers, and films via rapidly expanding solutions of polymers and other materials in SCFs.

B. Supercritical Fluid Processing Methods

Applications of SCFs in materials processing have received considerable attention since the mid-1980s. A number of reviews on the subject have appeared in the literature (30–51). One area of focus has been particle formation. Among the most widely investigated processing techniques are those involving precipitation processes in supercritical solutions known as SAS (*S*upercritical *A*nti*S*olvent) (32,52–54) and RESS (*R*apid *E*xpansion of *S*upercritical *S*olutions) (33,55–57). SAS generally refers to the precipitation for particle formation in a compressed fluid at supercritical as well as subcritical conditions. The process is also called PCA (*P*recipitation with *C*ompressed *A*ntisolvent) or GAS (*G*as *A*nti*S*olvent). As with any precipitation process, the antisolvent can be added to the solution (normal-addition precipitation) or the solution can be added to the antisolvent (reverse-addition precipitation). A typical SAS apparatus for particle preparation is illustrated in Figure 2. The method requires that the supercritical antisolvent be miscible with the solution solvent and that the solute be insoluble in the supercritical antisolvent. In the normal-addition SAS, a solute is dissolved in a liquid solvent, and a supercritical antisolvent is added to the solution in a partially filled, closed container that is initially at ambient pressure. With the addition of the supercritical antisolvent, both the volume of the solution/antisolvent mixture and the pressure of the closed container increase. The decrease in solubility of the solute with increasing antisolvent fraction in the mixture results in precipitation of the solute. The precipitate is then washed with the antisolvent to yield the desired particles. The size and size distribution of the particles are dependent on the selection of the solution/antisolvent system, the solution concentration, the relative solution and antisolvent quantities, the rate of the antisolvent addition, and the degree of mixing (52). In the reverse-addition SAS, a liquid solution is sprayed through a nozzle into a supercritical antisolvent. The rapid diffusion of the solvent from the solution droplets sprayed into the bulk SCF results in the solute precipitation. The precipitate is then washed with the antisolvent and filtered to obtain the desired particles.

The SAS methods have been used for preparing a variety of particles and fine powders from proteins, pharmaceuticals, pigments, polymers, and even explosives. For example, Debenedetti and coworkers used a continuous-flow, supercritical antisolvent process to prepare fine powders of trypsin, lysozyme, and insulin proteins (58–60). In the preparation a protein solution in dimethylsulfoxide (DMSO) was sprayed through a small orifice into supercritical CO_2. The particles had diameters ranging from 1 to 5 μm. The biological activity of the micrometer-sized powders was nearly the same as that of the starting materials. The method has also been used in the processing of pharmaceutically important compounds, such as salmeterol xinafoate (61), sulfathiazole (62), and methylprednisolone and hydrocortisone acetate (41). Kitamura et al. used the

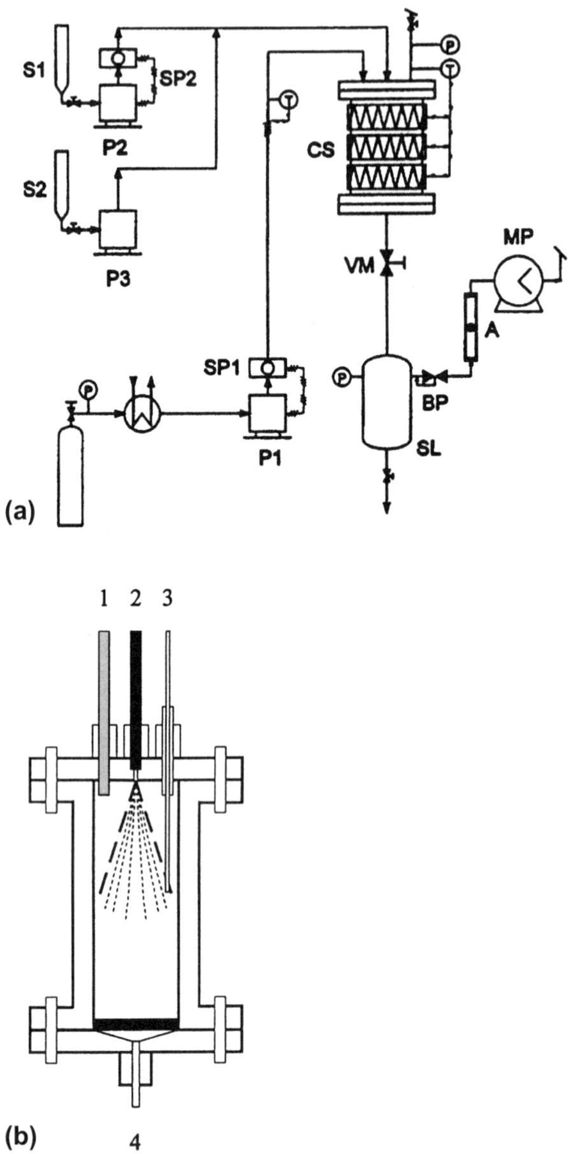

Figure 2 (a) Schematic representation of an SAS apparatus. P1, P2, and P3, high-pressure pumps; SP1 and SP2, pressure dampeners; S1 and S2, liquid solution supplies; CS, precipitation vessel; VM, micrometering valve; BP, back-pressure regulator; SL, liquid separator; A, calibrated rotameter; MP, wet test meter. (b) Schematic representation of the precipitation chamber. 1, Supercritical CO_2 inlet; 2, liquid solution inlet; 3, pressure and temperature measurements; 4, precipitator outlet. (From Ref. 13.)

stepwise addition of supercritical antisolvent to an ethanol solution of sulfathiazole to control the nucleation and particle growth processes (62). Sulfathiazole crystals in a wide size range (10–100 μm up to 2–6 mm) were obtained by varying the timing and supercritical antisolvent addition. Increasing the pressure of the initially added antisolvent resulted in faster nucleation and smaller particles. Pigments were prepared via SAS precipitation by Gao et al. (63). Particles of several commercial pigments were obtained using acetone solutions and compressed CO_2 as an antisolvent. The particle sizes were on the order of 1 μm. Larger particles could be achieved by decreasing the pressure or increasing the temperature of the antisolvent or by increasing the expansion nozzle size. For polymers, Johnston and coworkers used antisolvent precipitation to prepare both microparticles and fibers of polystyrene (64–67). In the reverse-addition mode, toluene solution of the polymer was sprayed into compressed CO_2 (67). The particle size was adjusted in the range of 100 nm to 20 μm by varying the CO_2 density and temperature, with higher CO_2 density and lower temperature resulting in smaller particles (Figure 3). The product morphology (particles vs. fibers) was found to be dependent on the polymer–solution concentration, with a higher concentration (1–5 wt %) promoting the formation of fibers. When a small amount of CO_2 was added to the polymer solution before expansion, porous particles and fibers were obtained (66). The application of SAS to the processing of explosives was described by Gallagher et al. (44). Fine crystals (<200 μm) were obtained from cyclotrimethylenetrinitramine (RDX) using subcritical or supercritical CO_2 as an antisolvent with solutions of a variety of solvents (44).

A significant advantage of the SAS and related SCF processing methods over a liquid solution–based technique is the ability to prepare dry powders in a single step (44). The preparation of dry powders at low temperature is particularly important for pharmaceuticals and protein samples as well as other materials that are thermally labile or shock sensitive. Because many compounds of interest have higher solubility in liquid solvents than in low-temperature SCFs, the SAS method generally allows higher throughputs than the RESS method.

The RESS process differs from the SAS process in that in RESS the solute is dissolved in an SCF and then the solution is rapidly expanded through a small nozzle or orifice into a region of lower pressure (30–33). The rapid reduction in pressure—and, thus, density—results in rapid precipitation of the solute. Experimentally, the supercritical solution can be generated either by heating and pressurizing a solution from room temperature or by continuously extracting the solute using an extraction column (30). The room-temperature method allows the expansion to be performed at a known constant concentration, whereas the extraction method is useful for solutes that are insoluble or sparingly soluble in the solvent at lower temperatures. In the latter method, the temperature of the extraction column may be the same as or different from the temperature at which

Figure 3 SEM micrographs of polystyrene spheres formed by spraying 1% polystyrene solution in toluene through a 100-µm nozzle into liquid CO_2 (density of 0.86 g/cm^3) at (a) 10°C and 45.8 bar and (b) 30°C and 133.7 bar. (From Ref. 67.)

the expansion is carried out. The extraction temperature and flow rate may also be used to control the solution concentration. The RESS process is driven by the decrease in pressure, which can propagate at speeds up to the speed of sound in the expansion nozzle. Because solubilities in SCFs can be up to a million times higher than those under ideal gas conditions, the rapid expansion from supercritical pressure to ambient pressure results in extremely high supersaturation—and, consequently, homogeneous nucleation of the solute—leading to narrow size distributions in the products. The RESS technique can be used to obtain several product morphologies, including particles, fibers, and films from a variety of inorganic, organic, and polymeric materials.

C. RESS for Particles, Fibers, and Films

The pioneering investigation in the use of RESS for particle production was conducted by Krukonis (55). In 1984, Krukonis reported the preparation of small particles and fibers via RESS from several classes of materials, including aluminum isopropoxide, dodecanolactam, polypropylene, β-estradiol, ferrocene, navy blue dye, and soybean lecithin (55). Since that time, many investigations have been focused on RESS and its application to materials preparation and processing. Some of the more representative investigations and the associated key results are summarized here.

1. Inorganic Materials

Following the report of Krukonis on the formation of micrometer-sized inorganic particles via RESS, Smith and coworkers carried out a series of RESS experiments aimed at not only processing of various materials but also evaluation of the effects of processing conditions on product morphology (particle, fiber, or film), size, and distribution (56,57,68–72). In the processing of inorganic oxides, supercritical water was used as a solvent. For example, silica particles less than 100–500 nm in diameter were prepared via the rapid expansion of a supercritical water solution at 470°C using a nozzle of a 60-μm inner diameter (Figure 4) (69,71,73,74). However, the product morphology was found to be sensitive to the RESS parameters, especially the preexpansion fluid temperature. In RESS with supercritical water solution, a small reduction in the temperature from 470°C to 450°C resulted in the formation of silica films rather than particles (70). The dramatic effects of temperature on the product morphology were rationalized in terms of the formation and agglomeration of fluid droplets in the expansion jet (70,72). Although the particle size could be manipulated to some extent by varying the preexpansion solution concentration (69,70), the inorganic particles obtained from RESS were generally in the submicrometer domain.

Another inorganic particle that has been prepared via RESS is iron oxide (Fe_2O_3) (75). The preparation involved rapid expansion of a $Fe(NO_3)_3$ solution in supercritical water. The expansion was at 500°C and 100 MPa through 50- to 200-μm-diameter orifices into an evacuated chamber. The Fe_2O_3 particles thus obtained were small and exhibited exceptional reactivities. In addition to inorganic oxides, several neutral metal carbonyls (chromium hexacarbonyl, dimanganese decacarbonyl, and triiron dodecacarbonyl) were processed via RESS to form micrometer-sized particles (76). The solubility of these compounds allowed the use of supercritical CO_2 in the RESS processing.

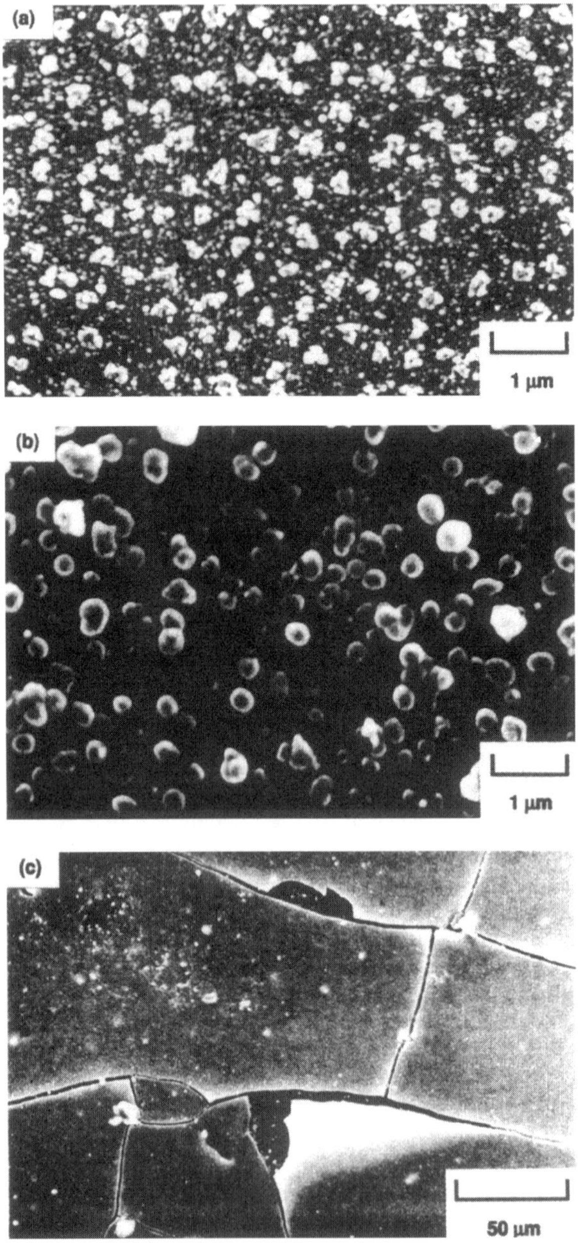

Figure 4 Silica products prepared using the RESS process. (a) and (b) SiO_2 powders produced using different concentrations of silica in the supercritical water solution, and (c) an SiO_2 film produced using a two-phase RESS expansion. (From Ref. 73.)

RESS was also used as an alternative to the conventional chemical vapor deposition (CVD) process in the preparation of thin films (77–79). A general procedure involved rapid expansion of an SCF solution onto a heated substrate in a vacuum chamber. For example, palladium thin films were prepared via the rapid expansion of a bis-(2,2,7-trimethyl-3,5-octanedionato)palladium(II) solution in supercritical pentane at 205°C and 1200 psia onto a silicon substrate heated to 515–740°C (77,78). High-quality InP films were obtained by a similar procedure (79).

2. Polymeric Materials

RESS has been particularly popular for the processing of polymeric materials. According to the original report by Krukonis (55), rapid expansion of a polypropylene solution in supercritical propylene resulted in the formation of fiber-like particles. The report marked the beginning of an extensive discussion on the issue of particles vs. fibers involving the use of RESS with polymers.

Smith and coworkers reported the preparation of both particles and fibers via RESS from a variety of polymers (57,68–72,80). For example, a solution of polystyrene (molecular weight about 300,000 and melting point about 170°C) in supercritical pentane was rapidly expanded at 350°C and 170 bar using a 25-μm nozzle; the result was spherical particles with an average diameter of 20 μm (68). Other polymers, including polypropylene, poly(carbosilane) (an important precursor for silicon carbide), poly(phenylsulfone), poly(methyl methacrylate), and cellulose acetate, were processed into micrometer-sized particles via RESS in a similar fashion (68). However, when the preexpansion temperature of the supercritical pentane solution was lowered from 350°C to 200°C, polystyrene fibers (100–1000 μm in length and 1 μm in diameter) were obtained.

Among the RESS parameters, the preexpansion solution temperature apparently had the most significant impact on product morphology (particle vs. fiber). When the preexpansion temperature was near or far from (significantly higher or lower than) the melting point of the polymer, fibers or particles were produced from RESS, respectively. The intermediate preexpansion temperatures resulted in mixtures of fibers and particles (68,73). The results showed that the formation of fibers from RESS was somewhat more difficult, requiring more specific processing conditions (the preexpansion temperature being near the polymer melting point). On the other hand, the polymer-solution concentration was found to have only a minor effect on product morphology (68,73).

Smith and coworkers also examined several technical issues concerning the production of fibers via RESS. For example, they found that solvent-free fibers could be produced under the appropriate expansion conditions and that the fractionation of the polymer sample due to selective solvation of lower molecular weight species could be avoided, yielding products with the same properties as those of the starting polymer sample. They also found that the fiber diameters were generally smaller than those of the expansion nozzles and uniform along the fiber length and that the diameters were affected by the nozzle size (72).

Lele and Shine reported the preparation of fibers and particles of polycaprolactone, poly(methyl methacrylate), and a block copolymer styrene-methyl methacrylate via RESS (81,82). Chlorodifluoromethane was used as solvent. The emphasis of the study was an evaluation of the effects of RESS processing conditions on product morphology (fibers vs. particles). They reported that the formation of fibers was promoted by a high polymer concentration, a high preexpansion temperature, a low preexpansion pressure, and a low nozzle length/diameter ratio (Figure 5). They explained their findings on the basis of solubility data, stating that the physical location in the nozzle at which the onset of solute precipitation occurs during the expansion process determines the product morphology. For example, precipitation occurring farther upstream in the expansion nozzle results in fiber formation. According to Lele and Shine, the dynamics of the precipitation process plays an important role. While the density reduction associated with the rapid expansion process is on the order of microseconds, the time scale of the precipitation process can vary from several seconds to a fraction of a microsecond, depending on where the precipitation occurs in the expansion process (70,81). Precipitation in the entrance region of the capillary nozzle probably results in fibers, whereas precipitation inside the nozzle results in particles.

The production of polymer fibers at temperatures below melting or glass transition temperatures was explained by Lele and Shine in terms of fluid-fluid phase separation (81,82). Cloud point curves were used to determine the onset of the liquid-liquid phase separation in polymer solutions. Solidification or vitrification occurs when the glass transition temperature or melting temperature reaches the local process temperature. By correlating the process time scale (time between the initiation of phase separation and vitrification), Lele and Shine found that the processes that crossed the cloud point curve before the nozzle entrance resulted in fibers and those that crossed the curve inside the expansion nozzle resulted in particles. Thus, they proposed that the formation of large polymer particles via RESS reported by others could be attributed to a slow process in the preheating zone (82). They also noted that those observations were all based on experiments in which the solutions were significantly heated prior to expansion. If the solute was first extracted at a lower temperature and then heated at a constant pressure, the decrease in density could result in the solute precipitating in the preheating zone and the formation of large particles (82).

Similar effects of RESS processing conditions on product morphology were observed by Mawson et al. in their preparation of fibers and particles of a crystalline fluoropolymer, poly(1,1,2,2-tetrahydroperfluorodecyl acrylate) via the rapid expansion of supercritical CO_2 solution (83). Shown in Figure 6 are

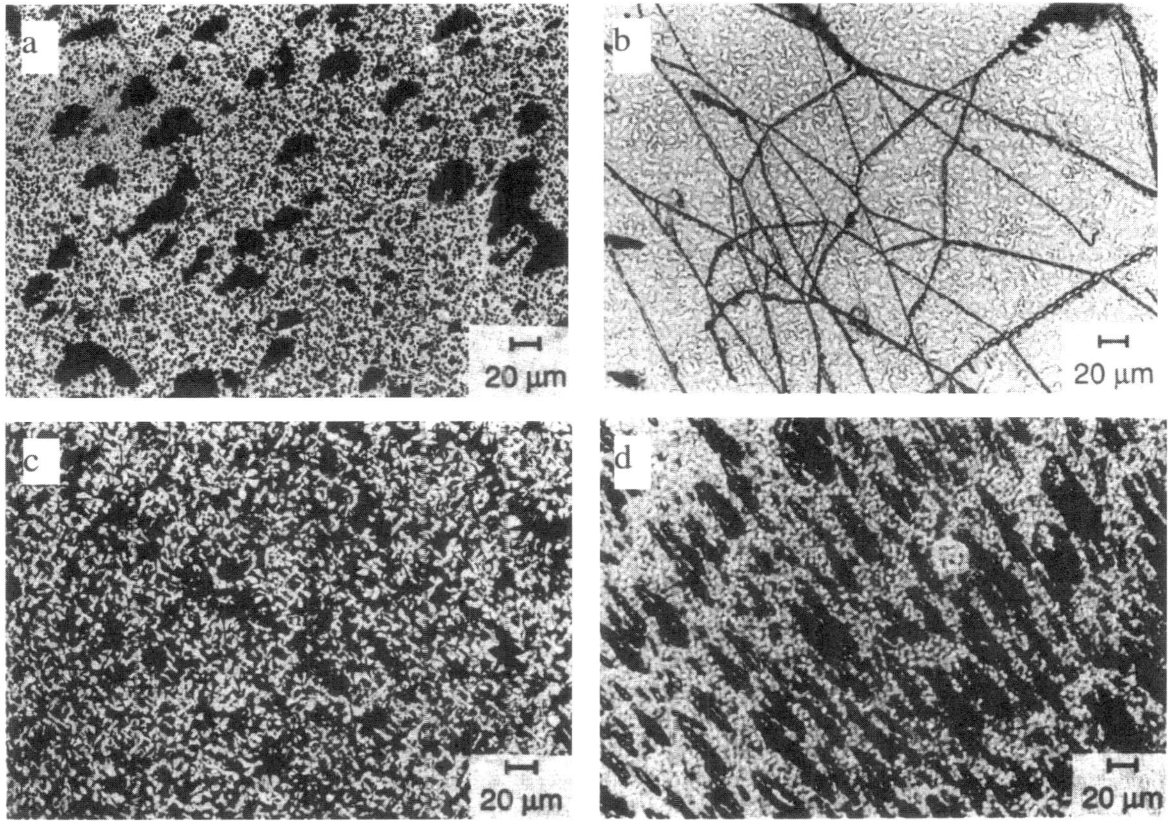

Figure 5 Optical photomicrographs of polycaprolactone precipitated from chlorodifluoromethane at 13.8 MPa and (a) 90°C (50-μm capillary nozzle), (b) 110°C (50-μm capillary nozzle), (c) 145°C (50-μm capillary nozzle), and (d) 110°C (30-μm capillary nozzle). (From Ref. 81.)

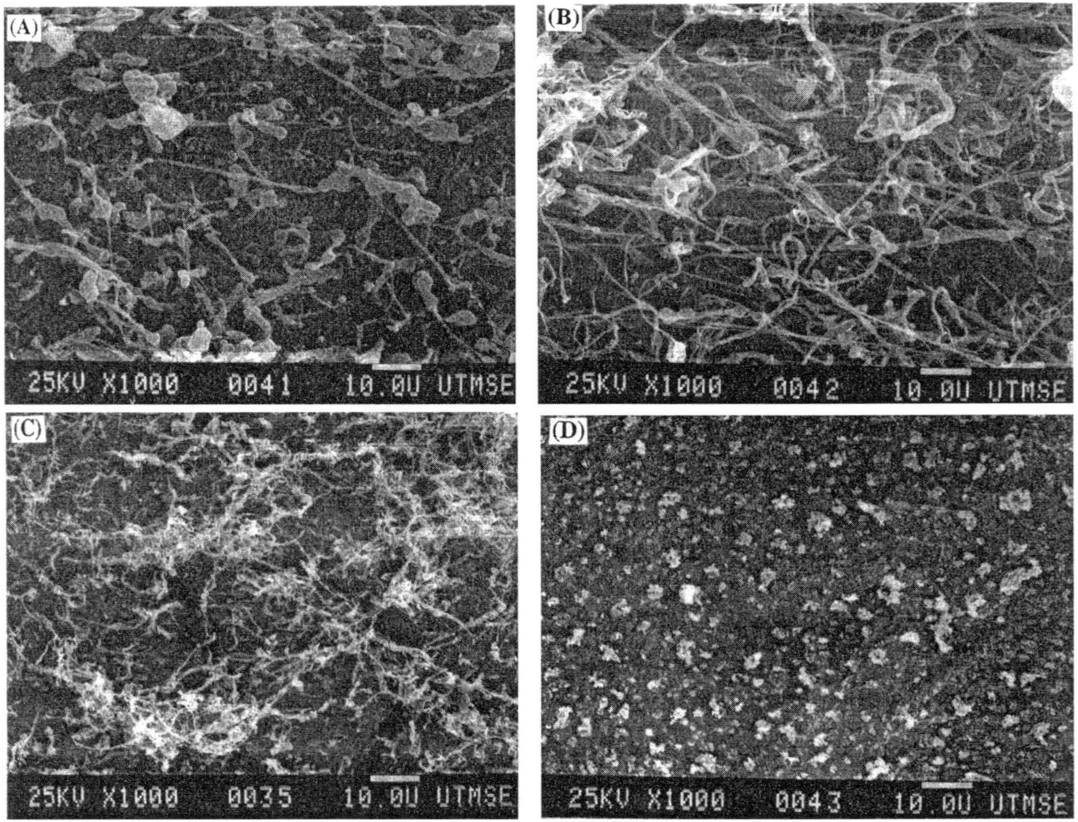

Figure 6 Poly(1,1,2,2-tetrahydroperfluorodecyl acrylate) microparticles collected from the RESS of a 0.5 wt % poly(1,1,2,2-tetrahydroperfluorodecyl acrylate)/CO_2 solution sprayed through a 50-μm capillary at pre-expansion temperatures of (A) 105°C, (B) 85°C, (C) 65°C, and (D) 45°C. (From Ref. 83.)

the poly(1,1,2,2-tetrahydroperflurodecyl acrylate) microparticles collected from the RESS at different preexpansion temperatures. Small polymer fibers with diameters as small as 0.3 μm were obtained.

Aniedobe and Thies found that in addition to the processing conditions discussed by Lele and Shine (81,82), concentration is an important parameter for the formation of fibers rather than particles via RESS (84). In their investigation cellulose acetate was dissolved in methanol at elevated temperatures (105–125°C) to form homogeneous solutions of high concentrations (14 wt %). The rapid expansion with a preexpansion temperature of 240°C and a pressure of 80–100 bar yielded cellulose-acetate continuous fibers that were 2–3 μm in diameter, with each individual fiber strand consisting of several entwined fibers (Figure 7). An interesting aspect of the investigation was the constant concentration expansion; methanol was used to push the solution from the dissolution chamber into the preheating zone in a plug-flow manner. After an expansion of several minutes, a mixing of the solution with fresh methanol resulted in the formation of both fibers and particles. As the solution was continuously diluted via mixing with fresh methanol, the product morphology underwent a transition from fibers to particles and, eventually, to films. It was thus concluded that concentration had significant effect on the product morphology (84).

In addition to particles and fibers, poly(methyl methacrylate), poly(ethyl methacrylate), and poly(L-lactic acid) films were prepared via RESS (61). The required condition for films is that the process path cross the vapor–liquid equilibrium line for the solute before the onset of precipitation (cloud point curve)—a condition similar to that for the preparation of inorganic oxide films (71). By controlling the processing conditions for inducing rapid precipitation in the nozzle, Lele and Shine were able to prepare a well-mixed (homogeneous on a scale estimated to be less than 30 nm) composite of two otherwise immiscible polymers: poly(methyl methacrylate) and poly(ethyl methacrylate) (81,82). The composite has a single glass transition temperature that is intermediate between those of the two components. However, a separation of the two phases can be induced at higher temperatures via thermal annealing, which results in two glass transition temperatures equal to those of the two components.

3. Organic Materials and Pharmaceuticals

The use of RESS in the processing of pharmaceuticals is driven, in large part, by the requirement for particles of a narrow size distribution. This requirement with regard to pharmaceutical particles is related to issues such as reproducibility and uniform solvation of drugs in the body. In the search for more favorable processing conditions through evaluations of various experimental parameters, organic materials such as polyaromatics are often used as models for pharmaceuticals. For example, Debenedetti and coworkers used naphthalene in supercritical CO_2

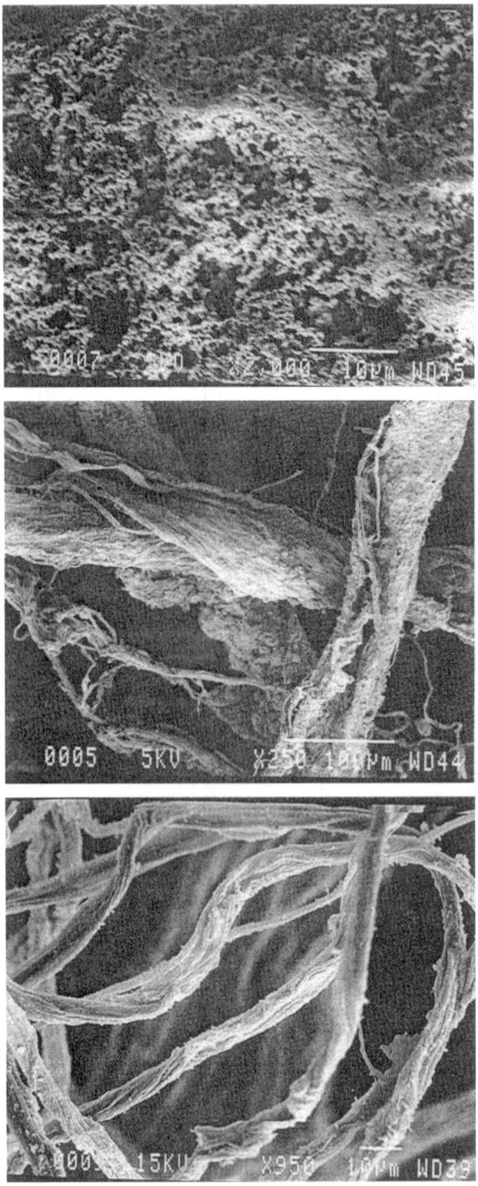

Figure 7 Agglomerated cellulose acetate (CA) powders produced from the rapid expansion of dilute (<1 wt %) CA/methanol solutions at 240°C and 81 bar (top); and CA fibers produced from the rapid expansion of methanol/CA solutions at 240°C and 82 bar with concentrated (14–15 wt %) CA/methanol solutions (middle) and more dilute CA/methanol solutions (bottom). (From Ref. 84.)

to investigate the dependence of product morphology on RESS processing conditions such as solution concentration and pre- and postexpansion temperatures and pressures (Figure 8) (85).

Generally, particle size increases with preexpansion temperature. For example, an increase in the preexpansion temperature from 110°C to 170°C resulted in an increase in particle size from 8–75 to 38–225 μm (85). The effect is also solution-concentration dependent. At a lower concentration the same change in the preexpansion temperature resulted in only a slight increase in particle size from 2–35 to 4–51 μm. Similarly, the product morphology is dependent on the postexpansion temperature, with a lower temperature favoring smaller particles and a narrower particle size distribution. On the other hand, the effect of the postexpansion pressure is insignificant, with only a slight decrease in particle size being observed with increasing postexpansion pressure.

In the same study, Debenedetti and coworkers evaluated the effect of naphthalene concentration on particle size (85). For two samples prepared under the same conditions, an increase in naphthalene mole fraction from 2.6% to 4.44% resulted in a decrease in particle size from 30–135 to 6–32 μm. The concentration effect was rationalized such that a higher naphthalene concentration should lead to a higher supersaturation ratio in the expansion jet and, therefore, a higher nucleation rate. Theoretically, the particle volume is inversely proportional to the

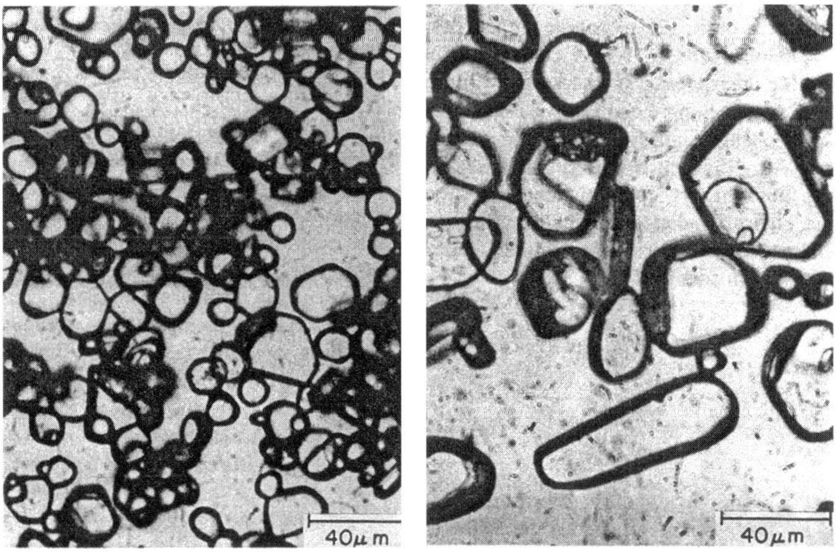

Figure 8 Photomicrographs of powders obtained with preexpansion naphthalene mole fractions of (left) 5.73% and (right) 1.88%. (From Ref. 85.)

nucleation rate, which means that a larger number of smaller particles is formed with a more concentrated naphthalene solution.

Several other groups used naphthalene to investigate the effects of RESS processing parameters (86–88). Liu and Nagahama examined the effects of pre- and postexpansion temperatures, postexpansion pressure, naphthalene concentration, and orifice diameter (86). An increase in the preexpansion temperature from 87°C to 120°C resulted in larger particle size and a broader size distribution, whereas a decrease in the postexpansion chamber temperature from 45°C to 15°C resulted in a smaller particle size and a narrower size distribution. Postexpansion chamber pressure had a negligible effect when varied in the 0.1- to 4.36-MPa range. For the naphthalene concentration, a larger particle size was obtained with a higher concentration. These results are in general agreement with those of Debenedetti and coworkers (85). In addition, the orifice diameter was found to have little effect on particle properties (86). Liu and Nagahama also broadened the scope of their investigation to include naphthalene-phenanthrene mixtures (88). An interesting finding was the dependence of morphology on particle composition for the naphthalene-phenanthrene composite particles. A systematic change from plate-like particles for pure naphthalene to needle-like particles for pure phenanthrene was observed. The investigation also included a study of the anthracene-phenanthrene system.

The effects of RESS processing conditions on product morphology were also evaluated using other materials and systems. For example, the salicylic acid-CO_2 system was examined by Reverchon et al. under various conditions (89). The primary product morphology was needles that were 5–170 μm in length and 1–10 μm in diameter. Similar to the effect of naphthalene concentration on particle size discussed above, a higher salicylic acid concentration favored smaller needles, causing decreases in both length and diameter. Pre- and postexpansion temperatures affected particle size and size distribution, respectively. An increase in the preexpansion temperature from 100°C to 140°C resulted in an increase in the needle length from 5–15 μm to 30–170 μm and the needle diameter from 1 μm to 6–8 μm. The increase in needle size with increasing preexpansion temperature was accompanied by a broadening in the size distribution for both needle length and diameter. On the other hand, a change in the postexpansion chamber temperature from 30°C to −10°C resulted in gradual decreases in the needle length and diameter and eventually a transition from needle-like to particle-like products; these decreases were also accompanied by decreased size distributions for both needle length and diameter. Shown in Figure 9 are optical micrographs of salicylic acid crystals obtained via RESS at 10°C and 0°C expansion chamber temperatures. These results agree well with those of Debenedetti and coworkers concerning the effects of RESS processing parameters on naphthalene particles (85). It appears that some tuning of particle (needle) sizes and size distributions may be achieved by varying the processing

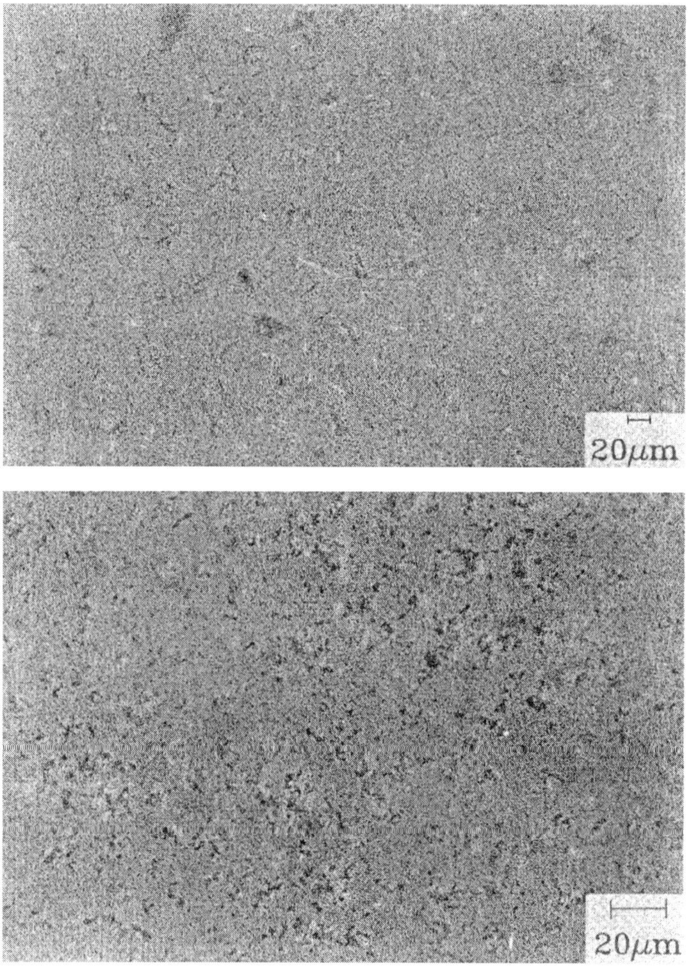

Figure 9 Optical photomicrographs of salicylic acid crystals obtained by RESS with (top) expansion chamber temperature = 10°C and (bottom) expansion chamber temperature = 0°C. (From Ref. 89.)

parameters. Reverchon and coworkers also investigated the RESS processing of the antimicotic compound griseofulvin using trifluoromethane and found that the preexpansion temperature had the most significant effect on product morphology (90,91). The RESS process at a preexpansion temperature of 60°C produced griseofulvin needles 13–36 μm in length and 1.1 μm in diameter, but the same process at a preexpansion temperature of 150°C yielded griseofulvin particles ~1 μm in diameter.

Similarly, Alessi et al. investigated the processing of the steroid drugs progesterone and medroxyprogesterone acetate via RESS with CO_2 and evaluated the effects of processing conditions such as solution concentration, postexpansion temperature and pressure, and nozzle diameter (92). For progesterone, an increase in the preexpansion solution concentration from 45.5 to 135 μM resulted in a decrease in the average particle size from 7.5 to 6 μm. Changes in the postexpansion temperature and pressure had only minor effect on particle properties, with a slight decrease in particle size for a temperature increase of 40–60°C and a more significant decrease in particle size (average 9.1–7.5 μm) for a pressure decrease of 50 – 1 bar. However, while the particle size distribution was negligibly affected by the change in postexpansion pressure, the product morphology was significantly changed. The materials prepared at a higher postexpansion pressure of 50 bar were needle-like, with a dendritic structure, whereas those prepared at a lower pressure of 1 bar were more particle-like (Figure 10). The effect of nozzle diameter was more pronounced, with a decrease in average particle size from 7.4 to 4.1 μm with a change in nozzle diameter from 100 to 30 μm.

Many other studies have been reported on the RESS processing of organic materials and pharmaceuticals and on the effects of processing conditions. Griscik et al. used the RESS method with CO_2 in the processing of octacosane (93). Ohgaki et al. reported the RESS processing of sigmasterol in CO_2 (94). Chang and Randolph prepared submicrometer-sized β-carotene particles via RESS with supercritical ethane, ethylene, and ethylene with 1.5% toluene as cosolvent (95). Their attempts to use CO_2 as a solvent were unsuccessful because of reactions between the solute and solvent. Interestingly, the expansion of β-carotene solution in ethylene at 70°C and 306 atm produced particles of an average size of 1 μm and a size distribution of 36%, whereas the expansion under the same preexpansion conditions into an aqueous gelatin solution yielded particles with an average size of 0.3 μm and a size distribution of 34% (95).

Domingo et al. investigated the use of porous frits as alternatives to capillary nozzles in the RESS processing of benzoic acid, salicylic acid, aspirin, and phenanthrene in supercritical CO_2 (96,97). Under their experimental conditions, expansions with porous frits resulted primarily in particles, whereas those with capillary nozzles resulted in more needle-like products (Figure 11). The particles obtained with frit nozzles were generally smaller than those obtained with capillary nozzles. An excessive clogging problem experienced with the frit nozzles made it impossible to use these nozzles in the processing of aspirin (96).

4. Particles of Mixtures and Composites

The RESS method has also been used in the preparation of multicomponent particles. One example was particles of KI-containing composites, including

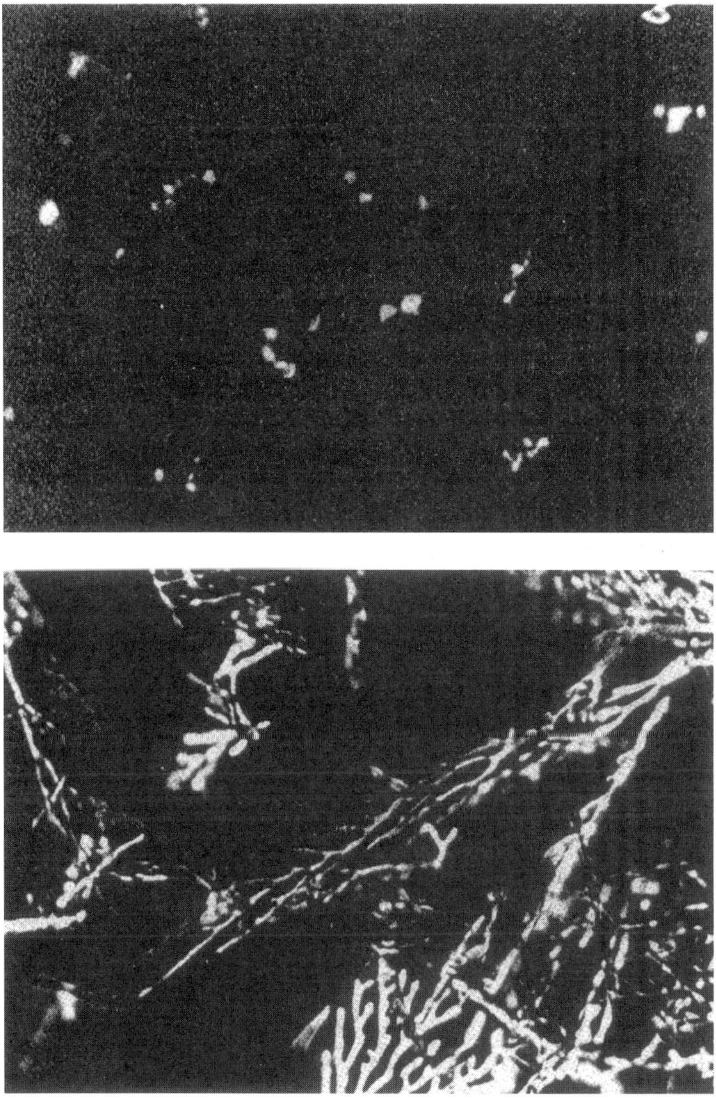

Figure 10 Effects of preexpansion and postexpansion conditions on particle size and size distribution. (top) Particles obtained by RESS with saturation pressure = 150 bar, saturation temperature = 60°C, postexpansion pressure = 1 bar, and postexpansion temperature = 40°C. (bottom) Particles obtained by RESS with saturation pressure = 150 bar, saturation temperature = 60°C, postexpansion pressure = 50 bar, and postexpansion temperature = 40°C. (From Ref. 92.)

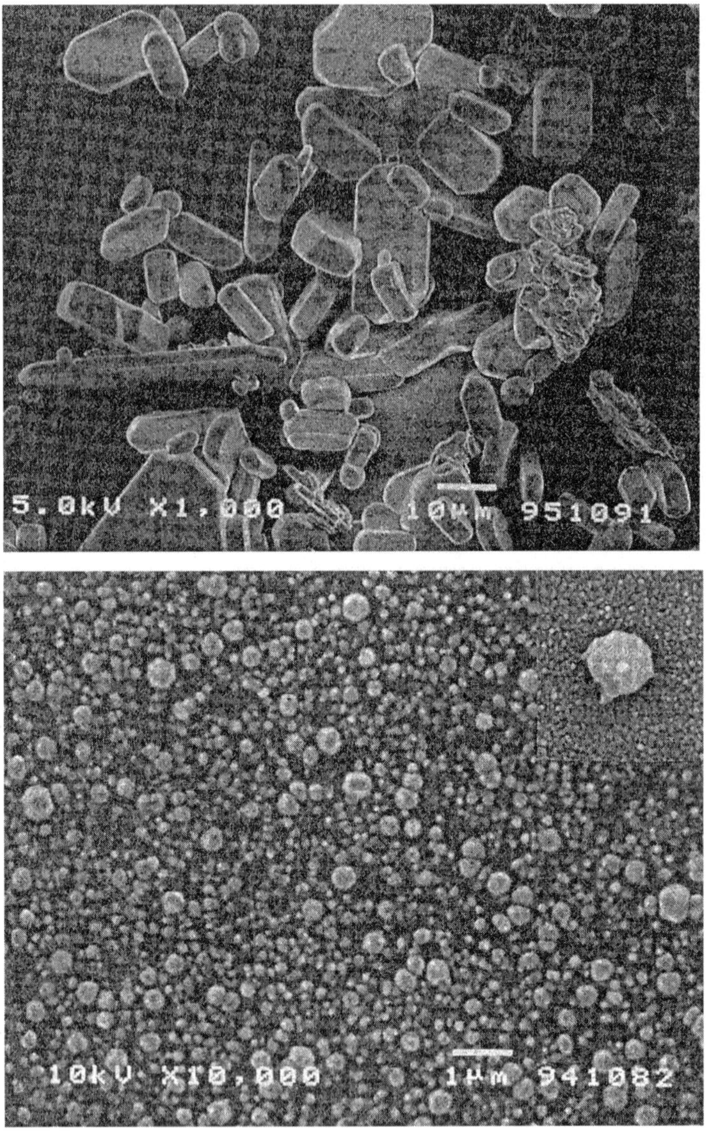

Figure 11 SEM photographs of benzoic acid crystals obtained using (top) a capillary nozzle and (bottom) a 0.5-μm pore size frit. (From Ref. 96.)

KI-SiO_2 and KI-poly(vinyl chloride) (69,71,73). For KI-SiO_2, a KI solution (1–10 mM) was pumped through the extraction region containing SiO_2 at 385°C, followed by rapid expansion at 580 bar and 500°C to yield agglomerates (~20 μm) of small particles. The presence of both KI and SiO_2 in agglomerated particles was confirmed via x-ray fluorescence characterization. Similarly, KI-poly(vinyl chloride) composite spheres of about 7 μm in diameter were prepared using a KI solution in ethanol to dissolve poly(vinyl chloride) in the extraction zone at 250°C, followed by rapid expansion at 170 bar and 350°C via a nozzle of 25 μm inner diameter. X-ray fluorescence was again used in the characterization of the composite particles.

An important feature of the RESS method is that it allows the preparation of composite particles in which components with very different properties are intimately mixed. One of the most promising applications of such composite materials is controlled drug release using polymer-based particles. For example, Tom and Debenedetti reported RESS processing of several polyhydroxy acids into particles for potential applications in controlled release of pharmaceuticals (98). Benedetti et al. reported the preparation of micrometer-sized particles of the biocompatible polymer hyaluronic acid benzylic ester via RESS and compared the results with those obtained using the SAS method (99). Debenedetti et al. successfully prepared poly(D,L-lactic acid) microparticles encapsulating lovastatin (an anticholesterol drug) (100). In the preparation the polymer and drug were extracted simultaneously using supercritical CO_2 at 200 bar and 55°C, followed by rapid expansion of the solution at 75–80°C to produce polymer particles containing lovastatin needles. The lovastatin concentration was estimated to be about 20%.

For controlled drug release applications, the ideal product morphology is uniform spherical drug particles homogeneously distributed throughout bioerodible polymer particles. In an attempt to examine the morphological issue, pyrene was used as a model "drug" for encapsulation in poly(L-lactic acid) microparticles via RESS (101,102). Pyrene was selected for its well-understood fluorescence properties, which allow easy characterization using fluorescence microscopy. In the preparation two extraction columns were used in parallel, with poly(D,L-lactic acid) being extracted with a CO_2-$CHClF_2$ mixture at 200 bar and 55°C in one column and pyrene being extracted with pure CO_2 at 200 bar and 65°C in the other column. The pyrene concentration was adjusted to the desired level through dilution with CO_2. The two solutions were mixed and heated to the preexpansion temperature. The rapid expansion of the final solution through a 50-μm capillary nozzle with a length/diameter ratio of 200 produced poly(D,L-lactic acid) microspheres that were uniformly dispersed with pyrene (101). Fluorescence was more significant at a higher pyrene concentration (Figure 12).

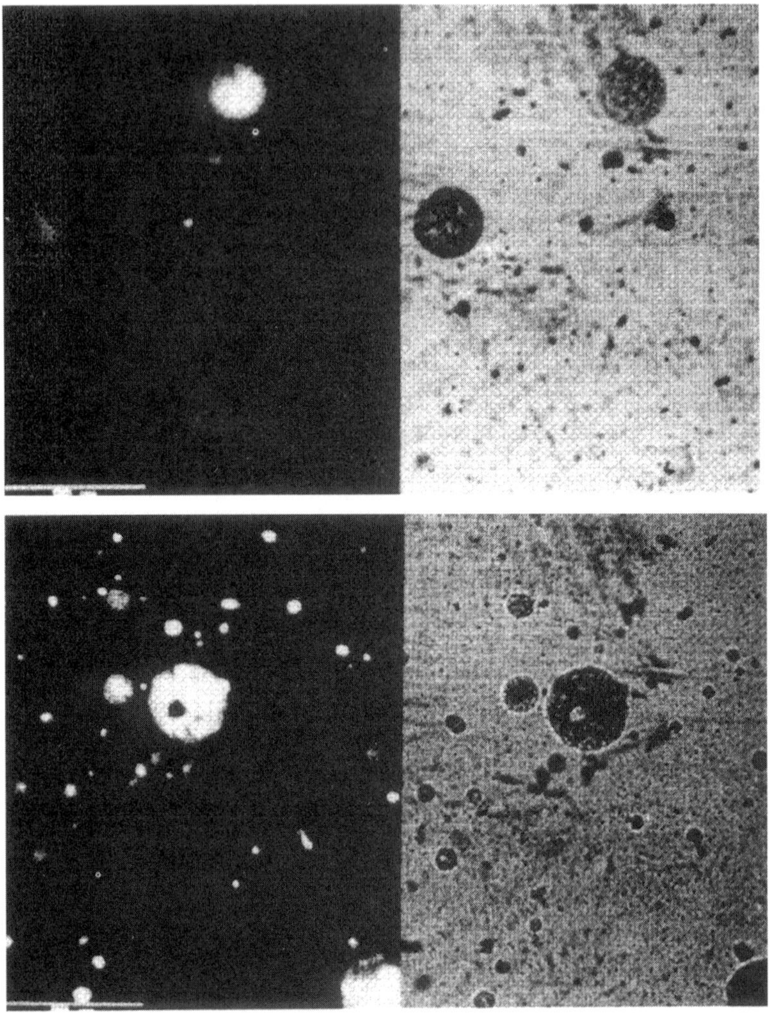

Figure 12 Fluorescence (left) and transmission (right) images of poly(L-lactic acid)-pyrene microspheres precipitated at pyrene concentrations in CO_2 of (top) ~0.0013 wt % and (bottom) >0.002 wt %. (From Ref. 101.)

Kim et al. also used poly(L-lactic acid) microparticles to encapsulate naproxen (6-methoxy-α-methyl-2-naphthaleneacetic acid) (103). The two solutes were extracted simultaneously with supercritical CO_2 using an extraction column. The rapid expansion of the supercritical solution at 190 bar and 114°C through a capillary nozzle (50 μm inner diameter) yielded poly(L-lactic acid) particles (10–90 μm in diameter) embedded with smaller naproxen particles (2–5 μm in diameter).

RESS is a versatile processing technique for a variety of materials, ranging from inorganic and organic compounds to thermally sensitive proteins and pharmaceuticals and from polymers to composites. Product morphology can be varied widely under a variety of processing conditions. The properties of the products can also be controlled to some extent. An additional important feature of the RESS method is the ability to prepare nonequilibrium mixtures and composites for different combinations of materials. Significant efforts in the modeling and simulation of the supercritical processes have provided an understanding of the relationships between processing conditions and product properties (32–34,81,82,104–109).

An emerging application of RESS and related methods is the preparation and processing of nanoparticles and other nanoscale materials and systems. This application will be discussed in detail, following a review of the relevant (more "conventional") preparation methods for nanomaterials.

II. METHODS FOR NANOPARTICLES AND RELATED MATERIALS

The development of methods for the preparation of nanoparticles has received considerable attention. The ideal method would be one that is versatile, applicable to a wide range of materials, readily implemented, inexpensive, and scalable, and one that would allow control of particle size, size distribution, shape, crystallinity, and particle surface properties. Although no single technique meets these objectives, methods have been developed for both generalized and specific applications. Here we provide an overview of methods for the preparation and processing of nanoparticles and other nanomaterials that are comparable to those using SCFs.

A. Solution Methods

Among the simplest techniques for the preparation of nanoparticles are solution methods. Examples of these methods include simple precipitation reactions in room-temperature solutions, chemical reduction of metal salts, controlled release of reactants, and thermal or photoinduced decomposition of specially designed precursors.

1. Mixing and Precipitation

Metal-sulfide nanoparticles have been investigated extensively. For their preparation a simple method involves mixing aqueous solutions of the metal ion and

Na_2S or purging H_2S gas through the metal-ion solution. For example, Bard and coworkers reported the preparation of CdS nanoparticles via the precipitation reaction of Cd^{2+} in acetonitrile with H_2S gas at room and lower temperatures (110). The sizes of the particles were generally 15–40 nm, and the samples prepared at lower temperatures exhibited less aggregation effects. Similarly, Murakoshi et al. reported the preparation of CdS nanoparticles via the same precipitation reaction (111). The Cd^{2+} source was a solution in dimethylformamide (DMF), and the S^{2-} source was either Na_2S solution in methanol or H_2S gas. The particles thus obtained were 18–42 nm in diameter, and the use of H_2S was found to produce larger particles and broader size distributions.

The aggregation of particles is a significant problem with the simple precipitation method. For stable suspensions of nanoparticles without unwanted aggregation and particle growth, surface capping or stabilization agents are widely used with this method. In the presence of thiophenolate or polymers as capping agents, the formation and properties of nanoparticles are dependent on the competition between precipitation reaction (particle growth) and particle surface capping in the solution.

Wang and coworkers prepared CdS nanoparticles by mixing a Cd^{2+} solution with a second solution that contained both S^{2-} and the capping agent PhS^- (112). The CdS particle sizes were varied systematically in the 15- to 35-nm range through changes in the S^{2-}/PhS^- ratio (Table 2 and Figure 13). Well-defined shoulders in the observed absorption spectra of the nanoparticle samples indicated relatively narrow particle size distributions. Similarly, Nosaka et al. prepared CdS nanoparticles via simple mixing of aqueous $CdCl_2$ and Na_2S solutions in the presence of different capping agents and protecting polymers (113). The particle sizes were estimated from the spectral positions of the plasmon absorption band for the nanoparticles.

A significant modification to the simple mixing of solutions is the slow and controlled release of one of the reactants. For example, Ohtaki et al. prepared CdS nanoparticles in nonaqueous solvents, where one of the reactants (S^{2-}) was made available slowly and uniformly via the controlled hydrolysis of P_2S_5 (114). The CdS particles were, on average, 6 nm in diameter, had a size distribution standard deviation of 1.2 nm, and formed stable suspensions under the protection of polymers (Figure 14). A similar strategy was used by Meisel and coworkers (115) and, more recently, by Yin et al. (116) in the preparation of CdS nanoparticles, where the slow release of S^{2-} was achieved through the use of pulse radiolysis.

2. Chemical Reduction

Chemical reduction of a salt solution has been widely done in the preparation of metal nanoparticles. Among the commonly employed reducing agents are

Table 2 Analytical Results for CdS/SPh Clusters Isolated as Powders (112)

S/SPh used	Cd (%)	S (%)	C (%)	H (%)	Empirical formula	Cd S C_6H_5	S/SPh calc.	X-ray diam (Å)
0.33[a]	39.25	20.02	38.62	2.60	$CdS_{0.26}(SPh)_{1.48}(HSPh)_{0.06}$	1:1.80:1.54	0.17	<15
0.5[a]	42.29	9.90	35.53	2.52	$CdS_{0.34}(SPh)_{1.31}(HSPh)_{0.00}$	1:1.65:1.31	0.26	<15
0.75[a]	47.68	20.40	29.31	2.19	$CdS_{0.54}(SPh)_{0.92}(HSPh)_{0.04}$	1:1.50:0.96	0.56	~20
1.17[a]	58.92	21.23	18.06	1.43	$CdS_{0.78}(SPh)_{0.44}(HSPh)_{0.04}$	1:1.26:0.48	1.62	~25
2.0[a]	61.38	21.45	15.22	1.20	$CdS_{0.84}(SPh)_{0.32}(HSPh)_{0.07}$	1:1.23:0.39	2.15	~30
4.5[a]	66.45	22.18	9.68	0.85	$CdS_{0.92}(SPh)_{0.16}(HSPh)_{0.09}$	1:1.17:0.25	3.70	~35
0.5[b]	45.89	21.80	26.94	2.04	$CdS_{0.75}(SPh)_{0.50}(HSPh)_{0.42}$	1:1.67:0.92	0.81	<15
0.75[b]	50.65	23.03	24.04	1.97	$CdS_{0.86}(SPh)_{0.28}(HSPh)_{0.46}$	1:1.60:0.74	1.16	~20

[a]Materials prepared in methanol solvent.
[b]Materials prepared in mixed solvent with acetonitrile.

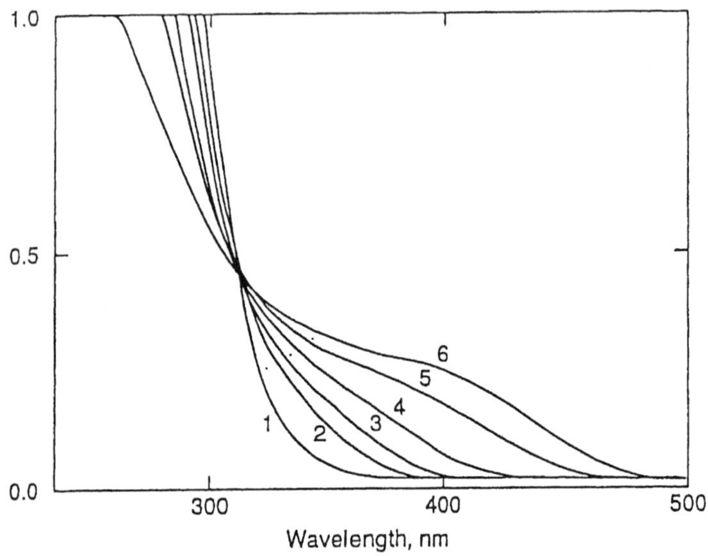

Figure 13 Absorption spectra of CdS/SPh clusters in acetonitrile solution with [Cd] = 4×10^{-3} M in 1 mm pathlength cells. (1) $(NMe_4)_4$-$(Cd_{10}S_4SPh_{16})$; (2) idealized S/SPh ratio used in preparation, 0.33; (3) 0.5; (4) 0.75; (5) 1.17; and (6) 2.0. (From Ref. 112.)

alcohols, hydrazine, borohydrides, aluminum hydrides, Grignard reagents, alkali metals, alkalides, and electrides. Radical anions that are generated chemically, radiolytically, and photolytically have also been used as strong reducing agents.

For the preparation of nanoparticles, the metal salts of the elements Pd, Au, and Pt can be easily reduced by alcohols (117). Numerous alcohols at various temperatures have been used in such reduction reactions. Yonezawa and Toshima applied the same reduction reaction for the preparation of bimetallic nanoparticles (118).

Most metals require reducing agents that are stronger than alcohols. For the elements Cu, Ni, Co, and Fe, Klabunde and coworkers attempted the preparation of nanoparticles using $NaBH_4$ as a reducing agent in a variety of solvents (119–124). The corresponding metal, metal boride, or metal-oxide particles obtained were dependent on the choice of solvent and experimental conditions. For example, reduction of Co^{3+} in diglyme produced Co metal particles, whereas the reaction in water yielded Co_2B particles. In addition to metal borides formed under certain experimental conditions, a small amount of boron impurity was found in all metal particle samples. Thus, this method is not applicable to the preparation of highly pure metal nanoparticles. The size of the particles was found to be dependent on the element and reaction conditions. According to

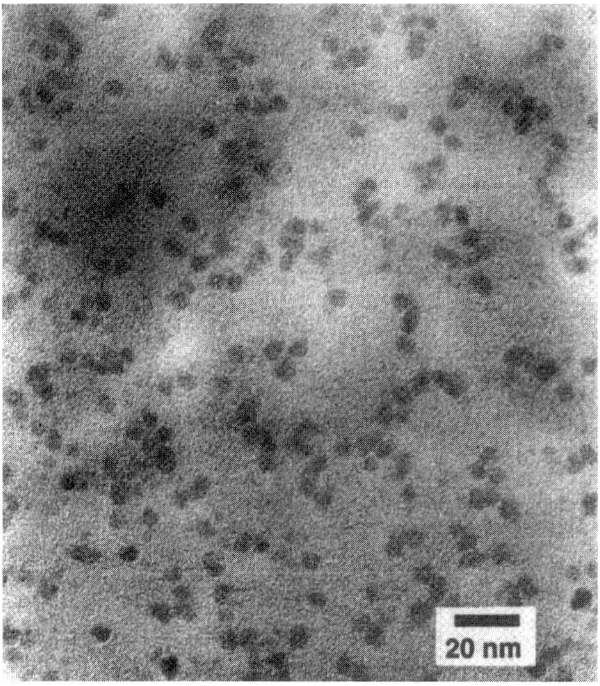

Figure 14 TEM image of CdS nanoparticles prepared by decomposition of P_2S_5 in absolute ethanol in the presence of PVP, $[Cd^{2+}] = [S^{2-}] = 1$ mmol dm^{-3}. (From Ref. 114.)

results from powder x-ray diffraction characterization, the average particle sizes were estimated to be generally less than 30 nm. The particles thus produced were found to be amorphous in most cases. The particle size distributions, although not determined, were probably very broad.

Using a similar strategy, Bönnemann et al. employed various hydrotriorganoborates (in particular, $LiBEt_3H$ and $NaBEt_3H$) as reducing agents to prepare numerous nanoscale metal powders in THF solutions (125–128). The average size of these metal particles was generally in the 1- to 5-nm range. Although many of the metal particle samples from the reduction with hydrotriorganoborates contained boron impurities (typically at a level near 1%), metal borides were not found. One advantage of hydrotriorganoborates is their ability to reduce ions of some reductive metals (iron, for example)—difficult or impossible with milder reducing agents such as hydrazine and $NaBH_4$. However, the high reactivity of hydrotriorganoborates can create problems. These reducing agents react with air and water as well as other functional groups, such as alcohols, which limits the selection of solvent systems and requires the use of highly pure solvents under air-free conditions.

Aluminum-based reducing agents—$LiAlH_4$ and Et_2AlH in particular—have also been employed. For example, Dawdle et al. prepared Ni particles using Et_2AlH as a reducing agent and triphenylphosphine as a stabilizing agent (129). In the preparation the reducing agent was added to a diethyl ether solution of $Ni(acac)_2$ at $-40°C$ and then slowly warmed to room temperature. During the slow warming process, a change in solution color occurred, accompanied by the formation of Ni nanoparticles as a black precipitate. According to TEM results (Figure 15), these particles were narrowly distributed in size, with most particles containing 19–21 atomic layers in the ⟨111⟩ lattice plane; this corresponds to

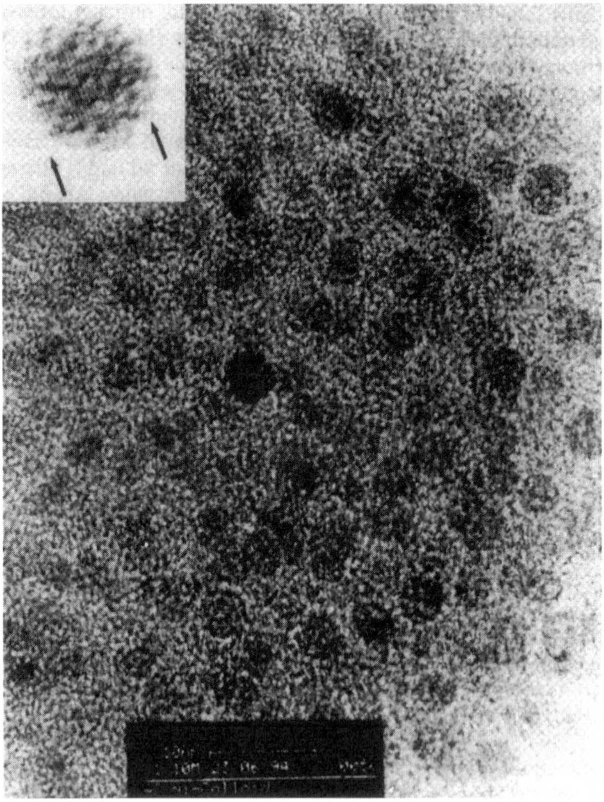

Figure 15 TEM image of nickel colloid. Some of the spherical particles show 19, 20 or 21 (111) layers, corresponding to 3.9, 4.1, or 4.3 nm diameter. Most of the particles have diameters in between the size range 3.9–4.3 nm. The insert shows a magnified image of a single colloidal particle consisting of 21 atomic layers. (From Ref. 129.)

particle sizes of 3.9 and 4.3 nm, respectively. A possible cause of the narrow size distribution was the low initial temperature of −40°C, which allowed a thorough mixing of the reagents under unreactive conditions.

Alkalides and electrides have been used as reducing agents for the preparation of metal nanoparticles under low-temperature conditions. For example, nanoparticles of numerous elements (Au, Pt, Cu, Te, Ni, Fe, Zn, Ga, Si, Mo, W, In, Sn, and Sb) and alloys (Au-Cu, Au-Zn, Cu-Te, and Zn-Te) were obtained with the reduction at −50°C (130,131). The average particle diameter was generally 2–15 nm. The low temperature did not aid the narrowing of particle size distribution in this case probably because the reduction reaction occurred immediately upon mixing of the reactants.

Photochemical reduction is an effective method for the preparation of nanoscale metal particles in solution. For example, silver nanoparticles were obtained via photolysis in an aqueous solution of silver nitrate containing poly(*N*-vinylpyrrolidone) (PVP) polymer as a stabilizing agent (132). The silver particles thus obtained had average sizes ranging from 10 to 22 nm and size distribution standard deviations of 3.2–6.3 nm. The average particle size and size distribution were found to be dependent on the photoirradiation time and the protective polymer PVP concentration. In addition, photochemical reduction has been widely used in the coating of nanoparticles with metal elements. A classical example is the preparation of semiconductor-metal composites such as TiO_2 nanoparticles photochemically coated with Pt and other metal elements (133–135).

Many metal nanoparticles can be prepared by several different reduction methods. For example, Mayer and Mark prepared Ag, Pd, Au, and Pt nanoparticles in solution using alcohols, KBH_4, and photochemical reduction, with amphiphilic diblock copolymers as protection agents (136). The particle sizes were in the 1- to 7-nm range, depending on preparation conditions.

The solution-based methods have some obvious advantages and disadvantages. These methods are relatively simple and easily coupled with a variety of strategies. For the chemical reduction of metal salts in solution, for example, the method is easily applicable to various metals and allows some tuning of particle sizes, but the particle size distribution is generally broad. A significant improvement in particle size distribution can be achieved in special types of solution methods, such as the sonochemical decomposition of precursors in solution (discussed in the following section). On the other hand, the use of protection agents provides some improvements to the solution-based methods. A special class of protection agents that offer the most significant benefits is surfactants (as discussed in detail in Sec. II.C).

B. Sonochemical Method

Sonochemical synthesis of nanoparticles has been based primarily on the sonolysis of precursors in solution. Suslick et al. used the sonolysis of iron pentacarbonyl under an inert argon atmosphere to prepare amorphous iron powders (137–139). The solution of the iron precursor for ultrasonic irradiation also contained a stabilizing agent, either PVP polymer or oleic acid (138). The size of the particles prepared in the presence of PVP was in the 3- to 8-nm range. However, the particles prepared in the presence of oleic acid also contained iron oxide and had an average particle size of 8 nm and a narrower size distribution than that of the PVP-stabilized particles. These iron particle samples also contained small fractions of carbon and oxygen impurities, probably due to the decomposition of the solvent or carbon monoxide. Suslick et al. also prepared nanoparticles of Co and Fe-Co alloys with different compositions via sonolysis of the respective precursors (139). The alloy particles were 10–20 nm in diameter. Characterization with energy-dispersive spectrometry (EDS) showed that the alloy particles were solid solution in nature, with the components in the particles being homogeneously distributed on the scale of a few nanometers.

The sonochemical method can also be used to prepare complicated nanoscale mixtures. For example, Gonsalves et al. used the sonochemical decomposition of organometallic precursors to prepare powders of M50-type steel (4% Cr, 4.5% Mo, 1% V, 0.8% C, with the balance being Fe) (140).

Gibson and Putzer prepared Co platelet particles by sonicating an aqueous Co^{2+} solution in the presence of hydrazine (141). The Co particles thus obtained were essentially hexagonal, about 100 nm wide, and 15 nm thick (Figure 16). An interesting issue with these particles was the shape-property relationships, such as the potential effects of shape anisotropy on the magnetic properties of the particles.

The sonochemical method has been particularly useful in the preparation of metal nanoparticles through decomposing organometallic precursors, especially for metals that are difficult to reduce chemically. This technique has also been used successfully in the preparation of nanoparticles of other materials, including metal carbides and sulfides (142–144). For example, the sonication of molybdenum and tungsten carbonyls resulted in nanoparticles of metal carbides (142,143). In the case of Mo_2C, for example, the particles were found to be somewhat small, with the average size of the particles being about 2 nm in diameter. The same method was recently extended to the preparation of MoS_2 nanoparticles by sonolysis in a slurry of molybdenum hexacarbonyl and sulfur in isodurane under argon protection (144). Results of x-ray diffraction and TEM measurements showed that the sample contained aggregates ($\sim$15 nm) of nanoparticles (1.6 nm in diameter). Shown in Figure 17 are SEM images of sonochemically and conventionally prepared MoS_2.

Another important feature of the sonochemical method is that nanoparticle formation is largely independent of the presence of other species in the solution. For example, sonolysis of a precursor solution in the presence of alumina or silica support resulted in the formation of nanoparticles on the support surface

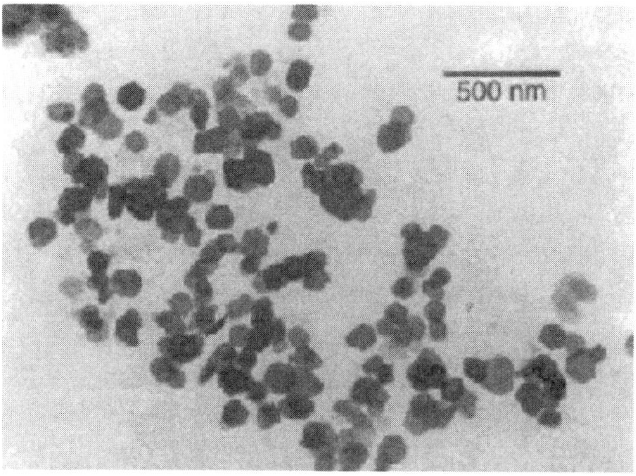

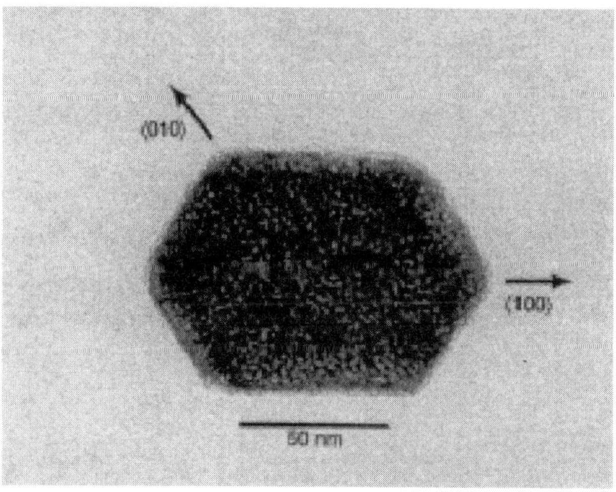

Figure 16 (top) A group of cobalt nanoclusters. The particles, which adopt a hexagonal or centered trigonal unit cell, are preferentially deposited as oriented (001) plates. (bottom) A single cobalt nanocluster. The orientations of the ⟨100⟩ and ⟨010⟩ zone axes were determined by SAD and TEM. (From Ref. 141.)

(145). The nanoparticles prepared in the presence and absence of the supports had essentially the same size.

The advantage of the sonochemistry-based methods is that they produce nanoparticles with a relatively narrow size distribution. One of the limitations of these methods is that the nanoscale metals produced are generally amorphous

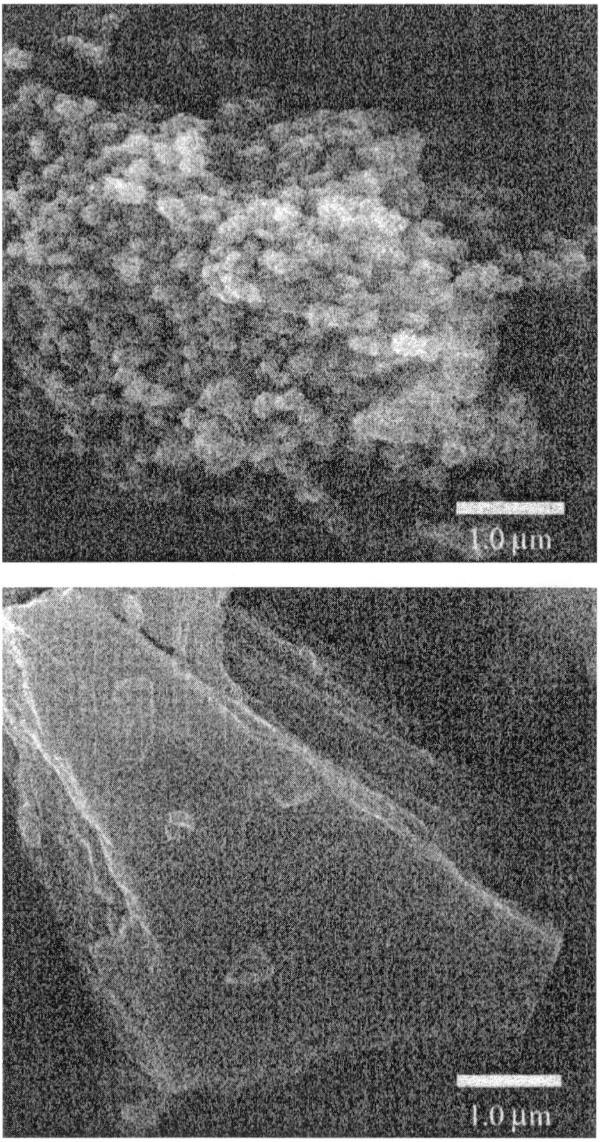

Figure 17 SEM micrographs of (top) sonochemically and (bottom) conventionally prepared MoS_2. (From Ref. 144.)

and that the products often contain carbon and oxygen impurities. Interest in sonochemical preparation of nanoparticles from other materials is increasing.

C. Microemulsion Methods

A versatile method for the preparation of nanoparticles from many different classes of materials involves the use of micellar structures as templates (146–203). The templating effect allows a certain degree of control over the size of the particles produced and even over variations in the product shape (from spherical to rod-like). In general, the size distribution of nanoparticles thus obtained is relatively narrow.

Micelles and reverse (or inverse) micelles are microemulsions that are typically stable and macroscopically homogeneous. To facilitate the microscopic phase separation, a micellar system consists of at least three components: two liquid phases that are immiscible and an amphiphilic surfactant. Microemulsions of normal micelles consist of spherical oil droplets in the bulk water phase. On the other hand, microemulsions of reverse micelles, consist of water droplets in the bulk oil (organic solvent) phase. The size of these droplets can be varied systematically by changing the oil/surfactant and water/surfactant ratios for normal and reverse micelles, respectively. The nanoscale cavities in the micellar structures serve as templates for the synthesis of semiconductor and metal nanoparticles. In these microemulsion systems, the contents of different micellar cavities undergo exchange when the micelles collide, creating opportunities for mixing and reactions between the reactants in the different cavities. Since the core in reverse micelles is water, it enables the dissolution of metal salts and other ionic species. Thus, when metal salts are incorporated in one reverse micelle emulsion and sulfide anions are incorporated in another, mixing of the two results in the formation of metal-sulfide nanoparticles in the nanoscale reverse-micellar cavities. If the second reverse-micelle emulsion contains a reducing agent rather than sulfide anions, mixing results in the formation of metal nanoparticles.

1. Microemulsions Under Ambient Conditions

Many researchers, most notably the Pileni group (146–186,190,191,197–201), have reported on the use of micelle emulsions for preparing nanoparticles of numerous materials, including sulfides, oxides, chlorides, metals and alloys, and polymers. For example, Pileni and coworkers used the technique to prepare small (4-nm) CdS nanoparticles (146,174,181). Similarly, Lianos and Thomas prepared CdS nanoparticles by simply mixing two reverse-micelle emulsions containing Cd^{2+} and S^{2-} ions (188). Absorption and emission properties of the colloidal CdS suspension were studied, and quenching of emission by Cu^{2+} and methylviologen (MV^{2+}) cations was evaluated. Recently, Haram et al. reported the preparation of CuS nanoparticles using reverse micelles containing Cu^{2+} and thiourea (189). The adsorption of Cu_2S nanoparticles on the wall of TX-100 inverse micelles is illustrated in Figure 18. Absorption spectroscopy was used to monitor reaction progress and sample stability.

Pérez and Quintela reported the synthesis of iron oxide nanoparticles using reverse micelles (187). Two reverse-micellar emulsions were prepared, with one containing an acidic solution of $FeSO_4$ and $FeCl_3$ salts as the aqueous core and the other containing cyclohexylamine. Mixing of the two emulsions at 65°C resulted in the formation of Fe_2O_3 nanoparticles. Characterization of the particles using x-ray diffraction and transmission electron microscopy (TEM) showed the particle size to be in the 2- to 7-nm range. These results were supported by those from magnetization measurements, which also showed that the standard deviation of the log-normal volume distributions was typically about 0.5 nm. Bagwe and Khilar used reverse micelles to prepare AgCl nanoparticles and evaluated the effects of the continuous phase, the surfactant, and the water/surfactant mole ratio on particle size (202). Tojo et al. also investigated the kinetics of particle formation in microemulsions (203).

The choice of surfactant plays a critical role in the formation of micelles. A widely used surfactant for reverse micelles is sodium bis(ethylhexyl)sulfosuccinate (AOT). More effective control of the amount of metal salts in the hydrophilic cavities of reverse micelles was achieved through a clever strategy developed by Pileni and coworkers to functionalize surfactants with the metal cations under consideration (153,164,168,179,199). As an example, for AOT, Na^+ was replaced with the metal cation of choice. The typical strategy of using a mixture of the parent and functionalized surfactant was applied in the preparation of a wide variety of nanoscale particles and related nanomaterials (153,164,168,179,199). For example, silver nanoparticles were prepared using the Ag^+-substituted AOT (AgAOT). One reverse-micelle emulsion was prepared using a 30% AgAOT–70% AOT mixture in isooctane and the other contained the

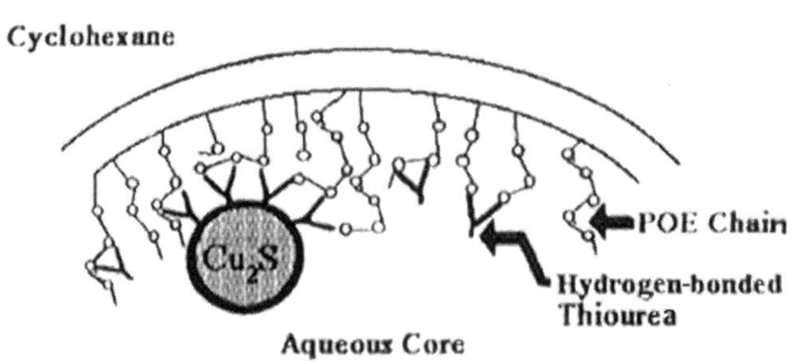

Figure 18 Schematic representation of the adsorption of Cu_2S nanoparticles on the inner wall of TX-100 inverse micelles. (From Ref. 189.)

reducing agent $NaBH_4$ or N_2H_4 (153,163,177). Mixing the two reverse-micelle emulsions resulted in chemical reduction and the formation of silver nanoparticles, with the average particle diameter being 2.5–7.5 nm and size distribution standard deviations being 1–3 nm. Variation in particle size was achieved by varying the water/surfactant ratio in the microemulsions (Figure 19). The particles obtained from the reduction with $NaBH_4$ had a narrower size distribution than those obtained from the reduction with N_2H_4 under the same conditions. Similarly, AgAOT was used in the construction of reverse micelles for the preparation of Ag_2S nanoparticles (151,161,166,200,201). The particles were capped by dodecanethiol by the addition of capping agent to the micelle mixture following particle formation. The Ag_2S nanoparticles thus obtained were 2–10 nm in diameter, with a polydispersity of 14%.

Copper nanoparticles were produced using both reverse and normal micelles (148,158,164,168,170,173,176,179). Functionalized AOT was used in the construction of reverse micelles. With $NaBH_4$ or N_2H_4 as a reducing agent, copper nanoparticles, cylindrical rods, and a mixture of particles and rods were prepared under various experimental conditions. The spherical particles were 1–12 nm in diameter and had a narrow size distribution, and the cylindrical rods had a typical aspect ratio (length/width) of more than 2, with length of 12–25 nm and diameter of 6.6–8.2 nm.

Dodecyl sulfate (DS) salts represent another series of metal-functionalized surfactants. For the preparation of copper nanoparticles using normal micelles, $Cu(DS)_2$ was used as a surfactant (160,168). Water was the continuous phase, and $NaBH_4$ was the reducing agent. The size and shape of the nanoparticles were manipulated by controlling the $Cu(DS)_2$ concentration. Elongated rod-like networks with an aspect ratio (length/width) up to 10 were produced at $Cu(DS)_2$ concentrations below the critical micelle concentration (CMC). At higher concentrations above the CMC, mixtures of elongated and spherical particles were produced (Figure 20). The relative populations of rods and particles were dependent on the $Cu(DS)_2$ concentration, with a higher concentration favoring more particles. At even higher $Cu(DS)_2$ concentrations, spherical particles became the primary product, and the particles became smaller, with average particle size in the 2- to 7-nm range. The size and shape distributions of the particles from the normal micelles were generally broader than those from the reverse micelles discussed above.

Normal micelles of $Fe(DS)_2$ and $Co(DS)_2$ salts were used to prepare nanoparticles of cobalt ferrite (162,172,184–186). The preparation involved stirring the emulsion for 2 h, which resulted in a magnetic precipitate of nanoparticles. The precipitate could be redispersed to form an aqueous suspension. The particle sizes were varied from 2 to 5 nm by changing the $Fe(DS)_2$ concentration, with a higher $Fe(DS)_2$ concentration favoring larger particles. TEM characterization showed that the particles had polydispersities in the 23–37% range.

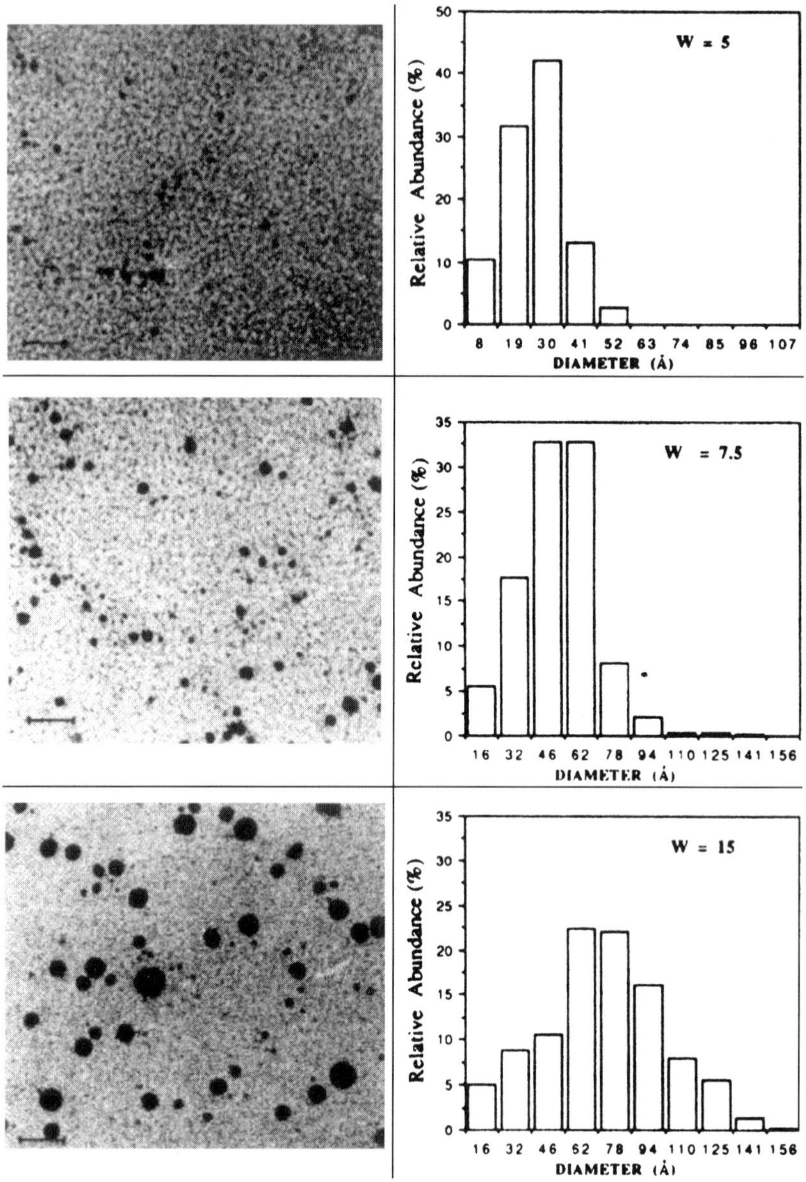

Figure 19 Electron microscopy and size distribution of silver crystallites synthesized in AgAOT-NaAOT-water-isooctane reverse micelles at various water contents. [AgAOT] = 3×10^{-2} M, [NaAOT] = 7×10^{-2} M, [$NaBH_4$] = 2.5×10^{-4} M, and $W_0 = 5$ (top), 7.5 (middle), and 50 (bottom). The marks represent 20.8 nm. (From Ref. 177.)

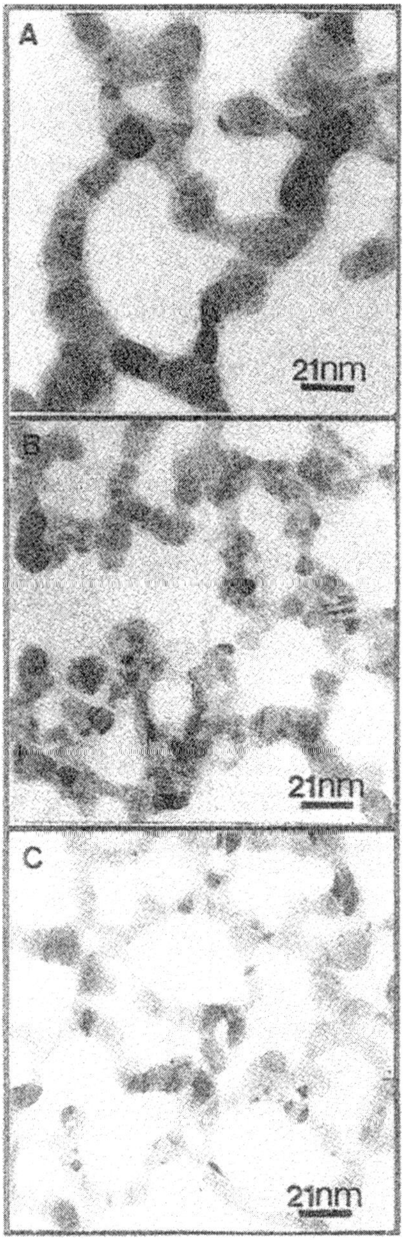

Figure 20 TEM patterns of colloidal copper dispersions prepared in a pure copper dodecyl sulfate solution. $[Cu(DS)_2] = 5 \times 10^{-4}$ M (A), 7.5×10^{-4} M (B), and 1.2×10^{-3} M (C). $[NaBH_4] = 2 \times [Cu(DS)_2]$. (From Ref. 160.)

Similarly, nanoparticles of Fe-Cu-B alloys were prepared by mixing reverse micelles of Fe- and Cu-functionalized AOTs with an aqueous $NaBH_4$ solution (157). For a sample prepared using the reverse micelles of 30% Fe(II)- and 70% Cu-functionalized AOTs, the TEM results showed that the particles were polydispersed and aggregated. Composition of Fe and Cu in the particles, as determined by energy-dispersive spectrometry (probe size 35 nm), varied widely from area to area. The significant variation in composition might be a result of the different redox potentials of the two metal ions. The inability of $NaBH_4$ to completely reduce Fe^{2+} to Fe^0 (known to be the case for the bulk reduction of Fe^{2+} in room-temperature solutions) could be responsible for the incorporation of boron in the product and also for the sample heterogeneity.

Other multiple-component systems prepared in microemulsions include ternary semiconductors. For example, reverse micelles were used to prepare $Cd_yZn_{1-y}S$ and $Cd_{1-y}Mn_yS$ nanoparticles (190,191). In the preparation process either three separate reverse-micelle emulsions (each containing one component) or two emulsions (one containing two metals and the other containing the anion) were mixed. The particles, on average, were 2–4 nm in diameter. Overall compositions of the three components in the particles were determined by energy-dispersive spectrometry, although probably the composition varied from particle to particle. The variation might be more significant for those particles obtained from mixing of the three emulsions.

The coating of nanoscale semiconductor particles by a different semiconductor has also been accomplished using the reverse-micelle approach. For example, Han et al. reported the preparation of Ag_2S-coated CdS nanoparticles (194). The nanoparticles were prepared first by mixing reverse-micelle emulsions containing the respective ions; these particles were subsequently coated by adding an $AgNO_3$ solution to the nanoparticle suspension. The cadmium on the CdS nanoparticle surface was displaced by silver, resulting in the formation of an outer Ag_2S shell around a CdS core.

The strategy for the core-shell particles has been applied to the preparation of nanoscale composites that contain homogeneously dispersed semiconductor nanoparticles. For example, Chang et al. prepared colloidal silica-CdS nanocomposites of various structural morphologies under different experimental conditions (195). The method was based on a somewhat difficult combination of the controlled hydrolysis of tetraethyl orthosilicate (TEOS) with the formation of CdS nanoparticles in a reverse-micelle emulsion. The mixing sequence and timing for the TEOS, $Cd(NO_3)_2$, and $(NH_4)_2S$ microemulsions determined the product morphology. One of the products was SiO_2 spheres of 40–300 nm diameter that were homogeneously dispersed with small CdS nanoparticles of 2.5 nm diameter. Other products resulting from different sets of mixing sequences and timing included CdS surface patches on SiO_2 spheres, CdS/SiO_2 and SiO_2/CdS core/shell spheres, CdS/SiO_2 multilayer particles, and SiO_2 spheres with CdS surface inclusions. In another variation, CdS was removed from the surface of SiO_2 spheres for the preparation of highly nanoporous silica particles.

Polymeric nanocomposites were prepared via the same principle. For example, Shiojiri et al. used reverse micelles to prepare CdS, ZnS, TiO_2, and AgI nanoparticles and incorporated them into polyurea via in situ polymerization of hexamethylene diisocyanate (Figure 21) (192). The size of the nanoparticles embedded in the composite was estimated from absorption measurements. The composite was evaluated for photocatalytic properties. Similarly, Premachandran

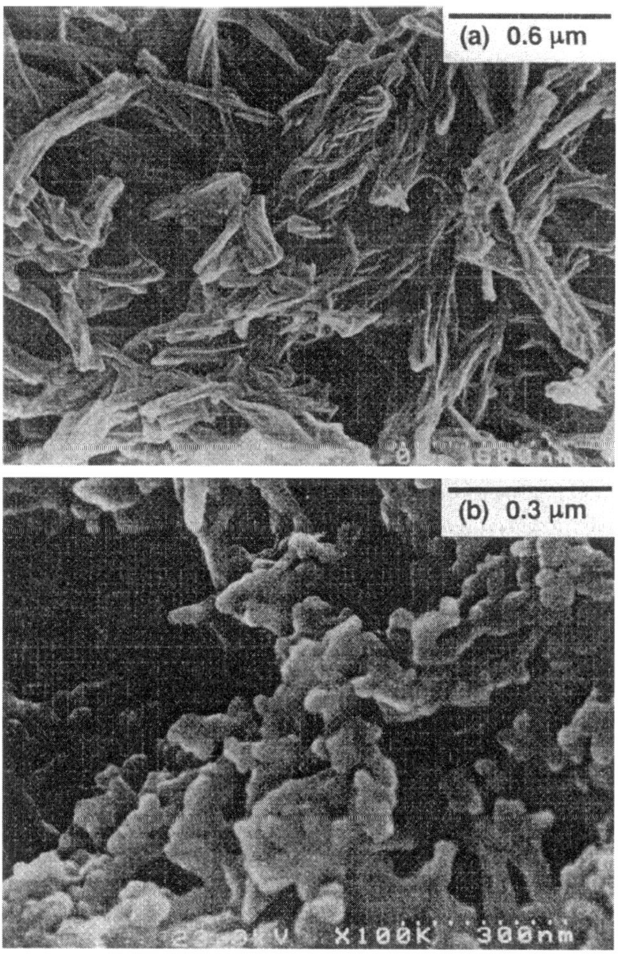

Figure 21 SEM images for (a) metal sulfide-free polyurea particles and (b) CdS-polyurea prepared in a reverse micellar system. (From Ref. 192.)

et al. used reverse micelles to prepare CdS-polymer nanocomposites (193). The preparation involved mixing two separate reverse-micelle emulsions—one containing CdS nanoparticles and the other thiol-containing polymers. Upon drying, the solid was washed with isooctane to remove the surfactant and any unbound CdS, yielding the polymer-capped CdS nanoparticles.

Pileni and coworkers also used reverse micelles to prepare polymer nanoparticles (169). In the preparation, reverse micelles of dodecyldimethylammonium methacrylate in a water-toluene mixture were mixed with the radical initiator AIBN, followed by UV irradiation for 7 h. The polymerization yields were estimated to be 50–70%. The polymer could not be separated from monomers, presumably because of the low molecular weight. The latex particles thus obtained were about 2 nm in diameter. At higher water/surfactant ratios, the size distribution of the particles increased, but the average diameter remained about 2 nm.

The nanoscale particles and related species prepared in microemulsions can be used as precursors for more complex nanostructures and nanomaterials. For example, Pileni and coworkers used Ag and capped Ag_2S nanoparticles of narrow size distribution for self-assembly into two-dimensional monolayers and three-dimensional superstructures (153,154,161,163,197–201).

2. The "Hot Soup"

Alivisatos and coworkers have developed a relatively simple "hot soup" method for the production of primarily semiconductor nanoparticles or nanocrystals (204–210,214,215,218). Shown in Figure 22 is the experimental apparatus. The method is based on the pyrolysis of organometallic reagents after rapid injection into a hot coordinating solvent (ligand). The high-temperature solution of reactants with the ligand facilitates the formation of nearly monodispersed nanoscale semiconductors. Among the ligands commonly used with this method are trioctylphosphine (TOP) and trioctylphosphine oxide (TOPO). Numerous nanocrystals from semiconductors such as CdS, CdSe, CdTe, InP, and InAs have been synthesized by this method (211,213–219). In the synthesis of CdSe, for example, dimethyl cadmium and selenium powder were dissolved in tributylphosphine or TOP, followed by rapid injection of this solution into hot TOPO (340–360°C) in a flask under vigorous stirring to produce ligand-coated CdSe nanoparticles. The nanoparticles thus produced had a narrow size distribution.

Mechanically, the method involves two steps: (A) a temporally discrete nucleation resulting from an abrupt supersaturation upon reagent injection, followed by (B) slower controlled particle growth on the existing nuclei (211). Kinetic control of particle growth plays a crucial role in determining the average particle size and size distribution. A continuous adjustment of the monomer concentration is required for an optimal growth sequence. When a higher monomer

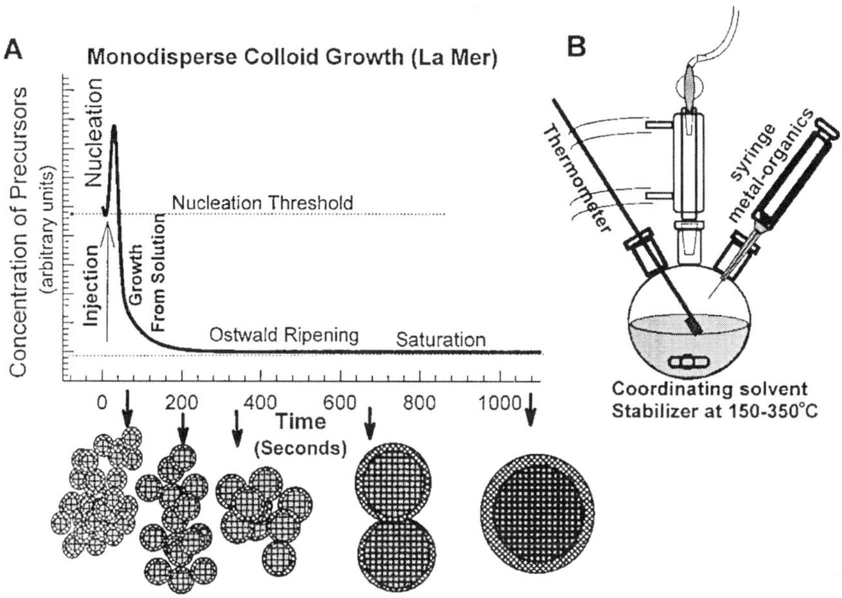

Figure 22 The "hot soup" method for the preparation of nanocrystals. (From Ref. 211.)

concentration is used, smaller nanocrystals grow more rapidly than larger ones, thus enabling control of size distribution and preparation of nearly monodispersed nanocrystals. The properties of the organometallic precursor and the coordinating solvent can also greatly influence particle growth; they can be used to manipulate particle size, shape, and size distribution (211,218,221). For example, in the synthesis of ZnSe nanocrystals, TOPO as a coordinating solvent hinders particle growth because of the strong interactions between TOPO and the Zn precursor, whereas a coordinating solvent, such as alkylamine, enhances particle growth significantly (220). Many other solvents have been used as ligands, including alkylphosphites, alkylphosphates, pyridines, and furans; however, a mixture of alkylphosphine and alkylphosphine oxide remains the optimum choice for production of semiconductor nanocrystals (211).

Peng et al. recently reported the use of the hot-soup method to produce CdSe quantum rods (222). Changes in several experimental parameters were found to affect properties of the quantum rods such as size and aspect ratio. These parameters included ligand selection, reaction time, injection and growth temperatures, and number of injections. Controlled variations in these parameters could also be employed to change the product morphology from rod-like to nearly spherical.

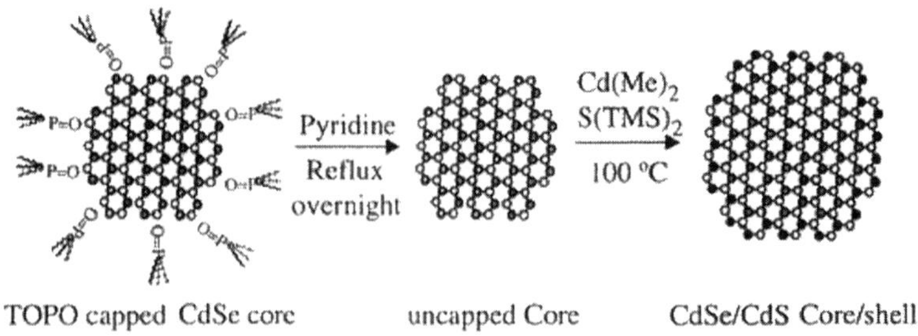

Figure 23 A scheme for the synthesis of CdSe/CdS core/shell nanocrystals. (From Ref. 225.)

The hot-soup method has also been applied to the production of semiconductor core/shell nanostructures, where a semiconductor nanocrystal is coated with a second semiconductor of wider bandgap. Shown in Figure 23 is a scheme for the synthesis of CdSe/CdS core/shell nanocrystals. These nanostructures were synthesized via a high-temperature route in a mixture of TOP and TOPO (223,224) and via a low-temperature route in pyridine (225). Among the semiconductor core/shell nanomaterials produced were CdSe/ZnS, CdSe/CdS, InAs/InP, InAs/CdSe, InAs/ZnSe, and InAs/ZnS (223,225–228). In these structures, the shell type and thickness allow further control of the optical, electronic, and other properties of semiconductor nanocrystals. For example, the shell may be used to passivate the imperfect surface of the core semiconductor, resulting in significantly improved luminescence efficiency.

3. Microemulsions in Supercritical Fluids

Microemulsions of surfactants in SCFs have emerged as effective supercritical solvent systems for the preparation and processing of materials. Microemulsions are particularly useful in the case of SCFs such as CO_2 that are environmentally benign but of relatively low solvent strength. However, because most conventional surfactants contain no CO_2-philic moieties, microemulsions in CO_2 require special surfactants such as fluorinated amphiphilic molecules (229–235). Other commonly used SCFs are simple hydrocarbons, which are compatible with conventional surfactants such as AOT. A unique advantage of SCF-based microemulsions is that properties of the microemulsions may be varied via changes in the pressure and temperature of the SCF (Figure 24) (236–244).

Microemulsions in SCFs have been used for the preparation of nanoparticles from several classes of materials (56,245–248). In an early publication (56), Matson et al. reported the production of nanometer- to micrometer-sized $Al(OH)_3$ particles through the use of AOT-based reverse micelles in supercritical propane. The experiment involved a simple precipitation reaction of aqueous

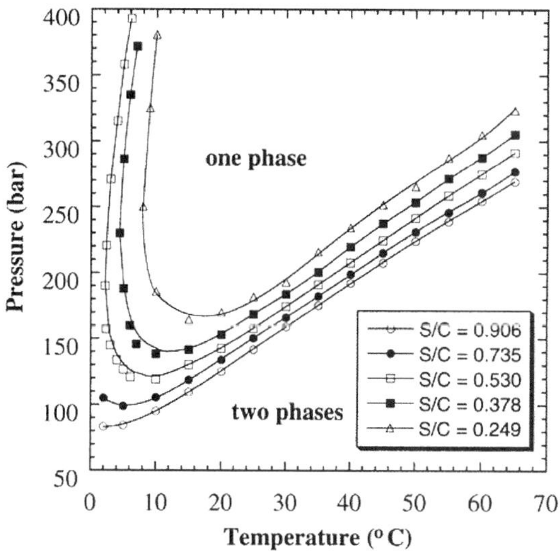

Figure 24 Phase behavior of the water/PFPE-NH_4/CO_2 system at $W_0 = 12.5$ vs. the surfactant-to-CO_2 ratio (S/C). Two distinct phase transitions are clearly evident, one at higher temperatures as the pressure is reduced at constant temperature or the temperature is increased at constant pressure (P_L or T_U) and one at lower temperatures as the temperature is reduced at constant pressure (T_L). (From Ref. 241.)

$Al(NO_3)_3$ in the reverse-micellar core with NH_3. More recently, Cason and Roberts prepared copper nanoparticles via the hydrazine reduction of copper ions in reverse micelles based on the Cu^{2+}-substituted AOT, $Cu(AOT)_2$, as surfactant in subcritical and supercritical ethane (245). The particle size was estimated to be 9–10 nm on the basis of in situ UV-vis absorption spectral analysis and the characterization of the recovered particles. The recovery involved a transfer from the microemulsion in high-pressure propane to a liquid-micelle emulsion. Results of TEM analysis of the recovered copper particles indicated agglomeration of the particles during recovery.

Recently, the use of microemulsions in supercritical CO_2 to produce nanoparticles has received considerable attention (240,246–249). Since conventional hydrocarbon surfactants for oil/water systems often exhibit low solubilities in CO_2 and are therefore incapable of solubilizing a significant amount of water (250,251), surfactants with fluorinated tails have been used for the formation of water-in-CO_2 microemulsions (251–254). Perfluoropolyether (PFPE) is a popular and commercially available fluorinated surfactant. Wai and coworkers prepared silver nanoparticles via chemical reduction in a microemulsion of water in supercritical CO_2 with PFPE (247). A slow flocculation of the nanoparticles was observed over approximately 1 h (Figure 25). The phenomenon was attributed to the low viscosity and low dielectric constant of the fluid, which probably facilitated a higher collision frequency and efficient exchange between the droplets and a stronger attractive interaction between the nanoparticles. Particle size was estimated to be 5–15 nm through in situ UV-vis absorption spectral analysis and TEM characterization of the recovered sample. Wai and coworkers also reported

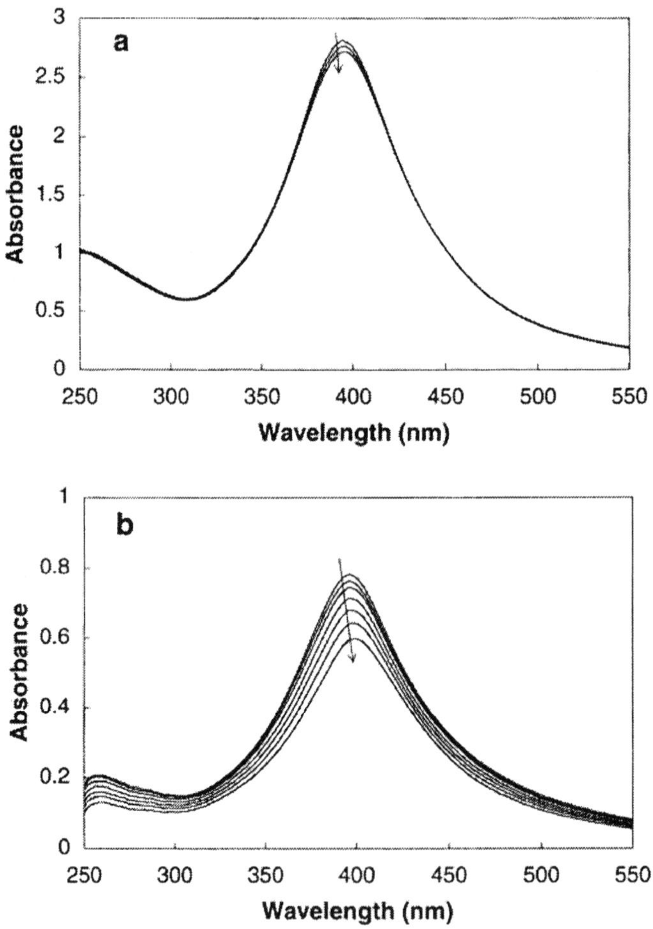

Figure 25 Series of UV-vis spectra of Ag nanoparticles dispersed in microemulsions formed in (a) liquid CO_2 (25°C and 300 bar) and (b) supercritical CO_2 (35°C and 367 bar) collected at intervals of 20 and 10 min, respectively. Spectra were acquired at evenly spaced intervals over the course of an hour after their formation. (From Ref. 247.)

the synthesis of silver-halide nanoparticles in the same type of microemulsion in supercritical CO_2 (248). The experiment involved the mixing of two separate microemulsions containing the reactants Ag^+ and X^-. The determining step in the formation of the silver-halide nanoparticles was the intermicellar exchange process.

Johnston and coworkers reported the preparation of nanoscale CdS particles in a PFPE-based microemulsion in supercritical CO_2 for evaluating the effects of the water/surfactant ratio (W_0) of the microemulsion on the nanoparticle properties (249). They found that the particle size increased significantly with an increase in the W_0 value, from which a correlation between average nanocrystal radius and water-core radius was established (Figure 26). Recently, Johnston and coworkers also prepared silver nanocrystals coated with fluorinated ligands (240). These coated nanocrystals could be dispersed in CO_2 at moderate pressure and temperature.

The examples discussed have demonstrated that SCF systems can be used successfully in the preparation of nanomaterials. A number of advantages associated with the SCF-based methods can be envisaged. For example, the tunable properties of SCFs through pressure and/or temperature changes will provide more possibilities to manipulate the nanoparticles produced. For potential applications the preparation of nanoparticles in situ in an SCF system may be coupled with other processes, such as rapid expansion for nanoscale coating and patterning. Another rationale for the use of SCF systems in the production of nanoscale materials is to take advantage of the templating effect associated with the rapid-expansion process, as discussed in detail in the next section.

III. RESS-INTO-SOLVENT (RESOLV) METHOD FOR NANOMATERIALS

As discussed in previous sections, the RESS method has been widely used in particle production. In conventional RESS processing, a supercritical solution is typically expanded into ambient air (31,71,73,82–84). However, theoretical calculations by Weber et al. have provided evidence that agglomeration processes are responsible for the growth of particles beyond the nanoscale (255). Therefore, Sun and coworkers made a significant modification to the conventional RESS process by expanding the supercritical solution into a liquid rather than a gas in an effort to eliminate agglomeration. Their modified process is called RESOLV (*R*apid *E*xpansion of Supercritical Solution into Liquid *SOLV*ent). It is typically coupled with a reacting system in which one reactant is dissolved in the supercritical solution and another is dissolved in the liquid-receiving solution. For example, Cd^{2+} ions in a sprayed supercritical ammonia solution react with S^{2-} in the receiving liquid-ethanol solution to form CdS nanoparticles (256). The

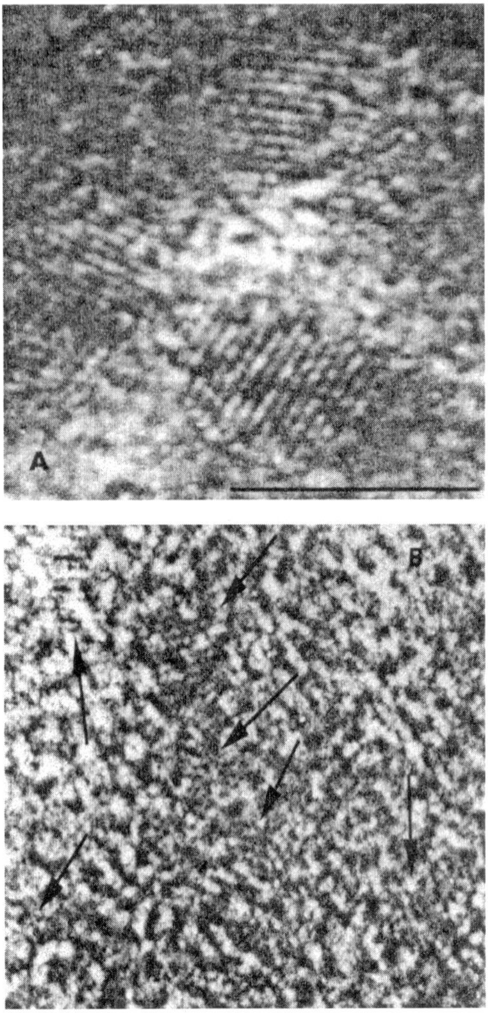

Figure 26 TEM images of CdS confirming the particle sizes of (A) $W_0 = 10$ and (B) $W_0 = 5$ obtained from the absorption spectra. The lattice fringes can be clearly seen for the top image. The arrows indicate the location of the lattice fringes in the bottom image. Scale = 5 nm (applies to both TEM images). (From Ref. 249.)

typical RESOLV apparatus for the preparation of nanoscale particles illustrated in Figure 27 consists of a syringe pump for pressure generation and pressure maintenance during the rapid-expansion process and a gauge for monitoring system pressure. The heating unit is a cylindrical solid copper block of high heat capacity in a tube furnace. The copper block is wrapped in a coil of stainless

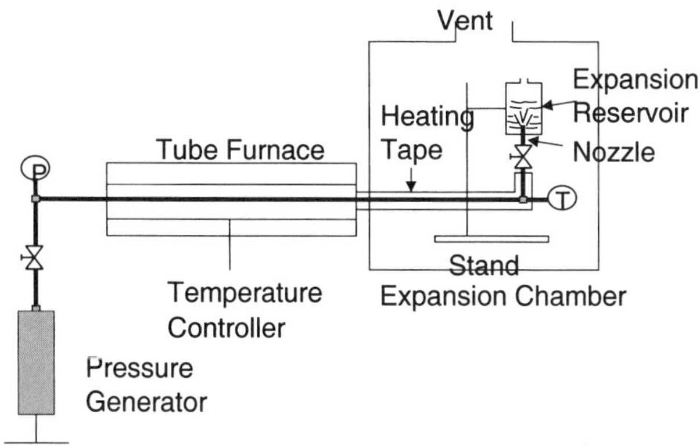

Figure 27 Experimental setup for the preparation of nanoparticles via RESOLV.

steel tubing and inserted tightly into a stainless steel tube to ensure close contact between the tubing coil and the copper block and, thus, efficient heat transfer. The copper block-tubing coil assembly is preheated to a set temperature before each experiment. For fluids with a high critical temperature, the syringe pump can be preheated to ensure that the solution reaches the designated temperature at the end of the tubing coil and becomes thermally equilibrated before rapid expansion. The expansion nozzle is typically a fused-silica capillary held within stainless steel tubing, which is inserted into the expansion chamber containing the ambient solution; micro-orifices can also be used for the rapid expansion. Sun and coworkers applied the RESOLV method to the preparation of nanoscale particles from a variety of elements and materials (256–263). The nanoparticles obtained via RESOLV are typically in the size domain of a few nanometers, with relatively narrow size distributions.

A. Nanoscale Metals

Metal nanoparticles have been prepared by coupling RESOLV with chemical reduction. For nanoscale silver particles, for example, a typical procedure involved the preparation of a homogeneous ammonia solution of $AgNO_3$ in a syringe pump. The solution was then heated to the desired temperature in the heating zone (Figure 27). Finally, the supercritical solution of $AgNO_3$ in ammonia at 160°C and 4000 psia was rapidly expanded through the expansion nozzle into the expansion chamber, which contained a room-temperature ethanol solution of hydrazine, to produce Ag-metal nanoparticles (257,260,261). Under the protection of a stabilizing agent, the nanocrystalline Ag particles formed a stable suspension in ethanol, which was visually indistinguishable from a typical colored homogeneous solution. As shown in Figure 28, the UV-vis spectrum of the stable suspension contains an intense absorption band that peaks at about 410 nm, characteristic of the plasmon absorption. A TEM image of the Ag particles is shown in Figure 29, from which an average Ag particle size of 5.6 nm and a size distribution of 0.78 nm are obtained (257). Shown in Figure 30 is the powder x-ray diffraction pattern of the nanocrystalline Ag particles in the solid state, which corresponds to the pattern for bulk face–center–cubic Ag metal.

When other reducing agents were used with RESOLV in the preparation of Ag nanoparticles, the particle properties differed somewhat. For reduction with $NaBH_4$, the Ag nanoparticles produced were smaller (Figure 29, average particle size of about 3.2 nm) than those obtained with hydrazine reduction, but the particle size distribution was similarly narrow (0.6 nm). As compared in Figure 28, the plasmon absorption bands of the Ag particles prepared with different reducing agents exhibit significant differences, which can hardly be attributed entirely to the difference in average particle size (257,263).

The same procedure was used to prepare Cu nanoparticles, except that a supercritical ammonia solution of $Cu(NO_3)_2$ was used in the rapid expansion (260). The nanoparticles were characterized using x-ray powder diffraction (Figure 31) and TEM.

Nickel nanoparticles were also prepared via RESOLV with $NaBH_4$ reduction (258). The preparation involved rapid expansion of a $NiCl_2$ solution in near-critical ethanol at 230°C into a room-temperature solution of $NaBH_4$ in DMF. The DMF solution, which also contained PVP polymer for particle protection, was deoxygenated before the expansion to avoid oxidation of the

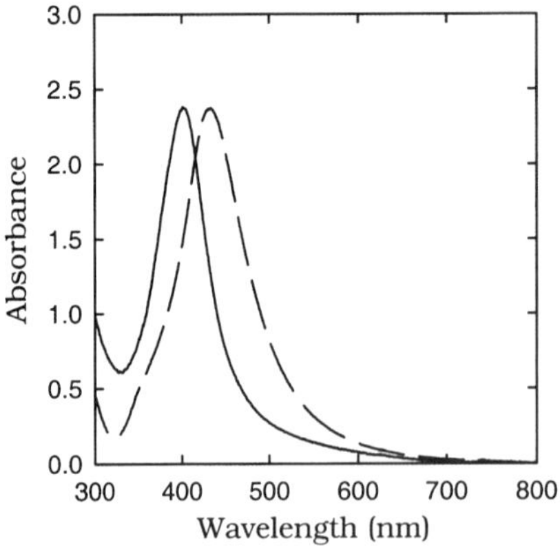

Figure 28 Absorption spectra of Ag nanoparticles prepared via RESOLV with N_2H_4 (dashed line) and $NaBH_4$ (solid line) reduction.

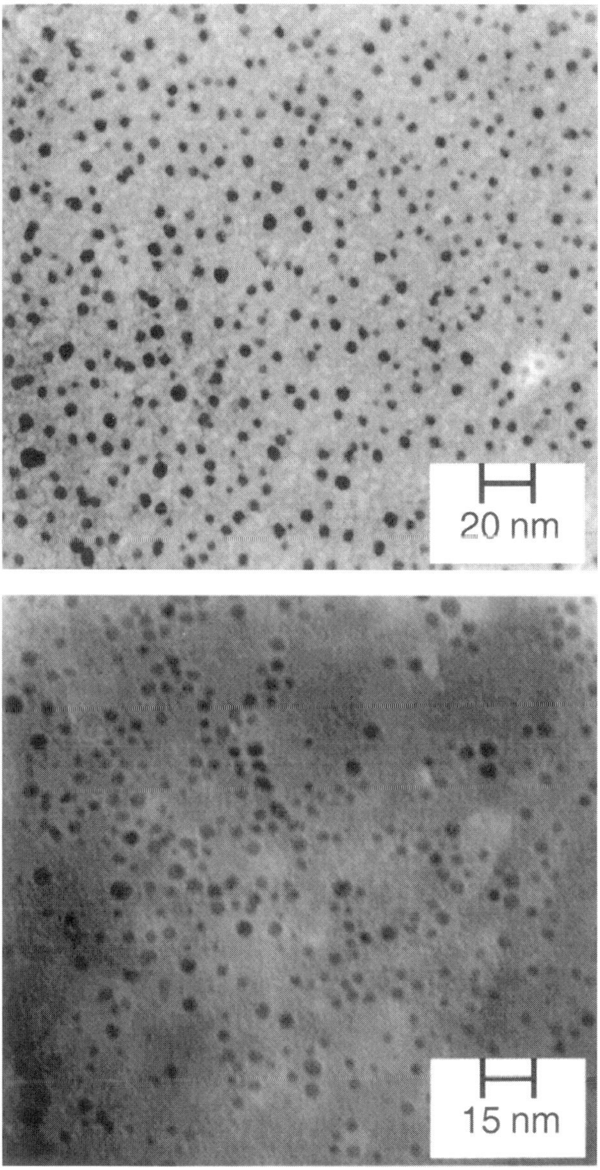

Figure 29 TEM images of the Ag nanoparticles prepared via RESOLV with (top) N_2H_4 and (bottom) $NaBH_4$ reduction.

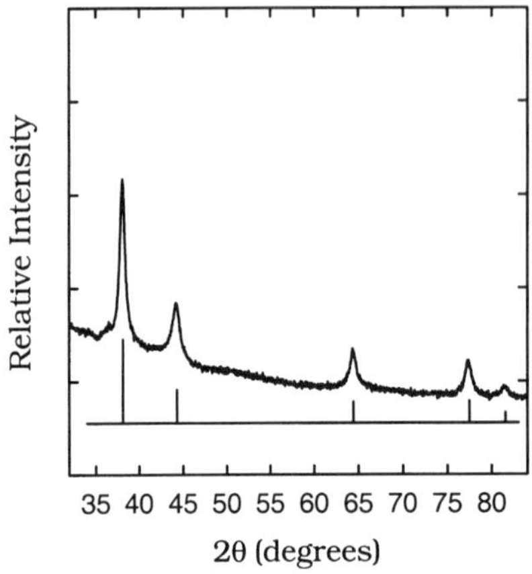

Figure 30 X-ray powder diffraction pattern of the Ag nanoparticles prepared via RESOLV with N_2H_4 reduction. The pattern for bulk Ag in the JCPDS database is also shown for comparison.

Ni particles produced. The expansion was at a constant pressure of about 3000 psia through a fused-silica capillary nozzle with an inner diameter of 77 μm. These Ni nanoparticles had an average particle size of 5.8 nm and a size distribution standard deviation of 0.54 nm, according to the TEM image (Figure 32). The particles were largely amorphous, exhibiting an extremely diffuse x-ray

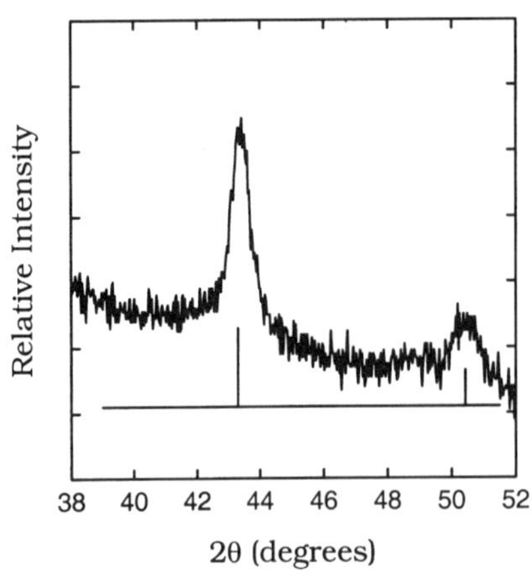

Figure 31 X-ray powder diffraction pattern of the Cu nanoparticles prepared via RESOLV with $NaBH_4$ reduction. The pattern for bulk Cu in the JCPDS database is also shown for comparison.

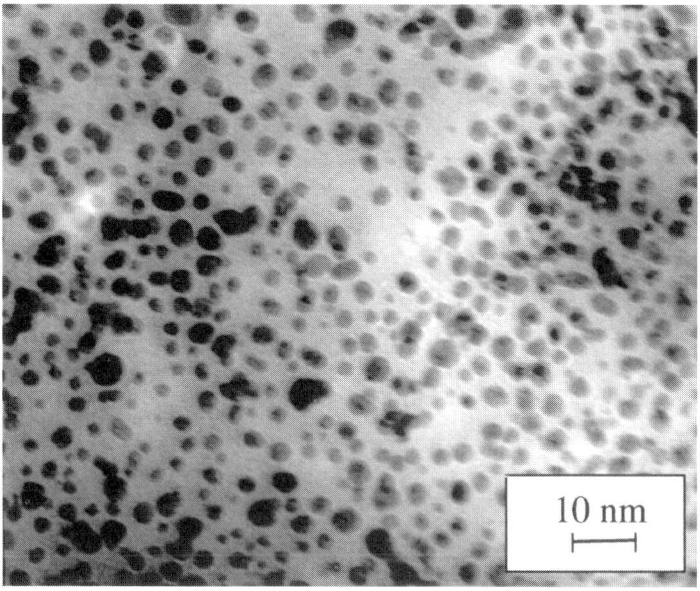

Figure 32 TEM image of the PVP-protected nickel nanoparticles prepared via RESOLV. (From Ref. 258.)

diffraction pattern. However, the crystallinity was significantly increased via thermal annealing at a high temperature, allowing characterization by powder x-ray diffraction (Figure 33).

Nanoscale Co particles were prepared and characterized in a similar fashion (258). These nanoparticles were also amorphous; thus, thermal annealing was necessary for characterization by powder x-ray diffraction (Figure 33).

In the application of the RESOLV chemical reduction method the extreme case was represented by the preparation of nanoscale iron particles. Because $NaBH_4$ was incapable of reducing Fe^{2+} to Fe^0, a stronger reducing agent such as $LiB(C_2H_5)_3H$ was required. However, because of the sensitivities of $LiB(C_2H_5)_3H$ toward water, alcohol, and many other organic functional groups, special experimental conditions were required. For example, anhydrous tetrahydrofuran (THF) distilled over sodium wires was used as the solvent for both the supercritical solution and the receiving ambient solution; and carefully dried polyethylene oxide polymer was used as the particle stabilization agent. The Fe nanoparticles obtained from the RESOLV chemical reduction process were found to have an average particle size of 7.6 nm and a size distribution standard deviation of 1.4 nm (258). The powder x-ray diffraction results for the as-prepared Fe nanoparticles are shown in Figure 33.

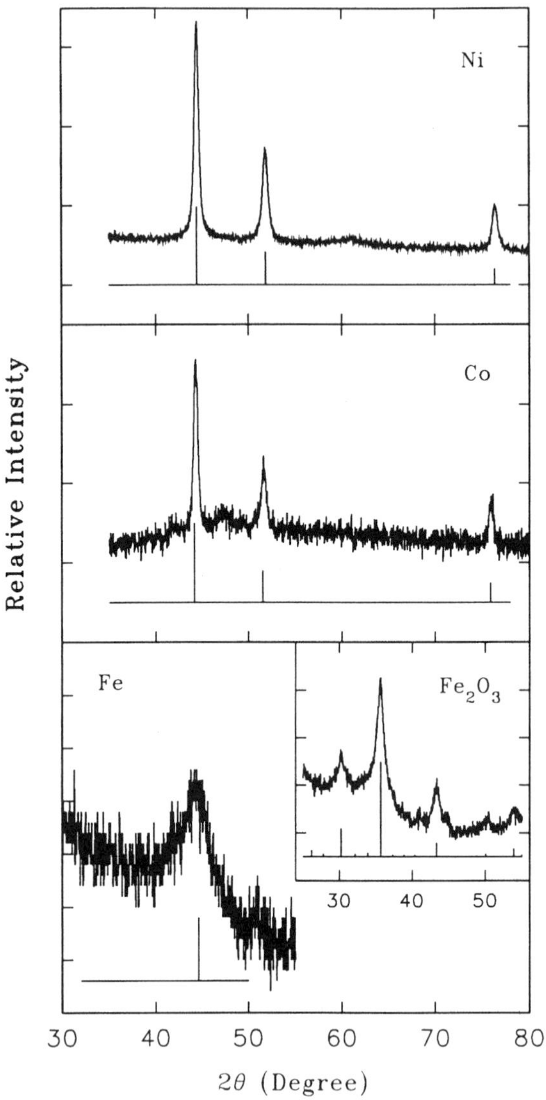

Figure 33 Results from X-ray powder diffraction measurements of the iron nanoparticles and heat-treated nickel, cobalt, and Fe_2O_3 nanoparticles. (From Ref. 258.)

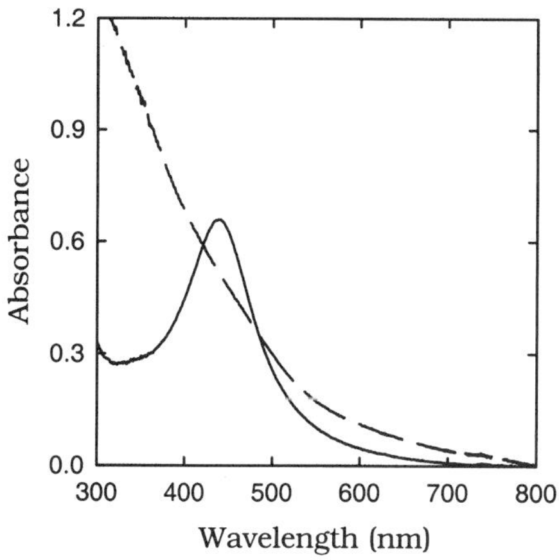

Figure 34 UV-Vis absorption spectra of the Ag nanoparticles in PVP polymer-protected suspension prepared via RESOLV. Solid line, as-prepared; dashed line, after dialysis against freshwater.

The nanoscale metal particles obtained via RESOLV were sensitive to oxidation both in stabilized suspensions and in the solid state. In particular, the Fe nanoparticles required storage under rigorously air-free conditions. Exposing a solid sample of the Fe nanoparticles to air in a sudden fashion resulted in fireworks-like flames. Under more controlled conditions, oxidation of the Fe nanoparticles produced iron oxide (Fe_2O_3) nanoparticles, which remained amorphous. The powder x-ray diffraction pattern of the thermally annealed Fe_2O_3 nanoparticles is also shown in Figure 33.

A similar oxidation process was observed for the other metal nanoparticles produced via RESOLV. For example, when a PVP polymer–stabilized suspension of Ag nanoparticles was purified via dialysis against freshwater, the UV-vis absorption spectrum was significantly altered (Figure 34). Gradual disappearance of the plasmon absorption band was probably due to the oxidation of the Ag nanoparticles (263). Similarly, oxidation of Cu nanoparticles in a suspension was evidenced by the suspension color changing gradually from dark yellow to blue (263).

B. Nanoscale Semiconductors

Among the most extensively studied quantum dots are semiconductor sulfide nanoparticles, with a tremendous amount of experimental and theoretical results being reported in the literature (264–271). These systems were used by Sun and coworkers in the development of the RESOLV method for nanoscale materials.

CdS nanoparticles were prepared via RESOLV by rapidly expanding a supercritical ammonia solution of $Cd(NO_3)_2$ into a room-temperature aqueous or ethanol solution of Na_2S (256,260). The CdS nanoparticles produced in the process formed a yellowish suspension under the protection of PVP polymer. The absorption spectrum of the suspension was typical of the quantum-confined CdS, with a shoulder at about 370 nm (Figure 35). The nanoparticles were characterized using x-ray powder diffraction and TEM, from which an average particle size of about 3.3 nm was estimated (256).

A more systematic investigation on the preparation of lead sulfide (PbS) nanoparticles via RESOLV was conducted (259). The procedure began with the preparation of a homogeneous solution of $Pb(NO_3)_2$ in liquid ammonia in the syringe pump (Figure 27). The solution was then heated to the preset temperature when passed through the long tubing coils in the heating unit. Finally, the supercritical ammonia/$Pb(NO_3)_2$ solution at 160°C and 3500 psia was rapidly expanded through a fused-silica capillary nozzle of 50 μm inner diameter into the expansion chamber, which contained a room-temperature solution of Na_2S

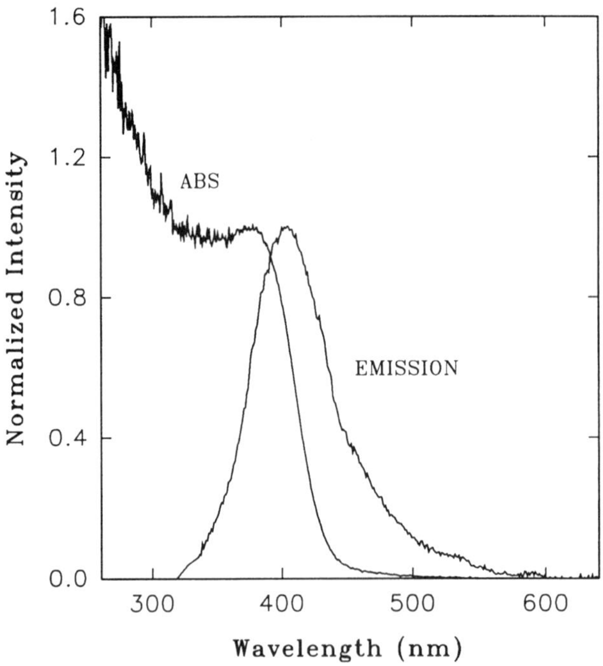

Figure 35 The UV-Vis diffuse reflectance spectrum and the luminescence spectrum (295 nm excitation) of CdS nanoparticles obtained from the RESOLV method with the supercritical ammonia solution expansion rate of 12 mL/min. (From Ref. 256.)

in ethanol, to produce PbS nanoparticles. The room-temperature solution in the expansion chamber also contained PVP polymer (6 mg/mL) for stabilizing the suspension of the PbS nanoparticles produced. The UV-vis absorption spectrum of the PbS nanoparticles in an ethanol suspension showed absorption onset at about 700 nm, with a clear shoulder at about 550 nm (Figure 36). According to a rough empirical correlation between particle size and absorption spectral position (271), the size of the PbS nanoparticles was estimated to be in the 2.5- to 4-nm range. More accurately, TEM analysis of the PbS nanoparticles (Figure 37) yielded an average particle size of 4 nm and a size distribution standard

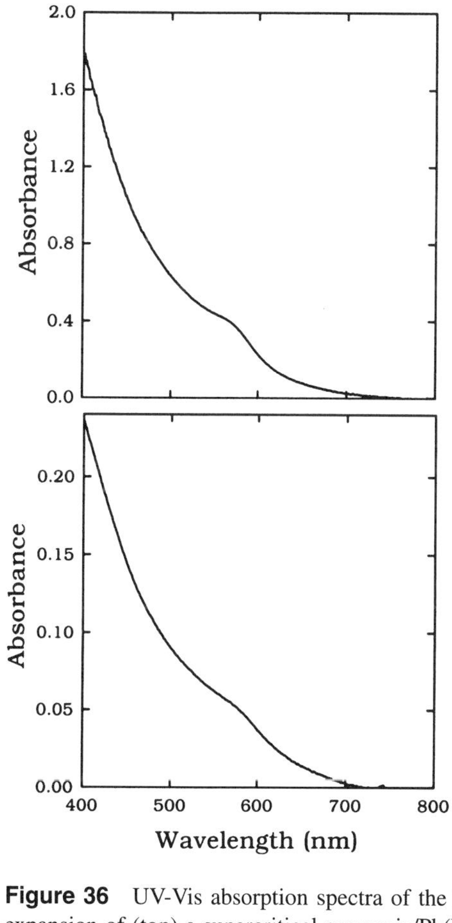

Figure 36 UV-Vis absorption spectra of the PbS nanoparticles obtained via the rapid expansion of (top) a supercritical ammonia/$Pb(NO_3)_2$ solution and (bottom) a supercritical acetone/$Pb(NO_3)_2$ solution into a room-temperature solution of Na_2S in ethanol. (From Ref. 259.)

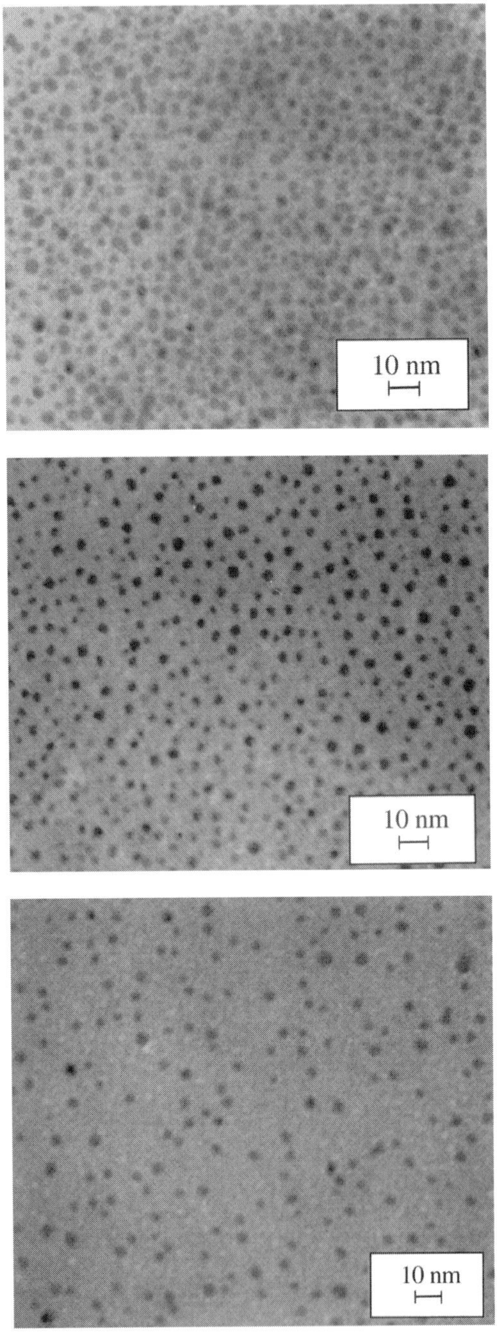
10 nm
10 nm
10 nm

deviation of 0.7 nm. On the other hand, the average crystal size of the PbS nanoparticles was estimated from the peak broadening in the x-ray diffraction pattern (Figure 38) in terms of the Debeye-Scherer equation (273).

$$D = K\lambda/(\beta \cos\theta) \tag{1}$$

where D is the average particle diameter in Å, β is the corrected band broadening (FWHM), K is a constant related to the crystallite shape and the manner in which D and β are defined, λ is the x-ray wavelength, and θ is the diffraction angle. The average size of 2.7 nm calculated in this way is significantly smaller than that obtained from TEM analysis, which is probably due to the PbS nanoparticles being partially amorphous (259).

PbS is a semiconductor that exhibits extreme quantum confinement effects, with the absorption spectral properties being highly sensitive to the physical parameters of the nanoparticles. Thus, the effects of the experimental parameters in RESOLV on the PbS nanoparticles produced can be evaluated in a systematic fashion. For example, PbS nanoparticles were prepared by the rapid expansion of a supercritical ammonia/$Pb(NO_3)_2$ solution at various temperatures. In the experiments all parameters [$Pb(NO_3)_2$ concentration, expansion nozzle size, and concentration of Na_2S and PVP polymer in the room-temperature ethanol solution] were held constant, except the temperature of the ammonia/$Pb(NO_3)_2$ solution for rapid expansion. The absorption spectra of the two PbS-nanoparticle samples obtained by rapid expansions at 160°C and 130°C were quite similar. The similarity in absorption properties probably reflected the fact that the PbS nanoparticles obtained at the two different preexpansion temperatures had similar particle sizes, which was supported by the similar x-ray powder diffraction patterns of the nanoparticle samples.

The effects of expansion nozzle size on the properties of the PbS nanoparticles were evaluated (259). The inner diameter of the capillary nozzle was varied (19, 50, and 77 μm), while other experimental parameters were held constant. Since the operating pressure was maintained near the designated value (3500 psia) during the RESS process, a smaller nozzle size resulted in a slower expansion. However, despite the significant difference in the rate of supercritical solution expansion, PbS nanoparticles obtained were similar. For the three PbS-nanoparticle samples obtained via RESOLV using different expansion nozzle sizes, the UV-vis absorption spectra were similar, indicative of their similar average particle sizes. This was confirmed by results from TEM analyses of these

---

Figure 37 TEM images of the PbS nanoparticles obtained via RESOLV with the rapid expansion of (top) supercritical ammonia/$Pb(NO_3)_2$ solution into ethanol/Na_2S solution, (middle) supercritical acetone/$Pb(NO_3)_2$ solution into ethanol/Na_2S solution, and (bottom) supercritical methanol/Na_2S solution into ethanol/$Pb(NO_3)_2$ solution. (From Ref. 259.)

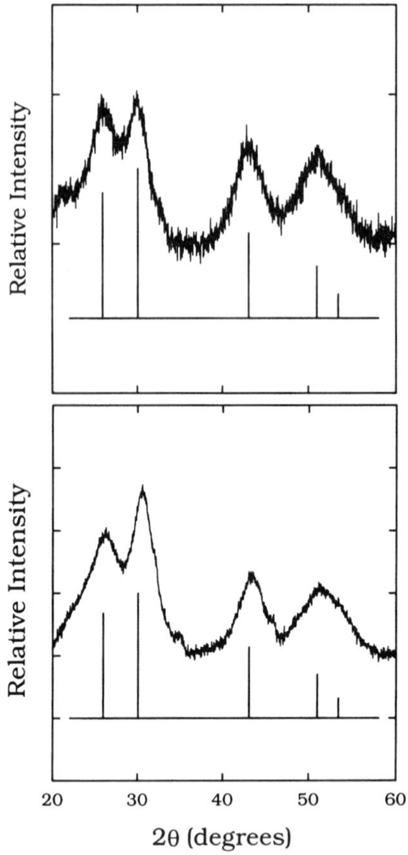

Figure 38 X-ray powder diffraction patterns of the PbS nanoparticles obtained via RESOLV with the rapid expansion of (top) supercritical ammonia/$Pb(NO_3)_2$ solution and (bottom) supercritical methanol/$Pb(NO_3)_2$ solution into ethanol/Na_2S solution. The pattern for bulk PbS in the JCPDS database is also shown for comparison. (From Ref. 259.)

samples (Table 3), which showed the PbS nanoparticles to have an average particle size of about 4 nm (259). Apparently, the change in expansion nozzle size only marginally effected the properties of the PbS nanoparticles produced.

The properties of the PbS nanoparticles produced via RESOLV were also found to be insensitive to selection of the supercritical solvents. For example, methanol was used in place of ammonia for preparation of a homogeneous methanol solution of $Pb(NO_3)_2$ in the syringe pump. After the solution was heated to the preset temperature in the heating unit of the experimental apparatus (Figure 27), the supercritical methanol/$Pb(NO_3)_2$ solution at 250°C and

Table 3 A Comparison of PbS Nanoparticles Produced Under Different Conditions (nm) (259)

	19-μm nozzle		50-μm nozzle		77-μm nozzle	
Solvent	*D*	σ	*D*	σ	*D*	σ
Ammonia	3.9	0.65	3.9	0.7	4.3	0.8
Methanol	4.1	0.5	3.9	0.6	4.2	0.8
Acetone	4.5	0.7	4.1	0.8	4.6	1

The average particle size in diameter (*D*, nm) and the size distribution standard deviation (σ, nm) were determined via TEM analyses.

3500 psia was rapidly expanded through a fused-silica capillary nozzle into the expansion chamber, which contained a room-temperature solution of Na_2S and PVP polymer in ethanol. The PbS nanoparticles obtained were characterized, and the results were analyzed for comparison with those of the PbS nanoparticles produced using the ammonia expansion. The absorption spectrum of the PbS nanoparticles obtained with the methanol expansion was similar to that of the particles obtained with the ammonia expansion; the x-ray powder diffraction patterns were also similar (Figure 38), as well as the TEM results for the two expansions (Table 3). In addition to methanol, acetone was used with the RESOLV method. The formation of PbS nanoparticles also resulted from the rapid expansion of a supercritical acetone/$Pb(NO_3)_2$ solution at 250°C and 3500 psia into a room-temperature ethanol solution of Na_2S. According to UV-vis absorption (Figure 36) and TEM (Figure 37) results, the PbS nanoparticles thus produced were similar to those obtained with the ammonia and methanol expansions (Table 3). Finally, methanol and acetone expansions with different nozzle sizes resulted in similar PbS nanoparticles (Table 3).

Sun and coworkers also demonstrated that the properties of the PbS nanoparticles were independent of which ion solution was used in the rapid expansion. For example, the rapid expansion of a supercritical methanol/Na_2S solution into a room-temperature solution of $Pb(NO_3)_2$ in ethanol resulted in the formation of PbS nanoparticles. The UV-vis absorption, x-ray, and TEM (Figure 38) results showed that the PbS nanoparticles thus produced were similar to those obtained via the rapid expansion of a supercritical ammonia/$Pb(NO_3)_2$ solution into a Na_2S solution.

The fact that similar PbS nanoparticles were obtained under a variety of operational conditions points to a useful feature of the RESOLV method; it allows the production of nanoscale semiconductors in a consistent and reproducible fashion. The negative aspect of the insensitivity to changes in experimental parameters is that this limits the ability to manipulate the nanoparticle properties by varying the RESS processing conditions. However, it should also be recognized that the changes in experimental parameters investigated by Sun and coworkers were relatively slight, and thus the results did not necessarily rule out the possibility of producing nanoparticles of different properties through more dramatic changes in the RESOLV processing parameters.

Mechanistically, the RESOLV method for producing PbS nanoparticles has been discussed in the same frame of reference as that for the conventional RESS production of micrometer-sized (submicrometer-sized, in some cases) particles (31,71,73,82–84). According to Sun and coworkers, a possible scenario is that the formation of PbS nanoparticles takes place through the reaction of S^{2-} species with the Pb^{2+} species in nanoscopic "solute droplets" that are produced in the rapid expansion process. With conventional RESS, the solute $Pb(NO_3)_2$ rapidly precipitates out of the supercritical solution to form initially species that may be considered as "solute droplets"; then these nanoscopic species form micrometer- or submicrometer-sized particles via agglomeration (255). However, the receiving liquid in RESOLV may temporarily stabilize the initially formed nanoscopic solute droplets and allow their capture by the reactant species (S^{2-}) in the same receiving solution to form PbS nanoparticles, which are subsequently stabilized by the protection agent (259).

Other nanoscale semiconductors that have been produced via RESOLV include ZnS (263), CdSe (263), and Ag_2S (257,260,263). The experimental conditions used for producing these semiconductor nanoparticles were generally similar to those for CdS and PbS discussed above. These nanoparticles were also characterized using UV-vis absorption, x-ray powder diffraction, and TEM methods. For example, shown in Figure 39 is the absorption spectrum of the ZnS nanoparticles in an ethanol suspension, which exhibits a significant quantum-confinement effect (263). An x-ray powder diffraction pattern and TEM images of the Ag_2S nanoparticles obtained via RESOLV are shown in Figures 40 and 41, respectively (257,260,263).

C. RESOLV with Microemulsions in CO_2

Since CO_2 is widely considered to be the desirable SCF because of its environmentally benign characteristics and ambient critical temperature, the use of supercritical CO_2 for the preparation and processing of nanomaterials has naturally received considerable attention. However, as discussed in previous sections, the poor solubility of most solutes in supercritical CO_2 represents a major limitation. Surfactants containing both CO_2-soluble and hydrophilic moieties are often added to CO_2 to form reverse micelles. Such water-in-CO_2 microemulsions offer a convenient means of dissolving hydrophilic compounds, as demonstrated by the in situ methods for the preparation of nanoparticles (240,247-249). For the production of nanoscale metals and semiconductors via RESOLV, on the other

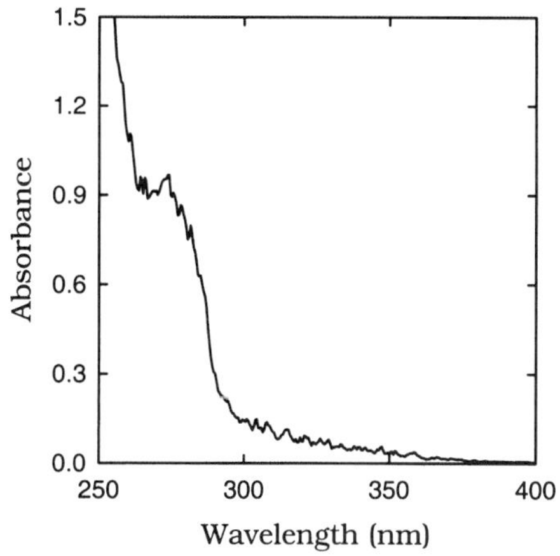

Figure 39 The absorption spectrum of the ZnS nanoparticles in PVP polymer–protected suspension prepared via RESOLV.

hand, the examples discussed above were based on the use of high-temperature supercritical solvents such as ammonia and THF. In an effort to replace the organic solvents with CO_2-based systems for RESOLV at ambient temperatures, Sun and coworkers investigated the use of water-in-CO_2 microemulsions to dissolve metal ions and the subsequent preparation of nanoparticles via RESOLV.

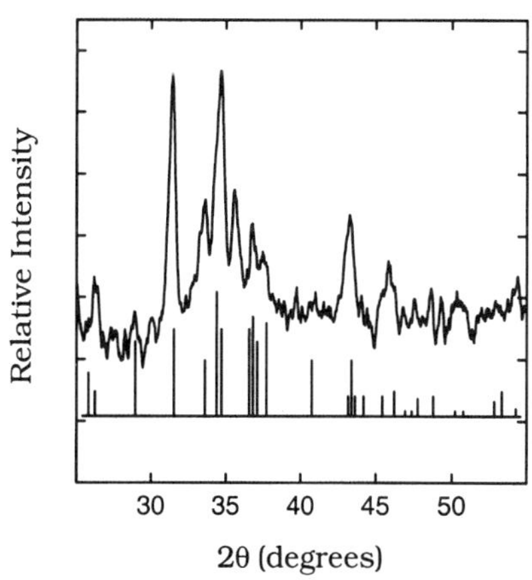

Figure 40 X-ray powder diffraction pattern of the Ag_2S nanoparticles obtained via RESOLV with the rapid expansion of supercritical ammonia/$AgNO_3$ solution into ethanol/Na_2S solution. The pattern for bulk Ag_2S in the JCPDS database is also shown for comparison. (From Ref. 260.)

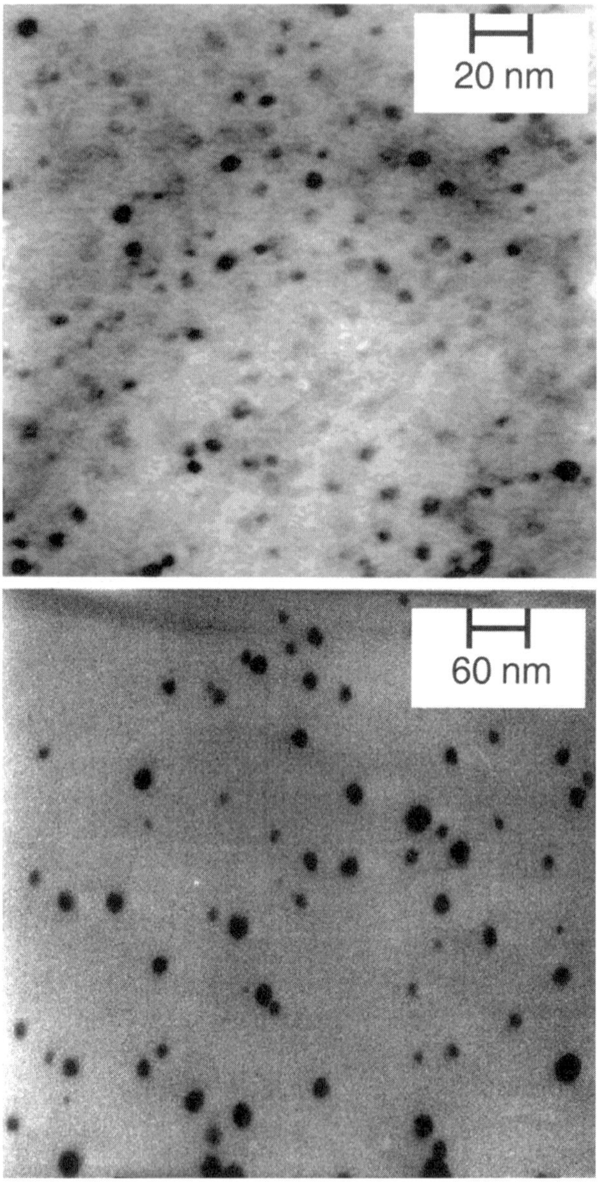

Figure 41 TEM images of the Ag_2S nanoparticles prepared via RESOLV with the rapid expansion of supercritical ammonia/$AgNO_3$ solution into (top) ethanol/Na_2S solution and (bottom) water/Na_2S solution and with (top) PVP polymer and (bottom) bovine serum albumin protein as stabilization agents, respectively.

They compared the results with those obtained from the use of organic solvents (262) and found that the properties of the nanoparticles produced were dependent on the preexpansion microemulsion conditions (260,262).

1. Dissolution of Metal Ions in Microemulsions

Sun and coworkers investigated the dissolution of metal salts in the reverse-micellar emulsions in CO_2, including a systematic evaluation of the experimental parameters such as surfactant concentration, water content, and salt concentration (274). $Cu(NO_3)_2$ was used in the investigation as an absorption spectroscopic probe because of the distinctive blue color of Cu^{2+} (strong red near-infrared absorption) in the aqueous phase. Shown in Figure 42 is a comparison of the absorption spectra of Cu^{2+} in an ambient aqueous solution vs. a water-in-CO_2 microemulsion stabilized by the surfactant perfluoropolyether ammonium carboxylate (PFPE-NH_4) (236,244). The absorption spectrum obtained from the microemulsion is blue shifted. According to McCleskey and coworkers, the blue shift might be attributed to interactions between Cu^{2+} and the hydrophilic head group of PFPE-NH_4 and also to a decrease in the polarity of water in the reverse-micellar core (275).

At a constant W_0 value of 4, the observed absorbance increases linearly with $Cu(NO_3)_2$ concentration in the water core (Figure 43) up to the maximum solubility of about 3.5 M (274). Since the reverse micelles are homogeneously distributed in the continuous-phase CO_2 in the high-pressure optical cell, the $Cu(NO_3)_2$ concentration with respect to the cell volume can be used

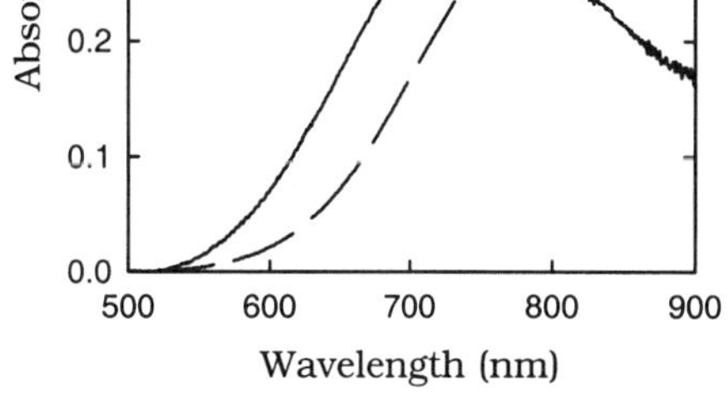

Figure 42 Absorption spectra of Cu^{2+} in an aqueous solution (solid line) and in the aqueous core of a water-in-CO_2 microemulsion (dashed line). (From Ref. 274.)

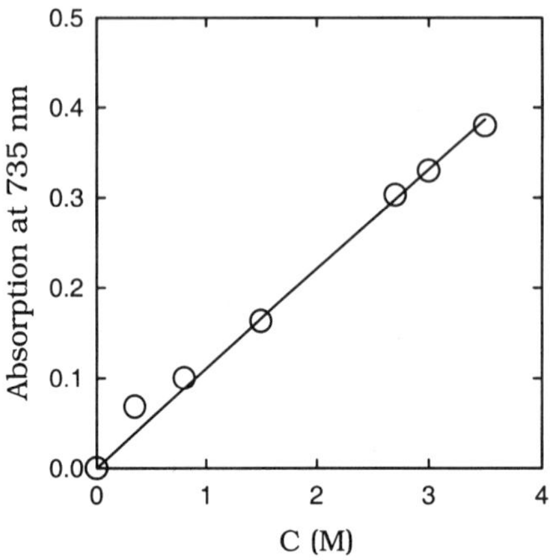

Figure 43 The absorbance at the spectral maximum as a function of the Cu^{2+} concentration in the aqueous core of a water-in-CO_2 microemulsion ($W_0 = 12$). (From Ref. 274.)

to calculate the molar absorptivity of Cu^{2+} in the water-in-CO_2 microemulsion. The molar absorptivity value thus obtained for the absorption band maximum is 25.8 M^{-1} cm^{-1}, which is almost three times larger than the value of 7.8 M^{-1} cm^{-1} obtained in an ambient aqueous $Cu(NO_3)_2$ solution (274).

The W_0 value of the water-in-CO_2 microemulsion was varied in two different ways: by changing the water content at a constant surfactant concentration and by changing the surfactant concentration at a constant water content (274). For a fixed water content, the $Cu(NO_3)_2$ concentration was also fixed. Thus, a decrease in the surfactant concentration resulted in an increase in the W_0 value. The absorption spectra of Cu^{2+} in the microemulsions were similar up to a W_0 value of 15, as shown in Figure 44. However, for larger W_0 values, the observed absorption spectra became significantly weaker due to some Cu^{2+} solution being excluded from the micelles because the surfactant concentration was insufficient to support the large micelles.

On the other hand, when the surfactant concentration was fixed, an increase in the water content resulted in a corresponding increase in the W_0 value. Since the $Cu(NO_3)_2$ concentration in the aqueous phase was also fixed, the increase in water content resulted in a proportional increase in the observed absorbance, as shown in Figure 45. However, the linear relationship ceased at a threshold W_0 value because of, again, insufficient surfactant concentration to support stable microemulsions. The threshold W_0 value was apparently dependent on the $Cu(NO_3)_2$ concentration in the aqueous phase; a lower $Cu(NO_3)_2$ concentration resulted in a significantly higher threshold W_0 value (Figure 45).

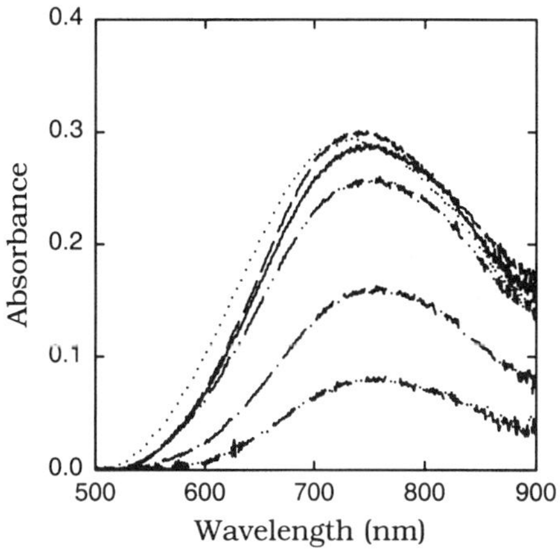

Figure 44 Absorption spectra of Cu^{2+} in water-in-CO_2 microemulsions containing the same amount of aqueous Cu^{2+} solution (2.7 M) but with different surfactant PFPE concentrations and, consequently, different W_0 values. (W_0 = 2, ——; 4, – –; 7, ·······; 15, – ·· –; 37, –· –; 70, – ·· –). (From Ref. 274.)

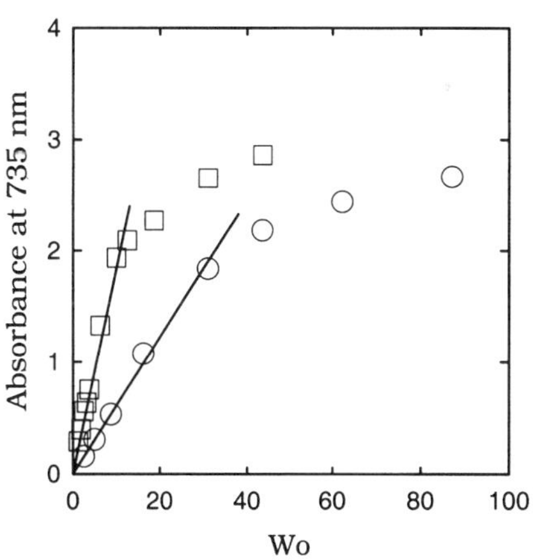

Figure 45 The absorbance at the spectral maximum as a function of the W_0 value – constant PFPE concentration but varying amounts of aqueous Cu^{2+} solution with concentrations of 0.8 M (circle) and 2.7 M (square). (From Ref. 274.)

These results are comparable to those for the formation of reverse micelles in CO_2 without salts in the water core. According to Johnston and coworkers, the threshold W_0 value is 20.7 for the water-in-CO_2 microemulsion to remain single phase at 45°C and 158.1 bar (236).

2. RESOLV with Microemulsions for Nanoparticles

Sun and coworkers investigated the production of nanoscale metal particles via RESOLV with water-in-CO_2 microemulsions (262,263). Ag was selected as a model system because the nanoparticles had been prepared via RESOLV with ammonia and because the high crystallinity of the nanoparticles would enable easier characterization using x-ray powder diffraction. The experiment involved the solubilization of $AgNO_3$ in a water-in-CO_2 microemulsion ($W_0 = 5$) in a syringe pump. The reverse-micellar solution of aqueous-$AgNO_3$/PFPE-NH_4 in CO_2 was rapidly expanded via a 50-μm fused-silica capillary nozzle into a room-temperature solution of $NaBH_4$ in ethanol for chemical reduction. The Ag nanoparticles produced in the rapid expansion/chemical reduction were protected from agglomeration in the presence of PVP polymer (5 mg/mL) and formed a stable suspension. The absorption spectrum of the suspension showed the characteristic surface plasmon absorption of Ag nanoparticles (276,277), and the x-ray powder diffraction pattern was typical of nanoscale Ag particles. TEM analysis of the as-prepared nanoparticles (Figure 46) yielded an average particle size of 7.8 nm and a size distribution standard deviation of 2.5 nm. These results are similar to those obtained via RESOLV with supercritical ammonia solution. However, the CO_2-based system has obvious advantages, including the environmentally benign nature of the solvent and the near-ambient system temperature. The latter is important in RESOLV processing of some metal salts that decompose at high temperatures. In addition, the preexpansion conditions for the reverse micelles may be used to influence the properties of the nanoparticles produced.

Sun and coworkers found that the average size of the Ag nanoparticles was dependent on the W_0 value of the reverse-micellar solution in supercritical CO_2. For example, reducing the amount of surfactant PFPE-NH_4 in CO_2 increased the W_0 value of the microemulsion from 5 to 20, with the $AgNO_3$ concentration in the aqueous core being kept constant. The rapid expansion of such an aqueous-$AgNO_3$/PFPE-NH_4/CO_2 system of a higher W_0 value of 20 into a room-temperature solution of $NaBH_4$ in ethanol again yielded Ag nanoparticles. While the absorption spectrum was similar to that of the suspension obtained via RESOLV with the microemulsion having a W_0 value of 5, TEM analysis revealed significantly larger particles (Figure 46), with an average particle size of 14.5 nm and a size distribution standard deviation of 5.8 nm. The trend was validated by results for the microemulsion having an intermediate W_0 value.

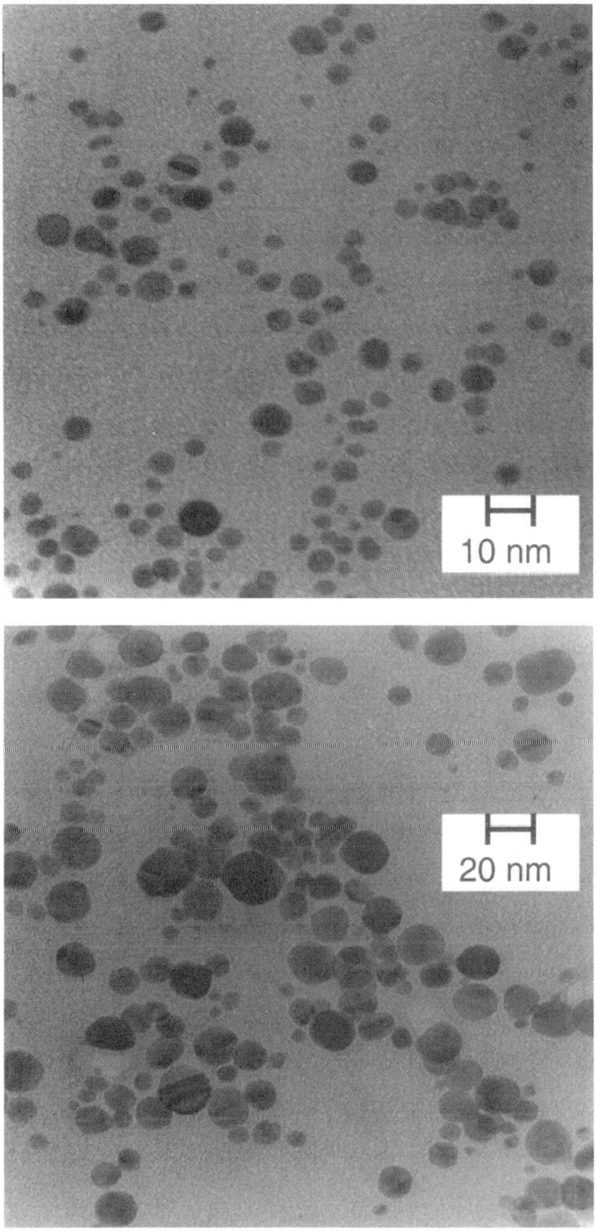

Figure 46 TEM images of the Ag nanoparticles prepared via RESOLV with the water-in-CO_2 microemulsions of W_0 equal to 5 (top) and 20 (bottom). (From Ref. 262.)

The Ag nanoparticles prepared via RESOLV with the microemulsion having a W_0 value of 12 had an average particle size of 10.4 nm and a size distribution standard deviation of 3.8 nm (263).

The W_0 value determines the average size of the reverse micelles and, consequently, the water and salt content of the microemulsion in CO_2. According to Zielinski et al., PFPE-NH_4-based water-in-CO_2 microemulsions having W_0 values of 11 and 31 contained water droplets with average radii of about 2 and 3.5 nm, respectively; and the droplets and droplet structure were negligibly affected by experimental parameters such as system pressure (254). Based on those results, Sun and coworkers estimated that the average size of the reverse-micellar core for the microemulsions used in their study was <2 and 2–3.5 nm for W_0 values of 5 and 20, respectively (262). Even with the surfactant molecules included, the average micellar diameter should still be significantly smaller than the expansion nozzle diameter, which would allow the micelles to survive, at least to some extent, the rapid expansion process. In this context, Sun and coworkers proposed that the reverse micelles might be playing a more significant role than that of dissolving metal ions in CO_2. The micelles might, in fact, be present throughout the RESOLV process, directly affecting the properties of the nanoparticles produced. The observed larger Ag nanoparticle sizes were attributed to agglomeration in the postexpansion chemical reduction under ambient conditions.

The same microemulsions in CO_2 were used to dissolve other metal ions, such as $Cu(NO_3)_2$ and $Pb(NO_3)_2$, for the preparation of nanoparticles (Cu and PbS) via RESOLV (263). The nanoparticles produced were similar to those obtained via RESOLV with ammonia as the supercritical solvent.

According to Sun and coworkers, the nanoparticles obtained via RESOLV with the rapid expansion of water-in-CO_2 microemulsions generally had a broader size distribution than those with ammonia as the supercritical solvent. As a result, opportunities were explored for optimization of the processing conditions to narrow the nanoparticle size distribution, and the results were encouraging. For example, the Ag nanoparticles obtained by rapid expansion of a microemulsion ($W_0 = 12$) into a more basic room-temperature ethanol solution of $NaBH_4$ exhibited a significantly better defined surface plasmon absorption band (Figure 47), indicative of a narrower particle size distribution (177,276). The narrower size distribution was confirmed by TEM analysis (Figure 48) (262,263).

The use of water-in-CO_2 microemulsions with RESOLV for nanoparticle production serves as an alternative to the in situ reaction method reported by Wai, Fulton, and coworkers (247,248) and by Johnston and coworkers (240,249). However, a significant feature of the RESOLV method is that the formation of nanoparticles occurs under ambient conditions outside the SCF chamber, avoiding difficulties associated with the in situ reaction method regarding collection of the as-prepared nanoparticles. In addition, as discussed in previous sections,

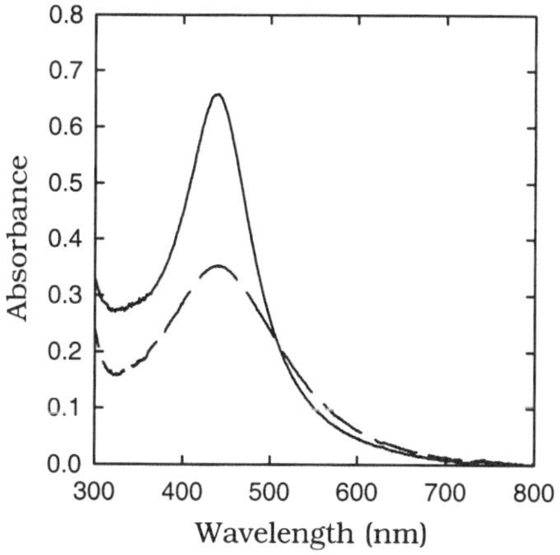

Figure 47 UV-Vis absorption spectra of the Ag nanoparticles obtained via rapidly expanding the aqueous/$AgNO_3$-in-CO_2 microemulsion ($W_0 = 12$) into room-temperature $NaBH_4$-ethanol solutions of pH equal to 8 (dashed line) and 10 (solid line). (From Ref. 262.)

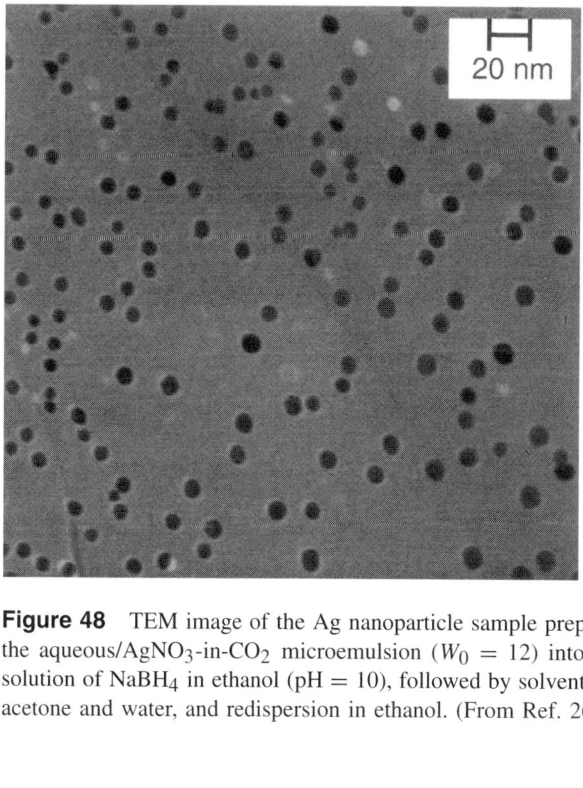

Figure 48 TEM image of the Ag nanoparticle sample prepared via rapidly expanding the aqueous/$AgNO_3$-in-CO_2 microemulsion ($W_0 = 12$) into a room-temperature basic solution of $NaBH_4$ in ethanol (pH = 10), followed by solvent evaporation, washing with acetone and water, and redispersion in ethanol. (From Ref. 262.)

rapid expansion of SCF is associated with several popular methods of materials preparation and fabrication. The study of Sun and coworkers is a valuable example of the use of microemulsions in the rapid expansion–based methods.

D. Optical Properties and Related Applications

The nanoscale metal and semiconductor particles produced via RESOLV typically form stable suspensions under the protection of a stabilization agent such as PVP or polyethylene oxide polymer. These suspensions have permitted the study of optical and other properties of nanoparticles under more controllable and reproducible conditions.

1. Luminescence

Sun and coworkers reported that the CdS nanoparticles produced via RESOLV with supercritical ammonia exhibited interesting luminescence properties. The luminescence spectrum contained an intense exciton emission band at about 400 nm (Figure 35), which is typical of suspended nanoscale CdS particles (256). However, the spectrum showed essentially no surface defect emissions in the long-wavelength region. For CdS nanoparticles prepared by other methods, surface defects such as dangling bonds and vacancies are typically treated by chemical passivation techniques, which results in the disappearance of defect luminescence and the appearance of strong exciton emission (264). Ammonia has been used as an effective passivation agent for several nanoscale semiconductors; however, for the CdS nanoparticles generated in other methods, surface defect emissions are significant—even with ammonia passivation (278). Thus, the absence of defect luminescence with the CdS nanoparticles obtained by the rapid expansion of supercritical ammonia may be due to some special surface passivation effects; such effects warrant further investigation.

2. Optical Limiting

Materials that exhibit optical-limiting responses are often called optical limiters (279–285). An ideal optical limiter has linear transmittance at low incident light fluences but becomes opaque at high incident light fluences. Among the widely investigated materials for optical-limiting applications are organic dyes such as metallophthalocyanines and porphyrins, fullerenes, and nanomaterials, including carbon black and carbon nanotube suspensions. Mechanistically, optical-limiting organic dyes and fullerenes are generally considered to be nonlinear absorbers (or reverse saturable absorbers), whereas carbon black suspensions undergo dramatic changes in transmittance due to laser irradiation–induced nonlinear scattering. However, mechanistic descriptions of the nonlinear absorption and nonlinear scattering processes in these materials are still subjects of debate, especially in view of the recent results concerning the concentration and medium dependencies in the optical-limiting responses of many nonlinear absorbers (286).

Sun and coworkers reported that Ag-containing nanoparticles produced via RESOLV exhibited excellent optical-limiting responses toward nanosecond laser pulses at 532 nm (257,287). These nanoparticles formed stable transparent suspensions in the presence of PVP polymer, which appeared to be indistinguishable from typical homogeneous solutions; thus, nanoparticle suspensions of high optical quality allowed quantitative optical-limiting measurements and also direct comparison of results with those from organic optical limiters. The physical and structural parameters of the nanoparticles used in the optical-limiting measurements are given in Table 4. Shown in Figures 49 and 50 are typical optical-limiting responses of Ag_2S and Ag nanoparticles, respectively, in stable ethanol suspensions to 5-ns laser pulses at 532 nm. These strong optical limiters, even at 90% linear transmittance, are significantly more effective than the benchmark limiters (C_{60} in toluene solution and chloroaluminum phthalocyanine in DMF solution) at the same linear transmittance (257). In a comparison of the Ag-containing nanomaterials, the Ag_2S nanoparticles were found to be more effective optical limiters than Ag nanoparticles when examined in similarly stable transparent suspensions.

The role of the Ag became obvious when the optical-limiting results for the Ag-containing nanomaterials were compared with those for other nanoparticles, including nanoscale CdS, PbS, and Ni particles in stable suspensions (257). The nanoparticles that contained no Ag were found to be considerably weaker optical limiters. For example, the Ni nanoparticles in a stable transparent suspension exhibited only marginal optical-limiting response to 5-ns laser pulses at 532 nm (Figure 50) (257).

Table 4 Physical and Structural Parameters of the Metal and Metal Sulfide Nanoparticles for Optical Limiting Measurements (257)

Particle	Supercritical solution	RT solution	Stabilization agent	X-ray diffraction	TEM Size (nm)	TEM σ (nm)[a]
Ag_2S	Ammonia	Ethanol	PVP	Monoclinic	7.3	1.7
CdS	Ammonia	Water	Gelatin	Cubic	~5	—
PbS	Methanol	Methanol	PVP	Cubic	6.6	1.0
Ag	Ammonia	Ethanol	PVP	Cubic	5.6	0.78
Ni	Ethanol	DMF	PVP	Cubic	5.8	0.54

[a]Size distribution standard deviation.

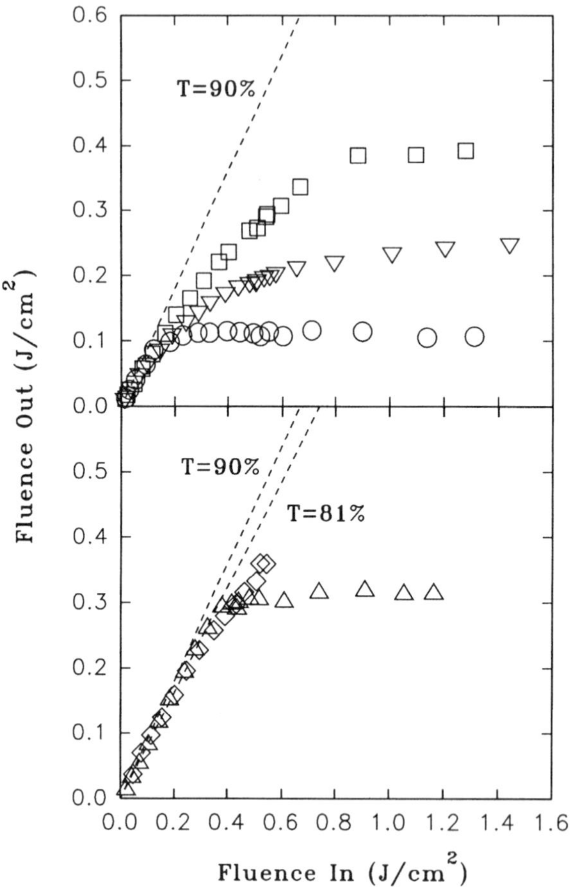

Figure 49 Optical limiting responses of the nanocrystalline Ag_2S particles in a PVP polymer–stabilized ethanol suspension (○) of 90% linear transmittance at 532 nm are compared with those of C_{60} in toluene (□) and chloroaluminum phthalocyanine in DMF (▽) of the same linear transmittance and those of the CdS nanoparticle suspension (◇) of 81% linear transmittance and the PbS nanoparticle suspension (△) of 90% linear transmittance. (From Ref. 257.)

Recently, Sun and coworkers evaluated the optical-limiting properties of Ag nanoparticles produced via RESOLV with supercritical ammonia as opposed to water-in-CO_2 microemulsion and with hydrazine reduction as opposed to $NaBH_4$ reduction (287). The nanoparticles obtained with the rapid expansion of a water-in-CO_2 microemulsion had a significantly broader particle size distribution than those with the supercritical ammonia solution; and the nanoparticles obtained from the hydrazine reduction were on average 50% larger than those

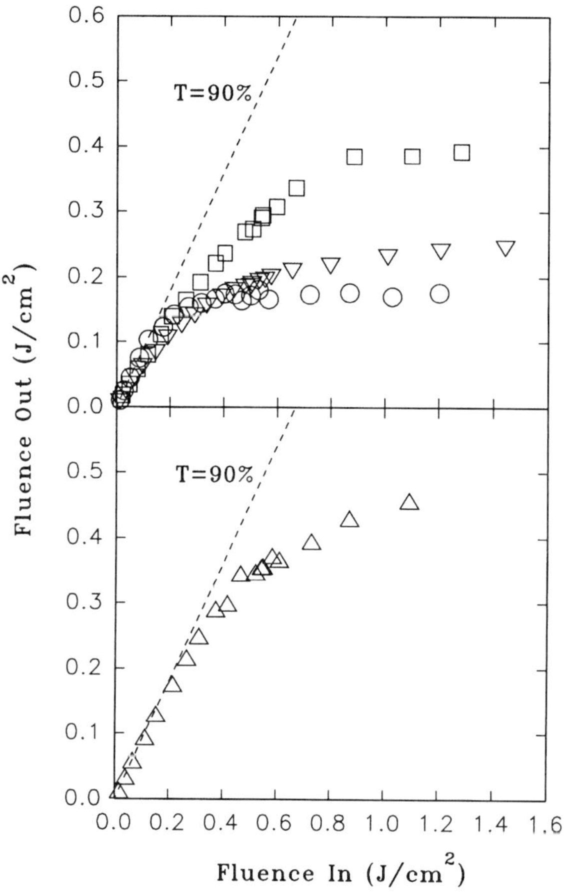

Figure 50 Optical limiting responses of the nanocrystalline Ag metal particles in PVP polymer–stabilized ethanol suspension (○) of 90% linear transmittance at 532 nm are compared with those of C_{60} in toluene (□) and chloroaluminum phthalocyanine in DMF (▽) and the Ni metal nanoparticles in DMF suspension (△) of the same linear transmittance. (From Ref. 257.)

from the $NaBH_4$ reduction. However, the optical-limiting responses of all of these nanoparticles in PVP polymer–stabilized suspensions were found to be similar (Figure 51).

3. Polymeric Nanocomposite Films

Adding appropriate polymers to the solution-like suspensions, Sun and coworkers prepared polymer films containing homogeneously dispersed nanoparticles

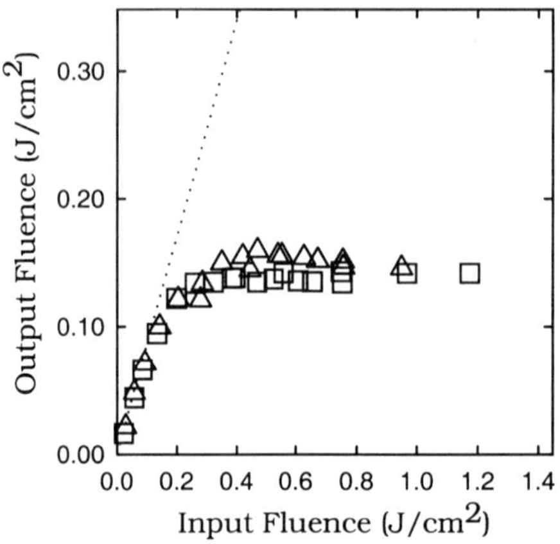

Figure 51 Optical limiting responses of the Ag_2S nanoparticles prepared via RESOLV with the rapid expansion of a supercritical/ammonia solution (narrow particle size distribution) (□) and a water-in-CO_2 microemulsion (broader particle size distribution) (△).

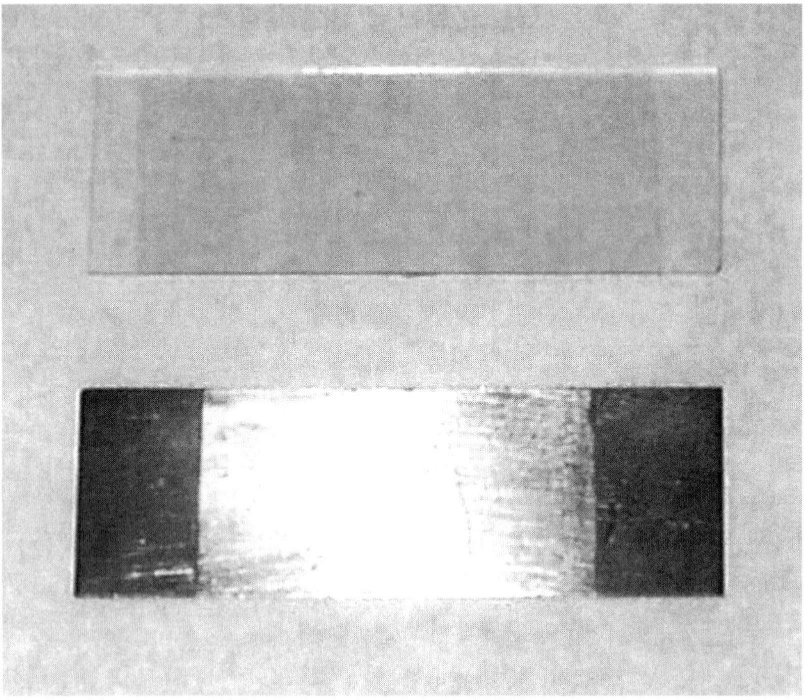

Figure 52 Photoconductive PVK-PbS nanoparticle composite thin films on a glass slide (top) and a copper substrate (bottom).

through the use of wet-casting methods (288). For example, a suspension of nanoscale PbS particles was mixed with poly(vinylcarbazole) to form a highly viscous polymer blend, which was then spin-cast into a thin polymer film (Figure 52). The transparent nanocomposite film of PbS nanoparticles embedded homogeneously in a poly(vinylcarbazole) matrix was found to have interesting photoconductive properties (260,288).

Polymer-nanoparticle composite materials have a wide range of existing and potential applications. Since the RESOLV method allows considerable flexibility in the selection of stabilization agents, including the use of the matrix polymer itself as a stabilization agent, polymer-nanoparticle composite films may be prepared that are free from foreign substances. These "clean" nanocomposite materials are particularly useful in biomedical applications.

In conclusion, SCF technology has been widely applied to the synthesis, processing, and fabrication of various materials. However, the use of SCF technology in the development of nanoscale materials still represents an emerging and progressive research area. In comparison with other more conventional techniques, the SCF methods not only serve as alternatives but also offer unique advantages and opportunities. We expect that the preparation and processing of nanoparticles and other nanomaterials by SCF technology will continue to receive significant attention and undergo broader based advances.

ACKNOWLEDGMENTS

We thank M. Whitaker and D. Elgin for assistance in the preparation of the manuscript. This work was made possible by the support of the Department of Energy under Contracts DE-FG02-00ER45859 (Y.-P.S.) and DE-AC07-99ID13727 (H.W.R.), the National Science Foundation under Grants CHE-9729756 and EPS-9977797 and through the Center for Advanced Engineering Fibers and Films (Y.-P.S), and the Air Force Office of Scientific Research and Dr. J. Tishkoff (C.E.B.).

REFERENCES

1. CE Bunker, HW Rollins, Y-P Sun. Chapter 1 of this book; and the references cited therein.
2. RW Shaw, TB Brill, AA Clifford, CA Eckert, EU Franck. Chem Eng News 69:26, 1991.
3. (a) PE Savage, S Gopalan, TI Mizan, CJ Martino, EE Brock. AIChE J 41:1723, 1995; (b) T Clifford, K Bartle. Chem Ind 12:449, 1996.
4. D Andrew, BT Des Islet, A Margaritis, AC Weedon. J Am Chem Soc 117:6132, 1995.
5. BJ Hrnjez, AJ Mehta, MA Fox, KP Johnston. J Am Chem Soc 111:2662, 1989.

6. Y Kimura, Y Yoshimura, M Nakahara. J Chem Phys 90:5679, 1989.
7. Y Kimura, Y Yoshimura. J Chem Phys 96:3085, 1992.
8. RD Weinstein, AR Renslo, RL Danheiser, JG Harris, JW Tester. J Phys Chem 100:12337, 1996.
9. NS Isaacs, NJ Keating. J Chem Soc Chem Commun 876, 1992.
10. CE Bunker, HW Rollins, JR Gord, Y-P Sun. J Org Chem 62:7324, 1997.
11. H Ksibi, P Subra, Y Garrabos. Adv Powder Technol 6:25, 1995.
12. P Subra, P Testin. Powder Technol 103:2, 1999; and references cited therein.
13. M Weber, MC Thies. Chapter 11 of this book.
14. MJ Kamlet, JL-M Abbound, RW Taft. J Am Chem Soc 98:6027, 1977.
15. MJ Kamlet, TN Hall, J Boykin, RW Taft. J Org Chem 44:2599, 1979.
16. DC Dong, MA Winnik. Photochem Photobiol 35:17, 1982.
17. DC Dong, MA Winnik. Can J Chem 62:2560, 1984.
18. W Rettig. Angew Chem Int Ed Eng 25:971, 1986.
19. Y-P Sun, CE Bunker, NB Hamilton. Chem Phys Lett 210:111, 1993.
20. Y-P Sun, CE Bunker. J Phys Chem 99:13786, 1995.
21. Y-P Sun, CE Bunker. Ber Bunsen-Ges Phys Chem 99:976, 1995.
22. JM Dobbs, JM Wong, KP Johnston J Chem Eng Data 31:303, 1986.
23. JM Dobbs, JM Wong, RJ Lahiere, KP Johnston. Ind Eng Chem Res 26:56, 1987.
24. JF Brennecke, DL Tomasko, J Peshkin, CA Eckert. Ind Eng Chem Res 29:1682, 1990.
25. JF Brennecke, CA Eckert. ACS Symp Series 14:406, 1989.
26. BJ Hrnjez, AJ Mehta, MA Fox, KP Johnston. J Am Chem Soc 111:2662, 1989.
27. CE Bunker, HW Rollins, JR Gord, Y-P Sun. J Org Chem 62:7324, 1997.
28. CE Bunker, Y-P Sun. J Am Chem Soc 117:10865, 1995.
29. CE Bunker, Y-P Sun, JR Gord. J Phys Chem A 101:9233, 1997.
30. MA McHugh, VJ Krukonis. Supercritical Fluid Extraction: Principles and Practice, 2 ed. Butterworth-Heinemann Series in Chemical Engineering, Butterworth-Heinemann, Stoneham, MA, 1994.
31. CA Eckert, BL Knutson, PG Debenedetti. Nature 383:313, 1996.
32. PG Debenedetti, JW Tom, X Kwauk, SD Yeo. Fluid Phase Equilib 82:311, 1993.
33. JW Tom, PG Debenedetti. J Aerosol Sci 22:555, 1991.
34. PG Debenedetti. AIChE J 36:1289, 1990.
35. RS Mohamed, DS Halverson, PG Debenedetti, RK Prudhomme. ACS Symp Ser 406:355, 1989.
36. X-Y Zheng, Y Arai, T Furuya. Trends Chem Eng 3:205, 1996.
37. JF Brennecke. Chem Ind 831, 1996.
38. F Cansell, B Chevalier, A Demourgues, J Etourneau, C Even, Y Garrabos, V Pessey, S Petit, A Tressaud, F Weill. J Mater Chem 9:67, 1999.
39. KA Larson, MA King. Biotech Prog 2:73, 1986.
40. M Sacchetti, MM Van Oort. Lung Biol Health Dis 94 (Inhalation Aerosols):337, 1996.
41. WJ Schmitt, MC Salada, GG Shook, SMI Speaker. AIChE J 41:2476, 1995.
42. B Subramaniam, RA Rajewski, K Snavely. J Pharm Sci 86:885, 1997.
43. JW Tom, GB Lim, PG Debenedetti, RK Prudhomme. ACS Symp Ser 514:238, 1993.

44. PM Gallagher, MP Coffey, VJ Krukonis, WW Hillstrom. J Supercrit Fluids 5:130, 1992.
45. PM Gallagher, MP Coffey, VJ Krukonis, N Klasutis. ACS Symp Ser 406:334, 1989.
46. KP Johnston, JML Penninger, eds. Supercritical Fluid Science and Technology, American Chemical Society, Washington, DC, 1989.
47. TG Squires, ME Paulaitis, eds. Supercritical Fluids, Chemical and Engineering Principals, American Chemical Society, Washington, DC, 1987.
48. FV Bright, MEP McNally, eds. Supercritical Fluid Technology: Theoretical and Applied Approaches to Analytical Chemistry, American Chemical Society, Washington, DC, 1992.
49. E Kiran, JF Brennecke, eds. Supercritical Fluid Engineering Science: Fundamentals and Applications, American Chemical Society, Washington, DC, 1993.
50. KW Hutchenson, NR Foster, eds. Innovations in Supercritical Fluids: Science and Technology, American Chemical Society, Washington, DC, 1995.
51. PR von Rohr, C Treep, eds. High Pressure Chemical Engineering, Elsevier, Amsterdam, 1996.
52. E Reverchon. J Supercrit Fluids 15:1, 1999.
53. DJ Dixon, G Lunabarcenas, KP Johnston. Polymer 35:3998, 1994.
54. DJ Dixon, KP Johnston, RA Bodmeier. AIChE J 39:127, 1993.
55. VJ Krukonis. Presented at the AIChE Annual Meeting, 1984.
56. DW Matson, JL Fulton, RD Smith. Mater Lett 6:31, 1987.
57. DW Matson, RC Petersen, RD Smith. Adv Ceram 21:109, 1987.
58. MA Winters, BL Knutson, PG Debenedetti, HG Sparks, TM Przybycien, CL Stevenson, SJ Prestrelski. J Pharm Sci 85:586, 1996.
59. SD Yeo, GB Lim, PG Debenedetti, H Bernstein. Biotechnol Bioeng 41:341, 1993.
60. SD Yeo, PG Debenedetti, SY Patro, TM Przybycien. J Pharm Sci 83:1651, 1994.
61. P York, M Hanna. Respir Drug Delivery V, Program Proc., (Int. Symp.), 5th:231, 1996.
62. M Kitamura, M Yamamoto, Y Yoshinaga, H Masuoka. J Cryst Growth 178:378, 1997.
63. Y Gao, TK Mulenda, Y-F Shi, W-K Yuan. J Supercrit Fluids 13:369, 1998.
64. S Mawson, S Kanakia, KP Johnston. Polymer 38:2957, 1997.
65. S Mawson, MZ Yates, ML O'Neill, KP Johnston. Langmuir 13:1519, 1997.
66. DJ Dixon, G Lunabarcenas, KP Johnston. Polymer 35:3998, 1994.
67. DJ Dixon, KP Johnston, RA Bodmeier. AIChE J 39:127, 1993.
68. RC Petersen, DW Matson, RD Smith. Polym Eng Sci 27:1693, 1987.
69. DW Matson, JL Fulton, RC Petersen, RD Smith. Ind Eng Chem Res 26:2298, 1987.
70. DW Matson, RC Petersen, RD Smith. Mater Lett 4:429, 1986.
71. RC Petersen, DW Matson, RD Smith. J Am Chem Soc 108:2100, 1986.
72. RD Smith, JL Fulton, RC Petersen, AJ Kopriva, BW Wright. Anal Chem 58:2057, 1986.
73. DW Matson, KA Norton, RD Smith. Chemtech 19:480, 1989.
74. DW Matson, RD Smith. J Am Ceram Soc 72:871, 1989.

75. AA Burukhin, BR Churagulov, NN Oleynikov, YV Kolen'ko. Mater. Res. Soc. Symp. Proc. Vol. 520, Nanostructured Powders and Their Industrial Applications, 171, 1998.
76. JR Williams, AA Clifford, KD Bartle, TP Kee. Powder Technol 96:158, 1998.
77. BN Hansen, BM Hybertson, RM Barkley, RE Sievers. Chem Mater 4:749, 1992.
78. BM Hybertson, BN Hansen, RM Barkley, RE Sievers. Mater Res Bull 26:1127, 1991.
79. VK Popov, VN Bagratashvili, EN Antonov, DA Lemenovski. Thin Solid Films 279:66, 1996.
80. RC Petersen, DW Matson, RD Smith. Polym Prepr 27:261, 1986.
81. AK Lele, AD Shine. AIChE J 38:742, 1992.
82. AK Lele, AD Shine. Ind Eng Chem Res 33:1476, 1994.
83. S Mawson, KP Johnston, JR Combes, JM Desimone. Macromolecules 28:3182, 1995.
84. NE Aniedobe, MC Thies. Macromolecules 30:2792, 1997.
85. RS Mohamed, PG Debenedetti, RK Prudhomme. AIChE J 35:325, 1989.
86. GT Liu, K Nagahama. Ind Eng Chem Res 35:4626, 1996.
87. CY Tai, CS Cheng. Chem Res Eng Des 75:228, 1997.
88. G-T Liu, K Nagahama. J Chem Eng Jpn 30:293, 1997.
89. E Reverchon, G Donsi, D Gorgoglione. J Supercrit Fluids 6:241, 1993.
90. E Reverchon, G Dellaporta, R Taddeo, P Pallado, A Stassi. Ind Eng Chem Res 34:4087, 1995.
91. G Donsi, E Reverchon. Pharm Acta Helv 66:170, 1991.
92. P Alessi, A Cortesi, I Kikic, NR Foster, SJ Macnaughton, I Colombo. Ind Eng Chem Res 35:4718, 1996.
93. GJ Griscik, RW Rousseau, AS Teja. J Cryst Growth 155:112, 1995.
94. K Ohgaki, H Kobayashi, T Katayama, N Hirokawa. J Supercrit Fluids 3:103, 1990.
95. CJ Chang, AD Randolph. AIChE J 35:1876, 1989.
96. C Domingo, E Berends, GM Vanrosmalen. J Supercrit Fluids 10:39, 1997.
97. C Domingo, EM Berends, GM Vanrosmalen. J Cryst Growth 166:989, 1996.
98. JW Tom, PG Debenedetti. Biotech Prog 7:403, 1991.
99. L Benedetti, A Bertucco, P Pallado. Biotechnol Bioeng 53:232, 1997.
100. PG Debenedetti, JW Tom, SD Yeo, GB Lim. J Controlled Rel 24:27, 1993.
101. JW Tom, PG Debenedetti, R Jerome. J Supercrit Fluids 7:9, 1994.
102. BL Knutson, PG Debenedetti, JW Tom. Drugs Pharm Sci 77:89, 1996.
103. JH Kim, TE Paxton, DL Tomasko. Biotech Prog 12:650, 1996.
104. I Kikic, M Lora, A Bertucco. Ind Eng Chem Res 36:5507, 1997.
105. S Cocks, SK Wigley, MI Chicarelliroinson, RM Smith. J Chromatogr A, 697:115, 1995.
106. JK Kim, T Aihara. Int J Heat Mass Transfer 35:2515, 1992.
107. BJ Jurcik, JR Brock. J Phys Chem 97:323, 1993.
108. H Ksibi, C Tenaud, P Subra, Y Garrabos. Eur J Mech, B/Fluids 15:569, 1996.
109. GR Shaub, JF Brennecke, MJ Mccready. J Supercrit Fluids 8:318, 1995.
110. JM Zen, FRF Fan, G Chen, AJ Bard. Langmuir 5:1355, 1989.
111. K Murakoshi, H Hosokawa, M Saitoh, Y Wada, T Sakata, H Mori, M Satoh, S Yanagida. J Chem Soc, Faraday Trans 94:579, 1998.

112. N Herron, Y Wang, H Eckert. J Am Chem Soc 112:1322, 1990.
113. Y Nosaka, K Yamaguchi, H Miyama, H Hayashi. Chem Lett 605, 1988.
114. M Ohtaki, K Oda, K Eguchi, H Arai. Chem Commun 1209, 1996.
115. D Hayes, OI Micic, MT Nenadovic, V Swayambunathan, D Meisel. J Phys Chem 93:4603, 1989.
116. Y Yin, X Xu, Z Zhang. Chem Commun 1641, 1998.
117. G Cardenas, J Acuna. Colloid Polym Sci 275:442, 2001.
118. T Yonezawa, N Toshima. J Chem Soc, Faraday Trans 91:4111, 1995.
119. GN Glavee, KJ Klabunde, CM Sorensen, GC Hadjipanayis. Inorg Chem 34:28, 1995.
120. GN Glavee, KJ Klabunde, CM Sorensen, GC Hadjipanayis. Langmuir 10:4726, 1994.
121. GN Glavee, KJ Klabunde, CM Sorensen, GC Hadjipanayis. Inorg Chem 32:474, 1993.
122. GN Glavee, KJ Klabunde, CM Sorensen, GC Hadjipanayis. Langmuir 9:162, 1993.
123. GN Glavee, KJ Klabunde, CM Sorensen, GC Hadjapanayis. Langmuir 8:771, 1992.
124. L Yiping, GC Hadjipanayis, CM Sorensen, KJ Klabunde. J Appl Phys 69:5141, 1991.
125. H Bönnemann, W Brijoux, R Brinkmann, E Dinjus, R Fretzen, T Joussen, B Korall. J Mol Cat 74:323, 1992.
126. H Bönnemann, R Brinkmann, R Köppler, P Neiteler, J Richter. Adv Mater 4:804, 1992.
127. H Bönnemann, W Brijoux, R Brinkmann, R Fretzen, T Joussen, R Köppler, B Korall, P Neiteler, J Richter. J Mol Catal 86:129, 1994.
128. H Bönnemann, W Brijoux, T Joussen. Angew Chem Int Ed Engl 29:273, 1990.
129. A Duteil, G Schmid, W Meyer-Zaika. J Chem Soc Chem Comm 31, 1995.
130. KL Tsai, JL Dye. Chem Mater 5:540, 1993.
131. KL Tsai, JL Dye. J Am Chem Soc 113:1650, 1991.
132. HH Huang, XP Ni, GL Loy, CH Chew, KL Tan, FC Loh, JF Deng, GQ Xu. Langmuir 12:909, 1996.
133. H Einaga, S Futamura, T Ibusuki. Environ Sci Tech 35:1880, 2001.
134. K Shiba, H Hinode, M Wakihara. Reac Kinet Catal Lett 64:281, 1999.
135. I Yamanaka, K Nishikawa, K Otsuka. Chem Lett 6:368, 2001.
136. ABR Mayer, JE Mark. Colloid Polym Sci 275:333, 1997.
137. KS Suslick, SB Choe, AA Cichowlas, MW Grinstaff. Nature 353:414, 1991.
138. KS Suslick, M Fang, T Hyeon. J Am Chem Soc 118:11960, 1996.
139. KS Suslick, T Hyeon, MM Fang, AA Cichowlas. Mater Sci Eng A 204:186, 1995.
140. KE Gonsalves, SP Rangarajan, CC Law, CR Feng, G-M Chow, A Garcia-Ruiz. ACS Symp Ser 622:220, 1996.
141. CP Gibson, KJ Putzer. Science 267:1338, 1995.
142. TH Hyeon, MM Fang, KS Suslick. J Am Chem Soc 118:5492, 1996.
143. KS Suslick, TW Hyeon, MM Fang. Chem Mater 8:2172, 1996.
144. MM Mdleleni, T Hyeon, KS Suslick. J Am Chem Soc 120:6189, 1998.
145. (a) ZY Zhong, T Prozorov, I Felner, AJ Gedanken. Phys Chem B 103:947, 1999; (b) K Okitsu, S Nagaoka, S Tanabe, H Matsumoto, Y Mizukoshi, Y Nagata. Chem Lett 3:271, 1999.

146. TK Jain, F Billoudet, L Motte, I Lisiecki, MP Pileni. Prog Colloid Polym Sci 89:106, 1992.
147. L Motte, F Billoudet, J Cizeron, MP Pileni. Prog Colloid Polym Sci 98:189, 1995.
148. I Lisiecki, M Bjoerling, L Motte, B Ninham, MP Pileni. Langmuir 11:2385, 1995.
149. MP Pileni, I Lisiecki, L Motte, C Petit, J Cizeron, N Moumen, P Lixon. Prog Colloid Polym Sci 93:1, 1993.
150. MP Pileni. Cryst Res Technol 33:1155, 1998.
151. L Motte, MP Pileni. J Phys Chem B 102:4104, 1998.
152. N Duxin, N Brun, C Colliex, MP Pileni. Langmuir 14:1984, 1998.
153. MP Pileni, A Taleb, C Petit. J Disper Sci Tech 19:185, 1998.
154. A Taleb, C Petit, MP Pileni. J Phys Chem B 102:2214, 1998.
155. C Petit, A Taleb, MP Pileni. Adv Mater 10:259, 1998.
156. MP Pileni. Ber Bunsen-Ges Phys Chem 101:1578, 1997.
157. N Duxin, N Brun, P Bonville, C Colliex, MP Pileni. J Phys Chem B 101:8907, 1997.
158. I Lisiecki, F Billoudet, MP Pileni. J Mol Liq 72:251, 1997.
159. N Feltin, MP Pileni. Langmuir 13:3927, 1997.
160. MP Pileni. Langmuir 13:3266, 1997.
161. MP Pileni, L Motte, F Billoudet, J Mahrt, F Willig. Mater Lett 31:255, 1997.
162. MP Pileni, N Moumen, JF Hochepied, P Bonville, P Veillet. J Phys IV 7:505, 1997.
163. A Taleb, C Petit, MP Pileni. Chem Mater 9:950, 1997.
164. (a) J Tanori, MP Pileni. Langmuir 13:633, 1997; (b) J Tanori, MP Pileni. Langmuir 13:639, 1997.
165. C Petit, MP Pileni. J Magn Magn Mater 166:82, 1997.
166. MP Pileni, L Motte, F Billoudet, C Petit. Surface Rev Lett 3:1215, 1996.
167. L Motte, F Billoudet, MP Pileni. J Mater Sci 31:38, 1996.
168. I Lisiecki, F Billoudet, MP Pileni. J Phys Chem B 100:4160, 1996.
169. A Hammouda, T Gulik, MP Pileni. Langmuir 11:3656, 1995.
170. J Tanori, MP Pileni. Adv Mater 7:862, 1995.
171. J Tanori, N Duxin, C Petit, I Lisiecki, P Veillet, MP Pileni. Colloid Polym Sci 273:886, 1995.
172. N Moumen, P Veillet, MP Pileni. J Magn Magn Mater 149:67, 1995.
173. I Lisiecki, MP Pileni. J Phys Chem 99:5077, 1995.
174. C Petit, TK Jain, F Billoudet, MP Pileni. Langmuir 10:4446, 1994.
175. MP Pileni. Adv Colloid Interface Sci 46:139, 1993.
176. MP Pileni, I Lisiecki. Colloid Surface A 80:63, 1993.
177. C Petit, P Lixon, MP Pileni. J Phys Chem 97:12974, 1993.
178. MP Pileni. J Phys Chem 97:6961, 1993.
179. (a) C Petit, P Lixon, MP Pileni. Langmuir 7:2620, 1991; (b) I Lisiecki, MP Pileni. J Am Chem Soc 115:3887, 1993.
180. MP Pileni, I Lisiecki, L Motte, C Petit. Res Chem Intermed 17:101, 1992.
181. L Motte, C Petit, L Boulanger, P Lixon, MP Pileni. Langmuir 8:1049, 1992.
182. MP Pileni, L Motte, C Petit. Chem Mater 4:338, 1992.
183. C Petit, P Lixon, MP Pileni. J Phys Chem 94:1598, 1990.
184. N Moumen, P Bonville, MP Pileni. J Phys Chem 100:14410, 1996.

185. N Moumen, MP Pileni. Chem Mater 8:1128, 1996.
186. N Moumen, MP Pileni. J Phys Chem 100:1867, 1996.
187. LJA Pérez, MAL Quintela. J Phys Chem B 101:8045, 1997.
188. P Lianos, JK Thomas. Chem Phys Lett 125:299, 1986.
189. SK Haram, AR Mahadeshwar, SG Dixit. J Phys Chem 100:5868, 1996.
190. J Cizeron, MP Pileni. J Phys Chem B 101:8887, 1997.
191. L Levy, N Feltin, D Ingert, MP Pileni. J Phys Chem B 101:9153, 1997.
192. S Shiojiri, T Hirai, I Komasawa, I Komasawa. Chem Commun 1439, 1998.
193. R Premachandran, S Banerjee, VT John, GL McPherson, JA Akkara, DL Kaplan. Chem Mater 9:1342, 1997.
194. MY Han, W Huang, CH Chew, LM Gan, XJ Zhang, W Ji. J Phys Chem B 102: 1884, 1998.
195. S-Y Chang, L Liu, SA Asher. J Am Chem Soc 116:6739, 1994.
196. C-L Chang, HS Fogler. Langmuir 13:3295, 1997.
197. C Petit, A Taleb, MP Pileni. J Phys Chem B 103:1805, 1997.
198. MP Pileni. New J Chem 22:693, 1998.
199. MP Pileni. Supramol Sci 5:321, 1998.
200. L Motte, F Billoudet, D Thiaudiere, A Naudon, MP Pileni. J Physique III 7:517, 1997.
201. L Motte, F Billoudet, E Lacaze, J Douin, MP Pileni. J Phys Chem B 101:138, 1997.
202. RP Bagwe, KC Khilar. Langmuir 13:6432, 1997.
203. C Tojo, MC Blanco, F Rivadulla, MA Lopez-Quintela. Langmuir 13:1970, 1997.
204. JN Wickham, AB Herhold, AP Alivisatos. Phys Rev Lett 84:923, 2000.
205. XG Peng, TE Wilson, AP Alivisatos, PG Schultz. Angew Chem Int Ed Engl 36:145, 1997.
206. XG Peng, J Wickham, AP Alivisatos. J Am Chem Soc 120:5343, 1998.
207. M Bruchez, M Moronne, P Gin, S Weiss, AP Alivisatos. Science 281:201, 1998.
208. WU Huynh, XG Peng, AP Alivisatos. Adv Mat 11:923, 1999.
209. U Banin, M Bruchez, AP Alivisatos, T Ha, S Weiss, DS Chemla. J Chem Phys 110:1195, 1999.
210. J Rockenberger, EC Scher, AP Alivisatos. J Am Chem Soc 121:11595, 1999.
211. (a) CB Murray, DJ Norris, MG Bawendi. J Am Chem Soc 115:8706, 1993; (b) CB Murray, CR Kagan, MG Bawendi. Ann Rev Mater Sci 30:545, 2000.
212. M Brust, J Fink, D Bethell, DJ Schiffrin, CJ Kiely. Chem Soc Chem Comm 1655, 1995.
213. T Vossmeyer, L Katsikas, M Giersig, IG Popovic, K Diesner, A Chemseddine, A Eychmuller, H Weller. J Phys Chem 98:7665, 1994.
214. JEB Katari, VL Colvin, AP Alivisatos. J Phys Chem 98:4109, 1994.
215. XG Peng, J Wickham, AP Alivisatos. J Am Chem Soc 120:5343, 1998.
216. OI Micic, CJ Curtis, KM Jones, JR Sprague, AJ Nozik. J Phys Chem 98:4966, 1994.
217. AJ Nozik, OI Micic. Mater Res Soc Bull 23:24, 1998.
218. AA Guzelian, JEB Katari, AV Kadavanich, U Banin, K Hamad, E Juban, AP Alivisatos, RH Wolters, CC Arnold, JR Heath. J Phys Chem B 100:7212, 1996.

219. F Mikulec. Ph.D. dissertation, Massachusetts Institute of Technology, Cambridge, MA, 1999.
220. MA Hines, P Guyot-Sionnest. J Phys Chem B 102:3655, 1998.
221. AA Guzelian, U Banin, AV Kadavanich, X Peng, AP Alivisatos. Appl Phys Lett 69:1432, 1996.
222. XG Peng, L Manna, WD Yang, J Wickham, E Scher, A Kadavanich, AP Alivisatos. Nature 404:59, 2000.
223. MA Hines, P Gnyot Sionnest. J Phys Chem B 100:468, 1996.
224. BO Dabbousi, J RodriguezViejo, FV Mikulec, JR Heine, H Mattoussi, R Ober, KF Jensen, MG Bawendi. J Phys Chem B 101:9463, 1997.
225. XG Peng, MC Schlamp, AV Kadavanich, AP Alivisatos. J Am Chem Soc 119:7019, 1997.
226. YW Cao, U Banin. Angew Chem Int Ed Eng 38:3692, 1999.
227. YW Cao, U Banin. J Am Chem Soc 122:9692, 2000.
228. LO Brown, JE Hutchison. J Am Chem Soc 121:882, 1999.
229. AI Cooper, JD Londono, G Wignall, JB McClain, ET Samulski, JS Lin, A Dobrynin, M Rubinstein, ALC Burke, JMJ Frechet, JM DeSimone. Nature 389:368, 1997.
230. MA Quadir, R Snook, RG Gilbert, JM DeSimone. Macromolecules 30:6015, 1997.
231. E Buhler, AV Dobrynin, JM DeSimone, M Rubinstein. Macromolecules 31:7347, 1998.
232. F Triolo, A Triolo, R Triolo, JD Londono, GD Wignall, JB McClain, DE Betts, S Wells, ET Samulski, JM DeSimone. Langmuir 16:416, 2000.
233. KL Harrison, SRP da Rocha, MZ Yates, KP Johnston, D Canelas, JM DeSimone. Langmuir 14:6855, 1998.
234. ML O'Neill, MZ Yates, KL Harrison, KP Johnston, DA Canelas, DE Betts, JM DeSimone, SP Wilkinson. Macromolecules 30:5050, 1997.
235. MZ Yates, ML O'Neill, KP Johnston, S Webber, DA Canelas, DE Betts, JM DeSimone. Macromolecules 30:5060, 1997.
236. MP Heitz, C Carlier, J deGrazia, KL Harrison, KP Johnston, TW Randolph, FV Bright. J Phys Chem B 101:6707, 1997.
237. MP Heitz, FV Bright. Appl Spectrosc 50:732, 1996.
238. KP Johnston, KL Harrison, MJ Clarke, SM Howdle, MP Heitz, FV Bright, C Carlier, TW Randolph. Science 271:624, 1996.
239. J Zhang, T Bright. J Phys Chem 96:9068, 1992.
240. PS Shah, JD Holmes, RC Doty, KP Johnston, BA Korgel. J Am Chem Soc 122:4245, 2000.
241. CT Lee, P Bhargava, KP Johnston. J Phys Chem B 104:4448, 2000.
242. SRP da Rocha, KP Johnston. Langmuir 16:3690, 2000.
243. MZ Yates, ML O'Neill, KP Johnston, S Webber, DA Canelas, DE Betts, JM DeSimone. Macromolecules 30:5060, 1997.
244. MJ Clarke, KL Harrison, KP Johnston, SM Howdle. J Am Chem Soc 119:6399, 1997.
245. JP Cason, CB Roberts. J Phys Chem 104:1217, 2000.
246. JJ Watkins, TJ McCarthy. Chem Mater 7:1991, 1995.
247. M Ji, XY Chen, CM Wai, JL Fulton. J Am Chem Soc 121:2631, 1999.

248. H Ohde, JM Rodriguez, XR Ye, CM Wai. Chem Commun 23:2353, 2000.
249. JD Holmes, PA Bhargave, BA Korgel, KP Johnston. Langmuir 15:6613, 1999.
250. KA Consani, RD Smith. J Supercrit Fluids 3:51, 1990.
251. K Harrison, J Goveas, KP Johnston, EA O'Rear. Langmuir 10:3536, 1994.
252. KP Johnston, T Randolph, F Bright, SW Howdle. Science 272:1726, 1997.
253. J Estoe, BMH Gazalles, DC Steytler, JD Holmes, AR Pitt, TJ Wear, RK Heenan. Langmuir 13:6980, 1997.
254. RG Zielinski, SR Line, EW Kaler, N Rosov. Langmuir 13:3934, 1997.
255. (a) M Weber, LM Russel, PG Debendetti. Presented at the AIChE Annual Meeting, (Talk 115c), 1998; (b) M Weber, MC Thies. Private communication.
256. Y-P Sun, HW Rollins. Chem Phys Lett 288:585, 1998.
257. Y-P Sun, JE Riggs, HW Rollins, R Guduru. J Phys Chem B 103:77, 1999.
258. Y-P Sun, HW Rollins, R Guduru. Chem Mater 11:7, 1999.
259. Y-P Sun, R Guduru, F Lin, T Whiteside. Ind Eng Chem Res 39:4663, 2000.
260. HW Rollins. Ph.D. dissertation, Clemson University, Clemson, South Carolina, 1999.
261. JE Riggs. Ph.D. dissertation, Clemson University, Clemson, South Carolina, 2001.
262. Y-P Sun, P Atorngitijawat, MJ Meziani. Langmuir 17:5707, 2001.
263. Y-P Sun et al. Unpublished results.
264. Y Wang. Adv Photochem 19:179, 1995.
265. (a) G Schmid. Chem. Rev. 92:1709, 1992; (b) R Dagani. C&EN News Jan. 5:27, 1998; (c) A Hagfeldt, M Graetzel. Chem Rev 95:49, 1995; (d) MA Fox, MT Dulay. Chem Rev 93:341, 1993; (e) AL Linsebigler, G Lu, JT Yates. Chem Rev 95:735, 1995; (f) CP Gibson, KJ Putzer. Science 267:1338, 1995.
266. (a) PMS Ferreira, AB Timmons, MC Neves, P Dynarowicz, T Trindade. Thin Solid Films 389:272, 2001; (b) SG Hickey, DJ Riley, EJ Tull. J Phys Chem B 104:7623, 2000.
267. (a) JZ Zhang. J Phys Chem B 104:7239, 2000; (b) EPAM Bakkers, JJ Kelly, DJ Vanmaekelbergh. Electroanal Chem 482:48, 2000.
268. (a) WCW Chan, SM Nie. Science 281:2016, 1998; (b) B Ludolph, MA Malik, P O'Brien, N Revaprasadu. Chem Commun 17:1849, 1998.
269. (a) S Gorer, JA Ganske, JC Hemminger, RM Penner. J Am Chem Soc 120:9584, 1998; (b) U Winkler, D Eich, ZH Chen, R Fink, SK Kulkarni, E Umbach. Chem Phys Lett 306:95, 1999; (c) MC Brelle, JZ Zhang, L Nguyen, RK Mehra. J Phys Chem A 103:10194, 1999.
270. (a) JO Joswig, M Springborg, G Seifert. J Phys Chem B 104:2617, 2000; (b) H Yao, Y Takada, A Ito, N Kitamura. Polym J 31:1133, 1999.
271. Y Wang, A Suna, W Mahler, R Kasowski. J Chem Phys 87:7315, 1987.
272. (a) T Hirai, T Saito, I Komasawa. J Phys Chem B 104:11639, 2000; (b) T Hirai, T Watanabe, I Komasawa. J Phys Chem B 103:10120, 1999; (c) SW Haggata, DJ Cole Hamilton, JR Fryer. J Mater Chem 7:1969, 1997; (d) D Gallagher, WE Heady, JM Racz, RN Bhagava. J Mater Res 10:870, 1995.
273. HP Klug, LE Alexander. X-Ray Diffraction Procedures. John Wiley & Sons, New York, 1959.
274. MJ Meziani, Y-P Sun. Langmuir 2002 (in press).

275. MZ Yates, DL Apodaca, ML Campbell, ER Birnbaum, TM McCleskey. Chem Commun 25, 2001.
276. KP Clarle, W Schulze. Ber Bunsen-Ges Phys Chem 88:350, 1984.
277. PV Kamat, M Flumiani, GV Hartland. J Phys Chem B 102:3123, 1998.
278. Y Wang, A Suna, J McHugh, EF Hilinski, PA Lucas, RD Johnson. J Chem Phys 92:6927, 1990.
279. L Tutt, TF Boggess. Prog Quantum Electron 17:299, 1993.
280. JW Perry. In Nonlinear Optics of Organic Molecules and Polymers (HS Nalwa, S Miyata, eds.). CRC Press, Boca Raton, 1997, p. 813.
281. EW Van Stryland, DJ Hagan, T Xiz, AA Said. In: Nonlinear Optics of Organic Molecules and Polymers (HS Nalwa, S Miyata, eds.). CRC Press, Boca Raton, 1997, p. 841.
282. Y-P Sun, JE Riggs, Z Guo, HW Rollins. In: Optical and Electronic Properties of Fullerenes and Fullerene-Based Materials (J Shinar, ZV Vardeny, ZH Kafafi, eds.), Marcel Dekker, New York, 1999, p. 43.
283. Y-P Sun, JE Riggs. Intern Rev Phys Chem 18:43, 1999.
284. RC Hollins. Curr Opin Sol Mater Sci 4:189, 1999.
285. Y-P Sun, JE Riggs, KB Henbest, RBJ Martin. Nonlinear Opt Phys Mat 9:481, 2000.
286. (a) JE Riggs, Y-P Sun. J Phys Chem A 103:485, 1999; (b) JE Riggs, Y-P Sun. J Chem Phys 112:4221, 2000; (c) JE Riggs, DB Walker, DL Carrol, Y-P Sun. J Phys Chem B 104:7071, 2000.
287. MJ Meziani, RB Martin, Y-P Sun. Unpublished results.
288. Y-P Sun, et al. Unpublished results.

Index